Phycology is the study of algae, a ubiquitous and extremely important range of species ecologically because of the dependence of other species on their primary production. This third edition of *Phycology*, is designed to serve both as a textbook for a one-semester course and also as a reference on modern concepts in the study of algae. Algae are divided into four natural groups, and chapters on each group cover the cytology, morphology, physiology, life cycles, ecology and systematics, all presented in a manner that can be understood by readers with a basic knowledge of biology. The use of clear, concise drawings, as well as the special emphasis placed on those algae that are commonly covered in phycology courses, and encountered by students in marine and freshwater habitats, has led to the establishment of this book as a classic introductory text suitable for courses throughout the world.

ROBERT EDWARD LEE has had a long and varied career, teaching worldwide in countries such as South Africa and Iran, as well as at Harvard Medical School and Colorado State University, where he currently works in the Department of Anatomy and Neurobiology.

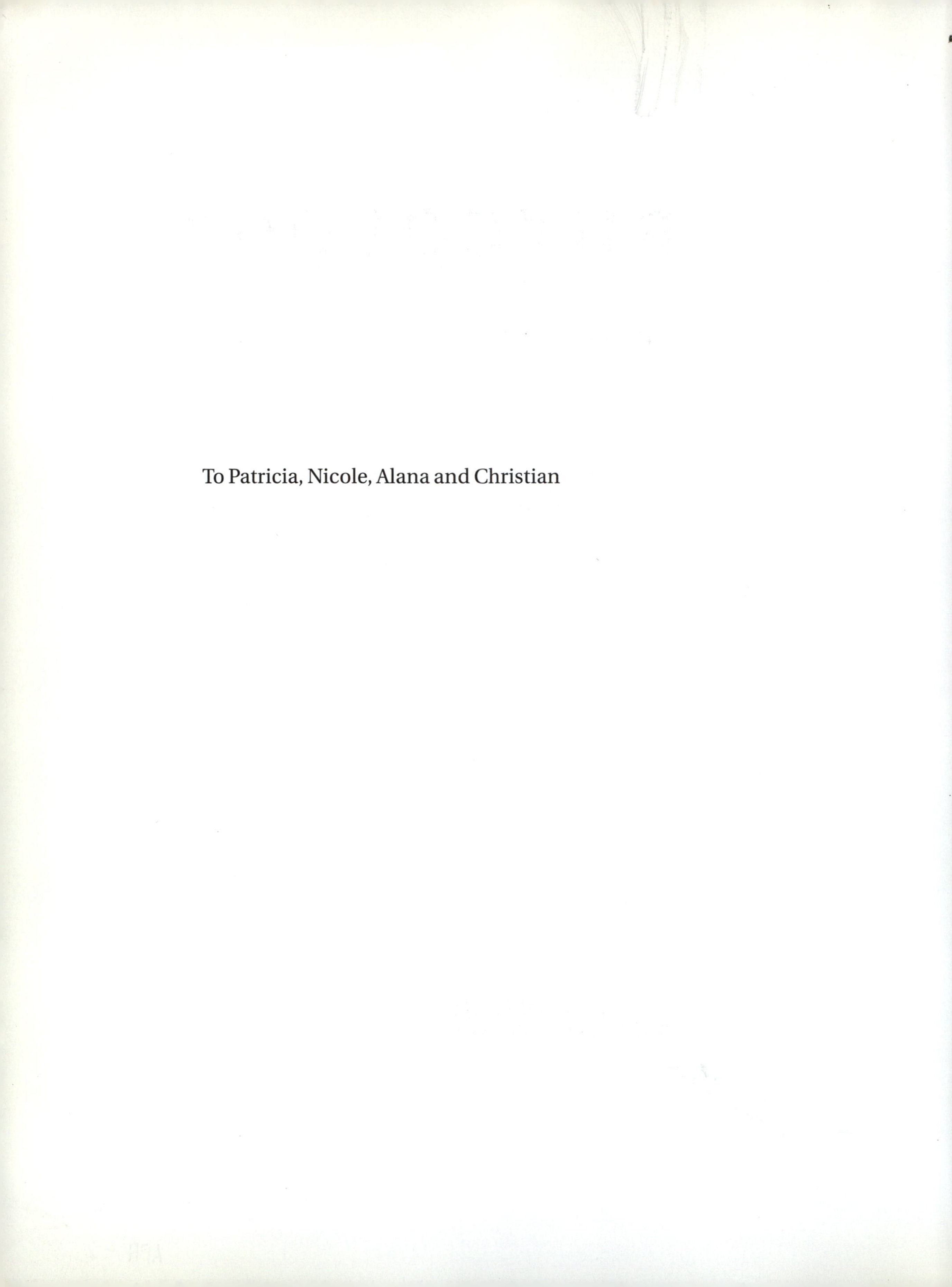

To Patricia, Nicole, Alana and Christian

PHYCOLOGY

THIRD EDITION

Robert Edward Lee

Colorado State University, USA

CAMBRIDGE
UNIVERSITY PRESS

PUBLISHED BY THE PRESS SYNDICATE OF THE UNIVERSITY OF CAMBRIDGE
The Pitt Building, Trumpington Street, Cambridge, United Kingdom

CAMBRIDGE UNIVERSITY PRESS
The Edinburgh Building, Cambridge CB2 2RU, UK www.cup.cam.ac.uk
40 West 20th Street, New York, NY 10011–4211, USA www.cup.org
10 Stamford Road, Oakleigh, Melbourne 3166, Australia
Ruiz de Alarcón 13, 28014 Madrid, Spain

First published 1980
Second edition 1989
Third edition 1999

Printed in the United Kingdom at the University Press, Cambridge

Typeface Utopia 9.5/13.5pt. *System* QuarkXPress® [SE]

A catalogue record for this book is available from the British Library

Library of Congress Cataloguing in Publication data

Lee, Robert Edward, 1942–
Phycology / Robert Edward Lee. – 3rd ed.
p. cm.
Includes bibliographical references and index.
ISBN 0 521 63090 8 (hardcover)
1. Algology. I. Title.
QK566.L44 1999
579.8–dc21 98–53255 CIP

ISBN 0 521 63090 8 hardback
ISBN 0 521 63883 6 paperback

CONTENTS

PREFACE

Professor Friedrich Oltmanns of the University of Freiburg in Germany. Professor Oltmanns was the author of the first comprehensive textbook of phycology *Morphologie und Biologie der Algen* published by G. Fischer in Jena, Germany, in 1905.

The third edition of *Phycology* contains some changes that reflect current thinking in the field. After two editions of referring to the prokaryotic algae as "blue-green algae," I have yielded to current thinking and use the term "cyanobacteria" in the third edition. I have dropped the chapter on the Prochlorophyta since investigations have shown that these algae arose a number of times within the cyanobacteria.

The last decade has been the decade of "nucleotide sequencing" with a large amount of valuable information being produced on the relationships of various groups of algae. The groups that have been most influenced by these studies are the heterokonts, prymnesiophytes and haptophytes, all of which have two membranes of chloroplast endoplasmic reticulum. As such, the reader will find these chapters to be those that have the most changes.

My main aim has always been to write a phycology textbook for a student taking a first course in the subject. I have been fortunate to be able to discuss such an approach with my colleague, Paul Kugrens, who actually teaches the phycology course at Colorado State University. After discussions with Paul, I have concentrated on fewer algae in the third edition and have deleted some of the information on algae that are usually not presented in a phycology course. The chapters on the red and green algae are most affected by this decision; both of these chapters are shorter than they were in previous editions.

Phycology was never intended to be a comprehensive treatise on the field of phycology. One of the more difficult things for me to do was to sift the many fine papers in the field and concentrate on what I thought would be of interest to a student taking a first course in phycology. My philosophy has been to consider in depth relatively few genera. I apologize to the many excellent investigators whose work I have not been able to cover because of space limitations.

I want to thank Paul Kugrens and Brec Clay, valued friends and colleagues, for taking the time to discuss various aspects of this edition and to improve the presentation of the material.

Robert Edward Lee
1999

PREFACE TO THE FIRST EDITION

It was that eccentric British soldier of fortune Col. Meinertzhagen, in his *Birds of Arabia*, who expressed the sentiment that prefaces should be kept short because few people ever read them. Accordingly, I would like to take a brief opportunity to express my gratitude to the people who offered encouragement and assistance during the preparation of this book. I would like to thank Adele Strauss Wolbarst, Robert Cnoops, Charmaine Slack, Sophia Skiordis, Caroline Mondel, Jill Keetley-Smith, Heather Edwards, Gail Arbeter, and the Lending Library at Boston Spa, England, for help while most of this manuscript was being prepared at the University of the Witwatersrand. For general encouragement while at Pahlavi (Shiraz) University and for providing assistance during the last turbulent and chaotic year of imperial rule in Iran, while the manuscript was being finished, I would like to thank Mark Gettner, Brian Coad, and Mumtaz Bokhari.

When photographs or drawings have been taken directly from the original material, this is indicated by stating in the legend that it is *from* the original work. Most of the drawings have been redrawn to suit my tastes, and these drawings are indicated by stating that the work is *after* the original. In some cases I have made drawings from photographs or have incorporated a number of drawings in one, in which case I state that the finished drawing is *adapted* from the original work or works.

I have used the metric system in this book, and the fine-structural illustrations are expressed in micrometers (μm) and nanometers (nm).

R.E.L.

PART I · Introduction

1 · Basic characteristics of the algae

Phycology or algology is the study of the algae. The word **phycology** is derived from the Greek word *phykos*, which means "seaweed." The term **algology**, described in Webster's dictionary as the study of the algae, has fallen out of favor because it resembles the term *algogenic* which means "producing pain." The algae are **thallophytes** (plants lacking roots, stems, and leaves) that have chlorophyll *a* as their primary photosynthetic pigment and lack a sterile covering of cells around the reproductive cells. This definition encompasses a number of plant forms that are not necessarily closely related, for example, the cyanobacteria which are closer in evolution to the bacteria than to the rest of the algae.

Algae most commonly occur in water, be it freshwater, marine, or brackish. However, they can also be found in almost every other environment on earth, from the algae growing in the snow of some American mountains to algae living in lichen associations on bare rocks, to unicellular algae in desert soils, to algae living in hot springs. In most habitats they function as the primary producers in the food chain, producing organic material from sunlight, carbon dioxide, and water. Besides forming the basic food source for these food chains, they also form the oxygen necessary for the metabolism of the consumer organisms. In such cases humans rarely directly consume the algae as such, but harvest organisms higher up in the food chain (i.e., fish, crustaceans, shellfish). Some algae, particularly the reds and browns, are harvested and eaten as a vegetable, or the mucilages are extracted from the thallus for use as gelling and thickening agents.

Structure of the algal cell

There are two basic types of cells in the algae, **prokaryotic** and **eukaryotic**. Prokaryotic cells lack membrane-bounded organelles (plastids, mitochondria, nuclei, Golgi bodies, and flagella) and occur in the Cyanophyta (Fig. 2.2). The prokaryotic structure of the Cyanophyta is described in detail in Chapters 2 and 3. The remainder of the algae are eukaryotic and have organelles.

A eukaryotic cell (Fig. 1.1) is often surrounded by a cell wall composed of

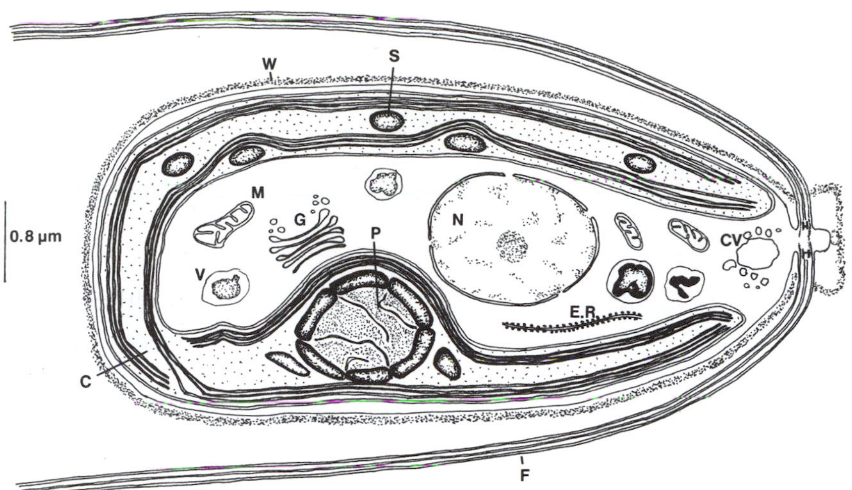

Fig. 1.1 A drawing of a cell of *Chlamydomonas* in the electron microscope showing the organelles present in a eukaryotic algal cell. (C) Chloroplast; (CV) contractile vacuole; (E.R.) endoplasmic reticulum; (F) flagella; (G) Golgi-body; (M) mitochondrion; (N) nucleus; (P) pyrenoid; (S) starch; (V) vacuole; (W) wall.

polysaccharides that are partially produced and secreted by the Golgi body. The plasma membrane (plasmalemma) surrounds the remaining part of the cell; this membrane is a living structure responsible for controlling the influx and outflow of substances in the protoplasm. Locomotory organs, the flagella, propel the cell through the medium by their beating. The flagella are enclosed in the plasma membrane and have a specific number and orientation of microtubules. The nucleus, which contains the genetic material of the cell, is surrounded by a double membrane with pores in it. The contents of the nucleus are a nucleolus, chromosomes, and the background material or karyolymph. The chloroplasts have membrane sacs called thylakoids that carry out the light reactions of photosynthesis. The thylakoids are embedded in the stroma where the dark reactions of carbon fixation take place. The stroma has small 70S ribosomes (see under "Ribosomes" later in this chapter), DNA, and in some cases the storage product. Chloroplasts are surrounded by the two membranes of the chloroplast envelope. Sometimes chloroplasts have a dense proteinaceous area, the pyrenoid, which is associated with storage-product formation. Double-membrane-bounded mitochondria have 70S ribosomes and DNA, and contain the respiratory apparatus. The Golgi body consists of a number of membrane sacs, called cisternae, stacked on top of one another. The Golgi body functions in the production and secretion of polysaccharides. The cytoplasm also contains large 80S ribosomes and lipid bodies.

FLAGELLA

Flagella consist of an **axoneme** of nine doublet microtubules surrounding two central microtubules, with all the microtubules surrounded by the

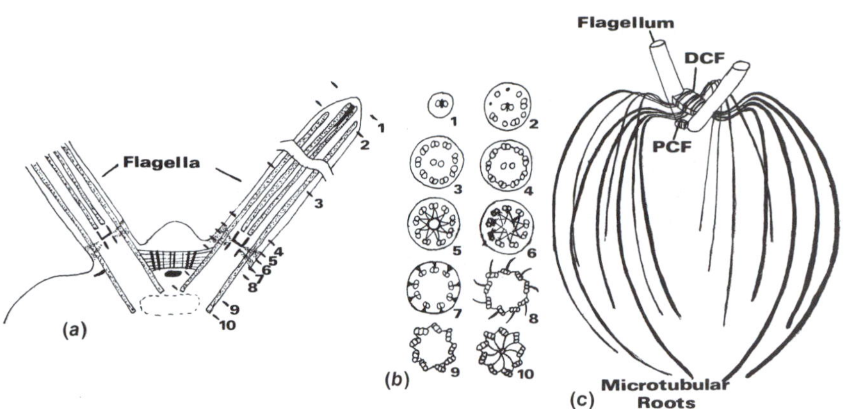

Fig. 1.2 The flagellar system in *Chlamydomonas*. (*a*) A diagrammatic drawing of a section of the flagellar system. The numbers refer to cross-sections of the flagellar system in (*b*). (*c*) Diagrammatic drawing of the whole flagellar apparatus. The two flagella are joined by the proximal connecting fiber (PCF) and distal connecting fiber (DCF). (After Ringo, 1967.)

plasma membrane of the cell (Fig. 1.2). On entering the cell body, the two central microtubules end at a dense plate, whereas the nine peripheral doublets continue into the cell, usually picking up an additional structure that transforms them into triplets.

There are also other structures between the microtubules in the basal region of the flagellum (**basal body**). Attached to the basal body there can be either microtubular roots or striated fibrillar roots. The former type of root consists of a group of microtubules running from the basal body into the protoplasm (Fig. 1.2(*c*)), whereas the latter consists of groups of fibers that have striations along their length. The gametes of the green seaweed *Ulva lactuca* (sea lettuce) illustrate both types of flagellar roots (Fig. 1.3) (Melkonian, 1980; Andersen et al., 1991). There are four **microtubular roots** composed of microtubules arranged in a cruciate pattern, and **fibrous roots** (**rhizoplasts**) composed of a bundle of filaments. There are two types of fibrous roots: (1) **system I fibrous roots** composed of 2 nm filaments cross-striated with a periodicity of approximately 30 nm and (2) **system II fibrous roots** composed of 4–8 nm filaments usually cross-striated with a periodicity greater than 80 nm. System I fibrous roots are non-contractile while system II fibrous roots are contractile when appropriately stimulated.

The flagellar membrane may have no **hairs** (**mastigonemes**) on its surface (**whiplash** or **acronematic** flagellum) or it may have hairs on its surface (**tinsel** or **hairy** or **pantonematic** or **Flimmergeissel**). There are two types of flagellar hairs: (1) **non-tubular flagellar hairs** less than 15 nm in diameter and (2) **tubular flagellar hairs** consisting of at least a hollow shaft (greater than 15 nm in diameter) often with one or more terminal filaments (Fig. 1.4) (Andersen et al., 1991). The flexible non-tubular flagellar hairs wrap around the flagellum increasing the surface area and efficiency of propulsion. The stiff "oarlike" tubular hairs, comparable to oars propelling a boat, reverse the thrust of the flagella. Non-tubular flagellar hairs are com-

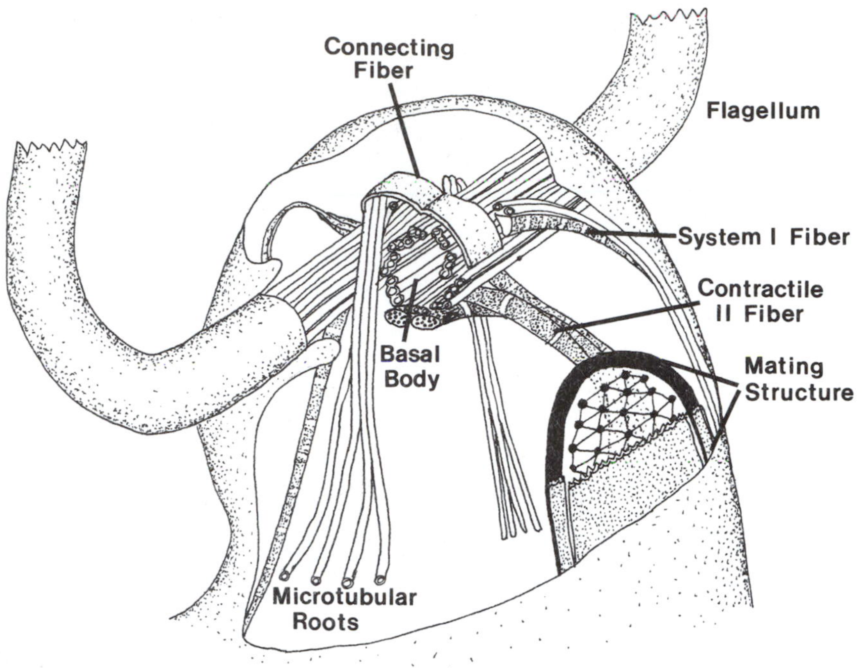

Fig. 1.3 Schematic three-dimensional reconstruction of the flagellar apparatus of a female gamete of *Ulva lactuca* showing the four cruciately arranged microtubular roots and the fibrous contractile roots associated with them. (Adapted from Melkonian, 1980.)

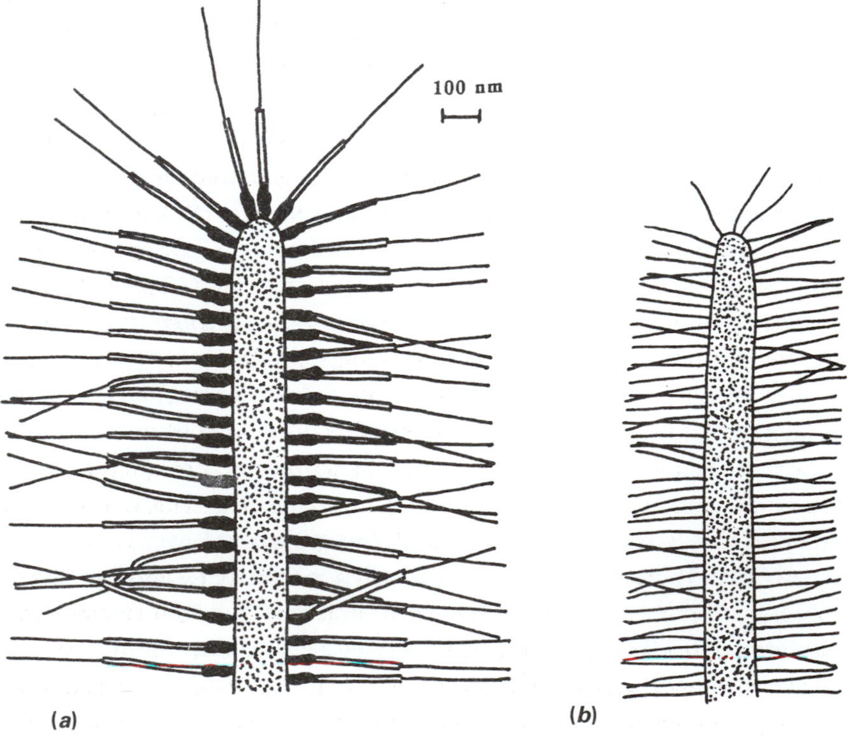

Fig. 1.4 Drawings of the types of hairs on algal flagella. (*a*) An example of tubular hairs on a flagellum (*Ascophyllum* sperm). Each hair is composed of a basal region attached to the flagellar membrane, the microtubular shaft, and a terminal fibril. (*b*) An example of non-tubular hairs on a flagellum (*Chlamydomonas* gamete). ((*a*) adapted from Bouck, 1969; (*b*) from Snell, 1976.)

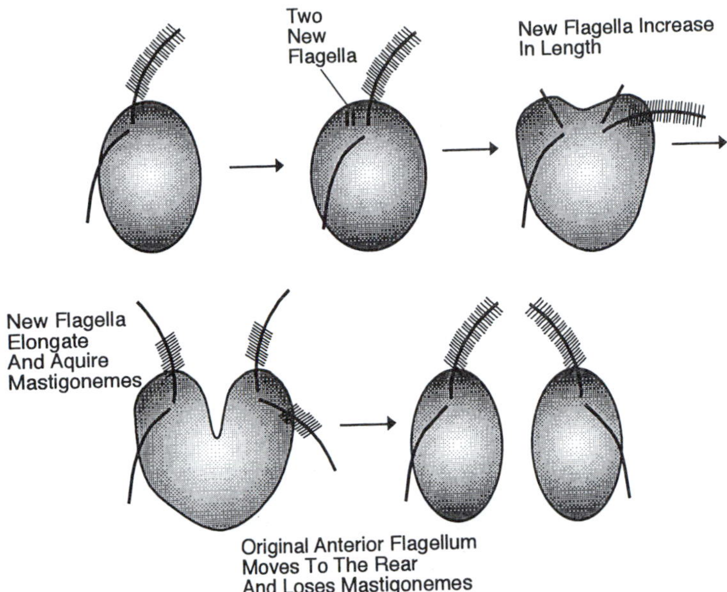

Fig. 1.5 The sequence of flagellar transformation during cell division.

posed of glycoproteins, whereas tubular flagellar hairs are composed of proteins and glycoproteins (Bouck et al., 1978). The non-tubular flagellar hairs are made up of solid fibrils 5–10 nm wide and 1–3 μm long. The tubular or tripartite hairs (Fig. 1.4(*a*)) are 20 nm wide and usually about 2 μm long, and are composed of three regions: (1) a tapering basal region 200 nm long attached to the flagellar membrane, (2) a microtubular shaft 1 μm long, and (3) a few 0.5-μm-long terminal filaments (Bouck, 1969, 1971). The bases of the hairs do not penetrate the flagellar membrane but are stuck to it. Development of the tubular hairs begins in the space between the inner and outer membrane of the nuclear envelope (perinuclear continuum) where the basal and microtubular regions are assembled. These then pass to the Golgi apparatus, where the terminal filaments are added. Finally the hairs are carried to the plasma membrane in Golgi vesicles, where they are discharged and attached to the flagellar membrane. Tripartite tubular hairs occur in the Heterokontophyta, Cryptophyta and some of the primitive Chlorophyta. The Chlorophyta, Euglenophyta and Dinophyta have non-tubular flagellar hairs (Moestrup, 1982). In addition to hairs, a number of different scale types occur on the surface of the flagella. These will be discussed in the chapters on the individual algal groups.

Flagella progress through a set of developmental cycles during cell division (Fig. 1.5). A biflagellate cell with an anterior flagellum covered with tubular hairs (tinsel flagellum), and a posterior smooth flagellum (whiplash flagellum), will be used as an example. Before the onset of cell division, two new flagella appear next to the anterior flagellum. These two new flagella

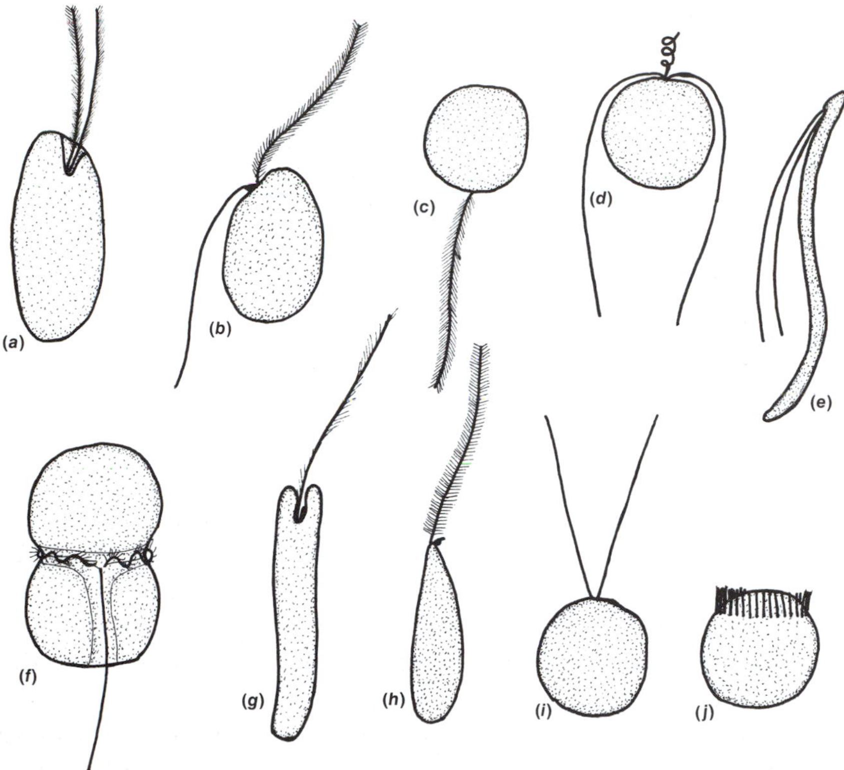

Fig. 1.6 The shape of eukaryotic motile algal cells and their flagella. There are a number of modifications in structure that are not included here. (*a*) Cryptophyta; (*b*) Xanthophyceae, Raphidophyceae, Chrysophyceae and Phaeophyceae of the Heterokontophyta; (*c*) Bacillariophyceae of the Heterokontophyta; (*d*) Prymnesiophyta; (*e*) Chlorophyta; (*f*) Dinophyta; (*g*) Euglenophyta; (*h*) Eustigmatophyceae of the Heterokontophyta; (*i,j*) Chlorophyta.

elongate while the original anterior flagellum moves toward the posterior of the cell and loses its tubular hairs, to become the posterior smooth flagellum of one of the daughter cells. The two new flagella at the anterior end of the cell acquire tubular hairs and become the tinsel flagella of the daughter cells. Thus, each daughter cell has one new anterior tinsel flagellum, and one posterior smooth whiplash flagellum that was originally a flagellum in the parent cell (Beech and Wetherbee, 1990; Melkonian et al., 1987).

Algal motile cells can have different arrangements of flagella (Fig. 1.6). If the flagella are of equal length, they are called **isokont** flagella; if they are of unequal length, they are called **anisokont flagella**; and if they form a ring at one end of the cell, they are called **stephanokont** flagella. **Heterokont** refers to an organism with a hairy and a smooth flagellum (Moestrup, 1982).

CELL WALLS AND MUCILAGES

In general, algal cell walls are made up of two components: (1) the fibrillar component, which forms the skeleton of the wall, and (2) the amorphous component, which forms a matrix within which the fibrillar component is embedded.

Fig. 1.7 Structural units of alginic acid, fucoidin, and agarose. (After Percival and McDowell, 1967.)

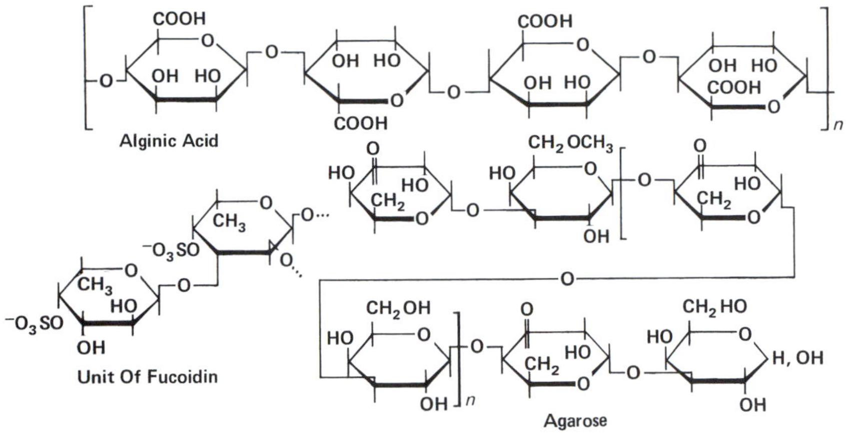

The most common type of fibrillar component is **cellulose**, a polymer of 1,4 linked β-D-glucose. Cellulose is replaced by a **mannan**, a polymer of 1,4 linked β-D-mannose, in some siphonaceous greens and in *Porphyra* and *Bangia* in the Rhodophyta. In some siphonaceous green algae and some Rhodophyta (*Porphyra, Rhodochorton, Laurencia,* and *Rhodymenia*), fibrillar **xylans** of different polymers occur.

The amorphous mucilaginous components occur in the greatest amounts in the Phaeophyceae and Rhodophyta, the polysaccharides of which are commercially exploited. **Alginic acid** (Fig. 1.7) is a polymer composed mostly of β-1,4 linked D-mannuronic acid residues with variable amounts of L-guluronic acid. Alginic acid is present in the intercellular spaces and cell walls of the Phaeophyceae. **Fucoidin** (Fig. 1.7) also occurs in the Phaeophyceae and is a polymer of α-1,2, α-1,3, and α-1,4 linked residues of L-fucose sulfated at C-4. In the Rhodophyta the amorphous component of the wall is composed of galactans or polymers of galactose, which are alternately β-1,3 and β-1,4 linked. These galactans include **agar** (made up of agaropectin and agarose, Fig. 1.7), **carrageenan**, **porphyran**, **furcelleran**, and **funoran**. The amorphous polysaccharides of the Chlorophyta are more complex, containing residues of D-galactose, L-arabinose, D-xylose, D-glucuronic acid, and L-rhamnose.

The cell walls of the Cyanophyta are more complicated and are similar to those of bacteria. They are considered in detail in the chapter on the Cyanophyta.

PLASTIDS

The basic type of plastid in the algae is a **chloroplast**, a plastid capable of photosynthesis. **Chromoplast** is synonymous with chloroplast; in the older

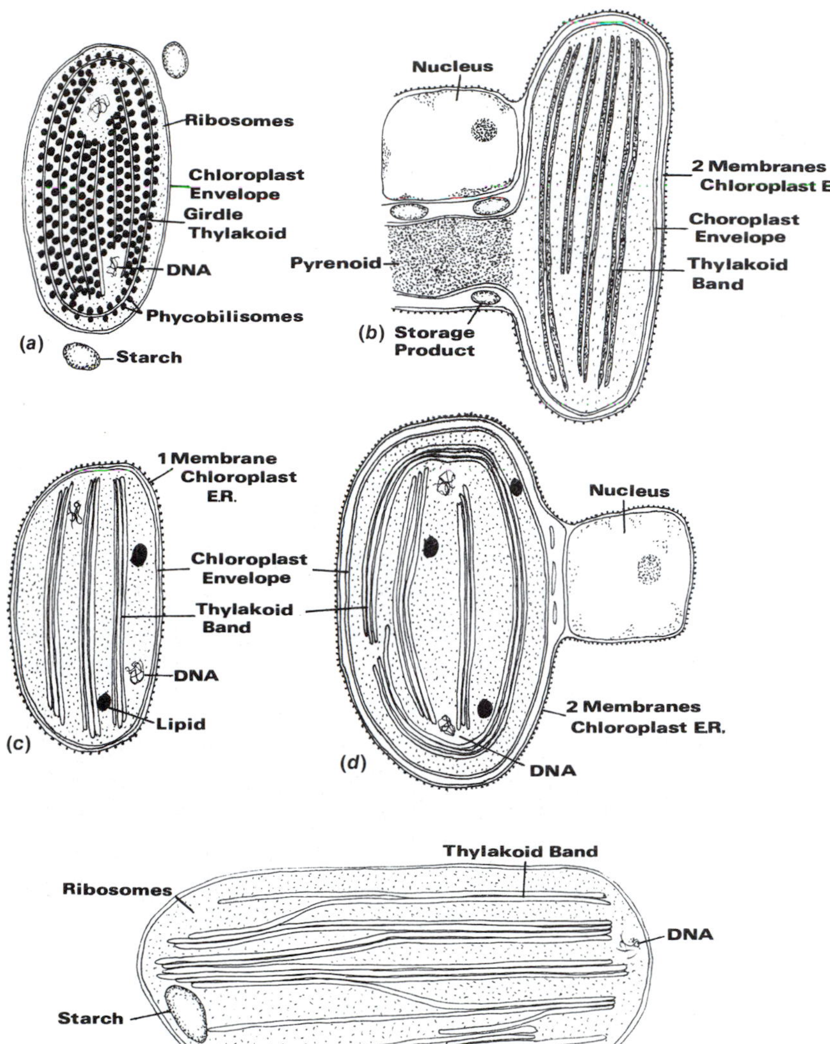

Fig. 1.8 Types of chloroplast structure in eukaryotic algae. (*a*) One thylakoid per band, no chloroplast endoplasmic reticulum (Rhodophyta). (*b*) Two thylakoids per band, two membranes of chloroplast E.R. (Cryptophyta). (*c*) Three thylakoids per band, one membrane of chloroplast E.R. (Dinophyta, Euglenophyta). (*d*) Three thylakoids per band, two membranes of chloroplast E.R. (Prymnesiophyta and Heterokontophyta). (*e*) Two to six thylakoids per band, no chloroplast E.R. (Chlorophyta).

literature a chloroplast that has a color other than green is often called a chromoplast. A **proplastid** is a reduced plastid with few if any thylakoids. A proplastid will usually develop into a chloroplast although in some heterotrophic algae it remains a proplastid. A **leucoplast** or **amyloplast** is a colorless plastid that has become adapted for the accumulation of storage product.

In the Rhodophyta and Chlorophyta, the chloroplasts are bounded by the double membrane of the chloroplast envelope (Fig. 1.8(*a*),(*e*)). In the other eukaryotic algae, the chloroplast envelope is surrounded by one of two membranes of **chloroplast endoplasmic reticulum** (chloroplast E.R.),

Table 1.1 *Characteristics of algal chloroplasts and storage products*

Algal class	Chl.	Phycobilins	Carotenoids	Thyl. band[a]	Memb. CER[b]	Stor. prod.[c]
Cyanophyta	a	C-Phycocyanin, C-phycoerythrin, allophycocyanin, phycoerythrocyanin	β-Carotene, zeaxanthin, echinenone, canthaxanthin, mutatochrome, antheraxanthin, β-cryptoxanthin, myxoxanthophyll, aphanizophyll, oscillaxanthin	1	–	–
Euglenophyta	a,b		β-Carotene, γ-carotene, diadinoxanthin, diatoxanthin, neoxanthin, β-cryptoxanthin (and its 5′, 6′-monoepoxide), echinenone, 3-hydroxyechinenone, astaxanthin ester	3	1	O
Dinophyta	a,c_2		β-Carotene, peridinin, diadinoxanthin, diatoxanthin, dinoxanthin	3	1	O
Cryptophyta	a,c_2	3 Phycoerythrins, 3 phycocyanins	α-Carotene, β-carotene, alloxanthin, crocoxanthin, monadoxanthin	2	2	B
Raphidophyceae	a,c		β-Carotene, lutein epoxide, antheraxanthin	3	2	O
Chrysophyceae	a,c_1,c_2		β-Carotene, fucoxanthin, diatoxanthin, diadinoxanthin, echinenone	3	2	O
Prymnesiophyceae	a,c_1,c_2		β-Carotene, fucoxanthin, diatoxanthin, diadinoxanthin, echinenone	3	2	O
Bacillariophyceae	a,c_1,c_2		β-Carotene, α-carotene, fucoxanthin, diatoxanthin, diadinoxanthin, neoxanthin	3	2	O
Xanthophyceae	a,c		β-Carotene, diadinoxanthin, diatoxanthin, heteroxanthin, vaucheriaxanthin ester, neoxanthin, β-cryptoxanthin 5′,6′-monoepoxide and 5,6′,5′,6′-diepoxide	3	2	O

Table 1.1 (*cont.*)

Algal class	Chl.	Phycobilins	Carotenoids	Thyl. band[a]	Memb. CER[b]	Stor. prod.[c]
Eustigmatophyceae	*a*		β-Carotene, violaxanthin, diatoxanthin, heteroxanthin, vaucheriaxanthin ester, neoxanthin, β-cryptoxanthin 5′,6′-monoepoxide and 5,6′,5′,6′-diepoxide	3	2	O
Phaeophyceae	a, c_1, c_2		β-Carotene, fucoxanthin, violaxanthin, zeaxanthin, antheraxanthin, mutatochrome	3	2	O
Rhodophyta	*a,d*	R-Phycocyanin, R-phycoerythrin, C-phycocyanin, C-allophycocyanin, C-phycoerythrin, b-phycoerythrin, B-phycoerythrin	β-Carotene, zeaxanthin, antheraxanthin, β-cryptoxanthin, lutein, neoxanthin	1	0	O
Chlorophyta	*a,b*		β-Carotene, lutein, violaxanthin, zeaxanthin, antheraxanthin, neoxanthin, β-cryptoxanthin, lutein-5,6-epoxide, loroxanthin, pyrenoxanthin, echinenone, canthaxanthin, 3-hydroxyechinenone ester, adonirubin ester, adonixanthin ester, crustaxanthin ester, astaxanthin ester, phoenicopterone	3–6	0	I

Notes

[a] Thylakoids per band.

[b] Number of membranes of chloroplast endoplasmic reticulum.

[c] Place where storage product is found: O, outside chloroplast; I, inside chloroplast; B, between chloroplast envelope and chloroplast endoplasmic reticulum.

Source: Bisalputra (1974); Ragan (1981).

which has ribosomes attached to the outer face of the membrane adjacent to the cytoplasm. In the Euglenophyta and Dinophyta, there is one membrane of chloroplast E.R. (Fig. 1.8(*c*); see also Table 1.1). In the Cryptophyta, Prymnesiophyta, and Heterokontophyta there are two membranes of chloroplast E.R., with the outer membrane of the chloroplast E.R. commonly being continuous with the outer membrane of the nuclear envelope, especially if the chloroplast number is low (see Fig. 1.8(*b*),(*d*)). Between the chloroplast E.R. and chloroplast envelope is a space that contains tubules, ribosomes, and, in the Cryptophyta, reserve product and a nucleomorph. The arrangement of the membranes of chloroplast E.R. might be the remains of an earlier endosymbiotic association (Lee, 1977) (see Figs. 1.32 and 1.34).

The basic structure of the photosynthetic apparatus in a plastid consists of a series of flattened membranous vesicles called **thylakoids** or **discs**, and a surrounding matrix or **stroma**. The thylakoids contain the chlorophylls and are the sites of the photochemical reactions; carbon dioxide fixation occurs in the stroma. The thylakoids can be free from one another or grouped to form **thylakoid bands**. In the Cyanophyta and Rhodophyta (Fig. 1.8(*a*)), the thylakoids are usually free from one another, with **phycobilisomes** (containing the phycobiliproteins) on the surface of the thylakoids. The phycobilisomes on the surface of one thylakoid alternate with those on the surface of an adjacent thylakoid. The phycobilisomes appear as 35-nm granules when phycoerythin predominates, or as discs when phycocyanin predominates. In the more primitive members of the Rhodophyta the thylakoids terminate close to the chloroplast envelope, whereas in advanced members of the Rhodophyta peripheral thylakoids are present, which enclose the rest of the thylakoids. In the Cryptophyta, the chloroplasts contain bands of two thylakoids (Fig. 1.8(*b*)); the phycobiliproteins are dispersed within the thylakoids. In the Euglenophyta, and Heterokontophyta the thylakoids are grouped in bands of three with a girdle or peripheral band running parallel to the chloroplast envelope. In the Dinophyta, Prymnesiophyta, and Eustigmatophyceae, the thylakoids are also in bands of three, but there is no girdle band (Fig. 1.8(*c*),(*d*)). In the Chlorophyta, the thylakoids occur in bands of two to six, with thylakoids running from one band to the next. The above grouping of algal thylakoids into bands occurs under normal growth conditions. Abnormal growth conditions commonly cause lumping of thylakoids and other variations in structure.

A pyrenoid (Fig. 1.8(*b*)) is a differentiated region within the chloroplast that is denser than the surrounding stroma that may or may not be traversed by thylakoids. A pyrenoid is frequently associated with storage products. Pyrenoids occur within every class of algae and within a class are considered to a primitive evolutionary characteristic.

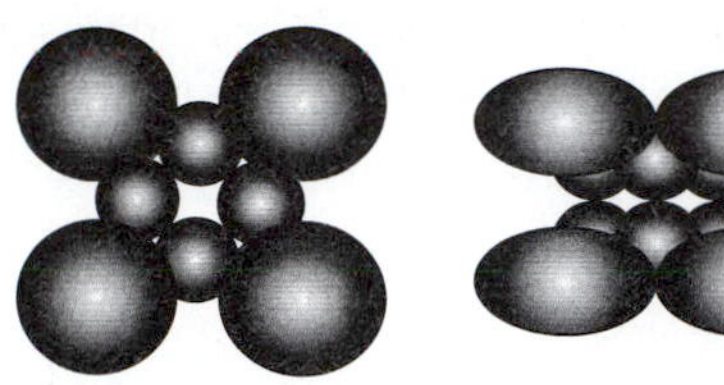

Fig. 1.9 The structure of Form I of ribulose-1,5-bisphosphate carboxylase showing the eight large subunits and the eight small subunits. (Adapted from Baker et al., 1977.)

Pyrenoids contain ribulose-1,5-bisphosphate carboxylase (Rubisco), the enzyme that fixes carbon dioxide (Rawat et al., 1996; Süss et al., 1995). Ribulose-1,5-bisphosphate carboxylase exists in two forms, both of which have essentially the same active site (Haygood, 1996; Kellogg and Juliano, 1997):

1 Form I occurs in some bacteria, the cyanobacteria, in all green plants and non-green plants. Form I is composed of eight large subunits and eight small subunits (Fig. 1.9) and has a high affinity for CO_2 and a low catalytic efficiency (low rate of CO_2 fixation).
2 Form II occurs in some eubacteria and in the dinoflagellates and is composed of two large subunits. Form II has a low affinity for CO_2 and a high catalytic efficiency.

The common ancestor of all ribulose-1,5-bisphosphate carboxylase was probably similar to Form II and was adapted to the anaerobic conditions and high CO_2 concentrations prevailing in the ancient earth (Haygood, 1996). Form I evolved as the earth's atmosphere became oxygenated, and CO_2 concentration declined and with it the need for a greater affinity for CO_2. The greater affinity for CO_2 in Form I, however, came at the price of reduced catalytic efficiency.

Chloroplasts commonly contain small (30–100 nm), spherical lipid droplets between their thylakoids (Fig. 1.8(*c*),(*d*)). These lipid droplets serve as a pool of lipid reserve for the synthesis and growth of lipoprotein membranes within the chloroplast. Many motile algae have groups of tightly packed carotenoid lipid globules that constitute an orange-red **eyespot** or **stigma**. These globules shade the **photoreceptor**, an area of the plasma membrane or chloroplast envelope containing specialized molecules (Melkonian and Robenek, 1980). For example, the photoreceptor in motile cells of the green alga *Chlamydomonas* consists of a **chromophore** (colored substance) linked to a protein that is embedded in the plasma membrane. The chromophore is 11-*cis*-retinol (Fig. 1.10), a rhodopsin that functions as a photoreceptor in animals (Foster et al., 1984). As the alga swims helically through the medium, the photoreceptor is shaded for varying periods of time by the eyespot (normally a group of lipid droplets near the photo-

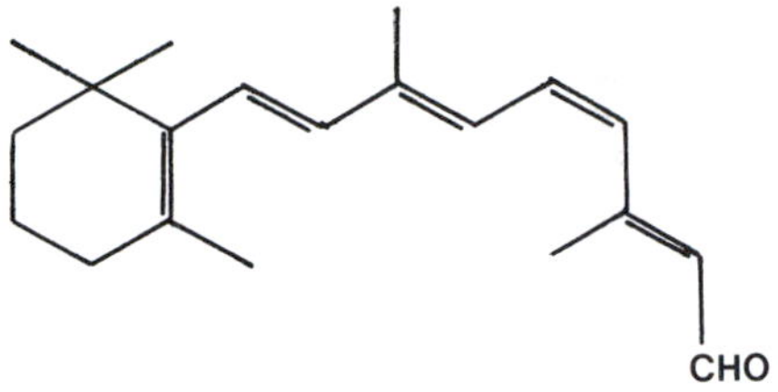

Fig. 1.10 The structure of 11-*cis*-retinol, the photoreceptor in *Chlamydomonas*.

receptor), depending on the orientation of the alga to the light. The shading results in a change in the membrane potential on the plasma membrane by a photoinductive phenomenon involving rhodopsin. The change in membrane potential causes an influx of calcium ions into the cell, which affects the beating of the flagella and the direction of cell movement (Kamiya and Witman, 1984). A motile cell may swim *toward* (**positive phototaxis**) or *away from* (**negative phototaxis**) light. At high light intensities most zoospores are negatively phototactic, whereas at low light intensities they are positively phototactic.

In the Chlorophyta, Cryptophyta and most of the Heterokontophyta, the eyespots occur as lipid droplets in the chloroplast. In the Euglenophyta, Eustigmatophyceae, and Dinophyta, the eyespot occurs as a group of membrane-bounded lipid droplets, free of the chloroplast.

Most chloroplasts contain **prokaryotic DNA** in an area of the chloroplast devoid of 70S ribosomes (Fig. 1.11). The individual DNA microfibrils are circular, are attached to the chloroplast membranes, and lack basic proteins (**histones**). The algae can be divided into two general groups according to the distribution of DNA in the plastids (Coleman, 1985). In the first group, the clumps of DNA (**nucleoids**) are scattered throughout the plastids. This group includes the Cryptophyta, Dinophyta, Prymnesiophyta, Eustigmatophyceae, Rhodophyta, and Chlorophyta. In the second group, the DNA occurs in a ring just within the girdle lamella. This group includes the Chrysophyceae, Bacillariophyceae, Raphidophyceae and Xanthophyceae (with the exception of *Vaucheria* and three genera known to lack girdle lamellae – *Bumilleria*, *Bumilleriopsis*, and *Pseudobumilleriopsis*). The Euglenophyta fit into neither group, showing a variable distribution of chloroplast DNA.

The photosynthetic algae have chlorophyll in their chloroplasts. **Chlorophyll** is composed of a porphyrin-ring system that is very similar to that of hemogloblin but has a magnesium atom instead of an iron atom (Fig. 1.12). The algae have four types of chlorophyll, *a*, *b*, *c* (c_1 and c_2), and *d*. Chlorophyll *a* is the primary photosynthetic pigment (the light receptor in photosystem I of the light reaction) in all photosynthetic algae and ranges from 0.3% to 3.0% of the dry weight. Chlorophyll *a* is insoluble in water and

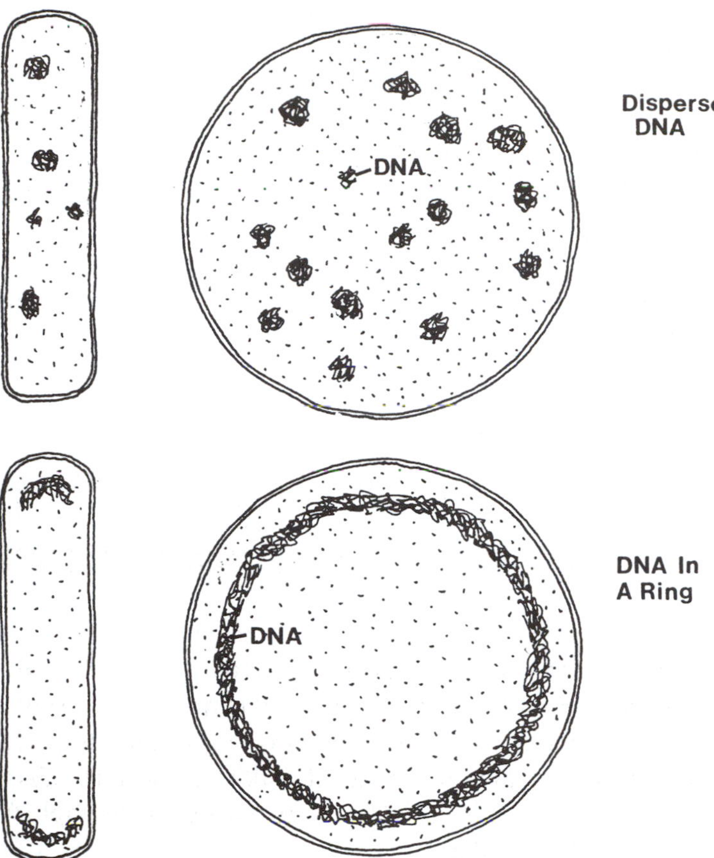

Fig. 1.11 Semi-diagrammatic drawing of the two types of distribution of DNA in algal chloroplasts. Side and face views of the plastids are drawn. (Adapted from Coleman, 1985.)

petroleum ether but soluble in alcohol, diethyl ether, benzene, and acetone. The pigment has two main absorption bands in vitro, one band in the red light region at 663 nm and the other at 430 nm (Fig. 1.13).

Whereas chlorophyll *a* is found in all photosynthetic algae, the other algal chlorophylls have a more limited distribution and function as accessory photosynthetic pigments. Chlorophyll *b* is found in the Euglenophyta and Chlorophyta (Fig. 1.12). Chlorophyll *b* functions photosynthetically as a light-harvesting pigment transferring absorbed light energy to chlorophyll *a*. The ratio of chlorophyll *a* to chlorophyll *b* varies from 2:1 to 3:1. The solubility characteristics of chlorophyll *a* are similar to chlorophyll *b*, and in vitro chlorophyll *b* has two main absorption maxima in acetone or methanol, one at 645 nm and the other at 435 nm (Fig. 1.13).

Chlorophyll *c* (Fig. 1.12) is found in the Dinophyta, Cryptophyta, and most of the Heterokontophyta. Chlorophyll *c* has two spectrally different components: chlorophyll c_1 and c_2. Chlorophyll c_2 is always present, but chlorophyll c_1 is absent in the Dinophyta and Cryptophyta. The ratio of

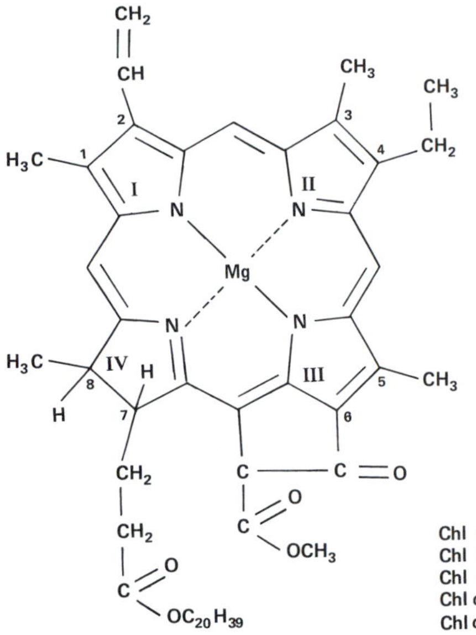

Fig. 1.12 The structure of the chlorophylls. (From Meeks, 1974.)

Chl *a*: as shown in structure
Chl *b*: II-3 = CHO
Chl *d*: I -2 = CHO
Chl c_1: IV-7 = CH=CHCOOH; double bond at IV-7, 8
Chl c_2: IV-7 = CH=CHCOOH; double bond at IV-7, 8
II-4 = $CH{=}CH_2$

chlorophyll *a* to chlorophyll *c* ranges from 1.2:2 to 5.5:1. Chlorophyll *c* probably functions as an accessory pigment to photosystem II. The pigment is soluble in ether, acetone, methanol, and ethyl acetate, but is insoluble in water and petroleum ether. Extracted chlorophyll c_1 has main absorption maxima at 634, 583, and 440 nm in methanol, whereas chlorophyll c_2 has maxima at 635, 586, and 452 nm.

Chlorophyll *d* (Fig. 1.12) is a minor component in the extracts of many Rhodophyta. It does not occur in the simpler red algae and has not been shown to exist in all higher red algae. The photosynthetic function is unknown. It is soluble in ether, acetone, alcohol, and benzene and very slightly soluble in petroleum ether. It has three main absorption bands at 696, 456, and 400 nm (Fig. 1.13).

Carotenoids are yellow, orange, or red pigments that usually occur inside the plastid but may be outside in certain cases. In general, naturally occurring carotenoids can be divided into two classes: (1) oxygen-free hydrocarbons, the **carotenes**; and (2) their oxygenated derivatives, the **xanthophylls.** The most widespread carotene in the algae is β-carotene (Fig. 1.14). There are a large number of different xanthophylls, with the Chlorophyta having xanthophylls that most closely resemble those in higher plants. Fucoxanthin (Fig. 1.14) is the principal xanthophyll in the golden-brown algae (Chrysophyceae, Bacillariophyceae, Prymnesiophyceae, and Phaeophyceae), giving these algae their characteristic color. Like the chlorophylls, the carotenoids are soluble in alcohols, benzene, and acetone but insoluble in water.

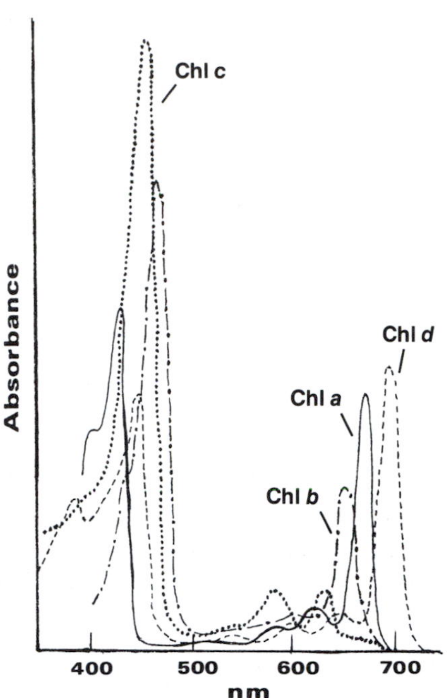

Fig. 1.13 The absorption spectra of chlorophylls *a*, *b*, *c*, and *d*.

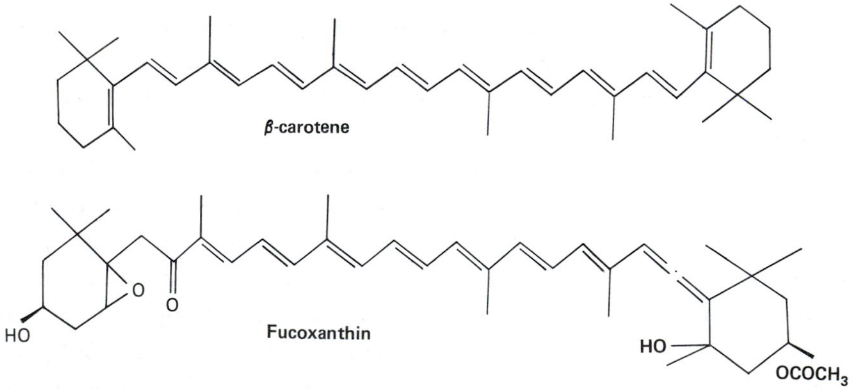

β-carotene

Fucoxanthin

HO

O

O

HO

$OCOCH_3$

Fig. 1.14 The structure of β-carotene and fucoxanthin.

Phycobiliproteins are water-soluble blue or red pigments located *on* (Cyanophyta, Rhodophyta) or *inside* (Cryptophyta) thylakoids of algal chloroplasts (Glazer, 1982). They are described as **chromoproteins** (colored proteins) in which the **prosthetic group** (non-protein part of the molecule) or **chromophore** is a tetrapyrole (bile pigment) known as **phycobilin.** The prosthetic group is tightly bound by covalent linkages to its **apoprotein** (protein part of the molecule) (see Fig. 1.15). Because it is difficult to separate the pigment from the apoprotein, the term **phycobiliprotein** is used. There are two different apoproteins, α and β, which together form the basic unit of the phycobiliproteins. To either α or β are attached the colored

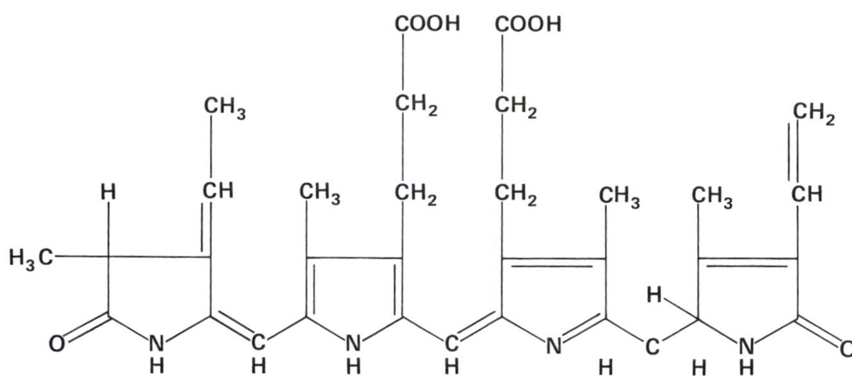

Fig. 1.15 The structure of phycoerythrobilin.

chromophores. The major "blue" chromophore occurring in **phycocyanin** and **allophycocyanin** is **phycocyanobilin**, and the major "red" chromophore occurring in **phycoerythrin** is **phycoerythrobilin** (Fig. 1.15).

The general classification of phycobiliproteins is based on their absorption spectra. There are three types of phycoerythrin: R-phycoerythrin and B-phycoerythrin in the Rhodophyta, and C-phycoerythrin in the Cyanophyta. There are also three types of phycocyanin: R-phycocyanin from the Rhodophyta and C-phycocyanin and allophycocyanin from the Cyanophyta. In addition, in the Cryptophyta there are three spectral types of phycoerythrin and three spectral types of phycocyanin.

The basic subunit of a phycobilisome consists of the apoproteins α and β, each of which is attached to a chromophore (Grossman et al., 1993) (Fig. 1.16). In the core of the phycobilisome, α and β are attached to allophycocyanin. In the outer rods, α and β are attached to phycocyanin or phycoerythrin. The α, β molecules are assembled into hexamers (α_1, β_1) cylindrical in shape. The hexamers that make up the core of the phycobilisome are assembled in pairs, with the hexamers of the rods radiating from the core. The hexamers are joined together by linker polypeptides. The linker polypeptides are basic whereas the hexamers are acidic; this suggests that electrostatic interactions are important in assembling phycobiliproteins. There are high-molecular-weight polypeptides that anchor the phycobilisome to the area of the thylakoid membrane that contains the reaction center and associated chlorophylls.

The pathway of energy transfer (Glazer et al., 1985) is

phycoerythrin ($\lambda_{max}=565$) or phycoerythrocyanin ($\lambda_{max}=568$)	$\rightarrow$ phycocyanin $\rightarrow$ ($\lambda_{max}=620$–638)	allophycocyanin $\rightarrow$ ($\lambda_{max}=650$)	allophycocyanin B ($\lambda_{max}=670$) or high-molecular-weight polypeptide ($\lambda_{max}=665$)	$\rightarrow$ chlorophyll *a*

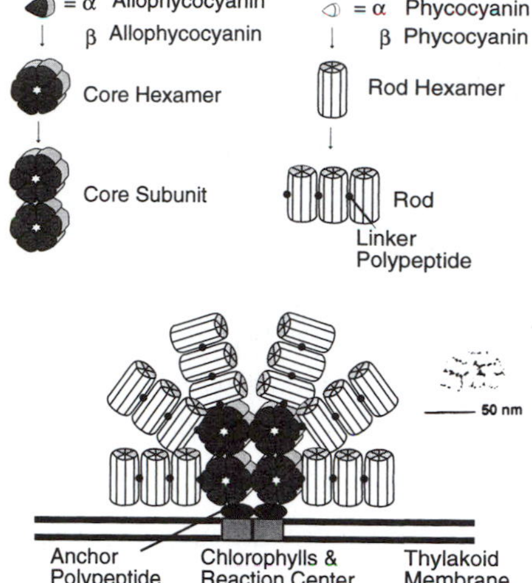

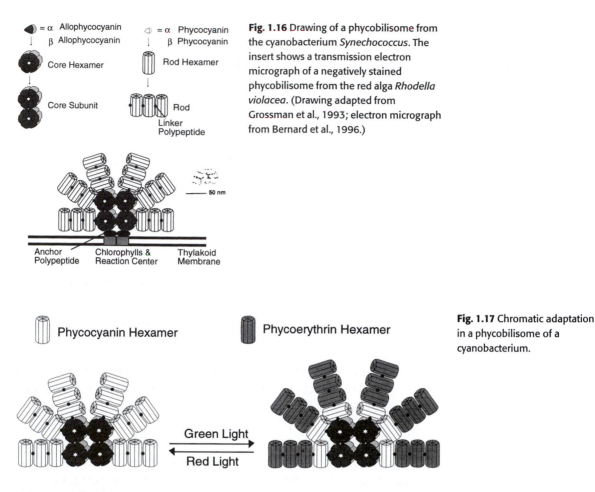

Fig. 1.16 Drawing of a phycobilisome from the cyanobacterium *Synechococcus*. The insert shows a transmission electron micrograph of a negatively stained phycobilisome from the red alga *Rhodella violacea*. (Drawing adapted from Grossman et al., 1993; electron micrograph from Bernard et al., 1996.)

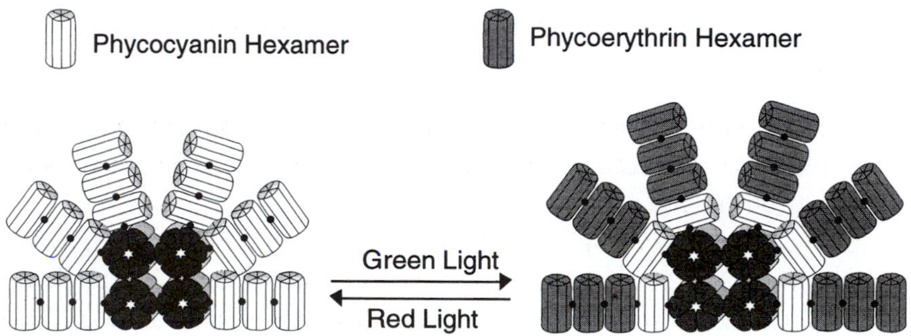

Fig. 1.17 Chromatic adaptation in a phycobilisome of a cyanobacterium.

In intact cells, the overall efficiency of energy transfer from the phycobilisome to chlorophyll *a* in the thylakoids exceeds 90% (Porter et al., 1978).

Chromatic adapters change their pigment components under different light wavelengths (Fig. 1.17). For example, the cyanobacterium *Synechocystis* grown in green light produces phycoerythrin (red in color), phycocyanin (blue) and allophycocyanin (blue-green) in a molar ratio of about 2:2:1; when it is grown in red light, the ratio is about 0.4:2:1. The phycobilisome structure changes appropriately, with the peripheral rods having more phycoerythrin hexamers under green light, and less phycocyanin hexamers. The allophycocyanin core hexamers stay the same.

Depriving cells of nitrogen results in an ordered degradation of phycobilisomes (Fig. 1.18). There is a progressive degradation of hexamer rod and linker polypeptides followed by the core peptides. New phycobilisomes are rapidly synthesized on the addition of nitrogen to the medium. Phycobilisomes are, thus, an important source of internal nitrogen and

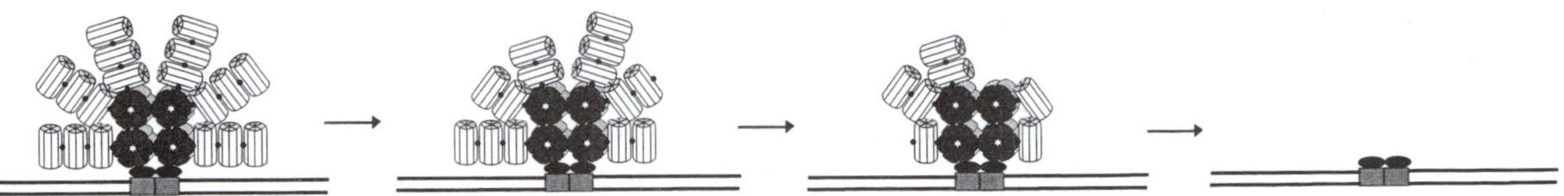

Fig. 1.18 Phycobilisome breakdown under conditions of nitrogen deprivation. (Adapted from Grossman et al., 1993.)

offer the algae that have phycobilisomes (cyanobacteria, cryptophytes and red algae) an important ecological advantage in the open ocean, which is predominantly nitrogen limited (Vergara and Niell, 1993).

The photosynthetically active pigments of algae are gathered in discrete pigment–protein complexes which can be divided functionally into two groups (Grossman et al., 1990):

1 the **photochemical reaction center** containing chlorophyll *a*, where light energy is converted into chemical energy.
2 the **light-harvesting complexes** that serve as antennae to collect and transfer available light energy to the reaction center.

The light-harvesting complexes use different antennae pigment complexes to capture light energy.

1 Green algae and higher plants use chlorophyll *a*/*b* binding proteins that are integral parts of the thylakoid membrane as three membrane-spanning helices.
2 Brown and golden-brown algae (diatoms, chrysophytes, dinoflagellates, brown algae and related groups) use a fucoxanthin chlorophyll *a*/*c* complex that is an integral part of the thylakoid membrane. The ratio of fucoxanthin to chlorophyll in this complex is approximately 2:1 and the characteristic brown or golden-brown color of these algae is due to the high level of fucoxanthin in these cells. Due to chlorophyll *c* and special xanthophylls, these organisms are especially suited to harvest blue and green light, which are the most abundant at increasing ocean depths. This light-harvesting complex also is composed of three membrane-spanning helices and is closely related to the light-harvesting complex in the first group (Caron et al., 1996).
3 Cyanobacteria, cryptophytes and red algae use the phycobilisome as the major light-harvesting complex.

MITOCHONDRIA AND PEROXISOMES

There are two types of mitochondria in algal cells (Leipe et al., 1994). Mitochondria with **flat lamellae cristae** occur in the red algae, green algae, euglenoids and cryptophytes (Fig. 1.19). These algae have either phycobili-

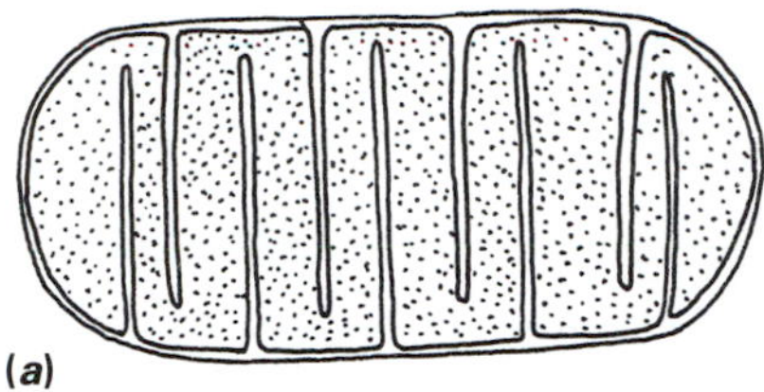

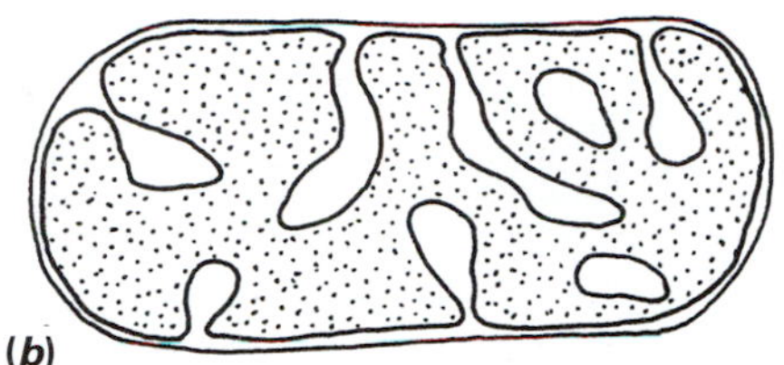

Fig. 1.19 Drawings of the two types of mitochondria that occur in the algae: (*a*) Mitochondrion with flat lamellar cristae and (*b*) Mitochondrion with tubular cristae.

somes or chlorophylls *a* and *b* (Stewart and Mattox, 1984). Mitochondria with **tubular cristae** occur in heterokonts and haptophytes, algae that do not have phycobiliproteins or chlorophylls *a* and *b*.

Glycolate, the major substrate of photorespiration, can be broken down by either **glycolate dehydrogenase** in the mitochondria, or by **glycolate oxidase** in **peroxisomes**, single membrane-bounded bodies in the cytoplasm (for the reactions, see the chapter on Chlorophyta). The distribution of the two enzymes is as follows (Betsche et al., 1992; Iwamoto et al., 1996):

1 Glycolate dehydrogenase occurs in the cyanobacteria, cryptophytes, euglenoids, diatoms and the green algae with the exception of the Charophyceae.
2 Glycolate oxidase occurs in the glaucophytes, red algae, brown algae, the Charophyceae in the green algae and higher plants.

STORAGE PRODUCTS

The storage products that occur in the algae are as follows:

High-molecular-weight compounds

1 α-1,4 Linked glucans
  a **Floridean starch** (Fig. 1.20): This substance occurs in the Rhodophyta and is similar to the amylopectin of higher plants. It stains red-violet with iodine, giving a color similar to that of the stain reaction of animal glycogen. Floridean starch occurs as bowl-shaped grains from 0.5 to 25 μm in size outside of the chloroplast.
  b **Myxophycean starch:** Found in the Cyanophyta, myxophycean starch has a similar structure to glycogen. This reserve product occurs as granules (α-granules), the shape varying between species from rod-shaped granules to 25-nm particles to elongate 31- to 67-nm bodies.
  c **Starch:** In the Chlorophyta, starch is composed of amylose and amylopectin. It occurs inside the chloroplast in the form of starch grains (Fig. 1.8(*e*)). In the Cryptophyta, starch has an unusually high content

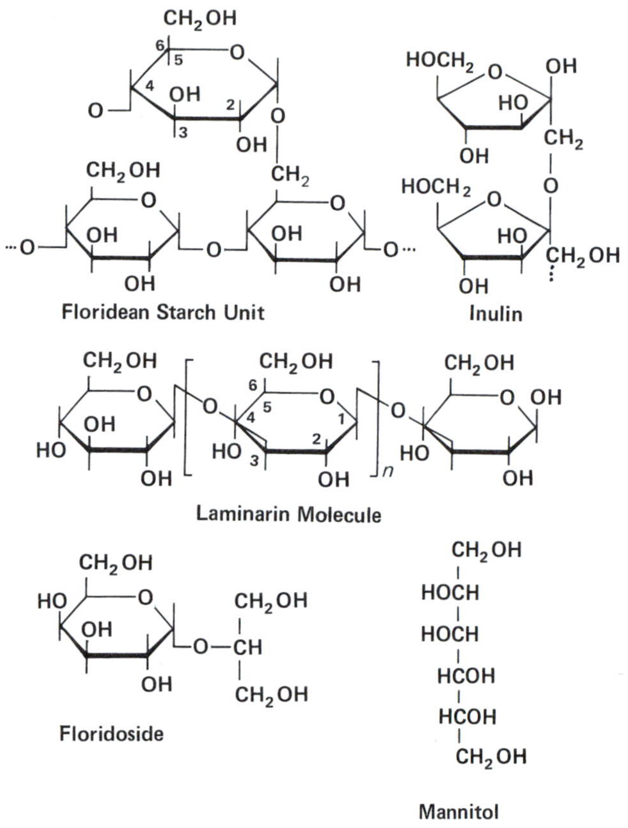

Fig. 1.20 The structure of floridean starch, inulin, laminarin, and floridoside. (After Percival and McDowell, 1967.)

Floridean Starch Unit

Inulin

Laminarin Molecule

Floridoside

Mannitol

of amylose (45%) and is found as grains between the chloroplast envelope and the chloroplast E.R. (Fig. 1.8(*b*)). In the Dinophyta also, starch occurs in the cytoplasm outside of the chloroplast, but its structure is not known.

2 β-1,3 Linked glucans

a **Laminarin** (Fig. 1.20): In the Phaeophyceae, laminarin consists of a related group of predominantly β-1,3 linked glucans containing 16 to 31 residues. Variation in the molecule is introduced by the number of $1 \rightarrow 6$ linkages, the degree of branching, and the occurrence of a terminal mannitol molecule. The presence of a high proportion of C-6 interresidue linkages and of branch points seems to determine the solubility of the polysaccharide in cold water: The greater the number of linkages, the higher the solubility. Laminarin occurs as an oil-like liquid outside of the chloroplasts, commonly in a vesicle surrounding the pyrenoid.

b **Chrysolaminarin (leucosin):** In the Chrysophyceae, Prymnesiophyta, and Bacillariophyceae, chrysolaminarin consists of β-1,3 linked D-glucose residues with two $1 \rightarrow 6$ glycosidic bonds per molecule. Chrysolaminarin occurs in vesicles outside of the chloroplast and has more glucose residues per molecule than laminarin.

c **Paramylon:** In the Euglenophyta, Xanthophyceae, and Prymnesiophyta (*Pavlova mesolychnon*), paramylon occurs as water-soluble, single-membrane-bounded inclusions of various shapes and dimensions outside of the chloroplast. Paramylon consists solely of β-1,3 linked glucose residues, and the molecule is about as large as that of chrysolaminarin.

3 **Frutosans:** *Acetabularia* (Chlorophyta) has an inulin-like storage product consisting of a series of 1,2 linked fructose units terminated by a glucose end group (Fig. 1.20). Fructosans also occur in storage products of the Cladophorales (Chlorophyta).

Low-molecular-weight compounds

1 **Sugars:** Chlorophyta and Euglenophyta form sucrose as a reserve product; trehalose is found in the Cyanophyta and at low levels in the Rhodophyta.
2 **Glycosides:** The glycerol glycosides, floridoside (Fig. 1.20) and isofloridoside, are widely distributed in the Rhodophyta.
3 **Polyols:** Mannitol is universally present in the Phaeophyceae. It is also present in lower green algae, where it replaces sucrose as a photosynthetic product. Free glycerol occurs widely in the algae and is an important photosynthetic product in several zooxanthellae (endosymbiotic algae in animals) and in some marine Volvocales, especially *Dunaliella.*

NUCLEUS

As in other eukaryotic plants, the nucleus in the algae is surrounded by a double-membrane nuclear envelope and contains DNA. There are, however, two basic types of nuclei in the eukaryotic algae. Evans (1974) has stated, "As in dinoflagellates, the mitotic process in euglenoids differs considerably from that encountered in higher plants." Thus it is possible to divide the algae into two types in regard to nuclear division (Lee, 1977): (1) that occurring in the Dinophyta and Euglenophyta, and (2) that occurring in the rest of the eukaryotic algae. The Dinophyta and Euglenophyta have the following "mesokaryotic" nuclear characteristics: (1) chromosomes condensed throughout the mitotic cycle; (2) a persistent nucleolus (**endosome**), which does not disperse during prophase and which divides by pinching in two; (3) large nuclei; (4) chromosomes that are attached to the nuclear membrane and not to spindle microtubules inside the nucleus; (5) an intact nuclear membrane during the whole mitotic cycle. In the

Dinophyta, there are few basic proteins (histones) associated with the DNA.

In the remainder of the eukaryotic algae, (1) the chromosomes condense at prophase and disperse during telophase; (2) the nucleolus disperses during prophase and condenses during telophase; (3) the nuclei are small; (4) the chromosomes are attached to spindle microtubules; and (5) the nuclear membrane may disperse or remain whole during nuclear division.

CONTRACTILE VACUOLES

The ability of algal cells to adjust to changes in the salinity of the medium is an important aspect of the physiology of these cells. In cells with walls, this osmoregulation is accomplished with the aid of turgor pressure, whereas in naked cells it is accomplished by means of contractile vacuoles and/or regulation of the solutes present in the cells. In the latter case, cells increase the internal concentration of osmotically active molecules and ions when the concentration of dissolved solutes increases in the external medium. Likewise, the internal concentration of such molecules decreases when the concentration of dissolved salts in the external medium decreases.

Most algal flagellates have two contractile vacuoles in the anterior end of a cell (Fig. 1.1). A contractile vacuole will fill with an aqueous solution (**diastole**) and then expel the solution outside of the cell and contract (**systole**). The contractile vacuole rhythmically repeats this procedure. If there are two contractile vacuoles, they usually fill and empty alternately. Contractile vacuoles occur more frequently in freshwater than marine algae, a phenomenon that gives credence to the theory that the contractile vacuoles maintain a water balance in the cells. The algal cells in freshwater have a higher concentration of dissolved substances in their protoplasm than in the surrounding medium so that there is a net increase of water in the cells. The contractile vacuoles act to expel this excess water. An alternate theory on the function of the contractile vacuoles is that they remove waste products from the cells. The Dinophyta have a structure similar to a contractile vacuole, called a pusule, which may have a similar function but is more complex.

The contractile vacuoles of the Cryptophyta are characteristic of the algae (Fig. 1.21). In the Cryptophyta, the contractile vacuole occurs in a fixed anterior position next to the flagellar depression (Patterson and Hausmann, 1981). At the beginning of the filling phase (**diastole**), there is no distinct contractile vacuole, only a region filled with small (*ca.* 0.5-μm diameter) contributory vacuoles. These vacuoles fuse to form a large irregular vacuole which subsequently rounds up. The contributory vacuoles destined to form the next contractile vacuole now appear around the rounded

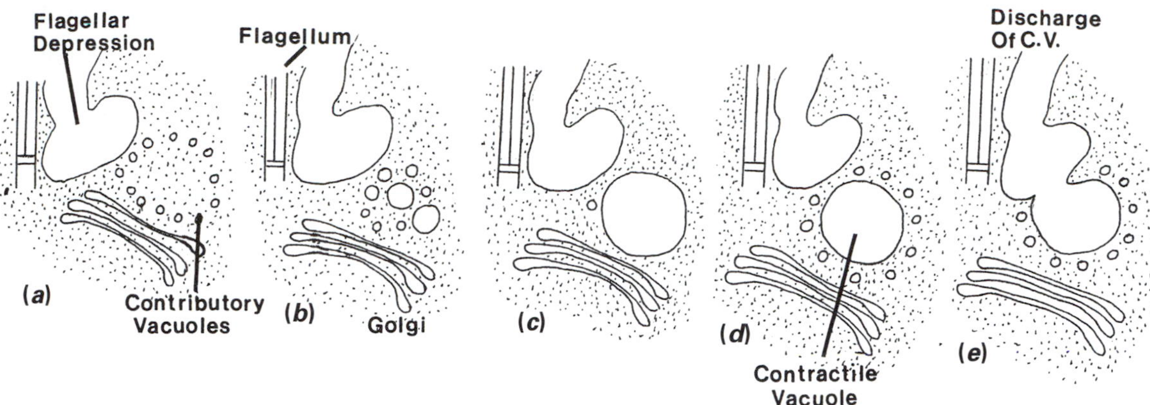

Fig. 1.21 Semi-diagrammatic illustration of the behavior of the contractile vacuole (c. v.) complex during filling and discharge in the Cryptophyta. (Adapted from Patterson and Hausmann, 1981.)

contractile vacuole. The contractile vacuole fuses with the plasma membrane of the flagellar pocket and discharges its contents outside the cell. The area of the plasma membrane that fuses with the contractile vacuole does not have a periplast (specialized plates within the plasma membrane). This area is, instead, bounded by microtubules. The membrane of the contractile vacuole is recovered by the cell as small vesicles with an electron-dense coat, and the membrane components are reutilized by the cell. These vesicles plus the contractile vacuole occur in the **spongiome** or area around the contractile vacuole. In freshwater algae the contractile vacuole cycle lasts for 4 to 16 seconds, whereas in marine species the cycle can last for up to 40 seconds.

Algal flagellates use a combination of contractile vacuoles and osmoregulation to control the water content of their cells. In the chrysophyte *Poterioochromonas malhamensis (Ochromonas malhamensis)*, the internal level of isofloridoside (*O*-α-D-galactopyranosyl-1$\rightarrow$1-glycerol) is proportional to the external osmotic value as long as the external solute concentration exceeds 75 mOsm (Wessel and Robinson, 1979). Below this external solute concentration, the influx of water into the cytoplasm is counterbalanced by means of the contractile vacuoles (Kauss, 1974).

RIBOSOMES

Ribosomes fall into two general classes, based on their sedimentation coefficients in an ultracentrifuge (expressed in Svedberg units or S): (1) the smaller "70S" prokaryotic type of ribosome found in bacteria, Cyanophyta, chloroplasts, and mitochondria; and (2) the "80S" eukaryotic type of ribosome found in the cytoplasm of eukaryotic cells outside the chloroplasts and mitochondria.

Table 1.2 *Types of nutrition found in the algae*

Type of nutrition	Principle source of energy for growth	Principal source of carbon for growth
Autotrophic		
Photoautotrophic	Light	Carbon dioxide
Chemoautotrophic	Oxidation of organic compounds	Carbon dioxide
Heterotrophic		
Photoheterotrophic	Light	Organic compounds
Chemoheterotrophic	Oxidation of organic compounds	Organic compounds

Nutrition

Algae can be either **autotrophic** (**lithotrophic** or **holophytic**) or **heterotrophic** (**organotrophic**) (Table 1.2). If they are **autotrophic**, they use inorganic compounds as a source of carbon. Autotrophs can be **photoautotrophic** (**photolithotrophic**), using light as a source of energy, or **chemoautotrophic** (**chemolithotrophic**), oxidizing inorganic compounds for energy. If they are **heterotrophic**, the algae use organic compounds for growth. Heterotrophs can be **photoheterotrophs** (**photoorganotrophs**), using light as a source of energy, or **chemoheterotrophs** (**chemoorganotrophs**), oxidizing organic compounds for energy. Heterotrophic algae may be **phagocytotic** (**holozoic**), absorbing food particles whole into food vesicles for digestion, or they may be **osmotrophic**, absorbing nutrients in a soluble form through the plasma membrane. If the algae live heterotrophically on dead material, they are **saprophytic**; if they live off a live host, they are **parasitic**. Some algae, particularly the flagellates, are **auxotrophic**, requiring a small amount of an organic compound, but not as an energy source. These algae usually require a vitamin. Some photosynthetic algae are **mixotrophic** (**facultatively heterotrophic**), capable of also using organic compounds supplied in the medium.

Rhythms

Circadian (literally, "about a day") or **diel rhythms** are oscillations in the physiology, morphology, or activity of an organism occurring in approximately 24-hour periods under natural illumination; they continue to be manifested under constant conditions of light and temperature (Sweeney, 1979). Light "times" the circadian rhythms. An exposure to light during the "day" part of the cycle has no effect; one in the early "night" causes the rhythm to jump ahead.

In the marine environment, rhythms entrained by the tides are not circadian, although they resemble them closely, but the phase is not changed by light pulses. **Tidal rhythms** show periods of slightly more than 24 hours (usually 24.8 hours) under natural illumination, so they are synchronized with the tidal cycles rather than the solar environmental cycles.

Endogenous **circannual rhythms** ("about a year") are adjusted to the actual time of the year by means of the annual course of day lengths and are considered to be a mechanism for anticipating a favorable season, be it for growth or reproduction (Lüning and tom Dieck, 1989; Molenaar et al., 1996). Kain (1989) has called these algae **season anticipators** and compared them to **season responders** whose growth and/or reproduction are controlled by the environmental conditions existing at the time.

Range of structure in the algae

MOTILE HABIT

The motile unicell is commonly a more or less spherical, oblong, or pear-shaped body, approximately circular in cross-section. The chloroplasts, when present, occupy the posterior region or lie along the sides, whereas the nucleus is frequently near the middle of the cell. A cell wall may surround the plasma membrane, or there may be nothing outside of the plasma membrane. Some flagellates have a periplast (Cryptophyta), an amphiesma (Dinophyta), or a pellicle (Euglenophyta) of strengthening material inside the plasma membrane. Some flagellates settle on a substrate and exhibit creeping movement with the protrusion of blunt **pseudopodia**. In some cases, these phases have long delicate cytoplasmic protrusions called **rhizopodia**.

Evolution from the motile unicell has resulted in the formation of the motile colony with varying numbers of unicells aggregated together, often within a mucilaginous envelope.

PALMELLOID AND DENDROID HABIT

In this type of organism, motility has been lost in the vegetative phase and occurs only during reproduction. In some genera, the non-motile vegetative phase is short-lasting, as during reproduction in *Chlamydomonas* (Chlorophyta) (see Fig. 5.42) when the cells settle down and divide. In a large number of algae this palmelloid habit is permanent, with the reproductive cells alone being motile (e.g., *Phaeocystis*, Prymnesiophyta). Dendroid colonies are a variant in which mucilage is produced locally, generally at the base of the cell, resulting in cells with a basal mucilage stalk (e.g., *Dinobryon*, Chrysophyta) (see Figs. 9.6 and 9.12).

COCCOID HABIT

Many motile unicells come to rest and withdraw their flagella before division of the protoplast to form a new generation of motile unicells. A prolongation of the non-motile phase, with a shortening of the swarming period, led to the evolution of a non-motile unicell that forms swarmers only at times of reproduction. This is known as the coccoid habit. A further step in this evolutionary line resulted in a complete loss of motility, even in the reproductive cells.

FILAMENTOUS HABIT

Here cell division occurs in one plane, with the products of cell division remaining attached to each other to form a filament. Filaments can be **uniseriate** (composed of a single row of cells) or **multiseriate** (composed of more than one row of cells). A modification occurs in which there is cell division in two planes to form a sheet of cells. Another variation of this type occurs when most cell divisions are in one or two planes and a few cell divisions are in the other one or two planes. The result is a true **parenchymatous thallus**. A **pseudoparenchymatous thallus** occurs when filaments are compacted together to give a structure that appears to be parenchymatous.

HETEROTRICHOUS HABIT

This is the most highly evolved type of habit, with the thallus consisting of two different parts: (1) a prostrate creeping system anchoring the thallus to the substrate, and (2) a projecting or erect system composed of usually branched filaments.

SIPHONACEOUS HABIT

Here enlargement of the plant body occurs with multiplication of organelles but without septation. This **siphonaceous** or **coenocytic** structure is multinucleate and has a large number of chloroplasts.

Reproductive cells

In the algae there are some general terms used in referring to reproductive cells. **Spores** are cells that germinate without fusing, to form new individuals. **Aplanospores** are non-motile spores. **Hypnospores** are aplanospores with a greatly thickened cell wall. **Gametes** are cells that fuse to form a **zygote**. Gametes can be motile (**planogametes**) or non-motile (**aplanogametes**). A **swarmer** is a general term for any motile cell. **Akinetes** are vegetative cells that have developed into a spore-like stage with very thick walls and abundant food reserves. Unlike aplanospores, akinetes always have additional wall layers around the protoplast fused with the parent wall. Akinetes are resistant to unfavorable environmental conditions.

ASEXUAL OR VEGETATIVE REPRODUCTION

In this type of reproduction there is no fusion of gametes to form a zygote. There are two basic types, fragmentation and zoospores. In **fragmentation**, a thallus breaks up, and each part grows to form a new thallus. **Zoospores** are naked swarmers than can be formed in ordinary vegetative cells or in specialized cells called **sporangia**. Zoospores are formed by **zoosporogenesis**. Many zoospores have eyespots and/or photoreceptors and show a phototactic response.

SEXUAL REPRODUCTION

Here there is a fusion of gametes to yield a zygote. In **homothallic** or **monoecious** species, gametes of different mating types are formed on the same plant, whereas in **heterothallic** or **dioecious** species, the gametes of different strains are formed on different plants. There are several types of sexual reproduction, depending on the structure and function of the gametes. Gametes are formed by **gametogenesis**.

In **isogamy**, the fusion of morphologically and physiologically similar gametes occurs. Because the two gametes that fuse to form a zygote look and behave the same way, it is not possible to call them male and female. The gametes are therefore referred to as plus (+) and minus (−). In **anisogamy**, the motile gamates are structurally and/or morphologically different. The larger gamete is the female, and the smaller is the male. The female gamete is usually more sluggish and is not motile for as long as the male. In **oogamy**, there is the fusion of a large non-motile egg or ovum with a smaller motile sperm (except in the Rhodophyta, where the spermatia are non-motile). The eggs are formed within an **oogonium**, and the sperm within an **antheridium**. The egg usually has one or more chloroplasts and many food reserves, whereas the sperm is commonly colorless with a reduced chloroplast and few, if any, food reserves. In some algae, **chemotaxis** occurs with the mature egg releasing a chemical substance that attracts the sperm to it. Oogamy is regarded as the most advanced type of sexual reproduction, and isogamy is the most primitive type.

After fertilization some zygotes will accumulate food reserves and a yellowish-red oil, and form a thick wall. Then called **zygospores** or **oospores**, they are able to withstand prolonged desiccation, germinating under favorable conditions. In some cases, gametes will not fuse to form a zygote and are able to germinate **parthenogenetically** to form a new plant like the parent.

Life histories

Five basic types of life history occur in the algae:

1 Predominantly diploid life history, with meiosis occurring before the formation of gametes. Thus the gametes are the only haploid part of the life cycle and fuse to form the diploid zygote (*Melosira*, Fig. 13.35; *Fucus*, Fig. 17.38).
2 Predominantly haploid life history, with meiosis occurring when the zygote germinates. Thus the zygote is the only diploid part of the life history (*Oedogonium*, Fig. 5.66; *Chara*, Fig. 5.23).
3 **Isomorphic** (**homologous**) alternation of generations, consisting of the alternation of haploid (**gametophytic**) plants bearing gametes with structurally identical diploid (**sporophytic**) plants bearing spores (*Ectocarpus*, Fig. 17.5; *Dictyota*, Fig. 17.35).
4 **Heteromorphic** (**antithetic**) alternation of generations, consisting of the alternation of small haploid plants bearing gametes with large diploid plants bearing spores (*Desmarestia*, Fig. 17.11; *Laminaria*, Fig. 17.19), or of large haploid plants alternating with smaller diploid plants (*Cutleria*, Fig. 17.13).
5 **Triphasic** life cycle in the red algae, consisting of a haploid gametophyte, a diploid carposporophyte, and a diploid tetrasporophyte (Fig. 4.7).

Algae and their environments

Some algae are predominantly **terrestrial**, occurring in the soil (e.g., many of the Chlorococcales, Chlorophyta), but generally most algae live in bodies of water from as small as a puddle to as large as an ocean. Within an aquatic environment an alga grows as either benthos, periphyton, or phytoplankton. **Benthos** are those organisms that grow on the bottom of a body of water. If the alga is attached to the surface of a rock, it is **lithophytic**. If the alga bores into, and lives inside, a rock (usually a limestone), it is **endolithic**. Algae living on the surface of mud or sand are **epipelic. Periphyton** are organisms attached to submerged vegetation. The periphyton is **epiphytic** if attached to the surface of an aquatic plant and **endophytic** if living inside the other plant. If an alga is attached to the surface of an animal, it is **epizoic**, and if living inside of an animal, it is **endozoic. Phytoplankton** (from the Greek *planktos*, meaning "to wander") are those plants that float aimlessly or swim too feebly to maintain a constant position against a water current. If the phytoplankton is large enough to be caught in a fine plankton net, it is called **net** or **macroplankton.** If it is too small to be caught in a net (smaller than 200 μm), it is called **microplankton.** The microplankton are from 20 to 200 μm, the **nannoplankton** from 10 to 20 μm, the **ultraplankton** from 2 to 10 μm, and the **picoplankton** from 0.2 to 2 μm. There is some looseness in the utilization of the above terms by different phycologists.

Cells in the picoplankton size range are adapted to planktonic life in that they sink extremely slowly: A spherical cell 10 μm in diameter sinks at a rate of 25 cm per day whereas a cell 1 μm in diameter sinks at a rate of only 0.25 cm per day. These rates are insignificant because of the turbulent motion of the water. An additional advantage is that the reduction in cell size increases the ability of the cell to take up nutrients. The reason is that the surface: volume ratio increases greatly as the cell volume becomes smaller (Fogg, 1986a). In the sea, picophytoplankton make up most of the phytoplankton. In seawater, there are commonly between 10^3 and 10^5 cells of picophytoplankton per milliliter, whereas in freshwater the number of cells is an order of magnitude higher.

Vertically, bodies of water can be divided into a **euphotic** (**photic**) and an **aphotic** zone (see Fig. 1.25). In the euphotic zone, there is sufficient light penetration that photosynthesis is greater than respiration when measured over a 24-hour period. The **compensation depth** (where photosynthesis equals respiration over a 24-hour period) divides the euphotic from the aphotic zone.

Algae show different tolerances to temperature. **Thermophiles** are adapted to live at elevated temperatures. Thermophilic cyanobacteria are adapted to live in hot springs up to 75 °C. **Psychrophilic** algae are adapted to cold-water environments with a temperature optima less than 15 °C. Marine diatoms in the Arctic ocean are psychrophiles, living in a stable environment that is always cold. **Psychrotrophs** are tolerant of cold temperatures but have a temperature optimum above 15 °C. Cyanobacteria that comprise the major component of the freshwater flora in the late summer in the Arctic and Antarctic are psychrotrophs. They are able to survive the cold winter conditions, but only grow in the warm summer environments (Tang et al., 1997).

Algae also show different tolerances to pH. **Acidophilic algae** are able to live in water with a pH as low as 2.0. These algae include the red alga *Cyanidium caldarium* and the green algae *Dunaliella acidophila* and species of *Chlamydomonas*. Acidophilic algae have one or more of the following mechanisms (Tatswzawa et al., 1996):

1 A cell-surface barrier that is extremely impermeable to protons (H+), thereby helping to keep the protoplasm at a more neutral pH.
2 A plasma membrane that has a higher concentration of saturated fatty acids, resulting in a decrease in the fluidity of the plasma membrane.
3 Accumulation of glycerol derivatives in the protoplasm in order to resist the osmotic imbalance caused by high concentrations of H_2SO_4.

Alkaliphiles are algae that grow in water with a high pH. The cyanobacterium *Spirulina platensis* is an obligate alkaliphile that thrives in extreme alkaline habitats, growing optimally at pH 9.0 to 10.0. Even in media at pH 11.5, growth rates are 80% of optimum. The pH of the protoplasm is less than the pH of the environment. The pH of the protoplasm is 8.0 when the cells are in a medium at pH 10.0. Alkaliphiles require sodium in order to maintain a low protoplasmic pH. Deprivation of sodium results in a rapid lysis of cells (Schlesinger et al., 1996).

TERRESTRIAL ENVIRONMENT

Some of the terrestrial algae are freshwater algae that live in the water held in moist soils, and thus are essentially the same as the algae found in freshwaters. Other terrestrial algae, mostly cyanobacteria and green algae (Bell, 1993), are adapted to rock (lithic environments). These lithic algae can be classified into four broad groups (Friedmann and Galun, 1974): (1) **epilithic**, or living on exposed rocks; (2) **chasmolithic**, or living in rock fissures; (3) **cryptoendolithic**, or living inside the rock; and (4) **sublithic**, or living on the undersurface of translucent stones embedded in soil.

Algae have the capability to live and grow even in desert environments. The presence of algae and lichens beneath the surface of porous rocks has been reported for the ice-free deserts of Victoria Land (in Antarctica) and the hot deserts of Israel, northern Mexico, and the southwestern United States (Friedmann, 1982; Bell et al., 1986). The rocks of cold deserts are dominated by chasmoendoliths living in rock cracks and fissures in coastal locations and by cryptoendoliths living in rocks in inland locations (Tschermak-Woess and Friedmann, 1984). In the cold deserts, a number of

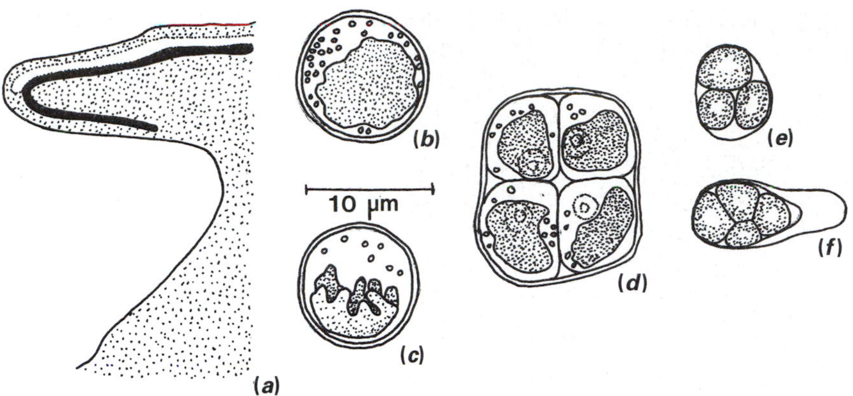

Fig. 1.22 (*a*) Diagrammatic cross-section of an overhanging sandstone rock colonized by cryptoendolithic microorganisms in the Antarctic desert. The upper rock face is colonized by lichens and *Hemichloris antarctica* (white area); the lower rock face is colonized only by *Hemichloris antarctica* (black area). (*b*)–(*d*) *Hemichloris antarctica*. Cell showing the chloroplast in top (*b*) and side (*c*) views. (*d*) Four autospores. (e)–(*f*) *Chroococcidiopsis kashaii*. (*a*)–(*d*) after Tschermak-Woess and Friedmann, 1984.)

different algae occur, such as the green alga *Hemichloris antarctica* (Fig. 1.22(*b*)–(*d*)) and cyanobacteria genera such as *Chroococcidiopsis* (Fig. 1.22(*e*),(*f*)) and *Gloeocapsa*. In contrast, the hot deserts are usually colonized by coccoid cyanobacteria in the genus *Chroococcidiopsis*. The endolithic algae occur beneath the surface of light-coloured rocks in a discrete band that follows the surface contours of the rock (Fig. 1.22(*a*)) (Bell et al., 1986). Examination of the fractured rocks shows the free-living algal colonies attached to the rock crystals. The extremes of temperature, scouring winds, and moisture stress, which make the rock surface uninhabitable, are mitigated within the rock (McKay and Friedmann, 1985).

Cryptogamic crusts (**microbiotic crusts**) cover extensive portions of the arid and semi-arid regions of the world (Johansen, 1993). These crusts consist of water-stable surface soil aggregates held together by algae, fungi, lichens and mosses. Cryptogamic crusts are dominated by cyanobacteria, although green algae, xanthophytes and diatoms occur in these crusts. The cryptogamic crusts are more common in alkaline soils, probably because cyanobacteria do not grow well in acidic conditions. The lack of vegetation makes desert soil easily erodible and the cryptogamic crusts are important in stabilizing these soils. Destruction of cryptogamic crusts by livestock and vehicular traffic may contribute to desertification of semi-arid areas.

FRESHWATER ENVIRONMENT

In the freshwater environment there are two basic type of communities: lakes and ponds, or **lentic** communities; and rivers and streams, or **lotic** communities. Lakes and ponds are divided into different zones; (1) The **supralittoral** zone is above the edge of standing water and is exposed to wave action and splash during windy periods. As a result of wave action and the subsequent abrasive effect of the sand and the drying of the zone, algae

Fig. 1.23 Scanning electron micrograph of periphyton. The dominant taxon is the diatom *Synedra ulna*. Bar = 100 μm. (From Steinman and McIntire, 1986.)

are sparse here. (2) The **littoral** zone extends from the water's edge to a depth of about 6 m. The outer edge of this zone is determined by the maximum depth of the rooted vegetation. This is the most productive area of the lake in terms of biomass with a large amount of periphytic algae present, particularly diatoms and desmids. (3) The **sublittoral** zone extends down to the compensation depth. (4) The **profundal** zone occurs beneath the compensation depth and contains resting spores of photosynthetic algae in addition to colorless heterotrophic algal cells.

Lakes can be classified by the amount of life they can support. This is expressed as **productivity** (change in biomass over a period of time) or as **standing crop** (amount of biomass present at a specific time). **Oligotrophic** lakes are relatively deep lakes characterized by low productivity. Usually there is a relatively narrow littoral and sublittoral zone but an extensive profundal area not subject to severe oxygen depletion. There are a large number of algal species present, mostly desmids and diatoms, but very few cells of each species; thus the total standing crop is low. **Eutrophic** lakes are shallower older lakes with a highly productive littoral and sublittoral region but with a much reduced profundal region. Here there are few algal species present, but there is a large representation for each species so that the total standing crop of algae is high. The algae that are present are usually cyanobacteria and green algae.

The influence of the host plant in determining the epiphytes (Fig. 1.23)

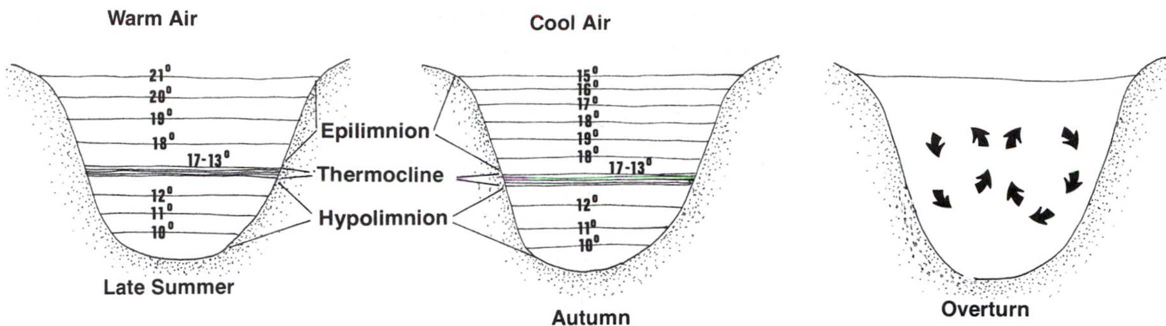

Fig. 1.24 The development of thermal stratification in a lake during the summer and its destruction by the autumn overturn. The temperature is expressed in degrees Celsius.

attached to it is greatest in infertile lakes and is of less importance as the water becomes more fertile. In infertile water, the availability to rooted aquatic macrophytes of nutrients in the sediments and the subsequent transfer of these nutrients by macrophytes to periphyton are important (Eminson and Moss, 1980). When there are abundant nutrients in the water, the transfer of nutrients from the host plant becomes insignificant and the number and type of epiphytes will reflect the conditions of the water. The periphyton is disadvantageous to its host because the periphyton absorbs much of the light that would otherwise reach the plant surface, an important factor in freshwaters that are less than clear. Some aquatic plants and filamentous algae have evolved mechanisms (e.g., slimy surfaces) that discourage periphyton attachment. Other macrophytes have a high rate of production of new leaves, with the periphyton being lost as the older leaves are shed from the macrophyte.

Thermal stratification of lakes occurs when the surface waters are warmed by radiation from the sun (Fig. 1.24). The surface water is then warmer and lighter than the colder and denser water beneath it. In lakes the upper warmer water is called the **epilimnion**, and the bottom denser water is the **hypolimnion**. These two layers are separated by the **thermocline**, a layer of water that has a sharp temperature drop within a few meters' depth. In a typical temperate lake, the water temperature may drop from 21 °C to 10 °C over a 3-m increase in depth (from 10 to 13 m). A thermally stratified lake is usually a lake low in productivity because the euphotic zone is in the epilimnion, and photosynthesis by plants soon uses up all of the available nutrients. If any of the organisms in the epilimnion die, they fall into the hypolimnion and decay. The nutrients from the decay are then locked in the hypolimnion and not available for photosynthesis because there is no exchange of water between the two layers. An exchange of water will occur only when there is an **overturn**. During an overturn the surface water becomes cooler and heavier than the water below it, and sinks, causing the end of the thermal stratification. Nutrient-rich water is then brought to the surface, and, if temperature and light are favorable, a large growth or bloom of phytoplankton occurs. In temperature lakes there are usually two overturns a year, once in the fall and once in the spring. In the fall, the surface water is cooled by the cold air, and the cool surface water sinks beneath the lower warmer water. In the spring, the melting ice leaves the water at 4 °C (and therefore of maximum density) on the surface, which again sinks, causing an overturn.

Most of the algae in streams and rivers (lotic environment) are those that are able to attach themselves to a stable bottom substrate. There is little phytoplankton in swiftly moving streams because the phytoplankton is quickly washed downstream, although in larger slowly moving rivers there can be a considerable amount of phytoplankton. The algae that are common in smaller streams and rivers are attached diatoms, cyanobacteria, and green algae (particularly *Stigeoclonium* and *Cladophora*).

MARINE ENVIRONMENT

In the ocean, there are two basic marine environments (Fig. 1.25), which are divided as follows:

1 **Pelagic** environment, consisting of the ocean water itself.
    (a) **Neritic** province, extending from the high tide mark to a depth of 200 m, which is usually also the outer limit of the continental shelf.
    (b) **Oceanic** province, consisting of those waters with a depth greater than 200 m.
        (i) **Euphotic** or **epipelagic** zone, from the surface of the water to a depth of 200 m. Photosynthesis occurs in the upper portion of this zone.
        (ii) **Aphotic** zone, at depths greater than 200 m. There is insufficient light at this depth to support photosynthesis.
2 **Benthic** environment, consisting of the ocean bottom.
    (a) **Littoral** province, extending from high tide to low tide.
    (b) **Deep-sea** province, including all the ocean bottom beneath the low-tide mark.
        (i) **Sublittoral** zone, from low tide to a depth of 200 m (approximately the end of the continental shelf). This is where the large red and brown seaweeds are found. The depth at which the seaweeds are found will vary according to the turbidity of the water and the resulting penetration of light into the water column. The algal population living at the greatest depth recorded consists of crustose coralline red algae living at a depth of 268 m in ocean waters off the coast of Belize (Littler et al., 1985).

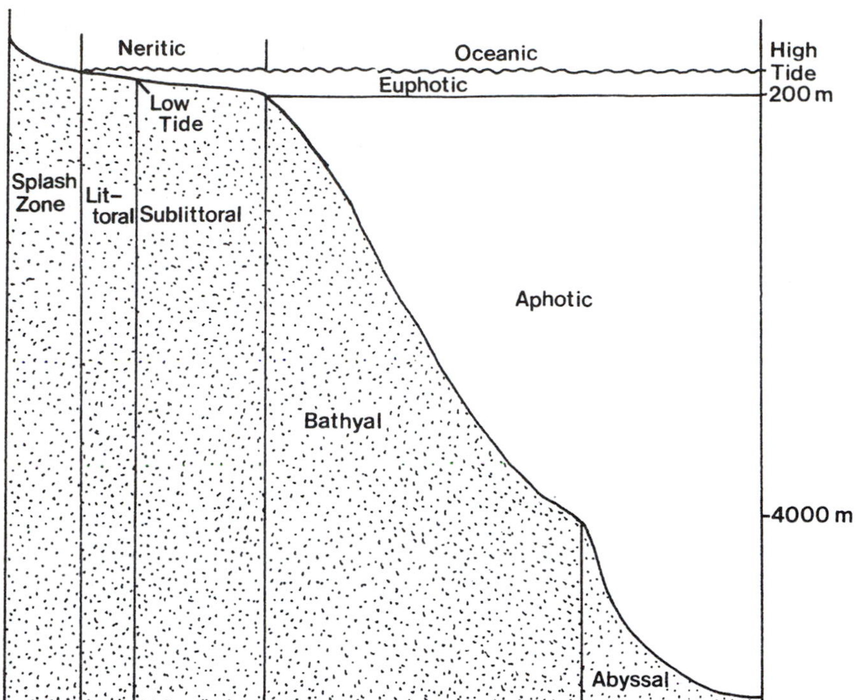

Fig. 1.25 Divisions of the marine environment.

(ii) **Bathyal** zone, extending from a depth of 200 to 4000 m, corresponding to the continental slope, the geomorphic province beyond the continental shelf.

(iii) **Abyssal** zone, at depths greater than 4000 m. It includes over 80% of the benthic environment.

Thermal stratification also occurs in the open ocean, resulting in areas of nutrient deficiencies and low phytoplankton growth. The large phytoplankton of the ocean, such as the dinoflagellate *Pyrocystis*, and the diatoms *Ethmodiscus* and *Rhizosolenia*, migrate vertically in the water column. These phytoplankton migrate down from the euphotic zone, which is poor in nutrients, to deeper water that is richer in nutrients, and then migrate back up into the euphotic zone (Villareal and Lipschultz, 1995).

Upwellings are oceanic areas where colder bottom waters rich in nutrients are brought up to the surface. The resulting combination of nutrients plus light results in large growths of phytoplankton. There are two different causes of upwelling. The first, common in waters around the Antarctic, is an ice shelf that cools the adjacent seawater, causing it to sink. This water is carried away to the northeast by the current. Water that replaces it from the northwest is warmer, even at great depths. This water rises above the colder

Antarctic water, causing a gigantic upwelling area. In the second situation, a surface current of coastal water suddenly turns away from shore, moving out to sea, causing a vertical movement of colder bottom waters to replace the surface water. The sudden change of surface current can be due to offshore winds or to the earth's rotation.

Most of the total photosynthetic capacity of the oceans of the world is carried out by picophytoplankton, most of which are tiny blue-green cyanobacteria (Fogg, 1986b; Glover et al., 1986). In the open ocean, the production of organic matter by phytoplankton ranges from 1 to 20 g of dry organic matter per cubic meter per day (Ryther, 1959; Eppley, 1982; Adey and Goertemiller, 1987).

Polar marine environments in the Arctic and Antarctic regions are characterized by extreme variation in irradiance and seasonal changes of day length, accompanied by low water temperatures in the range of −1.8 °C to 2 °C (Kirst, 1995). During sea-ice formation, phytoplankton are scavenged from the water column and incorporated into sea ice. Many species tolerate incorporation and form distinct sea-ice assemblages. Indeed, approximately a quarter of all primary productivity in the Southern Ocean occurs in sea ice (Arrigo et al., 1997). During the seasonal cycle of ice formation, growth of the ice sheet, and melting, the ice algal communities are subjected to large gradients in irradiance, temperature and salinity. In the process of formation, the growing ice crystals exclude salts which are combined as brine into a system of pockets and channels. The algae must be able to acclimate to a wide range in salinity of up to five times seawater, down to highly diluted seawater during ice melt. The temperature fluctuates at the same time from + 2 °C to − 15 °C.

There is a striking difference between Antarctic and Arctic regions in the amount of annual and perennial ice coverage. About 90% of the Arctic sea-ice is multiyear ice while most of the Antarctic sea ice is seasonal. In the Antarctic, the annual advance and retreat of the sea ice is a major physical determinant in the polar biota.

The greatest formation of algae is in the bottom 20 cm of sea ice in both Antarctic and Arctic (Kirst, 1995). Under sea-ice habitats present special problems to algal inhabitants because ice, and the snow above, strongly absorb photons in the red wavelengths and transmit blue and green photons (maximum 470–480 nm). Blue-green light emerging from algal-free sea ice, 2 meters thick, may be reduced to 10% or less of that at the surface (Robinson et al., 1995). Algae growing at the bottom of sea ice develop extremely high fucoxanthin to chlorophyll *a* ratios that result in increased absorption of the light passing through the sea ice. Specimens from many classes of algae are present in sea ice, although diatoms usually predominate, with pennate specimens being more common than centric (Kirst, 1995). The diatoms that are present tend to be compact and small, such as *Fragilariopsis curta*.

ESTUARINE ENVIRONMENT

An **estuary** is a semi-enclosed coastal body of water that has a free connection with the open sea. The estuary is strongly affected by tidal action, and within it seawater is mixed and diluted with freshwater from land drainage. Organisms living in estuaries must have wide temperature and salinity tolerance (i.e., be **eurythermic** and **euryhaline**) because of the wide variation of these factors in an estuary. This is in contrast to organisms living in the open ocean, which are usually **stenothermic** and **stenohaline**. Estuaries usually have few multicellular algae because of the instability of the muddy bottoms and variations in temperature and salinity. The **interstitial water**, or that within the muddy bottoms, undergoes relatively small variations in temperature and salinity, and some algae have adapted to this habitat by burrowing down into the mud during high tide and rising to the surface during low tide.

Light

The quantity and quality of light largely determine the type and occurrence of algae. As far as photosynthesis and photomorphogenesis of algae are concerned, there are three important components of the radiant energy of the sun: (1) near ultraviolet, with wavelengths from 290 to 380 nm; (2) visible light, with wavelengths from 380 to 760 nm; and (3) infrared, with wavelengths from 760 to 900 nm. Photosynthesis in most algae occurs in the presence of light with wavelengths of 300 to 700 nm. Photomorphogenesis of algae may involve a phytochrome which absorbs in the far red (700 to 760 nm) and part of the infrared (up to 800 nm) (Lüning, 1981).

Radiant energy is measured as the radiant energy (watts, W) incident upon a square meter per minute. One unit of radiant energy is equal to one joule per second ($1\ \mathrm{W} = 1\ \mathrm{J\ s^{-1}}$). Many meterological observations use calories per square centimeter per minute. The conversion to radiant energy is

$$1\ \mathrm{cal\ cm^{-2}\ min} = 698\ \mathrm{W\ m^{-2}}$$

Instead of measuring irradiance in terms of energy, the number of quanta (photons in the visible range) is often measured as microeinsteins per square meter per second:

$$1\ \mu\mathrm{E\ m^{-2}\ s^{-1}} = 6.02 \times 10^{17}\ \mathrm{quanta\ m^{2}\ s^{-1}}$$

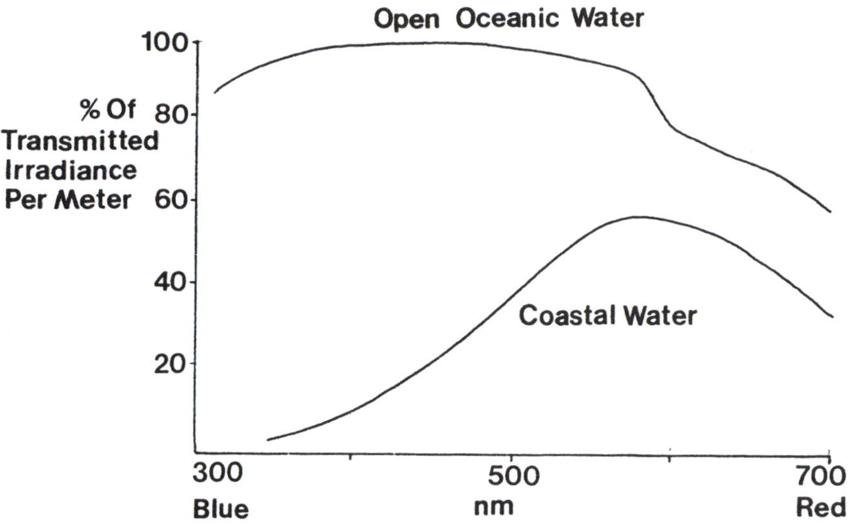

Fig. 1.26 Percentage of irradiance per meter depth of water for different wavelengths.

Microeinsteins perceived per square meter per second measures the number of quanta absorbed and not their energy. Thus this standard is more easily used for photochemical work (Lüning, 1981).

In water, light is attenuated (Fig. 1.26) by (1) absorption and (2) scattering (Grossman et al., 1990). **Absorption** of longer, low-energy, red wavelengths is greatest in water, with shorter, higher-energy wavelengths being least absorbed. Dissolved salts have little influence on light absorption. In many freshwater and coastal seawaters, shorter blue wavelengths are absorbed strongly by dissolved organic compounds which are produced by the decay of vegetation or are liberated by phytoplankton or brown algae. **Scattering** of light by water molecules and particulate matter is greatest for the shorter wavelengths. Scattering results in increased absorption of the light because of the irregular zigzag path the light has to travel to penetrate to a certain water depth.

The lowest normal limit of perennial benthic seaweeds (usually crustose coralline red algae) occurs at a depth that receives a total irradiance of about 10 E m^{-2} $year^{-1}$ (Lüning, 1981). This is equivalent to about 0.05% to 0.1% of the surface irradiance (Fig. 1.27). At Helgoland, off the coast of Germany, this corresponds to a depth of 15 m, whereas in the clearer waters of Jamaica and the Bahamas, it corresponds to a depth of 175 m. Different algae require different amounts of light. The deepest growing perennial kelps (brown algae) require more light, at least 70 E m^{-2} $year^{-1}$, in order to build up their thallus. The euphotic zone for phytoplankton occurs from the surface to a water depth that receives 1% of surface irradiance, which is about 110 m in clear oceanic water and 6 m in more turbid coastal waters.

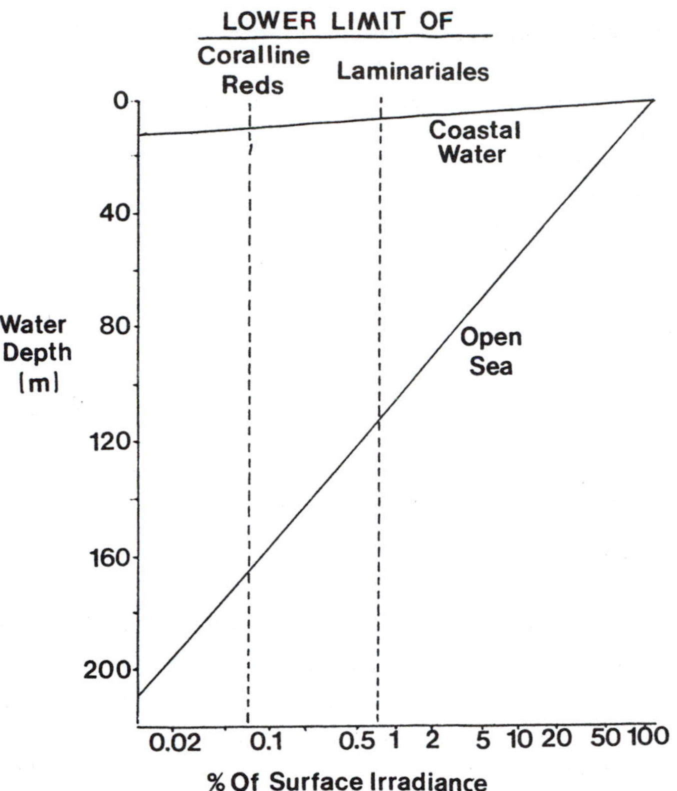

Fig. 1.27 Percentage of surface downward irradiance (350 to 700 nm) in relation to water depth and type of water. (Adapted from Lüning, 1981.)

The amount of light necessary to saturate photosynthesis varies with the position of the alga. Those seaweeds in the upper littoral zone have their photosynthesis saturated at a much higher light intensity than do deep-water seaweeds. Phytoplankton growing near the surface of the ocean have their photosynthesis saturated at levels similar (500 $\mu E\ m^{-2}\ s^{-1}$) to those of intertidal seaweeds, whereas phytoplankton from the bottom of the euphotic zone have their photosynthesis saturated around 200 $\mu E\ m^{-2}\ s^{-1}$. Growth is usually light-saturated at considerably lower irradiances than is photosynthesis (Table 1.3). Adult thalli of intertidal seaweeds (e.g., *Fucus*) have growth saturated at 150 to 250 $\mu E\ m^{-2}\ s^{-1}$ whereas the more bulky seaweeds of the upper sublittoral zone (e.g., *Laminaria, Chondrus, Codium*) are saturated at 30 to 100 $\mu E\ m^{-2}\ s^{-1}$. Growth saturates at light intensities that are different from those at which photosynthesis saturates, because processes other than photosynthesis become limiting to growth at high rates of photosynthesis.

Beginning at sunrise and ending at sunset, the irradiance from the sun at one place is approximately a parabola, barring atmospheric disturbances

Table 1.3 *Light levels required for saturation of growth rates in some adult seaweeds*

Intertidal species	
Fucus vesiculosus	150[a]
Fucus serratus	150
Gigartina stellata	170
Ascophyllum nodosum	250
Pelvetia canaliculata	250
Sublittoral species	
Laminaria saccharina	70
Chondrus crispus	94
Codium fragile	28
Plenosporium squarrulosum	11

Notes:
[a]In $\mu E\ m^{-2}\ s^{-1}$.
Source: Lüning (1981).

(Fig. 1.28). The rate of photosynthesis in an alga will follow the irradiance parabola if the radiant energy never saturates photosynthesis. However, if the radiant energy is greater than that required to saturate photosynthesis, the rate of photosynthesis will be independent of the radiant energy. In this situation, the rate of photosynthesis is predictable: There is a morning maximum, an afternoon depression, and a late afternoon recovery (Fig. 1.28). As much as 70% of the daily photosynthesis can occur in the first half of the day (Ramus and Rosenburg, 1980; Ramus, 1981). The afternoon depression is due to a number of factors acting in concert, including photorespiration, an increase in dark respiration, and daily rhythms of photosynthetic capacity.

The release of certain chlorofluorohydrocarbons from refrigerators and air-conditioners has resulted in destruction of ozone in the stratosphere where most of the ozone occurs. Ozone absorbs large amounts of ultraviolet-B radiation (280–320 nm), but does not significantly absorb ultraviolet-A radiation (320–400 nm) or photosynthetically active radiation (400–700 nm) (Franklin and Forster, 1997). During the last decade, about 7% of the ozone has been lost in the Northern Hemisphere, and about 11% in the Southern Hemisphere. This reduction in the ozone has resulted in an increase in the amount of ultraviolet-B radiation reaching the earth. Ultraviolet-B radiation damages DNA and RNA, the reaction center of photosystem II, and the carboxylating enzyme ribulose bisphosphate carboxylase/oxygenase (Rubisco) (Nilawati et al., 1997). The most significant damage to the ozone layer has occurred during late spring and early

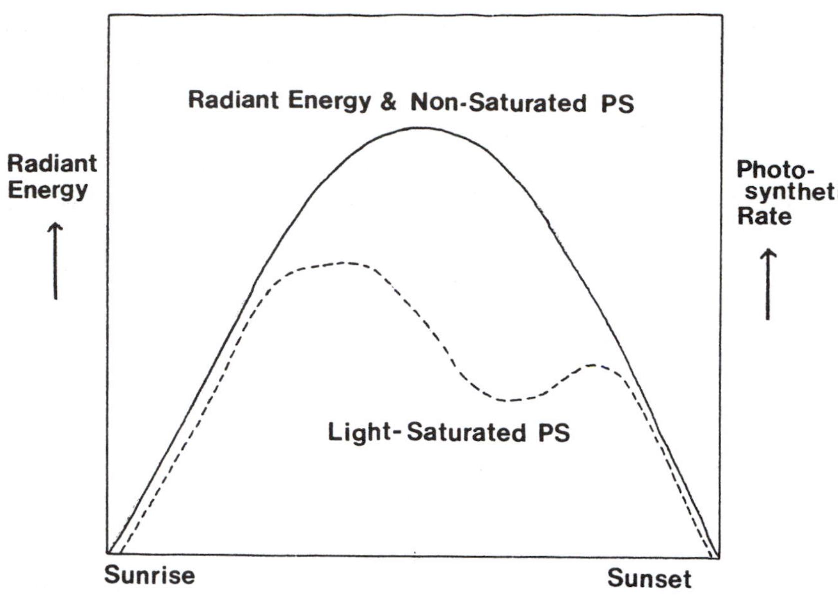

Fig. 1.28 Generalized graph of the amount of radiant energy striking a seaweed and the rate of photosynthesis under conditions when light is saturating and non-saturating. (After Ramus, 1981.)

summer in the Southern Hemisphere when an ozone hole appears. This is the time that there is a bloom of phytoplankton in the Southern Ocean. Recent measurements have shown that carbon fixation rates by phytoplankton in the ocean under the ozone hole are 10% less than normal (Schofield et al., 1995; Xiong et al., 1997).

Algae in general have four strategies to avoid or mitigate ultraviolet-B radiation (Quesada and Vincent, 1997):

1 The production of compounds such as flavonoids or sheath pigments that absorb maximally in the 300 nm range.
2 Migration into habitats with reduced light exposure such as sinking deeper into the water column.
3 Production of quenching agents such as carotenoids or superoxide dismutase that neutralize the highly reactive oxygen species produced by ultraviolet light.
4 Repair mechanisms such as DNA photorepair and ultraviolet-A-blue light guided repair of the photosynthetic apparatus.

Toxic algae

Algae can be harmful in two basic ways (Falconer, 1993; Hallegraeff, 1993; Lassus et al., 1995; Smayda and Shimizu, 1993; Yasumoto and Murata, 1993):

1 **Producing large populations in the aquatic environment** – Large growths of some algae (e.g., the diatom *Chaetoceros* or the prymnesio-

phyte *Chrysochromulina*) can clog the gills of fish and can be particularly a problem in aquaculture systems. Anoxic conditions, resulting in fish kills, can occur at the end of blooms of other algae (e.g., green algae) as the algae die and decompose.

2 **Production of toxins** Some algae produce toxins that sicken and kill other organisms that prey on these algae. Indeed, this probably was the reason that these algae were selected for in the evolutionary process since it reduced predation by grazers (Gilbert, 1996). Filter-feeding shellfish can accumulate large quantities of these toxins as they filter the algae out of the water. Consumption of the shellfish by man, birds and animals results in sickness and death.

The groups of algae and the toxins that they produce are outlined below. More information can be found in the chapters on these algae.

Cyanophyceae (cyanobacteria)

- **Neurotoxins** that block transmission of signal from neuron to neuron. These neurotoxins include the alkaloids (nitrogen-containing compounds) anatoxin and saxitoxin (Fig. 1.29).
- **Hepatotoxins** that are inhibitors of protein phosphatases 1 and 2A. These hepatotoxins include microcystin and nodularin (Fig. 1.29).

Dinophyceae (dinoflagellates)

- **Diarrhetic shellfish poisoning** caused by okadaic acid, macrolide toxins and yessotoxins which are inhibitors of protein phosphatases.
- **Ciguatera fish poisoning** caused by ciguatoxin and maitotoxins.
- **Paralytic shellfish poisoning** caused by saxitoxins.

Bacillariophyceae (diatoms)

- **Amnesic shellfish poisoning** caused by the neurotoxin domoic acid (Fig. 1.29).

Raphidophyceae (chloromonads)

- Unidentified toxins that have produced fish kills in Japan.

Prymnesiophyceae (haptophytes)

- Unidentified toxins that have produced fish kills in Scandinavia and Israel.

The activity of man has resulted in increased eutrophication of the earth's waters and an increase in the occurrence of toxic blooms (Hallegraeff, 1993). The most promising strategy to control blooms of marine phytoplankton is to spread flocculents, such as clay, that scavenge algal cells from seawater and carry them to bottom sediments (Anderson, 1997). Clay is a non-toxic, naturally occurring material that doesn't effect fish and bottom-dwelling organisms.

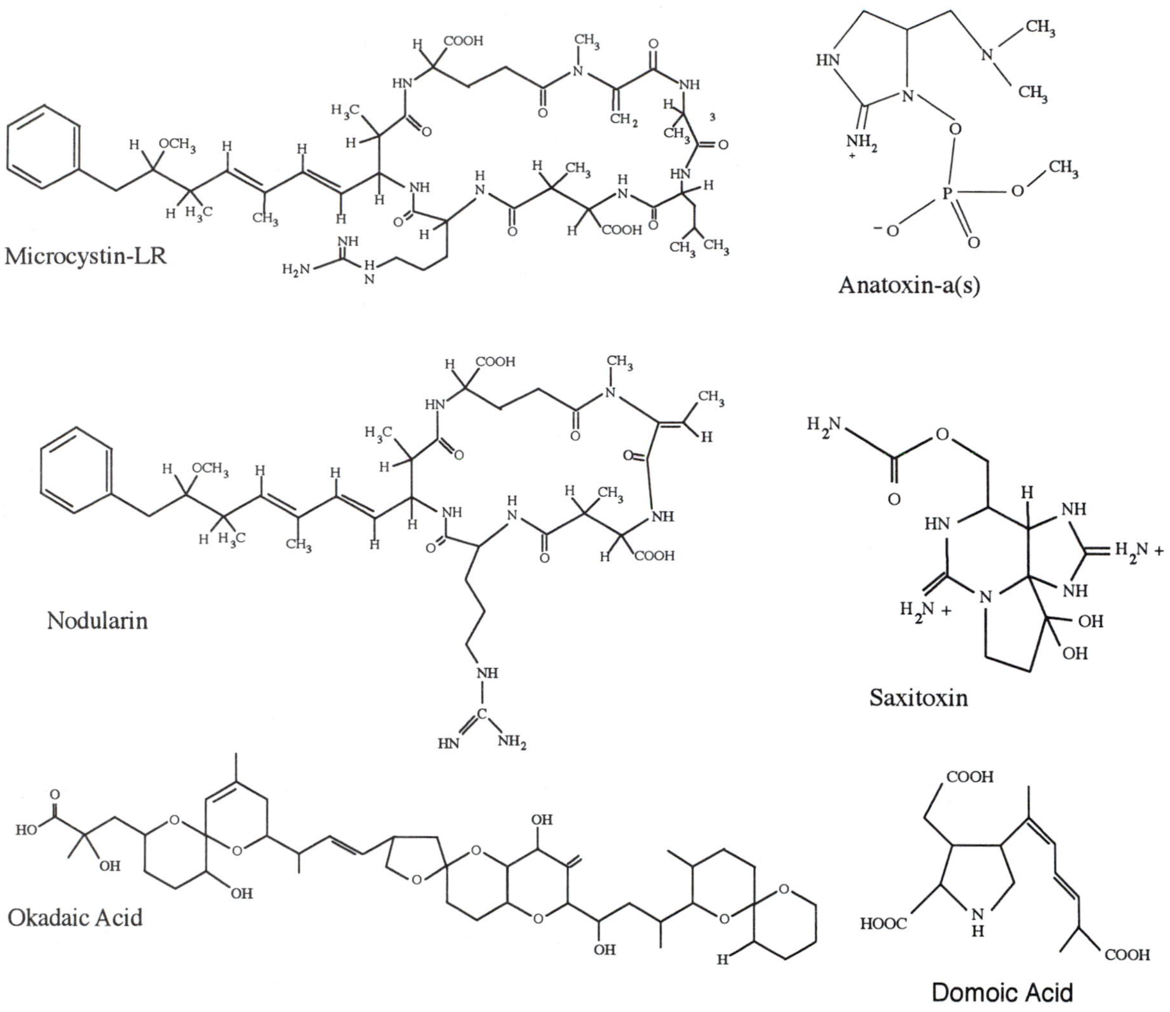

Microcystin-LR

Anatoxin-a(s)

Nodularin

Saxitoxin

Okadaic Acid

Domoic Acid

Fig. 1.29 The chemical structure of some algal toxins.

The grand experiment

Man's activities have resulted in an increase in CO_2 in the atmosphere and potential warming of the atmosphere of the earth due to the "greenhouse effect." This increase in CO_2 in the atmosphere can be addressed in one of two ways, either by a reduction in the burning of fossil fuels or by removing the CO_2 from the atmosphere.

John Martin of the Moss Landing Marine Laboratory in California put forth the hypothesis that iron availability limits phytoplankton production in nutrient-rich seas. He further suggested that it might be possible to fertilize the Southern Ocean (which has an abundance of unused nutrients) with iron, increase photosynthesis by plankton, and increase the flux of CO_2 from the atmosphere to the deep ocean, which contains 60 times more CO_2 than the atmosphere.

> I first said this more or less facetiously at a Journal Club lecture at Woods Hole Oceanographic Institute in July 1988. I estimated that with 300 000 tons of Fe, the Southern Ocean phytoplankton could bloom and remove two billion tons of carbon dioxide. Putting on my best Dr Strangelove accent, I suggested that with half a ship load of Fe, I could give you an ice age. Chisholm and Morel (1991)

Historically, there has been variation in the CO_2 concentration in the atmosphere. Man's activities have resulted in an increase in CO_2 concentration today to almost 350 parts per million (ppm). This is an increase from 200 ppm during the last ice age (glacial maximum, 18 000 years ago). The glacial minimum that preceded this, however, had an atmospheric concentration of 280 ppm, approximately the same as in 1900. The decrease in CO_2 concentration during the last glacial maximum is explained as follows. There was a fivefold increase in the arid areas of the earth, along with a 1.5 times increase in the winds. These two factors resulted in a 50-fold increase in the airborne dust particles. Since iron is the fourth most-common element on earth, the airborne dust contained a significant amount of iron, much of which was deposited in the oceans. This resulted in a threefold increase in photosynthesis and a decrease in CO_2 to around 200 ppm. This produced a decrease in the greenhouse effect, a cooling of the earth and an ice age (Martin, 1990).

The suggestion of adding iron to the Southern Ocean to reduce atmospheric CO_2 triggered a debate about whether we should engage in intentional large-scale intervention with the earth's natural biogeochemical cycles. Martin put forward the following:

> One could argue that this really is not such a new or weighty issue. After all, we have already changed dramatically the landscape of terrestrial ecosystems, we have converted forests to croplands, croplands to deserts, rivers to lakes and deserts to greenbelts. So why is there such a big fuss about the prospect of spreading some iron around the ocean? If we are inadvertently, but knowly, changing the chemistry of the atmosphere through fossil fuel burning, why should we not change it purposely through iron fertilization or some other scheme? Chisholm and Morel (1991)

Ultimately it was decided to try a test of the hypothesis. John Martin orchestrated the scientific and logistical planning for a large-scale iron enrichment experiment in the open ocean, although his untimely death from cancer in 1993 prevented him from seeing the outcome. In mid-November 1993, the RV *Columbus Iselin* arrived 500 miles south of the Galapagos

Islands with 480 kg of iron. The iron was pumped into the propellar wash as the vessel steamed to and fro across an 8×8 km field over 24 hours, raising the iron concentration from about 0.05 mM to about 4 mM. Water samples were taken and monitored for phytoplankton and nutrients, while a P-3 Orion airplane optically scanned the water for changes in phytoplankton pigments. The results showed that there was an increase in phytoplankton, although the increase was not as great as was predicted from laboratory cultures. This was probably due to increased grazing by zooplankton, which showed a 50% increase over the period (Wells, 1994). The experiment was repeated in 1995 (Coale et al., 1996) and again enrichment of photoplankton was obtained.

The experiment showed that it is possible to remove CO_2 from the atmosphere by adding Fe to the Southern Ocean. The political realities of performing Fe enrichment in quantities high enough to significantly effect the atmospheric CO_2 concentration is on hold as the debate over the increase in CO_2 in the atmosphere, and global warming, continues.

Gene sequencing and algal systematics

Specific sequences of nucleotides in DNA of the cell code for cell constituents. It is possible to isolate DNA from cells, multiply certain DNA segments and determine the nucleotide sequences of that DNA. Each species has differences in the nucleotides that make up the DNA and differences in nucleotides that can be used to produce an evolutionary history of the cell. The DNA nucleotides that are most commonly sequenced to produce phylogenies are those of ribosomal DNAs (rDNAs). These rDNA nucleotides make up the genes that code for the rRNAs. rRNAs occur in ribosomes and there are three types of ribosomes, each made up of a large and small subunit:

1 **Prokaryotic ribosomes.** The large 70S subunit contains 5S and 23S rRNAs as well as 34 ribosomal proteins. The small 30S subunit contains a single 16S rRNA and 21 proteins.
2 **Eukaryotic ribosomes** (Fig. 1.30). The large 60S subunit contains 28S, 5.8S and 5S rRNAs and 49 proteins. The small 40S subunit contains 18S RNA and 33 proteins.
3 **Mitochondrial ribosomes.** These are similar, although not the same, as prokaryotic ribosomes. They are not used to produce algal phylogenies, mostly because mitochondria have been transferred between eukaryotic hosts and, therefore, do not reflect the evolutionary history of the organism (Stiller and Hall, 1997).

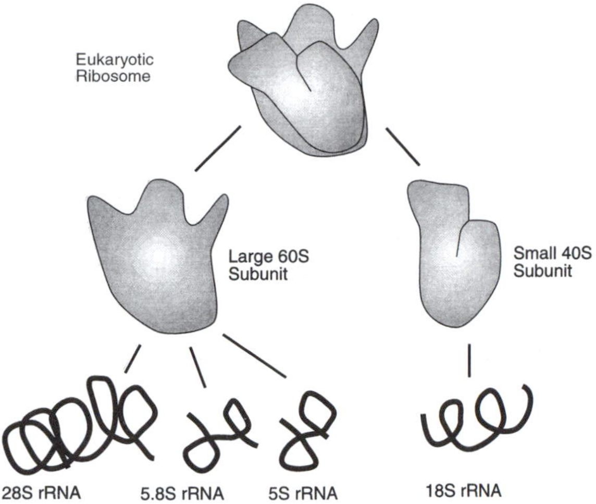

Fig. 1.30 The components of a eukaryotic ribosome.

The **rDNA** for the **18S rRNA** of the small subunit of the eukaryotic ribosome is the form of rDNA usually sequenced to determine the phylogeny of eukaryotic organisms. The nucleotides coding for all of the ribosomal subunits are encompassed within a single operon and transcribed by a single RNA polymerase (Kawai et al., 1997). The procedure for determining the nucleotide sequences is available in any basic biochemistry book.

The **rDNA** for the **5S rRNA** has been also used in phylogeny studies. Although less nucleotides are in the rDNA coding for 5S rRNA, making it easier to sequence, the data have been suspect because of large deviation in the nucleotides (Ragan, 1994). The DNA coding for other molecules, such as ribulose bisphosphate carboxylase/oxygenase (Freshwater et al., 1994; Fujiwara et al., 1994) and actin (Bhattacharya and Ehlting, 1995), have also been used in determining phylogeny.

Gene sequencing has been the most active field of phycological systematics in the last decade and has provided important new information on the relationships between algae. However, as stated by Manhart and McCourt (1992):

> . . . molecular data are not a magic bullet for species problems. They are data, no more, no less. Some molecular data are informative, and others are misleading. Molecular data are fraught with many of the same difficulties as morphological data . . .

Classification

There are four distinct groups within the algae. The first contains the only prokaryotic algae, the Cyanophyta (cyanobacteria), characterized by their prokaryotic cell organization.

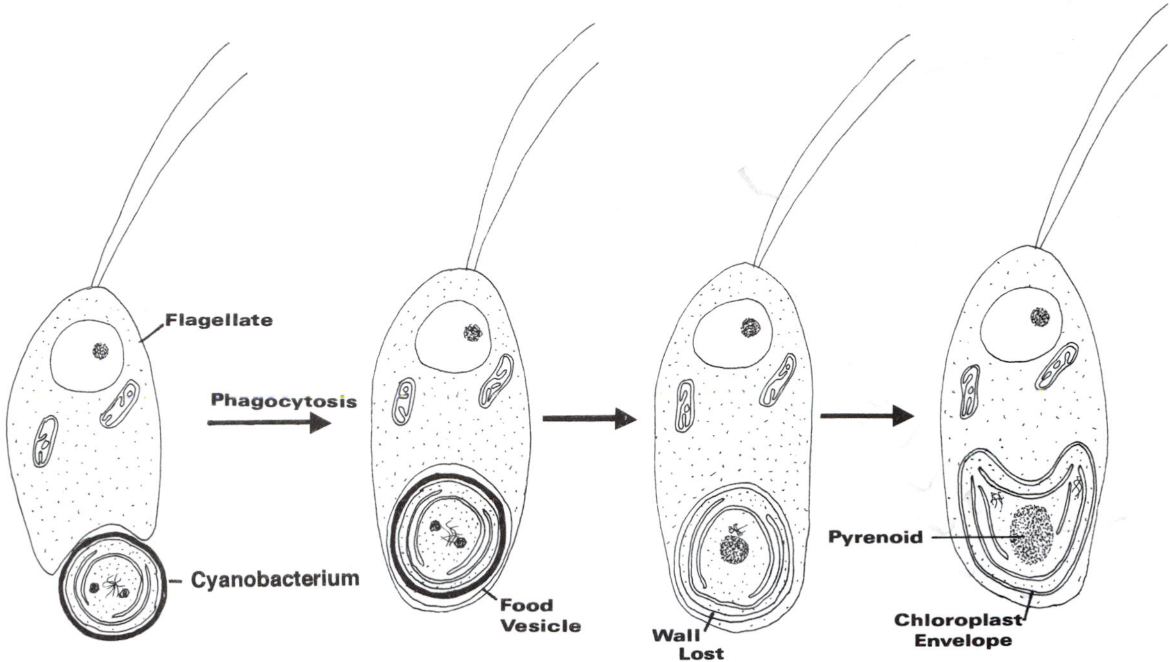

Fig. 1.31 Diagrammatic representation of the uptake of a prokaryotic alga by a protozoan into a food vesicle. This resulted in the establishment of an endosymbiosis between the prokaryotic alga and the protozoan. Through evolution, the endosymbiotic prokaryotic alga evolved into a chloroplast surrounded by two membranes of the chloroplast envelope.

The second group contains those eukaryotic algae with chloroplasts surrounded only by the chloroplast envelope, with no chloroplast endoplasmic reticulum. These algae evolved through an evolutionary event that involved the capture of a cyanobacterium by an aerobic phagocytic protozoan containing mitochondria and peroxisomes (Palmer, 1997) (Fig. 1.31). Normally the cyanobacterium would have been digested as a source of food by the heterotrophic protozoan. However, for some reason, the prokaryotic algal cell was retained in the host cytoplasm as an endosymbiont. Such an endosymbiosis was of benefit to the host protozoan because the host received some of the photosynthate from the endosymbiont. It was also of benefit to the endosymbiont, because the endosymbiont resided in the more stable and protected environment of the cytoplasm of the host. Eventually in the process of evolution, the plasma membrane of the endosymbiont became the inner membrane of the chloroplast envelope and the food vesicle membrane of the host became the outer membrane of the chloroplast envelope. The Glaucophyta represent an intermediate stage in this process, in which the endosymbiotic alga has not completely evolved into a chloroplast. The algae in the Rhodophyta (red algae) and Chlorophyta (green algae) represent the completion of this evolutionary pathway into mature chloroplasts. It appears that the chloroplast evolved only once through a single evolutionary line, and that all chloroplasts are derived from

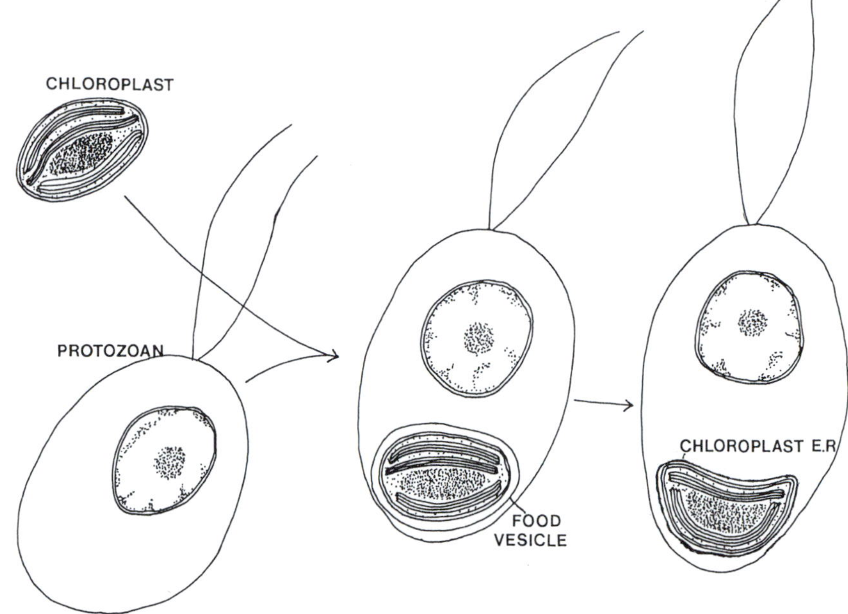

Fig. 1.32 Drawing illustrating the probable sequence of evolutionary events that led to a single membrane of chloroplast endoplasmic reticulum surrounding a chloroplast. Initially, a chloroplast was taken up by a phagocytotic protozoan into a food vesicle. An endosymbiosis resulted, with the food vesicle membrane eventually evolving into a single membrane of chloroplast endoplasmic reticulum.

this line (Bhattacharya and Medlin, 1995; Douglas and Murphy, 1994; Lee, 1972).

The third evolutionary group contains the Euglenophyta (euglenoids) and Dinophyta (dinoflagellates), algae that have the chloroplast surrounded by one membrane of chloroplast endoplasmic reticulum. This evolutionary pathway, originally outlined by Lee (1977), resulted when a chloroplast from a eukaryotic alga was taken up into a food vesicle by a phagocytotic protozoan (Fig. 1.32). Normally the protozoan would have digested the chloroplast as a source of food. However, in this case, the chloroplast was retained in the cytoplasm of the protozoa as an endosymbiont. The host protozoan benefited from the association by receiving photosynthate from the endosymbiont chloroplast. The endosymbiont chloroplast benefited by the stable environment created by the cytoplasm of the host. Eventually the food vesicle membrane of the host became the single membrane of chloroplast endoplasmic reticulum surrounding the chloroplast in the Euglenophyta and Dinophyta.

It appears that algae with chloroplast endoplasmic reticulum were selected for in evolution because of their ability to outcompete other algae in environments which are in low dissolved CO_2 (Lee and Kugrens, 1999). Before explaining the mechanism by which these algae are able to outcompete, it is necessary to understand the equilibria governing the distribution of carbon species in water.

Carbon occurs in water as dissolved inorganic carbon (DIC) which is composed of HCO_3^-, CO_3^{-2} and CO_2.

$$H_2O + CO_2 \rightleftharpoons H_2CO_3 \rightleftharpoons H^+ + HCO_3^- \rightleftharpoons 2H^+ + CO_3^-$$

Seawater in equilibrium with the atmosphere at pH 8.2 and 25 °C contains about 2200 μm of dissolved inorganic carbon. At this pH, 10 μm occurs as CO_2, 200 μm occurs as CO_3^{-2} with the remainder occurring as HCO_3^-. As the pH is lowered, more of the dissolved inorganic carbon occurs as CO_2. At pH 1 virtually all the dissolved inorganic carbon occurs as CO_2, with HCO_3^- completely absent.

Carbon dioxide is the form of carbon that is fixed by the carboxylating enzyme in photosynthesis, ribulose 1,5-bisphosphate carboxylase/oxygenase (Rubisco) (Falkowski and Raven, 1998). Many microalgae have developed a CO_2 concentrating mechanism (CCM) that concentrates CO_2 inside cells to a level several times higher than the CO_2 in the external medium (Fridlyand, 1997; Sukenik et al., 1997). The CO_2 concentrating mechanisms pumps bicarbonate (HCO_3^-) outside the cell through the plasma membrane and chloroplast membrane into the chloroplast. Once inside the chloroplast, the dissolved inorganic carbon stays mostly as HCO_3^- because the stroma is alkaline. A significant amount of the HCO_3^- passes into the lumen of the thylakoids, where in illuminated chloroplasts, the pH is about 5 (Fridlyand, 1997). The enzyme carbonic anhydrase is attached to the thylakoid membranes (Raven, 1997). In the lumen of the thylakoids, carbonic anhydrase converts the HCO_3^- to CO_2 at a rate hundreds of times faster than the non-enzymatic conversion.

$$HCO_3^- + H^+ \rightarrow CO_2 + H_2O$$

The result is a CO_2 concentration that is ten times higher than HCO_3^- in the lumen of the thylakoids (Raven, 1997).

Efflux of the CO_2 from the thylakoid lumen to the chloroplast stroma suppresses Rubisco oxygenase activity and stimulates carboxylase activity in the stroma (or pyrenoid if one is present). Through carboxylation by Rubisco, the CO_2 enters into the carbon reduction cycle. Thus the evolutionary goal of the carbon concentrating mechanism is to increase the concentration of CO_2 at the location of Rubisco.

Once generated by carbonic anhydrase, the CO_2 rapidly diffuses away from the thylakoid lumen and out of the chloroplast, reducing the effectiveness of the carbon concentrating mechanism. Algae with chloroplast E.R. have limited the egress of CO_2 by the following mechanism (Fig. 1.33). The volume between the chloroplast envelope and chloroplast

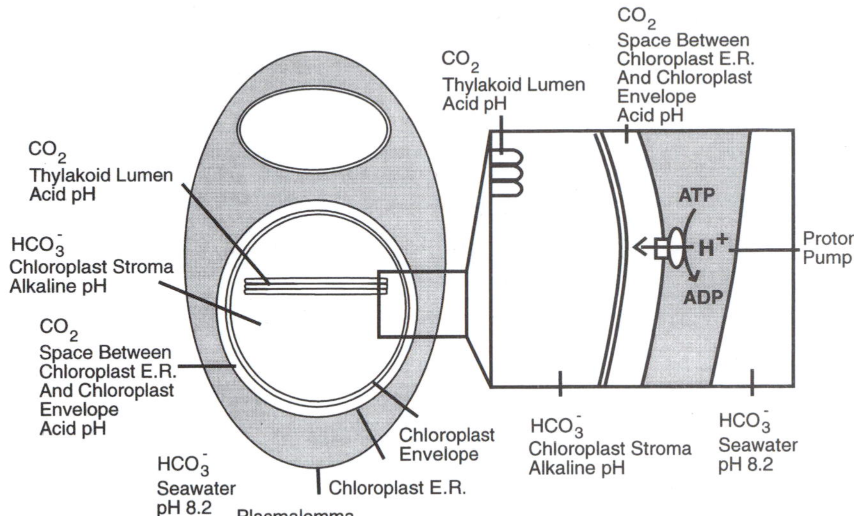

Fig. 1.33 Diagrammatic representation of the compartments present in an alga with one membrane of chloroplast endoplasmic reticulum. The pH and predominant form of dissolved inorganic carbon are indicated for each compartment.

endoplasmic reticulum most likely has an acid pH since the membrane of chloroplast E.R. next to the chloroplast envelope was the vacuolar membrane in the original endosymbiosis leading to the establishment of chloroplast E.R. (Lee, 1977). An acidic pH in the volume between the chloroplast envelope and the adjacent membrane of chloroplast E.R. is therefore probable since the pH of digestive vacuoles of phagocytic protozoa is usually acidic when tested for acid phosphatase (optimal pH 5). An acid pH in the volume between the chloroplast envelope and chloroplast E.R. would contain a relatively high concentration of CO_2, and a relatively low concentration of HCO_3^-, because of the effect of pH on the balance between the two entities. The chloroplast membranes are relatively permeable to CO_2 so the CO_2 in this space would freely move into the chloroplast stroma. A relatively high concentration of CO_2 would, therefore, be maintained in the chloroplast with CO_2 entering both from the acidic thylakoid lumen and the acidic space between the chloroplast E.R. and chloroplast envelope. This would provide algae with chloroplast E.R. an abundant supply of CO_2 for photosynthesis, particularly in the marine environment where at pH 8.2, the dissolved CO_2 concentration is particularly low. In such an environment, algae with chloroplast endoplasmic reticulum had a competitive advantage and were selected for in evolution.

The fourth evolutionary group contains those algae with two membranes of chloroplast endoplasmic reticulum (chloroplast E.R.) surrounding the chloroplast. Algae with two membranes of chloroplast endoplasmic

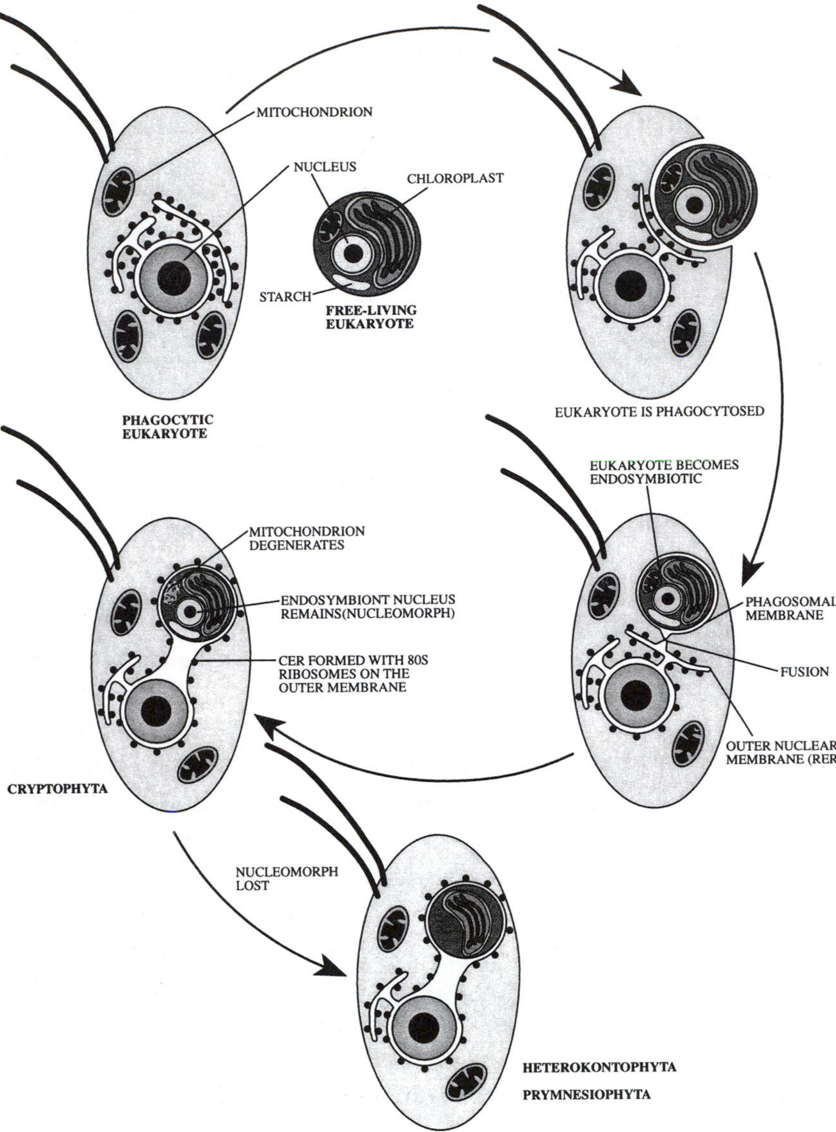

Fig. 1.34 The sequence of events that led to the formation of two membranes of chloroplast endoplasmic reticulum through a secondary endosymbiosis. (Drawing provided by Brec Clay.)

reticulum have the inner membrane of chloroplast E.R. surrounding the chloroplast envelope. The outer membrane of chloropast E.R. is continuous with the outer membrane of the nuclear envelope and has ribosomes on the outer surface (Fig. 1.34).

The algae with two membranes of chloroplast E.R. evolved by a secondary endosymbiosis. In the **secondary endosymbiosis** (Fig. 1.34) (Lee, 1977), a phagocytic protozoan took up a eukaryotic photosynthetic alga into a food vesicle. Instead of being phagocytosed by the protozoan, the photosynthetic alga became established as an endosymbiote within the food vesicle of the protozoan. The endosymbiotic photosynthetic alga benefited by the acidic environment in the food vesicle that kept much of the inorganic carbon in the form of carbon dioxide, the form needed by ribulose bisphosphate/carboxylase for carbon. The host benefited by receiving some of the photosynthate from the endosymbiotic alga. The food vesicle membrane eventually fused with the endoplasmic reticulum of the host protozoan, resulting in ribosomes on the outer surface of this membrane, which became the outer membrane of the chloroplast E.R. Through evolution, ATP production and other functions of the endosymbiont's mitochondrion were taken over by the mitochondria of the protozoan host, and the mitochondria of the endosymbiont were lost. The host nucleus also took over some of the genetic control of the endosymbiont, with a reduction in the size and function of the nucleus of the endosymbiont. The resulting cytology is characteristic of the extant cryptophytes which have a nucleomorph representing the degraded endosymbiont nucleus, as well as starch produced in what remains the endosymbiont cytoplasm.

The type of chloroplast E.R. that exists in the Heterokontophyta and the Prymnesiophyta resulted from further reduction. The nucleomorph was completely lost and storage product formation was taken over by the host. The resulting cell had two membranes of chloroplast envelope surrounding the chloroplast. Outside of this was the inner membrane of chloroplast E.R. that was the remains of the plasma membrane of the endosymbiont. Outside of this was the outer membrane of chloroplast E.R. which was the remains of the food vesicle membrane of the host.

Although the above evolutionary scheme is discussed in one sequence, it is probable that two membranes of chloroplast E.R. evolved at least twice, with one line going to the Cryptophyta, and the second (or more) line leading to the Heterokontophyta and Prymnesiophyta (Bhattacharya and Medlin, 1995; van der Auwera and deWachter, 1997).

The standard botanical classification system is used in the systematics of the algae:

Phylum – phyta
  Class – phyceae
    Order – ales
      Family – aceae
        Genus
          Species

Group 1 Prokaryotic algae

Cyanophyta (cyanobacteria) (Chapter 2): chlorophyll *a*; phycobiliproteins.

Group 2 Eukaryotic algae with chloroplasts surrounded only by the two membranes of the chloroplast envelope.

Glaucophyta (Chapter 3): algae that represent an intermediate position in the evolution of chloroplasts; photosynthesis is carried out by modified endosymbiotic cyanobacteria.

Rhodophyta (red algae) (Chapter 4): chlorophylls *a* and *d*; phycobiliproteins; no flagellated cells; storage product is floridean starch.

Chlorophyta (green algae) (Chapter 5): chlorophylls *a* and *b*; storage product, starch, is found inside the chloroplast.

Group 3 Eukaryotic algae with chloroplasts surrounded by one membrane of chloroplast endoplasmic reticulum.

Euglenophyta (euglenoids) (Chapter 6): chlorophylls *a* and *b*; one flagellum with a spiraled row of fibrillar hairs; proteinaceous pellicle in strips under the plasma membrane; storage product is paramylon; characteristic type of cell division.

Dinophyta (dinoflagellates) (Chapter 7): mesokaryotic nucleus; chlorophylls *a* and c_1; cell commonly divided into an epicone and a hypocone by a girdle; helical transverse flagellum; thecal plates in vesicles under the plasma membrane.

Group 4 Eukaryotic algae with chloroplasts surrounded by two membranes of chloroplast endoplasmic reticulum.

Cryptophyta (cryptophytes) (Chapter 8): nucleomorph present between inner and outer membrane of chloroplast endoplasmic reticulum; starch is formed in grains between inner membrane of chloroplast endoplasmic reticulum and chloroplast envelope; chlorophylls *a* and *c*; phycobiliproteins; periplast is inside plasma membrane.

Heterokontophyta (heterokonts) (Chapters 9–17): anterior tinsel and posterior whiplash flagellum, chlorophylls *a* and *c*, fucoxanthin, storage product usually chrysolaminarin occurring in vesicles in cytoplasm.

- **Chrysophyceae** (golden-brown algae) (Chapter 9)
- **Synurophyceae** (Chapter 10)
- **Dictyochophyceae** (silicoflagellates) (Chapter 11)
- **Pelagophyceae** (Chapter 12)
- **Bacillariophyceae** (diatoms) (Chapter 13)
- **Raphidophyceae** (chloromonads) (Chapter 14)
- **Xanthophyceae** (yellow-green algae) (Chapter 15)
- **Eustigmatophyceae** (Chapter 16)
- **Phaeophyceae** (brown algae) (Chapter 17)

Prymnesiophyta (haptophytes) (Chapter 18): two whiplash flagella, haptonema present, chlorophylls *a* and *c*, fucoxanthin, scales common outside cell, storage product usually chrysolaminarin occurring in vesicles in cytoplasm.

The term **stramenopile** (straw-hair) has been used to include all protists with tubular flagellar hairs (van der Auwera and deWachter, 1997; Leipe et al., 1994; Patterson, 1989). In addition to the algae in the Heterokontophyta, the stramenopiles include the fungal oomycetes, hyphochytridomycetes, thraustrochytrids and the bicosoecids and labyrinthulids.

Data from molecular studies have shown that the red algae diverged first from the common line leading to higher plants (Bhattacharya and Medlin, 1995; Inagaki et al., 1997; Stiller and Hall, 1997). This was followed by divergence of the green algae and, then, multiple independent secondary endosymbioses evolving to those algae with chloroplast endoplasmic reticulum. The host phagocytic organisms leading to the euglenoids was probably a kinetoplastid, that leading to the dinoflagellates was probably an apicomplexan, and that leading to the photosynthetic cryptophytes and haptophytes was a colorless cryptophyte and haptophyte, respectively. The host organisms leading to the heterokonts have not been identified.

Algae and the fossil record

The cyanobacteria are the oldest group of algae with definite fossil remains in the form of stromatolites (see Chapter 2), dating back almost 3000 million years. When the cyanobacteria evolved, the atmosphere contained little or no oxygen and was composed primarily of methane (CH_4), ammonia (NH_3), and other reduced compounds. Photosynthesis by the cyanobacteria eventually built up the oxygen content of the atmosphere to what it is today (20%). Evolution of eukaryotic algae did not occur until about 700 to 800 million years ago, the algae first appearing in a form similar to the Glaucophyta of today, with endosymbiotic cyanobacteria instead of chloroplasts (see Chapter 3). It is difficult to fix this date exactly because these first algae were composed of soft tissues and would not have been preserved. In order to appear in the fossil record, algae would usually have to be large or to have some calcified ($CaCO_3$) or silicified (SiO_2) structures, which are preserved in sedimentary rocks. The appearance of fossil members of the algal classes in the geological timetable is presented in Table 1.4. This table does not purport to show when the algal groups first evolved, but shows only where fossil specimens appear in the geological timetable. The fossil members of each of the algal classes are discussed in the chapter on the particular class.

Table 1.4 *First appearance of algae in the geological time scale*

Era	Period	Epoch	Millions of years ago	First appearance of algal fossil
Cenozoic	Quaternary	Holocene		
		Pleistocene	1.8	
	Tertiary	Pliocene	5.5	
		Miocene	25.0	Xanthophyta
		Oligocene	36.0	
		Eocene	53.5	Euglenophyta
		Palaeocene	65.0	
Mesozoic	Cretaceous		135	Chrysophyta
	Jurassic		191–205	
	Triassic		235–245	Bacillariophyta
Paleozoic	Permian		275–290	Prymnesiophyta
	Carboniferous		360–380	
	Devonian		405–430	
	Silurian		435–460	Dinophyta, Charophyta
	Ordovician		500–530	
	Cambrian		570–610	
Proterozoic	Precambrian		3000	Cyanophyta, Rhodophyta, Chlorophyta

Source: Dill et al. (1986); Loeblich (1974); Medlin et al. (1997); Young et al. (1994).

Fig. 1.35 *Left:* Paul Kugrens, *centre:* the author (with his son Christian) and *right:* Jeremy Pickett-Heaps FRS in Dr Pickett-Heaps laboratory at the University of Colorado in Boulder, Colorado.

Paul Kugrens Born June 29, 1942, in Jelgava, Latvia. Dr Kugrens received his BS (1965) and MS (1967) from the University of Nebraska. He received his PhD from the University of California, Berkeley, in 1971. From 1971 to the present, he has been in the Department of Botany at Colorado State University. Dr Kugren's early research interests were primarily in the red algae. His later research involves mostly the flagellates, primarily the Cryptophyta.

Robert Edward Lee Born September 15, 1942, in Worcester, Massachusetts. Dr Lee received his BS from Cornell University in 1964. From 1964 to 1966, he was a platoon leader with the First Infantry Division of the United States Army in the Republic of Vietnam. In 1971 he received his PhD from the University of Massachusetts. He was subsequently employed from 1971 to 1977 at the University of the Witwatersrand, Johannesburg, Republic of South Africa; from 1977 to 1979 in the Department of Biology at Shiraz (Pahlavi) University in Shiraz, Iran; from 1979 to 1981 at the Schepens Eye Research Institute and the Harvard Medical School in Boston, Massachusetts; and from 1981 to the present in the Department of Anatomy and Neurobiology of the College of Veterinary Medicine at Colorado State University in Fort Collins, Colorado, where he teaches veterinary neurology and veterinary anatomy, and dabbles in phycology.

Jeremy Pickett-Heaps FRS Born June 5 1940. Dr Pickett-Heaps did his undergraduate and graduate work at Cambridge University, receiving his PhD in 1965. From 1965 to 1970, he was a Research Fellow and then a Fellow at the National University in Canberra, Australia. From 1970 to 1988, he was in the Department of Molecular, Cellular and Developmental Biology at the University of Colorado in Boulder, Colorado. In 1988, he accepted the Chair in the Botany Department at the University of Melbourne, Australia. In his early investigations, Dr Pickett-Heaps worked on the formation of cell walls, mostly in higher plants. Later, he turned to algae, where his research laid the foundation for the current classification of the green algae. Lately, his research has focused on morphogenesis in heterokont algae.

References

Adey, W. H., and Goertemiller, T. (1987). Coral reef algal turfs: Master producers in nutrient poor seas. *Phycologia* 26:374–86.

Anderson, D. M. (1997). Turning back the harmful red tide. *Nature* 388:513–14.

Andersen, R. A., Barr, D. J. S., Lynn, D. H., Melkonian, M., Moestrup, O., and Sleigh, M. A. (1991). Terminology and nomenclature of the cytoskeletal elements associated with the flagellar/ciliary apparatus in protists. *Protoplasma* 164:1–8.

Arrigo, K. R., Worthen, D. L., Ligotta, M. P., Dixon, P., and Diekmann, G. (1997). Primary production in Antarctic sea ice. *Science* 276:394–7.

Baker, T. S., Suh, S. W., and Eisenberg, D. (1977). Structure of ribulose-1,5-bisphosphate carboxylase-oxygenase: Form III crystals. *Proc. Natl. Acad. Sci., USA* 74:1037–41.

Beech, P. L., and Wetherbee, R. (1990). Direct observations on flagellar transformation in *Mallomonas splendens* (Synurophyceae). *J. Phycol.* 26:90–5.

Bell, R. A. (1993). Cryptoendolithic algae of hot semiarid lands and deserts. *J. Phycol.* 29:133–9.

Bell, R. A., Athey, P. V., and Sommerfeld, M. R. (1986). Cryptoendolithic algal communities of the Colorado plateau. *J. Phycol.* 22:429–35.

Bernard, C., Etienne, A-L., and Thomas, J. C. (1996). Synthesis and binding of phycoerythrin and its associated linkers to the phycobilisome in *Rhodella violacea* (Rhodophyta): compared effects of high light and translation inhibitors. *J. Phycol.* 32:265–71.

Betsche, T., Schaller, D., and Melkonian, M. (1992). Identification and characterization of glycolate oxidase and related enzymes from the endocynotic alga *Cyanophora paradoxa* and from pea leaves. *Plant Physiol.* 98:887–93.

Bhattacharya, D., and Ehlting, J. (1995). Actin coding regions: gene family evolution and use as a phylogenetic marker. *Arch. Protistenkd.* 145:155–64.

Bhattacharya, D., and Medlin, L. (1995). The phylogeny of plastids: a review based on comparisons of small-subunit ribosomal RNA coding regions. *J. Phycol.,* 31:489–98.

Bisalputra, T. (1974). Plastids. In *Algal Physiology and Biochemistry,* ed. W. D. P. Stewart, pp. 124–60. Berkeley: Univ. Calif. Press.

Bouck, G. B. (1969). Extracellular microtubules. The origin, structure, and attachment of flagellar hairs in *Fucus* and *Ascophyllum* antherozoids. *J. Cell Biol.* 40:446–60.

Bouck, G. B. (1971). The structure, origin, isolation, and composition of the tubular mastigonemes of the *Ochromonas* flagellum. *J. Cell Biol.* 50:362–84.

Bouck, G. B. (1972). Architecture and assembly of mastigonemes. *Adv. Cell Mol. Biol.* 2:237–71.

Bouck, G. B., Rogalski, A., and Valaitis, A. (1978). Surface organization and composition of *Euglena.* II. Flagellar mastigonemes. *J. Cell Biol.* 77:805–26.

Caron, L., Douady, D., Quinet-Szely, M., deGoër, S., and Berkaloff, C. (1996). Gene structure of a chlorophyll *a/c*-binding protein from a brown alga: Presence of an intron and phylogenetic implications. *J. Mol. Evol.* 43:270–80.

Chisholm, S. W., and Morel, F. M. M. (1991). What controls phytoplankton production in nutrient-rich areas of the open sea? *Limnol. Oceanog.* 36:1507–970.

Coale, K. H., Johnson, K. S., Fitzwater, S. E., Gordon, R. M., Tanner, S., Chavez, F. P., Ferioli, L., Sakamoto, C., Rodgers, P., Millero, F., Steinberg, P., Nightingale, P., Cooper, D., Cochlan, W. P., Landry, M. R., Constantinou, J., Rollwagen, G., Trasvina, A., and Kudela, R. (1996). A massive phytoplankton bloom induced by an ecosystem-scale iron fertilization experiment in the equatorial Pacific Ocean. *Nature (Lond.)* 383:495–501.

Coleman, A. W. (1985). Diversity of plastid DNA configuration among classes of eukaryote algae. *J. Phycol.* 21:1–16.

Dill, R. F., Shinn, E. A., Jones, A. T., Kelly, K., and Steinen, R. P. (1986). Giant subtidal stromatolites forming in normal salinity waters. *Nature* 324:55–9.

Douglas, S. E. W., and Murphy, C. A. (1994). Structural, transcriptional, and phylogenetic analyses of the atpB gene cluster from the plastid of *Cryptomonas* (Cryptophyceae). *J. Phycol.* 30:329–40.

Eminson, D., and Moss, B. (1980). The composition and ecology of periphyton communities in freshwaters. 1. The influence of host type and external environment on community composition. *Br. Phycol. J.* 15:429–46.

Eppley, R. (1982). The PRPOOS program: A study of plankton rate processes in oligotrophic oceans. *EOS* 63:522–3.

Evans, L. V. (1974). Cytoplasmic organelles. In *Algal Physiology and Biochemistry,* ed. W. D. P. Stewart, pp. 86–123. Berkeley: Univ. Calif. Press.

Falconer, I. R. (ed). (1993). *Algal Toxins in Seafood and Drinking Water.* New York: Academic Press. 341 pp.

Falkowski, P. G., and Raven, J. (1997). *Aquatic Photosynthesis.* Science. Oxford: Blackwell Science. 256 pp.

Fogg, G. E. (1986a). Picoplankton. *Proc. R. Soc.* [*B*] 228:1–30.

Fogg, G. E. (1986b). Light and ultraphytoplankton. *Nature* 319:96–7.

Foster, K. W., Saranak, J., Patel, N., Zarilli, G., Okabe, M., Kline, T., and Nakanishi, K. (1984). A rhodopsin is the functional receptor for phototaxis in the unicellular eukaryote *Chlamydomonas. Nature* 311:756–9.

Franklin, L. A., and Forster, R. M. (1997). The changing irradiance environment: consequences for marine macrophyte physiology, productivity and ecology. *Eur. J. Phycol.* 32:207–32.

Freshwater, D. W., Fredericq, S., Butler, B. S., Hommersand, M. H., and Chase, M. W. (1994). A gene phylogeny of the red algae (Rhodophyta) based on plastid *rbc*L. *Proc. Natl. Acad. Sci., USA,* 91:7281–5.

Fridlyand, L. E. (1997). Models of CO_2 concentrating mechanisms in microalgae taking into account cell and chloroplast structure. *BioSystems* 44:41–57.

Friedmann, E. I. (1982). Endolithic microorganisms in the Antarctic cold desert. *Science* 215: 1045–53.

Friedmann, E. I., and Galun, M. (1974). Desert algae, lichens, and fungi. In *Desert Biology, 2* (Brown, G. W., ed.), pp. 165–212. Academic Press, New York.

Fujiwara, S., Sawada, M., Someya, J., Minaka, N., Kawachi, M., and Inouye, I. (1994). Molecular phylogenetic analysis of *rbc*L in the Prymnesiophyta. *J. Phycol.* 30:863–71.

Gilbert, J. J. (1996). Effect of food availability on the response of planktonic rotifers to a toxic strain of the cyanobacterium *Anabaena flos-aquae. Limnol. Oceanog.* 41:1565–72.

Glazer, A. N. (1982). Phycobilisomes: Structure and dynamics. *Annu. Rev. Microbiol.* 36:173–98.

Glazer, A. N., Yeh, S. W., Webb, S. P., and Clark, J. H. (1985). Disk-to-disk transfer as the rate-limiting step for energy flow in phycobilisomes. *Science* 227:419–23.

Glover, H. E., Keller, M. D., and Guillard, R. R. L. (1986). Light quality and oceanic ultraplankters. *Nature* 319:142–3.

Grossman, A., Manodori, A., and Snyder, D. (1990). Light-harvesting proteins of diatoms: Their relationship to the chlorophyll *a/b* binding protein of higher plants and their mode of transport into plastids. *Mol. Gen. Genetics* 224:91–100.

Grossman, A. R., Schaffer, M. R., Chiang, G. G., and Collier, J. L. (1993). The phycobilisome, a light-harvesting complex response to environmental conditions. *Microbiol Rev.* 57:725–49.

Hallegraeff, G. M. (1993). A review of harmful algal blooms and their apparent global increase. *Phycologia* 32:79–99.

Haygood, M. G. (1996). The potential role of functional differences between Rubisco forms in governing expression in chemoautotrophic symbiosis. *Limnol. Oceanog.* 41:370–1.

Hoops, H. J., and Witman, G. B. (1985). Basal bodies and associated structures are not required for normal flagellar motion or phototaxis in the green alga *Chlorogonium elongatum. J. Cell Biol.* 100:297–309.

Inagaki, Y., Hayashi-Ishimara, Y., Ehara, M., Igarashi, I., and Ohama, T. (1997). Algae or protozoa: phylogenetic position of euglenophytes and dinoflagellates as inferred from mitochondrial sequences. *J. Mol. Evol.* 45: 295–300.

Iwamoto, K., Suzuki, K., and Ikawa, T. (1996). Purification and characterization of glycolate oxidase from the brown alga *Spatoglossum pacificum* (Phaeophyta). *J. Phycol.* 32:790–8.

Johansen, J. R. (1993). Cryptogamic crusts of semiarid and arid lands of North America. *J. Phycol.* 29:140–2.

Kain, J. M. (1989). The seasons in the subtidal. *Br. Phycol. J.* 24:203–15.

Kamiya, R., and Witman, G. B. (1984). Submicromolecular levels of calcium control the balance of beating between the two flagella in demembranated models of *Chlamydomonas. J. Cell Biol.* 98:97–107.

Kauss, H. (1974). Osmoregulation in *Ochromonas.* In *Membrane Transport in Plants,* ed. U. Zimmermann, and J. Daintz, pp. 90–94. Springer-Verlag, Berlin.

Kawai, H., Nakayama, T., Inouye, I., and Kato, A. (1997). Linkage of 5S ribosomal DNA to other rDNAs in the chromophytic algae and related taxa. *J. Phycol.* 33:505–11.

Kellogg, E. A., and Juliano, N. D. (1997). The structure and function of RuBisCo and their implications for systematic studies. *Amer. J. Bot.* 84:413–28.

Kirst, G. O. (1995). Ecophysiology of polar algae. *J. Phycol.* 31:181–91.

Lassus, P., Arzul, G., Erad-Le Denn, E., Gentien, P., and Marcaillou-Le Baut. (eds.) (1995). *Harmful Algal Blooms.* Paris: Lavoisier. 422 pp.

Lee, R. E. (1972). Origin of plastids and the phylogeny of the algae. *Nature* 237:44–6.

Lee, R. E. (1977). Evolution of algal flagellates with chloroplast endoplasmic reticulum from the ciliates. *S. Afr. J. Sci.* 73:179–82.

Lee, R. E., and Kugrens, P. (1999). The ecological advantage of chloroplast E.R. The ability to outcompete at low dissolved CO_2 concentrations. *Protist* 149:341–5.

Leipe, D. D., Wainright, P. O., Gunderson, J. H., Porter, D., Patterson, D. J., Valois, F., Himmerich, S., and Sogin, M. L. (1994). The stramenopiles from a molecular perspective: 16S-like rRNA sequences from *Labyrinthuloides minuta* and *Cafeteria roenbergensis. Phycologia* 33:369–77.

Littler, M. M., Littler, D. S., Blair, S. M., and Norris, J. N. (1985). Deepest known plant life discovered on an uncharted seamount. *Science* 227:57–9.

Loeblich, A. R. (1974). Protistan phylogeny as indicated by the fossil record. *Taxon.* 23:277–90.

Lüning, K. (1981). Light. In *The Biology of Seaweeds,* ed. C. S. Lobban and M. J. Wynne, pp. 326–55. Berkeley and Los Angeles: Univ. Calif. Press.

Lüning, K., and tom Dieck, I. (1989). Environmental triggers in algal seasonality. *Bot. Mar.* 32:389–97.

McKay, C. P., and Friedmann, E. I. (1985). The cryptoendolithic microbial environment in the Antarctic cold desert: Temperature variations in nature. *Polar Biol.* 4:19–25.

Manhart, J. R., and McCourt, R. M. (1992). Molecular data and species concepts in the algae. *J. Phycol.* 28:730–7.

Martin J. H. (1990). Glacial–interglacial CO_2 change. The iron hypothesis. *Paleoceanography.* 5:1–13.

Medlin, L. K., Kooistra, W. H. C. F., Gersonde, R., Sims, P. A., and Wellbrock, U. (1997). Is the origin of the diatoms related to the end-Permian mass extinction? *Nova Hedwigia* 65:1–11.

Meeks, J. C. (1974). Chlorophylls. In *Algal Physiology and Biochemistry,* ed. W. D. P. Stewart, pp. 161–75. Berkeley: Univ. Calif. Press.

Melkonian, M. (1980). Flagellar roots, mating structure and gametic fusion in the green alga *Ulva lactuca* (Ulvales). *J. Cell Sci.* 46:149–69.

Melkonian, M., and Robenek, H. (1980). Eyespot membranes in newly released zoospores of the green alga *Chlorosarcinopsis gelatinosa* (Chlorosarcinales) and their fate during zoospore settlement. *Protoplasma* 104:129–40.

Melkonian, M., Reize, I. B., and Preisig, H. R. (1987). Maturation of a flagellum/basal body requires more than one cell cycle in algal flagellates: studies on *Nephroselmis olivaea* (Prasinophyceae). In *Algal Development, Molecular and Cellular Aspects,* ed. W. Wiessner, D. G. Robinson, and R. C. Starr, pp. 102–13. Heidelberg: Springer.

Moestrup, Ø. (1982). Flagellar structure in algae: A review, with new observations particularly on the Chrysophyceae. Phaeophyceae (Fucophyceae), Euglenophyceae, and *Reckertia. Phycologia* 21:427–528.

Molenaar, F. J., Venekamp, L. A. H., and Breeman, A. M. (1996). Life-history regulation in the subtidal red alga *Calliblepharis ciliata. Eur. J. Phycol.* 31:241–7.

Nilawati, J., Greenberg, B. M., and Smith, R. E. H. (1997). Influence of ultraviolet radiation on growth and photosynthesis of two cold ocean diatoms. *J. Phycol.* 33:215–24.

Palmer, J. D. (1997). Organelle genomes: going, going, gone. *Science* 275:790–1.

Patterson, D. J. (1989). Stramenopiles: chromophytes from a protistan perspective. In *The Chromophyte Algae: Problems and Perspectives*, ed. J. C. Green, B. S. C. Leadbeater and W. L. Diver, pp. 357–79. Oxford: Clarendon Press.

Patterson, D. J., and Hausmann, K. (1981). The behavior of contractile vacuole complexes of cryptophycean flagellates. *Br. Phycol. J.* 16:429–39.

Percival, E., and McDowell, R. H. (1967). *Chemistry and Enzymology of Marine Algal Polysaccharides.* New York: Academic Press.

Porter, G., Tredwell, C. J., Searle, G. F. W., and Barber, J. (1978). Picosecond time-resolved energy transfer in *Porphyridium cruentum. Biochim. Biophys. Acta* 501:232–45.

Quesada, A., and Vincent, W. F. (1997). Strategies of adaptation by Antarctic cyanobacteria to ultraviolet radiation. *Eur. J. Phycol.* 32:335–42.

Ragan, M. A. (1981). Chemical constituents of seaweeds. In *The Biology of Seaweeds*, ed. C. S. Lobban and M. J. Wynne, pp. 589–626. Los Angeles and Berkeley: Univ. Calif. Press.

Ragan, M. A. (1994). 18S ribosomal DNA sequences indicate a monophyletic origin of Charophyceae. *J. Phycol.* 30:490–500.

Ramus, J. (1981). The capture and transduction of light energy. In *The Biology of Seaweeds*, ed. C. S. Lobban and M. J. Wynne, pp. 458–92. Berkeley and Los Angeles: Univ. Calif. Press.

Ramus, J., and Rosenberg, G. (1980). Diurnal photosynthetic performance of seaweeds measured under natural conditions. *Mar. Biol.* 56:21–8.

Raven, J. (1997). CO_2 concentrating mechanisms: a direct role for thylakoid lumen acidification? *Plant Cell and Environ.* 20:147–54.

Rawat, M., Henk, M. C., Lavigne, L. L., and Morney, J. V. (1996). *Chlamydomonas reinhardtii* mutants without ribulose-1,5-bisphosphate carboxylase/oxygenase lack a detectable pyrenoid. *Planta (Berl.)* 198:263–70.

Ringo, D. L. (1967). Flagellar motion and fine structure of the flagellar apparatus in *Chlamydomonas. J. Cell Biol.* 33:543–71.

Robinson, D. H., Arrigo, K. R., Iturriaga, R., and Sullivan, C. W. (1995). Microalgal light-harvesting in extreme low-light environments in McMurdo Sound, Antarctica. *J. Phycol.* 31:508–20.

Ryther, J. H. (1959). Potential productivity of the sea. *Science* 130:602–8.

Schlesinger, P., Belkin, S., and Boussiba, S. (1996). Sodium deprivation under alkaline conditions causes rapid death of the filamentous cyanobacterium *Spirulina platensis. J. Phycol.* 32:608–13.

Schofield, O., Kroon, B. M. A., and Prezelin, B. B. (1995). Impact of ultraviolet-B radiation on photosystem II activity and its relationship to the inhibition of carbon fixation rates for antarctic ice algae communities. *J. Phycol.* 31:703–15.

Smayda, T. J., and Shimizu, Y. (eds.) (1993). *Toxic Phytoplankton Blooms in the Sea.* Amsterdam: Elsevier Science Publishers. 423 pp.

Snell, W. J. (1976). Mating in *Chlamydomonas:* A system for the study of specific cell adhesion. I. Ultrastructural and electrophoretic analyses of flagellar surface components involved in adhesion. *J. Cell Biol.* 68:48–69.

Steinman, A. D., and McIntire, C. D. (1986). Effects of current velocity and light energy on the structure of periphyton assemblages in laboratory streams. *J. Phycol.* 22:352–61.

Stewart, K. D., and Mattox, K. R. (1984). The case for a polyphyletic origin of mitochondria: Morphological and molecular comparisons. *J. Mol. Evol.* 21:54–7.

Stiller, J. W., and Hall, B. D. (1997). The origin of red algae: implication for plastid evolution. *Proc. Natl. Acad. Sci. USA,* 94:4520–5.

Sukenik, A., Tchernov, D., Kaplan, A., Huertas, E., Lubian, L. M., and Livne, A. (1997). Uptake, efflux, and photosynthetic utilization of inorganic carbon by the marine eustigmatophyte *Nannochloropsis* sp. *J. Phycol.* 33:969–74.

Süss, K. H., Prokhorenko, I., and Adler, K. (1995). *In situ* association of Calvin cycle enzymes, ribulose-1,5-bisphosphate carboxylase/oxygenase activase, feredoxin-NADP + reductase, and nitrate reductase with thylakoid and pyrenoid membranes of *Chlamydomonas reinhardtii* chloroplast as revealed by immunoelectron microscopy. *Plant Physiol.* 107:1387–97.

Sweeney, B. M. (1979). Endogenous rhythms in the movement of plants. In *Encyclopedia of Plant Physiology* ed. W. Haupt and M. E. Feinleib, Vol. 7, pp. 71–93. Berlin: Springer-Verlag.

Tang, E. P. Y., Tremblay, R., and Vincent, W. F. (1997). Cyanobacterial dominance of polar freshwater ecosystem: are high-latitude mat-forms adapted to low temperature. *J. Phycol.* 33:171–81.

Tatsuzawa, H., Takizawa, E., Wada, M., and Yamamoto, Y. (1996). Fatty acid and lipid composition of the acidophilic green alga *Chlamydomonas* sp. *J. Phycol.* 32:598–601.

Tschermak-Woess, E., and Friedmann, E. I. (1984). *Hemichloris antarctica* gen. et sp. nov. (Chlorococcales, Chlorophyta), a cryptoendolithic alga from Antarctica. *Phycologia* 23:443–54.

van der Auwera, G., and deWachter, R. (1997). Complete large subunit ribosomal RNA sequences from the heterokont algae *Ochromonas danica, Nannochloropsis salina,* and *Tribonema aequale,* and phylogenetic analysis. *J. Mol. Evol.* 45:84–90.

Vergara, J. J., and Niell, F. X. (1993). Effects of nitrate availability and irradiance on internal nitrogen constituents in *Corallina elongata* (Rhodophyta). *J. Phycol.* 29:285–93.

Villareal, T. A., and Lipschultz, F. (1995). Internal nitrate concentrations in single cells of large phytoplankton from the Sargasso Sea. *J. Phycol.* 31:689–96.

Wells, M. L. (1994). Pumping iron in the Pacific. *Nature (London)* 368:295–6.

Wessel, D., and Robinson, D. G. (1979). Studies on the contractile vacuole of *Poterioochromonas malhamensis* Peterfi. I. The structure of the alveolate vesicles. *Eur. J. Cell Biol.* 19:60–6.

Xiong, F., Komenda, J., Kopecky, J., and Nedbal, L. (1997). Strategies of ultraviolet-B protection in microscopic algae. *Physiologia Plantarum* 100:378–88.

Yasumoto, T., and Murata, M. (1993). Marine toxins. *Chemical Review* 93:1897–909.

Young, J. R., Bown, P. R., and Burnett, J. A. (1994). Palaeontological perspectives. In *The Haptophyte Algae,* ed. J. C. Green and B. S. C. Leadbeater, Systematics Assn. Special Vol. 51, pp. 379–92. Oxford: Clarendon Press.

PART II · The prokaryotic algae

The cyanobacteria or blue-green algae form a natural group by virtue of being the only prokaryotic algae. Prokaryotic algae have an outer plasma membrane enclosing protoplasm containing photosynthetic thylakoids, 70S ribosomes, and DNA fibrils not enclosed within a separate membrane. Chlorophyll *a* is the main photosynthetic pigment, and oxygen is evolved during photosynthesis.

2 · Cyanobacteria

Cyanophyceae

The Cyanophyceae or **blue-green algae** are, today, usually referred to as the **cyanobacteria** (blue-green bacteria). The term cyanobacteria acknowledges that these prokaryotic algae are more closely related to the prokaryotic bacteria than to the eukaryotic algae.

The cyanobacteria are prokaryotic organisms containing chlorophyll *a*, phycobiliproteins, glycogen as a storage product, and cell walls consisting of amino sugars and amino acids.

Some of the cyanobacteria have chlorophyll *b*, which at one time was used to place these organisms into a separate group, the **Prochlorophyta**. It has since been shown that chlorophyll *b* evolved a number of different times within the cyanobacteria and these organisms have been again placed within the cyanobacteria (Palenik and Haselkorn, 1992; Urback, Robertson and Chisholm, 1992).

Morphology

The simplest morphology in the cyanobacteria is that of unicells, free living (Figs. 2.1, 2.7(*c*)) or enclosed within a mucilaginous envelope (Fig. 2.24(*a*),(*b*)). Subsequent evolution resulted in the formation of a row of cells called a **trichome** (Figs. 2.4, 2.6, 2.7 and 2.20). When the trichome is surrounded by a sheath, it is called a **filament** (Fig. 2.6(*a*),(*c*), 2.17, and 2.20 (*a*),(*d*)). It is possible to have more than one trichome in a filament (Fig. 2.17(*e*) and 2.24(*e*)). The most complex type of thallus in the cyanobacteria is the branched filament (Figs. 2.19(*a*) and 2.26(*a*)). Such a branched filament can be **uniseriate** (composed of a single row of cells) or **multiseriate** (composed of one or more rows of cells).

Cell wall and sheath

The cell walls of the cyanobacteria are composed of a number of layers. Only the inner two layers (LI and LII) (Fig. 2.1) are the same in all cyanobacteria. The structure of the wall outside of the LII depends on environmental

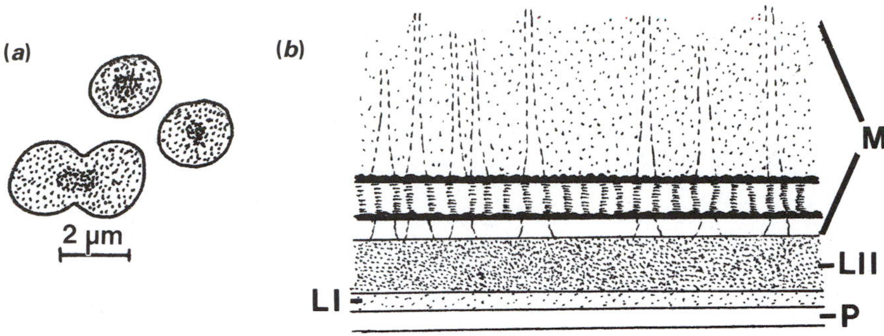

Fig. 2.1 *Synechocystis aquatilis*. (*a*). Drawing of whole cells. (*b*) Drawing of the structure of the plasmalemma (P), two wall layers (LI and LII), and mucilage (M). (After Smarda et al., 1979.)

conditions and the amount mucilage secreted (Smarda et al., 1979). The cell wall structure of the cyanobacteria is basically the same as that of Gram-negative bacteria, indicating a possible relationship between the two. The **mucilage layer** (**sheath** or **capsule**) outside of the cell wall is fibrillar. The sheath functions in protecting the cells from drying and is involved in gliding. Active growth seems necessary for sheath formation, a fact that may explain its sometimes poor development around spores and akinetes. The sheath of *Gloeothece* sp. is composed of neutral sugars and uronic acids including galactose, glucose, mannose, rhamnose, 2-*O*-methyl-D-xylose, glucuronic acid, and galacturonic acids (Weckesser et al., 1987). The sheath of *Gloeothece* contains only 2% protein and a trace of fatty acids and phosphate. Sheaths are often colored, with red sheaths found in algae from highly acid soils and blue sheaths characteristic of algae from basic soils (Drouet, 1978). Yellow and brown sheaths are common in specimens from habitats of high salt content, particularly after the algae dry out.

The major constituent of the cyanophycean cell wall is **peptidoglycan** (synonymous with **murein, glycopeptide, mucopeptide**), constituting up to 50% of the dry weight (Höcht et al., 1965). Although the structure of the peptidoglycan in the cyanobacteria has not been elucidated, it is probably similar to that of bacteria. Bacterial peptidoglycan is composed of polysaccharide chains of alternating *N*-acetylglucosamine and *N*-acetylmuramic acid residues. These polysaccharide chains are joined together by peptides consisting of four amino acids (among which is diaminopimelic acid, not found in eukaryotes).

Protoplasmic structure

Many of the protoplasmic structures found in the bacteria occur in the cyanobacteria. In the central protoplasm are the **circular fibrils of DNA** which are not associated with basic proteins (**histones**) (Figs. 2.2 and 2.3).

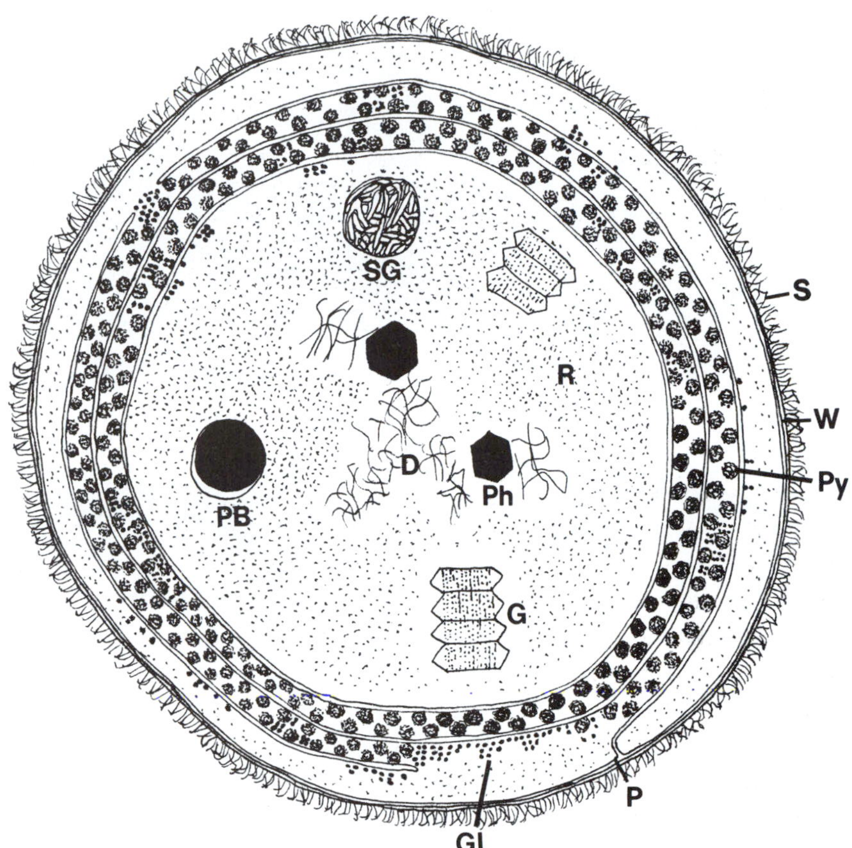

Fig. 2.2 Drawing of the fine-structural features of a cyanobacterial cell. (D) DNA fibrils; (G) gas vesicles; (Gl) glycogen granules; (P) plasmalemma; (PB) polyphosphate body; (Ph) polyhedral body; (Py) phycobilisomes; (R) ribosomes; (S) sheath; (SG) structured granules (cyanophycin granules); (W) wall.

The amount of DNA in unicellular cyanobacteria varies from 1.6×10^9 to 8.6×10^9 daltons. This is similar to the genome size in bacteria (1.0×10^9 to 3.6×10^9 daltons) and is larger than the genome size in mycoplasmas (0.4×10^9 to 0.5×10^9 daltons) (Herdman et al., 1979). The peripheral protoplasm is composed principally of **thylakoids** and their associated structures, the **phycobilisomes** (on the thylakoids, containing the phycobiliproteins) and **glycogen granules**. The **70S ribosomes** are dispersed throughout the cyanobacterial cell but are present in the highest density in the central region around the nucleoplasm (Allen, 1984).

Cyanophycin granules are large bodies composed of stored protein in the form of polypeptides, usually containing aspartic acid and arginine in the form of L-arginyl-poly(L-aspartic acid) (Lawry and Simon, 1982). Cyanophycin granules vary in appearance but usually seem to be full of convoluted membranes (Fig. 2.2). The amount of cyanophycin granule polypeptide varies with the algal growth cycle, being low in exponentially growing cells but high in cells of stationary phase cultures (Simon, 1973).

Carboxysomes or **polyhedral bodies** occur in the central part of cells

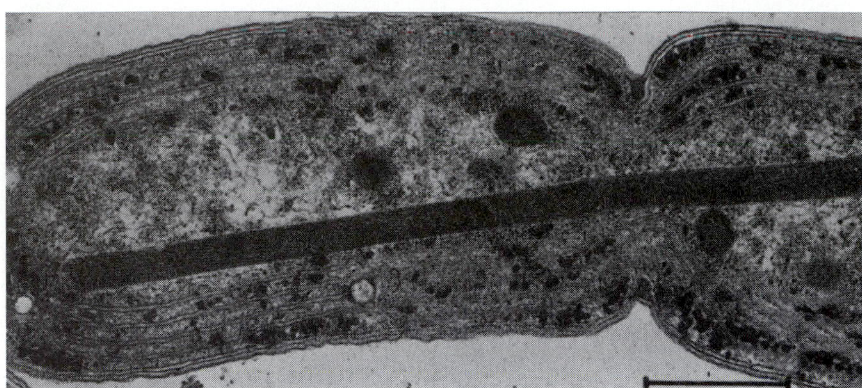

Fig. 2.3 Election micrograph of a dividing cell of *Anacystis nidulans* showing quadrilayered wall, thylakoids in peripheral cytoplasm, DNA microfibrils, ribosomes, and a long peripheral body in the central cytoplasm. Bar = 0.5 μm. (From Gantt and Conti, 1969.)

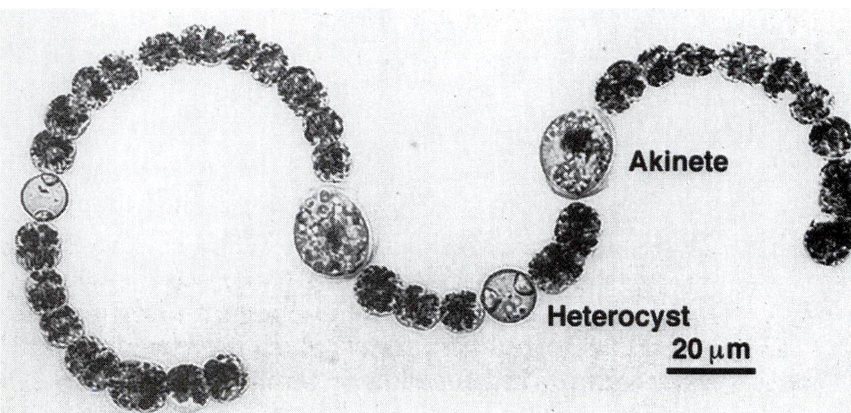

Fig. 2.4 Light micrograph of *Anabaena crassa* showing vegetative cells, akinetes and heterocysts. (From Li et al., 1997.)

(Figs. 2.2 and 2.3). Carboxysomes are similar to the carboxysomes in bacteria and contain the carbon dioxide-fixing enzyme ribulose-1,5-bisphosphate carboxylase/oxygenase (Skleryk et al., 1997). Cyanobacterial photosynthesis is similar to the C_3 type found in higher plants, where the initial carboxylation is carried out by ribulose-1,5-bisphosphate carboxylase/oxygenase. In the cyanobacteria, the enzyme occurs in both soluble and particulate forms. In the particulate form, the enzyme is located in carboxysomes. The amount of a cell occupied by carboxysomes increases as the inorganic carbon (HCO_3^-, CO_2) in the medium decreases (Turpin et al., 1984). Heterocysts (Fig. 2.4) lack ribulose-1,5-bisphosphate carboxylase/oxygenase and the ability to fix carbon dioxide. Heterocysts also lack carboxysomes (Winkenbach and Wolk, 1973).

Polyphosphate bodies (**metachromatic** or **volutin granules**) are spherical and appear similar to lipid bodies of eukaryotic cells in the electron microscope. Polyphosphate bodies contain stored phosphate, the bodies being absent in young growing cells or cells grown in a phosphate-deficient medium, but prominent in older cells (Tischer, 1957).

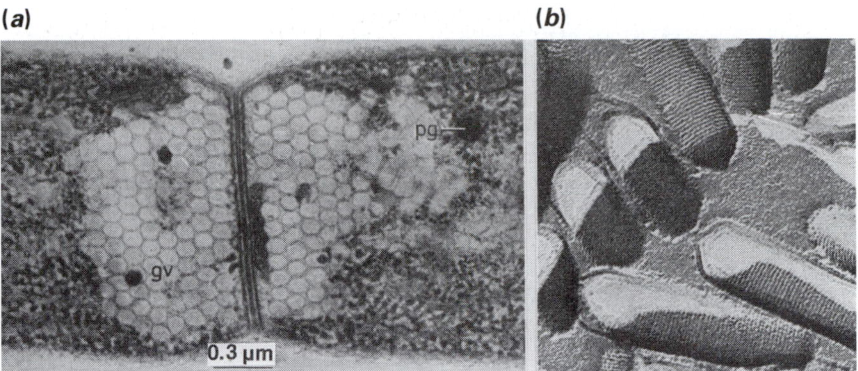

Fig. 2.5 (*a*) Electron micrograph of two cells of *Oscillatoria redekei* showing a cross wall separating areas of gas vacuoles (gv); (pg) lipid droplet. (From Whitton and Peat, 1969.) (*b*) Freeze-etch preparation of gas vacuoles. (From Jones and Jost, 1970.)

Polyglucan granules (**α-granules**) are common in the space between the thylakoids in actively photosynthesizing cells. These granules contain a carbohydrate, composed of 14 to 16 glucose molecules, that is similar to amylopectin (Hough et al., 1952; Frederick, 1951).

Gas vacuoles

A **gas vacuole** is composed of **gas vesicles**, or hollow cylindrical tubes with conical ends, in the cytoplasm of all orders of cyanobacteria except the Chamaesiphonales (Figs. 2.2 and 2.5) (Walsby, 1994; Oliver, 1994). Gas vesicles do not have true protein–lipid membranes, being composed exclusively of protein ribs or spirals arranged similarly to the hoops on a barrel. It is possible to collapse the gas vesicles by applying pressure to the cells, the collapsed vesicles having the two halves stuck together. The membrane of the gas vesicle is quite rigid, with the gas inside it at a pressure of 1 atm. The membrane is permeable to gases, allowing the contained gas to equilibrate with gases in the surrounding solution. The membrane must, however, be able to exclude water. It has been postulated that the inner surface must be **hydrophobic**, thereby preventing condensation on it of water droplets, and restraining, by surface tension, water creeping through the pores. At the same time these molecules must present a **hydrophilic** surface at the outer (water-facing) surface in order to minimize the interfacial tension, which would otherwise result in the collapse of the gas vacuole.

Cyanobacteria possessing gas vacuoles can be divided into two physiological-ecological groups. In the first group are those algae having vacuoles only at certain stages of their life cycle, or only in certain types of cells. In *Gloeotrichia ghosei* and in certain species of *Tolypothrix* and *Calothrix*, gas vesicles appear only in hormogonia. The hormogonia float when they are released, and it is possible that the buoyancy provided is of significance in

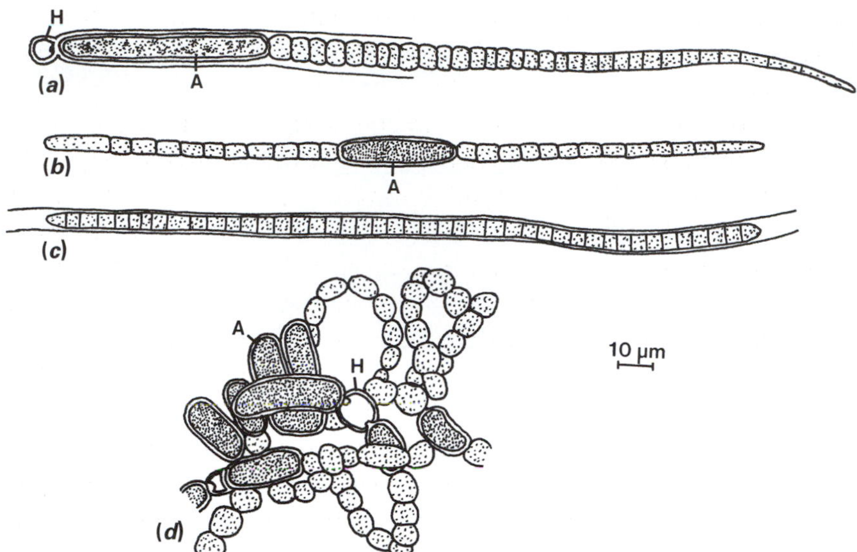

Fig. 2.6 (*a*) *Gloeotrichia echinulata*. (*b*) *Aphanizomenon flos-aquae*. (*c*) *Phormidium inundatum*. (*d*) *Anabaena flos-aquae*. (A) Akinete; (H) heterocyst. (After Prescott, 1962.)

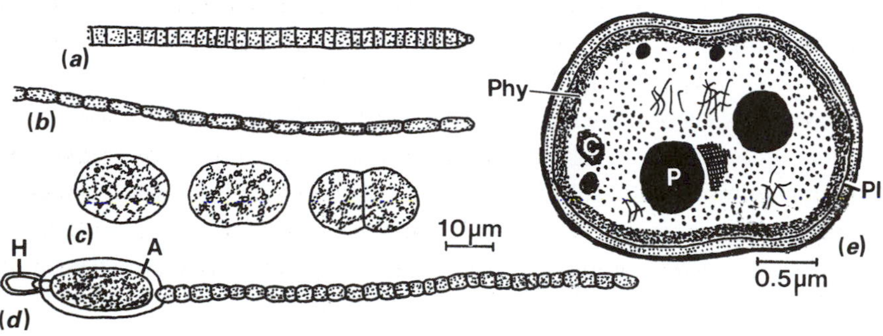

Fig. 2.7 (*a*) *Oscillatoria agardhii*. (*b*) *O. limnetica*. (*c*) *Synechococcus aeruginosus*. (*d*) *Cylindrospermum majus*. (A) akinete; (H) heterocyst. (*e*) Drawing of the fine structure of *Gloeobacter violaceus*. (C) Cyanophycin granule; (P) polyphosphate body; (Phy) probable layer of phycobiliproteins; (Pl) plasmalemma. (*c*, *d* after Prescott, 1962; e after Rippka et al., 1974.)

dispersal of these stages. The second group consists of planktonic cyanobacteria, including species of *Anabaena* (Figs. 2.4, 2.6(*d*) and 2.25(*b*)), *Gloeotrichia* (Fig. 2.6(*a*)), *Microcystis* (Fig. 2.24(*b*), *Aphanizomenon* (Fig. 2.6(*b*)), *Oscillatoria* (Fig. 2.7(*a*),(*b*)), *Trichodesmium* (Fig. 2.24(*g*)), and *Phormidium* (Figs. 2.6(*c*) and 2.24(*c*)). These algae derive positive buoyancy from their gas vesicles, and as a consequence form blooms floating near the water surface. The loss of buoyancy, and subsequent sinking of these algae in the water column, can be due to different factors. In *Anabaena flos-aquae* (Fig. 2.6(*d*)), the loss of buoyancy is caused by the loss of gas vesicles owing to increased turgor pressure, whereas in *Oscillatoria agardhii* (Fig. 2.7(*a*)), buoyancy is lost though the cessation of gas vesicle production and an increase in cell mass. In *Microcystis aeruginosa* (Fig. 2.24(*b*)), loss of buoyancy can be due to entrapment of whole colonies in a colloidal precipitate composed of organic material and iron salts. The colloidal precipitate is formed in certain lakes when dissolved iron in the anoxic water of the hypolimnion in stratified lakes becomes oxidized on mixing with aerated water of the epilimnion (Oliver et al., 1985).

There is a direct relationship between buoyancy and light quantity in nitrogen-fixing cyanobacteria such as *Anabaena flos-aquae* (Fig. 2.6*d*) (Spencer and King, 1985). The relationship is complex and also involves the concentration of ammonium ions (NH_4^+) in the water. Buoyancy in *Anabaena flos-aquae* increases under low irradiance (less than 10 $\mu E\ m^{-2}\ s^{-1}$), absence of NH_4^+, and low CO_2 concentrations. Such conditions occur in many stagnant eutrophic lakes during the summer. In these lakes, rapid growth of algae has depleted the NH_4^+ and CO_2. The water transmits little light because of the large standing crop of algae. Under these conditions, *A. flos-aquae* and other nitrogen-fixing cyanobacteria having gas vacuoles increase their buoyancy and rise close to the surface of the water. Here they are able to outcompete other algae because of their ability to fix nitrogen in water that has little available nitrogen. Cyanobacteria that do not fix nitrogen have reduced growth, and therefore reduced buoyancy, and sink in the water column. Once established, a bloom of buoyant, nitrogen-fixing, cyanobacteria tends to be self-perpetuating in that increased mass of the bloom maintains the reduced light and CO_2 levels required for maximum buoyancy.

Structures other than gas vesicles can cause significant variation in cell density, and therefore buoyancy (Konopka et al., 1987). Polyphosphate granules may have a density of 2 g cm^{-3} or greater, and glycogen (which is accumulated under high light intensities) has a density of about 1.5 cm^{-3}. Both have a higher density than water (1 g cm^{-3}) and can cause cells to sink (Booker and Walsby, 1981).

Pigments and photosynthesis

The major components of the photosynthetic light-harvesting system of the cyanobacteria are **chlorophyll** *a* in the thylakoid membrane, and the **phycobiliproteins**, which are water-soluble chromoproteins assembled into macromolecular aggregates (phycobilisomes) attached to the outer surface of the thylakoid membranes. Those cyanobacteria that were previously classified as Prochlorophyta (Palenik and Haselkorn, 1992; Urback et al., 1992), have **chlorophyll** *b*. The carotenoids of the Cyanophyceae differ from those of the eukaryotic algae in having **echineone** (4-keto-β-carotene) and **myxoxanthophyll**, which eukaryotic algae do not have; in lacking **lutein**, the major xanthophyll of chloroplasts; and in having much higher proportions of β-carotene than are found in eukaryotic algae (Goodwin, 1974).

The Cyanophyceae have four phycobiliproteins: **C-phycocyanin** (absorption maximum at a wavelength [λ] of 620 nm), **allophycocyanin** (λ_{max} at 650 nm), **C-phycoerythrin** (λ_{max} at 565 nm), and **phycoerythrocyanin** (λ_{max} at 568 nm). All cyanobacteria contain the first two, whereas C-phycoerythrin and phycoerythrocyanin occur only in some species. The phycobiliproteins of the cyanobacteria change in concentration in response to light quality and growth conditions. Cyanobacteria that produce the red phycoerythrin and the blue phycocyanin in white light, suppress phycoerythrin synthesis in red light and phycocyanin synthesis in green light (**complementary chromatic adaptation**; see Tandeau de Marsac, 1977).

In the evolution of the cyanobacteria, the thylakoids probably originated by invaginations of the plasmalemma; some cyanobacteria today have thylakoids that are continuous with the plasmalemma. An example of the primitive condition may be *Gloeobacter violaceus* (Fig. 2.7(*e*)), a unicellular cyanobacteria that lacks thylakoids but has chlorophyll *a*, carotenoids, and phycobiliproteins. In this alga the pigments, and presumably photosynthesis, are associated with the plasmalemma (Rippka et al., 1974).

Many cyanobacteria have the ability to photosynthesize under aerobic or anaerobic conditions. Under aerobic conditions, electrons for photosystem I are derived from photosystem II. Under anaerobic conditions, in the presence of sulfur, electrons are derived by the reduction of sulfur:

$$CO_2 + 2H_2S \xrightarrow[\text{chlorophyll}]{\text{light}} \underset{\text{sugar}}{(CH_2O)} + H_2O + S$$

These cyanobacteria are **facultative phototrophic anaerobes** and fill an important ecological niche in aquatic systems (Padan, 1979). Eukaryotic algae are restricted to photoaerobic habitats, whereas photosynthetic bacteria are restricted to photoanaerobic habitats. In habitats that fluctuate between the above conditions, cyanobacteria with facultative anaerobic photosynthesis have a clear selective advantage. An example of this is the Solar Lake, Elat, Israel, where in the winter high levels of sulfide are found in the anaerobic bottom layers of water of the thermally stratified lake. *Oscillatoria limnetica* (Fig. 2.7(*b*)) occurs in these highly anaerobic bottom layers, where the sulfide functions as an electron donor for photosynthesis. In the spring, the lake overturns, with all of the water becoming aerobic. *Oscillatoria limnetica* now activates aerobic photosystem II and carries out photosynthesis aerobically. Thus *O. limnetica*, by utilizing combined anoxygenic and oxygenic photosynthesis, is the dominant phototroph of the Solar Lake, with its fluctuating photoaerobic and photoanaerobic

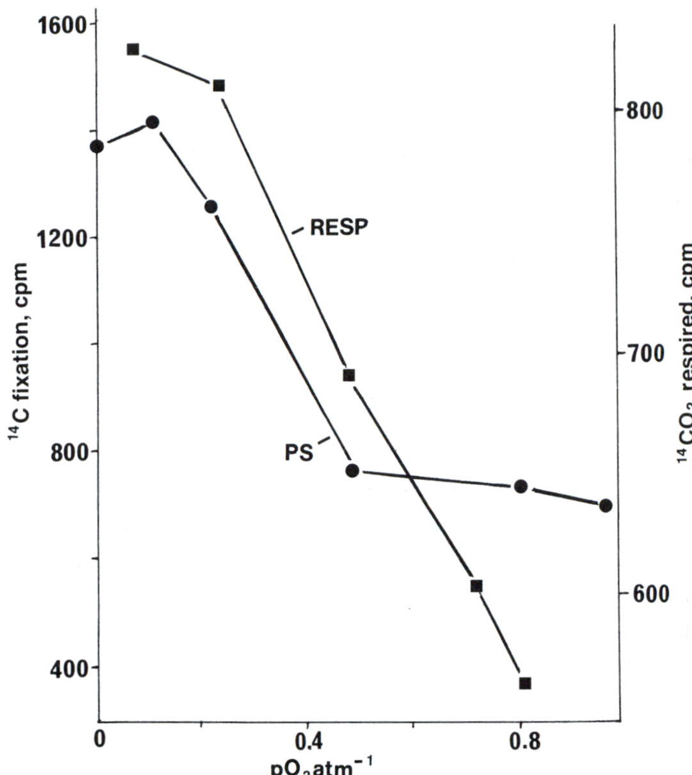

Fig. 2.8 Graph showing stimulation of photosynthesis (●–●) and respiration (■–■) by low concentrations of atmospheric O_2 in *Anabaena flos-aquae*. (After Stewart and Pearson, 1970.)

conditions. The interlinking position of the cyanobacteria in the phototropic world is compatible with the fact that they are among the oldest organisms, dating back to the Precambrian Period. Significantly, two of the sulfide-rich ecosystems containing high numbers of cyanobacteria – that is, hot sulfur springs and the marine littoral sediments – may represent old ecosystems that may predate the oxidized biosphere.

Photosynthesis in many cyanobacteria is stimulated by lowered oxygen concentration, the oxygen competing with carbon dioxide for the enzyme ribulose-1,5-bisphosphate carboxylase/oxygenase (Fig. 2.8) (Stewart and Pearson, 1970; Weller et al., 1975). This phenomenon probably reflects an adaptation to the absence of free oxygen in the atmosphere of Precambrian times when the cyanobacteria first evolved. After the evolution of the oxygen-evolving cyanobacteria, the oxygen in the atmosphere gradually built up, creating a protective ozone (O_3) layer in the atmosphere at the same time. The ozone layer removed most of the harmful ultraviolet radiation from the sun and allowed the evolution of more radiation-sensitive organisms. The cyanobacteria are relatively insensitive to radiation, having a system that repairs radiation damage (Bhattacharjee, 1977).

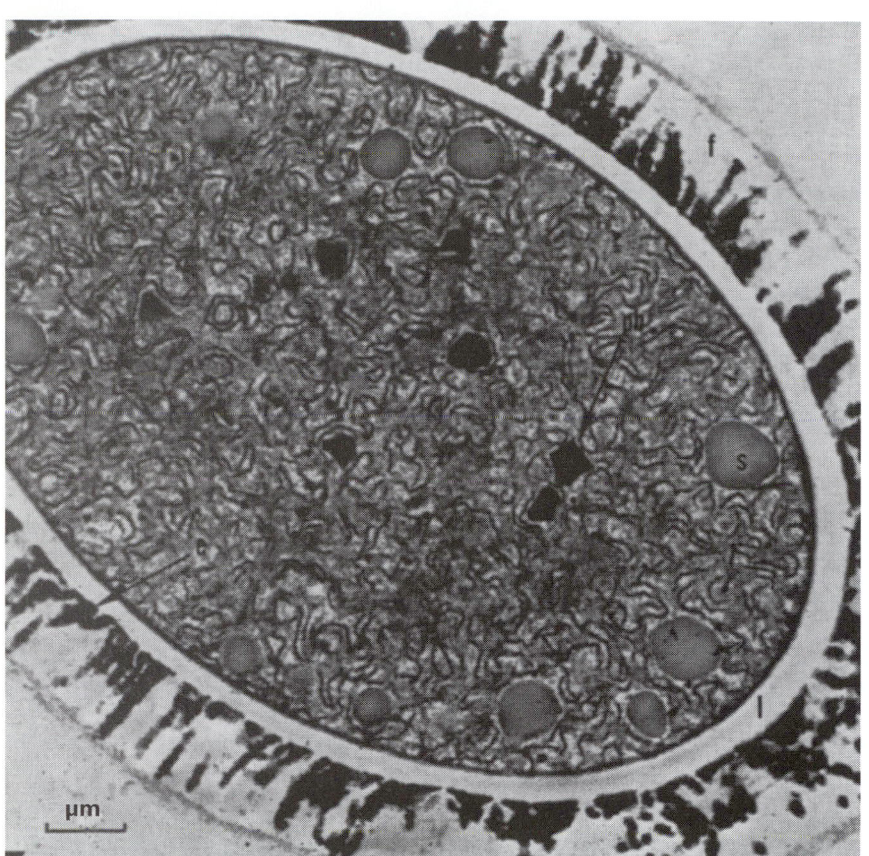

Fig. 2.9 Electron micrograph of a mature akinete of *Cylindrospermum* sp. with a thick layered wall (f,l) and cytoplasm full of proteinaceous structured granules (s), polyhedral bodies, and ribosomes. (From Clark and Jensen, 1969.)

Akinetes

Akinetes or **resting spores** are cells that are resistant to unfavorable environmental conditions. Akinetes occur in the orders Nostocales and Stigonematales. New filaments are produced by germination of akinetes, which can thus be considered to have a function similar to endospores in bacteria.

In *Aphanizomenon* (Fig. 2.6(*b*)) (Wildman et al., 1975), the development of akinetes from vegetative cells involves an increase in cell size, the gradual disappearance of gas vacuoles, and an increase in cytoplasmic density and number of ribosomes and cyanophycin granules. Akinetes of *Nostoc* lose 90% of their photosynthetic and respiratory capabilities, as compared to vegetative cells. This loss occurs even though there is little change in phycocyanin and chlorophyll, the main photosynthetic pigments (Chauvat et al., 1982). Mature akinetes are usually considerably larger than vegetative cells, contain protoplasm full of food reserves, and have a normal cell wall surrounded by a wide three-layered coat (Jensen and Clark, 1969; Cmiech et al.,

1986) (Figs. 2.4, 2.6, 2.7(*d*), and 2.9). Loss of flotation by an increase in cytoplasmic density causes filaments with akinetes to sink and overwinter in bottom sediments. In akinete germination, there is a reverse of the above events.

A wide range of physicochemical factors have been reported to stimulate akinete differentiation; for example, phosphate deficiency, low temperature, carbon limitation and reduction in the availability of light energy (Li et al., 1997, van Dok and Hart, 1995).

Heterocysts

Heterocysts are larger than vegetative cells and appear empty in the light microscope (whereas akinetes appear full of storage products) (Fig. 2.4, 2.6(*d*), 2.7(*d*), and 2.17(*a*),(*d*)). Heterocysts are photosynthetically inactive, they do not fix CO_2, nor do they produce O_2. They also exhibit a high rate of respiratory O_2 consumption and are surrounded by a thick, laminated cell wall that limits ingress of atmospheric gases, including O_2. The internal environment of heterocysts is, therefore, virtually anoxic, which is ideal for nitrogenase, a notoriously O_2 sensitive enzyme.

Heterocysts are formed from vegetative cells by the dissolution of storage granules, the deposition of a multilayered envelope outside of the cell wall, the breakdown of photosynthetic thylakoids, and the formation of new membranous structures (Fig. 2.10) (Kulasooriya et al., 1972). The frequency of heterocyst production is influenced by the concentration of molybdenum and combined nitrogen in the medium (molybdenum is a component of the nitrogen-fixing enzyme, **nitrogenase**). In *Anabaena cylindrica*, heterocyst frequency increases, whereas nitrogenase activity decreases, when the alga is grown under molybdenum-deficient conditions in a medium free of combined nitrogen. Molybdenum deficiency is not required for heterocyst formation. If *A. cylindrica* cells are grown in a medium free of combined nitrogen (ammonia, nitrate, or nitrite) and illuminated for 3 to 9 hours, the alga begins production of heterocysts (Fogg, 1944; Bradley and Carr, 1977). The formation of heterocysts is inversely related to the amount of nitrogen in the alga.

Apart from the exceptional cases of germination, heterocysts are unable to divide. Heterocysts have a limited period of physiological activity and appear to have a limited life. Senescent heterocysts undergo vacuolation and usually break off from the filament, causing fragmentation of the filament.

Heterocysts are dependent on a supply of substrates from adjacent vegetative cells through cytoplasmic connections (**microplasmodesmata**). These cytoplasmic connections probably convey nitrogen fixed in the

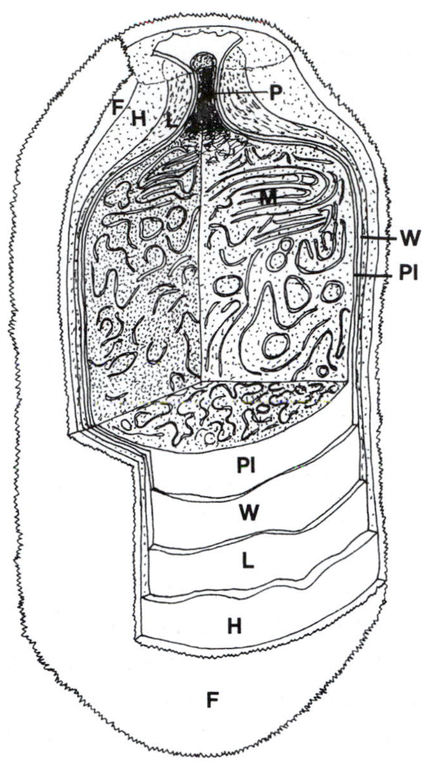

Fig. 2.10 Three-dimensional view of a heterocyst. The envelope has homogeneous (H), fibrous (F), and laminated (L) layers. (M) membranes; (P) pore channel; (Pl) plasmalemma; (W) cell wall. (After Lang and Fay, 1971.)

heterocyst to vegetative cells, whereas fixed carbon metabolites from vegetative cells are conducted to the heterocysts (Giddings and Staehelin, 1981). The transfer of metabolites into the heterocyst is necessary because the heterocyst is incapable of carbon fixation.

Nitrogen fixation

Cyanobacteria are **diazotrophs** (able to fix atmospheric nitrogen). In nitrogen fixation, N_2 from the atmosphere is fixed into ammonium in cyanobacterial cells using ATP as a source of energy and either molybdenum or vanadium as a cofactor (Attridge and Rowell, 1997) (Fig. 2.11). The ammonium is then incorporated into amino acids, such as glutamic acid and alanine. Nitrogen fixation occurs only in the prokaryotic bacteria and cyanobacteria, with cyanobacteria alone evolving oxygen during photosynthesis. This is important because the nitrogen-fixing enzyme, **nitrogenase**, is very sensitive to inactivation by oxygen. The cyanobacteria have evolved three different mechanisms for separating nitrogenase from oxygen:

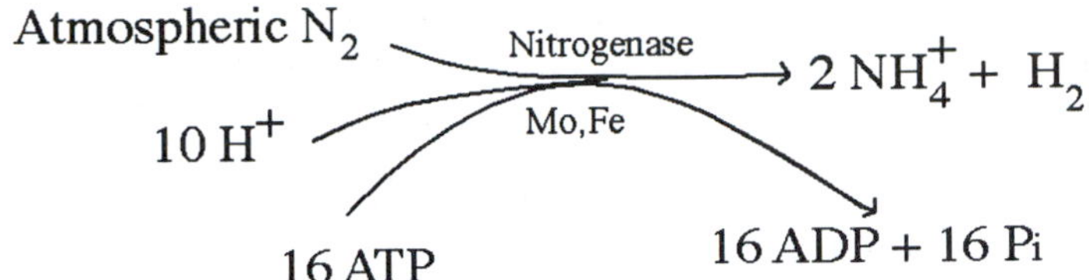

Fig. 2.11 Atmospheric nitrogen is fixed by cyanobacteria into ammonium which is incorporated into amino acids.

1 *Heterocystous cyanobacteria* Nitrogen fixation occurs in heterocysts in these cyanobacteria in the light. Heterocysts lack photosystem II of photosynthesis and, therefore, the ability to evolve oxygen. Yet by cyclic photophosphorylation, heterocysts can form the necessary ATP for nitrogen fixation. Heterocysts also contain a form of myoglobin called **cyanoglobin** that scavenges any oxygen present, preventing inhibition of nitrogenase (Potts et al., 1992). Under anaerobic conditions in an atmosphere of carbon dioxide and nitrogen, both the vegetative cells and heterocysts can fix nitrogen.

2 *Non-filamentous cyanobacteria that fix nitrogen in the dark but not in the light* These cyanobacteria fix nitrogen in the dark when photosynthesis is not producing nitrogenase-inhibiting oxygen. If *Synechococcus* (Fig. 2.7(*c*)) is grown under a 12-hour light: 12-hour dark cycle, most of the photosynthesis occurs during the dark period (Fig. 2.12). If the cells are subjected to continuous illumination, an endogenous timing cycle entrained by cell division continues to alternate the level of photosynthesis and nitrogen fixation (Mitsui et al., 1986; Chen et al., 1996). Nitrogenase activity peaks at the same time that it had in the previous dark period (Fig. 2.12). At the beginning of the next light period, oxygen is produced by photosynthesis and the nitrogenase is inactivated. New nitrogenase must be synthesized at the beginning of the next dark period for nitrogen fixation to occur.

3 *Trichodesmium* This is a filamentous, non-heterocystous cyanobacterium that fixes nitrogen under aerobic conditions only in the light (Bergman et al., 1997). The cycle of nitrogen fixation is entrained by an endogenous circadian clock (Capone et al., 1997). *Trichodesmium* produces colonies that are a major fraction of the planktonic flora of tropical seas (see p. 88) and are estimated to be responsible for fixing one quarter of the total nitrogen in the world's ocean (Bergman and Carpenter, 1991). Within the *Trichodesmium* filament, some cells are specialized to fix nitrogen, while others do not fix nitrogen. Cells that fix nitrogen are adjacent to one another. Cytologically, nitrogen-fixing cells have a denser thylakoid network with fewer gas vacuoles and cyanophycin granules (Fredriksson and Bergman, 1997).

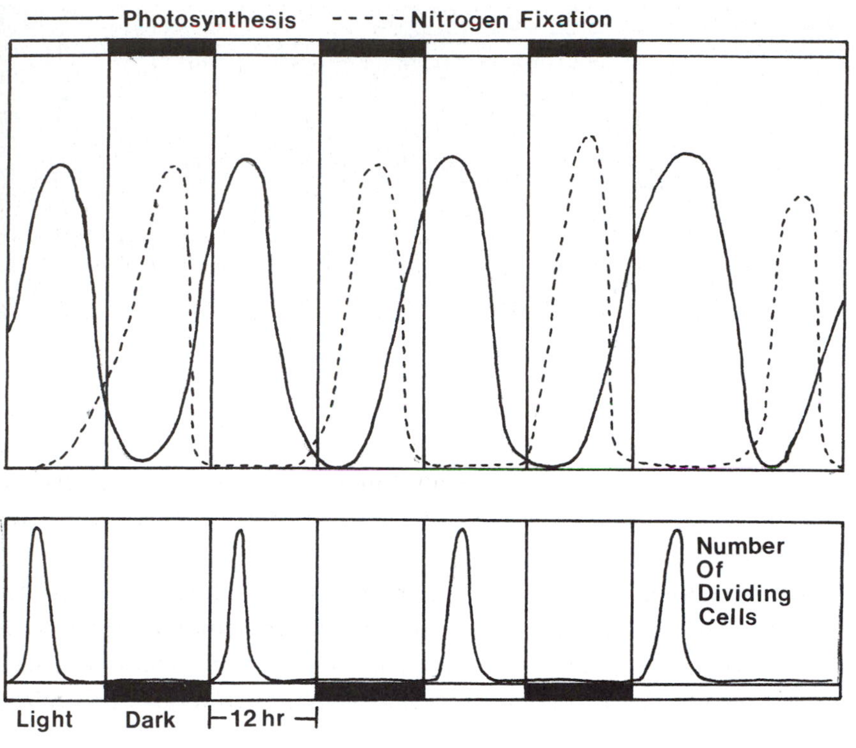

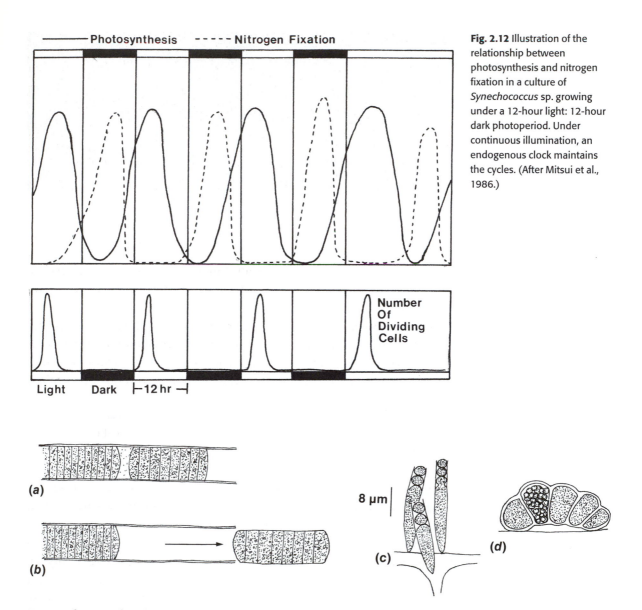

Fig. 2.12 Illustration of the relationship between photosynthesis and nitrogen fixation in a culture of *Synechococcus* sp. growing under a 12-hour light: 12-hour dark photoperiod. Under continuous illumination, an endogenous clock maintains the cycles. (After Mitsui et al., 1986.)

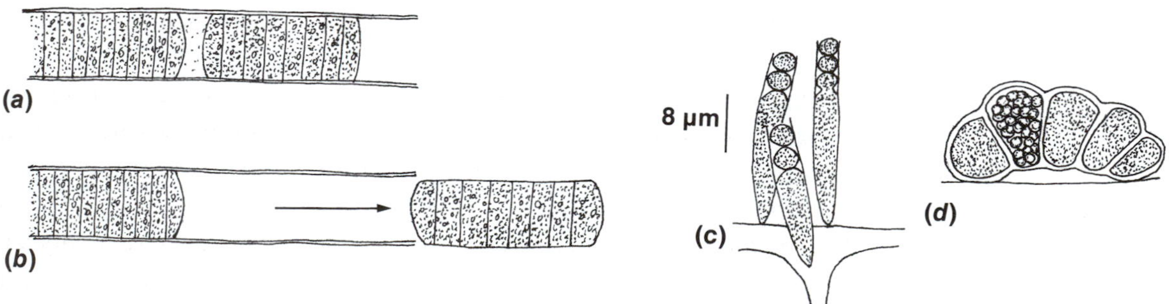

Fig. 2.13 (*a*), (*b*) Formation of a hormogonium in *Oscillatoria*. (*c*) Endospore formation in *Chamaesiphon incrustans*. (*d*) Endospore formation in *Dermocarpa pacifica*. (After Smith, 1950.)

Asexual reproduction

Asexual reproduction occurs by the formation of hormogonia or endospores (exospores are modified endospores) or by fragmentation of colonies (Fig. 2.13).

Hormogonia (or hormogones), which are characteristic of all truly filamentous cyanobacteria, are *short pieces of trichome that become detached from the parent filament and move away by gliding, eventually developing into a separate filament.* In simple filamentous algae such as

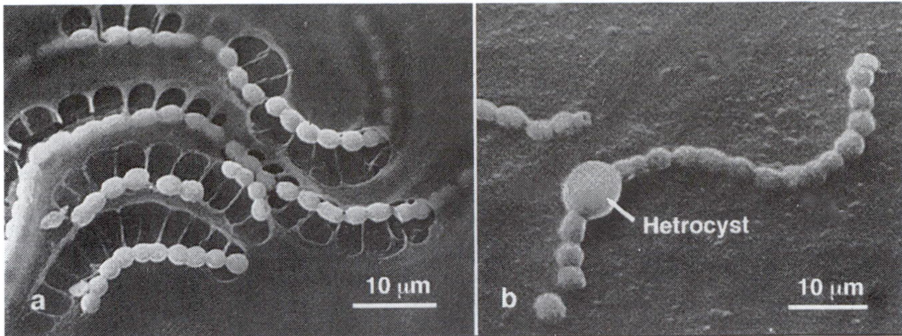

Fig. 2.14 Scanning electron micrographs of *Nostoc commune*. (From Hill et al., 1994.)

Oscillatoria and *Cylindrospermum*, the entire filament may break up (Fig. 2.13(*a*),(*b*)), whereas in others the hormogonia are produced at the tips of special branches. In some algae, specialized **separation discs** or **necridia** are involved in the breaking of the hormogone from the parent filament, whereas in others the filament just fractures.

Whereas akinetes are perennating structures, **endospores** are true reproductive cells formed in the cyanobacteria. These form by internal division of the protoplast, resulting in an irregular mass of spores (*Dermocarpa*) or a row of spores (*Chamaesiphon*) (sometimes called **exospores** or **budding**) (Fig. 2.13).

Growth and metabolism

In the cyanobacteria there are three nutritional types: (1) **facultative chemoheterotrophs**, or those organisms capable of growing in the dark on an organic carbon source and of growing phototrophically in the light (only a portion of the cyanobacteria exhibit this condition); (2) **obligate phototrophs**, or organisms that can grow only in the light on an inorganic medium (some of these are actually **auxotrophs**, requiring a small amount of an organic compound that is not used as a source of carbon, invariably meaning a vitamin); (3) **photoheterotrophs**, or those cells that are able to use organic compounds as a source of carbon in the light but not in the dark (Stanier, 1973).

Facultative chemoheterotrophs are able to grow in the dark on a very narrow range of substrates, being confined to glucose, fructose, and one or two disaccharides. This range of substrates is so small because the pentose phosphate pathway is the sole energy-yielding dissimilatory pathway. The tricarboxylic acid cycle lacks the enzymes α-ketoglutarate dehydrogenase and succinyl CoA synthetase, rendering it incomplete, and glycolysis likewise appears to be incomplete. Although the tricarboxylic acid cycle does not provide energy, it does provide carbon skeletons from the portions of it that are functional.

The pentose phosphate pathway does not operate in the *light* in the Cyanophyceae (although it is highly operational in the dark), being inhibited by ribulose-1,5-diphosphate, a product of light metabolism.

Even though some cyanobacteria grow in the dark, they grow very slowly. It is probable that the rate of chemoheterotrophic growth is always limited by the rate of dark ATP synthesis through oxidation of glucose-6-phosphate in the pentose phosphate pathway, a rate that is evidently not very great.

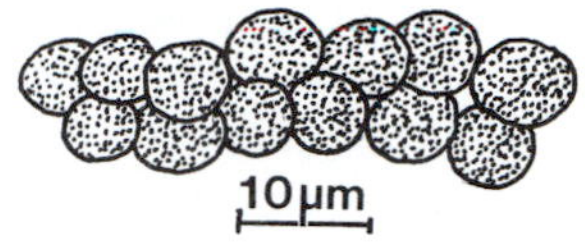

Fig. 2.15 *Chlorogloea fritschii.*

LACK OF FEEDBACK CONTROL OF ENZYME BIOSYNTHESIS

Cyanobacteria, such as *Chlorogloea fritschii* (Fig. 2.15), lack metabolic control of many pathways by repression and derepression of enzyme biosynthesis as occurs in many other organisms. In other organisms, as an end product of a pathway builds up to excess, the end product represses an enzyme involved in its synthesis, thereby directing precursors to other parts of the cell's metabolism, where they can be better used. When the amount of end product falls, the enzyme is derepressed, allowing manufacture of the end product again.

Cyanobacteria often release considerable quantities of nitrogenous and organic substances into the medium, with most of the compounds being excreted as peptides (Fogg, 1942, 1952; Walsby, 1974). The secretion of nitrogenous substances represents a considerable waste of potential metabolites which have been formed from carbon dioxide with an expenditure of ATP and reducing potential (e.g., NADH). Such an excretion, however, could well be a necessary consequence of ill-controlled amino acid biosynthesis. Unable accurately to adjust the synthesis of each amino acid to the needs of protein synthesis, the cells would inevitably have to synthesize an excess of amino acids to ensure that protein synthesis would proceed smoothly. This excess would then form the basis of the extracellular peptide material associated with the cyanobacteria.

The general lack of repression and derepression of enzyme synthesis in the cyanobacteria implies that the algal metabolic pathways developed before the control systems responsible for their regulation. The failure to develop transcriptional or regulatory control places the cyanobacteria under a selective disadvantage relative to other organisms possessing such metabolic regulators. They would be unable to divert biosynthetic capacity away from surplus synthesis to more necessary activity. Such a selective advantage would survive only in microorganisms less open to competition than most, or, in other words, in organisms in a particular ecological niche or in an environment with an abundant energy source. Autotrophic microorganisms such as the cyanobacteria fall into this category. Their source of energy (light) is at times more than sufficient, and in certain cases they occur in eccentric ecological situations that extremes of temperature or pH make relatively inhospitable.

Gliding and swimming

Some of the cyanobacteria are capable of movement by **gliding**, that is, *the active movement of an organism in contact with a solid substrate where there is neither a visible organ responsible for the movement nor a distinct change in the shape of the organism* (Jarosch, 1962). In gliding, the cells do not have the ability to steer in a particular direction, and in some species gliding is accompanied by rotation of the trichome. Gliding takes place within a mucilage sheath, with the sheath sticking to the substrate and being left behind by the advancing trichome. Regularly arranged fibrillar extensions of the protoplasm moving in waves against the sheath may be responsible for the propulsive force in gliding (Halfen and Castenholz, 1971).

A marine species of *Synechococcus* (Fig. 2.7(*c*)) has been isolated that is capable of **swimming** at speeds of 5 to 25 $\mu m\ s^{-1}$. The cells have no flagella or other organelles of motility. The swimming takes place without attachment to the substrate. The mechanism of swimming is unknown, but it is not related to gliding. **Chemotaxis** (*the movement of a whole cell in response to a concentration gradient of a chemical substance*) toward nutrient-enriched areas in the water may be the driving stimulus of the swimming (Waterbury et al., 1985).

Phototaxis is *an oriented movement either toward (positive phototaxis) or away from (negative phototaxis) a light source.* The cyanobacteria can be divided into two different classes on the basis of their phototactic response (Häder, 1987). The first group is represented by filaments of *Phormidium*, which show positive phototaxis but not negative phototaxis. The mechanism involves the front end of the filament receiving more light than the rear end, resulting in reorientation of the algal filament toward the light. Each cell is not capable of detecting the light direction, and the filament rotates around its axis during movement. In the second group, true steering occurs, with each cell of the filament able to detect the direction of the light. In *Anabaena*, the tip of the filament turns in the direction of the light source, and the filament does not rotate during movement. At high light intensities (27 $W\ m^{-2}$), the organism shows negative phototaxis and moves away from light.

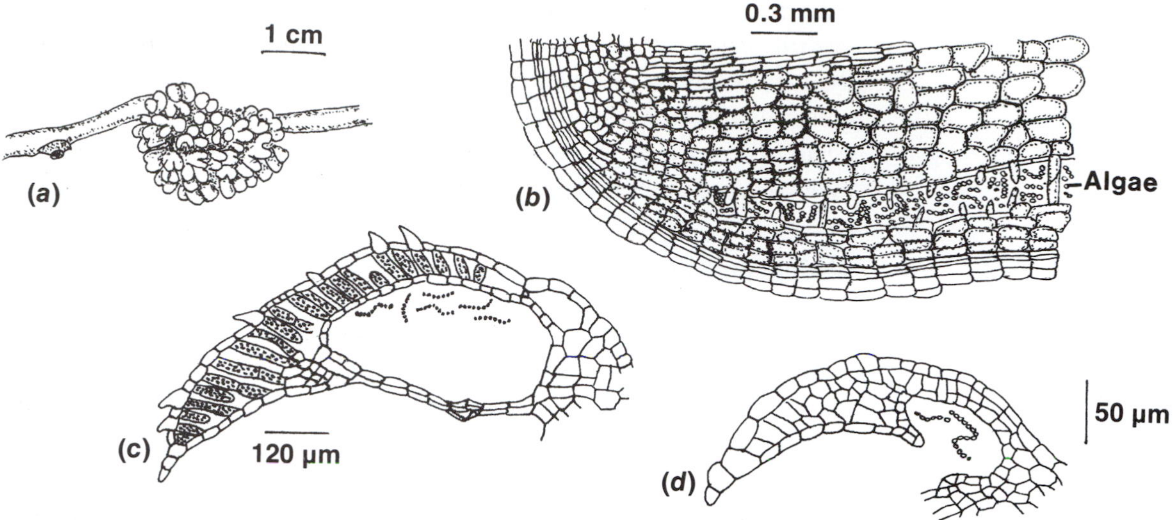

Fig. 2.16 (*a*) Root nodules of the cycad *Dioon spinulosum*. (*b*) Part of a longitudinal section of a nodule of *Encephalartos* with an algal zone. Longitudinal sections through the dorsal lobe of a mature (*c*) and an immature (*d*) leaf of *Azolla* illustrating filaments of *Anabaena* in leaf cavities. ((*a*), (*b*) after Spratt, 1915; (*c*), (*d*) after Smith, 1955.)

Symbiosis

The cyanobacteria occur in basically two types of associations, those in which the cyanobacterium is *extracellular* and those in which it is *intracellular*.

EXTRACELLULAR ASSOCIATIONS

The most common type of extracellular association is that with fungi to form lichens. In most lichens there is a single algal component (**phycobiont**) which is a green alga, a cyanobacterium, or, as in species of *Verrucaria*, a member of the Xanthophyceae. Cyanobacteria occur in about 8% of the species of lichens. Almost all of the fungal partners (**mycobionts**) are ascomycetes, but certain imperfect fungi and basidiomycetes also occur in lichens (Ahmadjian and Hale, 1973). Because the mycobionts frequently reproduce sexually whereas the phycobionts usually do not, the lichens are classified with the fungi. In the symbiotic association, the phycobiont fixes carbon in photosynthesis and liberates it as glucose, which the mycobiont converts into mannitol and assimilates. Approximately 40% of the photosynthate of the alga is secreted as glucose in the lichen association (Smith et al., 1970). Whereas the benefits to the fungus in the association are obvious, the benefits to the alga are probably limited to some protection against desiccation.

The water fern *Azolla* has cavities in the dorsal lobe of the leaf that are occupied by *Anabaena azollae* (Fig. 2.16(*c*),(*d*)). This cyanobacterium fixes nitrogen (Peters and Mayne, 1974), some of which is excreted into the cavity and taken up by the cells of the *Azolla*. It is possible to obtain algafree *Azolla* plants by treating them with antibiotic or subjecting them to growth conditions that allow the fern to outgrow the alga. The *Anabaena* is apparently unable to live outside of the host. The heterocyst frequency (compared to vegetative cells) in *A. azollae* is about 30%, which is very high, the highest for free-living *Anabaena* being about 8% (Hill, 1975). The host *Azolla* somehow modifies the surface properties of the *Anabaena*. The alga in the fern cavity has different surface antigens, as compared to when it is growing free in culture (Gates et al., 1980). Nitrogen fixation by the cyanobacterial symbionts of *Azolla* has for many years been utilized to advantage in the Far East; here the water fern has been used as a green manure in rice fields, where about 3 kg of atmospheric nitrogen per hectare per day is fixed by *Azolla* symbionts (Swaminathan, 1984, Canini, et al., 1992).

Marine sponges generally contain symbiotic cyanobacteria and/or bacteria. The endosymbionts maybe free-living in the mesophyl, aggregated in specialized vacuolated archaeocytes (cyanocytes), or contained in digestive vacuoles. The cyanobacteria are involved, at least partially, in the synthesis of the characteristic secondary metabolites of sponges which seem to deter grazing by fish through their toxicity or unpalatability (Green, 1977; Berthold et al., 1982).

Cycad roots are frequently infected with cyanobacteria that cause distortions of the roots (coralloid roots) (Fig. 2.16). In these roots the cyanobacteria occupy a clearly defined cortical zone midway between the pericycle and the epidermis. The cyanobacteria occur in the intercellular spaces and are surrounded by a sheath and a multilayered wall, whereas the adjacent cycad cells have a reduced chromosome number and secrete slime (Storey, 1968; Caiola, 1975). These cyanobacteria are able to fix nitrogen and contribute a portion of the fixed nitrogen to the cycad cells (Watanabe and Kiyohara, 1963).

INTRACELLULAR ASSOCIATIONS

Intracellular associations involving cyanobacteria are usually more specialized than extracellular associations, and it is not possible to culture the cyanobacteria away from the host. Pascher (1914) coined the term **cyanelle** for intracellular cyanobacteria in symbioses, and the term **cyanome** for the host cell. Cyanelles are present in a variety of cyanomes; such a variety that it is apparent that they are the result of a number of different symbiotic establishments and are not derived from a single event. It is probable that one of these endosymbiotic associations led to the development of chloroplasts in some of the algal groups. This line of evolution is considered under Glaucophyta (see beginning of Chapter 3).

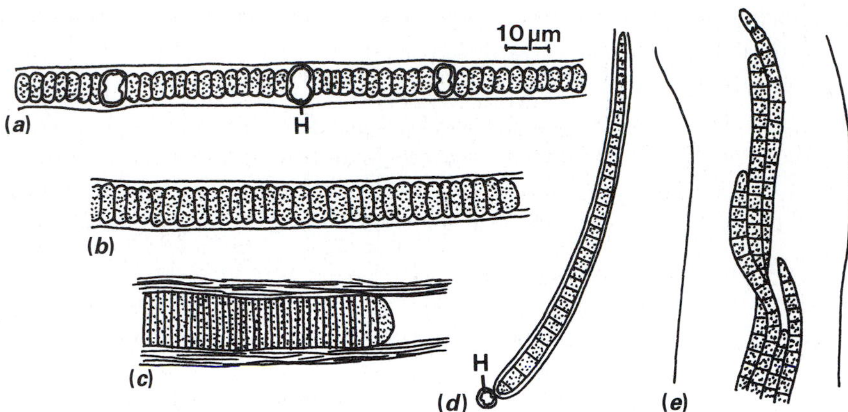

Fig. 2.17 (*a*) *Nodularia spumigena*. (*b*) *Lyngbya sordida*. (*c*) *L. majuscula*. (*d*) *Calothrix scopulorum*. (*e*) *Microcoleus chthonoplastes*. (H) Heterocyst. ((*b*), (*c*), (*e*), after Desikachary, 1959.)

Ecology of cyanobacteria

MARINE ENVIRONMENT

Littoral zone

Cyanobacteria in the littoral zone occur as a black encrusting film on rocks at the upper limit of the high-tide mark. This zone is sandwiched between the lower barnacle zone (the grazing of the molluscs in this zone limits the cyanobacteria) and a higher zone of maritime lichens such as *Verrucaria* and *Lichina*, the latter containing the cyanobacterium *Calothrix*, as the phycobiont. The brackish cyanobacterial zone is composed of various species of *Calothrix* (Fig. 2.17(*d*)), *Phormidium* (Figs. 2.6(*c*) and 2.24(*c*)), *Nodularia* (Fig. 2.17(*a*)), *Gloeothece* (Fig. 2.24(*a*)), and *Rivularia* (Fig. 2.25(*d*)) (Little, 1973). The vertical extent of this zone tends to be greater with increasing exposure of the shore; the greater the spray, the wider the zone. In most exposed areas it forms a thin adhering film, but in sheltered areas the growth may be much thicker, up to 5 mm. The nature of the rock is of some importance, with soft or granular rocks such as chalk and sandstones, respectively, sustaining the greatest growth. Most of the cyanobacteria in the littoral zone are nitrogen-fixing, and they make a significant contribution to the productivity of rocky shores and coral reefs (Mague and Holm-Hansen, 1975).

Ooids are spherical (0.2–2.0 mm in diameter), concentrically laminated, carbonate grains that form by carbonate accretion in agitated shallow tropical marine environments. *Hyella* spp. are cyanobacterial endoliths that bore into, and live in, ooids (Fig. 2.18) (Al-Thukair and Golubic, 1991). Extant *Hyella* spp. are similar to extinct *Eohyella* in 800 million-year old ooids (Fig. 2.18).

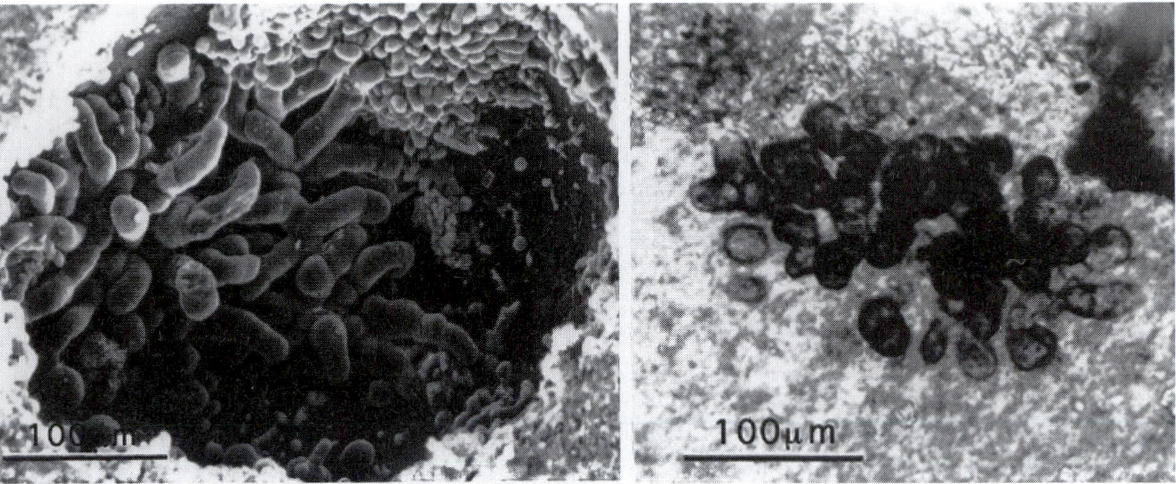

Fig. 2.18. *Left:* The endolith *Hyella immanis* inside a carbonate ooid sand grain. *Right:* The fossil endolith *Eohyella dichotoma* penetrating an ooid sand grain from the Late Proerozoic. (From Al-Thukair and Golubic, 1991.)

Open ocean cyanobacteria

In the open ocean, most of the total photosynthetic capacity is made up of **picophytoplankton** (*phytoplankton cells unable to pass through a filter with 2-µm-diameter holes*). The picophytoplankton is made up principally of tiny coccoid cyanobacteria at concentrations of around 10000 cells per milliliter (Glover et al., 1986). The coccoid cyanobacteria *Synechococcus* (Fig. 2.7(*c*)) and *Synechocystis* (Fig. 2.1) are the major organisms present. Although these cyanobacteria are small, they can be easily seen because their phycobiliproteins undergo autofluorescence in a fluorescence microscope. The picophytoplankton cyanobacteria have negligible sinking rates; thus they are ideally suited for planktonic life. The high surface:volume ratios, combined with steep diffusion gradients that are set up around small cells, allow them to take up nutrients at a high rate. Furthermore, a given amount of photosynthetic pigment dispersed in small cells absorbs more light than an equivalent amount of pigment packaged in large cells. Therefore, these cells grow best at low light intensities of less than 1/50 of full sunlight. The picophytoplanktonic cyanobacteria are more evident in nutrient-poor offshore waters where larger phytoplankton are less successful, because the larger phytoplankton are less able to utilize low concentrations of nutrients. Although the picophytoplanktonic cyanobacteria occur throughout the euphotic zone, they are concentrated at the bottom of the zone, not because they sediment out, but because they grow best under these conditions; they use the low irradiance efficiently to support growth and benefit from nutrients transported up from richer waters below. These algae contain large amounts of phycoerythrin that allows them to absorb the blue-green light penetrating into deep water (Fogg, 1986).

Cyanobacteria larger than picophytoplankton often form a significant part of oceanic phytoplankton. Massive development of filaments of the nitrogen-fixing *Trichodesmium* (Fig. 2.24(*g*)) (Carpenter et al., 1992) occurs in certain tropical waters. Each colony of *Trichodesmium* (sometimes considered a synonym for *Oscillatoria*) consists of a mass of filaments that secrete a flocculent mucilage which supports bacterial colonies; these in turn are fed on by different protozoa. The large surface area produced by the algal filaments forms a miniature ecosystem (Andersen, 1977). *Trichodesmium* is a major component of the Caribbean Sea plankton, comprising 60% of the total chlorophyll *a* in the upper 50 m and about 20% of the primary production. It is also an important source of nitrogen, fixing 1.3 mg of nitrogen per square meter per day (Carpenter and Price, 1977). The cells produce gas vacuoles which, under calm conditions, cause the cells to accumulate at the water surface, giving rise to the phenomenon known to sailors as "**sea sawdust**," or long orange or gray windrows of algae. One such bloom stretched 1600 km along the Queensland coast of Australia, extending from the shore to the Great Barrier Reef, and occupying an area of 52 000 km^2 (Ferguson-Wood, 1965).

Trichodesmium moves up in the water column by means of gas vesicles, and moves down in the water column by carbohydrate ballasting (Romans et al., 1994). The cells become progressively heavier from morning to evening as carbohydrates and polyphosphate bodies are produced. In the Caribbean, *Trichodesmium* cells have been found down to 200 m. The gas vesicles of this cyanobacterium are much stronger and more difficult to collapse than those found in any freshwater alga (Walsby, 1978). The gas vesicles are able to stand up to 20 atm of pressure, enabling *Trichodesmium* to rise from great depths.

FRESHWATER ENVIRONMENT

Freshwater blooms of cyanobacteria are common. Most freshwater blooms of cyanobacteria consist of *Microcystis* (Fig. 2.24(*b*)), *Anabaena* (Fig. 2.4, 2.6(*d*), 2.25(*b*)), *Aphanizomenon* (Fig. 2.6(*b*)), *Gloeotrichia* (Fig. 2.6(*a*)), *Lyngbya* (Fig. 2.17(*b*),(*c*)), or *Oscillatoria* (Fig. 2.7(*a*),(*b*)). Although they occur in lakes over the whole year, it is usually only in late summer and early autumn that they reach bloom proportions.

This because of (Tang et al., 1997):

1 the superior light-capturing abilities of cyanobacteria when self shading is the greatest.
2 their high affinity for nitrogen and phosphorus when nutrients limitation is most severe.

3 their ability to regulate their position in the water column by gas vacuoles to take advantage of areas richer in nutrients and/or light.
4 their higher temperature optima for growth and photosynthesis (greater than 20 °C).

Even in the Arctic and Antarctic, cyanobacteria dominate the algal flora in the late summer. These cyanobacteria are **psychrotrophs**, able to tolerate the severe conditions during the winter and then grow in the warmer summer months (as contrasted to **psychrophiles** that are able to grow at temperatures less than 15 °C (Tang et al., 1997).

Hot-spring cyanobacteria

Cyanobacteria are important in the colonization of non-acidic hot springs throughout the world. Some cyanobacteria have the ability to grow at temperatures as high as 70 to 73 °C, a much higher temperature tolerance than occurs in eukaryotic algae. In thermal environments above 56 to 60 °C, both photosynthetic and non-photosynthetic eukaryotic algae are always absent. In acid springs (pH less than 4), no cyanobacteria are present, and at temperatures above 56°C in these springs there are no photosynthetic organisms at all (Brock, 1973). *Mastigocladus laminosus* (Fig. 2.19(*a*)) is a cyanobacterium that occurs in thermal springs throughout the world, whereas other cyanobacteria such as *Synchococcus lividus* and *Oscillatoria terebriformis* have more restricted ranges (Castenholz, 1973). The cyanobacteria normally occur as mats mixed with flexibacteria, the cyanobacteria usually being more prevalent in the upper portion of the mat. The **thermophilic** cyanobacteria are especially adapted to live at elevated temperatures. *Synechococcus lividus* can grow at temperatures up to 73 °C but ceases to grow in culture when the temperature is lowered to 54 °C, with optimum growth occurring from 60 to 63 °C (Meeks and Castenholz, 1971); the cells will die if kept at 30 °C for 10 days. Enzymes isolated from thermal algae are more stable at higher temperatures than those from other organisms. For example, $NADPH_2$-cytochrome *c* reductase extracted from the thermal alga *Aphanocapsa thermalis* showed unimpaired activity after heating to 85 °C for 5 minutes, whereas that from *Anabaena cylindrica* was completely inactivated by a similar treatment.

TERRESTRIAL ENVIRONMENT

Terrestrial cyanobacteria play a major role as primary colonizers in the establishment of a soil flora and in the accumulation of humus. They do this in four main ways:

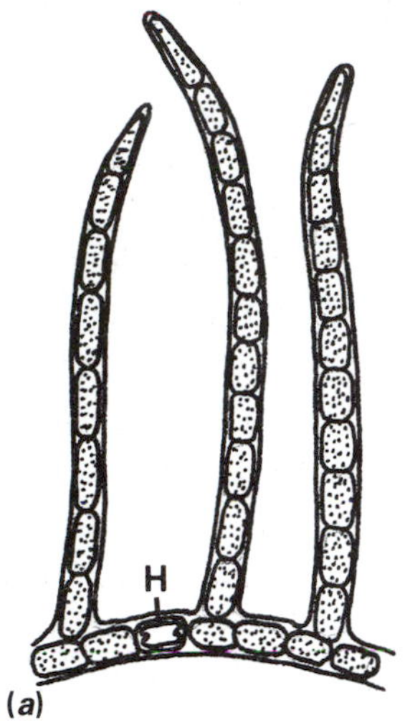

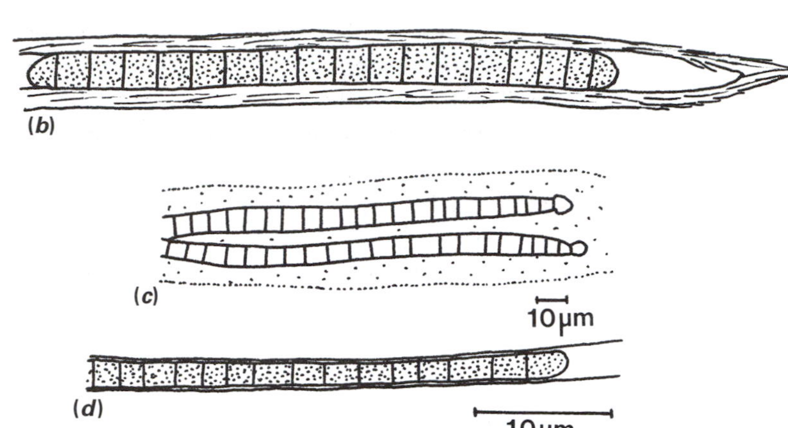

Fig. 2.19 (*a*) *Mastigocladus laminosus*. (H) Heterocyst. (*b*) *Porphyrosiphon notarisii*. (*c*) *Microcoleus vaginatus*. (*d*) *Plectonema notatum*. (*c*, *d* after Prescott, 1962.)

1 They bind sand and soil particles and prevent erosion. They do this with their gelatinous sheaths and by their growth pattern which produces closely intertwined ropelike bundles in and among soil particles. Genera that commonly perform this function are *Porphyrosiphon* (Fig. 2.19(*b*)), which covers large eroded areas in Brazil (Drouet, 1937), *Microcoleus* (Fig. 2.17(*e*)), *Plectonema* (Fig. 2.19(*d*)), *Schizothrix* and *Scytonema* (Fig. 2.26(*c*),(*d*)).
2 They help to maintain moisture in the soil. Booth (1941), in studies in Oklahoma, found that soil with an algal covering had a moisture content of 8.9% compared to 1.3% in the absence of algae.
3 They are important as contributors of combined nitrogen through nitrogen fixation. In grasslands, the soil surface between crowns of grasses may support extensive zones of cyanobacteria and lichens that include cyanobacteria as their phycobiont constituting up to 20% of the ground cover (Kapustka and DuBois, 1987).
4 It has been suggested that cyanobacteria assist higher plant growth by supplying growth substances.

Anhydrobiotics are organisms that can withstand the removal of the bulk of their intracellular water for extended periods of time. The cosmopolitan terrestrial cyanobacterium *Nostoc commune* is able to tolerate acute water stress and can survive in the air-dry state for many years. Approximately 0.1 g of blackened air-dried colonies becomes an olive-green rubbery mass of more than 20 g wet weight within 30 min of rehydration. In *Nostoc commune* (Figs. 2.14, 2.20(*b*)), rehydration rather than desiccation, appears to be the fatal event. To protect the cells during rehydration, a water stress protein and large amounts of the sugar trehalose are synthesized that stabilize the phospholipid bilayers of cellular membranes (Scherer and Potts, 1989; Potts, 1996).

Cyanobacteria comprise the dominant component of the soil photosynthetic community in hot and cold arid regions where higher plant vegetation is absent or restricted. Desert **cryptobiotic crusts** are initiated by the growth of cyanobacteria in the soil during episodic events of available moisture. The only cyanobacteria that are initial colonizers are those that have heterocysts, and are therefore able to fix nitrogen; and those cyanobacteria that produce **scytonemin**, a sunscreen that accumulates in the cyanobacterial sheaths and absorbs some of the strong sunlight in the near ultraviolet (370–384 nm) (Garcia-Pichel and Castenholz, 1991; Garcia-Pichel and Belnap, 1996). Mineral particles are entrapped in the network of algal filaments and the extracellular slime that the cyanobacteria produce. Subsequently, lichens, fungi and moss establish themselves in the crust, enriching and stabilizing the soil.

Nitrogen-fixing cyanobacteria make a major contribution to the fertility of paddy fields (Swaminathan, 1984). In many Eastern countries, peasant farmers do not fertilize their fields; the nitrogen is fixed by cyanobacteria, thus permitting a moderate harvest when in their absence there would be a poor one. There are about 100 million km^2 of paddy fields, of which some are in southern Europe and the United States, but about 95% are in India and the Far East. Usually rice is grown on land that is submerged under 10 cm or so of water for 60 to 90 days during the growing season, and then allowed to dry to facilitate harvest. The warm conditions demanded by the rice, the availability of nutrients, the reducing conditions in the soil, and the cyanobacterial ability to withstand desiccation all favor growth of cyanobacteria. Over 70% of algal species in Indian paddy soils are cyanobacteria (Pandey, 1965). In paddy soils in India there is a succession of cyanobacteria over the growing season of the rice (Singh, 1961). Early in the rainy season (starting at the end of June), the soil becomes covered with a thick patchy growth composed of a variety of algae. In July, the fields are flooded, and the top 20 cm of soil and algae are mixed to form a muddy suspension. The rice seedlings are transplanted, and after about a fortnight the soil and algae have settled. By the middle of September, there is an extensive brownish-yellow gelatinous growth on the soil, composed primarily of *Aulosira fertilissima* (Fig. 2.20(*a*)), the most important nitrogen-fixing species in the fields. This species is dominant for about 3 months and forms a papery growth on the soil surface when the fields are dry.

From the above discussion, it is plain that cyanobacteria occur in most environments on earth. There is, however, one important exception: cyanobacteria are usually absent in waters or soils with a pH below 5, and they are uncommon between pH 5 and 6 (Brock, 1973).

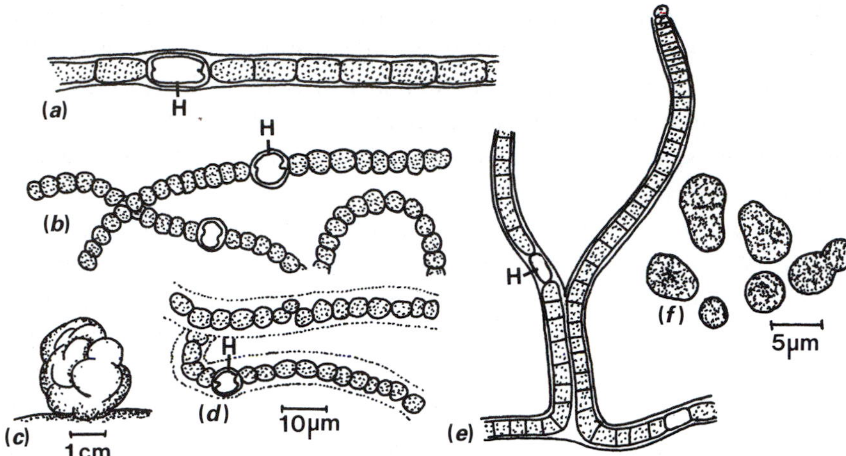

Fig. 2.20 (*a*) *Aulosira fertilissima*. (H) Heterocyst. (*b*) *Nostoc commune*. (*c,d*) *N. verrucosum*. (e) *Scytonema hofmanni*. (*f*) *Chamaesiphon* sp., showing formation of exospores. ((*a*), (e) after Desikachary, 1959; (*b*) after Prescott, 1962; (*f*) after Waterbury and Stanier, 1977.)

ADAPTION TO SILTING AND SALINITY

In salt marshes and mud flats, felts of cyanobacteria, particularly *Microcoleus chthonoplastes* (Fig. 2.17(*e*)) are important in stabilizing mud surfaces. Filaments of *M. chthonoplastes* are phototactic and chemotactic, resulting in migration to the surface of the mud at a rate of about 7 mm per 24 h. The movement of the filaments is an adaption that allows the cyanobacterium to survive during silting (Whale and Walsby, 1984). *M. chthonoplastes* is **euryhaline**, it is able to survive in a wide range of salinities in the estuarine environment by producing glycosylglycerol as an osmolyte to counteract the osmolarity of the surrounding environment. It also synthesizes trehalose to stabilize the phospholipid membrane bilayers of the cell (Karsten, 1996).

In **hypersaline** environments, such as the Great Salt Lake in Utah, the halotolerant and halophilic cyanobacteria accumulate the quaternary ammonium compounds glycine, betaine and glutamate betaine as osmoregulatory compounds.

Cyanotoxins

Some of the cyanobacteria produce toxins (cyanotoxins) (also see the section on algal toxins in the Introduction, p. 44). Physiologically, there are basically two types of cyanotoxins: neurotoxins and hepatotoxins (Carmichael, 1992; Bell and Cobb, 1994).

Neurotoxins – The neurotoxins are alkaloids (nitrogen-containing compounds of low molecular weight) that block transmission of the signal from neuron to neuron and neuron to muscle in animals and man. Symptoms include staggering, muscle twitching, gasping and convulsions. The neurotoxins can be fatal at high concentrations due to respiratory arrest caused by failure of the muscular diaphragm. The two neurotoxins produced by cyanobacteria are **anatoxin** and **saxitoxin** (Fig. 1.29). Anatoxins are synthesized by species of *Anabaena, Aphanizomenon, Oscillatoria* and *Trichodesmium* (Negri et al., 1997).

Hepatotoxins – The heptatotoxins are inhibitors of protein phosphatases 1 and 2A (Arment and Carmichael, 1996) and affect the animal by causing bleeding in the liver. Clinical signs include weakness, vomiting, diarrhea and cold extremities. Cyanobacteria produce two types of hepatotoxins, the **microcystins** and **nodularins** (Fig. 1.29), that are produced along a similar pathway (Rinehart et al., 1994). Microcystins are produced by species of *Microcystis, Anabaena, Nostoc, Nodularia* and *Oscillatoria* while the nodularins are produced by species of *Nodularia* (Kotak et al., 1995).

The cyanotoxins are mostly important in freshwaters where the cyanobacteria are ingested in drinking water by animals, with the algae dying and releasing their toxins in the intestinal tracts. The cyanotoxins are responsible for the loss of large numbers of stock each year throughout the world, usually in the warm summer months when the blooms of cyanobacteria are visible in the water. Seldom will man drink from such an unattractive water source. As such, poisoning of man by cyanotoxins is relatively rare.

The cyanotoxins function as anti-herbivore chemicals by inhibiting invertebrate grazers in the aquatic environment.

Many cyanobacteria produce **2-methylisoboneal** (**MIB**) and **geosmin**, both of which impart "earthy/musty" tastes and odors (Bufford et al., 1993). The presence of these taste/odor metabolites is a significant problem in large-scale municipal water supplies and in the aquaculture industry. Both MIB and geosmin are related to chlorophyll *a* and carotenoid synthesis and are associated with the photosynthetic lamella in the cells.

Cyanophages

Cyanophages are viruses that infect and commonly kill cyanobacteria (Suttle and Chan, 1994). Cyanophages can be extremely numerous, concentrations in excess of 100 000 viruses have been observed in surface seawater off the coast of Texas. Normally, however, cyanophages occur at concentrations of 1/10 that of the cyanobacterial host cells. Most cyanophages are actually resistant to attack by cyanophages with the cyanophage population being maintained by infection of the relatively rare cyanobacteria that are susceptible to infection. Cyanobacteria in the open ocean are more susceptible to infection than cyanobacteria from inshore waters. High temperature and phosphate limitation increase the probably of cyanobacterial infection by cyanophages (Wilson et al., 1996).

Secretion of antibiotics and siderophores

Some cyanobacteria, such as *Nostoc*, secrete antibiotics called **bacteriocins**, that kill related strains of the alga (Flores and Wolk, 1986). A **bacteriocin** is *a proteinaceous antibiotic that is active against procaryotic strains closely related to the organism that produces the antibiotic*. Other cyanobacteria secrete antibiotics that are active against a wide range of cyanobacterial and eukaryotic algae. *Scytonema hofmanni* (see Fig. 2.20(*e*)) produces such an antibiotic (Mason et al., 1982; Gleason and Paulson, 1984). This antibiotic, called **cyanobacterin**, is a chlorine-containing λ-lactone (Pignatello et all, 1983). All of these antibiotics probably play an active role in the survival of the producing organism by inhibiting growth of competing organisms.

The obligate requirement for iron, coupled with the low solubility of iron in many aquatic habitats, has led to the evolution of a mechanism for iron acqusition, at the cost of cellular energy and nutritional stores. A number of cyanobacteria release extracellular ferric-specific chelating agents ("**siderophores**") during periods of low iron availability. Siderophores function as extracellular ligands that aid in solubilization and assimilation of Fe^{3+} (Wilhelm et al., 1996).

Cyanobacteria are very sensitive to copper toxicity. Cyanobacteria can be controlled by adding 9.2 mg of copper sulfate per liter to impoundments without causing damage to higher plants or fish. Cyanobacteria react to copper in the water by secreting a compound that chelates (binds) copper, effectively removing copper ions from the water to a low enough level so the cyanobacteria are able to grow (Moffet and Brand, 1996).

Utilization of cyanobacteria as food

Cyanobacteria are used as human food and animal food supplements (Belay et al., 1996). In China, the *Spirulina* (Fig. 2.21, 2.24(*f*)) industry is supported by the State Science and Technology commission as a natural strategic program (Li and Qi, 1997). In 1996 there were more than 80 *Spirulina* factories with a total production of 400 tons of *Spirulina* dry powder and a total production area of 1 million square meters. Beside *Spirulina* pills and capsules, there are also pastries, blocks and *Spirulina*-filled chocolate blocks. In

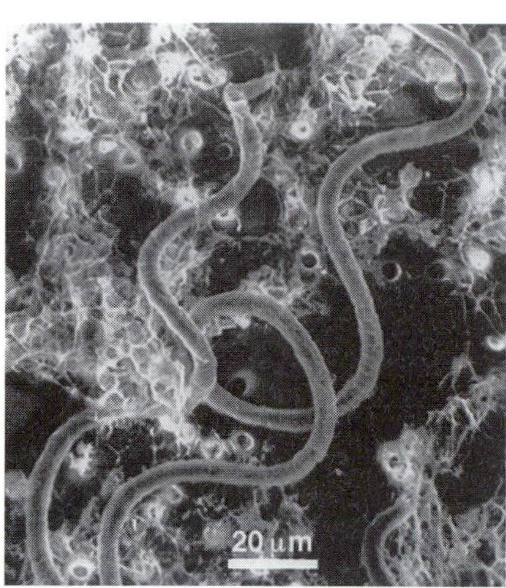

Fig. 2.21 Scanning electron micrograph of *Spirulina platensis*. (From El-Bestaway et al., 1996.)

Japan, *Aphanothece sacrum*, *Nostoc verrucosum* (Fig. 2.20(*c*),(*d*)) *N. commune* (Fig. 2.14, 2.20(*b*)) and *Brachytrichia* have been used as side dishes since ancient time. Bernal Diaz del Castillo, who accompanied Cortez to Mexico, described in 1521 how people living in the area of Mexico City "sell some small cakes made from a sort of ooze which they get out of the great lake, which curdles and from this they make bread having a flavor something like cheese."

Calcium carbonate and deposition and fossil record

Many species of cyanobacteria have calcium carbonate in the enveloping mucilage of the cells. In freshwater these algae usually grow in water where carbonate crystallizes out by nonbiological physicochemical mechanisms, and the crystals of calcite become trapped in the mucilage of the algae. Normally, only 1% to 2% of the calcium carbonate is actively precipitated by the cells (Pentecost, 1978). There are a large number of different forms of carbonate deposits attributed to the Cyanophyceae (for a review, see Golubić, 1973).

In the marine environment, cyanophycean depositions are the result of trapping and binding of the sediments as well as carbonate precipitation. **Stromatolithic** heads are probably the best known of these forms (Figs. 2.22, and 2.23). These heads are firmly gelatinous to almost cartilaginous in texture, predominantly hemispherical, and show fine concentric lamination. In the shallow subtidal waters of south Florida and the Bahamas, they are produced by a single species of *Schizothrix* (Monty, 1967). During the day there is growth of the algal filaments, resulting in the algae covering the

Fig. 2.22 Cyanobacterial stromatolites in the process of formation in Shark Bay, Western Australia. (From Logan, 1961.)

surface of the stomatolite head. During the night, growth ceases, and sediments accumulate on the surface of the head, forming sediment-rich laminae up to 100 μm thick. In the early part of the day, the algae penetrate through this deposited sediment, and grow a hyaline layer, 200 μm thick, with a low concentration of entrapped sediment. These alternating periods of growth and deposition give a laminated structure to the stromatolite (Fig. 2.23).

The production of laminae in stromatolites depends on fluctuations ultimately derived from the physical movements of the earth, sun, and moon, and requires some kind of rhythmicity that causes discontinuity in the accretionary process. The periodicity of laminae is due primarily to the daily photosynthetic cycle of the organisms in the stromatolites. In addition, stromatolites are **heliotropic** and grow toward sunlight (Awramik and Vanyo, 1986). This heliotropism allows the calculation of the extent of a year's deposition in a stromatolite. The yearly cycle of movement of the sun causes the sun to be higher in the sky in the summer and lower in the winter. The heliotropism of the stromatolites causes the stromatolites to grow in a sine waveform over the course of a year (Fig. 2.23). Counting the number of laminae in one sine wave has allowed paleontologists to calculate the number of days in a year for a geological period. This has allowed a continuous timing of a solar year back to 3500 million years ago (early Precambrian), when stromatolites first appeared in the fossil record. Such studies have shown that the solar year has varied considerably over this

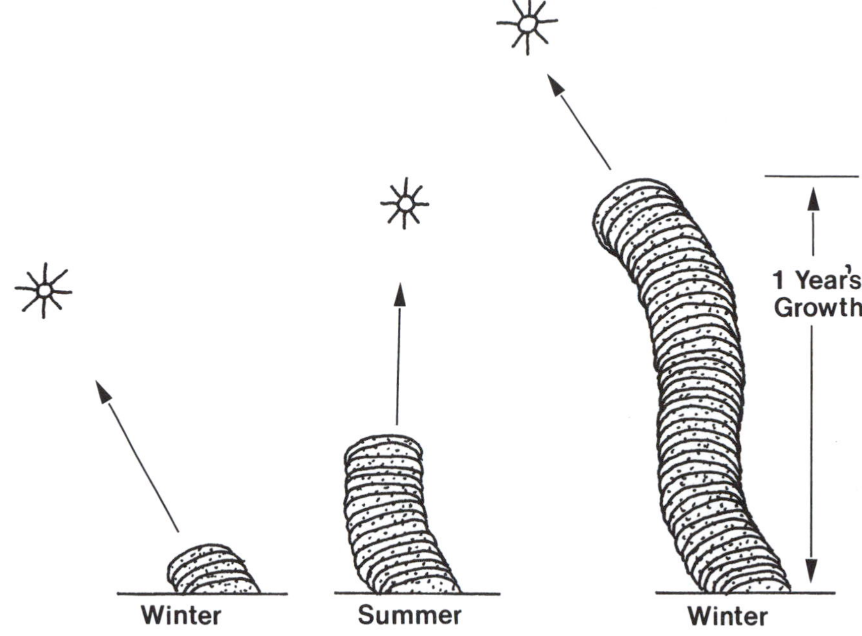

Fig. 2.23 Diagrammatic representation of the growth of a stromatolite over the period of a year. A year's growth is represented by an S-shaped curve.

period. For example, approximately 1000 million year ago, the solar year consisted of approximately 435 days (Vanyo and Awramik, 1985).

Up to 2000 million years ago, there were no grazing and boring organisms; thus stromatolites grew uncontested. Without competition, Precambrian stromatolites freely populated enormous areas and most likely grew in water down to a depth of 10 m. The occurrence and size of stromatolites declined dramatically after the evolution of grazing and boring organisms. Today stromatolites grow only in warm waters that are inhospitable to grazers and borers – waters such as the hypersaline waters in Shark Bay, Australia (Fig. 2.22), or in waters with a high tidal current that inhibit borers and grazers, such as those in the Bahamas where stromatolites grow to a height of 2 m (Dill et al., 1986).

Morphology and classification

The cyanobacteria include unicellular, colonial, and filamentous forms, with the most advanced type of morphological structure attained being the multiseriate branched filament. Culture studies of Cyanophyceae have made it clear that many of the described species are ecotypes of other species and therefore invalid. Species differentiation is frequently made on the basis of trichome diameter and sheath form (Geitler, 1932, Desikachary,

1959), both of which vary under environmental conditions. Heterocysts have been used as an important diagnostic feature in filamentous forms without taking into consideration the fact that the occurrence of heterocysts depends on the concentration of combined nitrogen as well as on the genotype. *Nostoc muscorum*, when grown in the dark on a medium containing glucose or sucrose, consists of coccoid cells (Lazaroff and Vishniac, 1961). After illumination at 1000 lux, these cells form the characteristic filaments of the species. Drouet (1968, 1978, 1981; Drouet and Daily, 1956) recognized the fact that the actual number of genotypes is only a fraction of the number of described species; however, this work has come under criticism (Stanier et al., 1971; Castenholz, 1992) for oversimplification without the necessary experimental evidence. It is clear, though, that there are far too many descriptions of species that are no more than ecotypes of a single organism, and that future taxonomic descriptions must be based on culture observations as well as descriptions from the field. DNA sequencing has not been used as much with cyanobacteria because of the need to use bacterial-free cultures (Bolch et al., 1996).

The Cyanophyceae can conveniently be divided into five orders (Fritsch, 1945). In this case, the filaments of the sulfur organisms in the Beggiatoaceae and Thiotrichaceae, which are probably colorless Cyanophyceae, are assigned to the bacteria. The following classification roughly agrees with that proposed by bacteriologists (Rippka et al., 1974).

Order 1 Chroococcales: single cells or cells loosely bound into gelatinous, irregular colonies

Order 2 Chamaesiphonales: epiphytes with a thallus showing polarity, multiplication by endospores or exospores

Order 3 Pleurocapsales: filaments without heterocysts, thallus divided into an erect and prostrate system (heterotrichy).

Order 4 Nostocales: filamentous thallus without true branching and with no heterotrichy

Order 5 Stigonematales: filamentous thallus with true branching and heterotrichy

CHROOCOCCALES

This order includes basically unicellular cyanobacteria which are held together in palmelloid colonies by mucilage. *Gloeothece* (Fig. 2.24(*a*)), *Microcystis* (Fig. 2.24(*b*)), *Synechococcus* (Fig. 2.7(*c*)), and *Synechocystis* (Fig. 2.1) are some of the genera in this order. Although the organisms in this order have a similar morphology, studies involving cell physiology, fatty acid composition, and DNA base composition have shown that the order is

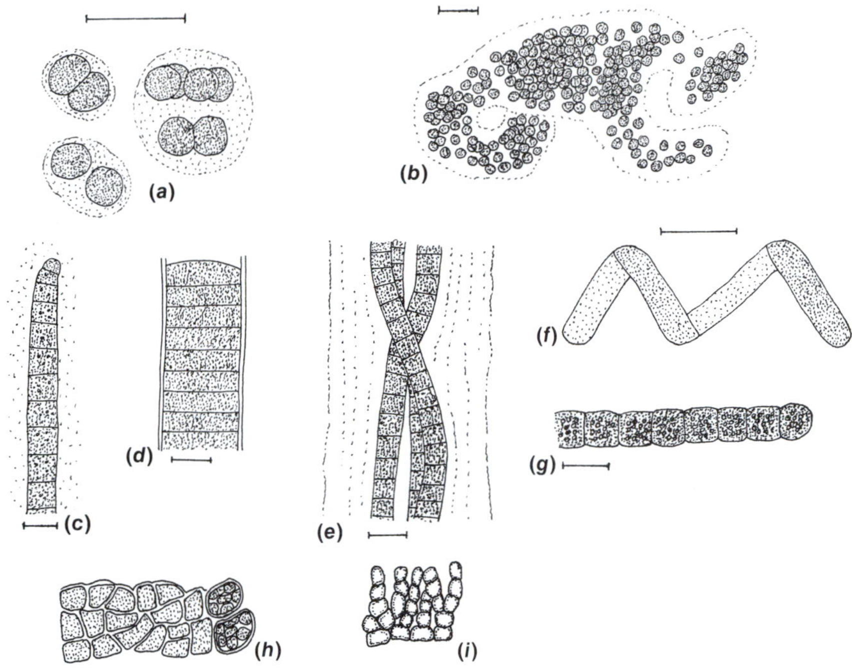

Fig. 2.24 (*a*) *Gloeothece magna*. (*b*) *Microcystis aeruginosa*. (*c*) *Phormidium autumnale*. (*d*) *Lyngbya birgei*. (*e*) *Hydrocoleus* sp. (*f*) *Spirulina major*. (*g*) *Trichodesmium lacustre*. (*h,i*) *Pleurocapsa minor*, surface view showing cells with endospores (*h*) and vertical section of thallus showing erect threads (*i*). Bar = 10 μm.

actually composed of several distinct and widely separated groups (Stanier et al., 1971).

CHAMAESIPHONALES

Epiphytic cyanobacteria with a polar-oriented thallus are placed in this order. Some genera have unicellular thalli, whereas others are multicellular. Reproduction is by **exospores** (**budding**) or **endospores** (**baeocytes**) (Rosowski et al., 1995). The order is more common in the sea than in freshwater.

Dermocarpa is primarily a marine genus, occurring in tide pools, and is composed of sessile cells that have one end broader than the other (Fig. 2.13(*d*)). Endospores are formed by repeated division of the protoplast in three planes, yielding four to a large number of endospores. *Chamaesiphon* is an example of a member of the order that reproduces by exospores (Figs. 2.13(*c*) and 2.20(*f*)).

PLEUROCAPSALES

In this order the thallus is composed of an erect and a prostrate system and is therefore heterotrichous. *Pleurocapsa* is a widely distributed lithophyte in

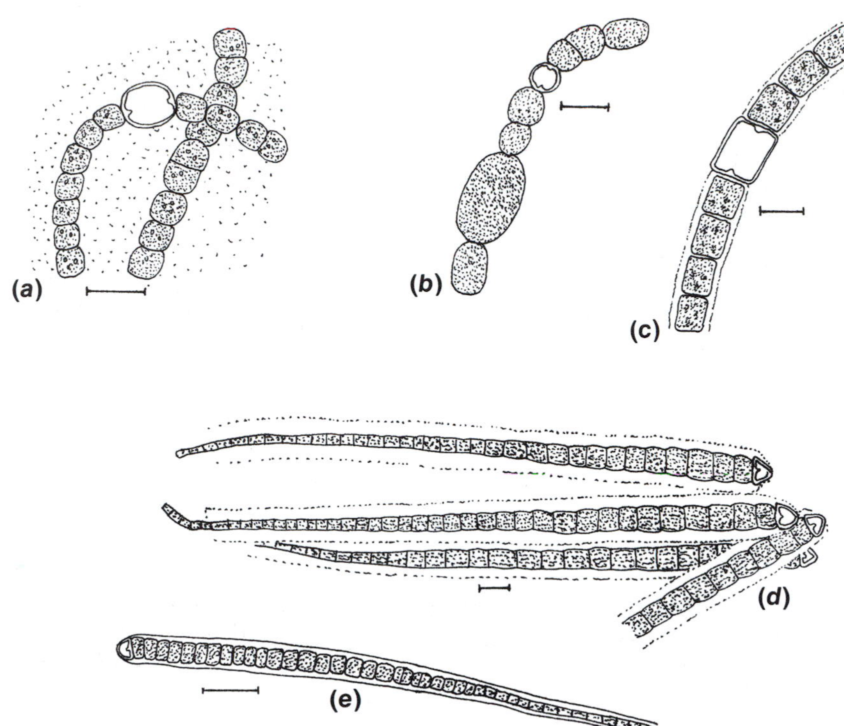

Fig. 2.25 (*a*) *Nostoc linckia*. (*b*) *Anabaena circinalis*. (c) *Aulosira implexa*. (*d*) *Rivularia dura*. (e) *Calothrix fusca*. Bar = 10 μm.

the marine and freshwater environment. Initially it consists of a filament creeping over the substrate. The filament branches and forms a pseudoparenchymatous disc. Development may stop here, or the prostrate basal disc cells may form erect threads (Fig. 2.24(*h*),(*i*)). Endospores are formed internally by division of a vegetative cell in three planes.

NOSTOCALES

These filamentous cyanobacteria without true branching make up by far the largest order in the class. The most common method of reproduction is by hormogonia. The classical method of classifying these algae relies heavily on the sheath characteristics, which unfortunately vary considerably, depending on environmental conditions. In *Lyngbya* (Fig. 2.24(*d*)), the sheath is discrete and contains a single trichome; in *Phormidium* (Fig. 2.24(*c*)), the sheaths are coalesced, whereas in *Hydrocoleus*, there are several trichomes within a single sheath (Fig. 2.24(*e*)). Trichome characters are also used in classification, as in *Spirulina* with its distinctive coiled trichome (Fig. 2.24(*f*)).

Heterocysts and akinetes occur in genera such as *Nostoc* (Fig. 2.14, 2.25(*a*)), *Anabaena* (Fig. 2.4, 2.25(*b*)), and *Aulosira* (Fig. 2.25(*c*)). In some

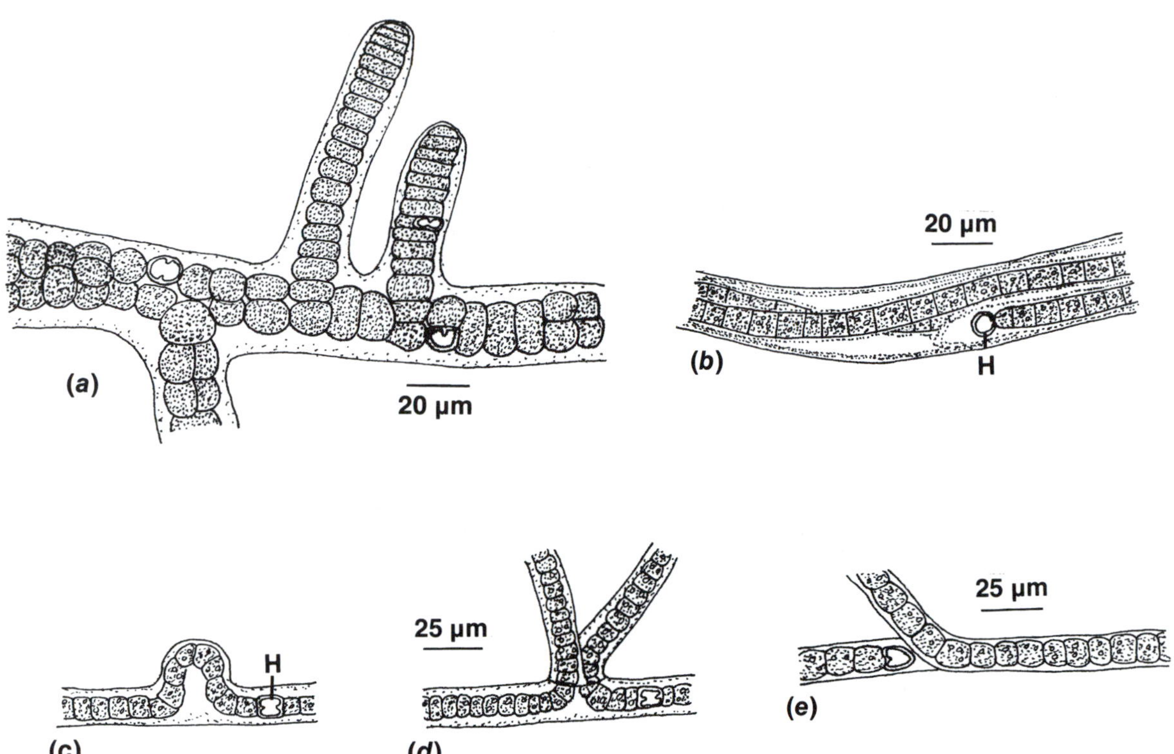

Fig. 2.26 (*a*) *Stigonema turfaceum* showing true branching. (*b*) *Desmonema wrangelii* with a number of trichomes in a sheath. (*c,d*) *Scytonema arcangelii* illustrating formation of false branches. (*e*) *Tolypothrix tenuis* with a single false branch. (H) heterocyst. (After Smith, 1950.)

genera there is a polarity: Heterocysts and/or akinetes are at the base, and a colorless hair is at the apex of the filament. A hair is a region of the trichome where the cells are narrow, elongated, highly vacuolate, and apparently colorless. *Calothrix* (Fig. 2.25(*e*)) is common in the littoral zone of the ocean, where it is attached to rocks. *Rivularia* (Fig. 2.25(*d*)) forms colonies in which the sheaths of one filament are confluent with others. The sheaths are usually heavily encrusted with lime and are very firm. *Rivularia* grows in freshwater submerged on stones and plants. The polarity of the thallus occurs only under conditions of low nitrogen concentration (Sinclair and Whitton, 1977). When these algae are grown in media containing nitrogenous compounds (e.g., NO_3^-, NH_3) heterocysts are not formed, and the trichomes are not tapered and lack colorless hairs at the apex. These algae then resemble species of genera such as *Oscillatoria.* Hairs can also be induced to form by growing the algae under conditions of phosphorus deficiency. Addition of phosphorus results in the loss of hairs and the formation of hormogonia (Livingstone and Whitton, 1983).

False branching occurs in some genera in this order. It results when the trichome lodges in the sheath, commonly in the area of a large heterocyst (Fig. 2.26(*c*)). Continued cell division in the trichome results in (1) a rupture of the sheath and (2) a break in the trichome that gives the appearance of a branch. The first type of false branching results in only one false branch protruding through the sheath, as in *Tolypothrix* (Fig. 2.26(*e*)). This type is due to the death of a cell or the formation of a separation disc or necridium (biconvex in shape because of pressure from adjacent cells; the necridia lyse as they mature). These weak points are normally next to a heterocyst and eventually result in a break of the trichome, with one of the broken ends of the trichome protruding through the sheath as a false branch. The second type of false branching results in the formation of two false branches, as in *Scytonema* (Fig. 2.26(*c*),(*d*)). Here a loop is formed that protrudes through the sheath. Eventually the loop breaks in the middle as cell division continues, resulting in two false branches. It is probable that the large heterocysts lodging in the sheath cause the immovability of the trichome in the sheath. Experimental evidence that this is so comes from experiments with algae grown in high combined-nitrogen levels, which inhibit both heterocyst formation and false branching.

STIGONEMATALES

These algae exhibit true branching with a tendency toward multiseriate thalli. The branching axis of *Stigonema* is partly or wholly multiseriate (Fig. 2.26(*a*)), occurring most frequently on wet rocks. Hormogonia are formed on the younger branches. The genera in this family are the most morphologically complex in the Cyanophyceae and are often differentiated into upright and prostrate filaments (heterotrichy).

Fig. 2.27 *Left:* Wayne W. Carmichael and *right:* G. E. Fogg.
Wayne W. Carmichael Born August 22, 1947 in Longview, Washington. Dr Carmichael received his BSc from Oregon State University in 1969, his MSc from the University of Alberta in 1972 and his PhD from the University of Alberta in 1974. From 1974 to 1976 he had a post-doctoral award

at the University of Alberta. Since 1976 he has been in the Department of Biological Sciences at Wright State University in Dayton University where he is Brage Golding Distinguished Professor of Research. Dr Carmichael is the leading researcher in the field of cyanobacterial toxins.

G. E. Fogg Born April 26, 1919. He was educated at Dulwich College, Queen Mary College, University of London (where he was a student of F. E. Fritsch) and St Johns College, Cambridge, where he received his PhD in 1943 and his ScD, in 1966. He was successively lecturer and reader in botany at University College, London from 1945 to 1960 (where he was strongly influenced by W. H. Pearsall); professor of botany at Westfield College, London, from 1960 to 1971; professor of marine biology in the University College of North Wales, from 1971 to 1985. He is now emeritus professor in the Department of Marine Biology at the University College of North Wales. In addition to his many journal publications in the cyanobacteria, he has written the following books: *The Metabolism of the Algae* (1953), *The Growth of Plants* (1963), *Algal Cultures and Phytoplankton Ecology* (1965), and *The Blue-Green Algae* (1973).

Fig. 2.28 Birgitte Bergman and Edward J. Carpenter.
Birgitte Bergman In 1977 she graduated in plant physiology at the University of Uppsala in Sweden and joined the Department of Botany at Stockholm University where she is a full professor.
Edward J. Carpenter Born 28 March 1942 in Buffalo, New York. In 1964 he received a BSc from the State University of New York at Fredonia, a MS and a PhD from North Carolina State University, the latter in 1970. From 1970 to 1975 Dr Carpenter was at the Woods Hole Oceanographic Institution. In 1975 he joined the Department of Marine Science at the State University of New York at Stony Brook where he is currently a full professor.

Dr Bergman and Dr Carpenter often collaborate on research involving marine bloom-forming cyanobacteria and nitrogen fixation by cyanobacteria.

References

Ahmadjian, V., and Hale, M. E. (1973). *The Lichens*. New York: Academic Press.

Allen, M. M. (1984). Cyanobacterial cell inclusions. *Annu. Rev. Microbiol.* 38:1–25.

Al-Thukair, A. A., and Golubic, S. (1991). New endolithic cyanobacteria from the Arabian Gulf. I. *Hyella immanis* sp. nov. *J. Phycol.* 27:766–80.

Anderson, O. R. (1977). Fine structure of a marine ameba associated with a blue-green alga in the Sargasso Sea. *J. Protozool.* 24:370–6.

Arment, A. R., and Carmichael, W. W. (1996). Evidence that microcystin is a thio template product. *J. Phycol.* 32:591–7.

Attridge, E. M., and Rowell, P. (1997). Growth, heterocyst differentiation and nitrogenase activity in the cyanobacteria *Anabaena variabilis* and *Anabaena cylindrica* in response to molybdenum and vanadium. *New Phytologist* 135:517–26.

Awramik, S. M., and Vanyo, J. P. (1986). Heliotropism in modern stromatolites. *Science* 231:1279–81.

Belay, A., Kato, T., and Ota, Y. (1996). *Spirulina* (*Arthrospira*): potential application as an animal feed supplement. *J. Appl. Phycol.* 8:303–11.

Bell, S. G., and Cobb, G. A. (1994). Cyanobacterial toxins and human health. *Rev. Med. Microbiol.* 5:256–64.

Bergman, B., and Carpenter, E. J. (1991). Nitrogenase confined to randomly distributed trichomes in the marine cyanobacterium *Trichodesmium thiebautii*. *J. Phycol.* 27:158–65.

Bergman, B., Gallon, J. R., Rai, A. N., and Stal, L. J. (1997). N_2 fixation by non-heterocystous cyanobacteria. *FEMS Microbiol. Rev.* 19:139–85.

Berthold, R. J., Borowitzka, M. A., and MacKay, M. A. (1982). The ultrastructure of *Oscillatoria spongeliae*, the blue-green algal endosymbiont of the sponge *Dysidea herbacea*. *Phycologia* 21:327–35.

Bhattacharjee, S. K. (1977). Unstable protein mediated ultraviolet light resistance in *Anacystis nidulans*. *Nature* 269:82–3.

Bolch, C. J. S., Blackburn, S. J., Neilan, B. A., and Grewe, P. M. (1996). Genetic characterization of strains of cyanobacteria using PCR-RFLP of the cpcBA integenic spacer and flanking regions. *J. Phycol.* 32:445–51.

Booker, M. J., and Walsby, A. E. (1981). Bloom formation and stratification by a planktonic blue-green alga in an experimental water column. *Br. Phycol. J.* 16:411–21.

Booth, W. E. (1941). Algae as pioneers in plant succession and their importance in erosion control. *Ecology* 22:38–46.

Bradley, S., and Carr, N. G. (1977). Heterocyst development in *Anabaena cylindrica*: The necessity for light as an initial trigger and sequential stages of commitment. *J. Gen. Microbiol.* 101:291–7.

Brock, T.D. (1973). Lower pH limit for the existence of blue-green algae: Evolutionary and ecological implications. *Science* 179:480–3.

Bufford, R. A., Seagull, R. W., Chung, S. Y., and Mille, D. F. (1993). Intracellular localization of the taste/odor metabolite 2-methylisoboneal in *Oscillatoria limosa* (Cyanophyta). *J. Phycol.* 29:91–5.

Caiola, M. G. (1975). A light and electron microscopic study of blue-green algae living in the coralloid roots of *Encephalartos altensteinii* and in culture. *Phycologia* 14:25–33.

Canini, A., Bergman, B., Civitareale, P., Rotilla, G., and Caiola, M. G. (1992). Localization of iron-superoxide dismutase in the cyanobiont of *Azolla filiculoides* Lam. *Protoplasma* 169:1–8.

Capone, D. G., Zehr, J. P., Paerl, H. W., Bergman, B., and Carpenter, E. J. (1997). *Trichodesmium*, a globally significant marine cyanobacterium. *Science* 275:1221–9.

Carmichael, W. W. (1992). Cyanobacteria secondary metabolites – the cyanotoxins. *J. Appl. Bacteriol.* 72:445–59.

Carpenter, E. J., and Price, C. C. (1977). Nitrogen fixation, distribution, and production of *Oscillatoria (Trichodesmium)* spp. in the western Sargasso and Caribbean seas. *Limnol. Oceanogr.* 22:60–72.

Carpenter, E. J., Capone, D. C., and Reuter, J. (eds.) (1992). *Marine Pelagic Cyanobacteria:* Trichodesmium *and other Diazotrophs*, Dordecht: Kluwer Academic Publishers. 312 p.

Castenholz, R. W. (1973). Ecology of blue-green algae in hot-springs. In *The Biology of the Blue-Green Algae*, ed. N. G. Carr and B. A. Whitton, pp. 379–414. Berkeley: Univ. Calif. Press.

Castenholz, R. W. (1992). Species usage, concept, and evolution in the cyanobacteria (blue-green algae). *J. Phycol.* 28:737–5.

Chauvat, F., Corre, B., Herdman, M., and Joset-Espardellier, F. (1982). Energetic and metabolic requirements for the germination of akinetes of the cyanobacterium *Nostoc* PCC7524. *Arch. Microbiol.* 133:44–9.

Chen, H. M., Chien, C-Y., and Huang, T-C. (1996). Regulation and molecular structure of a circadian oscillating protein located in the cell membrane of the prokaryote *Synechococcus* RF-1. *Planta* 199:520–7.

Clark, R. L., and Jensen, T. E. (1969). Ultrastructure of akinete development in a blue-green alga, *Cylindrospermum* sp. *Cytologia* 34:439–48.

Cmiech, H. A., Leedale, G. F., and Reynolds, C. S. (1986). Morphological and ultrastructural variability of planktonic Cyanophyceae in relation to seasonal periodicity. II. *Anabaena solitaria:* Vegetative cells, heterocysts, akinetes. *Br. Phycol. J.* 21:81–92.

Desikachary, T. V. (1959). *Cyanophyta.* Indian Council of Agricultural Research, New Delhi.

Dick, H., and Stewart, W. D. P. (1980). The occurrence of fimbriae on a N_2-fixing cyanobacterium that occurs in lichen symbiosis. *Arch. Microbiol.* 124:107–9.

Dill, R. F., Shinn, E. A., Jones, A. T., Kelly, K., and Steinen, R. P. (1986). Giant subtidal stromatolites forming in normal salinity waters. *Nature* 324:55–9.

Drouet, F. (1937). The Brazilian Myxophyceae. I. *Am. J. Bot.* 24:598–608.

Drouet, F. (1968). Revision of the classification of the Oscillatoriaceae. *Monogr. Acad. Nat. Sci. Philadeliphia* 15.

Drouet, F. (1978). Revision of the Nostocaceae with constricted trichomes. *Beihefte Nova Hedwigia* 57:1–258.

Drouet, F. (1981). Revision of the Stigonemataceae with a summary of the classification of the blue-green algae. *Beihefte Nova Hedwigia* 66:1–221.

Drouet, F., and Daily, W. (1956). Revision of the coccoid Myxophyceae. *Butler Univ. Bot. Stud.* 12.

El-Bestawy, E., Bellinger, E. G., and Sigee, D. C. (1996). Elemental composition of phytoplankton in a subtropical lake: X-ray microanalytical studies on the dominant algae *Spirulina platensis* (Cyanophyta) and *Cyclotella meneghiniana* (Bacillariophyceae). *Eur J. Phycol.* 31:157–66.

Ferguson-Wood, E. J. (1965). *Marine Microbial Ecology.* London: Chapman and Hall.

Flores, E., and Wolk, C. P. (1986). Production, by filamentous, nitrogen-fixing cyanobacteria, of a bacteriocin and of other antibiotics that kill related strains. *Arch. Microbiol.* 145:215–19.

Fogg, G. E. (1942). Studies on nitrogen fixation by blue-green algae. I. Nitrogen fixation by *Anabaena cylindrica* Lemm. *J. Exp. Biol.* 19:78–87.

Fogg, G. E. (1944). Growth and heterocyst production in *Anabaena cylindrica* Lemm. *New Phytol.* 43:164–75.

Fogg, G. E. (1952). The production of extracellular nitrogenous substances by a blue-green alga. *Proc. R. Soc.* [*B*] 139:372–9.

Fogg, G. E. (1986). Light and ultraphytoplankton. *Nature* 319:96.

Frederick, J. F. (1951). Preliminary studies on the synthesis of polysaccharides in the algae. *Physiol. Plant.* 4:621–6.

Fredricksson, C., and Bergman, B. (1997). Ultrastructural characterization of cells specialized for nitrogen function in a non-heterocystous cyanobacterium, *Trichodesmium* spp. *Protoplasma* 197:76–85.

Fritsch, F. E. (1945). *The Structure and Reproduction of the Algae,* Vol. 2. Cambridge: Cambridge University Press.

Gantt, E., and Conti, S. F. (1969). Ultrastructure of blue-green algae. *J. Bacteriol.* 97:1486–93.

Garcia-Pichel, F., and Castenholz, R. W. (1991). Characterization and biological implications of scytonemin, a cyanobacterial sheath pigment. *J. Phycol.* 27:395–409.

Garcia-Pichel, F., and Belnap, J. (1996). Microenvironments and microscale productivity of cyanobacterial desert crusts. *J. Phycol.* 32:774–82.

Gates, J. E., Fisher, R. W., Goggin, T. W., and Azrolan, N. I. (1980). Antigenic differences between *Anabaena azollae* fresh from the *Azolla* fern leaf cavity and free-living cyanobacteria. *Arch. Microbiol.* 128:126–9.

Geitler, L. (1932). Cyanophyceae. In *Krytogamenflora von Deutschland, Osterreich und der Schweiz*, ed. L. Rabenhorst, Vol. 14. Leipzig: Akademische Verlagsgesellschaft.

Giddings, T. H., and Staehelin, L. A. (1981). Observation of microplasmodesmata in both heterocyst-forming and non-heterocyst-forming filamentous cyanobacteria by freeze-fracture electron microscopy. *Arch. Microbiol.* 129:295–8.

Gleason, F. K., and Paulson, J. L. (1984). Site of action of the natural algicide, cyanobacterin, in the blue-green alga, *Synechococcus* sp. *Arch. Microbiol.* 138:273–7.

Glover, H. E., Keller, M. D., and Guillard, R. R. L. (1986). Light quality and oceanic ultraphytoplankters. *Nature* 319:142–3.

Golubić, S. (1973). The relationship between blue-green algae and carbonate deposits. In *The Biology of the Blue-Green Algae* (Carr, N. G., and Whitton, B. A., eds.), pp. 434–72. Univ. Calif. Press. Berkeley.

Goodwin, T. W. (1974). Carotenoids and biliproteins. In *Algal Physiology and Biochemistry*, ed. W. D. P. Stewart, pp. 176–205. Berkeley: Univ. Calif. Press.

Green, G. (1977). Ecology of toxicity in marine sponges. *Mar. Biol.* 40:207–15.

Häder, D. P. (1987). Photosensory behavior in procaryotes. *Microbiol. Rev.* 51:1–21.

Halfen, L. N., and Castenholz, R. W. (1971). Gliding motility in the blue-green alga, *Oscillatoria princeps*. *J. Phycol.* 7:133–45.

Herdman, M., Janvier, M., Rippka, R., and Stanier, R. Y. (1979). Genome size of cyanobacteria. *J. Gen. Microbiol.* 111:73–85.

Hill, D. J. (1975). The pattern of development of *Anabaena* in the *Azolla–Anabaena* symbiosis. *Planta* 122:179–84.

Hill, D. R., Peat, A., and Potts, M. (1994). Biochemistry and structure of the glycan secreted by dessication-tolerant *Nostoc commune* (Cyanobacteria). *Protoplasma* 182:126–48.

Höcht, H., Martin, H. H., and Kandler, O. (1965). Zur Kenntnis der chemischen Zusammensetzung der Zellwonc der Blaualgen. *Z. Pflanzenphysiol.* 53:39–57.

Hough, L., Jones, J. K. N., and Wadman, W. H. (1952). An investigation of the polysaccharide components of certain fresh-water algae. *J. Chem. Soc.* 3393–9.

Jarosch, R. (1962). Gliding. In *Physiology and Biochemistry of Algae*, ed. R. A. Lewin, pp. 573–81. Academic Press, New York and London.

Jensen, T. E., Clark, R. L. (1969). Cell wall and coat of the developing akinete of a *Cylindrospermum* species. *J. Bacteriol.* 97:1494–5.

Jones, J. D., and Jost, M. (1970). Isolation and chemical characterization of gas vacuole membranes from *Microcystis aeruginosa* Kuetz. emend Elenkin. *Arch. Mikrobiol.* 70:43–64.

Kapustka, L. A., and DuBois, J. D. (1987). Dinitrogen fixation by cyanobacteria and associative rhizosphere bacteria in the Arapaho prairie in the sand hills of Nebraska. *Am. J. Bot.* 74:107–13.

Karsten, U. (1996). Growth and organic osmolytes of geographically different isolates of *Microcoleus chthonoplastes* (cyanobacteria) from benthic microbial mats: response to salinity changes. *J. Phycol.* 32:501–6.

Konopka, A., Kromkamp, J., and Mur, L. R. (1987). Regulation of gas vesicle content and buoyancy in light- or phosphate-limited cultures of *Aphanizomenon flos-aquae* (Cyanophyta). *J. Phycol.* 23:70–8.

Kotak, B. G., Lam, A. K-Y., Prepas, E. E., Kenefick, S. L., and Hurdey, S.E. (1995). Variability of the hepatotoxin microcystin-LR in hypertrophic drinking water lakes. *J. Phycol.* 31:248–63.

Kulasooriya, S. A., Lang, N. J., and Fay, P. (1972). The heterocysts of blue-green algae. III. Differentiation and nitrogenase activity. *Proc. R. Soc. Lond.* [*B*] 181:199–209.

Lang, N. J., and Fay, P. (1971). The heterocysts of blue-green algae. II. Details of ultrastructure. *Proc. R. Soc. Lond.* [*B*] 178:193–203.

Lawry, N. H., and Simon, R. D. (1982). The normal and induced occurrence of cyanophycin inclusion bodies in several blue-green algae. *J. Phycol.* 18:391–9.

Lazaroff, N. and Vishniac, W. (1961). The effect of light on the development cycle of *Nostoc muscorum*, a filamentous blue-green alga. *J. Gen. Microbiol.* 25:365–74.

Li, D-M., and Qi, Y-Z. (1997). *Spirulina* industry in China: Present status and future prospects. *J. Appl. Phycol.* 9:25–8.

Li, R., Watanabe, M., and Watanabe, M. M. (1997). Akinete formation in planktonic *Anabaena* spp. (Cyanobacterium) by treatment with low temperature. *J. Phycol.* 33:576–84.

Little, M. G. (1973). The zonation of marine supra-littoral blue-green algae. *Br. Phycol. J.* 8:47–50.

Livingstone, D., and Whitton, B. A. (1983). Influence of phosphorus on morphology of *Calothrix parietina* (Cyanophyta) in culture. *Br. Phycol. J.* 18:29–38.

Logan, B. W. (1961). Cryptozoon and associate stromatolites from the Recent, Shark Bay, Western Australia. *J. Geol.* 69:517–33.

Mague, T. H., and Holm-Hansen, O. (1975). Nitrogen fixation on a coral reef. *Phycologia* 14:87–92.

Mason, C. P., Edwards, K. R., Carlson, R. E., Pignatello, J., Gleason, F. K., and Woods, J. M. (1982). Isolation of chlorine-containing antibiotic from the freshwater cyanobacterium *Scytonema hofmanni*. *Science* 215:400–2.

Meeks, J. C., and Castenholz, R. W. (1971). Growth and photosynthesis in an extreme thermophile, *Synechococcus lividus* (Cyanophyta). *Arch. Mikrobiol.* 78:25–41.

Mitsui, A., Kumazawa, S., Takashi, A., Ikemoto, H., Cao, S., and Arai, T. (1986). Strategy by which nitrogen-fixing unicellular cyanobacteria grow photoautotrophically. *Nature* 323:720–2.

Moffet, J. W., and Brand, L. E. (1996). Production of strong, extracellular Cu chelators by marine cyanobacteria in response to Cu stress. *Limnol. Oceanog.* 41:388–95.

Monty, C. L. V. (1967). Distribution and structure of recent stromatolite algal mats, eastern Andros Island, Bahamas. *Ann. Soc. Geol. Belg.* 90:55–99.

Negri, A. P., Jones, G. J., Blackburn, S. I., Oshima, Y., and Hideyuki, O. (1997). Effect of culture bloom development and of sample storage on paralytic shellfish poisons in the cyanobacterium *Anabaena circinalis*. *J. Phycol.* 33:26–35.

Oliver, R. L. (1994). Floating and sinking in gas-vacuolate cyanobacteria. *J. Phycol.* 30:161–73.

Oliver, R. L., Thomas, R. H., Reynolds, C. S., and Walsby, A. E. (1985). The sedimentation of buoyant *Microcystis* colonies caused by precipitation with an iron-containing colloid. *Proc. R. Soc. Lond.* [*B*] 223:511–28.

Padan, E. (1979). Facultative anoxygenic photosynthesis in cyanobacteria. *Annu. Rev. Plant Physiol.* 30:27–40.

Paerl, H. W. (1994). Spatial segregation of CO_2 fixation in *Trichodesmium* spp.: Linkage to N_2 fixation potential. *J. Phycol.* 30:790–9.

Palenik, B., and Haselkorn, R. (1992). Multiple evolutionary origins of prochlorophytes, the chlorophyll *b*-containing prokaryotes. *Nature (Lond.)* 355:26775–7.

Panday, D. C. (1965). A study of the algae from paddy soils of Ballia and Ghazipur districts of Uttar Pradesh, India. I. Cultural and ecological considerations. *Nova Hedwigia* 9:299–334.

Pascher, A. (1914). Über Symbiosen von Spaltpilzen und Flagellaten mit Blaualgen. *Ber. Dtsch. Bot. Ges.* 32:339–52.

Pentecost, A. (1978). Blue-green algae and freshwater carbonate deposits. *Proc. R. Soc. Lond.* [*B*] 200:43–61.

Peters, G. A., and Mayne, B. C. (1974). The *Azolla, Anabaena azollae* relationship. I. Initial characterization of the association. *Plant Physiol.* 53:813–19.

Pignatello, J. J., Porwoll, J., Carlson, R. E., Xavier, A., Gleason, F. K., and Wood, J. M. (1983). Structure of the antibiotic cyanobacterin, a chlorine containing γ-lactone from the freshwater cyanobacterium *Scytonema hofmanni. J. Org. Chem.* 48:4035–8.

Potts, M. (1996). The anhydrobiotic cyanobacterial cell. *Physiologia Plantarum* 97:788–94.

Potts, M., Angeloni, S. V., Ebel, R. E., and Bassam, D. (1992). Myogloblin in a cyanobacterium. *Science* 256:1690–2.

Prescott, G. W. (1962). *Algae of the Western Great Lakes Area.* W. C. Brown, Dubuque, Iowa.

Rinehart, K. L., Namikoshi, M., and Choi, B. W. (1994). Structure and biosynthesis of toxins from blue-green algae (cyanobacteria). *J. Appl. Phycol.* 6:159–76.

Rippka, R., Waterbury, J., and Cohen-Bazire, G. (1974). A cyanobacterium which lacks thylakoids. *Arch. Mikrobiol.* 100:419–36.

Romans, K. M., Carpenter, E. J., and Bergman, B. (1994). Buoyancy regulation in the colonial diazotrophic cyanobacterium *Trichodesmium tenue,* ultrastructure and storage of carbohydrate, polyphosphate, and nitrogen. *J. Phycol.* 30:935–42.

Rosowski, J. R., Bielik, I., and Lee, K. W. (1995). Observations on a Chamaesiphonaceous alga (Cyanophyceae) with special reference to exocyte formation. *J. Phycol.* 31:436–46.

Scherer, S., and Potts, M. (1989). Novel water stress protein from a dessication-tolerant cyanobacterium. *J. Biol. Chem.* 264:12546–53.

Skleryk, R. S., Tyrrell, P. N., and Espie, G. S. (1997). Photosynthesis and inorganic carbon acquisition in the cyanobacterium *Chlorogloeopsis* sp. ATCC 27193. *Physiologia Plantarum* 99:81–8.

Simon, R. D. (1973). The effect of chloramphenicol on the production of cyanophycin granule polypeptide in the blue-green alga *Anabaena cylindrica. Arch. Mikrobiol.* 92:115–22.

Sinclair, C., and Whitton, B. A. (1977). Influence of nitrogen source on morphology of Rivulariaceae (Cyanophyta). *J. Phycol.* 13:335–4.

Singh, R. N. (1961). *The Role of Blue-Green Algae in Nitrogen Economy of Indian Agriculture.* Indian Council for Agricultural Research. New Delhi, India.

Smarda, J., Caslavska, J., and Komarek, J. (1979). Cell wall structure of *Synechocystis aquatilis* (Cyanophyceae). *Arch. Hydrobiol.* [*Suppl.*] 56:154–65.

Smith, D. C, Muscatine, L., and Lewis, D. (1970). Carbohydrate movement from autotrophs to heterotrophs in parasitic and mutalistic symbiosis. *Biol. Rev.* 44:17–90.

Smith, G. M. (1950). *Freshwater Algae of the United States.* New York: McGraw-Hill.

Smith, G. M. (1955). *Cryptogamic Botany.* Vol. 1. New York: McGraw-Hill.

Spencer, C. N., and King, D. L. (1985). Interactions between light. NH_4^+, and CO_2 in buoyancy regulation of *Anabaena flos-aquae* (Cyanophyceae). *J. Phycol.* 21:194–9.

Spratt, E. R. (1915). The root nodules of the Cycadaceae. *Ann. Bot.* 29:619–26.

Stanier, R. Y. (1973). Autogrophy and heterotrophy in unicellular blue-green algae. In *The*

Biology of the Blue-Green Algae ed. N. G. Carr and B. A. Whitton, pp. 501–18. Berkeley: Univ. Calif. Press.

Stanier, R.Y., Kunisawa, R., Mandel, M., and Cohen-Bazire, G. (1971). Purification and properties of unicellular blue-green algae (order Chroococcales). *Bacteriol. Rev.* 35:171–205.

Stewart, W. D. P., and Pearson, H. W. (1970). Effects of aerobic and anaerobic conditions on growth and metabolism of blue-green algae. *Proc. R. Soc. Lond.* [*B*] 175:293–311.

Storey, W. B. (1968). Somatic reduction in cycads. *Science* 159:648–50.

Suttle, C. A., and Chan, A. M. (1994). Dynamics and distribution of cyanophages and their effect on marine *Synechococcus* spp. *Appl. Env. Microbiol.* 60:3167–74.

Swaminathan, M. S. (1984). Rice. *Sci. Am.* 250(1):80–93.

Tandeau de Marsac, N. (1977). Occurrence and nature of chromatic adaptation in cyanobacteria. *J. Bacteriol.* 130:82–91.

Tang, E. P. Y., Tremblay, R., and Vincent, W. F. (1997). Cyanobacterial dominance of polar freshwater ecosystems: are high-latitude mat-formers adapted to low temperature? *J. Phycol.* 33:171–81.

Tischer, I. (1957). Untersuchungen über die granulären Eihschlüsse und das Reduktions-Oxydations-Vermögen der Cyanophyceen. *Arch. Mikrobiol.* 27:400–28.

Turpin, D. H., Miller, A. G., and Canvin, D. T. (1984). Carboxysome content of *Synechococcus leopoliensis* (Cyanophyta) in response to organic carbon. *J. Phycol.* 20:249–53.

Urback, E., Robertson, D. L., and Chisholm, S. W. (1992). Multiple evolutionary origins of prochlorophytes within the cyanobacterial radiation. *Nature* (*Lond.*) 355:267–70.

Van Dok, W., and Hart, B. T. (1995). Akinete differentiation in *Anabaena circinalis* (Cyanophyta). *J. Phycol.* 32:557–65.

Vanyo, J. P., and Awramik, S. M. (1985). Stromatolites and earth–sun–moon dynamics. *Precambrian Res.* 29:121–42.

Walsby, A. E. (1974). The extracellular products of *Anabaena cylindrica* Lemm. I. Isolation of a macromolecular pigment–peptide complex and other components. *Br. Phycol. J.* 9:371–81.

Walsby, A. E. (1978). The properties and buoyancy-providing role of gas vacuoles in *Trichodesmium* Ehrenberg, *Br. Phycol. J.* 13:103–16.

Walsby, A. E. (1994). Gas vesicles. *Microbiol. Res.* 58:94–144.

Watanabe, A., and Kiyohara, T. (1963). Symbiotic blue-green algae of lichens, liverworts and cycads. In *Studies on Microalgae and Photosynthetic Bacteria*, ed. Japanese Soc. Plant Physiologists, pp. 189–96. Tokyo: University of Tokyo Press.

Waterbury, J., and Stanier, R. (1977). Two unicellular bacteria which reproduce by budding. *Arch. Microbiol.* 115:249–57.

Waterbury, J., Willey, J. M., Franks, D. G., Valois, F. W., and Watson, S. W. (1985). A cyanobacterium capable of swimming motility. *Science* 230:74–6.

Weckesser, J., Broll, C., Adhikary, S. P., and Jürgens, U. J. (1987). 2-*O*-methyl-D-xylose containing sheath in the cyanobacterium *Gloeothece* sp. PCC 6501. *Arch. Microbiol.* 147:300–3.

Weller, D., Doemel, W., and Brock, T. D. (1975). Requirement of low oxidation-reduction potential for photosynthesis in a blue-green alga (*Phormidium* sp.). *Arch. Mikrobiol.* 104:7–13.

Whale, G. F., and Walsby, A. E. (1984). Motility of the cyanobacterium *Microcoleus chthonoplastes* in mud. *Br. Phycol. J.* 19:117–23.

Whitton, B. A., and Peat, A. (1969). On *Oscillatoria redekei* Van Goor. *Arch. Mikrobiol.* 68:362–76.

Wildman, R. B., Loescher, J. H., and Winger, C. L. (1975). Development and germination of akinetes of *Aphanizomenon flos-aquae*. *J. Phycol.* 11:96–104.

Wilhelm, S. W., Maxwell, D. P., and Trick, C. G. (1996). Growth, iron requirements, and siderophore production in iron-limited *Synechococcus* PCC 7002. *Limnol. Oceanog.* 41:89–97.

Wilson, W. H., Carr, N. G., and Mann, N. H. (1996). The effect of phosphate status on the kinetics of cyanophage infection in the oceanic cyanobacterium *Synechococcus* sp. WH 7803. *J. Phycol.* 32:506–16.

Winkenbach, F., and Wolk, C. P. (1973). Activities of enzymes of the oxidative and the reductive pentose phosphate pathways in heterocysts of a blue-green alga. *Plant Physiol., Lancaster* 52:480–3.

Wyman, M., Gregory, R. P. F., and Carr, N. G. (1985). Novel role for phycoerythrin in a marine cyanobacterium, *Synechococcus* strain DC2. *Science* 230:818–20.

PART III · Evolution of the chloroplast

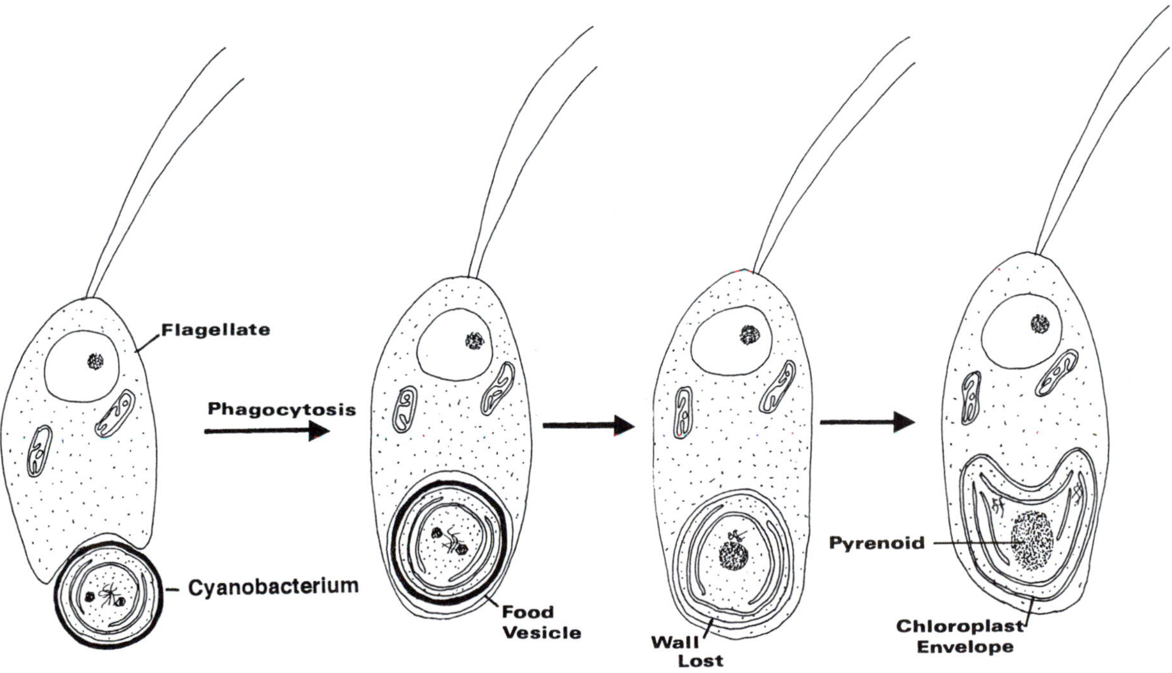

Fig. III.1 Diagrammatic representation of the uptake of a cyanobacterium by a protozoan into a food vesicle. This resulted in the establishment of an endosymbiosis between the cyanobacterium and the protozoan. Through evolution, the endosymbiotic cyanobacterium evolved into a chloroplast surrounded by two membranes of the chloroplast envelope.

The Rhodophyta (red algae) and Chlorophyta (green algae) form a natural group of algae in that they have chloroplasts surrounded by only the two membranes of the chloroplast envelope. The evolutionary event that led to the chloroplast occurred as follows (Fig. III.1). A phagocytotic protozoan took up a cyanobacterium into a food vesicle. Instead of being digested as a source of food, the cyanobacterium lived as an endosymbiont in the protozoan. This event benefited the protozoan because it received some of the photosynthate from the endosymbiotic alga, and it benefited the cyanobacterium because it received a protected stable environment. Through evolution the wall of the endosymbiotic cyanobacterium was lost. A mutation in the endosymbiont which resulted in a loss of the wall would have been selected for in evolution because it would have facilitated the transfer of compounds between the host and the endosymbiont. The food vesicle membrane of the phagocytotic host became the outer membrane of the chloroplast envelope. The plasma membrane of the cyanobacterium symbiont became the inner membrane of the chloroplast envelope. Rearrangement of the thylakoid membranes and evolution of polyhedral bodies into a pyrenoid completed the transition to a true chloroplast such as occurs in extant green and red algae.

3 · Glaucophyta

The Glaucophyta include those algae that have endosymbiotic cyanobacteria in the cytoplasm instead of chloroplasts. Because of the nature of their symbiotic association, they are thought to represent intermediates in the evolution of the chloroplast. The endosymbiotic theory of chloroplast evolution, first proposed by Mereschkowsky in 1905, is the one most widely accepted. According to this theory, a cyanobacterium was taken up by a phagocytic organism into a food vesicle. Normally the cyanobacterium would be digested by the flagellate, but by chance a mutation occurred, with the flagellate being unable to digest the cyanobacterium. This was probably a beneficial mutation because the cyanobacterium, by virtue of its lack of feedback inhibition, secreted considerable amounts of metabolites to the host flagellate. The flagellate in turn gave the cyanobacterium a protected environment, and the composite organism was probably able to live in an ecological niche where there were no photosynthetic organisms (i.e., a slightly acid body of water where free-living cyanobacteria do not grow; see Chapter 2). Pascher (1914) coined terms for this association; he called the endosymbiotic cyanobacteria **cyanelles**; the host, a **cyanome**; and the association between the two, a **syncyanosis**. In the original syncyanosis the cyanelle had a wall around it. Because the wall slowed the transfer of compounds from the cyanelle to the host and vice versa, any mutation that resulted in a loss of wall would have been beneficial and selected for in evolution. Most of the cyanelles in the Glaucophyta lack a wall and are surrounded by two membranes – the old food vesicle membrane of the cyanome and the plasma membrane of the cyanelle. As evolution progressed, these two membranes became the chloroplast envelope, the various inclusions of the cyanobacterium were lost (polyhedral bodies, cyanophycin granules, polyphosphate bodies), the cyanome cytoplasm took over formation of the storage product, and a pyrenoid was differentiated from proteinaceous bodies in the cyanelle. This endosymbiotic line of evolution probably led initially to the red algae.

There are a number of similarities between cyanobacteria and chloroplasts that support the endosymbiotic theory: (1) They are about the same size; (2) they evolve oxygen in photosynthesis (oxygen is not evolved in bacterial photosynthesis); (3) they have 70S ribosomes; (4) they have circular prokaryotic DNA without basic proteins; (5) they have chlorophyll *a* as the primary photosynthetic pigment.

The pigments of the Glaucophyta are similar to those of the Cyanophyceae: Both chlorophyll *a* and the phycobiliproteins are present; however, two of the cyanobacterial carotenoids, myxoxanthophyll and echinenone, are absent (Chapman, 1966).

Although similar to cyanobacteria, the cyanelles should be regarded as organelles rather than endosymbiotic prokaryotes (Betsche et al., 1992; Helmchen et al., 1995). The cyanelles have no respiratory electron transport chain (cyanobacteria have a respiratory electron transport chain) and the size of the genome is 5 to 10% of that of free-living cyanobacteria, and about the size of chloroplasts. Nevertheless, there are important differences between chloroplasts and cyanelles. The cyanelle genome codes for both subunits of ribulose bisphosphate carboxylase/oxygenase (Rubisco). In chloroplasts, the genome of the chloroplast encodes for the large subunit. Also, the cyanelles export glucose instead of the triose phosphates exported by chloroplasts.

The fact that in such syncyanoses one is dealing with composite organisms that exhibit features altogether new and no longer characteristic of either partner alone, led Skuja in 1954 to establish the phylum Glaucophyta. It must be appreciated that the organisms in the phylum represent a very old group, and that, when evolving, they were very plastic and undergoing a great deal of change in the attempt to reach the relatively stable level of a cell with a chloroplast. Such a dynamic group was formed consisting of a large number of organisms not well suited to compete with their more highly developed progeny. Such a situation led to the demise of many of the original members of the Glaucophyta, resulting in the existence today of few extant members of the group.

Cyanophora paradoxa is a freshwater flagellate with two cyanelles in the protoplasm, each cyanelle with a central dense body (Fig. 3.1(*a*)). Nitrate reduction, photosynthesis, and respiration in the cyanelles of *Cyanophora paradoxa* are similar to the corresponding processes of chloroplasts and dissimilar to those of cyanobacteria (Floener and Bothe, 1982; Floener et al., 1982). This fact is cited as evidence that the cyanelles of *Cyanophora paradoxa* are close to chloroplasts in evolution. However, the cyanelles of *Cyanophora paradoxa* are primitive in regard to where ribulose-1,5-bisphosphate carboxylase/oxygenase is produced. Ribulose-1,5-bisphosphate carboxylase/oxygenase, the carbon dioxide-fixing enzyme in photosynthesis, consists of 16 subunits, 8 large and 8 small. In higher plants the large subunits are encoded by DNA of the plastids, whereas the small subunits are encoded by nuclear DNA. In *Cyanophora paradoxa*, both sizes

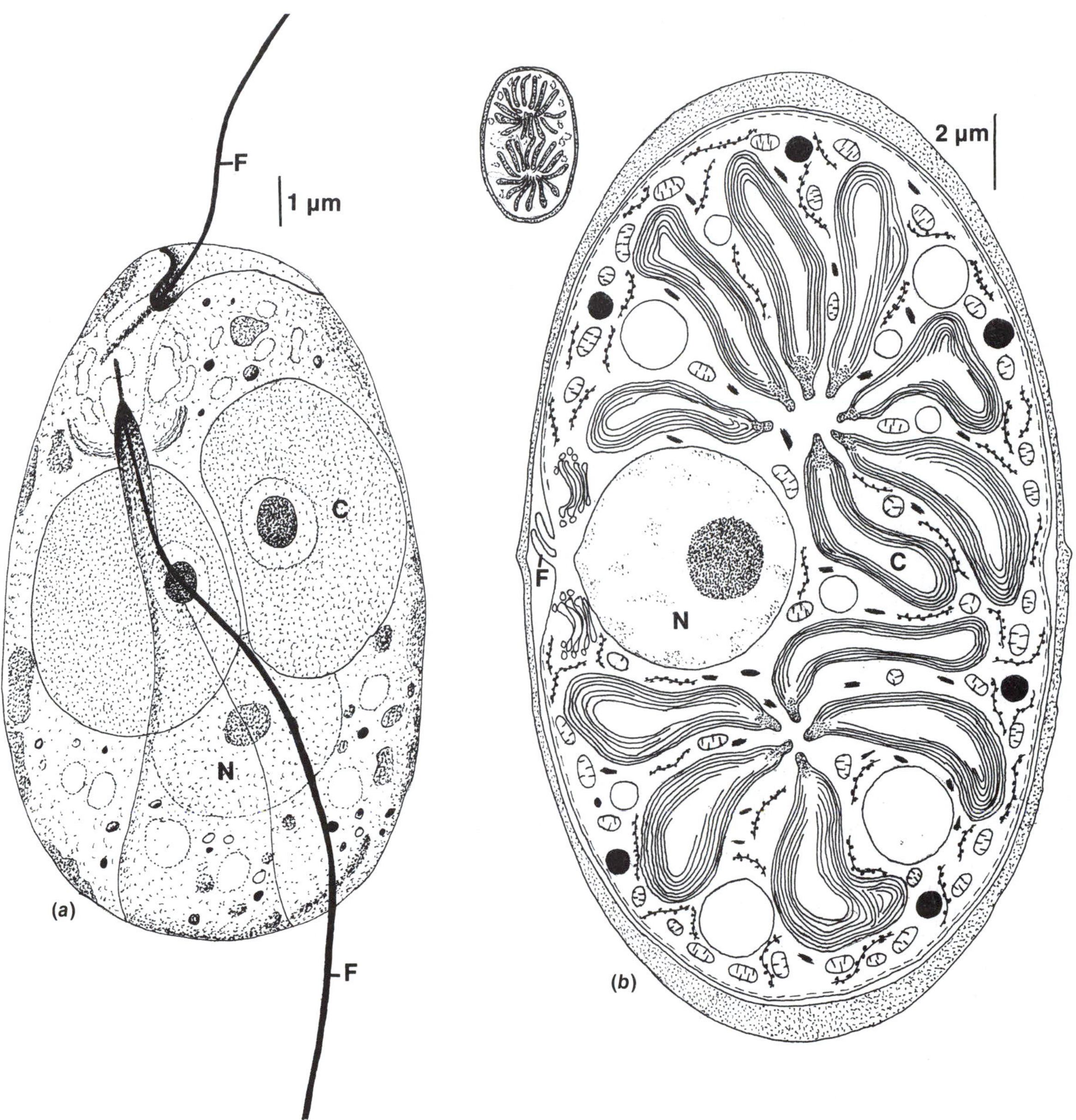

Fig. 3.1 (*a*) *Cyanophora paradoxa* with two cyanelles (C), nucleus (N), and

of subunits are encoded by cyanelle DNA, indicating that in this respect, the cyanelle has not advanced far in evolution toward a chloroplast (Heinhorst and Shively, 1983).

Glaucocystis is also a freshwater organism, found sparingly in soft-water lakes (lakes low in calcium). It has two groups of cyanelles, one on each side of the nucleus (Fig. 3.1(*b*)). The derivation of *Glaucocystis* from a biflagellate ancestor is evident from the two reduced flagella found inside the cell wall. Both of these organisms have starch formed in the cytoplasm, outside of the cyanelles, indicating that the host has accepted responsibility for the formation of the storage product.

There are other organisms that have endosymbiotic cyanobacteria that are not placed in the Glaucophyta because they represent evolutionary dead ends that did not lead to the evolution of chloroplasts. These organisms have cyanelles that still have a cell wall and are cytologically similar to cyanobacteria, such as the cyanelles of the fungus *Geosiphon* (Schnepf, 1964).

References

Betsche, T., Schaller, D., and Melkonian, M. (1992). Identification and characterization of glycolate oxidase and related enzymes from the endocyanotic algae *Cyanophora paradoxa* and from pea leaves. *Plant Physiol.* 98:887–93.

Chapman, D. J. (1966). Pigments of the symbiotic algae (cyanomes) of *Cyanophora paradoxa* and *Glaucocystis nostochinearum* and two Rhodophyceae, *Porphyridium aeruginosa* and *Asterocytis ramosa. Arch. Mikrobiol.* 55:17–25.

Floener, L., and Bothe, H. (1982). Metabolic activities in *Cyanophora paradoxa* and its cyanelles. II. Photosynthesis and respiration. *Planta* 156:78–83.

Floener, L., Danneberg, G., and Bothe, H. (1982). Metabolic activities in *Cyanophora paradoxa* and its cyanelles. I. The enzymes of assimilatory nitrate reduction. *Planta* 156:70–7.

Helmchen, T. A., Bhattacharya, D., and Melkonian, M. (1995). Analysis of ribosomal RNA sequences from glaucocystophyte organelles provide new insights into the evolutionary relationships of plastids. *J. Mol. Evol.* 41:203–10.

Heinhorst, S., and Shively, J. M. (1983). Encoding of both subunits of ribulose-1,5-bisphosphate carboxylase by organelle genome of *Cyanophora paradoxa. Nature* 304:373–4.

Mereschkowsky, C. (1905). Ueber Natur und Ursprung den Chromatophoren in Pflanzenreich. *Biol. Zentralbl.* 25:593–604.

Mignot, J. P., Joyon, L., and Pringsheim, E. G. (1969). Quelques particularités structurales de *Cyanophora paradoxa* Korsch., protozoaire flagellé. *J. Protozool.* 16:138–45.

Pascher, A. (1914). Uber Symbiosen von Spaltpilzen und Flagellaten. *Ber. Dtsch. Bot. Ges.* 32:339–52.

Schnepf, E. (1964). Zur Feinstruktur von *Geosiphon pyriforme.* Ein Versuch zur Deutung cytoplasmatischer Membranen und Kompartimente. *Arch. Mikrobiol.* 49:112–31.

Schnepf, E., Koch, W., and Deichgräber, G. (1966). Zur Cytologie und taxonomischen Einordnung von *Glaucocystis. Arch. Mikrobiol.* 55:149–74.

Skuja, H. (1954). Glaucophyta. In *Syllabus der Pflanzenfamilien,* 12 A. Engler, ed. H. Melchoir and E. Werdermann, I:56–7. Berlin: Borntraeger.

4 · Rhodophyta

Rhodophyceae

The Rhodophyceae, or **red algae**, comprise the only class in the division Rhodophyta. The Rhodophyceae are probably one of the oldest groups of eukaryotic algae. The red algae are most likely directly descended from a cyanome in the Glaucophyta (see Chapter 3). The Rhodophyceae lack flagellated cells, have chlorophylls *a* and *d*, phycobiliproteins, floridean starch as a storage product, and thylakoids occurring singly in the chloroplast.

A majority of the seaweeds are red algae, and there are more Rhodophyceae (about 4000 species) than all of the other major seaweed groups combined. Although marine red algae occur at all latitudes, there is a marked shift in their abundance from the equator to colder seas. There are few species in polar and subpolar regions where brown and green algae predominate, but in temperate and tropical regions they far outnumber these groups. The average size of the plants also differs according to geographical region. The larger species of fleshy red algae occur in cool-temperate areas, whereas in tropical seas the Rhodophyceae (except for massive calcareous forms) are mostly small, filamentous plants. The Rhodophyceae also have the ability to live at greater depths in the ocean than do members of the other algal classes. They live at depths as great as 200 m, an ability related to the function of their accessory pigments in photosynthesis. About 200 species of Rhodophyceae are found in freshwater, where they do not reach as great a size as the red seaweeds (Skuja, 1938). The majority of freshwater red algae occur in running waters of small to mid-sized streams (Sheath and Hambrook, 1988). Few red algae occur at currents of less than 30 cm s^{-1}. This fast flow probably favors red algae because loosely attached competitors are washed out and because of a constant replenishment of nutrients and gases.

Cell structure

The major features of a red algal cell (Fig. 4.1) are a chloroplast with one thylakoid per band and no chloroplast E.R., floridean starch grains in the cytoplasm outside the chloroplast, no flagella, pit connections between

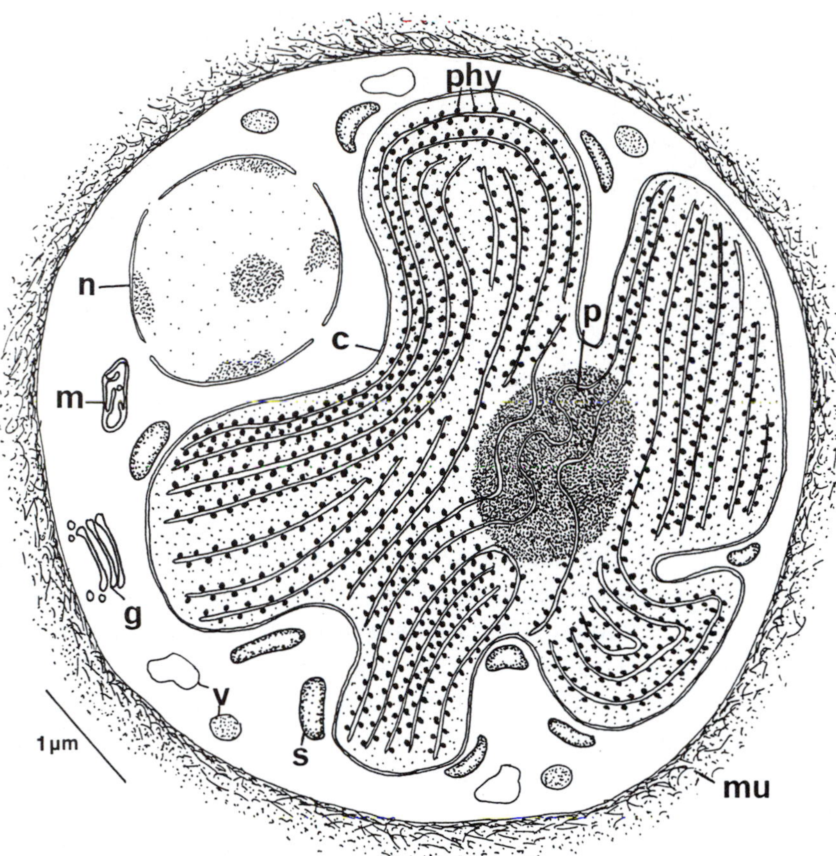

Fig. 4.1 Semi-diagrammatic drawing of a cell of *Porphyridium cruentum*. (c) Chloroplast; (g) Golgi; (m) mitochondrion; (mu) mucilage; (n) nucleus; (p) pyrenoid; (phy) phycobilisomes; (s) starch; (v) vesicle. (Adapted from Gantt and Conti, 1965.)

cells in filamentous genera, and a eukaryotic type of nucleus (Scott et al., 1980).

CELL WALLS

Cellulose forms the microfibrillar framework in most rhodophycean cell walls, although in the haploid phase of the Bangiales (*Bangia* and *Porphyra*) a β-1,3 linked xylan (polysaccharide composed of xylose residues) performs this function (Frei and Preston, 1964). Unicellular red algae have an amorphous matrix of sulphated polysaccharides without cellulose surrounding the cells (Arad et al., 1993). The amorphous polysaccharides or mucilages occur between the cellulose microfibrils in the rest of the red algae. The two largest groups of amorphous mucilages are the **agars** (Fig. 1.7) and the **carrageenans**. These mucilages may constitute up to 70% of the dry weight of the cell wall. Cuticles, composed mostly of protein, can occur outside the cell wall (Craigie et al., 1992).

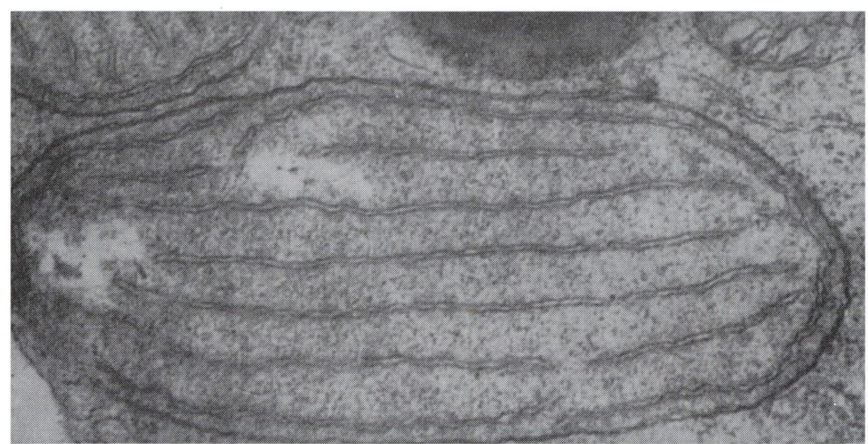

Fig. 4.2 Chloroplast in a carpospore of *Polysiphonia*. (From Tripodi, 1974.)

CHLOROPLASTS AND STORAGE PRODUCTS

Chloroplasts are usually stellate with a central pyrenoid in the morphologically simple Rhodophyceae (Fig. 4.1), whereas in the remainder of the Rhodophyceae they are commonly discoid (Fig. 4.2). In the Rhodophyceae with apical growth, the chloroplasts usually originate from small, colorless proplastids with few thylakoids in the apical cell (Lichtlé and Giraud, 1969). Chloroplasts are surrounded by the two membranes of chloroplast envelope with no chloroplast E.R. present (Figs. 4.1 and 4.2). Inside the chloroplasts thylakoids occur singly. DNA occurs as microfibrils, and the phycobilin pigments are localized into phycobilisomes on the surfaces of the thylakoids, a situation similar to that in the Cyanophyceae.

Chlorophylls *a* and *d* are in the chloroplasts. Chlorophyll *d* has been considered as an altered form of chlorophyll *a* in the past, but it is now accepted as a separate molecule (O'hEocha, 1971). Sagromsky (1964) claimed that it was present in plants in strongly illuminated habitats only. Of the carotenoids, zeaxanthin is found in greatest quantities.

The phycobiliproteins include R-phycocyanin, allophycocyanin, and three forms of phycoerythrin, the latter being present in the greatest amount, giving the algae their pinkish color. B-Phycoerythrin is present in the more primitive Rhodophyceae and has been found in *Porphyridium, Rhodosorus, Rhodochorton,* and *Smithora.* R-Phycoerythrin is found in almost all higher Rhodophyceae, and C-phycoerythrin has been found in *Porphyridium, Porphyra, Smithora, Grateloupia,* and *Polysiphonia.* The phycobiliproteins are in phycobilisomes on the surface of thylakoids (Fig. 4.1). If both phycoerythrin and phycocyanin are present, the phycobilisomes are spherical. If, however, only phycocyanin is present, then the phycobilisomes are discoid (Gantt, 1969).

Complementary chromatic adaptation occurs in the red algae. Orange and red light stimulates the production of long-wavelength absorbing phycocyanin, while green light stimulates the formation of short-wavelength absorbing phycocyanin (Sagert and Schubert, 1995). The color will vary, depending on where the plant grows; a species of red alga may be yellow-green when growing in the intertidal zone, whereas it is deep purple-red when growing 10 to 20 m beneath the water surface. The ratio of phycoerythrin to chlorophyll is much greater at low light intensities. As the intensity of light increases, the ratio of phycoerythrin to chlorophyll drops. This is reflected in the color of the plant and is the reason why many red algae turn green at higher light intensities (Calabrese, 1972). The greatest increase in the phycoerythrin:chlorophyll ratio occurs when the light intensity is dropped from 3000 to 400 lux. The ratio of phycoerythrin to phycocyanin to allophycocyanin stays about the same at different light intensities. There is little change in the thylakoid structure of the chloroplasts in different light regimes, but the number of phycobilisomes is much greater at low light intensities although the size of the individual phycobilisome remains constant (Waaland et al., 1974). This adaptation to greater amounts of phycoerythrin under low light intensities allows the Rhodophyceae to grow at greater depths than other algae because of the ability to utilize better the green and blue light that penetrates farthest into water (pink-red pigments absorb blue-green and reflect pink-red).

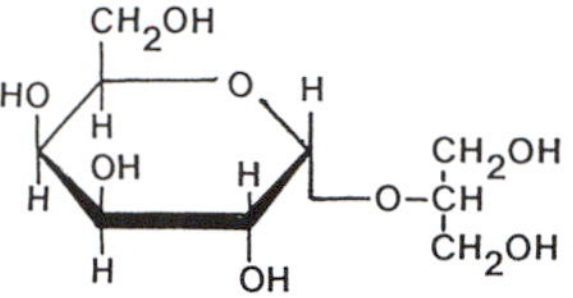

Fig. 4.3 The structure of floridoside.

Floridoside (O-α-D-galactopyranosyl-(1-2)-glycerol) Fig. 4.3) is the major product of photosynthesis in the red algae (Barrow et al., 1995). An isomeric form of floridoside, isofloridoside, occurs along with floridoside in the Bangiales. The concentration of floridoside increases in red algal cells as the salinity of the medium increases (Reed, 1985). This change in floridoside concentration is thought to compensate, at least in part, for the changes in external osmolarity, thereby preventing water from leaving the algal cells as the salinity increases. The levels of floridoside can be as high as 10% of the tissue dry-weight in some marine red-algal thalli. The first observable product of photosynthesis is phosphoglyceric acid, as is the case in higher plants. Floridoside appears after 30 seconds of illumination and after 2 hours floridoside is the major product of photosynthesis. Floridoside apparently has the same function as sucrose, the common product of photosynthesis in green algae and higher plants.

Floridean starch (Fig. 1.20) is the long-term storage product, occurring as grains in the cytoplasm outside of the chloroplast (Fig. 4.1). Floridean starch is similar to the amylopectin of higher plants, staining red-violet with iodine. In the more primitive Rhodophyceae the starch grains are clustered as a sheath around the pyrenoid of the chloroplast, whereas in the more advanced Rhodophyceae the starch grains are scattered in the cytoplasm (Hara, 1971; Lee, 1974).

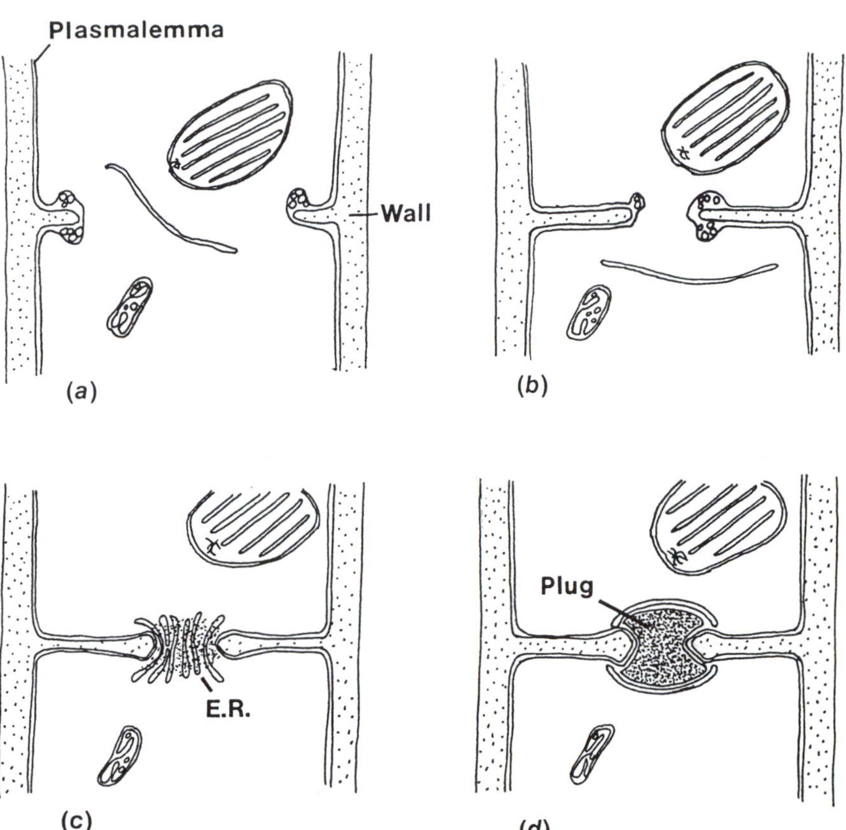

Fig. 4.4 Semi-diagrammatic drawing of the formation of a pit connection in a red alga: (*a*) The cross wall begins forming inward with the wall precursors found in vesicles derived from the cytoplasm: (*b*) the cross wall septum is complete, leaving an opening (aperture) in the center; (*c*) endoplasmic reticulum lies across the opening in the wall, and electron-dense material condenses in this area; (*d*) the pit connection is formed, consisting of a plug with the plasmalemma continuous from cell to cell. (Adapted from Ramus, 1969a, 1971; Lee, 1971.)

PIT CONNECTIONS

Pit connections occur between the cells in all of the orders except the Porphyridiales, and the haploid phase of the Bangiales. It has been pointed out that the term "pit connection" is inappropriate because the structure is neither a "pit" nor a "connection"; however, because the term has been used for so long, it is probably best to retain it. A pit connection consists of a proteinaceous plug core in between two thallus cells (Fig. 4.4, 4.5). Cap membranes separate the plug core from the adjacent cytoplasm. The cap membrane is continuous with the plasma membrane, which in turn is continuous from one cell to the next. On the inside of the cap membrane can be an inner layer, while on the outside of the cap membrane can be an outer cap layer (Pueschel, 1987). The structure of the pit connection can vary. The more primitive red algae, such as *Rhodochaete* and *Compsopogon*, lack cap membranes and cap layers, with only a plug core present. It has been postulated that this represents the ancestral condition (Pueschel, 1989).

There are two types of pit connections. **Primary pit connections** are formed between two cells during cell division. **Secondary pit connections**

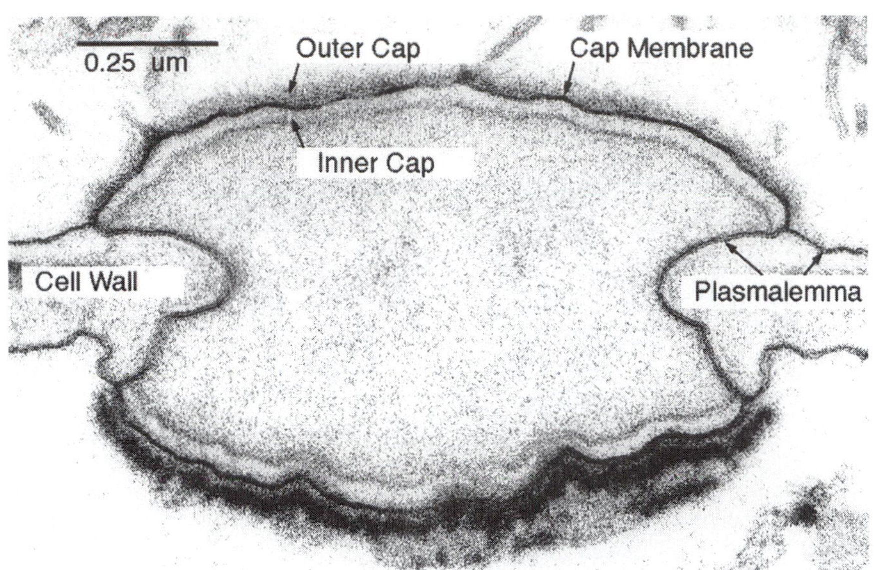

Fig. 4.5 A pit connection of *Palmaria mollis*. The plasma membrane is continuous from cell to cell. The cap membrane is continuous with the plasma membrane. The inner and outer cap layers are on each side of the cap membrane. (From Pueschel, 1987.)

result when two cells fuse. Both types of pit connections have the same structure (Kugrens and West, 1973). **Primary pit connections** are formed as follows (Fig. 4.4) (Ramus, 1969a): Soon after nuclear division, the cross wall grows inward from the lateral wall. When the cross wall is complete, there remains a hole (aperture) in the center through which the protoplasm of the two cells is continuous. A number of parallel vesicles traverse the hole, with electron-dense material condensing around the vesicles. Eventually the vesicles disappear, and the electron-dense material fills the hole. A membrane is formed around this material, producing a plug in the hole. The pit connection has been reported to contain proteins and polysaccharides (Pueschel and Trick, 1991; Ramus, 1971). The pit connection may function as a site of structural strength on the thallus (Kugrens and West, 1973). In some algae the plugs of the pit connections become dislodged from between the cells of a developing gonimoblast, leaving the protoplasm continuous between the cells and allowing the passage of metabolites to the developing reproductive cells (Turner and Evans, 1978).

Calcification

All members of the Corallinales and some of the Nemaliales (*Liagora; Galaxaura*) deposit $CaCO_3$ extracellularly in the cell walls. Anhydrous calcium carbonate occurs in two crystalline forms, **calcite** (rhomboidal) and **aragonite** (orthorhombic). The two forms differ markedly in specific gravity, hardness, and solubility. The Corallinales deposit $CaCO_3$ as calcite, whereas the calcified members of the Nemaliales deposit $CaCO_3$ as aragonite. In *Liagora cenomyce* (Nemaliales) the aragonite occurs as needle-like crystals in the wall, whereas in the Corallinales the calcite occurs as massive deposits (Borowitzka et al., 1974). Magnesium carbonate and strontium carbonate are commonly deposited along with calcium carbonate, there being less magnesium carbonate in aragonite than in calcite (Dixon, 1973). *Liagora* aragonite has about 1% magnesium carbonate, whereas in the Corallinaceae the calcite can contain 7% to 30% magnesium carbonate. Calcified walls of living cells probably have a mucilaginous component that shows the loss of Ca^{2+} into the medium (Pearse, 1972). If a calcified thallus is killed, the dispersal of the calcified wall is greatly accelerated.

The coralline algae thrive in rock pools and on rocky shores exposed to very strong wave action and swift tidal currents. The red algae that have the highest rates of calcification also have the highest rates of photosynthesis and are usually found in waters less than 20 m deep (Goreau, 1963). Calcification of the thallus occurs about two to three times more rapidly in the light than in the dark, although significant calcification does occur in the dark (Okazaki et al., 1970). The above observations have led to the theory that calcification may be linked to photosynthesis (Pearse, 1972). The most quoted theory on calcification is that calcium salts are precipitated from seawater by the alkalinity brought about by the extraction of carbon dioxide during photosynthesis, calcium carbonate being less soluble in alkaline waters than acid. The obvious and often mentioned drawback to this theory is that because all algae carry out photosynthesis, it is difficult to understand why they do not all calcify. Also the continued calcification of corallines in the dark is another argument against this theory.

Seawater is more or less saturated with respect to calcium carbonate, and the addition of either calcium or carbonate will cause the carbonate to precipitate. The concentration of CO_3^{2-} is related through a complex series of equilibria (Digby, 1977a,b):

$$CO_2 + H_2O \rightleftharpoons H_2CO_3 \rightleftharpoons H^+ + HCO_3^- \rightleftharpoons 2H^+ + CO_3^{2-}$$

then

$$CO_3^{2-} + Ca^{2+} \rightleftharpoons CaCO_3 \text{ (ppt.)}$$

The addition of acid will drive the reactions to the left and cause carbonate to dissolve, whereas the addition of base will drive the reactions to the right and form more carbonate. At the pH of seawater (8.4), almost all of the CO_2 in the water is in the form of bicarbonate ion, HCO_3^-. The addition of one equivalent of hydroxyl ions to seawater saturated with respect to calcium carbonate will precipitate one equivalent of calcium carbonate:

$$Ca^{2+} + HCO_3^- + OH^- \rightleftharpoons CaCO_3 \text{ (ppt.)} + H_2O$$

The fact that seawater is nearly saturated with calcium carbonate was demonstrated with seawater from the coast of Maine by Digby (1977a). By raising the pH of this seawater to 9.6, he caused precipitation of carbonates. Calcium carbonate precipitated first, being less soluble, followed by carbonate richer in magnesium.

Digby (1977b) proposed a theory of calcification of red algae based on raising the pH of the seawater immediately outside the cells, causing precipitation of carbonates as outlined above. The first process is the normal photosynthetic splitting of water:

$$H_2O \rightarrow \tfrac{1}{2}O_2 + 2H^+ + 2e^-$$

The oxygen then diffuses out of the cell. As mentioned above, in the sea most of the carbon dioxide is in the form of bicarbonate ions; these ions diffuse into the cells and receive the electrons freed initially by photosynthesis. The bicarbonate ions are then converted into carbonate ions and hydrogen according to the following reaction:

$$2HCO_3^- + 2e^- \rightleftharpoons 2H^+ + 2CO_3^{2-}$$

The carbonate ions diffuse out of the cell where they partially dissociate, forming bicarbonate and hydroxyl ions and thereby raising the pH:

$$2CO_3^{2-} + H_2O \rightleftharpoons 2HCO_3^- + 2OH^-$$

When saturation with regard to calcium and carbonate is reached by a rise in pH outside the cells, calcium carbonate precipitates on the walls:

$$2Ca^{2+} + 2CO_3^{2-} = 2CaCO_3 \text{ (ppt.)}$$

Continued precipitation of $CaCO_3$ results in the calcified wall of the Rhodophyceae. Although the above theory explains the mechanism of calcification, it does not explain why calcification is specific to certain red algae.

It has been theorized that calcification of red algal thalli evolved as a protection against grazing by organisms such as limpets, although it has also been pointed out that grazing is beneficial to the coralline algae in that the grazers remove epiphytes from the red algal thallus (Pueschel and Miller, 1996).

Reproductive structures

The Rhodophyceae have no flagellated cells or cells with any vistigial structure of flagellation such as basal bodies. In sexual reproduction **spermatia** are produced which are carried passively by water currents to the female

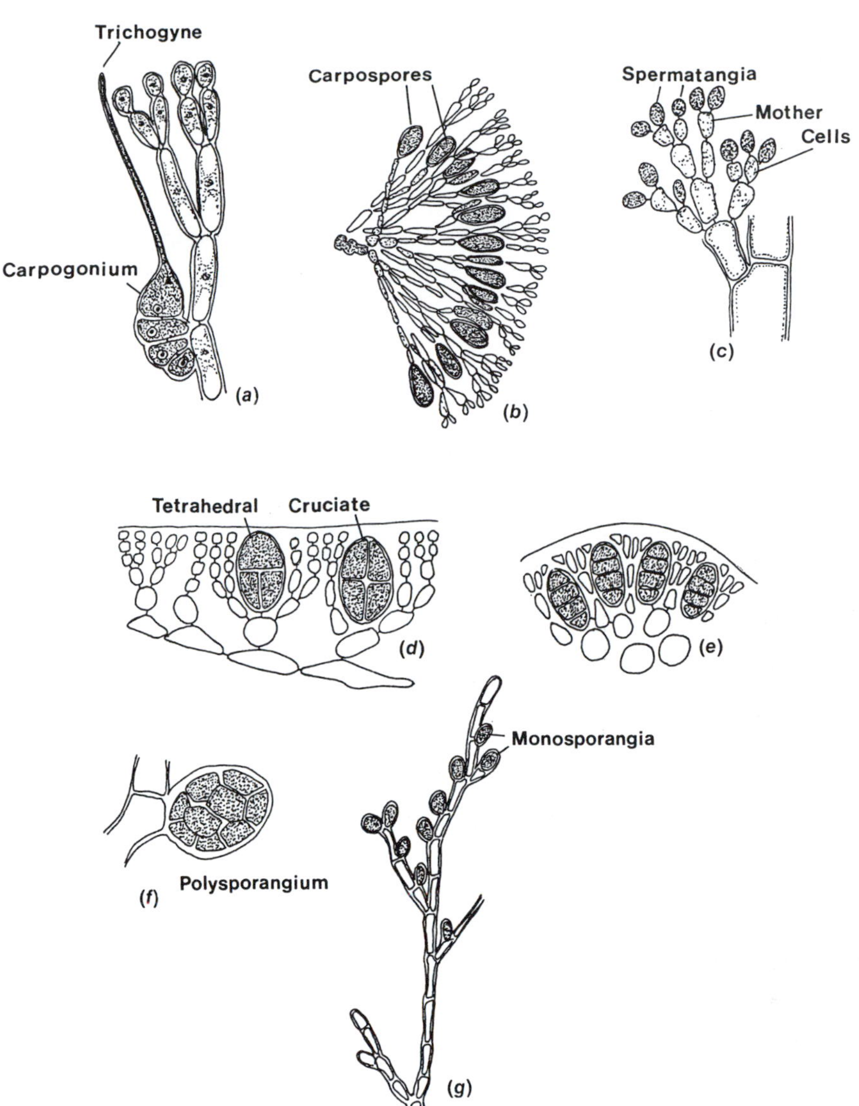

Fig. 4.6 (*a*) A filament of *Liagora viscida* with a carpogonial branch. (*b*) Gonimoblast filaments of *L. viscida* with carpospores. (*c*) A spermatangial branch of *Acrochaetium corymbiferum* with spermatangia and spermatangial mother cells. (*d*) Tetrahedral and cruciate tetrasporangia of *Nemastoma laingii*. (e) Zonate tetrasporangia of *Hypnea musciformis*. (*f*) Polysporangium of *Pleonosporium vancouverianum*. (*g*) Monosporangia of *Kylinia rhipidandra*. ((*a*), (*b*), (*c*), (e) after Kylin, 1930; (*f*) after Kylin, 1924; (*g*) after Kylin, 1928.)

organ, the **carpogonium** (Fig. 4.6(*a*)). The fertilized carpogonium produces **gonimoblast filaments** that form **carposporangia** and diploid **carpospores** (Fig. 4.6(*b*) and 4.7) at the tips. The carpospores produce the diploid **tetrasporophyte**, which subsequently gives rise to haploid tetraspores (Figs. 4.6(*d*),(*e*) and 4.7). Advanced red algae form chiefly tetrahedral tetrasporangia (Fig. 4.6(*d*)) with large spores, whereas less advanced groups generally form cruciate or zonate tetrasporangia (Fig. 4.6(*d*),(*e*)) with smaller spores (Ngan and Price, 1979). Tetraspores are generally larger than carpospores. The tetraspores complete the life cycle by germinating to form

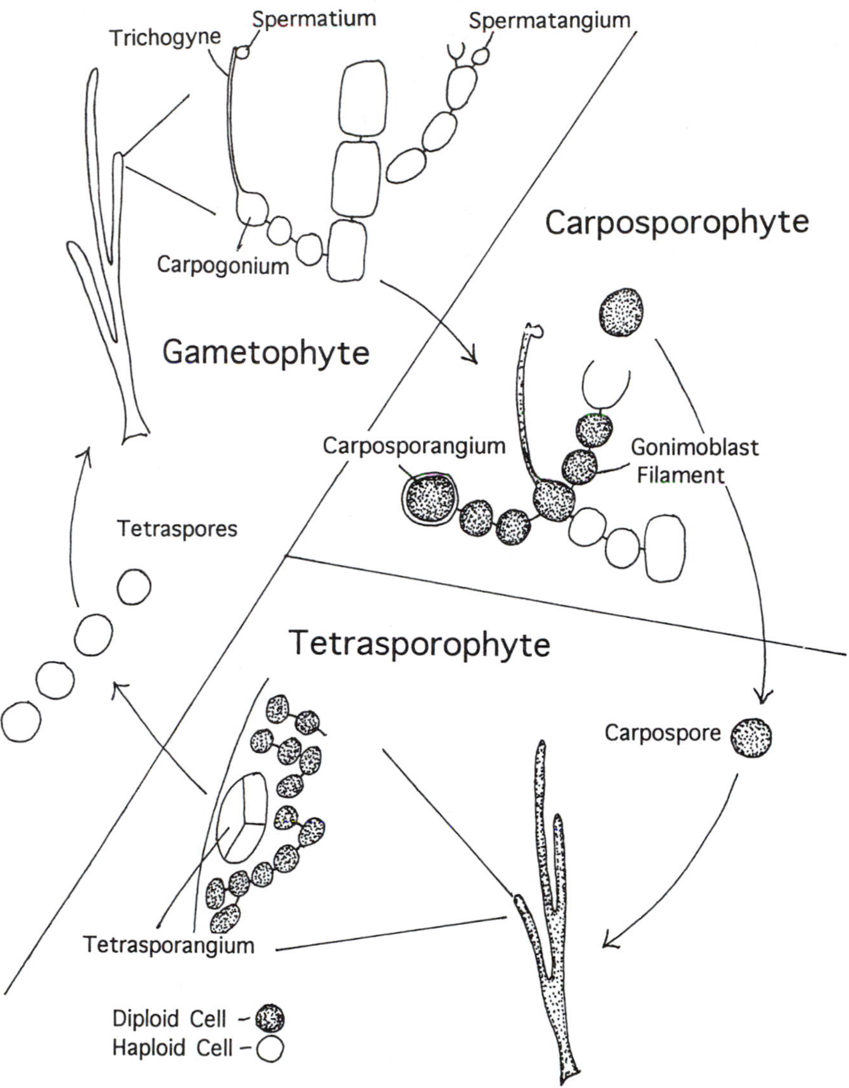

Fig. 4.7 Simplified life cycle of a typical red alga.

the gametophyte. Although this is the general life cycle of most Rhodophyceae, there are a number of modifications of it.

The postfertilization events vary from one order to another. The more advanced orders have **auxiliary cells** with which the fertilized carpogonium fuses to form a multinucleate fusion cell. Papenfuss (1966) recognizes two types of auxiliary cells, nutritive and generative. **Nutritive auxiliary cells** provide nutrients for the developing carposporophyte, whereas **generative auxiliary cells** give rise to gonimoblast filaments. The diploid tissue formed from the fertilized carpogonium forms the gonimoblast filaments. The gonimoblast filaments produce terminal carposporangia, which in turn form the carpospores. The carposporangia enlarge considerably during

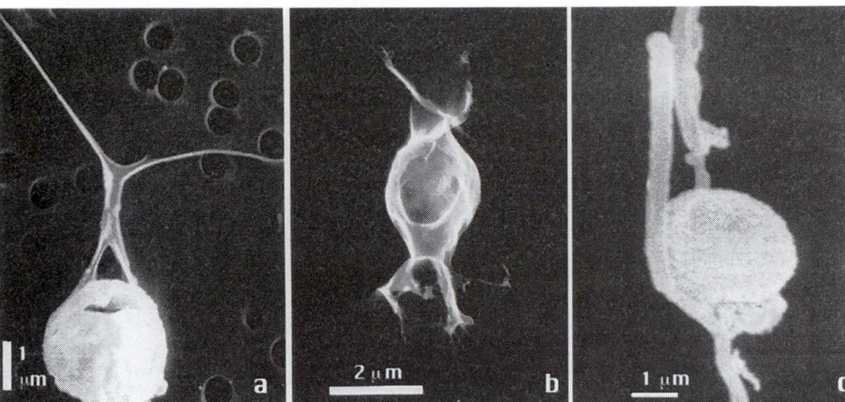

Fig. 4.8 Examples of spermatangial appendages. (*a*) Appendage of *Antithamnion nipponicum*. (*b*) Fimbriate cone-shaped appendages at each end of *Aglaothamnion neglectum*. (*c*) Fibrous strands attached to spermatia of *Tiffaniella snyderae*. ((*a*) from Kim and Fritz, 1993b; (*b*) from Magruder, 1984; (*c*) from Fetter and Neushul, 1981.)

their maturation because of the development of the chloroplasts and the vesicles containing wall precursors. The pit connection between the carposporangium and the gonimoblast breaks before release of the carpospore. Also during the development of the gonimoblast filaments, the pit connections between the older gonimoblast cells usually dissolve (Kugrens and West, 1972a, 1973, 1974).

CARPOGONIUM

The female organ, or carpogonium, consists of a dilated basal portion and a usually narrow gelatinous elongate tip, the **trichogyne**, which receives the male cells (Fig. 4.6(*a*)). Usually there are two nuclei in a carpogonium, one in the trichogyne, which degenerates soon after the carpogonium matures, and one in the basal part of the carpogonium, which functions as the female gamate nucleus. In most Rhodophyceae the carpogonium terminates a short, often branched, three- to four-celled lateral called the carpogonial branch. The cell from which the carpogonial branch arises is the **supporting cell**. The carpogonium and carpogonial branch are commonly colorless, although in some Nemaliales this is not true.

SPERMATIUM

The spermatia of the Rhodophyceae are spherical or oblong cells produced in **spermatangia**, a single spermatium being formed in a spermatangium and then released, leaving the empty spermatangium (Fig. 4.9). The spermatangia (Fig. 4.8) are formed on spermatangial mother cells (Fig. 4.6(*c*)). The young spermatangia frequently have a pronounced polar orientation, with the nucleus in the apical portion and one or more vacuoles in the basal portion (Scott and Dixon, 1973a). As the spermatangium ages, vacuoles

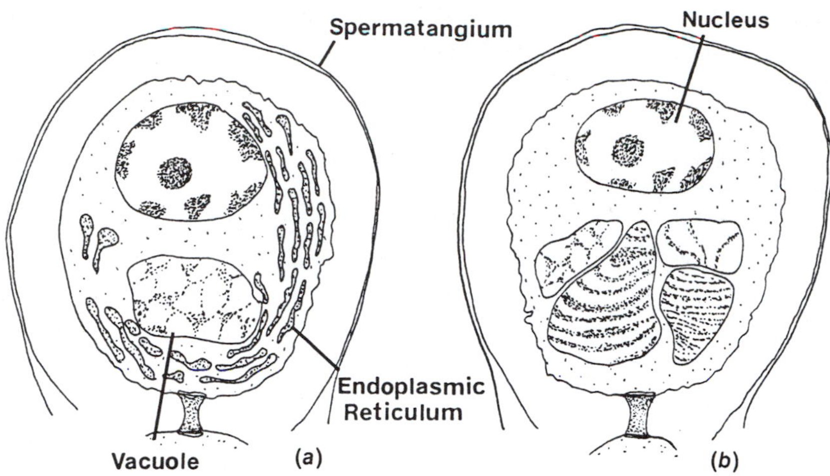

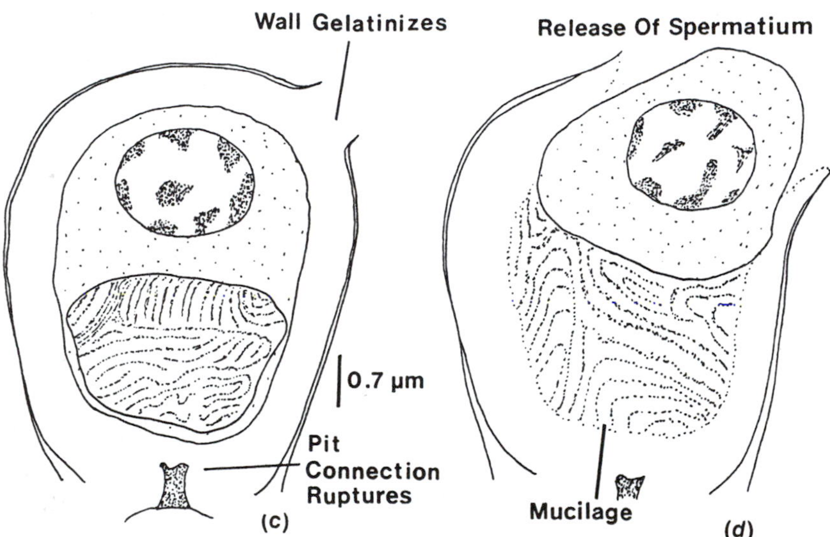

Fig. 4.9 A semi-diagrammatic drawing of the formation of a spermatium in the Rhodophyceae. (*a*) Young spermatium. (*b*) Formation of vacuoles containing fibrous material. (*c*) Fusion of vacuoles, breaking of the pit connection, and gelatinization of the wall of the spermatangium. (*d*) Extrusion of mucilage and release of the spermatium. (Adapted from Scott and Dixon, 1973a; Kugrens, 1974.)

form in the basal area. These vacuoles contain fibrous material (probably mucopolysaccharides) and make up half the volume of the spermatangium. Subsequently the vacuoles fuse to form one large vacuole. The spermatium is released by the gelatinization of the spermatangial wall near the apex and the concurrent release of the fibrous material in the basal vacuole. The fibrous material presumably swells and pushes the spermatium out of the spermatangium (Fig. 4.9). The fibrous material is sticky, and some of it adheres to the spermatium, thereby facilitating attachment to the trichogyne (Fig. 4.8(*c*)). During the development of the spermatium the pit connection with the spermatangium mother cell is severed. The mature spermatium is uninucleate, and wall-less but surrounded by mucilage, and may (Simon-Bichard-Bréaud, 1971; Peyrière, 1971) or may not (Kugrens and West, 1972a; Kugrens, 1974) contain functional chloroplasts.

FERTILIZATION

The spermatium usually is carried passively by water currents to the trichogyne of the carpogonium, although some spermatia can glide in a manner similar to the gliding in *Porphyridium*. At the point of contact the wall of the spermatium and the trichogyne dissolve, and the male nucleus moves into the carpogonium. Fusion of the male and female nuclei occurs in the basal portion of the carpogonium. The trichogyne will usually continue to grow until contact is made with a spermatium. After fertilization has occurred, the trichogyne becomes separated at its base from the rest of the carpogonium by the progressive thickening of the cell wall.

There is a relatively low statistical probability that the non-motile spermatia will be carried to receptive trichogynes of carpogonia. Species of red algae have evolved two mechanisms to increase this probability:

1 Some species have appendages on the spermatia that extend the reach of the spermatia five- to tenfold, resulting in a better chance of attaching to a receptive carpongium. The appendages are initially contained within vesicles with the spermatia and unfold when the spermatia are released (Fig. 4.8).
2 The outer cell walls of the spermatia differ in their carbohydrate composition from the cell walls of the vegetative cells. The spermatia have α-D-methyl mannose residues on their surface and these residues bind specifically to the trichogynes of the carpogonia. Thus, vegetative cells may be contacted by spermatia as the spermatia are moved by water currents, but the spermatia will only bind to the trichogynes of the carpogonia (Kim and Fritz, 1993a,b).

MEIOSPORANGIA AND MEIOSPORES

Tetrasporangia, polysporangia, and bisporangia formed on diploid plants are generally regarded as being the seat of meiosis although there are exceptions to this. The **tetrasporangia** form four tetraspores either in a row (zonate), crosswise (cruciate), or most commonly in a tetrad (tetrahedral). In the formation of tetraspores a wall is laid down inside the tetrasporangia by the protoplast, which has prominent dictyosomes (each dictyosome associated with a mitochondrion as is common in the Rhodophyceae; Kugrens and West, 1972b; Scott and Dixon, 1973b). The tetraspores are not joined by pit connections.

In some of the red algae it is possible to control tetrasporogenesis by varying the light period, but there is no general rule that can be applied to the response. In *Rhodochorton purpureum* (West, 1972; Dring and West, 1983) and *Acrochaetium asparagopsis* (Abdel-Rahman, 1982), tetrasporophytes produce tetrasporangia under short-day conditions. According to the physiological clock hypothesis, photoperiodism is controlled by an endogenous free-running **circadian** (*approximately 24-hour*) oscillation of some biochemical change. Each oscillation involves a regular alternation of two phases (each lasting approximately 12 hours) with a different sensitivity to light. The two phases are a **photophile** (light-loving) phase and a **skotophile** (dark-loving) phase. The initiation of a particular event depends on initiation or inhibition of metabolic changes of short-day or long-day organisms, respectively, by exposure to light at a particular point in the skotophile phase. In the case of *Rhodochorton purpureum* and *Acrochaetium asparagopsis*, light-breaks in the dark (skotophile) phase result in inhibition of tetrasporogenesis, whereas dark-breaks in the light (photophile) phase result in stimulation of tetrasporogenesis. Tetrasporogenesis in these algae is therefore a short-day phenomenon (Abdel-Rahman, 1982).

Polysporangia contain more than four spores, usually in multiples of four (Fig. 4.6(*f*)). Polysporangia probably evolved from tetrasporangia because polysporangia occur predominantly in the most advanced order, the Ceramiales. **Bisporangia** are probably reduced tetrasporangia in which cell division has resulted in two spores after meiosis instead of four.

ASEXUAL SPORES

Monosporangia (one spore per sporangium) and **parasporangia** (more than one spore per sporangium) produce asexual spores that re-form the parent thallus. Monosporangia can be formed by the release of a vegetative cell from the thallus (*Goniotrichum, Asterocystis*), or by the formation of sessile one-celled branches (*Kylinia*, Fig. 4.6(*g*)).

Secretory cells

Secretory cells (vesicular cells) occur in some Rhodophyceae (Fig. 4.10(*a*),(*b*)). These cells are colorless at maturity and commonly have a large central vacuole. The secretory cells in *Bonnemaisonia* are prominent and associated with high concentrations of iodine (Fig. 4.10(*a*)). The concentration of iodine can be high enough to produce a blue color in herbarium paper with starch as a filler (the chemical test for starch). Wolk (1968) has shown that if these algae are grown in an absence of bromine, then the secretory

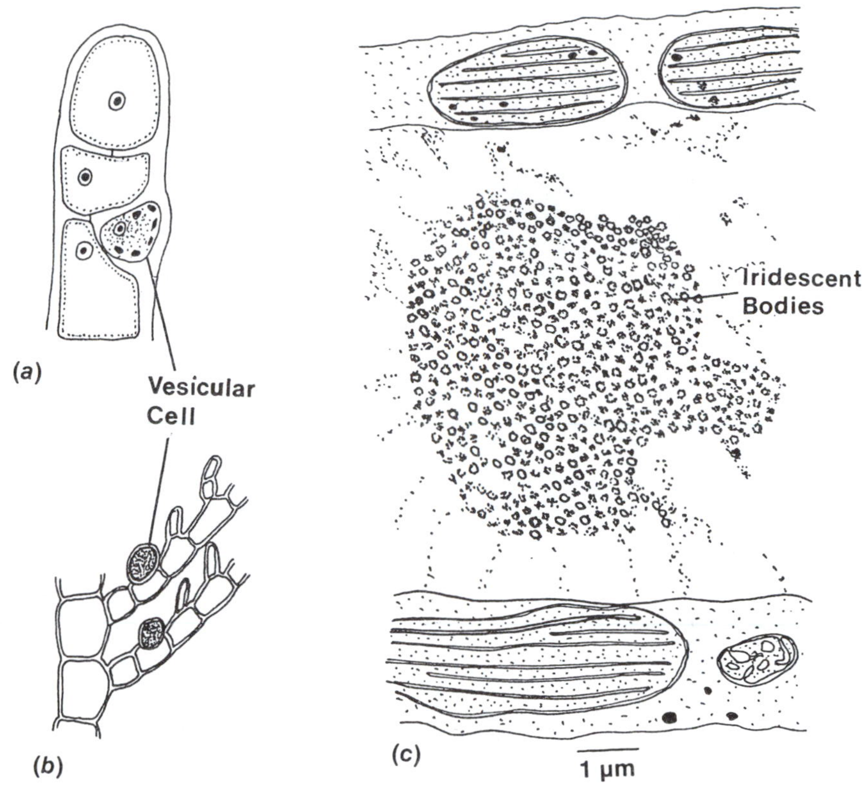

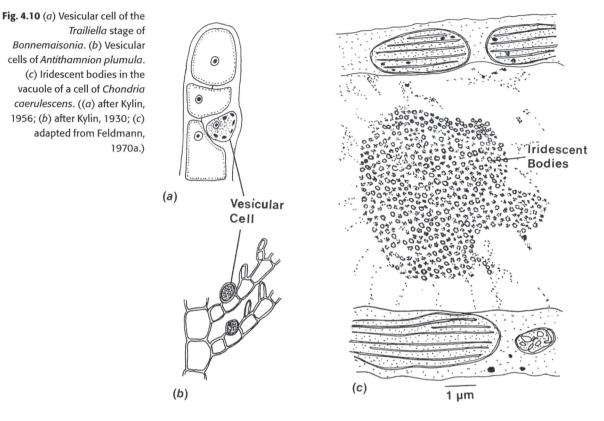

Fig. 4.10 (*a*) Vesicular cell of the *Trailiella* stage of *Bonnemaisonia*. (*b*) Vesicular cells of *Antithamnion plumula*. (*c*) Iridescent bodies in the vacuole of a cell of *Chondria caerulescens*. ((*a*) after Kylin, 1956; (*b*) after Kylin, 1930; (*c*) adapted from Feldmann, 1970a.)

cells are vestigial, lacking the normal large vacuole with its refractile contents. Bromine can also occur as granular deposits in mucilage, such as in the mucilage of the thallus medulla in *Thysanocladia densa* (Pallaghy et al., 1983) or in the cuticle as in *Polysiphonia nigrescens* (Peders'en et al., 1981).

Other types of secretory cells not associated with the formation of halogens occur. The cells are often called secretory cells, even though they are apparently not involved in secretion. In *Antithamnion* (Figs. 4.10(*b*) and 4.11(*a*)), these cells have a large central vacuole containing sulfated acidic polysaccharide (Young and West, 1979). In *Opuntiella californica*, there are "gland cells" with a large vacuole containing a homogeneous proteinaceous material (Young, 1979) (Fig. 4.11(*b*)). These "secretory cells" and "gland cells" may have compounds that act as deterrents to grazing, or they may accumulate special reserves for metabolic use.

Iridescence

The thalli of some Rhodophyceae show a marked blue or green iridescence when observed in reflected light. **Iridescence** is solely a physical

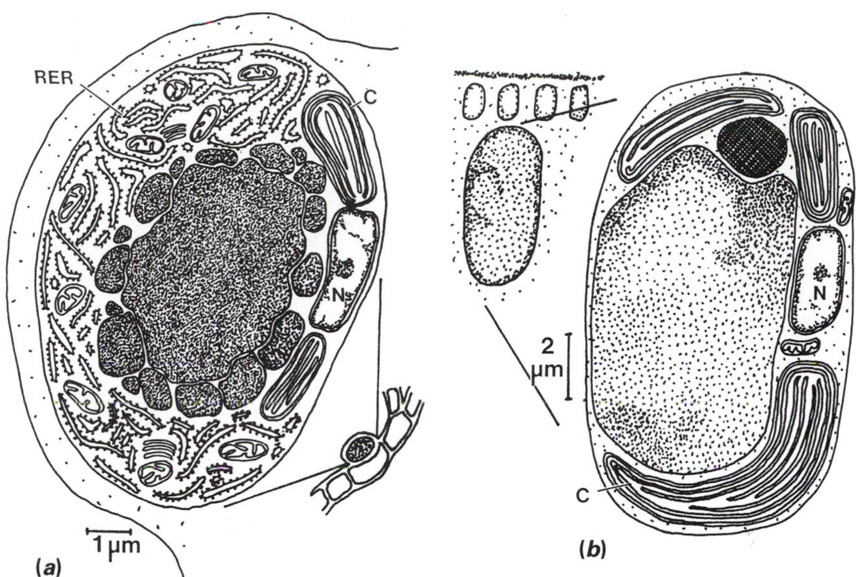

Fig. 4.11 (*a*) Semi-diagrammatic drawing of the fine structure of a vesicular cell of *Antithamnion*. The cell has a large central vacuole surrounded by protoplasm containing rough endoplasmic reticulum (RER), mitochondria, chloroplasts (C), and a nucleus (N). (*b*) Semi-diagrammatic drawing of the fine structure of a gland cell of *Opuntiella californica*. The cell has a large central vacuole containing chloroplasts (C), a nucleus (N), and mitochondira. ((*a*) after Young and West, 1979; (*b*) after Young, 1979.)

interference and is not related to any light-producing phenomena such as phosphorescence or bioluminescence (Gerwick and Lang, 1977). It results from the interference of light waves reflected from the surfaces of very thin multiple laminations separated by equally thin or thinner layers of material with a contrasting refractive index; the layers are uniform and produced by periodic secretion and deposition. Iridescence in the Rhodophyceae has been attributed to different causes by different investigators. Feldmann (1970a,b) found iridescent bodies in *Chondria* (Fig. 4.10(*c*)) and *Gastroclonium*, whereas Gerwick and Lang (1977) attributed the iridescence in *Iridaea* to a multilayered cuticle.

Epiphytes and parasites

Rhodophycean organisms range from autotrophic, independent plants to complete heterotrophic parasites. The spectrum includes non-obligate epiphytes (in the *Acrochaetium-Rhodochorton* complex), obligate epiphytes (*Polysiphonia lanosa* on *Ascophyllum*), semi-parasites that have some photosynthetic pigments (*Ceratocolax, Gonimophyllum*), and parasites with no coloration (*Harveyella, Holmsella*).

The association between the obligate epiphyte red alga *P. lanosa* and its brown alga host *Ascophyllum* has been well studied. After the spore of *P. lanosa* germinates on the host, the red alga sends down a rhizoid that digests its way into the host tissue by means of enzymatic digestion of the

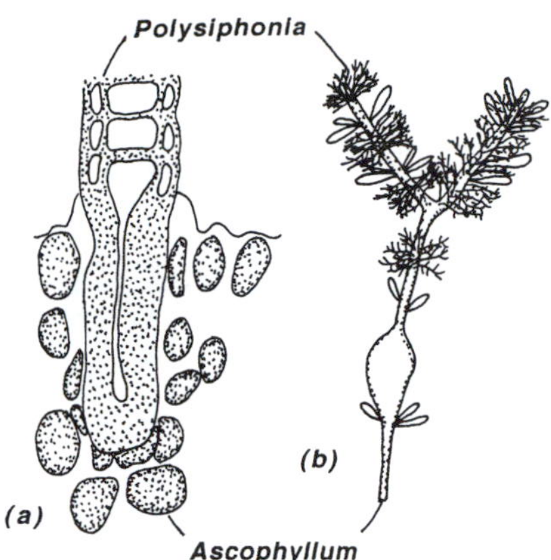

Fig. 4.12 (*a*) The rhizoid of *Polysiphonia lanosa* penetrating tissue of *Ascophyllum nodosum*. (*b*) *Polysiphonia* epiphytic on *Ascophyllum*. ((*a*) after Rawlence, 1972.)

host tissues. The enzymes are discharged from vesicles at the tip of the rhizoid. Once the rhizoid has established itself, intrusive cells form the basal parietal cells of the thallus (Fig. 4.12) (Rawlence, 1972). Although *P. lanosa* is an obligate epiphyte, there is no transfer of metabolites from the host to the epiphyte, the epiphyte manufacturing all of its own requirements through photosynthesis (Harlin and Craigie, 1975; Turner and Evans, 1978).

Parasitic red algae can be either adelphoparasites (*adelpho* = brother) or alloparasites (*allo* = other). **Adelphoparasites** are closely related to, or belong to the same family as their hosts and constitute 90% of parasitic red algae (Goff et al., 1996). **Alloparasites** are not closely related to their hosts. The parasitic habit apparently has been adapted more easily when the host is closely related to the parasite (adelphoparasites) than when it is not (alloparasites), partially because it is easier for the parasite to establish secondary pit connections with the host (and therefore transfer nutrients) if the host and parasite are related.

Choreocolax polysiphoniae is an example of a rhodophycean parasite (Fig. 4.13). The alga is a complete parasite and is interesting in that it is parasitic on *Polysiphonia fastigata*, which is itself epiphytic on *Ascophyllum*. Because *Choreocolax* is in the Gigartinales, and *Polysiphonia* is in the Ceramiales, this is a case of alloparasitism. *Choreocolax* consists of a more or less hemispherical white external portion made up of subdichotomously branched filaments enclosed and surrounded by gelatinous matter, and a mass of haustorial cells growing inside the host (Sturch, 1926). When the haustorial cells have reached a certain distance from the original point of infection, they frequently give rise to a second external cushion. Secondary pit connections are established between the haustorial internal filaments of the parasite and the larger cells of the host. Apically dividing filaments of the

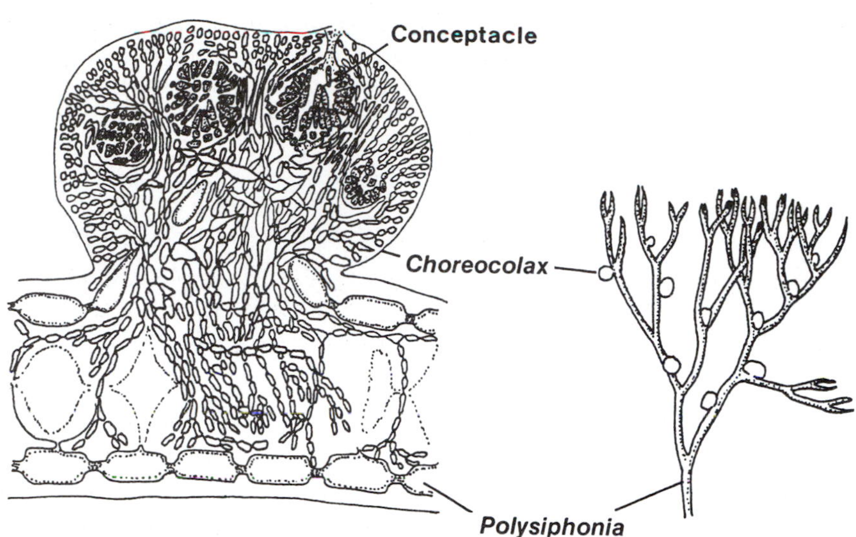

Fig. 4.13 Drawing of a section of *Choreocolax polysiphoniae* on *Polysiphonia*. The parasitic *Choreocolax* has carposporophytes in various stages of development. (After Sturch, 1926.)

parasite *Choreocolax* cells produce small conjunctor cells (Fig. 4.14) by the asymmetrical division of a cell. A conjunctor cell contains a highly condensed, small nucleus. The conjunctor cell fuses with an adjacent host *Polysiphonia* cell. The nucleus and cytoplasm of the conjunctor cell are incorporated into the host cell. The pit connection that originally connected the conjunctor cell and the sister *Choreocolax* cell now connects the host *Polysiphonia* cell and the parasite. This is often called a "secondary pit connection," even though it is actually an ordinary pit connection. Up to several hundred *Choreocolax* cells can fuse with a single host cell. The infected host *Polysiphonia* cell enlarges and the vacuole decreases in size with a concomitant increase in cytoplasmic contents (Goff and Coleman, 1984). The chloroplasts become scattered throughout the cytoplasm, instead of lying in the peripheral cytoplasm as they do in the non-infected cells. Although the parasite cells are colorless, they do contain very reduced plastids (Kugrens and West, 1973). In the association between the colorless parasite *Holmsella pachyderma* and the host *Gracilaria verrucosa*, both of which are Rhodophyceae, the main product of photosynthesis, floridoside, is transferred from the host to the parasite where it is accumulated as floridoside, mannitol, and starch (Evans et al., 1973).

Commercial utilization of red algal mucilages

The two most important polysaccharides derived from the Rhodophyceae are agar and carrageenan. **Agar** is defined pharmaceutically as a phycocolloid of red algal origin that is insoluble in cold water but readily soluble in

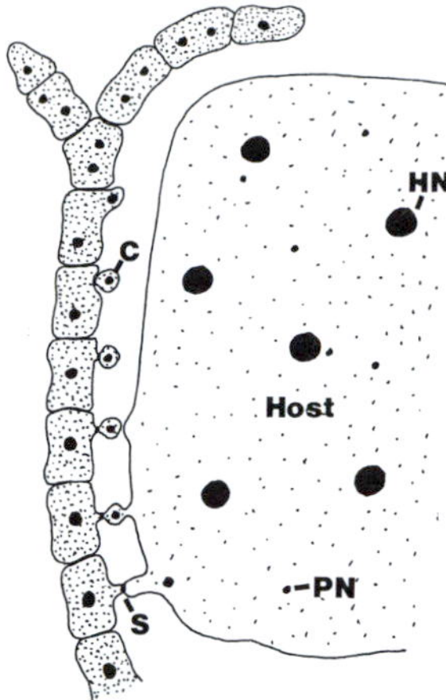

Fig. 4.14 A drawing illustrating the progressive formation of conjunctor cells by a filament of the parasitic red alga *Choreocolax* living on a host *Polysiphonia*. The parasite conjunctor cell fuses with the host *Polysiphonia* cell, leaving the pit connection of the conjunctor cell as a secondary pit connection. (C) Conjunctor cell; (HN) *Polysiphonia* host nucleus; (PN) *Choreocolax* parasite nucleus introduced into the host cell after fusion with the conjunctor cell of the parasite; (S) secondary pit connection. (After Goff and Coleman, 1984.)

hot water; a 1.5% solution is clear and forms a solid and elastic gel on cooling to 32 to 39 °C, not dissolving again at a temperature below 85 °C. Agar is composed of two polysaccharides, agarose (Fig. 1.7) and agaropectin.

Agar is obtained commercially from species of *Gelidium* and *Pterocladia* as well as from various other algae, such as *Acanthopeltis*, *Ahnfeltia*, and *Gracilaria* (Therkelsen, 1993). These algae are often loosely referred to as **agarophytes.** Commercial production of agar was a world monopoly of the Japanese for many years, and even in 1939 Japan was still the major producer. Wartime demands in areas deprived of Japanese agar led to the development of agar industries in many of the Allied countries, some of which have continued and prospered while others have declined or disappeared. The agarophytes are collected by diving, dragging, or raking them offshore at low tide. In the traditional processing procedure the plants are then cleaned and bleached in the sun, with several washings in freshwater used to facilitate bleaching. The material is boiled for several hours, and the extract is acidified. This extract is then frozen and thawed. On thawing, water flows from the agar, carrying impurities with it. The agar that remains is dried and marketed as flakes or cakes. The more modern method extracts the agar under pressure in autoclaves. The agar is decolorized and deodorized with activated charcoal, filtered under pressure, and evaporated under reduced pressure. Further purification by freezing is then undertaken.

The greatest use of agar is in association with food preparation and technology, and in the pharmaceutical industry. It is used for gelling and thickening purposes, particularly in the canning of fish and meat, reducing the undesirable effects of the can and providing some protection against shaking of the product in transit. It is also used in the manufacture of processed cheese, mayonnaise, puddings, creams, and jellies. Pharmaceutically agar is used as a laxative, but more frequently it serves as an inert carrier for drug products where slow release of the drug is required, as a stabilizer for emulsions, and as a constituent of cosmetic skin preparations, ointments, and lotions. The use of agar as a stiffening agent for growth media in bacteriology and mycology, which was its main use almost a century ago, is still responsible for a very considerable part of the demand.

Carrageenan is a phycocolloid similar to agar but with a higher ash content and requiring higher concentrations to form gels. It is composed of varying amounts of the principal components, κ-carrageenan and λ-carrageenans, both negatively charged high-molecular-weight polymers (Chiovitti et al., 1995; Therkelsen, 1993). κ-Carrageenan is distributed throughout the wall while λ-carrageenan is localized to the cuticle (Vreeland et al., 1992). κ-carrageenan precipitates selectively from a cold, dilute solution in the presence of potassium ions. It forms a gel when heated

Fig. 4.15 Cultivation of the carrageenan-containing *Kappophycus alvarezii* in an offshore area (*left*) and a shrimp pond (*right*) in Vietnam. (From Ohno et al., 1996.)

and cooled with potassium ions and is therefore the gelling component. λ-Carrageenan is the non-gelling component and is not precipitated or gelled by potassium. λ-Carrageenan contains galactose-2,6-disulfate, whereas κ-carrageenan contains 3,6-anhydro-D-galactose. It has been shown in *Chondrus crispus* and *Gigartina stellata* that the proportion of κ- and λ-carrageenan in the cell wall varies according to the ploidy of the plant. In the tetrasporophyte the amount of λ-carrageenan present is high as compared to the amount of κ-carrageenan, whereas just the opposite is true in the gametophyte (Chen et al., 1970). Such results may prove valuable in determining the ploidy of Rhodophyceae that have unknown life cycles.

Carrageenan is usually obtained from wild populations of Irish moss, the name for a mixture of *C. crispus* and the various species of *Gigartina*, particularly *G. stellata*. In the Philippines, *Eucheuma*, and in Vietnam, *Kappophycus* (Fig. 4.15) are extensively cultured as a source of carrageenan (Velasquez, 1972). Commercial extraction is similar to that for agar although carrageenan cannot be purified by freezing. The dried alga is washed with freshwater to reduce the salt content and then boiled with 2 to 4 parts of alga to 100 parts of water. The soluble carrageenan is separated from the insoluble residue in a centrifuge. Following filtration and some evaporation under vacuum, the carrageenan is dried on a rotary drier.

Carrageenans are used extensively for many of the same purposes as agar; however, because of their lower gel strength, carrageenans are used less for stiffening purposes than is agar, although for stabilization of emulsions in paints, cosmetics, and other pharmaceutical preparations carrageenans are preferred to agar. Also, for the stiffening of milk and dairy products, such as ice cream, carrageenans have supplanted agar completely in recent years, and it is in this area that demands for these products are the greatest. One particular use is for instant puddings, sauces, and creams, made possible by the gelling action, which does not require refrigeration.

Carrageenans inhibit human immunodeficiency virus (HIV) replication and reverse transcriptionase *in vitro* (in the test tube) (Bourgougnon et al., 1996). Replication of the HIV virus depends on interaction of a glycoprotein on the HIV virus envelope with a receptor on the target cells in the human body. The sulphated carrageenans prevent attachment of the HIV virus to the target cells. This occurs by the stronger negative R-O-SO_3^- groups on the carrageenan binding to a loop on the HIV molecule.

Classification

The Rhodophyta has a single class, the Rhodophyceae. In the past, the Rhodophyceae was divided into two subclasses, the Bangiophycidae and the Florideophycidae. The Bangiophycidae were supposed to lack pit connections, apical growth, and probably sexual reproduction, whereas the Florideophycidae had pit connections, apical growth, and sexual reproduction with a triphasic life cycle. The Bangiophycidae have since been found to have pit connections and apical growth in the *Conchocelis* filamentous stage of the Bangiaceae and probably in the Rhodochaetales; and there is sexual reproduction in the Bangiaceae and Rhodochaetales. In turn, the Florideophycidae do not necessarily have apical growth [intercalary growth occurs in the Corallinales (Dixon, 1973)], nor do they all have a triphasic life history (e.g., red algae in the Batrachospermales). For the above reasons, the two subclasses have been dropped in this treatment of the Rhodophyceae, as suggested by Gabrielson et al. (1985).

The classification of the more advanced orders of the red algae is based on complex characteristics of sexual reproduction. Harald Kylin, the father of red algal classification, placed many of the families into the superorders, such as the Gigartinales and Cryptonemiales, with the understanding that this might not be a natural grouping. One of the more active fields of phycology in the last decade has been the application of molecular techniques in a delineation of the evolutionary relationships of these algae (Freshwater et al., 1994; Ragan et al., 1994; Ragan and Gutell, 1995; Saunders and Kraft, 1996). While producing a more natural grouping of red algae, these excellent studies have produced an even more complex classification system, which is difficult to present to a student taking a first course in phycology, for which this book is directed. In writing the current edition of this book, the author has spent some time trying to decide how to present the red algal classification, and has decided that a presentation of all the more advanced orders would overwhelm the beginning student. As such, the author has selected those red algae that are commonly studied in phycology courses and/or are economically or ecologically unique.

Order 1 **Porphyridiales:** unicells, or multicellular Rhodophyceae that are held together by mucilaginous walls.
Order 2 **Rhodochaetales:** filamentous organisms with pit connections; gametes are formed singly by the mother cell.
Order 3 **Acrochaetiales:** algae with a uniseriate filamentous gametophyte and tetrasporophyte (if both are present).
Order 4 **Bangiales:** plants having a filamentous phase with pit connections and a macroscopic phase without pit connections.
Order 5 **Batrachospermales:** uniaxial (one apical cell per branch); gonimoblast usually develops from the carpogonium or hypogenous cell.
Order 6 **Nemaliales:** multiaxial (more than one apical cell per branch); usually the gonimoblast develops from the carpogonium or the hypogenous cell.
Order 7 **Corallinales:** heavily calcified algae with the reproductive organs in conceptacles.
Order 8 **Gelidiales:** fleshy agarophytes, carpogonial branch consisting of a single cell, the carpogonium, no differentiated auxiliary cells.
Order 9 **Gracilariales:** fleshy agarophytes, two-celled carpogonial branch, no auxiliary cells, or connecting cells.
Order 10 **Ceramiales:** relatively delicate or filamentous forms with an auxiliary cell cut off after fertilization and borne on the supporting cell of a four-celled carpogonial filament.

The first red alga was undoubtedly a unicell which subsequently evolved into a filamentous alga with pit connections between the cells. At first, the method of sexual reproduction was probably simple and may have been close to that in *Rhodochaete* (Fig. 4.19), where a single spermatium is cut off from the mother cell, with the female cell resembling a vegetative cell. From this filamentous alga there were most likely two basic lines of evolution. In the first, pit connections were lost in the macroscopic phase, leading to the Bangiales. In the second line, a more complex method of sexual reproduction was evolved in the form of the carposporophyte. The remaining Rhodophyceae fall into this line, which in turn was composed of a number of lines arising from an *Acrochaetium*-like ancestor.

The relationship between freshwater and marine Rhodophyceae, as well as their evolution, was discussed in an interesting paper by Skuja (1938). He believed that the Rhodophyceae are a very old group (as is borne out by their fossil record) that originated in shallow coastal waters of primitive seas poor in salt. Living in shallow water, these plants had no need of a large quantity of phycoerythrins to absorb the blue-green light present at greater depths of water. Consequently these primitive Rhodophyceae were not pinkish-red but blue-green in color. These plants are represented by the freshwater Rhodophyceae of today, which are predominantly blue-green in color, and found primarily in the more primitive orders such as the Porphyridiales, Bangiales, Acrochaetiales, and Nemaliales. Only later did the Rhodophyceae develop greater quantities of phycoerythrins, and a pinkish-red color, and penetrate into deeper waters where they attained their present state of development.

PORPHYRIDIALES

These algae are either unicells or filaments composed of cells embedded in thick mucilage. The order probably contains a diverse group phylogenetically. Some of the algae are probably very primitive (e.g., *Cyanidium*, *Cyanidioschyzon*), while others are probably carpospores, tetraspores or monospores of higher red algae that never germinated (e.g., *Rhodella*) (Ragan et al., 1994; Freshwater et al., 1994).

The algae in the Porphyridiales do not have cellulose microfibrils in their cell walls as do large rhodophytes (Arad et al., 1993). The unicells are differentiated by cytological characteristics. Thus *Porphyridium* (Fig. 4.1) has a single large stellate chloroplast with central pyrenoid, *Rhodosorus* (Fig. 4.16(*a*)) has a lobed chloroplast with a basal pyrenoid, and *Rhodella* has a stellate chloroplast with a central pyrenoid, but with a more dissected chloroplast than *Porphyridium* (Fig. 4.16(*b*)).

Porphyridium is a common alga on soil and damp walls where it forms several-layered blood-red mucilaginous strata. Even though it is a soil alga, most species grow best in marine liquid media, indicating that it is probably of brackish or marine origin. *Porphyridium* has the ability to glide over a substrate it is in contact with. Overhead illumination results in random movement, whereas unilateral light causes movement toward the light source (Sommerfield and Nichols, 1970). The positively phototactic cells move by the extrusion of mucilage in vesicles in one direction, which results in the formation of a mucilage stalk behind the cells (Lin et al., 1975). *Prophyridium* releases different amounts of polysaccharides, depending on the environmental conditions it is living under (Ramus and Robins, 1975). During the log phase of growth, large Golgi bodies form polysaccharides, which are stored in vesicles under the cell membrane. During the stationary phase of growth in culture, the polysaccharide is secreted outside the cell, giving rise to a capsule. This behavior in culture can be related to the survival of the cells in nature. The rapid log phase of growth is equivalent to a soil environment that is moist with available nutrients. Here the polysaccharides are stored inside the cell, and there is only a thin

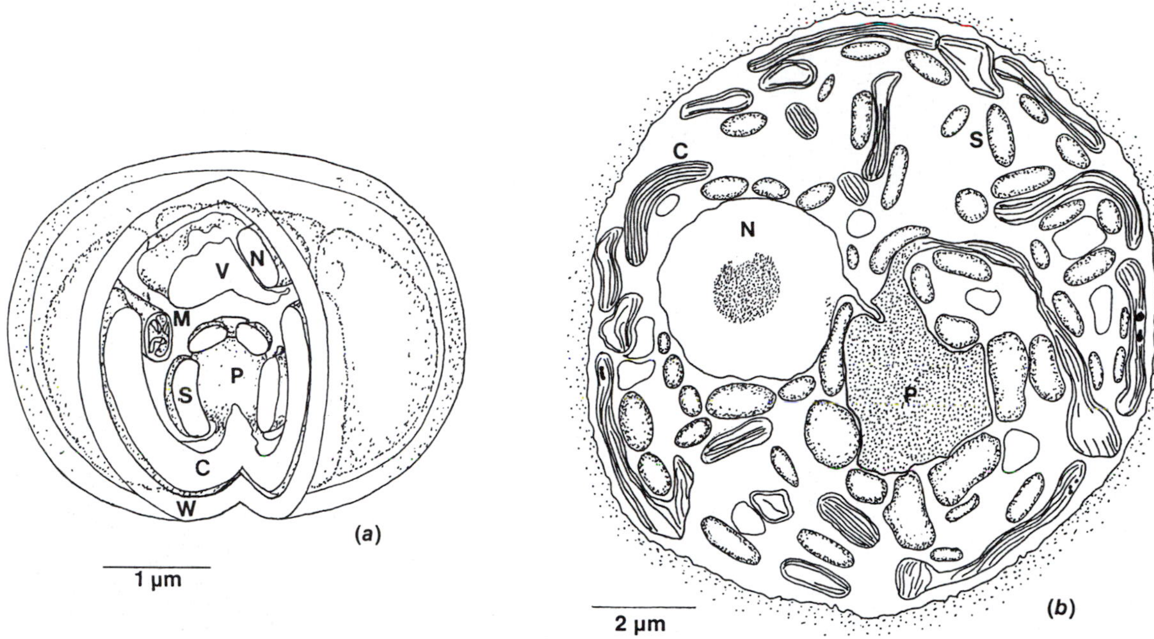

Fig. 4.16 (*a*) A diagrammatic drawing of a cell of *Rhodosorus marinus*. (*b*) A semi-diagrammatic drawing of a section through a cell of *Rhodella maculata*. (C) Chloroplast; (M) mitochondrion; (N) nucleus; (P) pyrenoid; (S) starch grain; (V) vacuole; (W) wall. ((*a*) after Giraud, 1962; (*b*) adapted from Evans, 1970.)

mucilage layer around the cell. The stationary phase of growth is equivalent to a soil environment that is drying out with nutrients becoming limiting, thereby causing a cessation of cell growth. Here the polysaccharides are released to the outside of the cell, where they form a capsule that enables the cell to withstand the desiccation that follows.

Cyanidium caldarium and *Cyanidioschyzon merolae* are probably the most primitive red algae (Seckbach, 1991). Each of these cells contains a single nucleus, mitochondrion and plastid (Fig. 4.17). They differ in that *Cyanidium* is round, has a cell wall and forms four endospores while *Cyanidioschyzon* is club-shaped has no cell wall and divides by binary fission (Ohta et al., 1997). *Cyanidioschyzon* has the smallest genome size (8–13 Mb) so far recorded in eukaryotes (Kuroiwa et al., 1994).

Cyanidium caldarium (Fig. 4.17) is the sole photosynthesizing organism in volcanic hot springs and in hot soils with a pH of less than 5 and average temperatures exceeding 40 °C (Doemel and Brock, 1971). *Cyanidium caldarium* shares with a variety of thermophilic cyanobacteria its preference for thermal habitats.

Also included in this order are algae that have cells joined together in thick mucilaginous filaments. *Goniotrichum* is a common marine epiphyte made up of branched mucilaginous filaments (Fig. 4.18(*b*)). *Goniotrichum* forms monospores simply by the release of a vegetative cell from a filament in photoperiods of over 12 hours of light (Fries, 1963). *Asterocytis* (Fig. 4.18(*a*)) exhibits what is probably an intermediate position in the evolution

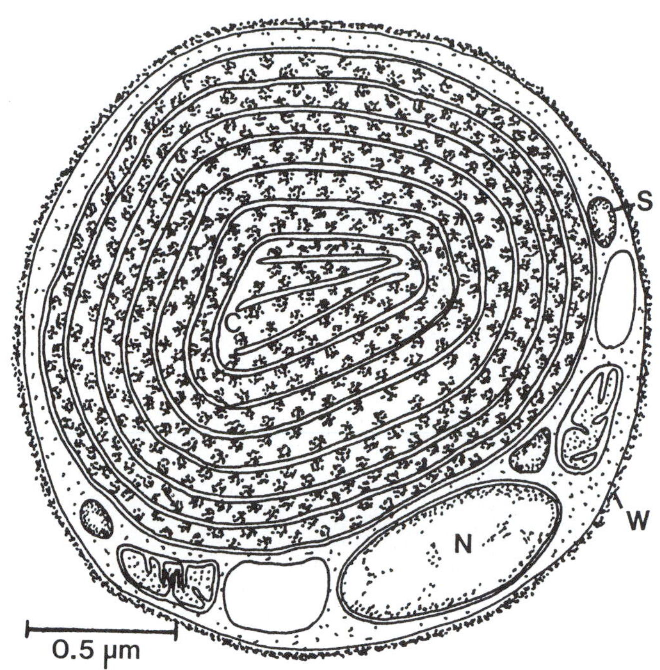

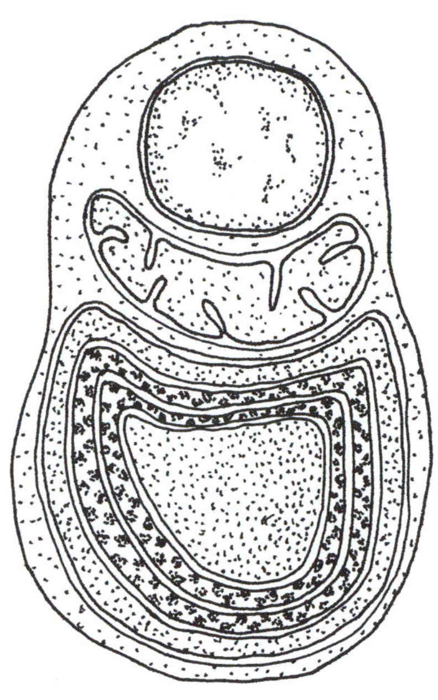

Fig. 4.17 *Left: Cyanidium caldarium. Right: Cyanidioschyzon merolae.* (C) Chloroplast; (M) mitochondrion; (N) nucleus; (S) Starch; (W) wall. (*Cyanidium* is after Seckbach and Ikan, 1972.)

of a red unicell into a mucilaginous filamentous alga. In normal seawater *Asterocytis* forms branched filaments, whereas in seawater of one-fourth strength the organism forms unicells, which were previously classified in the genus *Chroothece* (Lewin and Robertson, 1971).

RHODOCHAETALES

This order has only one genus, *Rhodochaete*, a relatively rare alga found in the Mediterranean and West Indies. The thallus consists of uniseriate filaments of elongate cells with infrequent branches. There are pit connections between the cells (Magne, 1960b), and each cell has several elongate chloroplasts. The gametophyte of *R. parvula* (with four chromosomes) cuts off a small spermatium by means of a curved wall (Fig. 4.19). The spermatium is released and passes to the female cell, which looks like a vegetative cell. The spermatium fuses with the female cell to form zygote, which subsequently cuts off a large carpospore in the same manner that the spermatium was produced (Magne, 1960a). The fate of the diploid carpospore is not known.

The simple method of reproduction in this genus makes it an attractive candidate for the most primitive multicellular red alga, or at least a close extant relative of the first multicellular red alga. The development of a carpogonium with a trichogyne and a simple spermatial branch from the reproductive structures in *Rhodochaete* would not have been difficult.

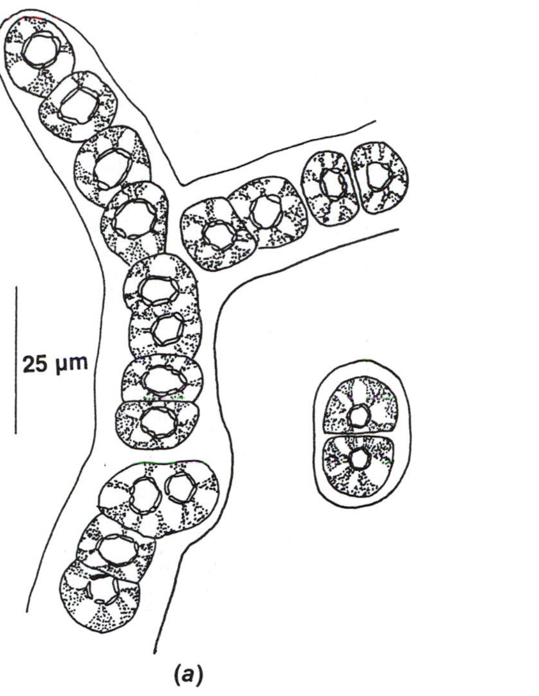

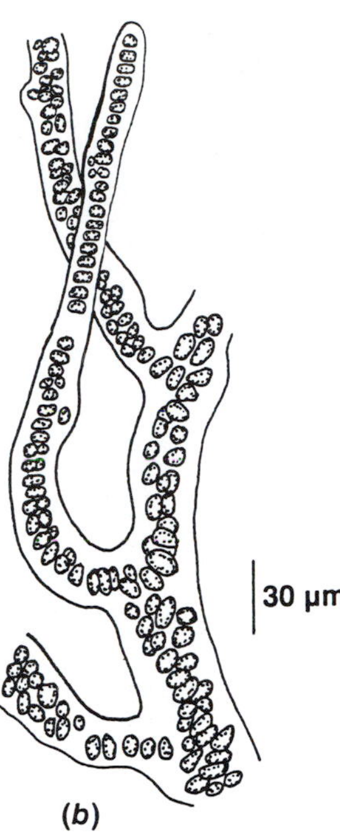

Fig. 4.18 (*a*) *Asterocytis* sp. in the filamentous and bicellular form. (*b*) *Goniotrichum alsidii*. ((*a*) after Belcher and Swale, 1960; (*b*) after Taylor, 1957.)

ACROCHAETIALES

Algae with uniseriate filaments are in this order (Chemin, 1937; Feldmann, 1953). There is a dispute as to how many genera are present in the order. Parke and Dixon (1976) recognize only the genus *Audouinella* as encompassing all the algae in the order. Papenfuss (1945, 1947), however, recognizes four major genera: (1) *Rhodochorton*, with each cell containing a few to many small discoid chloroplasts (Fig. 4.20); (2) *Acrochaetium*, with each cell having one parietal or laminate chloroplast; (3) *Audouinella*, with each cell having one or more spiral chloroplasts; (4) *Kylinia*, with each cell having one or more stellate chloroplasts (Fig. 4.6(*g*)).

Most of these algae are small epiphytes or endophytes. Some of them may still prove to be the alternate phase of more complex higher Rhodophyceae. *Rhodochorton investiens* can be used as an example of a triphasic life cycle (Fig. 4.20) (Swale and Belcher, 1963). Both the gametophyte and tetrasporophyte produce similar obovoid monospores in monosporangia arising just beneath a cross wall of the filament. The spore is liberated through the apex of the sporangial wall, which remains attached to the filament. After release, the monospores germinate without a resting phase to re-form the parent plant. The gametophyte is monoecious, with a filament ending in a cluster of spermatangia and a carpogonium being borne on the cell under the supporting cell of the spermatangia. The sper-

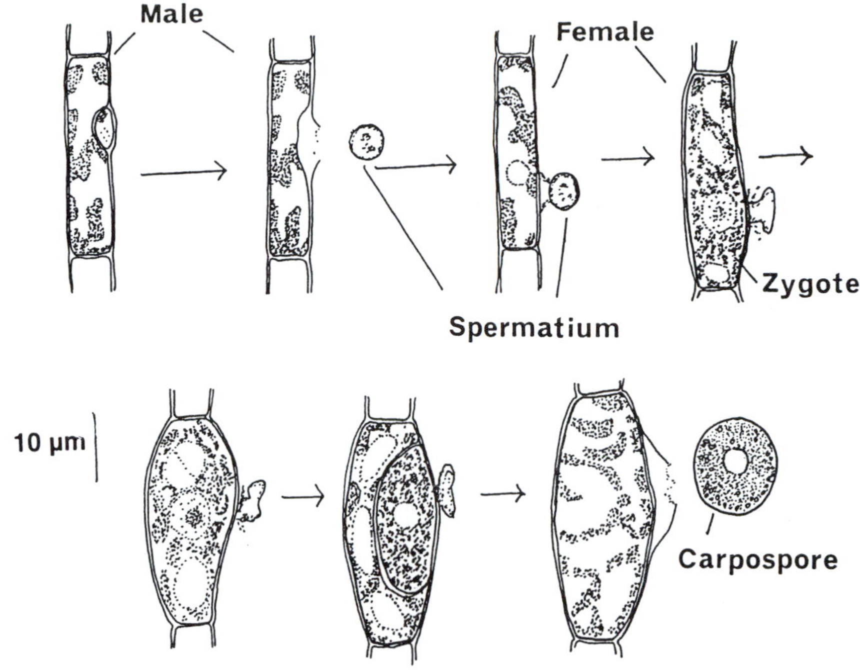

Fig. 4.19 Gametogenesis, fertilization, and carpospore formation in *Rhodochaete parvula*. (After Magne, 1960a).

matangia occur in groups of four to six, emerging from the enlarged and flattened distal end of a terminal cell. The carpogonia are sessile and appear in the position of branch cells. At the distal end of the carpogonium is a narrow trichogyne. After fertilization, the carpogonium becomes divided by three cross walls into a row of four cells. The first transverse wall develops below the trichogyne, the upper cell then elongating and dividing into three cells. This results in the trichogyne emerging from the second cell of the row. Two-celled gonimoblast filaments develop from each cell of the row, each gonimoblast filament producing two to three terminal carpospores. The carpospores germinate to form the tetrasporophyte, with larger cells of deeper color than those of the gametophyte. Tetrasporangia either are sessile or terminate a one-celled branch. The tetraspores germinate to produce the gametophyte, and thus complete the life cycle.

BANGIALES

The algae in this order show alternation of a haploid thallus stage having no pit connections, with a diploid filamentous *Conchocelis* state that has pit connections (Lee and Fultz, 1970; Kornmann, 1994).

Porphyra is an intertidal seaweed in the colder waters of the world. The thallus arises from a holdfast and is composed of a sheet of cells one to two layers thick. *Porphyra gardneri* (Fig. 4.21) is a foliose (leafy), monostromatic

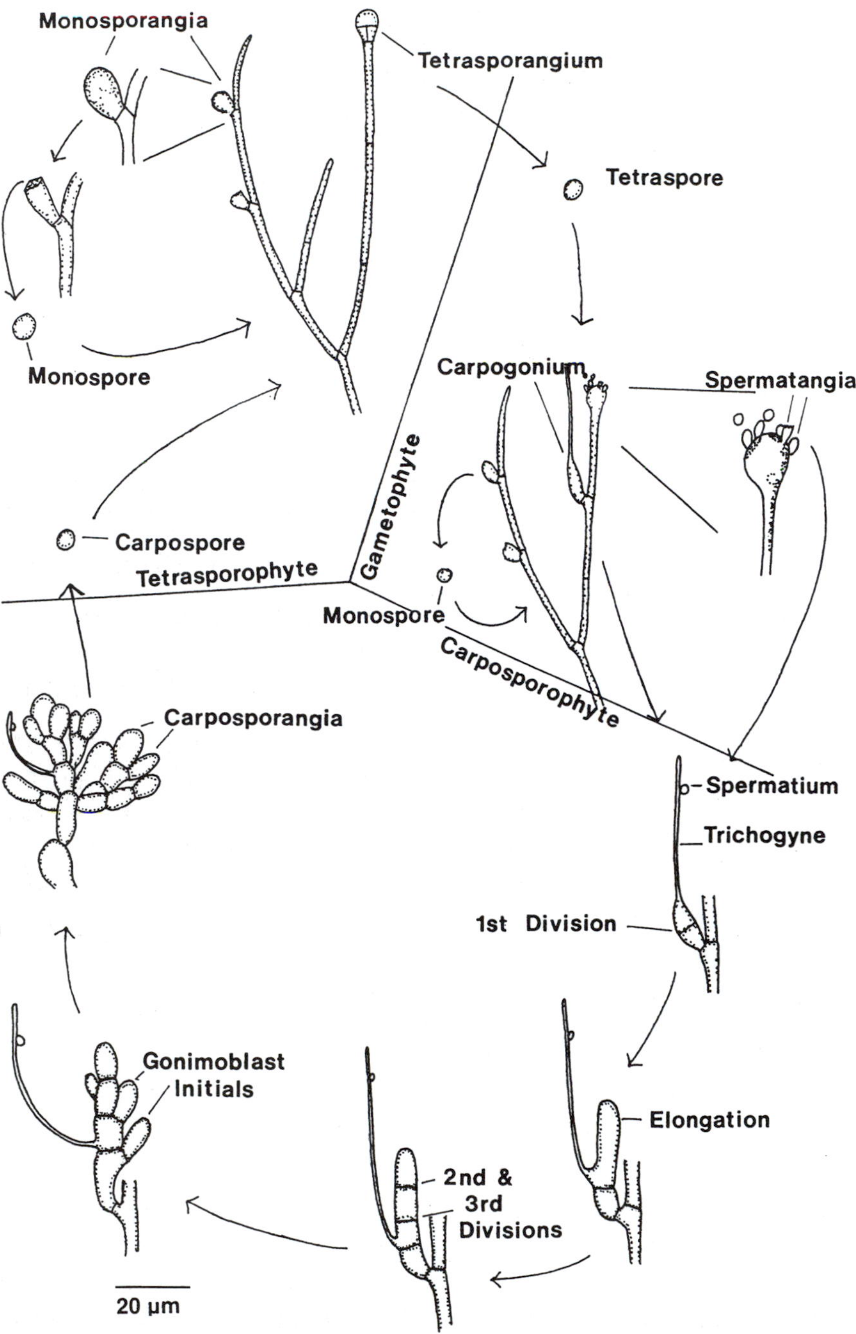

Fig. 4.20 The life cycle of *Rhodochorton investiens*. (Adapted from Swale and Belcher, 1963.)

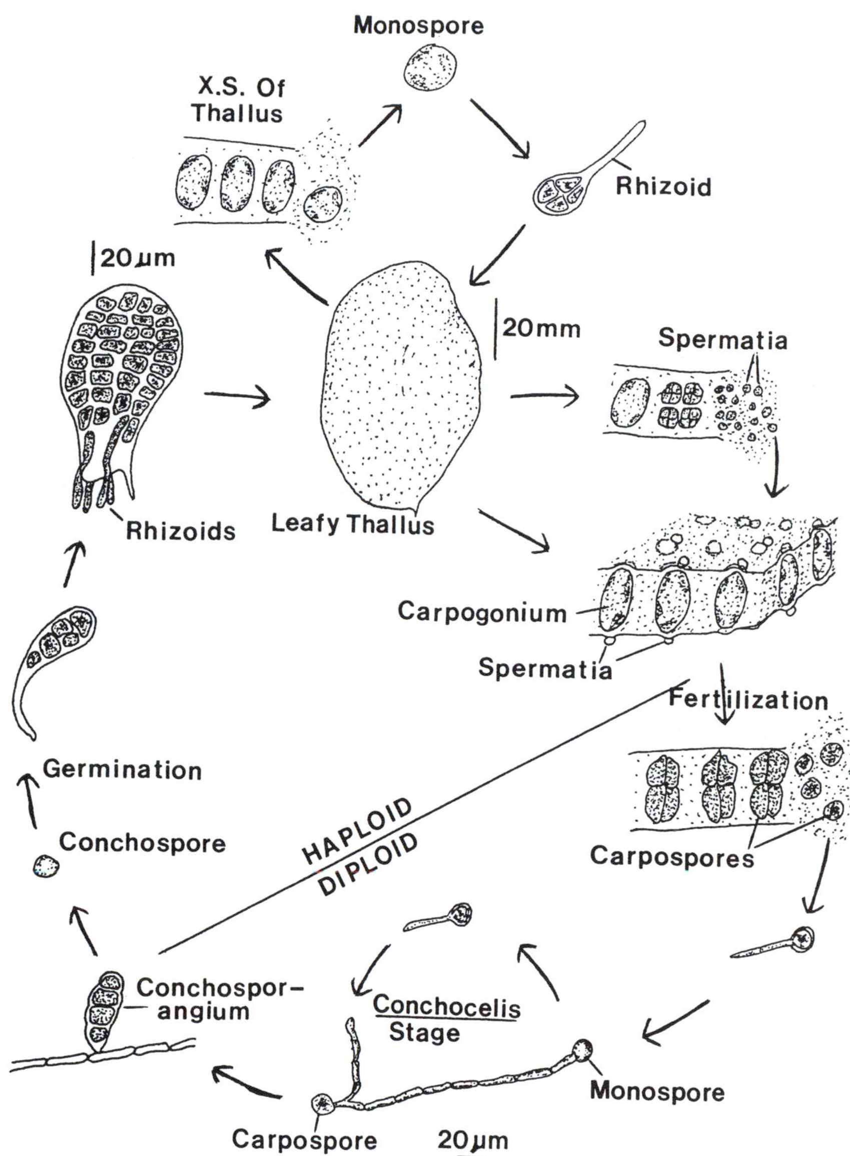

Fig. 4.21 The life cycle of *Porphyra gardneri*. (Adapted from Hawkes, 1978.)

(single layer of cells) alga found growing epiphytically on several members of brown algae in the Laminariales. In British Columbia, Canada, the host *Laminaria setchellii* has its blade worn back almost to the stipe by November. During December, a new *Laminaria* blade is rapidly produced. The first thalli of *Porphyra gardneri* appear epiphytically on the *Laminaria* at the end of February. **Asexual reproduction** occurs soon after *Porphyra gardneri* appears in February. The margins of the thallus break down and release single-celled **monospores**. After 1 or 2 days, the monospores germinate by sending out long rhizoids that anchor the monospores in the host

Laminaria tissue. From this, a new leafty thallus appears. Prolific monospore production results in a great increase of *Porphyra gardneri* during the spring months. **Sexual reproduction** begins during late April. Spermatium mother cells in the thallus divide to form 64 spermatia. The spermatia contain a degenerate chloroplast with only a few thylakoids. Vesicles containing a fibrous material are discharged by the spermatia just before spermatia liberation. The released spermatia are 3 to 5 μm in diameter, have no starch granules, and are surrounded only by the fibrous material from the released vesicles. The spermatia are carried to the carpogonia by water currents. Carpogonia (with a chromosome number of 4) differentiate from vegetative cells by the production of a swollen area of the cell wall, the **prototrichogyne**, directly above the carpogonium. In a monostromatic species, such as *Porphyra gardneri*, two prototrichogynes are produced by each carpogonium, one on each surface. In distromatic species, with two sheets of cells in the thallus, a single prototrichogyne is produced per carpogonium. A spermatium attaches to the prototrichogyne, a fertilization canal appears in the prototrichogyne, and the spermatial nucleus moves through the canal to fuse with the carpogonium. By the beginning of May, the fertilized carpogonia have divided to form two to four diploid carpospores (with a chromosome number of 8), 14 to 20 μm in diameter. Maximum carpospore production occurs during the months of June through August. The carpospores germinate in 2 to 3 days to produce the diploid *Conchocelis* stage (Hawkes, 1978). The *Conchocelis* stage is filamentous and commonly lives in shells of dead marine animals. Under long-day conditions, the *Conchocelis* stage differentiates monospores, which re-form the *Conchocelis* stage (Dixon and Richardson, 1970). Under short days the *Conchocelis* stage forms conchosporangia (fertile cell rows), each cell of which produces a conchospore. Conchospores are released from the conchosporangia under low-temperature conditions (about 5 °C) (Chen et al., 1970). The formation of the conchosporangia under short-day conditions is a true photoperiodic response because a light break in the middle of the dark period is inhibitory (Dring, 1967a). A functional phytochrome system is operative, with red light being the most effective in breaking the dark period (Dring, 1967b). This is one of the few demonstrations of a true photoperiodic response in the red algae, and it is unlikely that this phytochrome type of response occurs in sublittoral Rhodophyceae because far-red light penetrates to less than 1 m of seawater and red light no deeper than 10 m (Dixon and Richardson, 1970). On release, the conchospores germinate in a bipolar manner, forming a germling that grows into the thallus phase, completing the life cycle.

Porphyra perforata lives in the intertidal zone, where at low tide the plants are routinely exposed to air drying. As a result of evaporative water

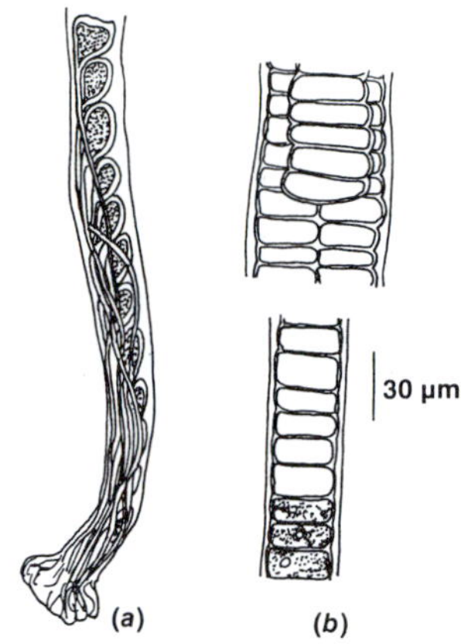

Fig. 4.22 *Bangia fuscopurpurea* showing (*a*) holdfast area and (*b*) uni- and multiseriate filaments. ((*a*) after Womersley, 1965; (*b*) after Taylor, 1957.)

loss, the salt concentration of extracellular water can increase up to 10 times above normal levels. During dessication at low tide, the alga can lose up to 90% of its fresh weight. Such dessication results in inhibition of photosynthesis. Some of the inhibition of photosynthesis is probably due to a decrease in electron flow between water and photosystem II because of a reduced concentration of water in the cells (Satoh et al., 1983).

Bangia forms upright threads that are at first uniseriate, the cells subsequently undergoing longitudinal division to form a multiseriate filament (Fig. 4.22). *Bangia* occurs in both marine and freshwater environments. It is possible to adapt freshwater *Bangia fuscopurpurea* to seawater by increasing the salinity by 10% of that of seawater every time the alga sporulates (den Hartog, 1971). If the thallus is moved directly from freshwater to seawater, the plant dies, illustrating that the spores have a better ability to adapt to changed salinity. Such an experiment shows the ease with which some of the smaller red algae can change from one habit to another. *Bangia* has a life cycle similar to that of *Porphyra* (Richardson, 1970; Sommerfeld and Nichols, 1973).

The diploid *Conchocelis* phase of *Porphyra* and *Bangia* differs chemically from the haploid thallus-phase (Liu et al., 1996). The *Conchocelis* phase has cellulose in the wall of the cells, whereas in the thallus phase, cellulose is absent and, as the structural polysaccharide, is replaced by a xylan (polysaccharide composed of xylose residues) (Gretz et al., 1980; Mukai et al., 1981). The galactans in the *Conchocelis* phase are also different from those in the thallus phase (Gretz et al., 1980). These chemical differences are in addition to the structural differences, particularly the occurrence of pit connections in the *Conchocelis* phase and their absence in the thallus phase.

The *Conchocelis* phase has been found as the fossil genus *Palaeoconchocelis starmachii* in the Upper Silurian of the Paleozoic (425 million years ago) (Campbell, 1980). The *Conchocelis* stage of *Bangia* and *Porphyra* is similar in appearance to, and probably has evolved from, one of the filamentous stages in the Acrochaetiales. The only place in the higher Rhodophyceae where pit connections do not occur (as is the case in the thallus stages of *Bangia* and *Porphyra*) is between the cells of the tetraspores (Kugrens and West, 1972b; Scott and Dixon, 1973b). If the tetrasporangium were to differentiate into a haploid multicellular thallus without pit connections, this might be the first step in the evolution of a plant such as *Bangia* or *Porphyra.*

The leafy thallus phase of *Porphyra* is eaten as a vegetable in the Far East and Nova Scotia (Canada). In Japan, *Porphyra* is eaten as a vegetable called *nori*; in Nova Scotia, it is called *laver*. *Porphyra* is cultivated on farms in China (Fig. 4.23) and Japan. In Japan, most of the *Porphyra* comes from

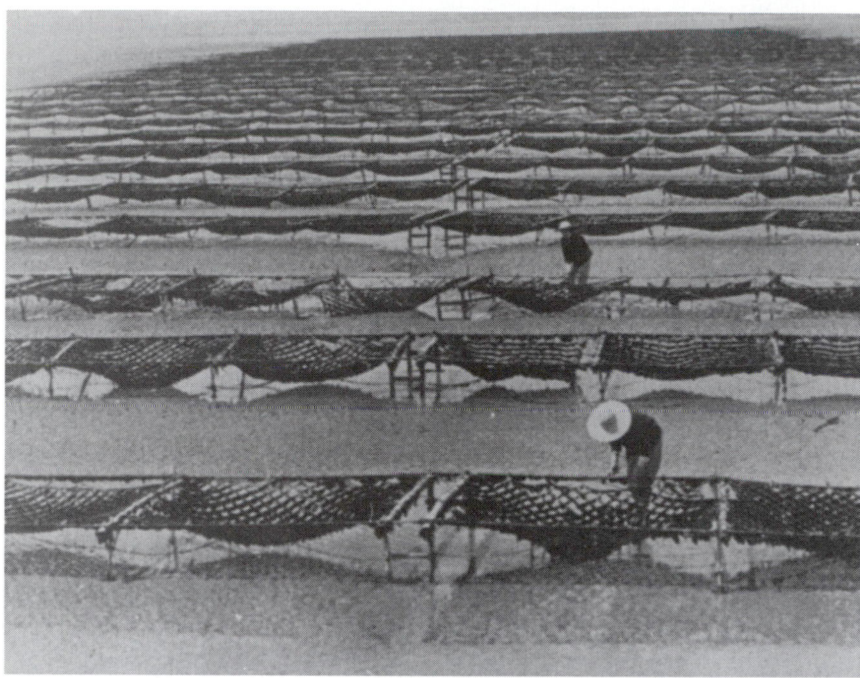

Fig. 4.23 Cultivation nets with *Porphyra* at low tide at Rudang, Jiangsu Province, People's Republic of China. (From Tseng, 1981.)

Porphyra farms in the shallow waters of such places as the Inland Sea and Tokyo Bay, although there is considerable collecting of plants from natural populations. *Porphyra* was first cultivated around 1700 in Tokyo Bay by placing bundles of bamboo or oak bushes (known as hibi) into the mud in early autumn, the usual procedure being to arrange the bundles in regular rows and at such a depth that the twigs were well covered by water at high tide. The modern method is to drive bamboo stakes into the mud in rows and then place netting between the stakes (Mumford and Miura, 1988). The *Conchocelis* stage growing in seashells releases the conchospores, which settle on the brush or nets and germinate to form the foliose *Porphyra* plant. From late November to March, the *Porphyra* plants (sometimes mixed with the green alga *Monostroma*) are harvested by a person in a narrow boat who picks or scrapes the plants off by hand. The *Porphyra* is brought to the factory, where it is washed and chopped into small fragments. These fragments are stirred in a vat, from which measured amounts of the mixture are dipped by means of a small wooden container and poured over a stiff porous mat. As the liquid drains away, the nori fragments are spread evenly over the mat, which is hung on outdoor bamboo racks to dry. The thin film of dry nori is removed as a sheet from the mat, folded, and packaged for market. The food value of nori or laver lies in its high protein content (25% to 30% of the dry weight), vitamins, and mineral salts, especially iodine. The vitamin C content is about 1½ times that of oranges per unit weight, and it is also rich in vitamin B. Humans digest about 75% of the protein and carbohydrate, and in this respect it is much better than other seaweeds.

Prior to the discovery of the alternate *Conchocelis* phase of *Porphyra* by Drew, the yield of *Porphyra* fluctuated sharply from one year to the next. Up to this time, the number of *Porphyra* plants formed depended on the production of conchospores by the *Conchocelis* phase. These fluctuations in the production of spores have been overcome by the artificial cultivation of the *Conchocelis* phase, usually on shells. The shells containing the *Conchocelis* phase are attached to the nets, or the nets are dipped into baths to which crushed shells have been added. The best settlement of spores occurs in waters of high nitrogen, that is, near sewage outflows. Although the production of nori increased until the early 1960s, there has been no increase in production since then, mostly because of the increasing pollution of the shallow waters in which *Porphyra* farming is carried out (Dixon, 1973).

BATRACHOSPERMALES

This order includes the uniaxial (each filament with a single apical cell) freshwater Rhodophyceae. The gonimoblasts usually arise from the fertilized carpogonium. No tetraspores are formed, and meiosis probably occurs when the diploid filamentous stage forms the thallus initials.

Batrachospermum (often called the "frog spawn" alga) is a freshwater alga that occurs in well-aerated, slow-moving streams. The gametophyte (Fig. 4.24) appears as delicate violet beads on a string. Each "bead" consists of whorls of branches arising at the cross walls of the elongated cells of the main axis. The gametophytes produce terminal carpogonia on short branches arising from the whorls of branches. Spherical spermatia are formed by small groups of antheridia at the tips of branches. The spermatia are carried to the carpogonium by water currents. After fertilization, the zygote cuts off gonimoblast initials which develop into gonimoblast filaments with terminal carposporangia. The carposporangia release diploid carpospores that germinate into filamentous prothalli. Monospores can be produced by the prothalli. The monospores germinate to re-form the parent plant. The prothalli also form erect filaments that elongate by apical growth. The apical cell of the erect filament cuts off three to five cells mitotically, then undergoes two meiotic divisions. The first meiotic division results in (1) a polar body and (2) a cell that divides again to form a second polar body and the apical cell of the haploid gametophyte. The macroscopic plant is thus composed of basal diploid cells on a haploid plant. The diploid portion of the plant has cells that are larger than those of the haploid portion (Hurdelbrink and Schwantes, 1972; von Stosch and Theil, 1979; Balakrishnan and Chaugule, 1980).

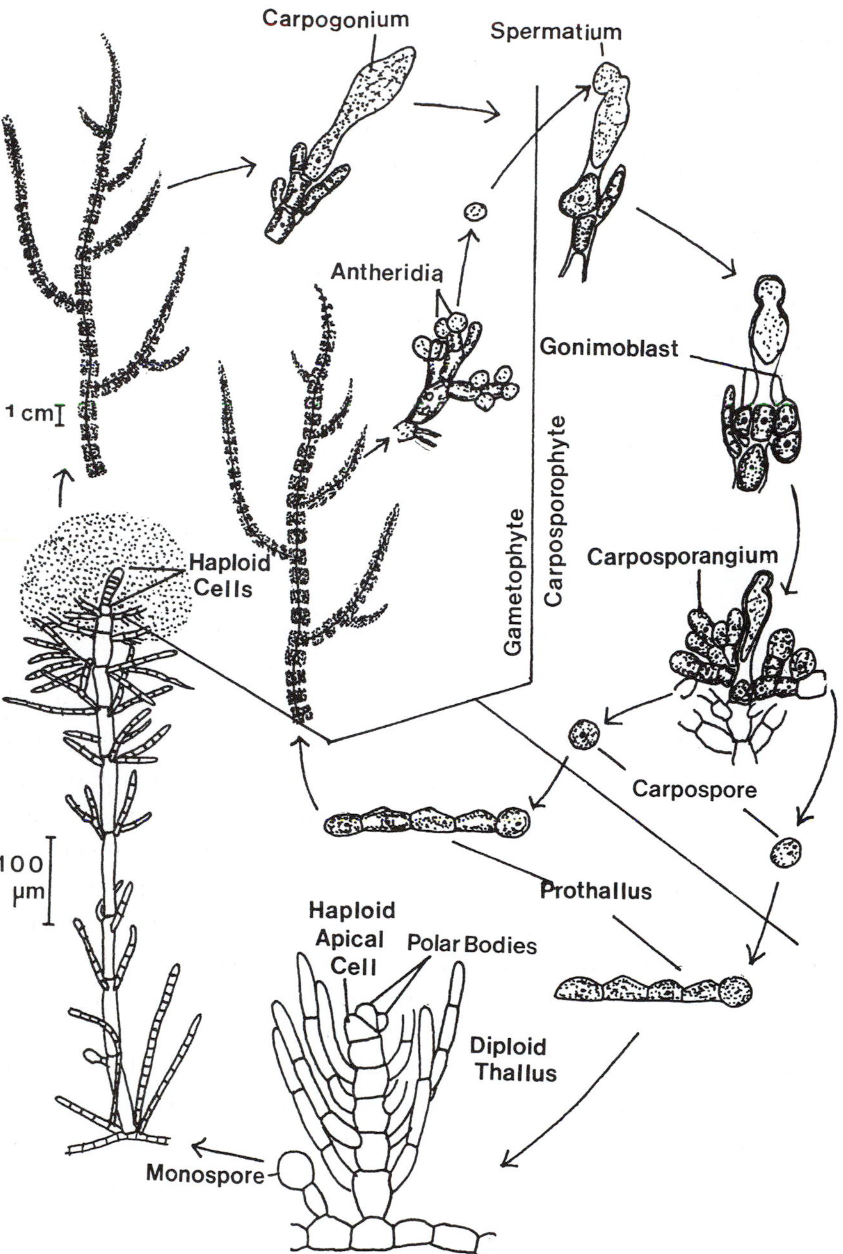

Fig. 4.24 The life cycle of *Batrachospermum* sp. (Adapted from Balakrishnan and Chaugule, 1980; von Stosch and Thiel, 1979.)

NEMALIALES

This order has multiaxial Rhodophyceae (with more than one apical cell), which usually have gonimoblasts developing from the carpogonium or hypogynous cell. There may be auxiliary cells present, but if there are, they are always nutritive auxiliary cells.

Nemalion is a common intertidal alga in north temperate seas. The thallus is a soft gelatinous cylinder reaching a length of 25 cm with a limited number of dichotomous branches (Fig. 4.25), being composed of a number of axial threads or filaments in the center and richly branched laterals around the periphery. The laterals arise from the axial threads and grow out horizontally on all parts of the thallus except for the tip where they radiate out vertically. The laterals are all of about the same length, and their tips intercalate so as to give the thallus an even surface. The central axial cells are colorless, whereas the peripheral laterals usually have a stellate chloroplast with a central pyrenoid.

In *Nemalion*, the plants are homothallic. The carpogonial branch consists of an ordinary lateral of four to seven cells (Fig. 4.25). The elongate trichogyne projects slightly beyond the surface of the thallus. Spermatangial branches are produced from the terminal cells of the laterals, and at the tip of the two- to four-celled spermatangial branch are formed three to four spermatangia. A spermatium is produced in the spermatangium and released, and passes to the trichogyne of the carpogonium where fertilization occurs. After fusion of the two gamete nuclei, the large zygote nucleus and the chloroplast divide into two. The carpogonium then divides transversely into two cells, the upper of which forms the gonimoblasts. The lower cell of the carpogonium gradually fuses with the **hypogynous cells** (those underneath the carpogonium), and eventually the upper carpogonial cell that has produced the gonimoblasts also fuses with these cells. These fusions probably have a nutritive function, providing the developing gonimoblasts and carposporangia with storage products. The gonimoblast threads hang downward, and each cell of the thread forms an upwardly curved two- to three-celled branchlet, the terminal cell of which enlarges to form the carposporangium. The carpospores give rise to a filamentous phase that produces tetraspores under short-day conditions (Cunningham and Guiry, 1989). The tetraspores produce filamentous gametophytes that form the erect axes under long-day conditions.

GELIDIALES

This is an order of uniaxial marine Rhodophyceae in which the carpogonial branch consists of a single cell, the carpogonium. After

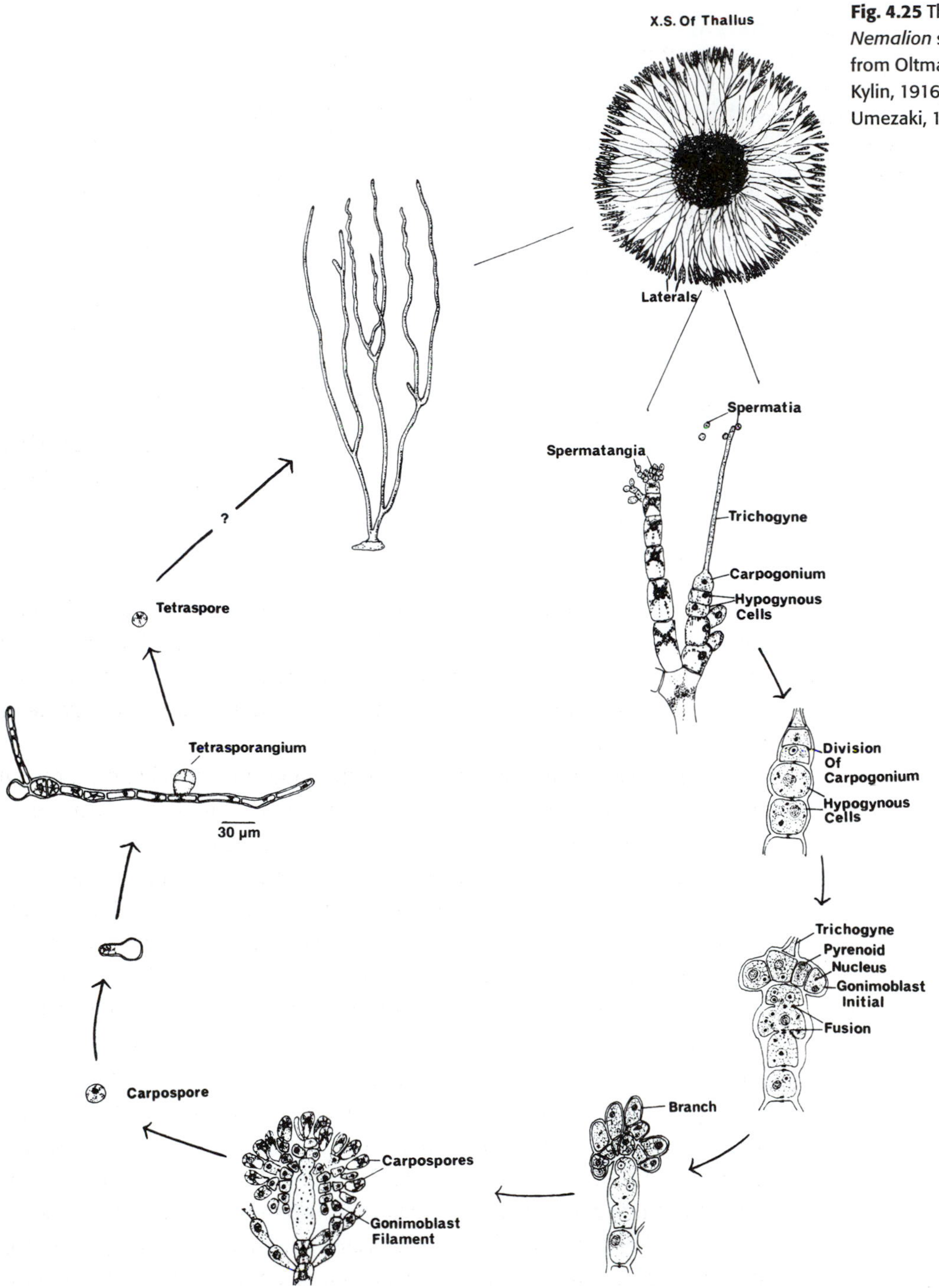

Fig. 4.25 The life cycle of *Nemalion* sp. (Adapted from Oltmanns, 1904; Kylin, 1916; Fries, 1967; Umezaki, 1967.)

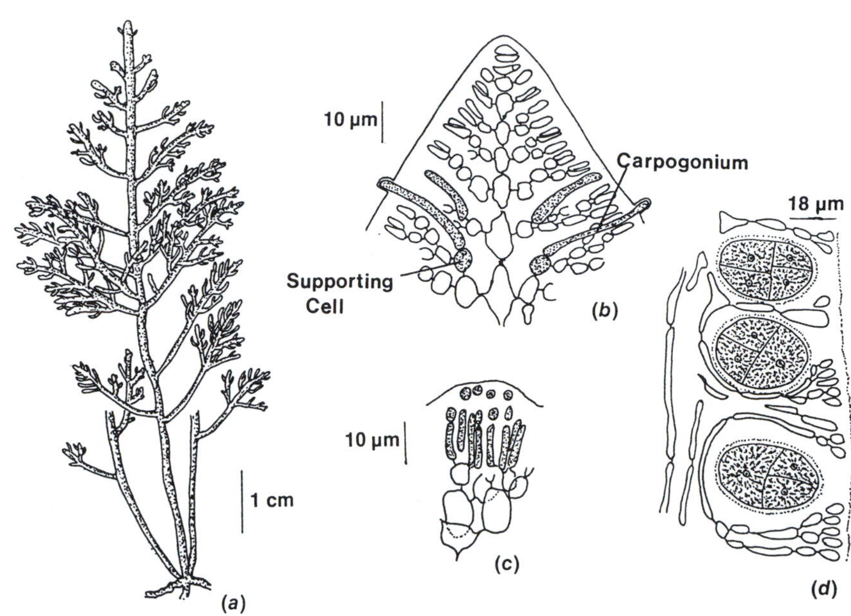

Fig. 4.26 *Gelidium cartilagineum* var. *robustum*. (*a*) Whole plant. (*b*) Section of fertile apex of filament showing apical cell, carpogonia, and supporting cells. (*c*) Section of spermatangial area. (*d*) Section of tetrasporangial area. ((*a*) after Smith, 1969; (*b*),(*c*) after Fan, 1961; (*d*) after Smith, 1938.)

fertilization, the carpogonium may fuse with the supporting cell and/or nutritive filaments. The tetrasporophyte and the gametophyte are macroscopic plants, although not necessarily similar. The plants are commonly used in the production of agar, with almost half of the world supply coming from members of this order (Lewis and Hanisak, 1996).

No member of the order has had its life cycle completed in culture, but it is presumed from plants collected in the field that there is a triphasic life cycle of gametophyte, tetrasporophyte, and carposporophyte.

The gametophyte and tetrasporophyte of *Gelidium* have a dome-shaped apical cell that cuts off daughter cells basipetally (Fig. 4.26). The daughter cells divide to form a thallus that soon loses its uniaxial nature in the mature parts. The carpogonia are usually formed on special **ramuli** (branches) with a deep apical notch, behind which there is a depression on both surfaces (Fig. 4.26(*b*) and 4.27(*a*),(*b*)). The carpogonia are produced in these depressions. The carpogonium is cut off from a cell beneath the surface of the thallus and has a long trichogyne that reaches to the outside of the thallus. The carpogonial branch thus consists of a single cell. Nutritive filaments are cut off from the cells at the base of each of the laterals in the fertile area. After fertilization, the carpogonium may fuse with the supporting cell and/or nutritive filaments, with the gonimoblast filaments and carposporangia developing from this fusion cell. Male plants are similar in morphology to female plants, with the spermatangial areas forming irregular patches on the thalli. The cortical cells of a fertile area elongate, fade in color, and become transformed into spermatangial mother cells (Fig. 4.26(*c*)). The col-

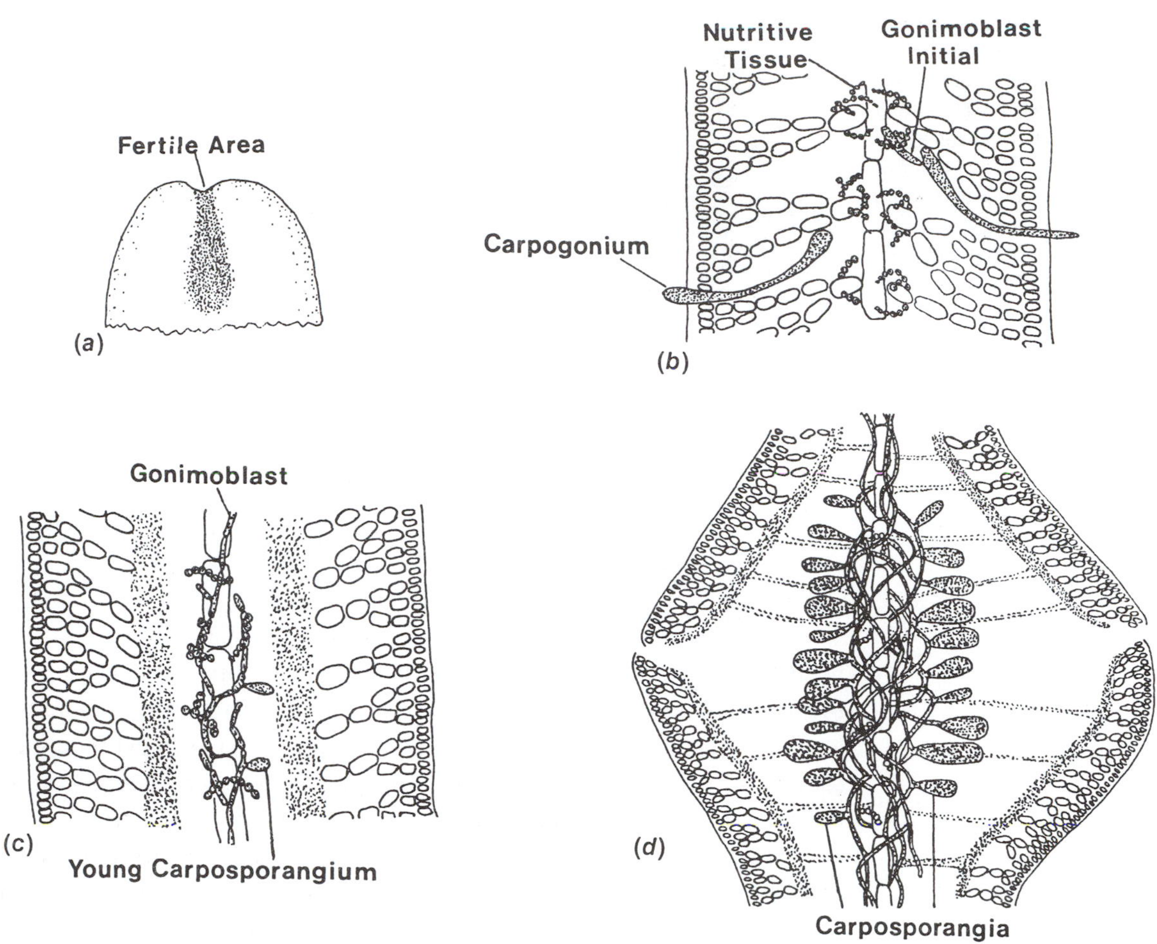

orless spermatangia are formed by transverse division of the mother cell. The tetrasporophyte forms tetrasporangia on the terminal portions or ramuli, the tetrasporangial mother cell being a cortical cell that is terminal on a lateral. It divides to produce four tetraspores in either a cruciate or a tetrahedral arrangement (Fig. 4.26(*d*)).

Fig. 4.27 *Gelidium cartilagineum*. (*a*) Apex of fertile thallus. (*b*) Longitudinal section of thallus showing carpogonium. (*c*) Gonimoblast producing young carposporangia. (*d*) Carposporophyte with mature carposporangia. ((*a*) after Kylin, 1928; (*b*)–(*d*) after Smith, 1938.)

CORALLINALES

The Corallinales is an order of heavily calcified red algae (Johansen, 1981; Silva and Johansen, 1986). Cytologically, the outer cap layer of the pit connections is large and dome-shaped (Pueschel and Trick, 1991). The order is characterized by having reproductive organs in conceptables (cavities that open to the thallus surface) opening to the exterior by one or more pores (Fig. 4.28(*d*)). In some genera, the tetrasporic conceptacles differ from sexual conceptacles in having numerous small pores in the roof rather than

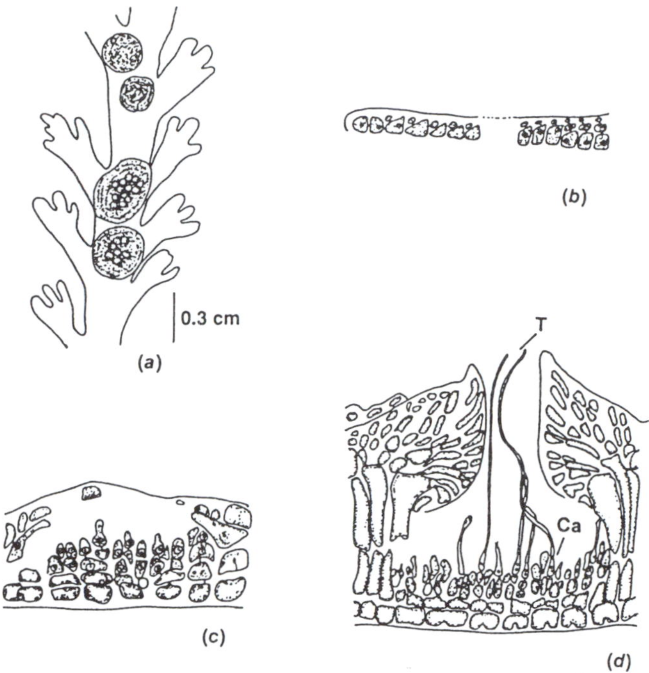

Fig. 4.28 (*a*) *Melobesia marginata* epiphytic on *Laurencia spectabilis*. *Melobesia lejolisii*: (*b*) section of sterile thallus showing cover cells and (*c*) section of immature fertile thallus. (*d*) *Melobesia limitata*: conceptacle with carpogonia (Ca) and trichogynes (T). ((*a*) after Smith, 1969; (*b*) after Suneson, 1937.)

a single pore. The sexual plants are usually dioecious, with marked differences between male and female conceptacles. Both male and female organs are borne in **nemathecia** (wart-like elevations of the surface containing many reproductive organs), which develop on the floor of the conceptacle. Spermatangia are formed abundantly from short filaments on the conceptacle floor. The female procarp consists of a two-celled carpogonial filament arising from a basal cell that functions as an auxiliary cell (Fig. 4.28(*d*)). The long trichogynes from the many carpogonia project through the conceptacular ostiole. After fertilization, a short ooblast from the carpogonium joins the auxiliary cell. All of the auxiliary cells of the conceptacle then fuse to form a large fusion or placental cell, from the margins of which issue the gonimoblast filaments with their carposporangia (Fig. 4.29(*b*)).

The thallus of the Corallinales is usually divided into two areas, the hypothallus and the perithallus. The **hypothallus** has relatively large cells and forms the basal part of the crustose plants (Fig. 4.29(*b*)) and the central part of the erect branches. Also, in case of injury the scar tissue that develops is hypothallus tissue. The **perithallus** has smaller cells and is located above the hypothallus in crustose forms and outside of the medullar hypothallus in branched forms.

The Corallinaceae of the Corallinales has two subfamilies. The crustose and nodular forms are in the Melobesioideae, and the articulated or jointed forms are in the Corallinoideae. In the Melobesiodeae, the simplest type of

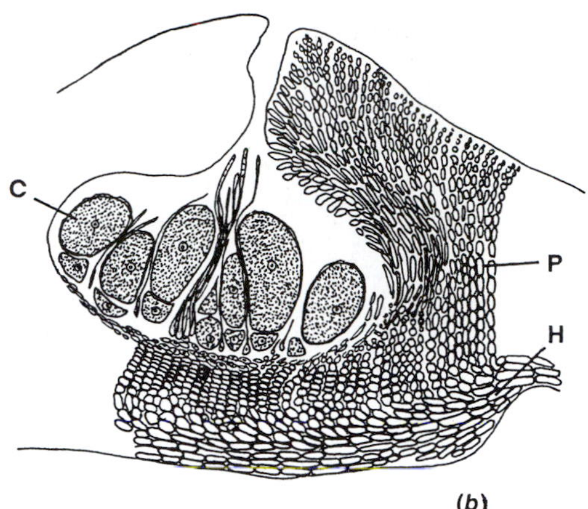

Fig. 4.29 (*a*) *Lithothamnion* sp. (*b*) *L. lenormandi*: drawing of a section of thallus with a hypothallus (H) and perithallus (P). The section includes a mature conceptacle with carposporangia (C). ((*a*) after Oltmanns, 1904; (*b*) after Suneson, 1943.)

thallus is in *Melobesia*, which has thin pink or red crusts that are widely distributed, especially as epiphytes on other algae and marine plants (Fig. 4.28(*a*)). The thallus consists of one to five layers of prostrate threads compacted to form a disc. A marked feature is the flat cover cells, which also occur in other Corallinaceae, forming the outer layer of cells (Fig. 4.28(*b*)). *Lithophyllum* and *Lithothamnion* are lithophytes that usually have considerably thicker crusts and sometimes nodules (Fig. 4.29).

In the Corallinoideae the plants are multiaxial, having a medulla of elongated cells and a cortex of shorter cells (Fig. 4.30). Calcification normally occurs only in the cell walls of the cortical cells. The plants are composed of a number of calicified segments, each segment joined by a non-calcified joint. The segments consist of calcified cortical and non-calcified medullary tissue, the arrangement of tissues giving the plants a certain amount of flexibility.

The crustose Corallinaceae occur in the intertidal zone, but only in areas that are not exposed to excessive drying, either on exposed rocks where they are kept moist by spray from breakers or in well-shaded areas. In some places they occur near the high-tide mark but are well covered by other algae. The sublittoral zone is a more favorable area for crustose algal growth, especially on reefs from the low-tide mark to a depth of 25 to 30 m. The depth and agitation of the water have a considerable influence on the growth form of the coralline algae. Crustose types are present at all depths, but the highly ramified or branching forms occur only near the surface, where they are most plentiful down to 30 m. In the crustose forms, the thickest crusts are formed in shallow waters; the crusts become thinner with depth (as a result of thinner hypothalli and smaller cells), probably as a result of slower growth. The crustose Corallinaceae are among the deepest-growing algae, down to 125 m in clear water. They are also among the

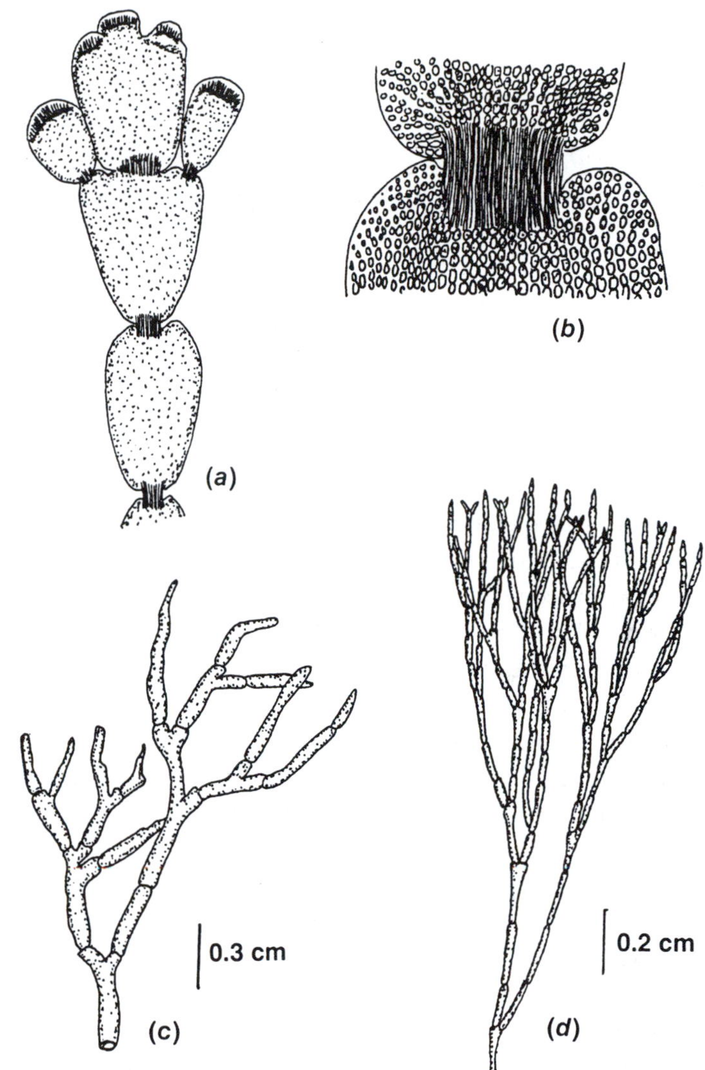

Fig. 4.30 (*a*) *Corallina* sp. (*b*) *Corallina* sp., showing non-calcified joints and calcified segments. (*c*) *Amphiroa rigida* var. *antillana*. (*d*) *Jania rubens*. (*a*),(*b*) after Oltmanns, 1904; (*c*),(*d*) after Taylor, 1957.)

longest-living, their life-span ranging from 10 to 50 years (Adey, 1970), probably as a result of their slow growth rate (0.3 to 3 mm year^{-1}). The light saturation for photosynthesis for red crustose corallines was found by Adey (1970) to be between 700 and 1000 lux, which is considerably lower than the 4000- to 10000-lux saturation intensities found for other Rhodophyceae (Kanwisher, 1966; Brown and Richardson, 1968). These light saturation values are probably related to the great depth at which the crustose corallines are able to live.

Maërl is composed of shallow, subtidal deposits of calcareous red algae belonging to the Corallinales. Maërl has been obtained commercially for many years by dredging from the coast of Brittany in France, off Falmouth

Harbor in England and Bantry Bay in Ireland. Maërl is placed on acidic soil to increase the pH of the soil for crops (Blunden et al., 1997).

The Corallinaceae form an important part of atolls and reefs (Dawson, 1966). The reefs are built up by the combined growth of coralline red algae (mostly species of *Porolithon*) and corals. When a reef first breaks the surface, the rigid, branched, brittle corals tend to break and fragment under severe surf action, whereas the massive coralline Rhodophyceae are unaffected by the pounding. In fact, the stronger the pounding, the faster they grow. The coralline reds thus grow into the breaking surf, developing upward and outward to form a rim slightly above sea level. While the corallines are forming the main framework of the reef, 90% of the reef comes to consist of sand and detritus cemented together by the plants and animals. Inside this rim of coralline Rhodophyceae, there is relative calm over the inner part of the reef. Within this inner area, carbonic acid, resulting from the solution of respiratory carbon dioxide produced by the plants and animals living there, tends to dissolve the solid calcareous materials. This dissolution of the inner part of the reef results in a central lagoon that stabilizes itself at a depth of 65 to 100 m.

The reef community is adapted to a low-stress environment characterized by the absence of significant seasonal change. The mean winter temperature of the water where the reefs grow is between 27 and 29 °C, and the difference between the monthly mean temperatures is 3 °C or less. The water is clear (so the penetration of light is at a maximum), agitated, and of normal salinity. Even under these ideal circumstances many reef organisms (e.g., corals) do not grow at depths greater than 20 m.

The **Solenoporaceae** is an extinct family that arose in the Ordovician Period of the Paleozoic Era and became extinct during the Jurassic Period. It is the family that gave rise in the Jurassic to the Corallinaceae. The Solenoporaceae were differentiated from the Corallinaceae as follows (Johnson, 1961): (1) The sporangia were external and non-calcified in the Solenoporaceae, whereas they are internal in the Corallinaceae. (2) The cells of the Solenoporaceae were larger and longer than those of the Corallinaceae. (3) Generally the Solenoporaceae did not have a differentiation of the thallus into a hypothallus and perithallus as the Corallinaceae do. During the Paleozoic, the numbers of Solenoporaceae were low, so that in the Silurian and Devonian limestone reefs, the animal corals were more important. This is in marked contrast to more recent reefs where the Corallinaceae are more important. The Solenoporaceae probably lived under conditions similar to those of the Corallinaceae today; thus the decline in Solenoporaceae during the Cretaceous Period, with the associated appearance and increase in the Corallinaceae, may well have been due to competition between the two groups of calcareous red algae.

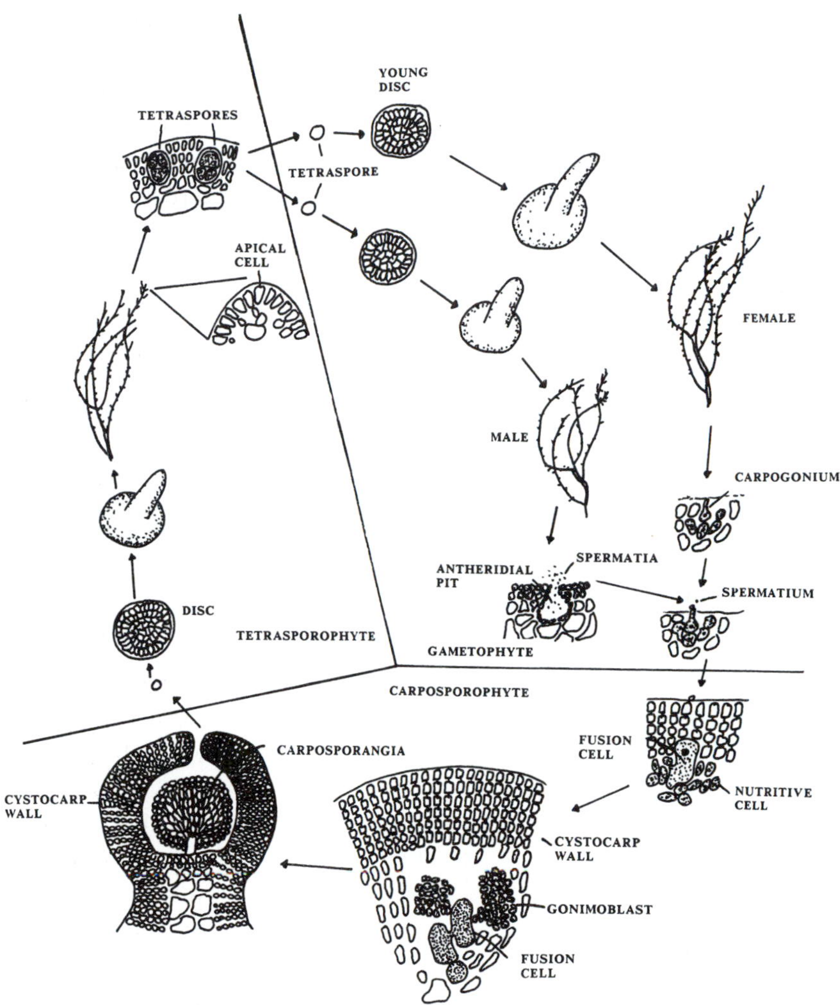

Fig. 4.31 The life cycle of *Gracilaria* spp. (Adapted from Kylin, 1930; Ogata et al., 1972).

GRACILARIALES

The Gracilariales are agarophytes that have a female reproductive system with a supporting cell of intercalary origin that bears a two-celled carpogonial branch flanked by two or more sterile branches (Fredericq and Hommersand, 1989).

The plants in the order are fleshy, having a tendency to be flattened or foliose with pseudoparenchymatous tissues that lack filamentous cells in the mature vegetative thallus. The principal genus in the family is *Gracilaria*, a widely distributed northern lithophyte found at low-tide level and below, with about 100 species. The dark-red thallus grows by means of a two-sided apical cell and has tapering branches (Fig. 4.31). There are large isodiametric cells in the medulla, with small cortical cells containing a number of ribbon-shaped chloroplasts. Unicellular hairs arise from enlarged peripheral cells that become multinucleate as they age.

The gametophytic plants are either male or female, and equal numbers of each are produced from tetraspores (Kain and Destombe, 1995). The male plants produce spermatia in antheridial pits over the surface of the thalli. The female plants form supporting cells from the outer layer of the large cells of the medulla (Kylin, 1930), the supporting cells producing the two-celled carpogonial branch and a number of laterals, a cell of which functions as the auxiliary cell. All of the cells of the procarp become multinucleate and develop into nutritive cells except for the carpogonium and the cell beneath it. After fertilization the carpogonium fuses with one of the nutritive cells that is acting as an auxiliary cell. Subsequently, this fusion cell fuses with the other multinucleate nutritive cells. At the same time, the cortical cells above the procarp divide to produce the cystocarp walls, the inner cells of which constitute nutritive cells. Gonimoblast initials are cut off from the fusion cell and develop into an inner sterile area that supports the outer carposporangia. The carposporangia ripen successively from the outside in. In some species, elongate cells radiate from the compact regions of the gonimoblast, penetrating the **pericarp** (cystocarp wall), and become connected with the cells of the pericarp.

The carpospores germinate to produce a parenchymatous disc that forms the tetrasporophyte as an erect protuberance. The tetrasporophyte is morphologically similar to the gametophyte and about the same size as the female gametophyte. Cruciate tetrasporangia are formed terminally on laterals in the cortex and are embedded in the thallus. The tetraspores germinate to form a parenchymatous disc that produces the gametophyte as an erect protuberance (Ogata et al., 1972).

Gracilaria is a major agarophyte, currently providing greater than half of the world's supply of agar. The cultivation of *Gracilaria*, both in the sea and in tanks, has been a principal factor in making this genus a source of agar-containing seaweeds (Lewis and Hanisak, 1996).

CERAMIALES

These plants have the auxiliary cell cut off after fertilization and borne on the supporting cell of the four-celled carpogonial filament. Most of the plants are relatively delicate filamentous or membranous forms.

In *Polysiphonia* the uninucleate, dome-shaped apical cell is polyploid and contains 64 to 128 times the amount of DNA in most of the mature cells in the alga (Goff and Coleman, 1986). Division of the cells derived from the apical cell is usually not accompanied by DNA replication; therefore, the

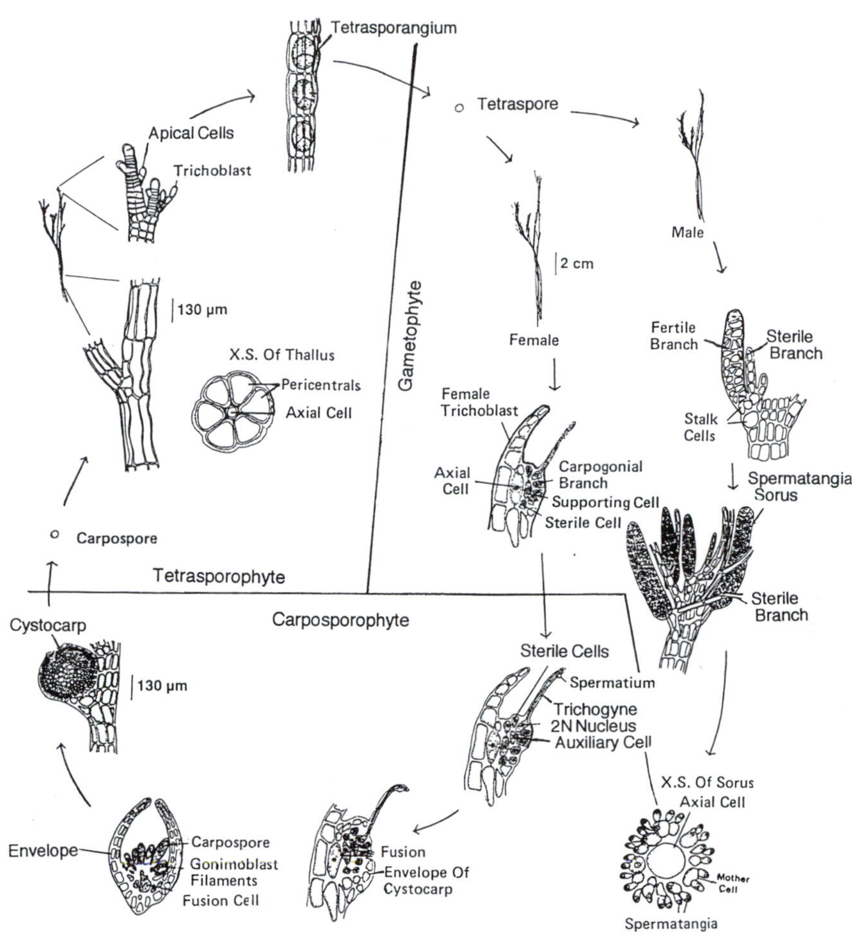

Fig. 4.32 The life cycle of *Polysiphonia* sp. (Adapted from Kylin, 1923; Edwards, 1970a.)

farther the daughter cell is from the apical cell, the lower the ploidy of the cell, until the ploidy number stabilizes at 1 *n*. The apical cell forms daughter cells that produce lateral branches before dividing longitudinally into central and pericentral cells (Fig. 4.32). The pericentrals are the same length as the axial cells. The lateral branches are of two kinds, the ordinary branches and the trichoblasts. The ordinary branches are polysiphonous with unlimited growth, similar to the main axis. The **trichoblasts** are uniseriate, usually colorless, and bear the sex organs. They commonly drop off in the older parts of the thallus.

Polysiphonia species occur either as lithophytes or as epiphytes on other algae. When growing on a solid substrate, some of the polysiphonous axes creep over the substratum to which they are firmly anchored by thick-walled, lobed rhizoids. When growing as an epiphyte on another alga, the rhizoid penetrates the host tissue (Fig. 4.12). The procarps are produced near the base of a trichoblast (Fig. 4.32). The two basal cells of the trichoblast become polysiphonous, and a procarp develops from a pericentral of the upper of the two basal cells. The pericentral (supporting cell) produces in succession a lateral sterile cell, the four-celled carpogonial branch, and, last, the second sterile cell. The sterile cells divide after fertilization and may have a nutritive role. The cells of the carpogonial branch (with the exception of the carpogonium) are commonly binucleate (Kylin, 1923). A large area of endoplasmic reticulum extends from one pit connection to the other pit connection in each cell of the carpogonial branch. This may be how the message of fertilization is transmitted down the carpogonial branch (Broadwater and Scott, 1982).

The male plants of *Polysiphonia* bear antheridial sori on a trichoblast consisting of a two-celled stalk surmounted by the fertile region. The upper stalk cell frequently bears a branch. Fertile regions become polysiphonous, and the pericentrals divide copiously to form a compact layer of mother cells, each of which gives rise to two or three antheridia (Kylin, 1923). After the spermatium fertilizes the trichogyne, the auxiliary cell is cut off from the supporting cell. The auxiliary cell then fuses with the carpogonium. The male nucleus fuses with the female in the carpogonium (Yamanouchi, 1906), and the diploid nucleus divides once. One of the diploid nuclei passes into the auxiliary cell, which subsequently fuses with the supporting cell. The fusion cell also fuses with the axial cell of the fertile segment. The gonimoblast initials are cut off from the fusion cell and develop into a number of gonimoblast filaments. The terminal cells of these filaments develop into pear-shaped carposporangia. In the meantime, the fusion cell unites with the gonimoblast initial and the fertile cells. The sterile outer envelope of the cystocarp originates from the other pericentrals of the axial cell that gave rise to the fertile pericentral that acted as the supporting cell. The young envelope consists of two lateral valves, composed of fused threads, which enclose the procarp like the shells of an oyster, the trichogyne alone projecting. After fertilization the two valves unite, and the envelope becomes two-layered. The carpospores form a tetrasporophyte, which is similar to the gametophyte, and which forms tetrahedrally arranged tetraspores in polysiphonous branches called **stichidia**. The tetraspores then germinate to form the gametophyte (Edwards, 1969).

Polysiphonia denudata completes its life history in 1.5 months in culture; thus the species probably has several life cycles each year (Edwards, 1970a). In some species of *Polysiphonia*, it is possible to influence stages of the life cycle by changing the photoperiod, but there appears to be no regularity among the different species (Edwards, 1970a,b).

Fig. 4.33 *Left:* Heinrichs Leonhards Skuja. *Right:* Harald Kylin. (Photograph of Skuja from Willen and Wingqvist, 1986; photograph of Kylin from *Die Gattung der Rhodophyceen*, CWK Gleerups Forlag, Lund, Sweden.)

Henrichs Leonhards Skuja Born September 8, 1892 in Majori, Rigas-Jurmula, Latvia; died July 19, 1972. During his youth, Dr Skuja lived close to the Latvian coast and took an early interest in aquatic plants and animals. During World War I, he lived in the Caucasus, where he was engaged in floristic studies on the Apsheron Peninsula. In 1922, he began his academic studies at the faculty of natural sciences at the University of Latvia, passed a *mag. rer. nat.* examination in 1929 and became *dr. rer. nat.* in 1943. In the autumn of 1944, Dr Skuja arrived in Sweden, where in 1947 he obtained a position as research professor (*laborator*) and in 1958 was awarded the degree of *doctor honoris causa* by the University of Uppsala. Besides his extensive publications on the algae, Dr Skuja also published in the areas of general botany, mycology, and lichenology.

Harald Kylin, 1879–1949 Dr Kylin was professor of botany at the University of Lund, Sweden. He contributed a large number of publications on his investigations of the red algae. The last of these was his book *Die Gattung der Rhodophyceen*, which was finished by his wife after his death.

References

Abdel-Rahman, M. H. (1982). The involvement of an endogenous circadian rhythm in photoperiodic timing in *Acrochaetium asparagopsis* (Rhodophyta, Acrochaetiales). *Br. Phycol. J.* 17:389–400.

Adey, W. H. (1970). The effects of light and temperature on growth rates in boreal-subarctic crustose corallines. *J. Phycol.* 6:269–76.

Arad, S., Dubinsky, O., and Simon, B. (1993). A modified cell wall mutant of the red microalga *Rhodella reticulata* resistant to the herbicide 2,6-dichlorobenzonitrile. *J. Phycol.* 29:309–13.

Balakrishnan, M. S., and Chaugule, B. (1980). Cytology and life history of *Batrachospermum mahabaleshwarensis*. *Cryptogamie: Algologie* 1:83–97.

Barrow, K. D., Karsten, U., King, R. J., and West, J. A. (1995). Floridoside in the genus *Laurencia* (Rhodomelaceae: Ceramiales) – a chemosystematic study. *Phycologia* 34:279–83.

Belcher, J. H., and Swale, E. M. F. (1960). Some British freshwater material of *Asterocystis*. *Br. Phycol. Bull.* 2:33–5.

Blunden, G., Campbell, S. A., Smith, J. R., Guiry, M. D., Hession, C. C., and Griffin, R. L. (1997). Chemical and physical characteristics of calcified red algal deposits known as maerl. *J. Applied Phycol.* 9:11–17.

Borowitzka, M. A., Larkum, A. W. D., and Nockolds, C. E. (1974). A scanning electron microscope study of the structure and organization of the calcium carbonate deposits of algae. *Phycologia* 13:195–203.

Bourgougnon, N., Lehaye, M., Quemener, B., Chermann, J-C., Rimbert, M., Cormaci, M., Furnari, G., and Kornprobat, J-M. (1996). Annual variation in composition and *in vitro* anti-HIV-1 activity of the sulfated glucuronogalactan from *Schizymenia dubyi* (Rhodophyta, Gigartinales). *J. Applied Phycol.* 8:1155–61.

Broadwater, S. T., and Scott, J. (1982). Ultrastructure of early development in the female reproductive system of *Polysiphonia harveyi* Bailey (Ceramiales, Rhodophyta). *J. Phycol.* 18:427–41.

Brown, T. E., and Richardson, F. L. (1968). The effect of growth and environment on the physiology of the algae: Light intensity. *J. Phycol.* 4:38–54.

Calabrese, G. (1972). Research on red algal pigments. 2. Pigments of *Petroglossum nicaeense* (Duby) Schotter (Rhodophyceae, Gigartinales) and their seasonal variations at different light intensities. *Phycologia* 11:141–6.

Campbell, S. E. (1980). *Palaeoconchocelis starmachii*, a carbonate boring micro-fossil from the Upper Silurian of Poland (425 millions years old): Implications for the evolution of the Bangiaceae (Rhodophyta). *Phycologia* 19:25–36.

Chemin, E. (1937). Le développement des spores chez les Rhodophycées. *Rev. Gen. Bot.* 49:205–34, 300–27, 353–74, 424–48, 478–536.

Chen, L. C-M., Edelstein, T., Ogata, E., and McLachlan, J. (1970). The life history of *Porphyra miniata*. *Can. J. Bot.* 48:385–9.

Chen, L. C-M., McLachlan, J., Neish, A. C., and Shacklock, P. F. (1973). The ratio of kappa- to lamda-carrageenan in nuclear phases of the Rhodophycean alga, *Chondrus crispus* and *Gigartina stellata*. *J. Mar. Biol. Assoc. UK* 53:11–16.

Chiovitti, A., Liao, M-L., Kraft, G. T., Munro, S. L. A., Craik, D. J., and Bacic, A. (1995). Cell wall polysaccharides from Australian red algae of the family Solieriaceae (Gigartinales, Rhodophyta): iota/kappa/beta-carrageenans from *Melanema dumosum*. *Phycologia* 34:522–7.

Cole, M., and Sheath, G. (1990). *Biology of the Red Algae*. Cambridge University Press, 517 pp.

Craigie, J. S., Correa, J. A., and Gordon, M. E. (1992). Cuticles from *Chondrus crispus*. *J. Phycol.* 28:777–86.

Cunningham, E. M., and Guiry, M. D. (1989). A circadian rhythm in the long-day photoperiodic induction of erect axis development in the marine red alga *Nemalion helminthoides*. *J. Phycol.* 25:705–12.

Dawson, E. Y. (1966). *Marine Botany*. New York: Holt, Rinehart and Winston.

den Hartog, C. (1971). The effect of the salinity tolerance of algae on their distribution, as exemplified by *Bangia*. *Proc. 7th Int. Seaweed Symp.*, pp. 274–6.

Digby, P. S. B. (1977a). Growth and calcification in the coralline algae, *Clathromorphum circumscriptum* and *Corallina officinalis*, and the significance of pH in relation to precipitation. *J. Mar. Biol. Assoc. UK* 57:1095–109.

Digby, P. S. B. (1977b). Photosynthesis and respiration in the coralline algae,

Clathromorphum circumscriptum and *Corallina officinalis* and the metabolic basis of calcification. *J. Mar. Biol. Assoc. UK* 57:1111–24.

Dixon, P. S. (1973). *Biology of the Rhodophyta,* University Reviews in Biology No. 4. Edinburgh: Oliver and Boyd.

Dixon, P. S., and Richardson, W. N. (1970). Growth and reproduction in red algae in relation to light and dark cycles. *Ann. N.Y. Acad. Sci.* 175:764–77.

Doemel, W. N., and Brock, T. D. (1971). The physiological ecology of *Cyanidium caldarium. J. Gen. Microbiol.* 67:17–32.

Dring, M. J. (1967a). Effects of daylength on growth and reproduction of the *Conchocelis*-phase of *Porphyra tenera. J. Mar. Assoc. UK* 47:501–10.

Dring, M. J. (1967b). Phytochrome in red alga, *Porphyra tenera. Nature* 215:1411–12.

Dring, M. J., and West, J. A. (1983). Photoperiodic control of tetrasporangium formation in the red alga *Rhodochorton purpureum. Planta* 159:143–50.

Edwards, P. (1969). The life history of *Callithamnion byssoides* in culture. *J. Phycol.* 5:266–8.

Edwards, P. (1970a). Field and cultural observations on the growth and reproduction of *Polysiphonia denudata* from Texas, *Br. Phycol. J.* 5:145–53.

Edwards, P. (1970b). Field and cultural studies on the seasonal periodicity of growth and reproduction of selected Texas benthic marine algae. *Contrib. Mar. Sci. Univ. Texas* 14:59–114.

Evans, L. V. (1970). Electron microscopical observations on a new red algal unicell, *Rhodella maculata* gen. nov., sp. nov., *Br. Phycol. J.* 5:1–13.

Evans, L. V., Callow, J. A., and Callow, M. E. (1973). Structural and physiological studies on the parasitic red alga *Holmsella. New Phytol.* 72:393–402.

Fan, K-C. (1961). Studies on *Hypneocolax,* with a discussion on the origin of parasitic red algae. *Nova Hedwigia* 3:119–28.

Feldmann, G. (1970a). Sur l'ultrastructure des corps irisants des *Chondria* (Rhodophycées). *C. R. Séances Acad. Sci. Paris* 270:945–50.

Feldmann, G. (1970b). Sur l'ultrastructure de l'appareil irisant du *Gastroclonium clavatum* (Roth.) Ardissonne (Rhodophyceae). *C. R. Séances Acad. Sci. Paris.* 270:1244–6.

Feldmann, J. (1953). L'évolution des organes femelles chez les Floridées. *Proc. 1st Int. Seaweed Symp.*, pp. 11–12.

Fetter, R., and Neushul, M. (1981). Studies on developing and released spermatia in the red alga, *Tiffaniella snyderae* (Rhodophyta). *J. Phycol.* 17:141–59.

Fredericq, S., and Hommersand, M. H. (1989). Proposal of the Gracilariales ord. nov. (Rhodophyta) based on an analysis of the reproductive development of *Gracilaria verrucosa. J. Phycol.* 25:213–19.

Frei, E., and Preston, R. D. (1964). Non-cellulosic structural polysaccharides in algal cell walls. II. Association of xylan and mannan in *Porphyra umbilicalis. Proc. R. Soc. Lond.* [*B*] 160:314–27.

Freshwater, D. W., Fredericq, S., Butler, B., Hommersand, M. H., and Chase, M. W. (1994). A gene phylogeny of the red algae (Rhodophyta) based on plastid *rbcL. Proc. Natl. Acad. Sci., USA* 91:7281–5.

Fries, L. (1963). On the cultivation of axenic red algae. *Physiol. Plant.* 16:695–708.

Fries, L. (1967). The sporophyte of *Nemalion multifidum* (Weber et J. Ag.). *Sven. Bot. Tidskr.* 61:457–62.

Gabrielson, P. W., Garbary, D. J., and Scagel, R. F. (1985). The nature of the ancestral red alga: Inferences from a cladistic analysis. *BioSystems* 18:335–46.

Gantt, E. (1969). Properties and ultrastructure of phycoerythrin from *Porphyridium cruentum. Plant Physiol.* 44:1629–38.

Gantt, E., and Conti, S. F. (1965). The ultrastructure of *Porphyridium cruentum*. *J. Cell Biol.* 26:365–81.

Gerwick, W. H., and Lang, N. J. (1977). Structural, chemical and ecological studies on iridescence in *Iridaea* (Rhodophyta). *J. Phycol.* 13:121–27.

Giraud, G. (1962). Les infrastructures de quelques algues et leur physiologie. *J. Microscopie* 1:251–64.

Goff, L. J., and Coleman, A. W. (1984). Transfer of nuclei from a parasite to a host. *Proc. Natl. Acad. Sci. USA* 81:5420–4.

Goff, L. J., and Coleman, A. W. (1986). A novel pattern of apical cell polyploidy, sequential polyploidy reduction and intercellular nuclear transfer in the red alga *Polysiphonia*. *Am. J. Bot.* 73:1109–30.

Goff, L. J., Moon, D. A., Nyvall, P., Stache, B., Mangin, K., and Zuccarello, G. (1996). The evolution of parasitism in the red algae: molecular comparisons of adelphoparasites and their hosts. *J. Phycol.* 32:297–312.

Goreau, T. F. (1963). Calcium carbonate deposition by coralline algae and coral in relation to their roles as reef builders. *Ann. N.Y. Acad. Sci.* 109:127–67.

Gretz, M. R., Aronson, J. M., and Sommerfeld, M. R. (1980). Cellulose in the cell walls of the Bangiophyceae (Rhodophyta). *Science* 207:779–81.

Gretz, M. R., McCandless, E. L., Aronson, J. M., and Sommerfeld, M. R. (1983). The galactan sulfates of the Conchocelis phases of *Porphyra leucostricta* and *Bangia atropurpurea (Rhodophyta)*. *J. Exp. Bot.* 34:705–11.

Hara, Y. (1971). An electron microscopic study on the chloroplasts of the Rhodophyta. *Proc. 7th Int. Seaweed Symp.*, pp. 153–8.

Harlin, M. M., and Craigie, J. S. (1975). The distribution of photosynthate in *Ascophyllum nodosum* as it relates to epiphytic *Polysiphonia lanosa*. *J. Phycol.* 11:109–13.

Hawkes, M. W. (1978). Sexual reproduction in *Porphyra gardneri* (Smith et Hollenberg) Hawkes (Bangiales, Rhodophyta). *Phycologia* 17:329–53.

Hurdelbrink, L., and Schwantes, H. O. (1972). Sur le cycle de développement de *Batrachospermum*. *Mem. Soc. Bot. Fr.*, 269–74.

Johansen, H. W. (1981). *Coralline Algae: A First Synthesis*. CRC Press, Boca Raton, Florida.

Johnson, J. (1961). *Limestone-Building Algae and Algal Limestones*. Colorado School of Mines.

Kain, J. M., and Destombe, C. (1995). A review of the life history, reproduction and phenology of *Gracilaria*. *J. Appl. Phycol.* 7:269–81.

Kanwisher, J. W. (1966). Photosynthesis and respiration in some seaweeds. In *Some Contemporary Studies in Marine Science* (Barnes, R., ed.), pp. 407–20. New York: Academic Press.

Kim, G.H., and Fritz, L. (1993a). Gamete recognition during fertilization in a red alga *Antithamnion nipponicum*. *Protoplasma* 174:69–73.

Kim, G. H., and Fritz, L. (1993b). Ultrastructure and cytochemistry of early spermatangial development in *Antithamnion nipponicum* (Ceramiaceae, Rhodophyta). *J. Phycol.* 29:797–805.

Kornmann, P. (1994). Life histories of monostomatic *Porphyra* species as a basis for taxonomy and classification. *Eur. J. Phycol.* 29:69–71.

Kugrens, P. (1974). Light and electron microscope studies on the development and liberation of *Janczewskia gardneri* Setch. spermatia (Rhodophyta). *Phycologia* 13:295–306.

Kugrens, P., and West, J. A. (1972a). Ultrastructure of spermatial development in the parasitic red alga *Levringiella gardneri* and *Erythrocystis saccata*. *J. Phycol.* 8:331–43.

Kugrens, P., and West, J. A. (1972b). Ultrastructure of tetrasporogenesis in the parasitic red alga *Levringiella gardneri* (Setchell) Kylin. *J. Phycol.* 8:370–83.

Kugrens, P., and West, J. A. (1973). The ultrastructure of an alloparasitic red alga *Choreocolax polysiphoniae*. *Phycologia* 12:175–86.

Kugrens, P., and West, J. A. (1974). The ultrastructure of carposporogenesis in the marine hemiparasitic red alga *Erythrocystis saccata*. *J. Phycol.* 10:139–47.

Kuroiwa, T., Kawazu, T., Takahashi, H., Suzuki, K., Ohta, N., and Kuroiwa, H. (1994). Comparison of ultrastructures between the ultra-small eukaryote *Cyanidioschyzon merolae* and *Cyanidium caldarum*. *Cytologia* 59:149–88.

Kylin, H. (1916). Über *Spermothamnion roseolum* (Ag.) Pringsh. und *Trailiella intricata* Batters, *Bot. Not.*, 83–92.

Kylin, H. (1923). Studien über die Entwicklungsgeschichte der Florideen. *Sven, Vet. Akad. Handl.* 63, No. 11.

Kylin, H. (1924). Bemerkungen über einige *Ceramium*-Arten. *Bot. Not.*, 443–52.

Kylin, H. (1928). Entwicklungsgeschichtliche Florideenstudien. *Lunds Univ. Arsskr. N.F. II* 24, No. 4.

Kylin, H. (1930). Uber die Entwicklungsgeschichtliche der Florideen. *Lunds Univ. Arsskr. N.F. II* 26, No. 6.

Kylin, H. (1956). *Die Gattungen der Rhodophyceen*. Gleerups, Lund.

Lee, R. E. (1971). The pit connections of some lower red algae: Ultrastructure and phylogenetic significance. *Br. Phycol. J.* 6:29–38.

Lee, R. E. (1974). Chloroplast structure and starch grain production as phylogenetic indicators in the lower Rhodophyceae. *Br. Phycol. J.* 9:291–5.

Lee, R. E., and Fultz, S. A. (1970). The ultrastructure of the *Conchocelis* stage of the marine red alga *Porphyra leucosticta*. *J. Phycol.* 6:22–28.

Lewin, R. A., and Robertson, J. A. (1971). Influence of salinity on the form of *Asterocytis* in pure culture. *J. Phycol.* 7:236–8.

Lewis, R. J., and Hanisak, M. D. (1996). Effects of phosphate and nitrate supply on productivity, agar content and physical properties of agar of *Gracilaria* strain G-16S. *J. Appl. Phycol.* 8:41–9.

Lichtlé, C., and Giraud, G. (1969). Etude ultrastructurale de la zone apicale du thalle du *Polysiphonia elongata* (Harv.) Rhodophyceé, Floridée. Evolution des plastes. *J. Microscopie* 8:867–87.

Lin, H., Sommerfeld, M. R., and Swafford, J. R. (1975). Light and electron microscope observations on motile cells of *Porphyridium purpureum* (Rhodophyta). *J. Phycol.* 11:452–7.

Liu, Y. L., Ross, N., Lanthier, P., and Reith, M. (1996). A gametophyte cell wall protein of the red alga *Porphyra purpurea* (Rhodophyta) contains four apparent polysaccharide-binding domains. *J. Phycol.* 32:995–1003.

Ludlow, C. J., and Park, R. B. (1969). Action spectra for photosystems I and II in formaldehyde fixed algae. *Plant Physiol.* 44:540–3.

Magne, F. (1960a). Sur l'existence d'une reproduction sexuée chez le *Rhodochaete parvula* Thuret. *C. R. Séances Acad. Sci. Paris* 251:1554–5.

Magne, F. (1960b). Le *Rhodochaeta parvula* Thuret (Bangiodée) et sa reproduction sexuée. *Cah. Biol. Mar.* 1:407–20.

Magruder, W. H. (1984). Specialized appendages on spermatia from the red alga *Aglaothamnion neglectum* (Ceramiales, Ceramiaceae) specifically bind with trichogynes. *J. Phycol.* 20:436–40.

Mukai, L. S., Craigie, J. S., and Brown, R. G. (1981). Chemical composition and structure of the cell walls of the *Conchocelis* and thallus phases of *Porphyra tenera* (Rhodophyceae). *J. Phycol.* 17:192–8.

Mumford, T. F., and Miura, A. (1988). *Porphyra* as food: cultivation and economics. In *Algae and Human Affairs*, ed. C. A. Lembi and J. P. Waaland, pp. 87–117. Cambridge, UK: Cambridge University Press.

Ngan, Y., and Price, I. R. (1979). Systematic significance of spore size in the Florideophycidae (Rhodophyta). *Br. Phycol. J.* 14:285–303.

Ogata, E., Matsui, T., and Nakamura, H. (1972). The life cycle of *Gracilaria verrucosa* (Rhodophyceae, Gigartinales) in vitro. *Phycologia* 11:75–80.

O'hEocha, C. (1971). Pigments of the red algae. *Oceanogr. Mar. Biol. Annu. Rev.* 9:61–82.

Ohno, M., Nang, H. Q., and Hirase, S. (1996). Cultivation and carrageenan yield and quality of *Kappaphycus alvarezii* in the waters of Vietnam. *J. Applied Phycol.* 8:431–7.

Ohta, N., Sato, N., Ueda, K., and Kuroiwa, T. (1997). Analysis of a plastid gene cluster reveals a close relationship between *Cyanidioschyzon* and *Cyanidium*. *J. Plant Res.* 110:235–45.

Okazaki, M., Ikawa, T., Furuya, K., Nisizawa, K., and Miwa, T. (1970). Studies on calcium carbonate deposition of a calcareous red alga *Serraticardia maxima*. *Bot. Mag. Tokyo* 83:193–201.

Oltmanns, F. (1904). *Morphologie und Biologie der Algen*, Bd. I. Jena.

Pallaghy, C. K., Minchinton, J., Kraft, G. T., and Wetherbee, R. (1983). Presence and distribution of bromine in *Thysanocladia densa* (Solieriaceae, Gigartinales), a marine red alga from the Great Barrier Reef. *J. Phycol.* 19:204–8.

Papenfuss, G. F. (1945). Review of the *Acrochaetium–Rhodochorton* complex of the red algae. *Univ. Calif. Publ. Bot.* 18:299–334.

Papenfuss, G. F. (1947). Further contributions toward an understanding of the *Acrochaetium–Rhodochorton* complex of the red algae. *Univ. Calif. Publ. Bot.* 18:433–47.

Papenfuss, G. F. (1966). A review of the present system of classification of the Florideophycidae. *Phycologia* 5:247–55.

Parke, M., and Dixon, P. S. (1976). Check-list of British marine algae – Third revision, *J. Mar. Biol. Assoc. UK* 56:527–94.

Pearse, V. B. (1972). Radioisotopic study of calcification in the articulated coralline alga *Bossiella orbigniana*. *J. Phycol.* 8:88–97.

Peders'en, M. E. E., Roomans, G. M., and v. Hofsten, A. (1981). Bromine in the cuticle of *Polysiphonia nigrescens:* Localization and content. *J. Phycol.* 17:105–8.

Peyrière, M. (1971). Etude infrastructurale des spermatocystes du *Griffithsia flosculosa* (Rhodophycée). *C. R. Séances Acad. Sci. Paris* 273:2071–4.

Pueschel, C. M. (1987). Absence of cap membrane as a characteristic of pit plugs in some red algal orders. *J. Phycol.* 23:150–6.

Pueschel, C. M. (1989). An expanded survey of the ultrastructure of red algal pit plugs. *J. Phycol.* 25:625–36.

Pueschel, C. M., and Trick, H. N. (1991). Unusual morphological and cytochemical features of pit plugs in *Clathromorphum circumscriptum* (Rhodophyta; Corallinales). *Br. Phycol. J.* 26:335–42.

Pueschel, C. M., and Miller, T. J. (1996). Reconsidering prey specializations in an algal-limpet grazing mutualism: Epithallial cell development in *Clathromorphum circumscriptum* (Rhodophyta, Corallinales). *J. Phycol.* 32:28–36.

Ragan, M. A., and Gutell, R. R. (1995). Are red algae plants? *Bot. L. Linn. Soc.* 118:81–105.

Ragan, M. A., Bird, C. J., Rice, E. L., Gutell, R. B., Murphy, C. A., and Singh, R. K. (1994). A molecular phylogeny of the marine red algae (Rhodphyta) based on the nuclear small-subunit rRNA gene. *Proc. Natl. Acad. Sci., USA* 91:7276–80.

Ramus, J. (1969a). Pit connection formation in the red alga *Pseudogloiophloea*. *J. Phycol.* 5:57–63.

Ramus, J. (1969b). Dimorphic pit connections in the red alga *Pseudogloiophloea*. *J. Cell Biol.* 41:340–5.

Ramus, J. (1971). Properties of septal plugs from the red alga *Griffithsia pacifica. Phycologia* 10:99–103.

Ramus, J., and Robins, D. M. (1975). The correlation of Golgi activity and polysaccharide secretion in *Porphyridium. J. Phycol.* 11:70–4.

Rawlence, D. J. (1972). An ultrastructural study of the relationship between rhizoids of *Polysiphonia lanosa* (L.) Tandy (Rhodophyceae) and the tissue of *Ascophyllum nodosum* (L.) Le Jolis (Phaeophyceae). *Phycologia* 11:279–90.

Reed, R. H. (1985). Osmoacclimation in *Bangia atropurpurea* (Rhodophyta, Bangiales): The osmotic role of floridoside. *Br. Phycol. J.* 20:211–18.

Richardson, N. (1970). Studies on the photobiology of *Bangia fuscopurpurea. J. Phycol.* 6:215–19.

Sagert, S., and Schubert, H. (1995). Acclimation of the photosynthetic apparatus of *Palmaria palmata* (Rhodophyta) to light qualities that preferentially excite photosystem I or photosystem II. *J. Phycol.* 31:547–54.

Sagromsky, H. (1964). Ist Chlorophyll *d* der Rotalgen Umwandlungsprodukt von Chlorophyll *a. Ber. Dtsch. Bot. Ges.* 77:323–6.

Satoh, K., Smith, C. M., and Fork, D. C. (1983). Effects of salinity on primary processes of photosynthesis in the red alga *Porphyra perforata. Plant Physiol.* 73:643–7.

Saunders, G. W., and Kraft, G. T. (1996). Small-subunit rRNA gene sequences from representatives of selected families of the Gigartinales and Rhodymeniales (Rhodophyta). 2. Recognition of the Halymeniales ord. nov. *Can. J. Bot.* 74:694–707.

Scott, J. L., and Dixon, P. S. (1973a). Ultrastructure of spermatium liberation in the marine red alga *Ptilota densa. J. Phycol.* 9:85–91.

Scott, J. L., and Dixon, P. S. (1973b). Ultrastructure of tetrasporogenesis in the marine red alga *Ptilota hypnoides. J. Phycol.* 9:29–46.

Scott, J., Bosco, C., Schornstein, K., and Thomas, J. (1980). Ultrastructure of cell division and reproductive differentiation of male plants in the Florideophycidae (Rhodophyta): Cell division in *Polysiphonia. J. Phycol.* 16:507–24.

Seckbach, J. (1991). Systematic problems with *Cyanidium caldarium* and *Galdiera sulphuraria* and their implications for molecular biology studies. *J. Phycol.* 27:794–6.

Seckbach, J., and Ikan, R. (1972). Sterols and chloroplast structure of *Cyanidium caldarium. Plant Physiol.* 49:457–9.

Sheath, R. G., and Hambrook, J. A. (1988). Mechanical adaptation to flow in freshwater algae. *J. Phycol.* 24:106–11.

Silva, P. C., and Johansen, H. W. (1986). A reappraisal of the order Corallinales (Rhodophyceae). *Br. Phycol. J.* 21:245–54.

Simon-Bichard-Bréaud, J. (1971). Un appareil cinétique dans les gamétocystes mâles d'une Rhodophycée: *Bonnemaisonia hamifera* Hariot. *C. R. Séances Acad. Sci. Paris* 273:1272–5.

Skuja, H. (1938). Comments on fresh-water Rhodophyceae. *Bot. Rev.* 4:665–76.

Smith, G. M. (1938). *Cryptogamic Botany,* Vol. 1. New York: McGraw-Hill.

Smith, G. M. (1969). *Marine Algae of the Monterey Peninsula, California.* 2nd ed. Stanford, Calif.: Stanford University Press.

Sommerfeld, M. R., and Nichols, H. W. (1970). Comparative studies in the genus *Porphyridium* Naeg. *J. Phycol.* 6:67–78.

Sommerfeld, M. R., and Nichols, H. W. (1973). The life cycle of *Bangia fuscopurpurea* in culture. I. Effects of temperature and photoperiod on the morphology and reproduction of the *Bangia* phase. *J. Phycol.* 9:205–10.

Sturch, H. H. (1926). *Choreocolax Polysiphoniae* Reinsch. *Ann. Bot.* 40:585–605.

Suneson, S. (1937). Studien über die Entwicklungsgeschichte der Corallinaceen. *Lunds Univ. Arsskr. N.F. II* 33, No. 2:1–132.

Suneson, S. (1943). The structure, life-history and taxonomy of the Swedish Corallinaceae, *Lunds Univ. Arsskr. N.F.* Aud. 2, Bd. 39.

Svedelius, N. (1942). Zytologisch-entwickungsgechichtliche Studien über *Galaxaura* eine diplobiontische Nemalionales-Gattung. *Nova Acta Regiae Soc. Sci. Up., Ser.* 4 13:5–154.

Swale, E. M. F., and Belcher, J. H. (1963). Morphological observations on wild and cultured material of *Rhodochorton investiens* (Lenormand) nov. comb. (*Balbiana investiens* (Lenorm.) Sirodot). *Ann. Bot.* 27:281–90.

Taylor, W. R. (1957). *Marine Algae of the Northeastern Coast of North America.* Ann Arbor: University of Michigan Press.

Therkelsen, G. H. (1993). Carrageenan. In *Industrial Gums: Polysaccharides and Their Derivatives.* 3rd. Ed., ed. R. L. Whistler and J. N. BeMiller, pp. 145–80. San Diego: Academie Press.

Tripodi, G. (1974). Ultrastructural changes during carpospore formation in the red alga *Polysiphonia. J. Submicrosc. Cytol.* 6:275–86.

Tseng, C. K. (1981). Commercial cultivation. In *The Biology Of Seaweeds* ed. C. S. Lobban, and M. J. Wynne. Berkeley and Los Angeles: Univ. Calif. Press.

Turner, C. H. C., and Evans, L. V. (1978). Translocation of photoassimilated ^{14}C in the red alga *Polysiphonia lanosa. Br. Phycol. J.* 13:51–5.

Umezaki, I. (1967). The tetrasporophyte of *Nemalion vermiculare* Suringar. *Rev. Algol.* 7:19–24.

Velasquez, G. T. (1972). Studies and utilization of the Philippine Marine Algae. *Proc. 7th Int. Seaweed Symp., Tokyo*, pp. 62–5.

von Stosch, H. A., and Theil, G. (1979). A new mode of life history in the freshwater red algal genus *Batrachospermum. Am. J. Bot.* 66:105–7.

Vreeland, V., Zablackis, E., and Laetsch, W. M. (1992). Monoclonal antibodies as molecular markers for the intracellular and cell wall distribution of carrageenan epitopes in *Kappaphycus* (Rhodophyta) during tissue development. *J. Phycol.* 28:328–42.

Waaland, J. R., Waaland, S. D., and Bates, G. (1974). Chloroplast structure and pigment composition in the red alga *Griffithsia pacifica:* Regulation by light intensity. *J. Phycol.* 10:193–9.

West, J. A. (1972). Environmental regulation of reproduction *Rhodochorton purpureum.* In *Contributions to the Systematics of Benthic Marine Algae of the North Pacific*, ed. I. A. Abbott, and M. Kurogi, pp. 213–30. Kobe: Jpn. Soc. Phycol.

Wolk, C. P. (1968). Role of bromine in the formation of the refractile inclusions of the vesicle cells of the Bonnemaisoniaceae (Rhodophyta). *Planta* 78:371–8.

Womersley, H. B. S. (1965). The Helminthocladiaceae (Rhodophyta) of Southern Australia. *Aust. J. Bot.* 13:451–87.

Yamanouchi, S. (1906). The life history of *Polysiphonia violacea. Bot. Gaz.* 42:401–49.

Young, D. N. (1979). Fine structure of the "gland cells" of the red alga, *Opuntiella californica* (Solieriaceae, Gigartinales). *Phycologia* 18:288–95.

Young, D. N., and West, J. A. (1979). Fine structure and histochemistry of vesicle cells of the red alga *Antithamnion defectum* (Ceramiaceae). *J. Phycol.* 15:49–57.

5 · Chlorophyta

The Chlorophyta, or **green algae**, have chlorophylls *a* and *b*, and form starch with the chloroplast, usually in association with a pyrenoid. The Chlorophyta thus differ from the rest of the eukaryotic algae in forming the storage product in the chloroplast instead of in the cytoplasm. No chloroplast endoplasmic reticulum occurs around the chloroplasts.

The Chlorophyta are primarily freshwater; only about 10% of the algae are marine, whereas 90% are freshwater (Smith, 1955). Some orders are predominantly marine (Caulerpales, Dasycladales, Siphonocladales), whereas others are predominantly freshwater (Ulotrichales, Coleochaetales) or exclusively freshwater (Oedogoniales, Zygnematales). The freshwater species have a cosmopolitan distribution, with few species endemic in a certain area. In the marine environment, the green algae in the warmer tropical and semi-tropical waters tend to be similar everywhere in the world. This is not true of the Chlorophyta in the colder marine waters; the waters of the Northern and Southern hemispheres have markedly different species. The warmer waters near the equator have acted as a geographical barrier for the evolution of new species and genera.

Cell structure

In the Chlorophyta, microtubular hairs do not occur on the flagella, although fibrillar hairs (*Chlamydomonas*, Fig. 1.4(*b*)) and Golgi-produced scales (*Pyramimonas*, Fig. 5.10) are present in some genera.

Cell walls usually have **cellulose** as the main structural polysaccharide, although **xylans** or **mannans** often replace cellulose in the Caulerpales (Huizing et al., 1979). The primitive algae in the Micromonadophyceae have extracellular scales, or a wall derived from interlacing scales, composed of acidic polysaccharides (Becker et al., 1996). Algae in the Volvocales have walls composed of glycoproteins (Goodenough and Heuser, 1985).

Chloroplast pigments are similar to those of higher plants; chlorophylls *a* and *b* are present, and the main carotenoid is **lutein**. The siphonaceous genera and *Tetraselmis* are the only green algae to have **siphonoxanthin** and its ester **siphonein**. In addition to the above chloroplast carotenoids,

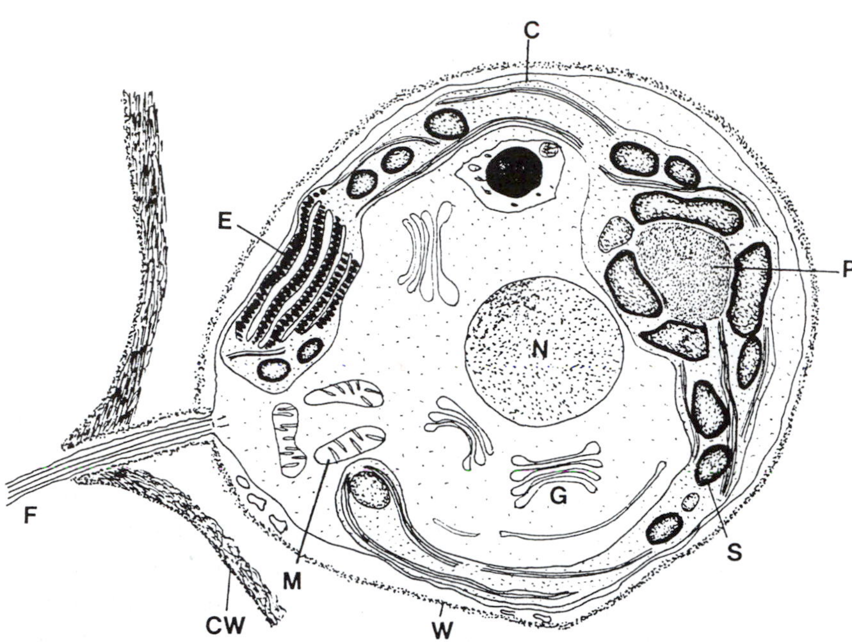

Fig. 5.1 Semi-diagrammatic drawing of a cell in a *Volvox* vegetative colony. The colony wall (CW) is distinct from the cell wall (W). (C) Chloroplast; (E) eyespot; (F) flagellum; (G) Golgi; (M) mitochondrion; (N) nucleus; (P) pyrenoid; (S) starch. (Adapted from Pickett-Heaps, 1970.)

accumulation of carotenoids occurs outside the chloroplast under conditions of nitrogen deficiency, coloring the alga orange or red. The carotenoid is usually β-carotene and/or a keto derivative (**hematochrome** is a general term for these carotenoids and/or their derivatives). Extra-plastidic carotenoids are often present in zygotes and sexual reproductive structures.

Chloroplasts are surrounded only by the double-membrane chloroplast envelope, with no chloroplast endoplasmic reticulum present (Fig. 5.1). The thylakoids are grouped into bands of three to five thylakoids without grana. In some of the siphonaceous genera (e.g., *Caulerpa*), amyloplasts containing starch grains and few thylakoids are found within the chloroplasts.

Starch is formed in the chloroplast, in association with a pyrenoid if one is present (Fig. 5.1). The starch is similar to that of higher plants and is composed of amylose and amylopectin. The photosynthetic pathways are similar to those of higher plants, many of these pathways first being worked out in green algae such as *Chlorella*.

Contractile vacuoles are present in vegetative cells of most Volvocales. Usually in biflagellate genera there are two contractile vacuoles at the base of the flagella. When there are two contractile vacuoles, they contract alternately with a rapid contraction and slow distention. The contractile vacuoles may control the water content of the cells where the protoplasm has a higher concentration of solutes than does the medium, leading to a total inflow of water that is compensated by the water pumped out by the contractile vacuoles. The contractile vacuoles may also function in removing wastes from the cells. Contractile vacuoles are sometimes called pulsating vacuoles because of their alternate filling and emptying action.

Phototaxis and eyespots

There are two types of phototactic movement in the Chlorophyta: movement by flagella and movement by the secretion of mucilage.

Most of the flagellated cells that show phototactic movement have an eyespot. In the Chlorophyta, the eyespot or stigma is always in the chloroplast, usually in the anterior portion near the flagella bases (Fig. 5.1). The eyespot consists of one to a number of layers of lipid droplets, usually in the stroma between the chloroplast envelope and the outermost band of thylakoids. The eyespot is usually colored orange-red from the carotenoids in the lipid droplets. The phototactic response varies with light intensity; Strasburger in 1878 (Bendix, 1960) observed that organisms with positive phototaxis at moderate light intensities exhibited negative phototaxis at very high light intensities. He also noted that at a given light intensity, temperature will have an effect on phototaxis. *Haematococcus* zoospores at a given light intensity will be negatively phototactic at 4 °C, positively phototactic at 16 to 18 °C, and very strongly positively phototactic at 35 °C. Similar results were obtained with *Ulothrix* and *Ulva*.

Green algae use a two-instant mechanism for perceiving light (i.e., successive measurements of light are performed by one receptor as the cell changes its position in relation to the light source) (Boscov and Feinleib, 1979). Such a mechanism can operate only if the cell frequently changes its position with respect to the light source. The photoreceptor then compares the light intensity at two different time intervals. The photoreceptor in *Chlamydomonas* is in the plasma membrane above the eyespot (Melkonian and Robenek, 1980b) and consists of a **chromophore** (colored substance) linked to a protein. The chromophore is **11-*cis*-retinol** (Fig. 5.2), a rhodopsin that functions as a photoreceptor in animals (Kroger and Hegemann, 1994; Schlicher et al., 1995). The eyespot acts as an interference filter by reflecting blue and green light back onto the photoreceptor in the plasma membrane (Kriemer and Melkonian, 1990). Different amounts of light are reflected onto the photoreceptor as the alga swims through the medium. This results in changes in membrane potential involving rhodopsin. Entry of calcium into the cell is affected by the membrane potential of the plasma membrane, and, in turn, the concentration of calcium ions in the cytoplasm affects the rate of beating of the flagella. The

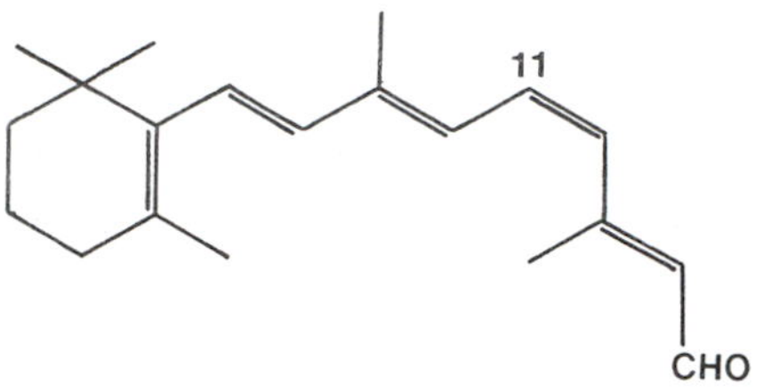

Fig. 5.2 The structure of 11-*cis*-retinol, part of the photoreceptor molecule in *Chlamydomonas*.

swimming direction of the cell is affected by the rate of beating, because at one concentration of calcium ions, each flagellum beats differently (Kamiya and Witman, 1984). Therefore, changing the cytoplasmic calcium concentration differentially changes the beat of each flagellum, causing the cell to swim in a different direction. Melkonian and Robenek (1980b), using freeze-fracture replicas, have shown that the plasma membrane and outer membrane of the chloroplast envelope have more particles in the areas of the membranes over the eyespot than in other areas. The portion of the plasma membrane overlying the eyespot of *Chlorosarcinopsis gelatinosa* has 8200 protein particles μm^{-2} in the half of the plasma membrane next to the cytoplasm (Melkonian and Robenek, 1980a). The plasma membrane not over the eyespot has only 2100 particles μm^{-2}. These protein particles in the membranes over the eyespot probably represent a part of the photoreceptor system because they disappear during flagellar retraction and before cell wall secretion, a time when the photoreceptor system would be of no use.

A second type of phototactic movement in the Chlorophyta uses secretion of mucilage. In 1848, Ralfs, in his monograph on desmids, described their movement to the surface of mud brought into the laboratory, and presumed this movement to be due to the stimulus of light. Braun (1851) noticed that young cells of *Penium curtum* quickly aligned their long axis and moved toward the light, accumulating on the lighted side of the vessel they were growing in. The movement in desmids is brought about by the extrusion of slime through cell wall pores in the apical part of the cell (Domozych et al., 1993; Nossag and Kasprik, 1993).

Green algae can also show geotactic responses to gravity. *Chlamydomonas* exhibits negative geotaxis by swimming against gravity. Such a feature would be selected for in evolution because when the algal cell (which is heavier than water) is confronted with darkness, it must move up to the surface in order to obtain light for growth and reproduction. In *Chlamydomonas*, negative geotaxis is an energy-dependent response that requires a horizontal swimming path of at least 200 μm because the normal geotactic orientation maneuvers require long gradual turns. The rate of geotaxis is steady but slow relative to the average swimming speed (Bean, 1977).

Asexual reproduction

There are a number of types of asexual reproduction, the simplest being **fragmentation** of colonies into two or more parts, each part becoming a new colony. **Zoosporogenesis** commonly occurs, usually induced by a change in the environment of the alga. In the Chlorophyta, **zoospores** are normally produced in vegetative cells (e.g., *Ulothrix*, Fig. 5.27), and only in a few cases are they formed in specialized sporangia (e.g., *Derbesia*, Fig. 5.34). Zoospores are usually formed in the younger parts of filaments, and the number of zoospores is generally a power of two in uninucleate genera. **Aplanospores** are non-flagellated and have a wall distinct from the parent cell wall (e.g., *Trebouxia*, Fig. 5.55). Aplanospores are considered to be abortive zoospores and have the ability to form a new plant on germination. **Autospores** are aplanospores that have the same shape as the parent cell, and are common in the Chlorococcales (e.g., *Chlorella*, Fig. 5.57). Autospores are usually formed in a multiple of two in the parent cell. **Coenobia** are colonies with a definite number of cells arranged in a specific manner (e.g., *Volvox*, Fig. 5.48). Genera with colonies arranged in coenobia form daughter colonies with a certain number of cells. In maturation of the daughter coenobia, there is enlargement but no division of the vegetative cells in the coenobia.

Sexual reproduction

Sexual reproduction in the Chlorophyceae may be isogamous, anisogamous, or oögamous, with the general line of evolution occurring in the same direction. Usually gametes are specialized cells and not vegetative cells, although in the one-celled Volvocales the latter can occur. If the species is isogamous or anisogamous, the gametes are usually not formed in specialized cells although in the oogamous species, gametes are normally formed in specialized gametangia (e.g., *Coleochaete*, Fig. 5.22). Whereas most Chlorophyta form motile flagellated gametes (zoogametes), in the Zygnematales aplanogametes or amoeboid gametes are formed.

In some of the Chlorophyta, gametogenesis is induced by environmental changes, whereas in others the presence of two sexually different strains is necessary. In the latter, vegetative cells of one sex secrete a substance that initiates sexual differentiation in competent cells of the opposite sex. Such a situation is common in the Volvocales (Starr, 1972; Kirk and Kirk, 1986) and is considered in more detail later. In *Oedogonium*, sex organs form without the complementary strain, but subsequent fertilization is under a complex hormonal control. In other genera, a chemotactic substance is sometimes produced by the egg that attracts the spermatozoids. This does not generally happen in isogamous species. In isogamous species, sexually different gametes meet at random and immediately adhere by means of an **agglutination** reaction. The agglutinative flagellar adhesion between gametes of different sex is designated as the **mating-type reaction**. Initially after mixing, the gametes of opposite sexes adhere by their flagella tips in clusters of up to 25 gametes. Soon the anterior ends of complementary gametes fuse, and the flagella free themselves. The motile zygote then swims for some time before settling and secreting a thick wall.

The mating-type substances (responsible for flagellar agglutination) are localized and function at the flagella tips. It is possible to isolate the mating-type substances that still have the ability to interact with the gametes of the opposite sex. When added to the opposite gamete type, they cause **iso-agglutination** (male gametes will clump with each other when a female mating-type substance is added to the culture). The mating-type substances are discussed in more detail for *Chlamydomonas*, which has been most intensively studied.

Soon after the gametes fuse (**syngamy**), meiosis is known to occur in the thick-walled zygotes of the Volvocales, Ulotrichales, Oedogoniales, Chlorococcales, and Zygnematales.

Classification

The four important classes in the Chlorophyta are the Micromonadophyceae, Charophyceae, Ulvophyceae, and the Chlorophyceae. The distinguishing characteristics of each group are outlined below and in Table 5.1; a more detailed explanation of the characteristics used to delineate the class follows the outline.

Class 1 **Micromonadophyceae:** scaly or naked flagellates with interzonal spindles that are persistent during cytokinesis; primitive green algae, some of which gave rise to the other classes in the Chlorophyta.

Class 2 **Charophyceae:** motile cells asymmetrical; two flagella attached in a lateral position in cell; flagellar root consisting of a broad band of microtubules and a second smaller microtubular root; multilayered structure (MLS) may be present; no rhizoplast; scales common outside of motile cells; persistent interzonal mitotic spindle in telophase; phragmoplast produces new cross walls after cell division; eyespots usually not present; glycolate broken down by glycolate oxidase; urea broken down by urease; predominantly freshwater; sexual reproduction involves the formation of a dormant zygote; meiosis occurs when the zygote germinates.

Class 3 **Ulvophyceae:** flagella attached at anterior end of cell; motile cells

Table 5.1 *Characteristics of the four classes of green algae*

	Micromonadophyceae	Charophyceae	Ulvophyceae	Chlorophyceae
Position of flagella in cell		Lateral	Anterior	Anterior
Microtubular root		Large band with smaller band	Four, cruciately arranged	Four, cruciately arranged
Rhizoplast	May be present	No	Common	Common
Multilayered structure	May be present	Yes	No	No
Covering on motile cells	Scales	Scales	Scales	Theca
Interzonal spindle	Persistent	Persistent	Persistent	Collapsing
New cross wall formation		Phragmoplast	Cleavage furrow	Phycoplast
Cellulose terminal complex		Rosettes	Linear row	Linear row
Eyespot		None	Common	Common
Glycolate degradation	Glycolate dehydrogenase	Glycolate oxidase	Glycolate dehydrogenase	Glycolate dehydrogenase
Urea degradation		Urease	Urease	Urea amidolyase
Cu/Zn superoxide dismutase		Present	Absent	Absent

have near-radial symmetry externally; flagella roots consist of four cruciately arranged microtubular roots and sometimes a rhizoplast; no multilayered structure (MLS); scales may be present on motile cells; persistent interzonal spindle in telophase; cleavage furrow produces the new cross wall in cell division; eyespots common; glycolate broken down by glycolate dehydrogenase; urea broken down by urease; predominantly marine; no dormant zygotes; alternation of generations common.

Class 4 **Chlorophyceae:** motile cells with radial or near-radial external symmetry; flagella attached at anterior end of cell; flagella roots consist of four cruciately arranged microtubular roots and sometimes a rhizoplast; no multilayered structure (MLS); theca common in motile cells; in telophase, the interzonal spindle collapses; phycoplast produces the new cross wall in cell division; eyespots common; glycolate breakdown by glycolate dehydrogenase; urea broken down by urea amidolyase; predominantly freshwater; zygote undergoes a dormant period; meiosis occurs when the zygote germinates.

POSITION OF FLAGELLA IN CELLS

In the **Charophyceae**, the flagella are attached in a lateral position in the cell (Fig. 5.3). In the **Ulvophyceae** and the **Chlorophyceae**, the flagella are attached at the anterior end of the cell.

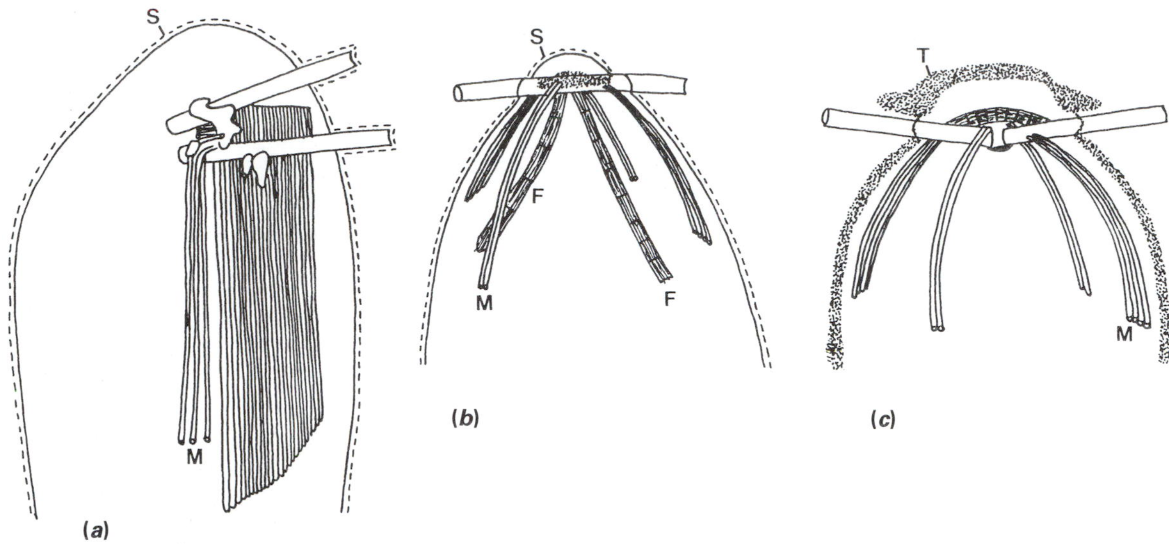

Fig. 5.3 Schematic drawings of the side view of swarmers produced by three of the classes of green algae. (*a*) Charophyceae: scaly cell with one large root (the MLS), one smaller root, and flagella extending at an angle from the point of insertion. (*b*) Ulvophyceae: four microtubular roots (two of each kind) in a cruciate arrangement, a pair of fibrous roots and a scaly covering over the cell. (*c*) Chlorophyceae: cruciate roots and the cell covered with a theca. (F) Fibrous roots; (M) microtubular root; (S) scales, (T) theca. (After Mattox and Stewart, 1984.)

FLAGELLAR ROOTS

Flagellar basal bodies are anchored in the protoplast by **microtubular roots** and/or **rhizoplasts (fibrous roots)** (Fig. 5.3) (Melkonian et al., 1988).

Microtubular roots

Microtubular roots consist of groups of 24-nm diameter microtubules that can have one of two basic configurations: (1) There can be a microtubular root consisting of a *large broad band of microtubules with a smaller second microtubular root* (**Charophyceae**), or (2) there can be *four groups of cruciately arranged microtubular roots* running from the basal bodies (**Ulvophyceae** and **Chlorophyceae**). The cruciately arranged microtubular roots have what is called an X-2-X-2 arrangement. This notation refers to the fact that two of the microtubular roots are usually composed of two microtubules, whereas the two other roots can have different numbers of microtubules in different organisms. Thus *Chlamydomonas moewusii* has a 4-2-4-2 arrangement, whereas motile cells of *Ulothrix* sp. have a 5-2-5-2 arrangement (Moestrup, 1978). One of the roots containing two microtubules is often linked to the outer membrane of the chloroplast envelope and is probably involved in phototaxis.

Rhizoplasts (fibrous roots)

A rhizoplast is usually a cylinder containing 5- to 10-nm-diameter filaments interrupted at approximately 80-nm intervals by bands of electron-dense

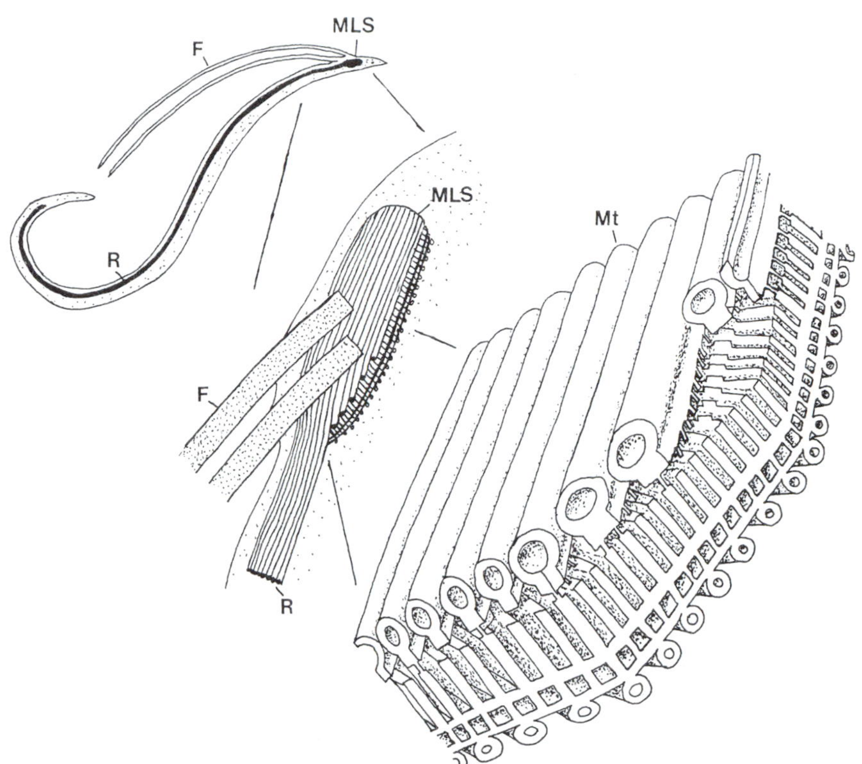

Fig. 5.4 Drawing of a green algal cell containing a multilayered structure. (F) Flagellum; (Mt) microtubule; (MLS) multilayered structure; (R) microtubular root. (Adapted from Carothers and Kreitner, 1967.)

material (in the electron microscope). A rhizoplast runs from the basal bodies posteriorly toward the nucleus. Rhizoplasts are contractile (Salisbury and Floyd, 1978), and the distance between the bands in the rhizoplast varies depending on the state of contraction of the rhizoplast. The size of the filaments in the rhizoplast is similar to that of actin–myosin filaments in animal muscle cells. The method of contraction of the rhizoplast may be similar to that of muscle. Rhizoplasts may be **present** in the **Micromonadophyceae**, **Chlorophyceae**, and **Ulvophyceae**, but are **absent** in the **Charophyceae**.

MULTILAYERED STRUCTURE (MLS)

A multilayered structure (MLS) (Fig. 5.4) consists of a more or less rectangular body attached to the anterior end of the single broad band of microtubules in the Charophyceae and in the spermatozoids of lower land plants. The MLS lies directly beneath the basal bodies of the flagella. The MLS consists of four layers. The layer closest to the plasma membrane contains the microtubules of the root. Under this are two electron-dense layers. The bottom-most layer is composed of small microtubules. An MLS

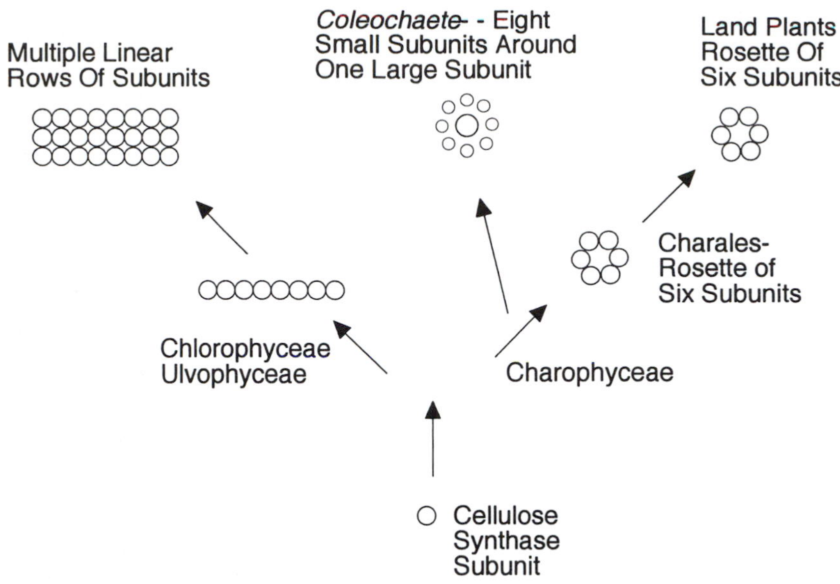

Fig. 5.5 Proteins of cellulose synthase in the plasma membrane have aggregated into terminal complexes along two phylogenetic lines in the Chlorophyta. Rosettes of cellulose synthase proteins have evolved in the Charophyceae, while aggregations of linear rows have evolved in the Chlorophyceae and Ulvophycease. (Adapted from Okuda and Brown, 1992.)

may be present in the **Micromonadophyceae** and **Charophyceae**, but is **absent** in the **Chlorophyceae** and **Ulvophyceae**.

OCCURRENCE OF SCALES OR A WALL ON THE MOTILE CELLS

Motile cells covered with scales may occur in the **Micromonadophyceae**, **Charophyceae**, and **Ulvophyceae**. The presence of scales on the motile cells is probably the primitive condition. As evolution progressed, the scales became interweaved along their edges so a coherent cell covering was formed, as in the genus *Tetraselmis* (Domozych et al., 1981; Mattox and Stewart, 1984). The end result of this evolution was the **theca** that covers the motile cells in the **Chlorophyceae**. This theca has a crystalline substructure and is composed of hydroxyproline-rich glycoproteins associated with various polysaccharides (Roberts, 1974; Miller, 1978; Deason, 1983). The theca in motile cells of the Chlorophyceae is thus not to be confused with the cell walls of non-motile stages of the more advanced Chlorophyta, which have **cellulose** as the main skeletal molecule.

Cellulose is produced by the enzyme **cellulose synthase** that occurs as proteins embedded in the plasma membrane of the cell (Fig. 5.5). Six to ten cellulose synthetase molecules are grouped into a single subunit. The subunits are, in turn, aggregated into **terminal complexes**. In the Chlorophyta, there are two different types of terminal complexes (Fig. 5.5) (Okuda and Mizura, 1993; Okuda et al., 1994):

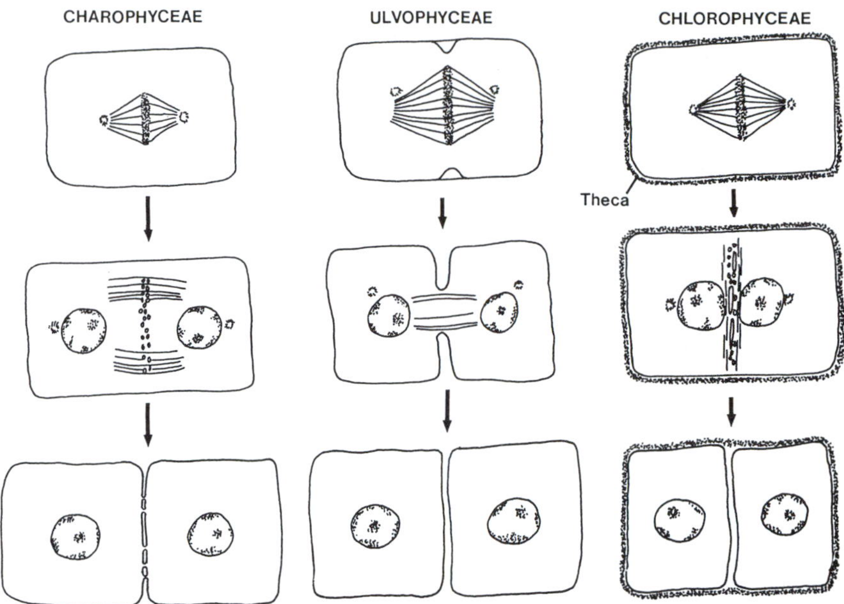

Fig. 5.6 Schematic drawing of the type of cell division in the Charophyceae, Ulvophyceae, and Chlorophyceae.

1 In the Charophyceae, terminal complexes have subunits aggregated into rosettes.
2 In the Chlorophyceae and Ulvophyceae, terminal complexes consist of linear rows of subunits.

CELL DIVISION

Two types of interzonal spindles occur in telophase cells of the Chlorophyta: the **persistent type** and the **collapsing type**. The new cell wall can be formed between the daughter cells by means of a **phragmoplast**, a **cleavage furrow**, or a **phycoplast**. Three of the classes in the Chlorophyta are characterized by the type of spindle in the class and the way the new cross wall is formed (Fig. 5.6).

Persistent interzonal spindle and phragmoplast (Charophyceae)

In the **Charophyceae**, the **microtubular spindle persists** even after the daughter nuclei have separated in telophase. The daughter nuclei are separated by the length of the persistent spindle while a new cross wall is formed by a **phragmoplast** in the more advanced members. Wall formation by a phragmoplast initially involves the production of vesicles by the dictyosomes. The vesicles contain the components of the new cross walls. The persistent spindle microtubules may function in guiding the vesicles to the area of the new cross wall which will separate the two daughter cells. The vesicles fuse, releasing their contents which form the new cross wall. Plasmodesmata are formed in the cross wall between the daughter cells where the persistent spindle microtubules traverse the cross wall. In some of the primitive Charophyceae, the new cross wall is formed by a phragmoplast in association with an infurrowing of the plasma membrane. In the more advanced Charophyceae (those in the Coleochaetales and Charales), the cross wall is formed only by a phragmoplast.

Persistent interzonal spindle and a cleavage furrow (Ulvophyceae)

In the **Ulvophyceae**, the **interzonal spindle persists** during telophase, holding the daughter cells apart while the new cross wall is formed by an **infurrowing of the plasma membrane**. As the plasma membrane furrows inward, dictyosome vesicles fuse with the plasma membrane, behind the infurrowing, producing the new cross wall.

Collapsing interzonal spindle and a cleavage furrow (Chlorophyceae)

In the **Chlorophyceae**, the **mitotic spindle collapses** after nuclear division. This results in the two daughter nuclei coming close together in telophase because there is no longer a persistent interzonal spindle to hold the nuclei apart. The position of the new cross walls becomes outlined by microtubules of the **phycoplast** that arise perpendicular to the former position of the spindle microtubules. Dictyosome vesicles fuse between the phycoplast microtubules, forming the new cross wall. The Chlorophyceae have motile cells with a cell wall (theca) and thus differ from the Ulvophyceae and Charophyceae. The phycoplast evolved in conjunction with the production of cell walls in these algae (Mattox and Stewart, 1977). During cell division in cells with a persistent spindle (in the Micromonadophyceae, Charophyceae, and Ulvophyceae), there is extensive elongation of the cell during anaphase. This cell elongation presents no problem in naked cells or cells covered with scales. However, in the Chlorophyceae the cells are not easily able to elongate in response to the anaphase elongation of the persistent spindle. The reason is that the Chlorophyceae have walls around the motile cells. Therefore, evolution of the phycoplast and collapsing spindle, which does not involve rapid elongation of the daughter cells, presents an evolutionary advantage.

GLYCOLATE DEGRADATION

Glycolate, the major substrate of photorespiration, is derived from phosphoglycolate, which is formed by the oxygenation of ribulose-1,5-diphosphate:

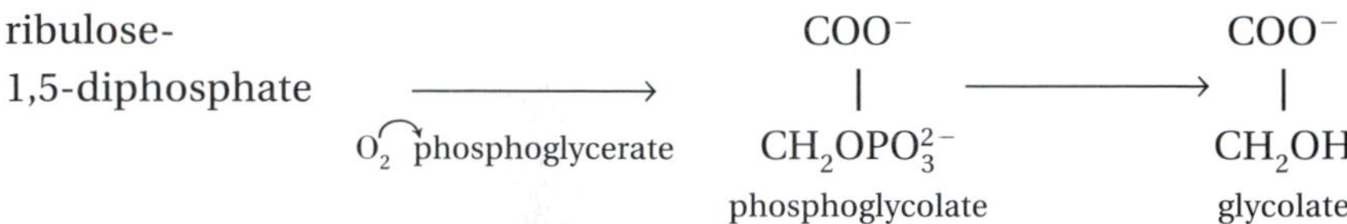

The glycolate is metabolized in microbodies called **peroxisomes**. In the peroxisomes, glycolate is oxidized to glyoxylate. The H_2O_2 produced in the reaction is cleaved by the enzyme **catalase** to H_2O and O_2:

$$\underset{\text{glycolate}}{\begin{matrix} COO^- \\ | \\ CH_2OH \end{matrix}} \xrightarrow{O_2 \curvearrowright H_2O_2} \underset{\text{glyoxylate}}{\begin{matrix} COO^- \\ | \\ C-H \\ \| \\ O \end{matrix}}$$

The above oxidation of glycolate can be catalyzed by the enzyme **glycolate dehydrogenase** or the enzyme **glycolate oxidase** (Gruber et al., 1974). In the **Charophyceae**, the reaction is catalyzed by **glycolate oxidase**, whereas in the **Chlorophyceae** and the **Ulvophyceae**, the reaction is catalyzed by **glycolate dehydrogenase** (Suzuki et al., 1991). Glycolate oxidase probably represents the primitive condition since *Cyanophora paradoxa* in the Glaucophyta catalyzes the reaction with this enzyme (Betsche et al., 1992).

UREA DEGRADATION

In the **Charophyceae**, **Ulvophyceae**, and **higher plants**, urea is broken down by the enzyme **urease**:

$$\underset{\text{urea}}{H_2N-\overset{\overset{\displaystyle O}{\|}}{C}-NH_2} + 2H_2O \xrightarrow[\text{urease}]{} 2NH_4^+ + CO_2$$

In the **Chlorophyceae**, urea is broken down by the enzyme **urea amidolyase** (Syrett and Al-Houty, 1984):

$$\underset{\text{urea}}{H_2N-\overset{\overset{\displaystyle O}{\|}}{C}-NH_2} + ATP \underset{\substack{HCO_3^- \\ \text{urea} \\ \text{amidolyase}}}{\overset{Mg^{2+}, K^+}{\rightleftharpoons}} 2NH_4^+ + HCO_3^- + ADP + P_i$$

SUPEROXIDE DISMUTASE

Superoxide dismutases (SOD) are a group of enzymes that catalyze the reaction:

$$O_2^- + O_2^- + 2H^+ \rightarrow H_2O_2 + O_2$$

Superoxide dismutases are important because they take highly reactive, potentially damaging oxygen radicals (O_2^-) and other toxic oxygen species (OH·, singlet oxygen) and convert them to less toxic moieties.

There are three forms of superoxide dismutases, named for the metal in their reaction centers: **iron SOD** (FeSOD), **manganese SOD** (MnSOD) and **copper-zinc SOD** (Cu/Zn SOD). The amino acid sequences and three-dimensional protein structure of FeSOD and MnSOD are very similar, suggesting a close evolutionary relationship, whereas those of Cu/Zn SOD are quite different, indicating very little, if any, evolutionary relationship to the other two (deJesus et al., 1989).

FeSOD and MnSOD occur in all classes of green algae and most land plants. Cu/Zn SOD, however, occurs only in the Charophyceae and land plants, reinforcing the concept that organisms similar to those in the Charophyceae evolved into land plants.

Cu/Zn SOD is the only superoxide dismutase that occurs in the cytosol. MnSOD occurs in the mitochondrion while FeSOD occurs in the chloroplast. de Jesus, Tabatabai and Chapman (1989) argue that green algae evolving to land plants would need superoxide dismutase in the cytosol to counteract oxygen diffusing in from the atmosphere, which would produce damaging radicals in the cytosol. The Charophycean lineage of green algae is the only one to have superoxide dismutase in the cytosol and is a major reason for the successful colonization of the land by descendants of the Charophyceae. The other lineages of green algae were unable to make the transition to the land because they had superoxide dismutase restricted to mitochondria, chloroplasts and peroxisomes and were not able to counteract the production of damaging radicals formed by oxygen diffusing in from the atmosphere.

Ancestral green flagellate

The green algae evolved from an ancestral green flagellate which had the following characteristics (Melkonian, 1979; Mattox and Stewart, 1984; O'Kelly and Floyd, 1984):

1 A cell with an **asymmetrical configuration**. If the cell was cut in half in any direction, the two halves were not the same.

2 The flagella and the cell body were covered with **scales**.
3 The **flagella were inserted subapically** and not right at the anterior end of the cell. The flagella were also probably inserted into a depression in the cell.
4 The flagella were anchored in the cell by a **single band of microtubules** and a **rhizoplast**.
5 A **multilayered structure** (MLS) was associated with the microtubular root.
6 The **interzonal spindle was persistent** during telophase.
7 Urea was degraded by **urease**.

The ancestral green flagellate was probably similar to some of the primitive green algae in the Micromonadophyceae (Kantz et al., 1990). The Charophyceae evolved more or less directly from the ancestral green flagellate to the land plants, with little change in basic cell structure (Fig. 5.7) (McCourt, 1995). The rhizoplast was the only major cytoplasmic structure lost in this evolutionary line to the Charophyceae. Within the Charophyceae, an alga similar to *Coleochaetae* probably evolved into the Bryophytes, while an alga similar to *Chara* or *Nitella* probably evolved to the remainder of the land plants (Okuda and Brown, 1992).

In the evolution of the Ulvophyceae and the Chlorophyceae, a cruciate root system developed and a multilayered structure (MLS) was lost. In the Ulvophyceae, cell walls evolved in non-motile stages. In the Chlorophyceae, evolution involved the aggregation of scales to form a theca and the collapse of the spindle before new cross walls were formed by a phycoplast.

Micromonadophyceae

The Micromonadophyceae contain primitive green flagellates, usually covered with scales composed of acidic polysaccharides (Becker et al., 1996). According to Mattox and Stewart (1984), these algae "exhibit primitive characteristics for many of the features used to define the other classes." This definition of the Micromonadophyceae does not result in a class with well-defined cytological and morphological features, other than the fact that as primitive green algae, members of the class are closely related to the ancestral green flagellate. Molecular data has shown that the algae placed in the class are closely related to each other (Daugbjerg et al., 1995).

Some phycologists take many of the organisms in the Micromonadophyceae and place them in a class called the Prasinophyceae, although molecular studies have not encouraged this view (Kim et al., 1994; Steinkötter et al., 1994).

Much of the interest in the Micromonadophyceae lies in the fact that

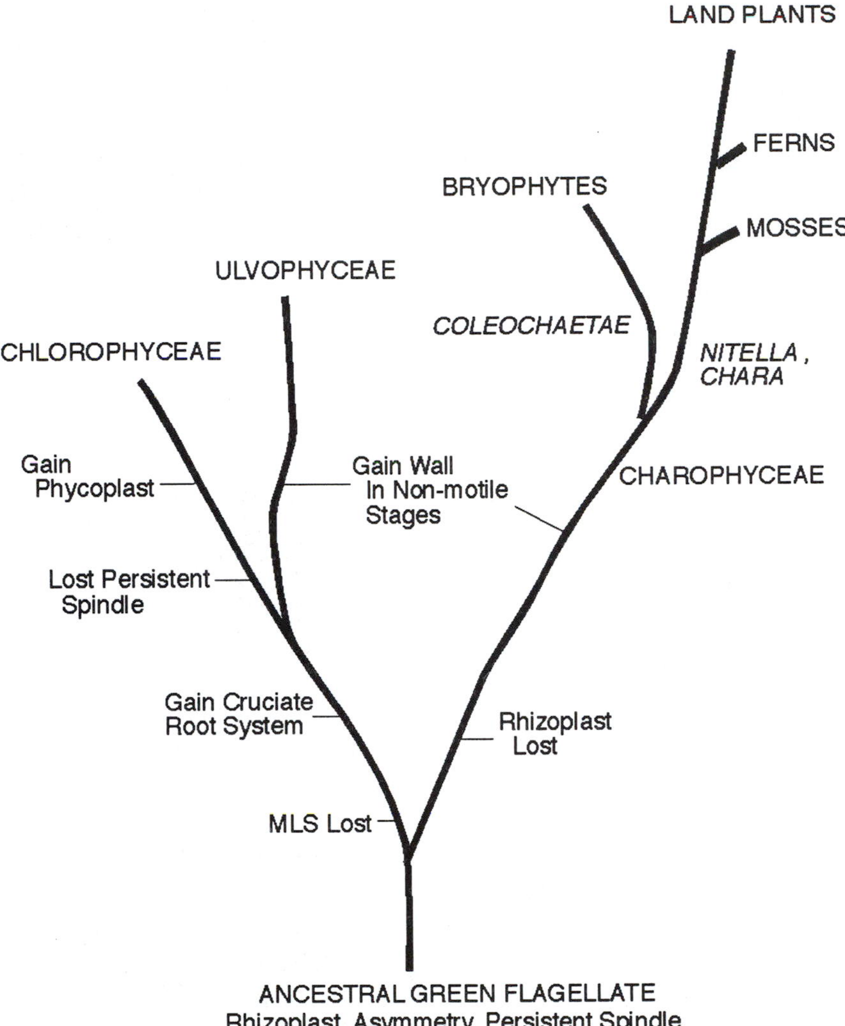

Fig. 5.7 Drawing of the evolution of the Bryophytes, land plants and other classes of green algae from an ancestral green flagellate similar to the flagellates that exist in the Micromonadophyceae today. (Modified from McCourt, 1995.)

they represent a pool of primitive genera that are probably similar to forms from which higher green algae sprang. *Mantoniella* (Fig. 5.8(*a*)) has characteristics that are considered to be the most primitive in the Micromonadophyceae and, therefore, the Chlorophyta. *Mantoniella* is a scaly green flagellate with two subapically attached flagella (one of the flagella is very short). There is one microtubular root composed of four microtubules and a second composed of two microtubules. Duplication of a microtubular root system like that in *Mantoniella* probably resulted in the cruciate root system of the Chlorophyceae.

Nephroselmis olivacea is bean shaped with two flagella of unequal length (Fig. 5.9). The cell contains a single cup-shaped chloroplast with a

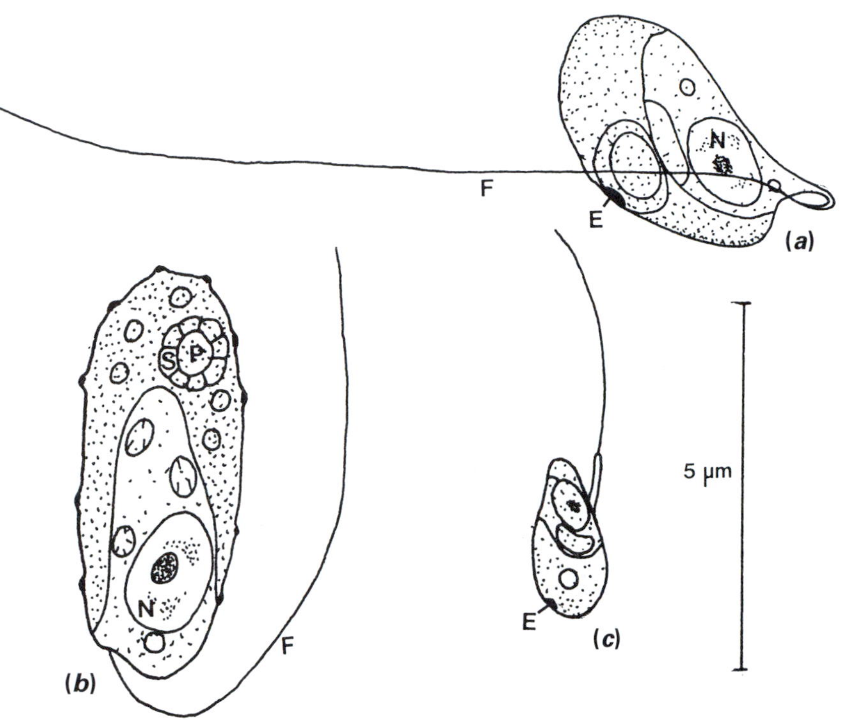

Fig. 5.8 Algal flagellates classified in the Micromonadophyceae. (*a*) *Mantoniella squamata*. The cell is covered with scales and has a second, very short flagellum, neither of which is shown. (*b*) *Pedinomonas tuberculata* (*c*) *Micromonas pusila*. (E) Eyespot; (F) flagellum; (N) nucleus; (P) pyrenoid; (S) starch. (After Manton and Parke, 1960; Barlow and Cattolico, 1980.)

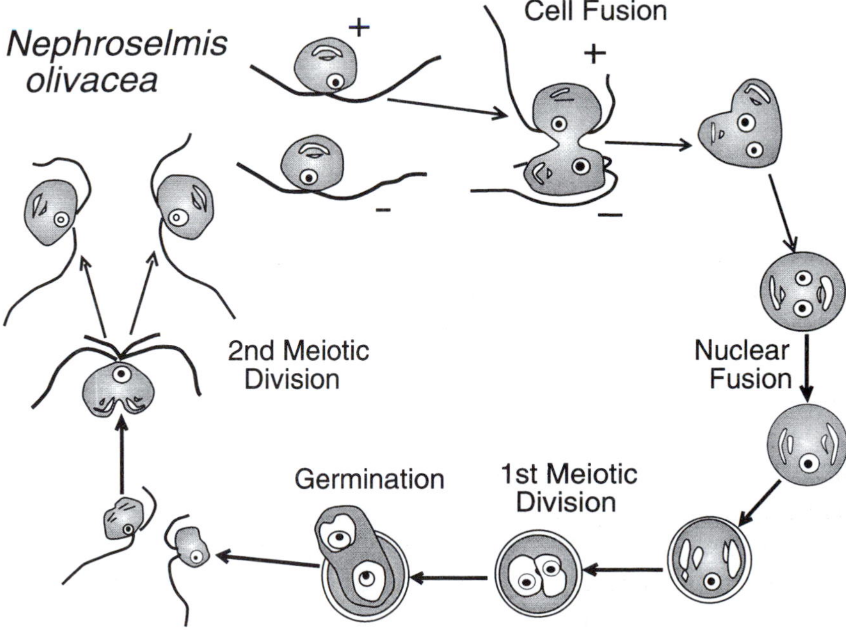

Fig. 5.9 The life cycle of *Nephroselmis olivacea*. (Adapted from Suda et al., 1989.)

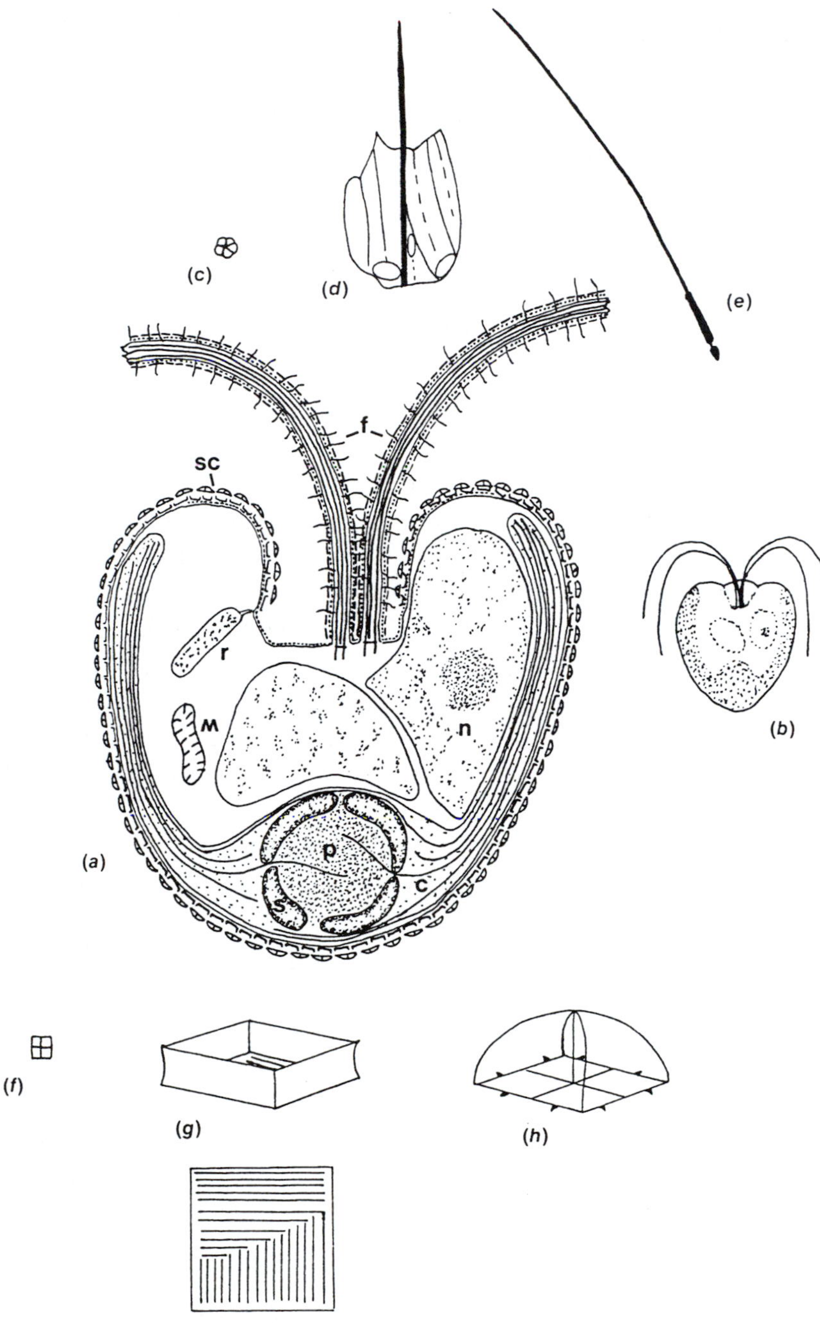

Fig. 5.10 *Pyramimonas obovata*. (*a*) Semi-diagrammatic drawing of cell showing the organelles, two of the four flagella, and the layers of scales on the body and the flagella. (*b*) Whole cell. (*c*) Inner flagellar scale. (*d*) Outer flagellar scale. (*e*) Flagellar hair. (*f*) Inner body scale. (*g*) Intermediate body scale, top and side view. (*h*) Outer body scale. (c) Chloroplast; (f) flagellum; (m) mitochondrion; (n) nucleus; (p) pyrenoid; (r) scale reservoir; (s) starch; (sc) scales. (After Belcher et al., 1974.)

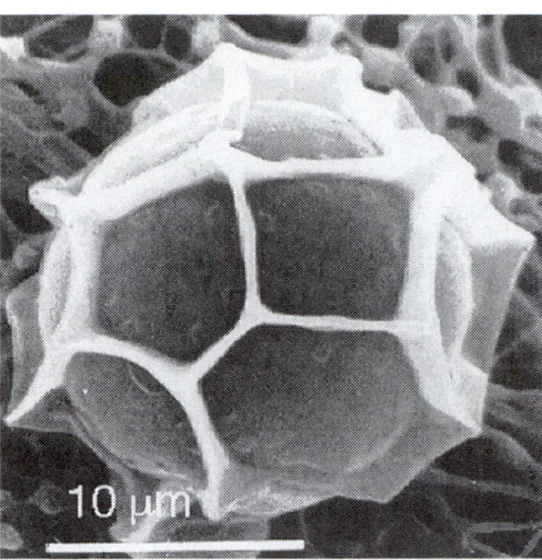

Fig. 5.11 A scanning electron micrograph of a phycoma of *Pterosperma cristatum*. Square, pentagonal or hexagonal compartments are former by wing-like protrusions of the cell wall. (From Inouye et al., 1990.)

single pyrenoid. The alga has a heterothallic sexual reproduction with plus and minus gametes that are morphologically similar. After fusion, the mature zygote germinates to produce two biflagellate daughter cells, each of which divides one further time (Suda et al., 1989).

Pyramimonas obovata is another alga in this class. The cells are heart-shaped (cordate), with four flagella arising from an anterior flat-bottomed depression (Fig. 5.10). There are three different layers of scales on the body and two layers on the flagellum (Belcher et al., 1974). The scales are formed in the dictyosomes, from which they are transported to the scale reservoir and then to the surface of the cell in vesicles.

Species of *Pyramimonas* have adapted to tidepools by means of a settling rhythm. Cells in tidepools anticipate the incoming tide and move down into the sand where they attach to sand grains by their flagella, before the first waves cover the tide pool. After the tide pool is exposed by the receding tide, the *Pyramimonas* cells move upward into the tide pool. The settling is controlled by an endogenous circadian oscillator and will continue in the laboratory for a few days after the cells are removed from the wild (Griffin and Aken, 1993).

Pterosperma is a genus that is closely related to *Pyramimonas. Pterosperma* produces walled cysts called *phycomata* (Fig. 5.11) that are similar to microfossils such as *Tasmanites* and *Cymatiosphaera* that have been reported from the Precambrian up to the Holocene (Inouye et al., 1990).

Tetraselmis (Fig. 5.12) is commonly found in marine waters. The cells are oval-shaped and surrounded by a theca that is formed by the coalescence of many small stellate scales that are produced in the Golgi apparatus. Four flagella are inserted in an apical pit in the cell. The flagella are covered with hairs and scales, and the flagella emerge from an opening in the theca. There is a cup-shaped chloroplast with a basal pyrenoid. A

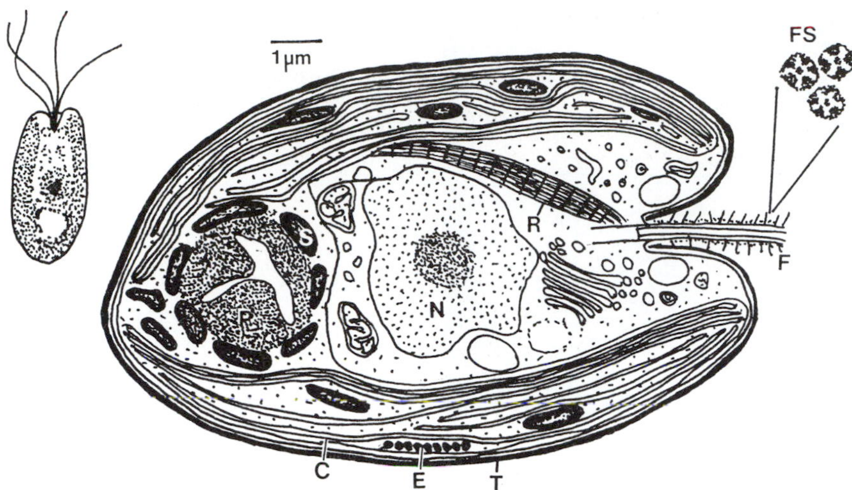

Fig. 5.12 Drawing of a cell of *Tetraselmis*. (C) Chloroplast; (E) eyespot; (F) flagellum; (FS) flagellar scales; (N) nucleus; (P) pyrenoid; (R) rhizoplast (only one of two is shown); (S) starch; (T) theca.

nucleus occurs in the center of the cell, and Golgi bodies are found anterior to the nucleus.

Tetraselmis produces stellate scales in the Golgi apparatus which fuse extracellularly to yield the theca (Manton and Parke, 1960). The theca is composed of neutral and acidic polysaccharides associated with certain amino acids (Becker et al., 1996), similar to the glycoprotein walls in volvocean flagellates (those green algae in the Volvocales) (Roberts, 1974) but unlike the cellulose cell walls in the Charophyceae. The typical volvocean flagellates and eventually the rest of the higher Chlorophyceae probably arose from a flagellate such as *Tetraselmis* (Domozych et al., 1981).

The genus *Tetraselmis* was previously divided into a number of genera including *Tetraselmis, Platymonas,* and *Prasinocladus* before it was recognized that all of the cells were similar. All of the organisms are now recognized as the single genus *Tetraselmis* with subgenera (Hori et al., 1982a,b):

Subgenus *Tetraselmis:* spherical pyrenoid into which narrow channels of cytoplasm extend.

Subgenus *Prasinocladia:* pyrenoid not penetrated by cytoplasm but by a single lobe of the nucleus so that the pyrenoid appears cup-shaped in the microscope.

Algae in the subgenus *Prasinocladia* grow as densely yellow-green tufts on rocks or shells in small tide pools near the high-tide mark (Fig. 5.13) (Chihara, 1963). The plants occur on most temperate coasts. In the mature attached vegetative state, the two to four protoplasts are at the apical area of the wall. Protoplasts are released as zoospores by escaping through a break in the apical area of the wall. The quadriflagellate zoospores swim actively

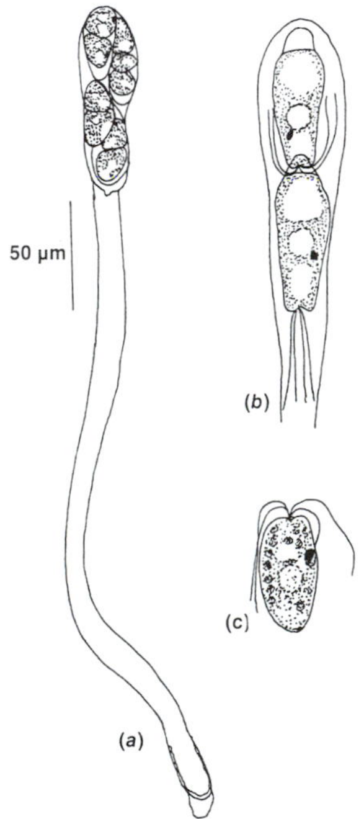

Fig. 5.13 *Tetraselmis* subgenus *Prasinocladia*. (*a*) Vegetative cells on a stalk. (*b*) Zoospores within a stalk. (*c*) Zoospore. (After Proskauer, 1950.)

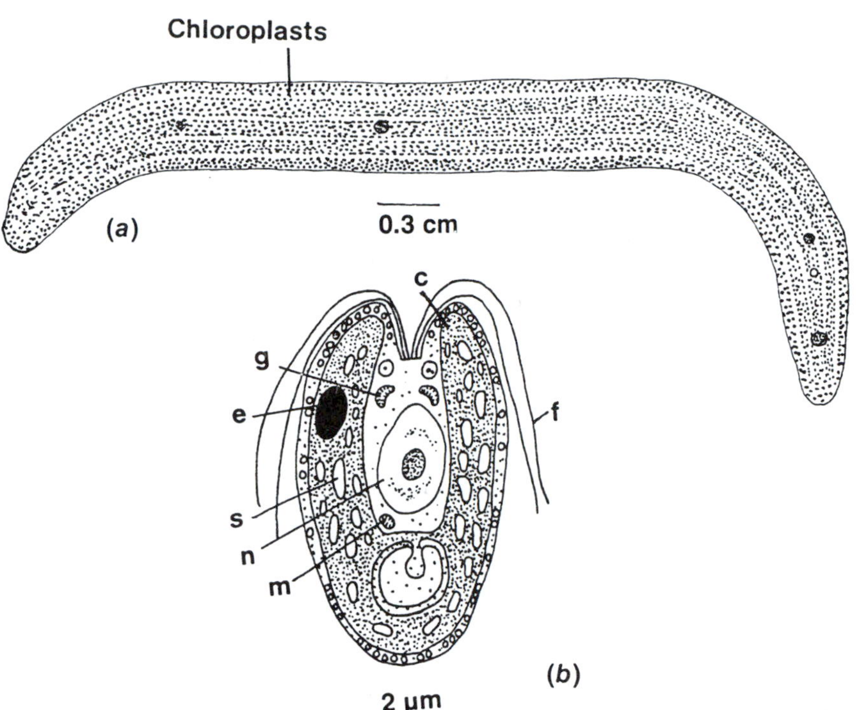

Fig. 5.14 (*a*) *Convoluta roscoffensis* flatworm with symbiotic green algae; (*b*) *Tetraselmis convolutae*, the algal symbiont of *Convoluta*. (c) Chloroplast; (e) eyespot; (f) flagella; (g) Golgi body; (m) mitochondrion; (n) nucleus; (s) starch. ((*a*) after Russell and Yonge, 1963; (*b*) after Parke and Manton, 1967.)

for a while, settle, lose their flagella, and secrete a wall. Germination occurs immediately after settling and first involves the secretion of an outer wall. The protoplast elongates with the deposition of an inner wall, causing a break in the outer wall; and, still surrounded by the elongating inner wall, the protoplast secretes additional wall material and positions itself at the top of the resulting tube. The protoplasts can divide into two cells to form new tubes. Any protoplast is capable of forming flagella and becoming a zoospore.

Convoluta is a small flatworm, several millimeters long, that lives on sandy marine shores (Fig. 5.14) (Russell and Yonge, 1963). It occurs in large colonies forming green patches on the yellow sand, appearing from beneath the sand immediately after the tide has left it and disappearing just before the tide returns. The green color of the flatworm is due to the presence in the animal of vast numbers of algae. The algae are not present in the egg, but there are algae adhering to the egg case that are ingested by the emerging worms. The alga that infects *Convoluta roscoffensis* is *Tetraselmis* (Fig. 5.14) (Oschman, 1966). These algae lie in the extracellular spaces between adjacent cells of the animal. The infecting algae undergo morphological alterations within the worms, losing their flagella, theca, and eyespot in that order. The *Tetraselmis* theca does not contain cellulose (Lewin, 1958), a possible explanation for the relative ease with which the worm can dissolve the theca and set up the symbiosis. Upon loss of the theca, the alga assumes an irregularly shaped form, with fingerlike processes of the algal cells penetrating between adjacent animal cells. *Convoluta* flatworms may contain algal cells of either the *Tetraselmis* or *Prasinocladia* subgenera, although one animal will usually contain cells of only one subgenus. In the United Kingdom, the flatworms having cells of the *Tetraselmis* subgenus are longer, are more likely to contain gametes, and lay more egg capsules containing more embryos than those containing cells of the *Prasinocladia* subgenes (Douglas, 1985). If the *Convoluta* eggs are not infected by *Tetraselmis*, they fail to develop properly and soon die. In early life, *Convoluta*, like other flatworms, feeds on smaller animals. As the flatworm gets older, it relies on photosynthate from the internal algae, and its digestive organs degenerate so that it is not able to feed like a normal animal. The *Convoluta*, however, needs a more varied diet than just that of algal photosynthate and begins after a time to feed upon the algae – to kill the geese that laid the golden eggs – so that the algae gradually disappear, the flatworm presenting the strange appearance of a green head and a white tail. Finally the flatworm dies of starvation, but not before it has laid a large number of eggs.

Charophyceae

This is the line of algal evolution that led to the development of land plants. The motile cells of the advanced members of the class are similar to the flagellated male gametes of the bryophytes and vascular cryptogams. The motile cells of the Charophyceae are asymmetrical and have two laterally or subapically inserted flagella. The microtubular root system contains a multilayered structure that is associated with a broad microtubular root and a second, smaller, microtubular root. Rhizoplasts are not present. The mitotic spindle is persistent during cytokinesis, and cell division occurs by means of a phragmoplast. No eyespots occur. Sexual reproduction results in the formation of a dormant zygote. Meiosis occurs when the zygote germinates. Glycolate is broken down by glycolate oxidase, whereas urea is broken down by urease. The algae in the class are found predominantly in freshwaters.

Classification

Within the Charophyceae are four important orders:

Order 1 **Klebsormidiales:** unbranches filaments without holdfasts; plasmodesmata absent; zoospores naked and released through a pore in the wall.

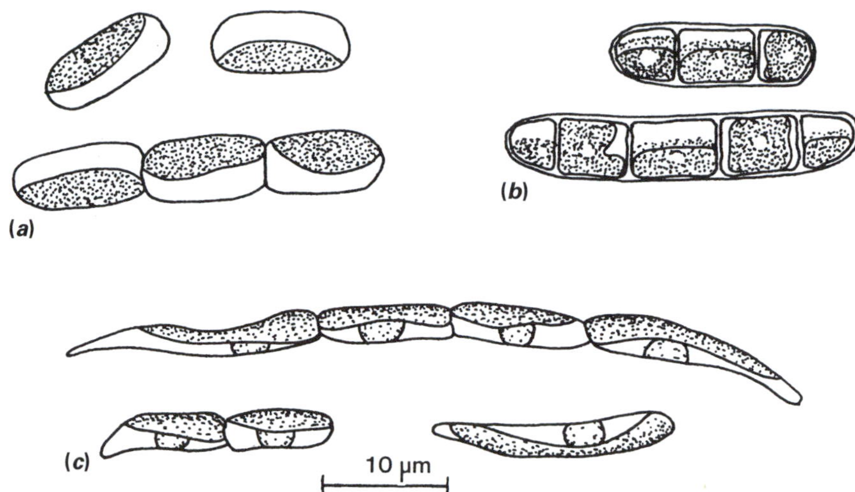

Fig. 5.15 (*a*) *Stichococcus bacillaris*. (*b*) *Klebsormidium* sp. (*c*) *Rhaphidonema nivale*. ((*c*) after Hoham, 1973.)

Order 2 **Zygnematales:** sexual reproduction by conjugation; unicells or unbranched filaments without holdfasts; plasmodesmata absent in filamentous forms; flagellated cells not produced.

Order 3 **Coleochaetales:** oogamous sexual reproduction; motile cells have a covering of scales, sheathed setae present; unicells, branched filaments or discoid thalli.

Order 4 **Charales:** oogamous sexual reproduction; sterile cells surround antheridia and oogonia; male gametes covered with scales; zoospores not produced; complex plant body with apical growth and differentiation into nodes and internodes; plasmodesmata present.

KLEBSORMIDIALES

The algae in this order are terrestrial or freshwater algae with unbranched filaments that do not have holdfasts. Each cell contains a single parietal chloroplast. There are no plasmodesmata between cells. The zoospores are naked and are released through a pore in the wall. Algae definitely placed in this order on the basis of their cytology include *Klebsormidium* (Pickett-Heaps, 1974), *Stichococcus* (Floyd et al., 1972), and *Raphidonema* (Pickett-Heaps, 1976) (Fig. 5.15).

Sexual reproduction is isogamous by biflagellate gametes. Reproduction and morphology of the algae in this order are similar to those in species of *Ulothrix*, with which some of the members of the Klebsormidiales were originally classified before their ultrastructural differences became apparent.

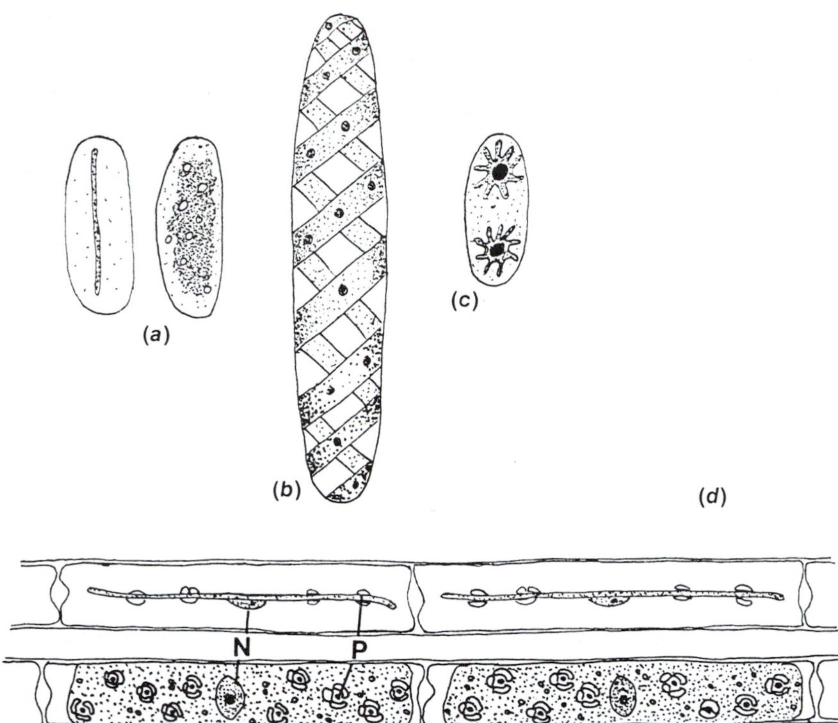

Fig. 5.16 (*a*) *Mesotaenium de greyi*, high- and low-intensity orientations of the chloroplast. (*b*) *Spirotaenia condensata*. (*c*) *Cylindrocystis brebissonii*. (*d*) *Mougeotia scalaris*, above-profile and below-surface views of the chloroplast. (N) Nucleus; (P) pyrenoid.

ZYGNEMATALES

The Zygnematales are freshwater algae that are unique among the Chlorophyta in having sexual reproduction by isogamous conjugation in which the gametes are non-flagellated. The union of the two gametes can be through a conjugation tube formed by the parent cells, or the gametes can move from their parent cells into the medium and fuse. The zygote or zygospore forms a cell wall and goes through a resting period before germinating meiotically. The life cycle is therefore primarily haploid, with the zygote representing the diploid generation. Zygnematalean algae, particularly *Mougeotia*, dominate freshwaters affected by acid precipitation (Graham et al., 1996).

There are basically three types of chloroplasts in the order: (1) spirally twisted bands extending the length of the cell (*Spirogyra, Spirotaenia*); (2) an axial plate extending the length of the cell (*Mougeotia, Mesotaenium*); and (3) two stellate chloroplasts next to each other (*Zygnema*). In those cells with flat axial chloroplasts there is a marked chloroplast orientation in response to light intensity. In *Mesotaenium* and *Mougeotia* (Fig. 5.16), under low light intensities the chloroplast presents a surface view to the light. When irradiated with high-intensity light or red light at low intensity, the chloroplast rotates to present an edge view. Actin microfilaments attached to the chloroplasts are directly responsible for movement of the chloroplast (Mineyuki et al., 1995) with a phytochrome system directing actin functioning.

Within the Zygnematales there are three families (some phycologists group the last two families into one family). Sequencing the large subunit of ribulose-1,5-bisphosphate carboxylase/oxygenase indicates that the order is paraphyletic (McCourt et al., 1995).

Family 1 Zygnemataceae: cylindrical cells united permanently into unbranched filaments; cell wall without pores.

Family 2 Mesotaeniaceae: basically non-filamentous; cell walls without pores; no new semi-cell formed in cell division.

Family 3 Desmidaceae: basically non-filamentous; cell walls with pores; new semicell formed in cell division.

Zygnemataceae

The cells of the Zygnemataceae are permanently united into unbranched filaments, and the cell walls lack pores. Union of the two aplanogametes is usually by the establishment of a conjugation tube between two cells.

Members of the Zygnemataceae are among the most common filamentous freshwater algae, favoring small stagnant bodies of water, but with a few found attached in the littoral zone of lakes (*Spirogyra adnata*) and in flowing water. They are especially abundant in the spring months, generally occurring as bright green free-floating masses, with some type of attaching organ present in young stages of several genera. Planktonic species of *Spirogyra* or *Mougeotia* often have twisted or spirally coiled threads. *Zygogonium* which lives next to thermal springs with acidic waters, is responsible for the dense purple mats that occur in the hot springs of the Yellowstone National Park, where it lives at temperatures of 21 to 30 °C and a pH of 2.4 to 3.1; the optimum temperature for photosynthesis is 25 °C, and the pH optimum is 1 to 5 (Lynn and Brock, 1969).

Spirogyra occurs primarily in the springtime because it tolerates high light intensities in cool water (Graham et al., 1995). *Spirogyra* has ribbon-shaped chloroplasts with a number of pyrenoids along the length of the chloroplast (Fig. 5.17). A nucleus is suspended in the center of the cell. Every cell in the filament except the basal one is capable of cell division. Asexual reproduction occurs by fragmentation of the filaments, whereas sexual reproduction is by conjugation, which is initiated by two filaments coming to lie next to one another, and being bound together in a layer of mucilage (Fig. 5.17). The cells of one filament put out papillae toward the cells of the opposite filament. Subsequently, papillae arise from opposite and

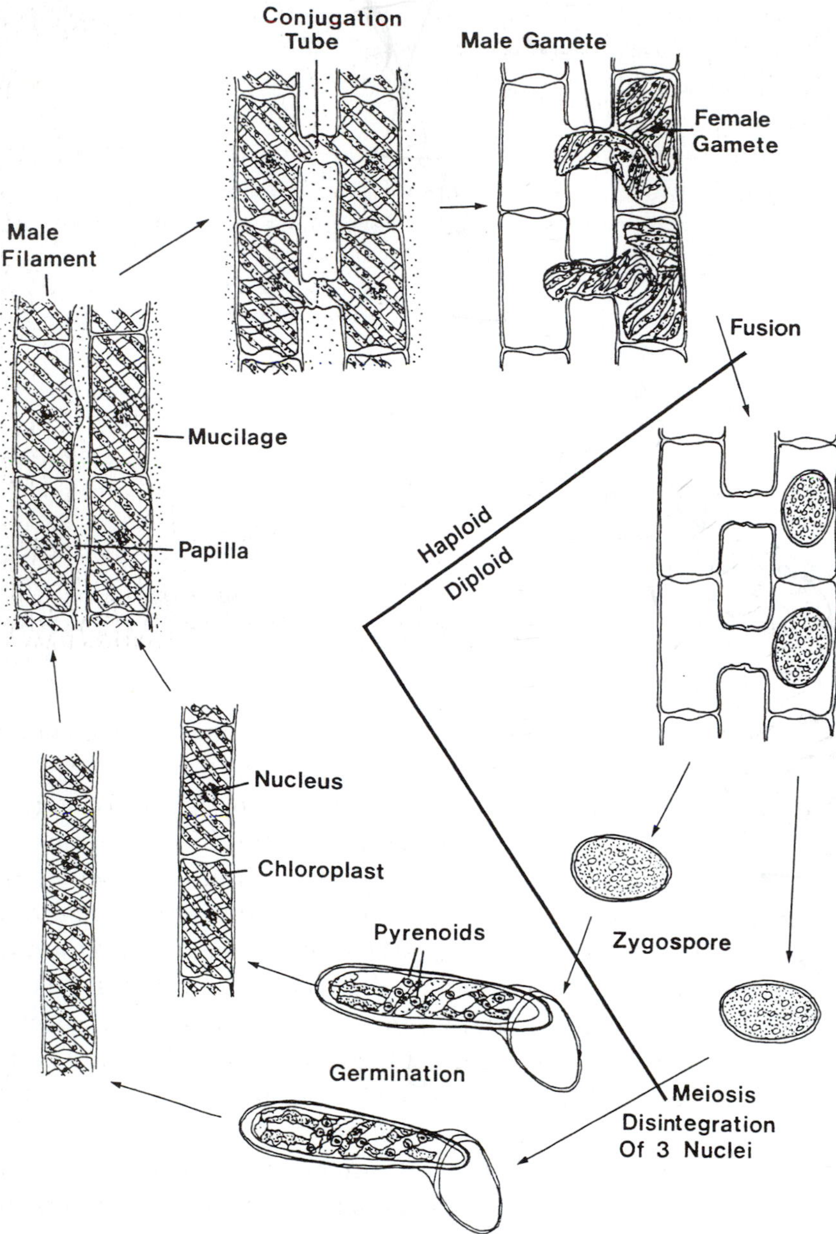

Fig. 5.17 The life cycle of *Spirogyra*.

corresponding points on the other filament so that the two papillae are in contact from the beginning of their formation. The two threads are then pushed apart as the papillae elongate. A hole is dissolved at the tips of the papillae so that the conjugation tube is continuous from one cell to the next. The male gamete (the protoplast that moves through the conjugation tube) contracts by the bursting of small contractile vacuoles within the cell mem-

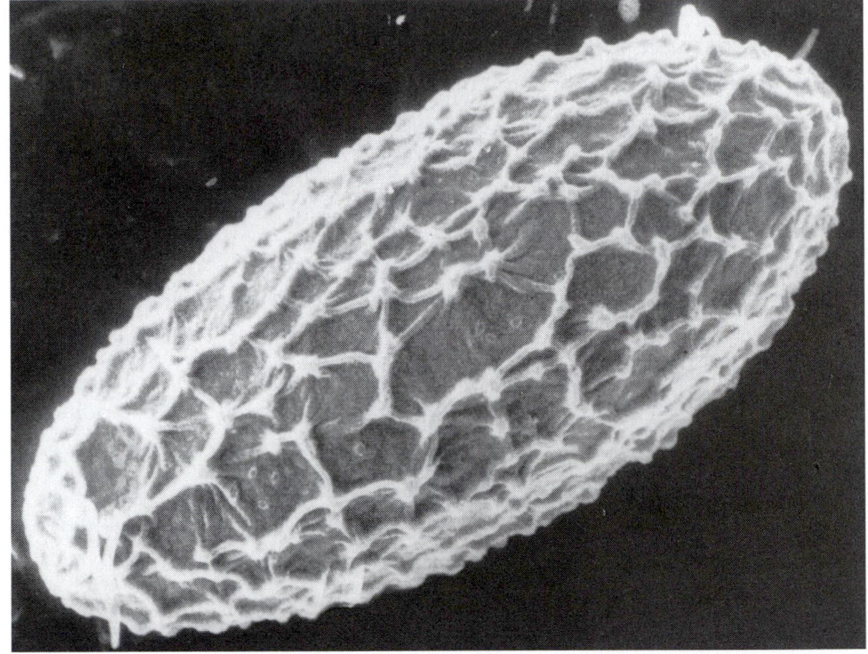

Fig. 5.18 Scanning electron micrograph of a zygote of *Spirogyra acanthopora* showing the sculpturing of the wall. (From Simons et al., 1982.)

brane, which is followed by a similar contraction of the female gamete as the male gamete moves through the conjugation tube. The male protoplast fuses with the female inside the parent wall of the female. The zygote (zygospore) secretes a three-layered wall around itself. The three layers of the wall from the outside to the inside are the **exospore**, **mesospore**, and **endospore**. The **exospore** is sometimes sculptured (Fig. 5.18) and contains cellulose and/or pectin. The **mesospore** is sometimes colored and contains sporopollenin. The **endospore** is thin and colorless, and contains cellulose and pectin (Simons et al., 1982). Ripening of the zygote is accompanied by the disappearance of chlorophyll and the conversion of the starch into a yellowish oil. During the resting period of the zygote (in nature till the following spring), the nucleus divides meiotically to form four haploid nuclei, three of which disintegrate. The zygote germinates by sending out a tubular growth that ruptures the outer two wall layers while the inner wall extends to accommodate the growth. The outgrowth then undergoes transverse divisions to form the first cells of the filament. The life cycle is thus primarily haploid, with the zygote being the only diploid cell. Conjugation is referred to as "physiologically anisogamous" because of the different behavior of morphologically similar gametes.

The above type of conjugation, which is called **scalariform** conjugation, occurs between two separate filaments. Another type of conjugation is **lateral** conjugation, which occurs between cells of the same filament. Here a conjugation tube is formed between adjacent cells, or in some cases the cross wall between adjacent cells simply dissolves. The members of this family would seem to be excellent tools for the study of fertilization, but this is not so because of the difficulty of inducing sexual reproduction at will.

Mesotaeniaceae

This family has organisms that are basically unicellular, even though in some cases they are joined together to form filaments, and the cells taper to their ends. The walls do not have pores. The Mesotaeniaceae are called the saccoderm desmids; unlike the placoderm desmids of the following family, they do not have a new semicell formed after cell division. The nucleus is central in the cell, and there are three types of chloroplast structure, which are similar to those in the Zygnemataceae. *Mesotaenium* (Fig. 5.16(*a*)) has a flat axial platelike chloroplast with one to several pyrenoids. *Cylindrocystis* has a pair of stellate axial chloroplasts, each with a massive central pyrenoid (Fig. 5.16(*c*)); and in several species of *Spirotaenia*, the chloroplast is a parietal spiral band (Fig. 5.16(*b*)). Sexual reproduction takes place by conjugation and is similar to that in the other two families of the order.

These organisms are indicative of water low in magnesium and calcium and therefore usually unpolluted water. They are common in upland pools and peat bogs; *Mesotaenium* and *Cylindrocystis* are found in wet soil.

Desmidaceae

These are the placoderm desmids. They have two distinct halves or semicells separated by a median constriction, the **sinus**, and joined by a connection zone, the **isthmus**. The cell walls have pores in them (Gerrath, 1969). The cells can be solitary, joined end to end in filamentous colonies. or united in amorphous colonies. Some of these plants have a relatively complex and attractive shape (Fig. 5.19). The taxonomy of many of these desmids is complicated by **polymorphism** (different forms within the same species), which often makes organisms of the same species appear as representative of different species (Sormus and Bicudo, 1974).

Cosmarium botrytis has a central nucleus and two chloroplasts in each semicell, each chloroplast with a central pyrenoid (Fig. 5.20). At the tip of each semicell are crystals of barium carbonate that move about irregularly (Brook, 1989). Mixing cells of opposite strains (*mt*+ and *mt*−) results in the cells moving about with each strain indifferent to the other. During this period the *mt*− cells release a proteinaceous pheromone that causes *mt*+ cells to produce gametes, likewise *mt*− cells are similarly induced to produce gametes by a pheromone (Fukumoto et al., 1997).

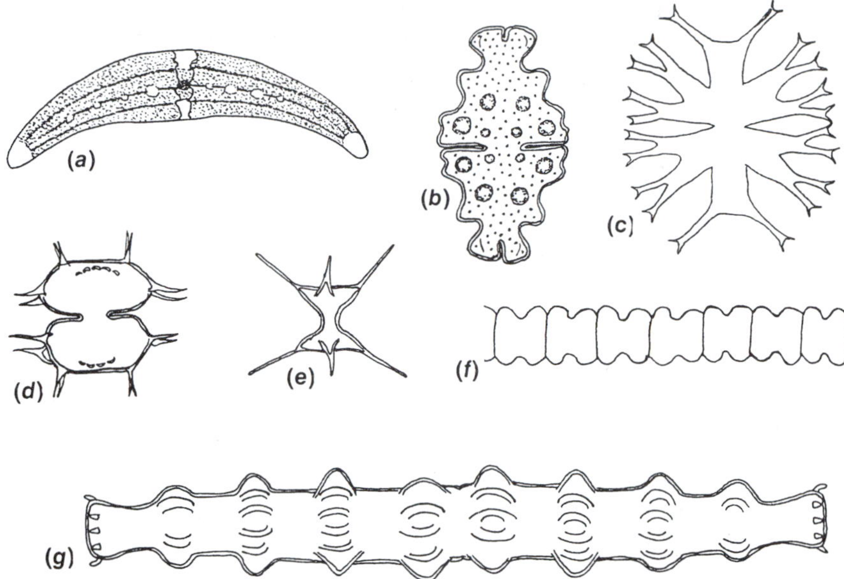

Fig. 5.19 (*a*) *Closterium moniliforme*. (*b*) *Euastrum affine*. (c) *Micrasterias radiata*. (*d*) *Xanthidium antilopaeum*. (*e*) *Staurastrum curvatum*. (*f*) *Spondylosium moniliforme*. (*g*) *Pleurotaenium nodosum*. (After Smith, 1950.)

After pairing, the cells move around each other and twist about until their broadest faces are in contact, and their longitudinal axes are at right angles. A sphere of mucilage is secreted around the conjugants and also between them, forcing the cells apart. The protoplast escapes from the gametes by the separation of the wall in the region of the isthmus (Starr, 1954). Within 4 to 7 minutes after making contact, the protoplasts fuse, forming an irregularly shaped zygote. The zygote then rounds off and forms a smooth-walled cell. The chromatophores arrange themselves at the periphery of the cytoplasm, and the ornamentation of the wall begins to appear about 1 hour after the fusion of the gametes. Spines develop on the wall, with the tips commonly divided into two or three parts. The zygote germinates meiotically, completing the life cycle.

The desmids are usually indicators of relatively unpolluted water, low in calcium and magnesium, with a slightly acidic pH. Normally a large number of species are present in such waters, without a single species comprising most of the population.

COLEOCHAETALES

The algae in the Coleochaetales are characterized by the presence of sheathed setae and by oogamous sexual reproduction. Small-subunit ribosomal RNA sequencing (Wilcox et al., 1993; Kranz et al., 1995) has indicated that the green algae in this order evolved into the bryophytes and lycopods. Like the bryophytes and lycopods, the algae in the Coleochaetales have

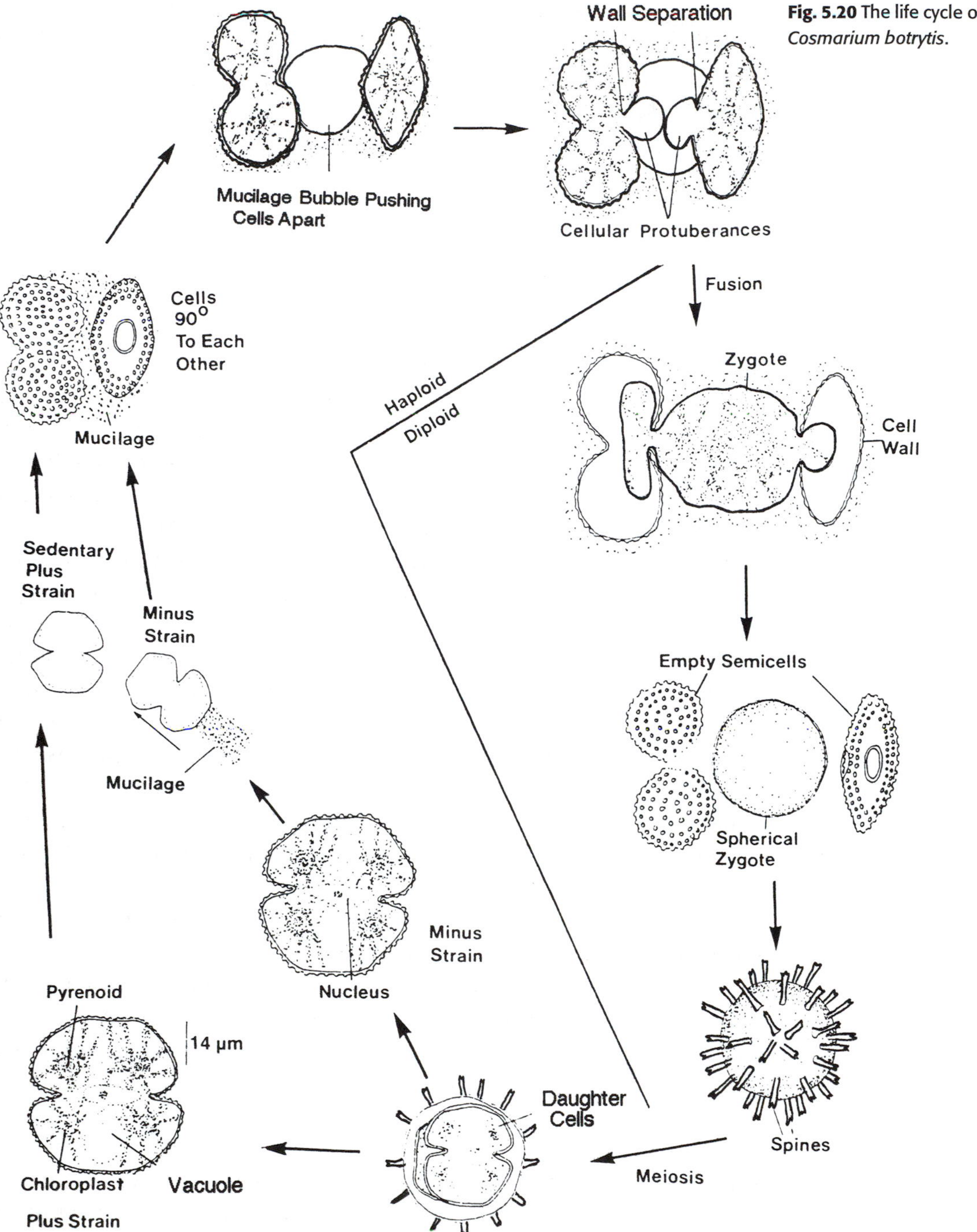

Fig. 5.20 The life cycle of *Cosmarium botrytis*.

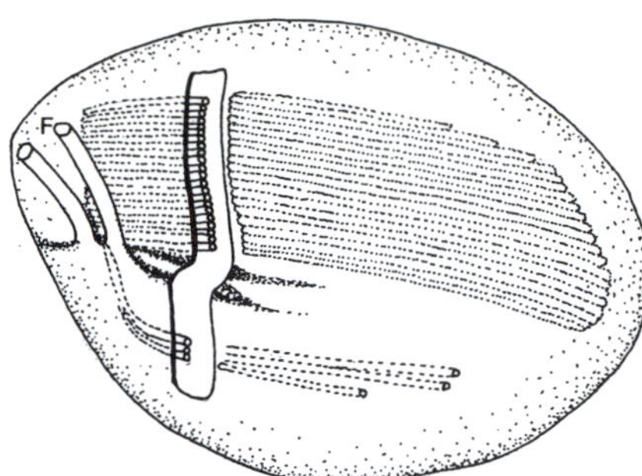

Fig. 5.21 Drawing showing the subapical flagella (F) in a zoospore of *Coleochaete*. The flagella are anchored in the cytoplasm by two microtubular roots. A broad microtubular root is associated with a multilayered structure, and a small microtubular root is composed of three microtubules. (After Sluiman, 1983.)

asymmetrical motile cells covered with scales, microtubular roots consisting of a large and small band (Sluiman, 1983), a persistent interzonal spindle and a phragmoplast at cytokinesis; glycolate is degraded by glycolate oxidase (Tolbert, 1976). In addition some of the species in the Coleochaetales have evolved a protective sheath around the oogonium, which is an advance toward the protected oogonia in the Bryophyta.

Most of the species of *Coleochaete* occur as epiphytes in the shallow littoral zone of freshwater lakes and may be observed by microscopic examination of macrophyte leaves or inorganic substrates, such as beer cans, soda bottles, or plastic bags discarded by fishermen (Graham, 1984). The thallus is composed of branching filaments, which are free in some species, whereas in others the branches are laterally opposed to form a pseudoparenchymatous disc. All members of the genus are characterized by uniquely sheathed hairs called **setae**. The base of setae is covered with a gelatinous material. The setae probably function as an antiherbivore defense, since broken setae exude a substance that repels potential predators (Marchant, 1977).

Zoospores and spermatozoids of *Coleochaetae* are asymmetrical, are covered with scales, and possess multilayered structures and microtubular roots similar to those of lower land plants (Graham and McBride, 1979; Sluiman, 1983). The flagella arise subapically and project to the side of the cell (Fig. 5.21). The flagella are anchored by a microtubular root system composed of a single broad band of microtubules associated with a multilayered structure and a second microtubular root.

Asexual reproduction is by biflagellate zoospores which are formed singly in a cell, most frequently in the spring. Initiation of zoosporogenesis is induced primarily by temperature. Northern temperate species undergo zoosporogenesis when brought to 20°C for a few days (Graham and Krantzfelder, 1986). Day length and irradiance have little effect on zoosporogenesis.

Sexual reproduction in *Coleochaete* is oogamous, and the plants can be

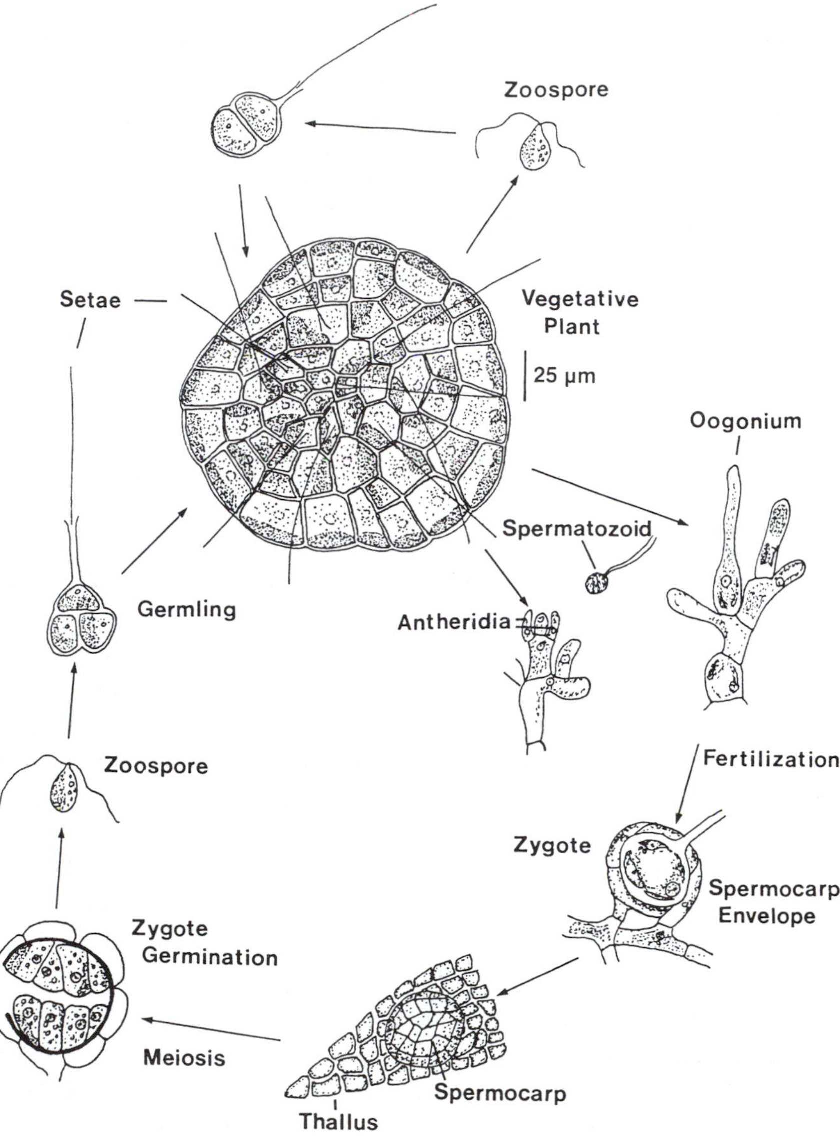

Fig. 5.22 The life cycle of *Coleochaete scutata*. (After Oltmanns, 1898; Smith, 1950.)

homothallic or heterothallic, depending on the species. Antheridia are borne at the tips of the branches, and each antheridium forms a single spermatozoid (Fig. 5.22). Oogonia are modified one-celled branches, and an oogonium is a flask-shaped structure with a long colorless neck, the trichogyne. At maturity the tip of the neck breaks down, with some colorless protoplasm being extruded and the basal protoplast rounding off to form a single egg. The spermatozoid swims into the oogonium and fertilizes the egg. The zygote remains in the oogonium, secretes a thick wall, and greatly increases in size. At the same time there is an upgrowth of branches from the cell below the oogonium and from neighbouring cells to form a pseudoparenchymatous layer enclosing the oogonium. The oogonium, with its enclosing sheath layer of cells, becomes reddish and is termed a **spermocarp**; the spermocarps remain dormant over the winter. The zygote germinates meiotically to form 8 to 32 biflagellate zoospores, which are liberated through a break in the spermocarp and zygote walls, and swim for a short time before settling and developing new thalli.

CHARALES

The green algae in the Charales have a complex plant body with apical growth and differentiation of the body into **nodes** and **internodes**. Reproduction is oogamous, with sterile cells surrounding the antheridia and oogonia. The male gametes have a cell covering of scales. No zoospores are formed. A phragmoplast develops during cell division, resulting in the formation of a cross wall with plasmodesmata. Land plants (embryophytes) probably evolved from algae similar to those in the Charales (Wilcox et al., 1993; McCourt et al., 1996).

The algae in the Charales are primarily freshwater, with only a few species occurring in brackish water. They are most common at the bottom of clear lakes, usually forming extensive growths. Many of the Charales are heavily calcified, with concentrations of plants on the bottom of lakes leading to the formation of **marl** ($CaCO_3$ and $MgCO_3$ deposits) and hence the common name of the Charales, **stoneworts**. Calcification in *Chara* and *Nitella* results from the precipitation of calcium carbonate ($CaCO_3$) in water high in Ca^{2+}: The localized OH^- efflux at certain areas of the internodal cell raises the local CO_3^{2-} ion concentration, leading to $CaCO_3$ supersaturation and precipitation (Lucas, 1979). In *Chara corallina*, rectangular crystals of calcite are deposited in bands on the surface of the cell wall of the cylindrical internodal cell. These bands correspond to localized alkaline (pH 9.5 to 10.5) regions on the cells. There is no organic material associated with the crystals, and they are formed entirely outside the cell wall (Borowitza, 1982; Wray, 1977).

The algae in the Charales have an axis divided into nodes and internodes (Fig. 5.23). Each **node** bears a whorl of branches composed of a number of cells that cease to grow after they have reached a certain length. The **internode** consists of a single long cell in *Nitella*. In some genera, such as *Chara*, the internodal cell is ensheathed (corticated) by a layer of vertically elongated cells of a much smaller diameter originating at the nodes (Fig. 5.23). Growth occurs by a dome-shaped apical cell, each derivative of the apical cell dividing transversely into two daughter cells. The upper daughter cell is the nodal initial, and the lower is the internodal initial. The **nodal initial** matures into the cells of the node, and the **internodal initial** matures into

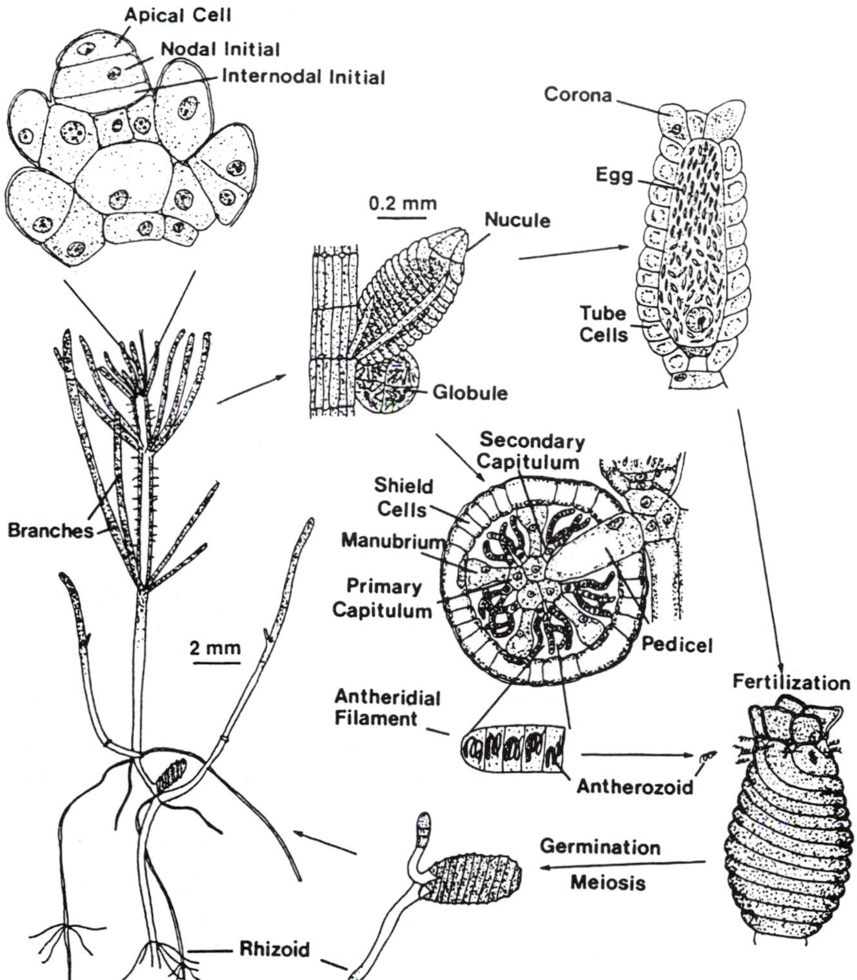

Fig. 5.23 The life cycle of *Chara* sp. (Adapted from Smith, 1955; Scagel et al., 1965.)

the single long cell of the internode (Smith, 1955). The axis is attached to the substratum by uniseriately branched rhizoids. The rhizoids are positively geotropic and grow downward. Near the rhizoid tip are a group of **statoliths** consisting of vacuoles containing crystallites of barium sulfate that may be associated with the response of the rhizoids to gravity (Sack, 1991; Hodick, 1994).

The uninucleate cells have many small ellipsoidal chloroplasts in longitudinal, spirally twisted, parallel rows. In the center of the cells is a single large vacuole. The large size of the internodal cell and its vacuole has made the cell a favorite tool of physiologists interested in the control of ion uptake in plants. The pH of the cell sap of *Chara* is controlled at about 5 as a result of a dynamic equilibrium between active influx and passive efflux of H^+ across the tonoplast. The H^+ transport system involves two proton (H^+) trans-

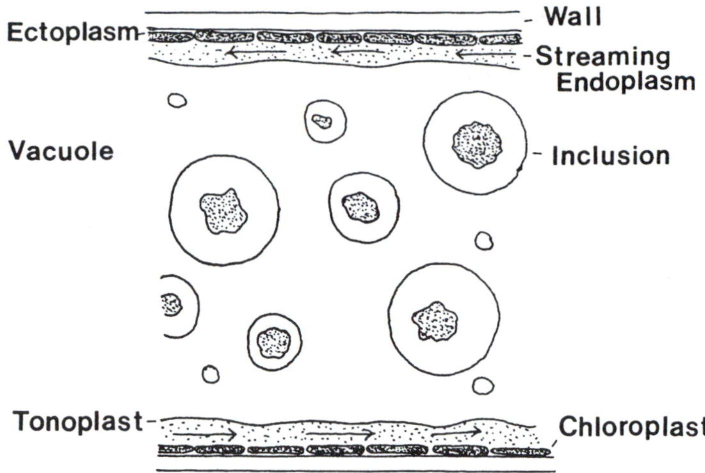

Fig. 5.24 Schematic view of part of an internodal cell of *Nitella*. The chloroplasts are stationary in the ectoderm, whereas the endoderm moves by cytoplasmic streaming. (After Kamitsubo, 1980.)

locating systems in the tonoplast (Shimmen and MacRobbie, 1987). **Cytoplasmic streaming** is particularly evident in these large internodal cells. The cytoplasm is divided into an inner **endoplasm** next to the vacuole, and an outer **ectoplasm** containing the chloroplasts (Fig. 5.24). Streaming endoplasm follows the orientation of the spirally arranged stationary chloroplasts in the ectoplasm at a rate of 60 $\mu m\ s^{-1}$ at 20 °C. Streaming is due to the interaction of actin microfilaments with cytoplasmic microtubules (Collings et al., 1996).

The Charophyta do not form zoospores, but there are specialized sexual bodies. Asexual reproduction may be effected by (1) star-shaped aggregates of cells developed from the lower nodes, called **amylum stars** because they are filled with starch; (2) **bulbils** developed on rhizoids; and (3) protonema-like outgrowths from a node.

The sexual organs are the **globule** (male) and **nucule** (female) (Fig. 5.23), the terms antheridium and oogonium not being appropriate because the sexual reproductive structures include both a sex organ and a multicellular sheath derived from cells beneath the sex organ (Smith, 1955). Globules and nucules are borne on the nodes, usually on the same plant. In *Chara*, the nucule is above the globule. The globule is attached to the axis by the pedicel cell and is surrounded by shield cells. At the end of the pedicel are the primary capitulum cells, to which are attached the manubrium cells. Each primary capitulum cell cuts off six secondary capitulum cells, which form the antheridal filaments (Maszewski and van Bel, 1996; Kwiatkowski, 1996). Each cell of an antheridial filament is an antheridium whose protoplast matures into a single antherozoid. The antherozoid is coiled in a compact helix of 2½ turns in the antheridium (Fig. 5.25). When the antherozoids are mature, the shield cells of the globule separate from one another and

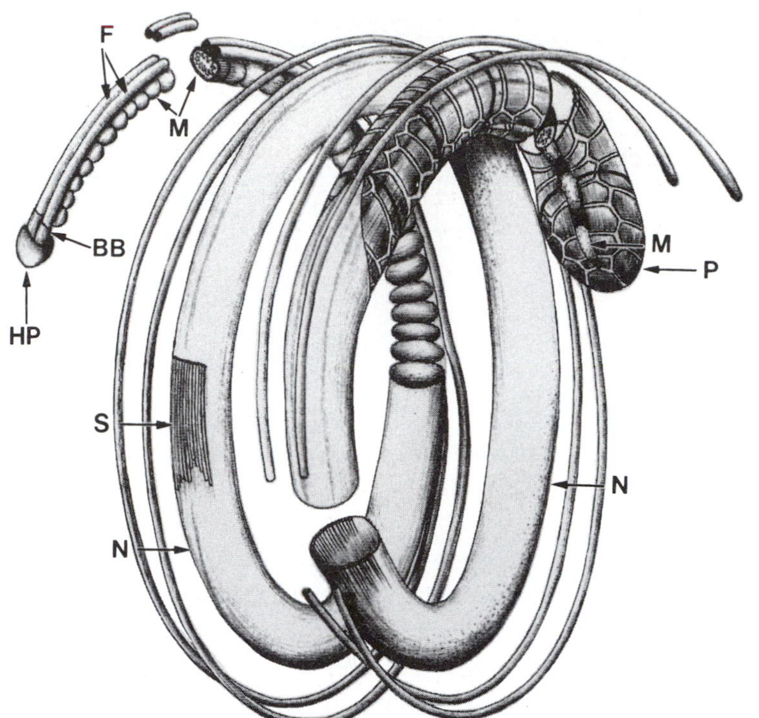

Fig. 5.25 Reconstruction of the mature spermatozoid of *Chara vulgaris*. BB-basal bodies, F-flagella, HP-head piece, M-mitochondrion, N-nucleus, P-plastid, S-spline of microtubules. (From Duncan et al., 1997.)

expose the antheridial filaments. Soon after the antheridial filaments are exposed, and antherozoids emerge backward through a pore in the cell wall.

Liberation of antherozoids generally takes place in the morning, and their swarming may continue until evening (Smith, 1955). The sperm have two somewhat unequal flagella attached subterminally near the anterior end of the cell (Fig. 5.25). There are three different regions within the body of the sperm: (1) the head region, consisting of a microtubular sheath over the mitochondria taking up one-fourth of the body; (2) the middle region, with microtubules covering the nucleus and taking up half of the body; and (3) the tail region, where the microtubules ensheath the plastids. Scales formed by the Golgi are on the outside of the flagella.

The nucule is supported by a pedicel cell, and in the center is the oogonium with its single egg. Five spirally twisted tube cells cover the oogonium except at the tip, where there are five corona cells. When the nucule is mature, the spirally twisted tube cells separate from one another just below the corona. Antherozoids swim through these openings to the oogonium, where they penetrate its gelatinized wall. The zygote secretes a thick wall, and the inner wall of the tube cells becomes thickened. Other portions of the walls of the sheath decay, leaving the persisting portions of walls of the tube cells projecting like the threads of a screw. The zygote (Fig. 5.26), with

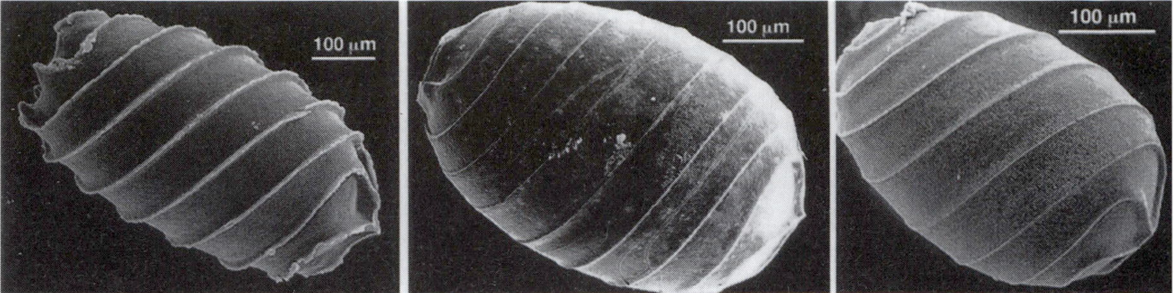

Fig. 5.26 Scanning electron micrographs of oospores of *Chara muelleri* (*left*) and *Chara fibrosa* (*middle, right*). (From Casanova, 1997.)

the surrounding remains of the sheath, falls to the bottom of the pool and there germinates after a period of weeks or months (Smith, 1955).

The zygote can be carried considerable distances attached to water fowl (Proctor, 1962). The zygote will not germinate in darkness and needs white or red light (Takatori and Imahori, 1971). Meiosis occurs in the first division of the zygote; thus the thallus is haploid, and the zygote is the only diploid part of the life history.

Because calcification occurs in most genera, the group is well represented as fossils, especially by the female fructification, which when calcified is termed a **gyrogonite**. The earliest fossils of Charales occur in the Uppermost Silurian. All of the extant Charales are placed in the Characeae, which date back to the Upper Carboniferous and which have the enveloping cells of the female fructification in a left-hand spiral. An extinct family, the Trochiliscaceae, also had a spirally twisted envelope, but it was twisted to the right.

Ulvophyceae

The motile cells of the Ulvophyceae have *apically attached flagella, near-radial symmetry externally, and a cruciate microtubular root system that is associated with a multilayered structure*. These characteristics differ from those of the Charophyceae but are similar to those of the Chlorophyceae. The Ulvophyceae, however, differ from the Chlorophyceae in having (1) *a persistent interzonal spindle that does not collapse at telophase, and* (2) *motile cells without a cell wall*.

The algae in the Ulvophyceae are predominantly marine although there are a number of freshwater species. All filamentous marine green algae or larger green seaweeds studied ultrastructurally have been shown to belong to the Ulvophyceae. The life cycle usually involves the alternation of a haploid thallus with a diploid thallus. The wide occurrence of alternation of generations in the Ulvophyceae might be due to the more stable marine environment fostering the evolution of a longer life cycle. Dormant zygotes are not known in the class. A number of genera in the Ulvophyceae produce swarmers with scales (Mattox and Stewart, 1973), indicating that the class had a scaly ancestor.

Classification

The following orders are placed in the Ulvophyceae. Cladistic analysis of nuclear-encoded rRNA sequence data have shown that the Ulotrichales and Ulvales form one group in the class, while the Caulerpales, Siphonocladales and Dasycladales form a second group (Zechman et al., 1990).

Order 1 **Ulotrichales:** uninculeate filamentous algae with a parietal chloroplast.

Order 2 **Ulvales:** uninucleate cells with a parietal chloroplast; thallus is a hollow cylinder or a sheet, one or two cells thick.

Order 3 **Cladophorales:** multinucleate filamentous algae with a parietal perforate or reticulate chloroplast.

Order 4 **Dasycladales:** thallus has radial symmetry composed of an erect axis bearing branches; thallus uninucleate but multinucleate just before reproduction; gametes formed in operculate cysts.

Order 5 **Caulerpales:** coenocytic algae lacking cellulose in the walls; siphonoxanthin and siphonein usually present.

Order 6 **Siphonocladales:** algae with segregative cell division; siphonoxanthin present.

ULOTRICHALES

Uninucleate filamentous green algae with a parietal chloroplast constitute this order.

Ulothrix (Fig. 5.27) is found in quiet or running freshwater and occasionally on wet rocks or soil. The thallus consists of unbranched filaments of indefinite length that are adfixed to the substratum by a special basal cell. All of the cells except the basal one are capable of cell division and forming zoospores or gametes. Species with narrow cells form 1, 2, or 4 quadriflagellate zoospores per cell, whereas those with broad cells form 2, 4, 8, 16 or 32 zoospores per cell. The zoospores have a conspicuous eyespot and are liberated through a pore in the side of the parent wall. Zoospores from species with narrow filaments are the same size, whereas those from broad-celled species form macro- and microzoospores that differ from each other in size, position of the eyespot, and length of the swarming period.

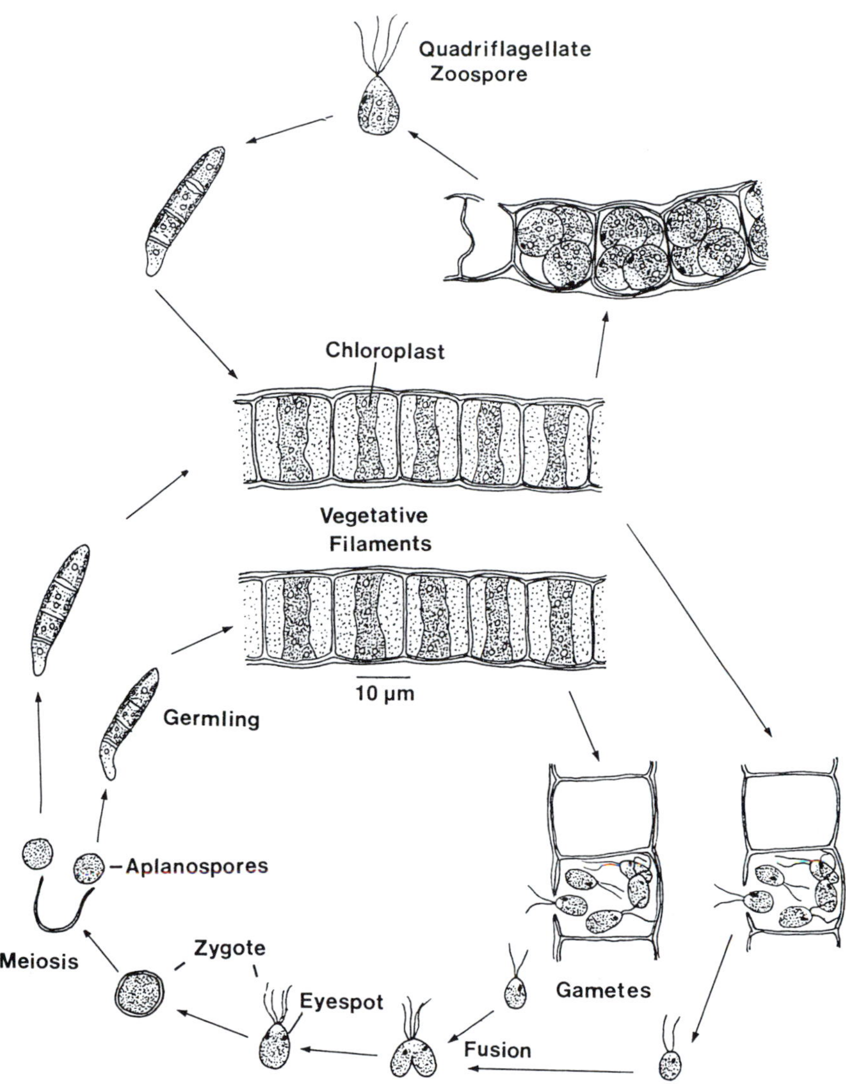

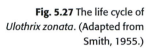

Fig. 5.27 The life cycle of *Ulothrix zonata*. (Adapted from Smith, 1955.)

Zoospores that are not discharged from the parent may secrete a wall and become thin-walled aplanospores. These later germinate to form a new filament.

Gametes of *Ulothrix* are formed in the same way as zoospores but are biflagellate. The gametes are of the same size, with fusion occurring only between gametes from different filaments; and there is never any parthenogenetic development of unfused gametes. The zygote remains for a while, settles, secretes a thick wall, and undergoes a resting period during which it accumulates a large amount of storage material. The first division of the zygote is meiotic, with the zygote forming 4 to 16 zoospores or aplanospores (Berger-Perrot et al., 1993).

In many northern lakes in the United States and Canada, *Ulothrix zonata* grows abundantly in early spring in shallow waters along rocky shorelines. *Ulothrix zonata* is dominant until the water temperature reaches 10 °C, when it disappears owing to massive conversion of the thallus to zoospores. At this time, *Cladophora glomerata* becomes the dominant attached alga. In culture, formation of the quadriflagellate zoospores of *Ulothrix zonata* occurs around 20 °C at relatively high light levels (520 μE m^{-2} s^{-1}) and photoperiods of either short-day (8 hours light: 16 hours dark) or long-day cycles (16 hours light: 8 hours dark). Zoospore formation is minimal at 5 °C, low irradiance (32.5 μE m^{-2} s^{-1}), and neutral day length (12 hours light: 12 hours dark) (Graham and Krantzfelder, 1986).

ULVALES

In nature, these plants have a thallus that is either an expanded sheet one or two cells thick (*Monostroma* and *Ulva*, respectively) or a hollow cylinder with a wall one cell in thickness (*Enteromorpha*). The normal morphology of these algae is lost if the plants are grown without the presence of bacteria. Under these conditions, *Ulva* and *Enteromorpha* develop into a pincushion-like colony, whereas *Monostroma* grows as a group of round cells with rhizoids (Provasoli and Pinter, 1980; Nakanishi et al., 1996). The addition of filtrates from bacterial, red algal, or brown algal cultures restores normal morphology.

Ulva fronds are composed of two layers of cells, with each cell having a large cup-shaped chloroplast toward the exterior of the cell (Fig. 5.28). The holdfast is formed by the cells of the thallus, sending down long slender filaments that coalesce to form the holdfast. The holdfast portion is perennial and proliferates new blades each spring. Cell division may occur anywhere in the thallus, but all divisions are in a plane perpendicular to the thallus surface.

The vegetative state of *Ulva mutabilis* is maintained by the blade cells excreting regulatory factors into the cell walls and into the environment. The production of one of these regulatory factors, a glycoprotein, decreases as the thallus matures. Eventually, the level of this regulatory factor decreases to the point where the regulatory factor is too low to maintain the vegetative state, and gametogenesis begins (Stratmann et al., 1996).

Ulva (Fig. 5.28) demonstrates an isomorphic alternation of generations, with the gametophyte forming biflagellate gametes and the sporophyte producing quadriflagellate zoospores. There is a periodic fruiting pattern in the genus, which is controlled by the lunar cycle. Gametes are released a few days before zoospores, with fruiting occurring during a series of neap tides (tide of minimum range occurring at the first and third quarters of the

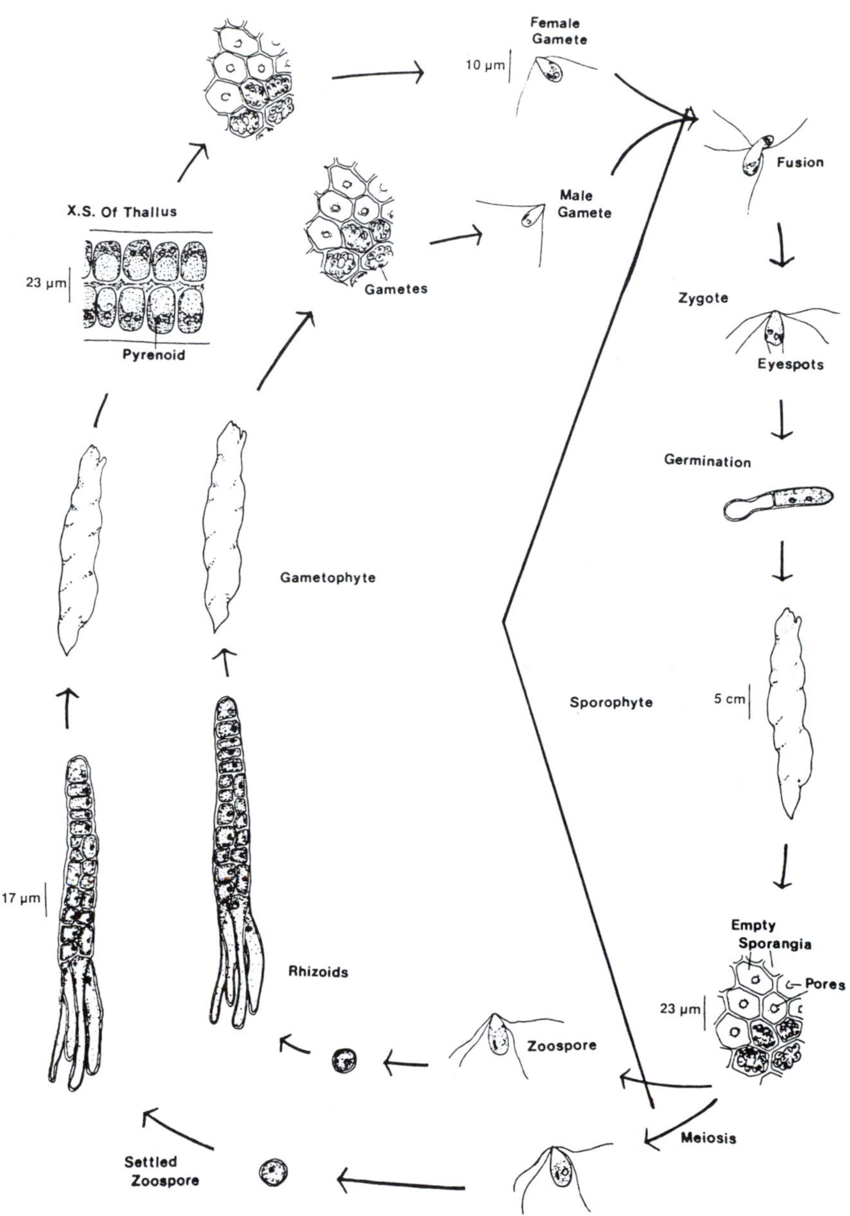

Fig. 5.28 The life cycle of *Ulva arasaki*. (Adapted from Chihara, 1969a.)

moon) in *U. pertusa* in Japan (Sawada, 1972) and during a series of spring tides (tide of maximum range during the new and full moon) in *U. lobata* on the Pacific Coast of North America (Smith, 1947). *Enteromorpha intestinalis* also shows a similar fortnightly periodicity at the beginning of a series of spring tides in England, except that here both gametes and zoospores are released at the same time (Fig. 5.29) (Christie and Evans, 1962). Although it is possible to correlate fruiting in the genus with tidal patterns, it is not the

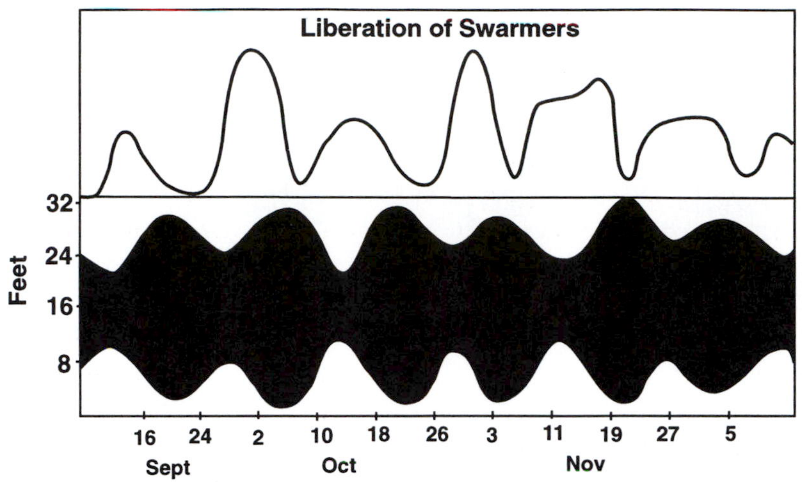

Fig. 5.29 Relationship between liberation of swarmers (gametes and zoospores) by *Enteromorpha intestinalis* and tidal amplitude. Maximum liberation of swarmers occurs 3 to 5 days before highest tide of each lunar period. (After Christie and Evans, 1962.)

tidal movements that induce fruiting and release of swarmers because the plants will exhibit a regular periodicity even when submerged. It is probably the amount of moonlight that the plants receive that gives the initial impetus to fruiting, although this may be timed by the tide as is the case with *Dictyota.*

Reproductive areas are formed near the margins of the fronds of *Ulva*, with the fertile portions changing from green to olive-green to brownish-green (5.28). Gametogenesis and sporogenesis (Melkonian, 1980a, 1980b) are similar and marked by a sharp reduction in photosynthetic capacity (Gulliksen et al., 1982). A cell divides to form 8 to 32 motile cells, with meiosis occurring in the formation of zoospores but not in the formation of gametes. Prior to the development of the motile cells, the mother cell forms a beaklike outgrowth extending to the thallus surface, through which the swarmers escape. The swarmers are released when the thallus is wet by the water of the incoming tide. The positively phototactic biflagellate gametes have a parietal chloroplast with a pyrenoid and eyespot. Haploid fronds are unisexual, so gametes from the same frond will not fuse. Gametes can be the same size, or one can be slightly larger than the other. The mixing of gametes of different strains results in the formation of cell clusters, joined by the tips of their flagella. Clusters separate almost immediately into mating pairs held together by their flagella (Bråten, 1971). The anterior ends of the gametes fuse within a few seconds, and the flagella separate, with the cells becoming negatively phototactic and swimming away from the light source. Within 3 minutes the cells jackknife and fuse laterally, with the quadriflagellate zygote remaining motile for a couple of minutes. The zygote settles, attaches to the substrate by its anterior flagellated end, and absorbs the flagella into the protoplasm. The zygote secretes a wall as soon as it settles, and nuclear fusion has occurred 30 minutes after the onset of copulation. The chloroplast from the plus gamete disintegrates (Bråten, 1973). In many ways the above process is similar to that in *Chlamydomonas* except that it occurs much more quickly.

Within a few days the zygote germinates, with mitotic division of the nucleus. After the first cell division one cell develops into a rhizoid, whereas the other eventually forms the blade. In some species it is possible to get a parthenogenetic development of gametes into a new plant.

Zoospores of *Ulva* are usually negatively phototactic, whereas gametes are positively phototactic. Upon fusion of gametes, phototaxis is reversed and the partially fused gamete pairs swim away from the light source. The total number of particles in the outer chloroplast envelope membrane over the large eyespot of *Ulva* zoospores is 11 300, whereas over the smaller eyespots of the female and male gametes there are only 5500 and 4900 particles, respectively. The lower number of particles in the gametes may be related to their positive phototaxis, especially because on fusion of the gametes, the total number of particles becomes 10 400, close to that of zoospores, with the fused gametes becoming negatively phototactic. The eyespot is on the same side as the mating structure in *Ulva* gametes, which means that on fusion the eyespots are positioned side by side and may be cooperating in shading the photoreceptor (Melkonian, 1980b).

Ulva is normally a marine genus although it can be found in brackish waters, particularly in estuaries. It normally grows on rocks in the middle to low intertidal zone, although the fronds are not situated at the same level throughout the year. During the colder months the plants grow mainly in the middle intertidal zone, covering wide vertical areas. In the warmer months the *Ulva* is lower in the intertidal zone and in a narrower band. Here the fronds are less exposed and subjected to less desiccation, which is more damaging to the plants in the high summer temperatures.

Ulva is an opportunistic alga, capable of rapid colonization and growth when conditions are favorable. This occurs primarily because of a rapid growth rate and the ability to take up and store nutrients available in pulsed supply. Because of its ability to quickly respond to enhanced nutrient supply, *Ulva* has proliferated in many areas that have received anthropogenic (related to man) nutrient enrichment. A feature of nuisance growths of *Ulva* in enclosed and semienclosed waters is that *Ulva* comprises a large proportion of drift plants, which may smother other benthic communities or be cast ashore where they decompose, causing considerable aesthetic nuisance (Hawes and Smith, 1995; Rivers and Peckol, 1995; Hernandez et al., 1997).

Ulva is commonly known as the sea lettuce or green laver, and has been eaten as a salad or used in soups, principally in Scotland. The chemical composition of dried *U. latuca* is 15% protein, 50% sugar and starch, less than 1% fat, and 11% water, making it usable as roughage in the human digestive system. In World War I, Phillipsen prepared a salad with *U. latuca*, *Enteromorpha*, and *Monostroma*, which he flavored with salad cream, vinegar, lemon, pepper, onions, and oil, and which he described as "wonderfully nice, slightly piquant and not inferior to the best garden salad" although one wonders after adding all of his condiments whether he was able to appreciate anything about seaweeds. The eminent French algologist Savaugeau prepared such a salad without condiments and said that "it was leathery and waxy in taste, and in spite of a good digestion I thought I would be ill" (Chapman, 1970).

CLADOPHORALES

The filamentous genera in this order have multinucleate cells, usually with a parietal or reticulate chloroplast. The filaments may be branched or unbranched. The reticulate chloroplast has pyrenoids at the intersections of the reticulum.

Cladophora and *Chaetomorpha* (Fig. 5.30), each with an isomorphic alternation of generations, are common members of this order. *Cladophora*, found in freshwater and marine habitats, may be the most ubiquitous macroalga in freshwaters worldwide (Dodds and Gudder, 1992). This filamentous alga can reach nuisance levels as a result of cultural eutrophication. *Cladophora* is predominantly benthic, and is often found in the region of unidirectional flow or in periodic wave action. In freshwater, it is a mid- to late-succession species. *Cladophora* is colonized by a wide variety of epiphytes because it offers a substrate that is anchored against flow disturbance.

DASYCLADALES

The plants in this order are all tropical and subtropical marine plants, most of them calcified. The members of the order are a clearly defined group having the following characteristics: (1) radial symmetry with an erect axis bearing branches; (2) uninucleate vegetative thallus, with a multinucleate condition developing just before reproduction; (3) gametes formed in operculate cysts within specialized gametangia.

The Dasycladales has a paleontological record that extends back to the Precambrian–Cambrian boundary (*ca.* 570 millions years ago) (Berger and Kaever, 1992). Of the 175 known fossil genera, only 11 are extant. The

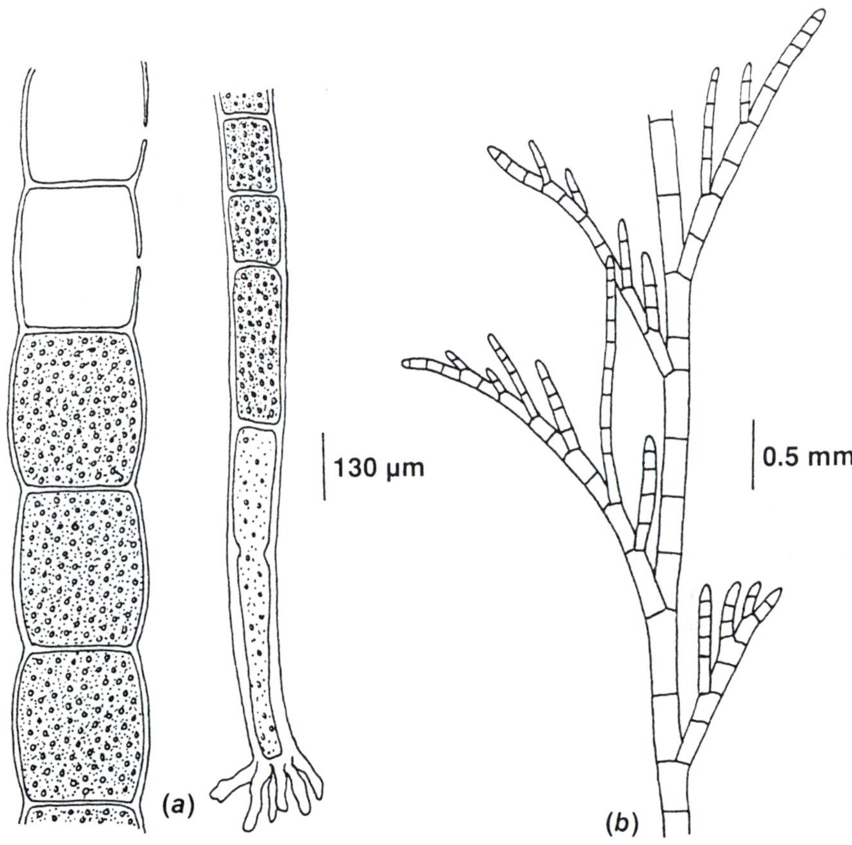

Fig. 5.30 (*a*) *Chaetomorpha aerea*. (*b*) *Cladophora microcladioides*. (After Smith, 1969.)

Dasycladales are in fact "living fossils." As defined by Stanley (1979), living fossils are organisms that include extant clades that have survived for long intervals of geological time at low numerical diversity and exhibit primitive morphological characteristics that have undergone little evolutionary change.

There are two families within the order, the first of which is extinct:

Family 1 Receptaculitaceae: laterals produced spirally on erect axis; relatively large plants; all extinct.

Family 2 Dasycladaceae: laterals produced in whorls on erect axis; relatively small plants; extinct and extant.

Receptaculitaceae

The plants in this order existed from the Lower Ordovician to the Permo-Carboniferous Period. They were non-septate marine dasycladaceous

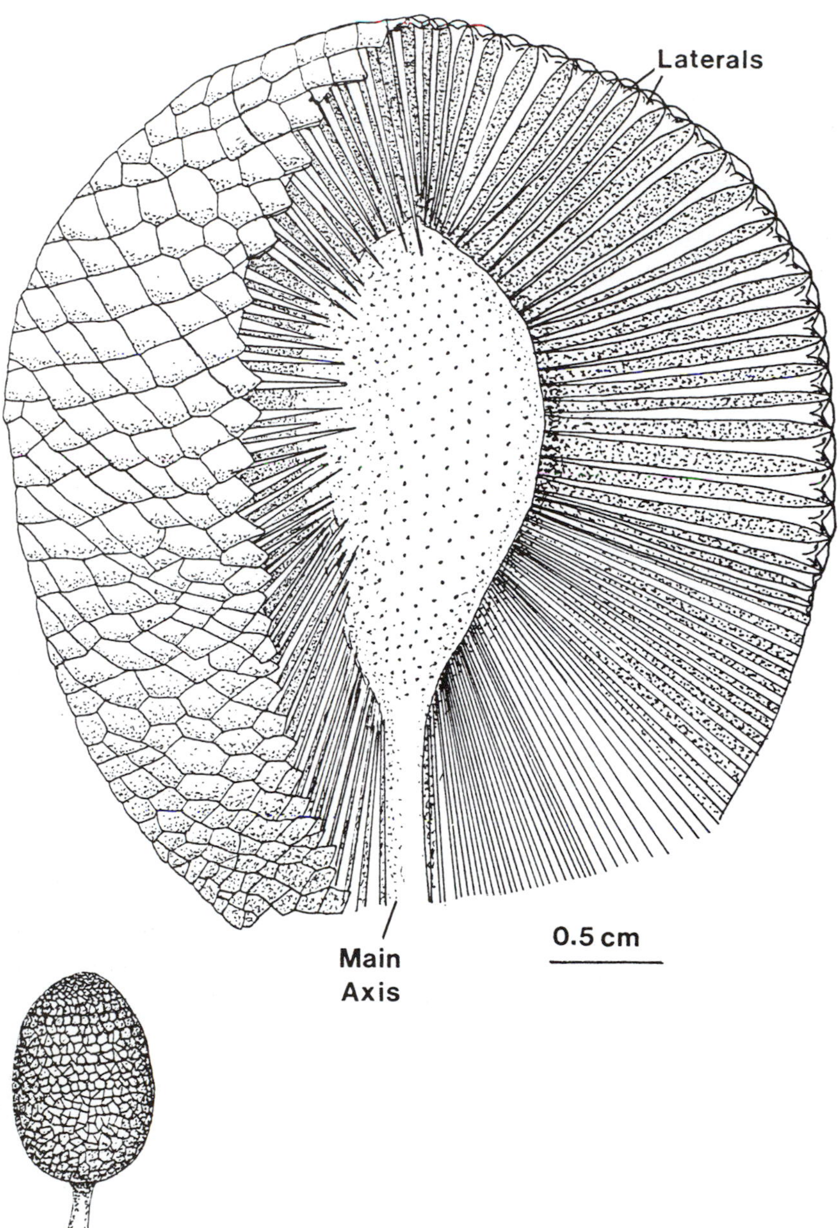

Fig. 5.31 Reconstruction of *Ischadites*, whole plant and cutaway drawing of the upper portion showing main axis, laterals, and lateral scars. (After Nitecki, 1971.)

algae shaped like a light bulb, with the upper portion consisting of a large number of spiral laterals attached to the main axis. They were calcified, with calcification being heavier around the periphery of the thallus. *Ischadites abbottae* was a Silurian alga that grew in shallow reef water. The thallus was globose, formed by spirally attached laterals with the largest laterals in the mid latitudes (Fig. 5.31). The upper laterals gradually became wider toward the periphery of the plant, then suddenly expanded into heads at the periphery. The main axis and the laterals were calcified (Nitecki, 1971).

Dasycladaceae

In the Dasycladaceae the plants have the laterals arranged in whorls on the main axis. There are about seven living genera, all marine and limited to warm waters. *Acetabularia* (Fig. 5.33) is the best known representative, consisting at maturity of a naked axis with usually a single gametangial disc at the apex. The axes of *Neomeris* and *Dasycladus* (Fig. 5.32), however, are enclosed by whorls of laterals that may form a fairly solid cortication.

The first fossils of this family appeared about the Middle Silurian, and some of the extant genera are well represented in the fossil record, *Neomeris* going back to the Cretaceous and *Acetabularia* to the Tertiary (Johnson, 1961).

The life cycle of *Acetabularia* is similar to that of other plants in the order. *Acetabularia* (mermaid's wineglass) is a warm-water alga found in shallow protected lagoons and on the borders of mangrove swamps growing on shells, coral fragments, and other algae. The thallus is calcified, with less calcification occurring in warm stagnant water. The young single-celled *Acetabularia* plant (Fig. 5.33) has two growing apices, one giving rise to the rhizoidal system that attaches the plants to the substrate, with the other growing apex becoming the erect thallus. As the apex of the axis grows, whorls of sterile hairs are produced just beneath the apex. Each of these whorls is eventually shed, leaving whorls of scars to mark their former attachment positions. During the vegetative growth of the thallus, the nucleus remains in one of the rhizoids.

A mature thallus forms a number of gametangial rays at the apex of the thallus, which can be joined or free from each other, depending on the species (Fig. 5.33). Near the base of each gametangial ray, at the tip of the thallus, is a coronal knob, which commonly bears sterile hairs. The coronal knobs together comprise the corona superior. Some species also have a corona inferior beneath the gametangial rays. Once the gametangial rays have reached full size, the primary nucleus in one of the rhizoids enlarges to

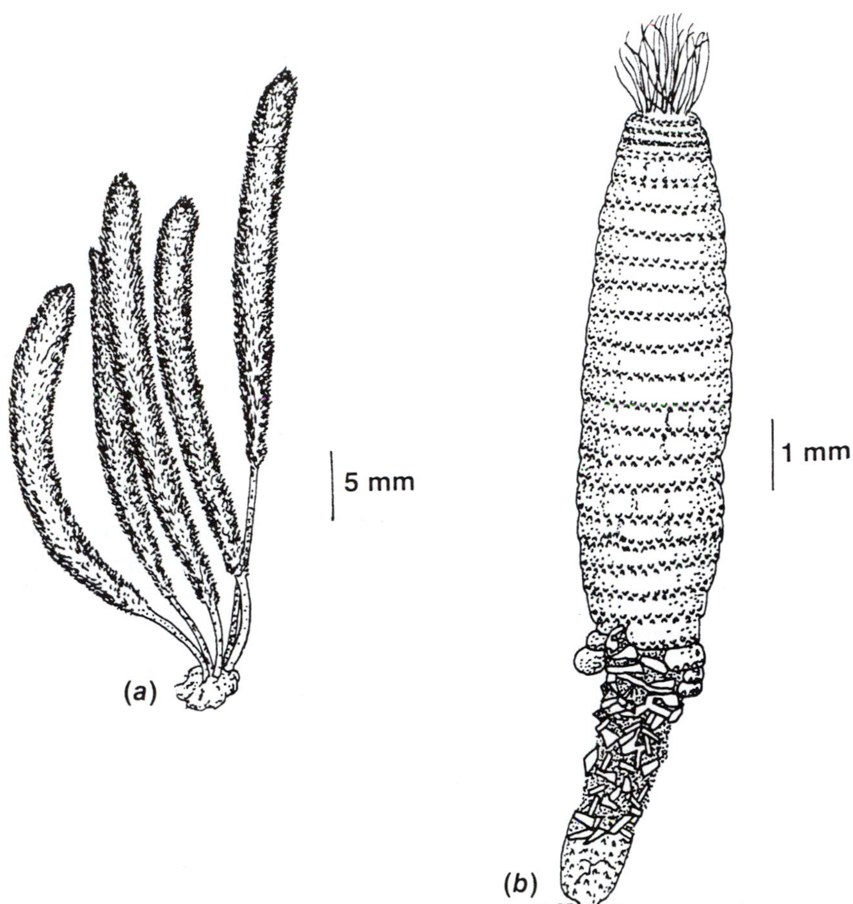

Fig. 5.32 (*a*) *Dasycladus vermicularis*. (*b*) *Neomeris annulata*. (After Taylor, 1960.)

about 20 times its original diameter. Thus nucleus then divides by meiosis (Koop, 1975a) into a large number of small secondary nuclei (Woodcock and Miller, 1973a,b), which are carried by cytoplasmic streaming along microtubules into the gametangial rays (Menzel, 1986). In the rays, each nucleus is held in place, a certain distance from other nuclei, by microtubules (Woodcock, 1971). The cytoplasm contracts around each nucleus, and a wall is formed, producing a resistant resting cyst. The cysts enlarge to many times their original size, a process that is accompanied by a number of nuclear divisions. The gametangial rays fall off, with a plug sealing the supporting part of the thallus (Menzel, 1980). The cysts of *Acetabularia mediterranea* are usually formed in the summer, are strongly calcified, and do not germinate until the following spring, requiring a resting period of 12 to 15 weeks (Koop, 1975b). At germination of the cyst, the protoplasm divides into a thousand or more pyriform biflagellate isogametes. The gametes are released through a lid in the cell wall. Gametes produced by

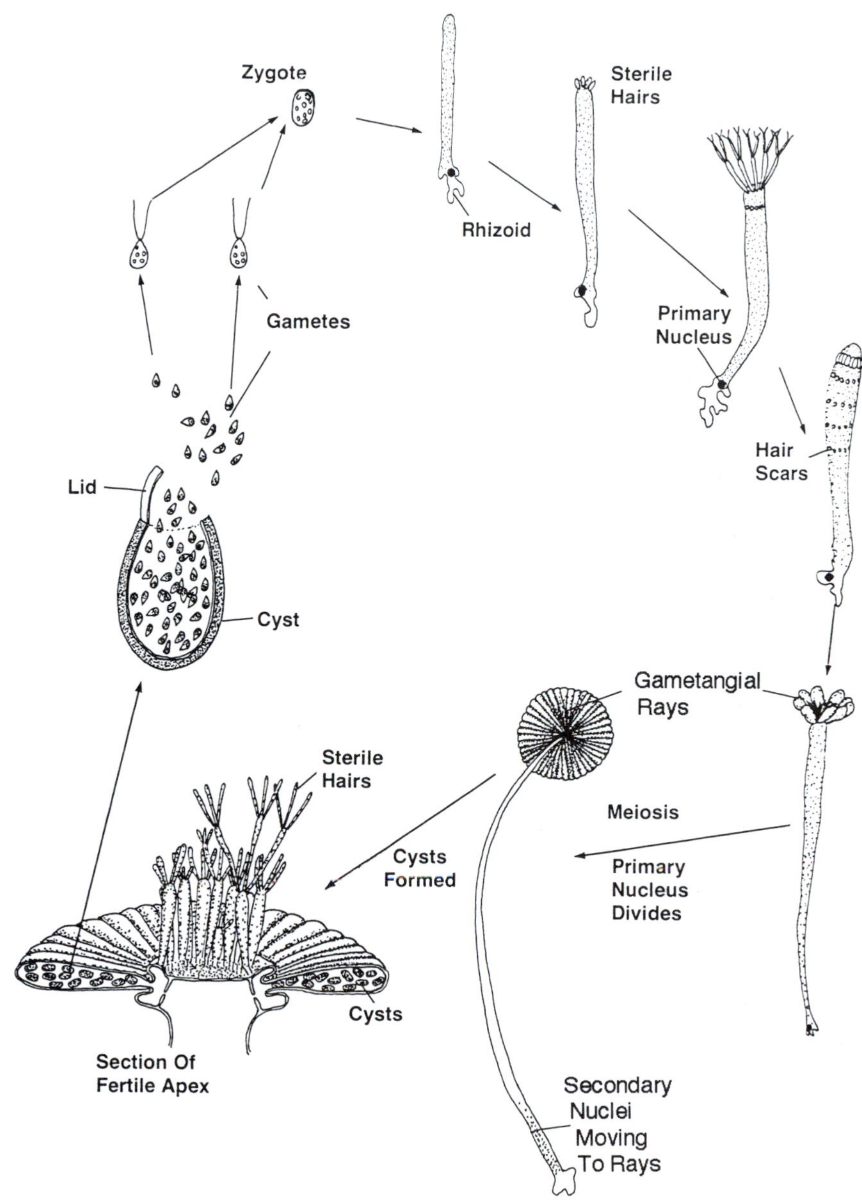

Fig. 5.33 The life cycle of *Acetabularia mediterranea*. (Adapted from Egerod, 1952; Smith, 1955.)

different cysts are morphologically similar, with gametes from a single cyst probably of the same sex. Parthenogenetic development of gametes has been noted. The zygote can germinate soon after fusing of gametes. In *Acetabularia calyculus*, gametes are of two sizes. After fusion to form a zygote, the chloroplasts of the smaller, male gametes are preferentially destroyed, leading to maternal inheritance of chloroplast genes (Kuroiwa, 1985; Kuroiwa, 1985).

CAULERPALES

This order contains the **coenocytic** or **siphonaceous** Chlorophyta. The **non-septate thallus** thus resembles a garden hose without any cross walls separating the usually large thallus, except during reproduction. The cells have numerous lens-shaped or fusiform-shaped chloroplasts and, in some cases, amyloplasts. Two carotenoids, **siphonoxanthin** and **siphonein**, not normally found in the Chlorophyta, occur in this order (with the exception of *Dichotomosiphon*, which has only siphonein; Kleinig, 1969). Cellulose is usually not a wall component and is replaced by a **β-1,3 linked xylan** or a **β-1,4 linked mannan** (Parker, 1970). The Caulerpales are marine algae and occur as seaweeds in the warmer oceans.

The most important families in the Caulerpales (or Siphonales or Codiales) are as follows:

Family 1 Derbesiaceae: stephanokontic zoospores; no amyloplasts; no oogamous reproduction.
Family 2 Codiaceae: only biflagellate swarmers; thallus basically filamentous; amyloplasts may be present; no oogamous reproduction.
Family 3 Caulerpaceae: only biflagellate swarmers; thallus composed of a stem bearing blades; amyloplasts present; no oogamous reproduction.
Family 4 Dichotomosiphonaceae: oogamous reproduction.

Derbesiaceae

Derbesia is a filamentous alga found in tropical and temperate waters growing on stones near the low-tide line or on larger algae. The life cycle of *Derbesia* (Kornmann, 1938; Feldmann and Feldmann, 1947; Feldmann, 1950) involves the alternation of this filamentous sporophyte with a bulbous vesicular gametophyte, the *Halicystis* stage, which was originally described as an independent plant. The *Halicystis* stage is found in deep water, usually epiphytic on a coralline alga. The filamentous *D. tenuissima* sporophyte (Zeigler and Kingsbury, 1964; Page and Kingsbury, 1968) has an interwoven basal portion that supports erect branched filaments, which are occasionally divided by pluglike septa (Fig. 5.34).

This filamentous sporophyte forms ellipsoidal sporangia as short lateral branches cut off by a septum from the main axis. The sporangia form zoospores meiotically (Neumann, 1969) with a whorl of flagella at one end. The zoospores swim for a while, settle, and germinate to produce

Fig. 5.34 The life cycle of *Derbesia*. (Adapted from Smith, 1955; Zeigler and Kingsbury, 1964).

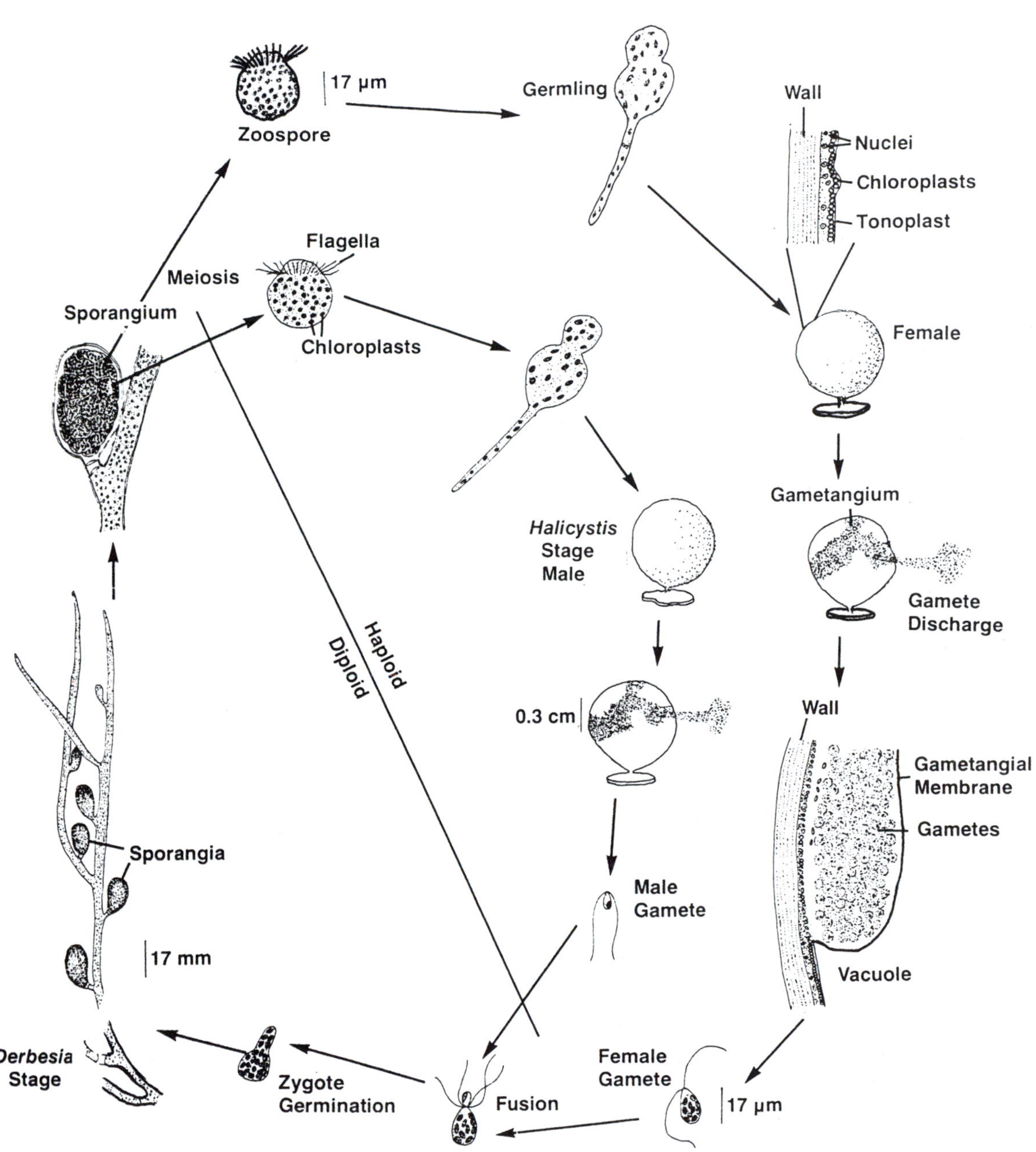

a filament that forms the vesicular *Halicystis* gametophyte. The *Halicystis* plants have nuclei arranged in an outer layer under the cell wall, with the chloroplasts in an inner layer next to the large central vacuole. The first visible stage of gamete formation is the migration of protoplasm to the plant apex; this protoplasm becomes separated by a membrane and is now the gametangium. The nuclei in the gametangium undergo synchronous divisions, with the protoplasm cleaving to form the uninucleate gametes.

Plants are heterothallic, producing only one type of gamete. The male plants have olive-green gametangia, whereas those of the female are blackish. Release of the gametes is induced by light, which causes an instantaneous increase in turgor pressure, rupturing the weakened pore area of the wall and causing a forcible expulsion of the gametes. Following release, the pore is sealed by the gametangial membrane (Wheeler and Page, 1974). Within 12 hours the cell is uniformly green, and it is capable of forming a second gametangium after 24 hours. The biflagellate gametes have one chloroplast in the male and 8 to 12 in the larger female (Roberts et al., 1981). The female gametes are less active than the male. Immediately after mixing, the male gametes surround the female. One of the males begins to fuse with the female, and the whole group of gametes sinks. The male and female nuclei in the zygote do not fuse. The other male gametes swim away from the zygote, which germinates immediately to form a new *Derbesia* plant. The coenocytic *Derbesia* is heterokaryotic, having nuclei derived from the male and female gametes existing separately in the protoplasm. Fusion of female and male nuclei (karyogamy) occurs in the sporangia, immediately before meiosis and the formation of zoospores (Eckhardt et al., 1986).

An endogenous rhythm controls gamete formation in the *Halicystis* stage of *D. tenuissima* (Page and Kingsbury, 1968). Gametogenesis has a basic period of 4 to 5 days in the laboratory, but in nature usually occurs in multiples of this figure, and is evidently timed by the tides. The rhythm is unaffected by changes in temperature or light intensity, indicating that it is an endogenous rhythm not directly linked to metabolic processes, such as photosynthesis. After induction of gametogenesis, about 7 hours of dark is necessary for maturation of the gametes, after which light causes their immediate release.

Bryopsis is a genus commonly found in quiet water of tide pools and other sheltered locations. The genus (Fig. 5.35(*d*)) usually has a main axis that supports lateral upright branches.

The cell walls of the gametophytic phases of *Derbesia tenuissima* and *Bryopsis plumosa* contain large amounts of xylans, whereas the walls of the sporophytes contain large amounts of mannans (Huizing et al., 1979).

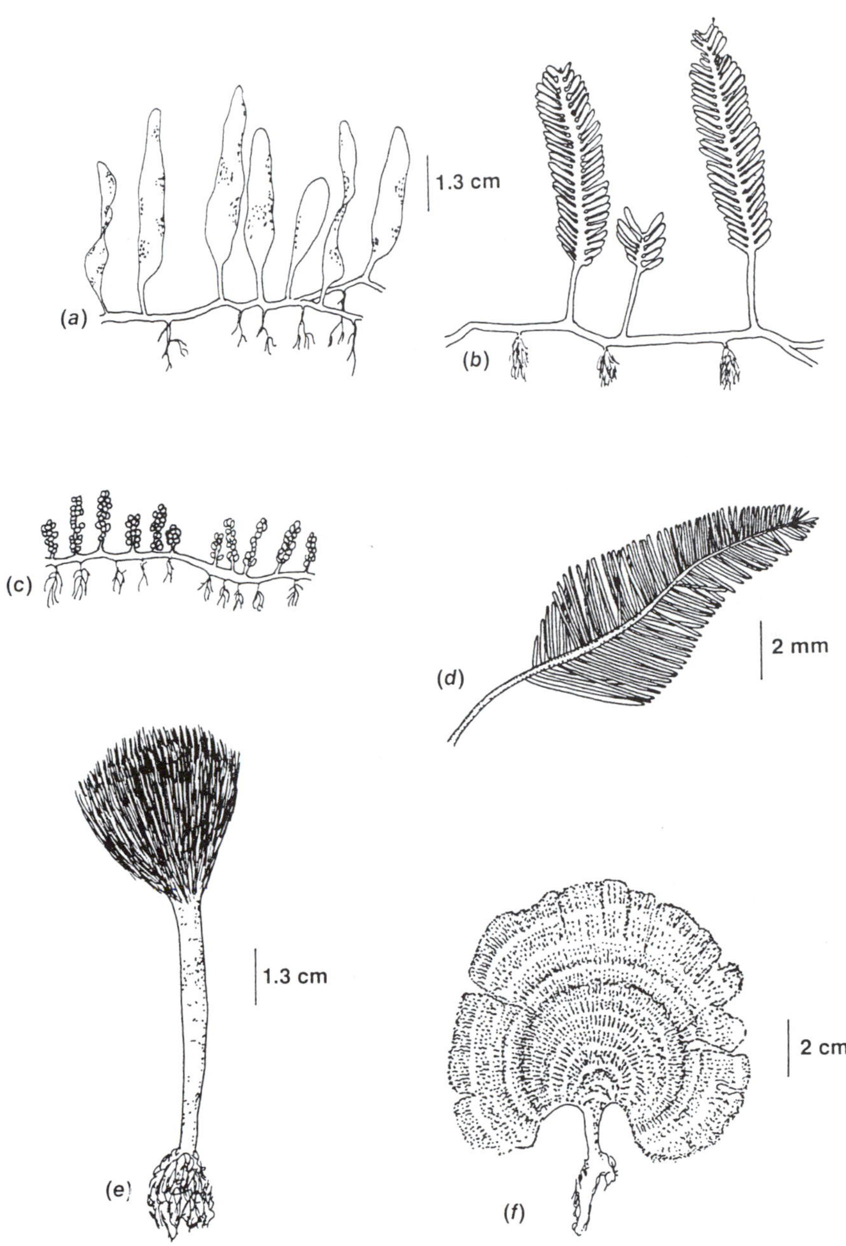

Fig. 5.35 (*a*) *Caulerpa prolifera*. (*b*) *Caulerpa floridana*. (*c*) *Caulerpa microphysa*. (*d*) *Bryopsis plumosa*. (e) *Penicillus capitatus*. (*f*) *Udotea conglutinata*. (After Taylor, 1960.)

Codiaceae

These algae differ from those in the Derbesiaceae in having biflagellate swarmers. The structure of the thallus is basically filamentous. The family can constitute a significant proportion of seaweed populations in tropical waters, including some attractive forms such as "mermaid's fan" or *Udotea* and "Neptune's shaving brush" or *Penicillus* (Fig. 5.35).

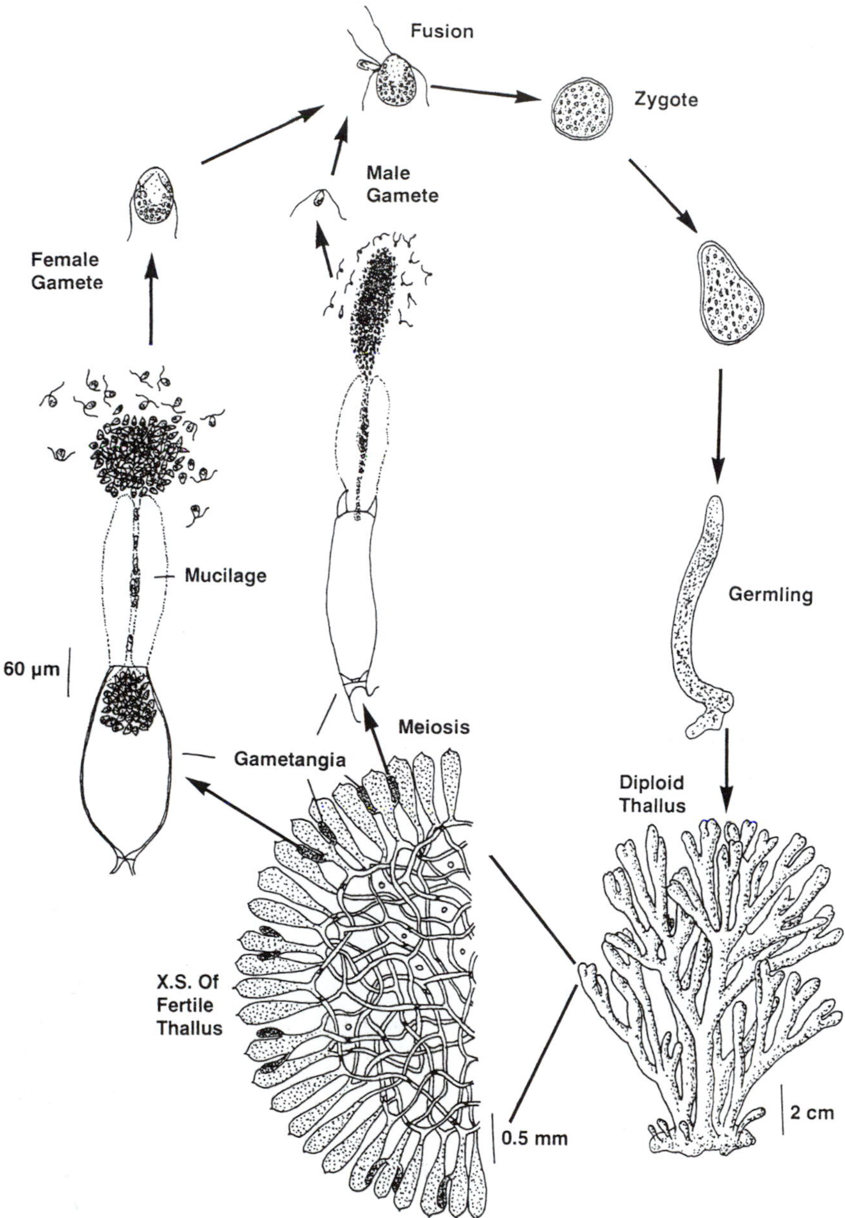

Fig. 5.36 The life cycle of *Codium* sp.

Codium (Fig. 5.36, 5.37) occurs from the low-tide mark up to 70 m depth in tropical and temperate marine waters. The genus was originally absent from much of the East Coast of North America. In 1957, *C. fragile* was found along the central Atlantic Coast (Bouck and Morgan, 1957), and it has subsequently spread as far north as Maine. It is probable that the alga was introduced on oysters transplanted from Europe. Since its introduction, *Codium*

Fig. 5.37 *Codium* sp. on a scallop.

has become a pest in oyster beds, attaching to oysters and causing them to be cast adrift during heavy storms (Fig. 5.37).

The thallus has a crustose prostrate portion that bears several cylindrical dichotomously branched shoots. The shoots have a central medulla composed of interwoven colorless filaments that give rise to inflated branchlets, the **utricles**, which surround the medulla. The utricles have a thick peripheral layer of cytoplasm around a large central vacuole. The discoid chloroplasts are in the outer part of the cytoplasm and the small nuclei in the interior. The colorless filaments of the medulla are divided in places by walls, especially near the base of the utricles. Dark-green female and brown male gametangia are produced from the utricles of the diploid thallus (Fig. 5.36). The gametes are formed meiotically and are released when the lidlike apical portion of a gametangium ruptures, extruding a gelatinous mass with a central canal through which the gametes move. The gametes initially lack flagella and are carried passively. After flagella extrusion, the gametes swim away, with the male gamete fusing with the side of a larger female gamete. The flagella from the male gamete are lost, and the flagella of the female gamete propel the zygote. After settling and flagella retraction, the zygote germinates immediately into a new *Codium* thallus. The gamete thus constitutes the only haploid structure in the life cycle.

Whole plants of *C. fragile* are able to fix nitrogen, owing to an association between the alga and a nitrogen-fixing bacterium (*Azotobacter*) on the surface of the alga (Head and Carpenter, 1975). The alga secretes 0.7 to 1.3 mg glucose per gram of dry weight of the alga per hour, or 16% to 31% of the carbon assimilated to the outside of the thallus. The bacterium uses the secreted glucose and in turn fixes the nitrogen. The nitrogen fixation occurs only under conditions of nitrogen deficiency and is probably an important factor in the growth of *Codium* in shallow bays under oligotrophic conditions.

Codium fragile shows a number of adaptations to its habitat. During winter months when the availability of dissolved inorganic nitrogen is at its highest in the water, *C. fragile* accumulates reserves of nitrogen which are utilized in times of relative nitrogen deficiency (Hanisak, 1979). The period of maximum carbon fixation, pigment content, and chloroplast size occurs during the early winter when competition from other algae is minimal and variation in tidal amplitude is decreased (Benson et al., 1983). In the summer, the physical environment is more extreme, owing to increased drying in the intertidal zone and increased competition from other algae. *Codium fragile* effectively retires from much of this competition by undergoing reproduction in the summer. This is accompanied by the development of frond hairs which may increase nutrient uptake.

Symbiotic associations between a number of molluscs and flatworms with chloroplasts of the Codiales are fairly common. Some molluscs (*Elysia, Tridachia, Placobranchus*) normally feed on siphonaceous Chlorophyceae such as *Codium* and *Caulerpa* by puncturing the cells and sucking out the contents. The chloroplasts are not always digested, and many chloroplasts lodge in the body of the animal and actively photosynthesize (Kawaguti and Yamasu, 1965; Trench et al., 1969, 1973a,b). In *E. viridis*, the chloroplasts can remain functional for a least 3 months when the animals are starved in the light. The rates of photosynthesis of chloroplasts intact in *Codium* and of chloroplasts in *Elysia* are of the same order (Trench et al., 1973b). Chloroplasts isolated from *Codium* release only 2% of their fixed carbon into the medium, mainly as glycolic acid. Isolated chloroplasts in animal homogenate release up to 40% of the fixed carbon, mostly as glucose with some glycolic acid. The animal cells obviously cause the symbiotic chloroplasts to release a large amount of their photosynthate, calculated to be at least 36% of the fixed carbon (Trench et al., 1973b).

The plants of *Halimeda* (Fig. 5.38) consists of calcified segments separated by more or less flexible little-calcified nodes. Coenocytic filaments

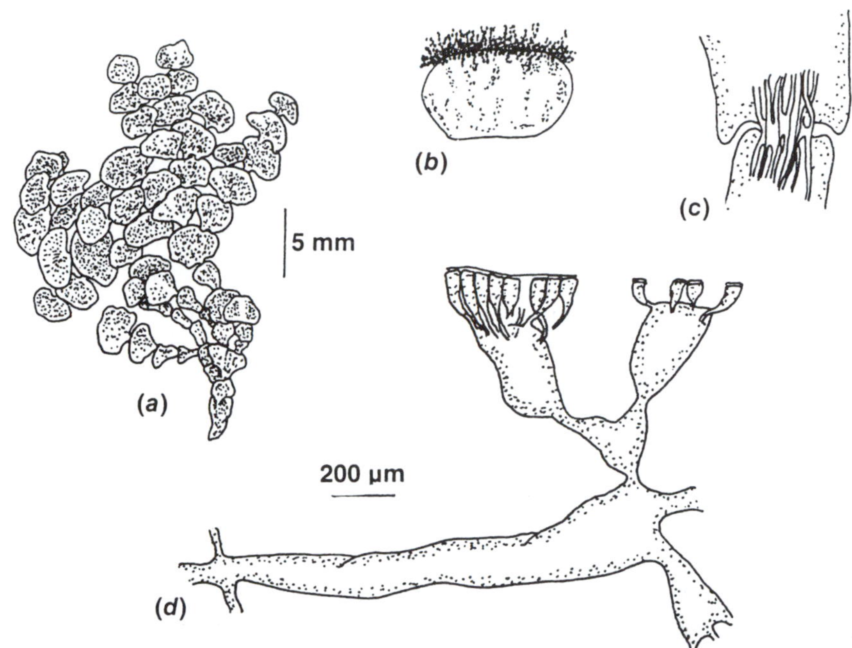

Fig. 5.38 (*a*) *Halimeda tuna*. (*b*) Fertile segment of *Halimeda*. (*c*) Nodal region of *Halimeda opuntia*. (*d*) Central filament of *Halimeda discoidea*, with lateral branches forming the outer utricles. ((*a*) after Taylor, 1960; (*c*),(*d*) after Egerod, 1952.)

make up the thallus. At the surface of each segment the filaments are inflated to form utricles which are closely appressed to one another in mature plants and form an unbroken surface separating the intercellular spaces from contact with the outside. The cell walls of the utricles are calcified. Occasionally, *Halimeda* plants bear clusters of beadlike reproductive structures on branched stalks arising from the surface of the segments. These form biflagellater swarmers, but it is not known whether they are gametes or zoospores.

Two types of plastids exist in *Halimeda* – amyloplasts and chloroplasts. Both types develop from proplastids (Borowitzka and Larkum, 1974). Actively growing plants of *Halimeda* may add a segment a day, the newly formed segment being non-calcified and white, containing only amyloplasts (Wilbur et al., 1969). At this stage the utricles have not completely closed, leaving intercellular spaces continuous with the outside medium. After the segment is 36 to 48 hours old, the utricles have closed with one another, and aragonite $CaCO_3$ crystals begin to appear on fibrous material outside the walls of the utricles. By the time that calcification has begun, the segment contains well-developed chloroplasts and is green. The older the segment becomes, the fewer amyloplasts and the more chloroplasts there are present.

Incorporation of radioactive ^{45}Ca is stimulated by light, with calcium incorporation into the thallus showing a diurnal rhythm, being greater during the day than during the night (Stark et al., 1969), even under constant illumination. This difference in calcification rates is reflected in the movement of chloroplasts to the periphery of the segments during the day and away from the periphery during the night. The calcification process appears to be a two-step process: First the ions are bound to the wall, with the consequent increase in their concentration, and then the $CaCO_3$ is precipitated. *Halimeda* and *Penicillus* have lower rates of $CaCO_3$ deposition than the calcified red algae (Goreau, 1963).

Halimeda is particularly successful at colonizing bottom habitats where ambient light intensities are from 10 to 20 times less than at the surface. The Halimedas are exceptional among the calcareous Chlorophyta in that they tend to be more heavily calcified in deep than in shallow water. Although this difference may be due to a decreased amount of organic matter rather than an increased calcification, the overall effect is opposite to that of other calcareous green algae such as *Penicillus*, *Udotea*, and *Rhipocephalus*, which are invariably less calcified in deep water. *Halimeda* often performs the major role in calcification in lagoons. Hoskin (1963) examined the sand of Alacran Reef, Mexico, and found that it consisted of 35% *Halimeda*, 29% coral, 8% other coralline algae, 8% mollusks, 6% foraminifera, 1% miscellaneous skeletal grains, 9% fecal pellets, and 4% aggregates by volume.

The family Codiaceae is one of the most important groups of rock-building algae, and, in the course of its long history, has been represented by a large number of genera (Johnson, 1961). In extant (living) plants the calcified genera usually have the outer portion calcified, whereas the inner portion is not. This type of calcification exists in fossil forms, resulting in good preservation in the outer part of the fossil, but with the structural features gradually fading toward the center. A section of *Paleocodium* from the lower Carboniferous (Fig. 5.39) shows a structure similar to that of *Codium* (Fig. 5.36). Restoration of a branch of *Ovulites margaritula* from the Eocene (Fig. 5.39) results in a structure similar to that of *Halimeda*.

Caulerpaceae

The coenocytic plants of this family have two types of plastids – chloroplasts and amyloplasts. *Caulerpa* (Fig. 5.35) is the only genus in the family and is a common inhabitant of intertidal and infratidal tropical and semitropical marine waters. The plants have a creeping green rhizome with rootlike colorless rhizoids and frondlike erect shoots. The erect shoots exhibit a considerable variation in morphology, many times resembling the blades of higher plants, after which some of the species are named. The thallus derives support from turgor pressure and from wall ingrowths, the **trabeculae**. The walls have a β-1,3 linked xylan as the main structural component.

Chloroplasts are prominent in the leaves and rhizome but totally lacking

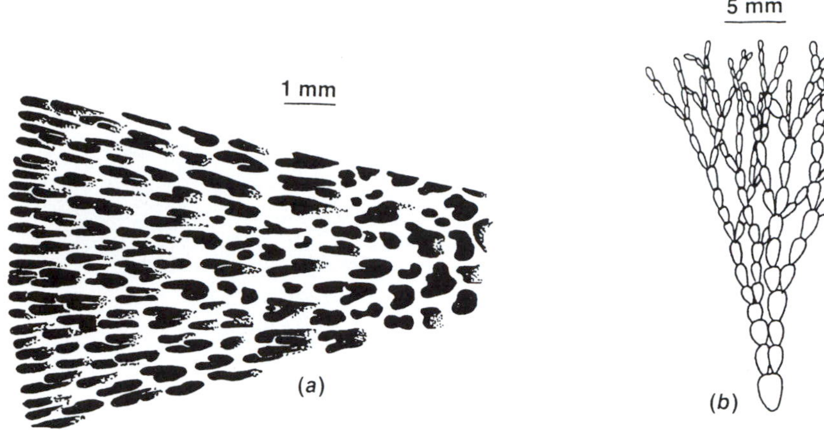

Fig. 5.39 Diagrammatic sketch of a section of the thallus of *Paleocodium* from the Lower Carboniferous, showing arrangement of branching filaments. (*b*) Restoration of a branch of *Ovulites margaritula* from the Eocene. ((*a*) after Johnson, 1961; (*b*) after Munier-Chalmas in Johnson, 1961.)

in the rhizoids and the extreme apex of the growing rhizome tip and growing blade. Amyloplast distribution is the reverse of chloroplast distribution, large numbers of amyloplasts being present in the rhizoids and blade tip with few amyloplasts in the rhizome and blades. There is a large central vacuole except at the growing tips. The vacuole contains material that quickly repairs damage to the coenocyte and prevents the loss of cell contents. After wounding, vacuolar contents are extruded to the surface of the wound by turgor pressure where they undergo sol-to-gel transformation, forming a gelatinous plug composed of carbohydrate (Goddard and Dawes, 1983; Menzel, 1988). Inside this gelatinous plug, there is the formation, for up to 11 hours, of an internal wound plug having a composition different from that of the original plug (Dreher et al., 1978). The cytoplasm retracts away from the wound site, and a large number of vesicles form between the cytoplasm and the surface of the internal wound plug. New wall synthesis begins and is complete 2 to 6 days from the time the thallus was wounded. After wounding, there is a decrease in the rate of photosynthesis and an increase in the rate of respiration, with both rates returning to those of unwounded tissues within 6 hours. Wounding depresses the rate of starch and sucrose synthesis, but increases the rate of β-1,3 linked glucan, lipid, and sulfated polysaccharide synthesis. These changes are consistent with the direct involvement of sulfated polysaccharides in the healing process (Hawthorne et al., 1981).

Mature regions of *Caulerpa* have two systems of protoplasmic streaming: large longitudinal streams in the vacuole and smaller streams oriented 45° to the blade axis in the peripheral cytoplasm. Bundles of microtubules are associated with the cytoplasmic streaming (Sabnis and Jacobs, 1967). The rate of streaming is relatively slow, 3–5 $\mu m\ s^{-1}$, as compared to 60 $\mu m\ s^{-1}$ for *Nitella*.

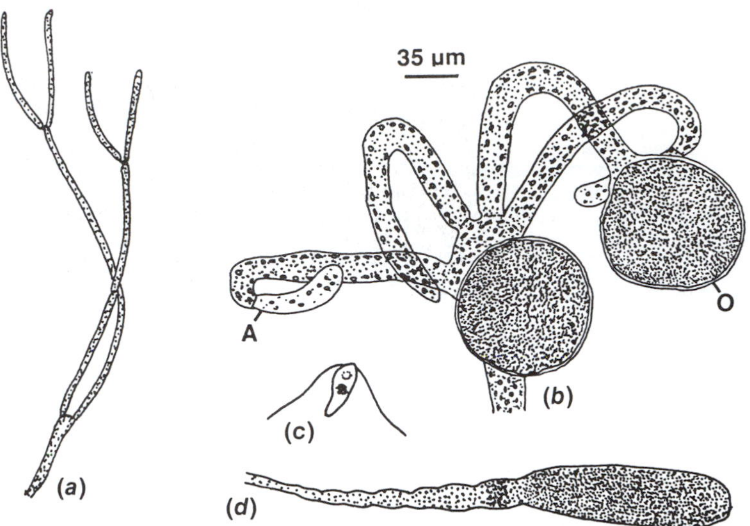

Fig. 5.40 *Dichotomosiphon tuberosus*. (*a*) Vegetative thallus. (*b*) Fertile plants with oogonia (O) and antheridia (A). (*c*) Spermatozoid. (*d*) Portion of thallus with akinetes. ((*b*) after Ernst, 1902; (*c*) after Moestrup and Hoffman, 1975; (*d*) after Smith, 1955.)

Asexual reproduction in *Caulerpa* is by fragmentation of the thallus or by abscission of proliferous shoots, with no zoospores produced, similar to that of other members of the Caulerpales.

In sexual reproduction, male and female gametangia are formed on the same frond by migration of the cytoplasm from the rhizomes into the fronds and subsequent cleavage to yield gametes (Goldstein and Morral, 1970). The large female gametes and smaller males are released in a greenish viscous fluid. After a few minutes the gametes agglutinate in groups of up to 50, followed by separation of pairs and the formation of zygotes. The development of *Caulerpa* has not been followed beyond zygote formation.

Dichotomosiphonaceae

Dichotomosiphon differs from the rest of the Caulerpales in having oogamous sexual reproduction. The thallus is a dichotomously branched tubular coenocyte bearing colorless rhizoids. Both lens-shaped chloroplasts and amyloplasts are present in the cytoplasm (Moestrup and Hoffman, 1973). Siphonein is present, but the related pigment siphonoxanthin is absent (Kleinig, 1969). The cell wall has a β-1,3 linked xylan as the main structural wall component (Maeda et al., 1966).

There are two species of *Dichotomosiphon:* the marine *D. pusillus* and the freshwater *D. tuberosus* (Fig. 5.40). The latter grows in lakes with an organic silty bottom, most of the thallus usually buried and only the tips of the branches above the silt. In deep water (more than 2 m), only asexual reproduction occurs, by the formation of akinetes in series at the ends of branches. The akinetes germinate directly to form new thalli (Ernst, 1902). In shallow water, reproduction is sexual, homothallic, and oogamous. Conical antheridia are produced at the tips of branches and separated from the rest of the thallus by a septum. The antheridia burst open explosively at the apex, releasing the biflagellate spermatozoids, which have a single reduced chloroplast but lack an eyespot (Moestrup and Hoffman, 1975). An oogonium is spherical, and just before fertilization develops a small beak-like opening at its apex.

SIPHONOCLADALES

These organisms have multicellular thalli, are wholly marine, and are usually tropical. The cells are multinucleate, with reticulate chloroplasts, and divide in a distinct manner known as **segregative cell division**. Most of the organisms have siphonoxanthin (except *Dictyosphaeria*) in addition to the normal pigments of the Chlorophyta (Kleinig, 1969). Sexual reproduction appears to be isogamous in most cases.

Siphonocladus tropicus initially has an undivided single-celled primary vesicle (Fig. 5.41). In segregative cell division, the continuous protoplast of the primary vesicle breaks into spherical masses of varying size that soon become surrounded by a wall and enlarge to fill the area within the expanding parent vesicle. After each segment has become firmly pressed against adjacent ones, it sends out a lateral protuberance, which constitutes a branch initial. The mature thallus consists of an erect axis with lateral branches (Egerod, 1952).

In *Valonia* (Fig. 5.41), a young plant consists of a bladderlike multinucleate primary cell. By segregative cell division small holdfast cells are formed so that the thallus consists of a large (3 cm) primary cell with a number of smaller rhizoidal cells. Segregative cell division begins with a differentiation of a dense, lens-shaped mass of protoplasm at a local area in the cell. A wall is formed around this dense mass, and the small, lens-shaped cell that has differentiated elongates into a holdfast cell. In some species of *Valonia*, other large primary cells are formed by segregative cell division. A mature cell has a conspicuous central vacuole surrounded by a relatively thick layer of cytoplasm containing an outer layer of polygonal chloroplasts with pyrenoids and an inner layer of nuclei. Because of the large size of the primary vesicle cell, the organism is a common tool for physiologists interested in characterizing the relationships between the vacuole and the protoplasm. It is relatively easy to remove the vacuolar contents from such a cell. The osmotic values of the vacuolar sap are 1 to 3 atm

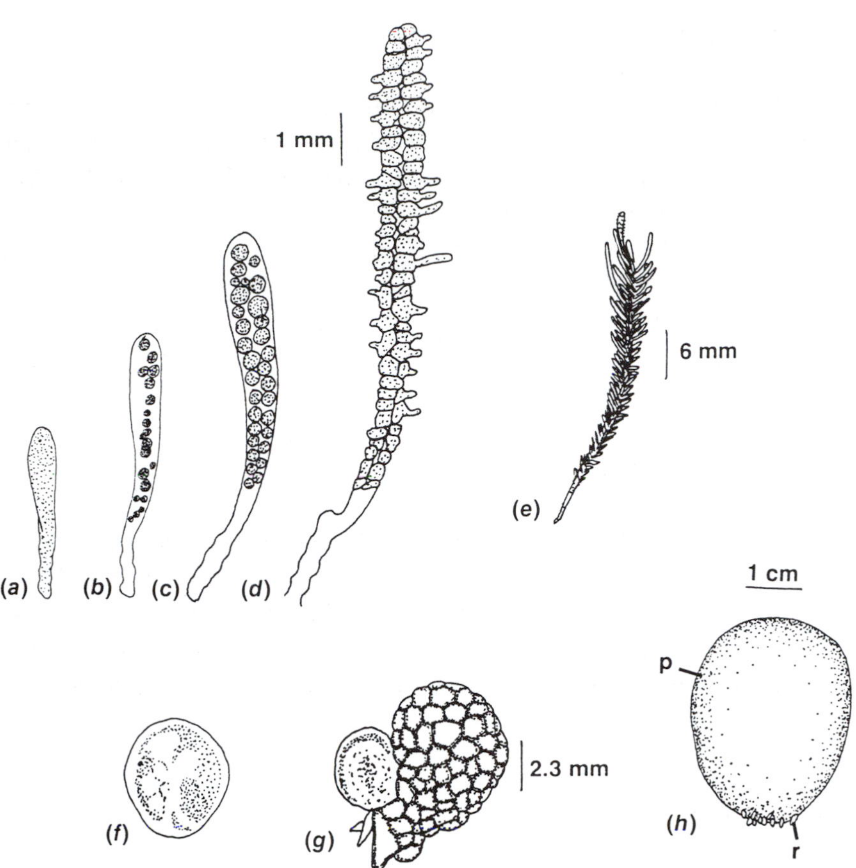

Fig. 5.41 Segregative cell division in *Siphonocladus tropicus*: (*a*) germling; (*b*) cytoplasm in spherical masses; (*c*) expansion of cytoplasmic masses; (*d*) lateral branches forming; (e) mature thallus. *Dictyosphaeria cavernosa*: (*f*) young aseptate vesicle; (*g*) secondary vesicle attached to primary vesicle. (*h*) *Valonia ventricosa*. (p) Primary vesicle cell; (r) rhizoidal cell. ((*f*),(*g*) after Egerod 1952; (*h*) after Taylor, 1960.)

higher than that of seawater, with the concentration of potassium in the vacuolar sap being 66 times higher than that in seawater (Gessner, 1967).

Any vegetative cell of *Valonia* (Fig. 5.41) can divide into biflagellate swarmers. The pyriform, uninucleate swarmers escape through pores in the cell wall. Although zygotes have been seen, fusion of gametes has not, but it is presumed that meiosis occurs in the cells immediately before the formation of swarmers.

When the large cells of *Valonia* or *Boergesenia forbesi* are either placed in double-concentration seawater or subjected to mechanical damage (pinching or probing), the protoplasm is transformed within 24 hours into numerous aplanospores. After release from the parent thallus, they germinate into new plants. The above mechanism appears to be an adaptation to ensure survival of the populations that otherwise might decline due to deaths of cells from injuries (Nawata et al., 1993).

Algae in the Siphonocladales are primarily plants of shallow warm oceanic water growing on stones and heavy shells.

Chlorophyceae

The distinguishing characteristics of the Chlorophyceae are the **theca outside of the cells** and a **collapsing telophase spindle** that brings the daughter cells close together, followed by **cell division by a phycoplast**. The flagellar root system is cruciate.

Some of the flagellates in the class do not have a theca, but these are assumed to have lost the theca in evolution because the cells have the other characteristics of the class. The Chlorophyceae are predominantly freshwater. The few unicellular, planktonic species that occur in coastal seawater are members of genera that have a much greater number of freshwater species, such as *Chlamydomonas*. The Chlorophyceae whose sexual reproduction is known produce a dormant zygote, with meiosis usually occurring when the zygote germinates.

The other characteristics of the Chlorophyceae were described earlier in this chapter and include motile cells with radial or near-radial external symmetry, flagella attached at the anterior end of the cell, the possibility of a rhizoplast, no multilayered structure, eyespots common, glycolate breakdown by glycolate dehydrogenase, and urea breakdown by urea amidolyase.

Classification

The Chlorophyceae are divided into the following important orders:

Order 1 **Volvocales:** vegetative cells flagellated and motile.

Order 2 **Tetrasporales:** non-filamentous colonies with immobile vegetative cells capable of cell division; pseudocilia may be present.

Order 3 **Schizogoniales:** foliose marine algae with stellate chloroplasts.

Order 4 **Chlorococcales:** unicellular or non-filamentous colonial algae; if colonial, then daughter colonies formed as coenobia; vegetative cells non-motile.

Order 5 **Sphaeropleales:** unbranched filaments with new walls formed inside the old filament walls, resulting in H-shaped wall pieces.

Order 6 **Chlorosarcinales:** daughter cells retained within parent cell wall; no plasmodesmata present.

Order 7 **Chaetophorales:** branched or unbranched filaments; plasmodesmata present.

Order 8 **Oedogoniales:** uninucleate filamentous freshwater algae with a unique type of cell division; motile spores and gametes with a whorl of flagella at one pole.

VOLVOCALES

In the Volvocales the vegetative cells are flagellated and motile. The algae can be unicellular or multicellular. If they are multicellular, then the number of cells in the vegetative colony is a multiple of two. Almost all of these organisms are freshwater, being abundant in waters high in nitrogenous compounds.

There are two important families in the order:

Family 1 **Chlamydomonadaceae:** unicellular algae.
Family 2 **Volvocaceae:** colonial algae formed into coenobia (colonies with a definite number of cells arranged in a specific manner).

Chlamydomonadaceae

In this family are all the unicellular Volvocales. All of the genera are uninucleate, and usually the chloroplast is cup-shaped.

Chlamydomonas is a unicellular, biflagellate organism with a cup-shaped, basal chloroplast containing a central pyrenoid (Fig. 5.42). The cells are uninucleate, with two contractile vacuoles at the base of the flagella, and there may, or may not, be an anterior eyespot in the chloroplast. A cell wall usually surrounds the protoplast. When growing on soil, or on agar in the laboratory, *Chlamydomonas* is non-motile and grows in gelatinous colonies. If these colonies are flooded, motile unicells are formed. The cells are widely distributed in freshwaters and in damp soils where they are commonly found in areas with high concentrations of nitrogen, such as farmyard soils.

Asexual reproduction begins by the *Chlamydomonas* cell coming to rest and usually retracting or discarding the flagella. The protoplast then divides into 2, 4, 8, or 16 daughter protoplasts within the parent wall (Fig. 5.42). These protoplasts develop walls and flagella, and are released on gelatinization of the parent wall, if growing in liquid medium.

Most species of *Chlamydomonas* exhibit isogamous sexual reproduction (Musgrave, 1993). In *C. moewusii* (*C. eugametos*), there is no structural difference between the gametes and the vegetative cells, and gametogenesis involves the production of agglutinins that cover the flagella (Demets et al., 1990). When gametes of a plus strain are mixed with those of a minus strain, the flagella of the opposite strains adhere because of the agglutinins on the flagella. Initially the gametes clump in groups of up to 50 cells with their flagella toward the center and with different numbers of plus and minus gametes in each clump. Eventually pairs of opposite gametes fuse at their anterior ends, the flagella become free, and the pair swims away from the clump. Before fusion of gametes, a hole is dissolved in the wall at the interior end of the gametes, and a fertilization tubule is pushed out through

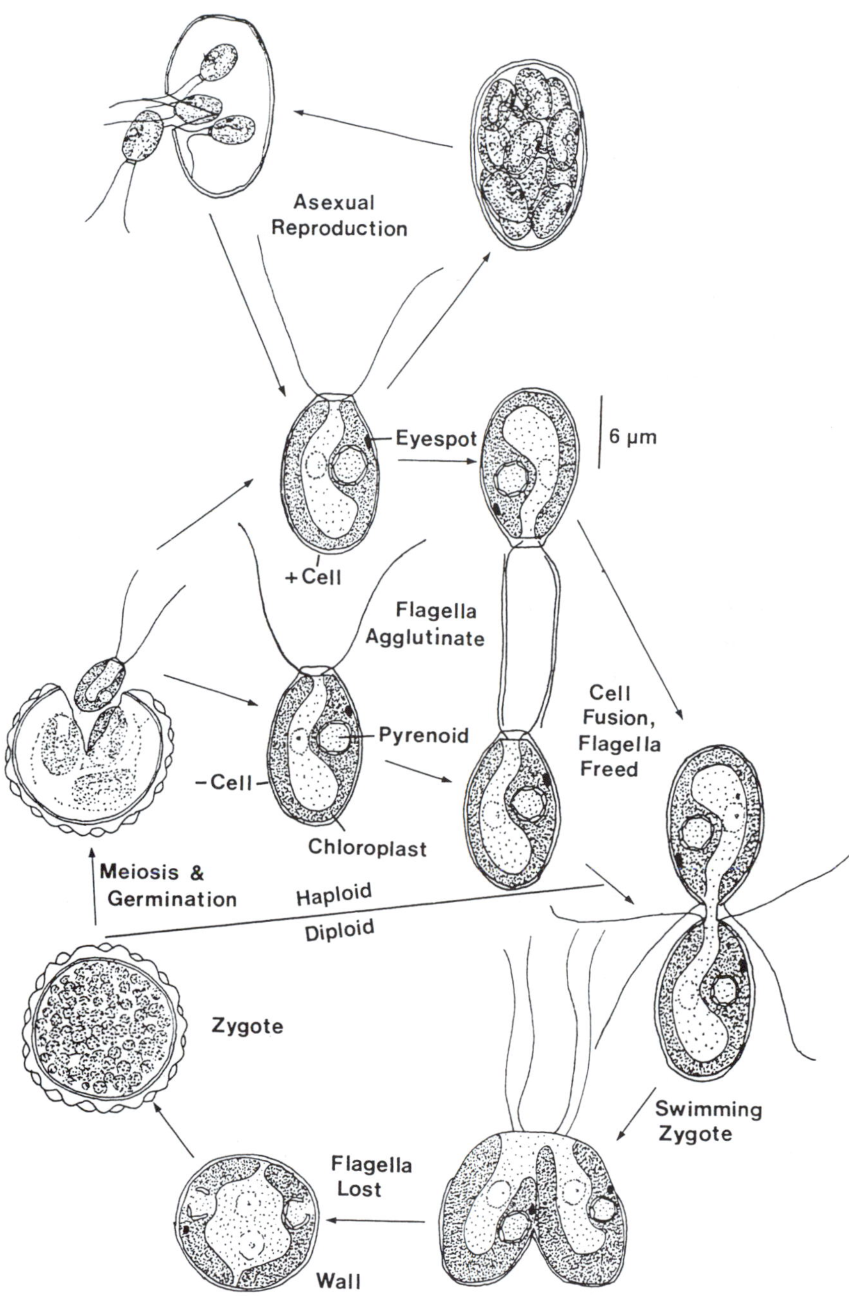

Fig. 5.42 The life cycle of *Chlamydomonas moewusii*. (Adapted from Brown et al., 1968.)

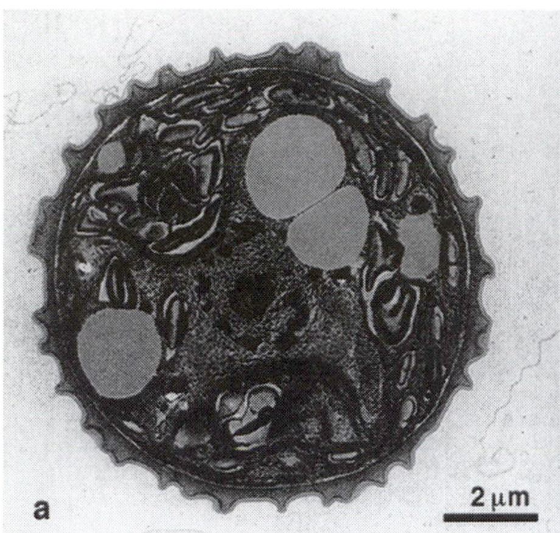

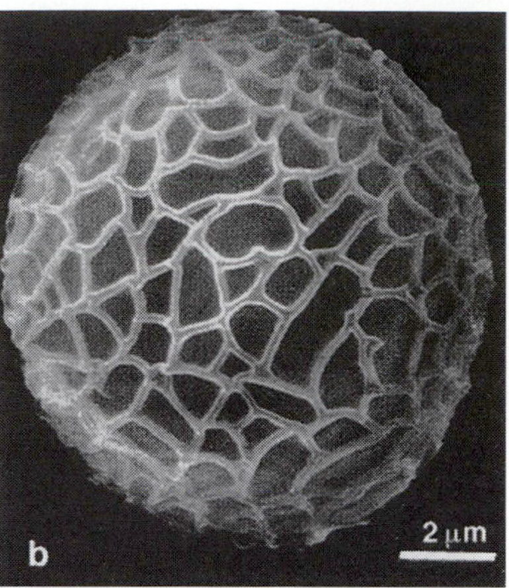

Fig. 5.43 Transmission and scanning electron micrographs of mature zygospores of *Chlamydomonas monoica*. (From Van Winkle-Swift and Rickoll, 1997.)

the hole. These tubules fuse at their tips, beginning fusion in the gametes. This connecting tubule shortens, causing the apical portion of the gametes to fuse, followed by jackknifing of the two gametes and lateral fusion. The quadriflagellate zygote swims for a while before settling down and forming the primary zygote wall. The nuclei fuse within 24 hours, and a secondary zygote wall with horns and spines is laid down. As the zygote matures, it may or may not enlarge, depending on the species, but it always accumulates large quantities of oils and starch, which may turn it reddish. Light and carbon dioxide are required for development of the zygote (Lewin, 1957). Zygote (zygospore) (Fig. 5.43) germination occurs in the dark; the inner layer of the zygospore is dissolved away and the sculptured exine splits open to liberate the motile zoospores. The zoospores are formed by meiosis. Usually 4 or 8 zoospores are released, but there can be as many as 16 or 32.

The sequences of events in *Chlamydomonas reinhardtii* is slightly different (Martin and Goodenough, 1975). Here gametogenesis is induced by placing the vegetative cells in nitrogen-free medium (Sager and Granick, 1954). In as little as 10 to 12 hours, gametes are differentiated. The mating reaction is initiated when mating-type "plus" (mt^+) and "minus" (mt^-) gametes are mixed together (Mesland et al., 1980; Snell, 1985). The resulting interactions can be divided into seven stages, all of which are completed within 30 seconds (Fig. 5.44).

1 The cells adhere to one another via *mt*-specific flagellar surface **agglutinins**, which are glycoproteins of extremely high molecular mass (Saito and Matsuda, 1984). The agglutinins cover the flagellar surface and are continuously lost and replaced from a large cytoplasmic pool (Snell and Moore, 1980; Saito et al., 1985).
2 Pairs of adhering cells move their agglutinins out to their respective flagellar tips, a "tipping" response that brings the cells close together.

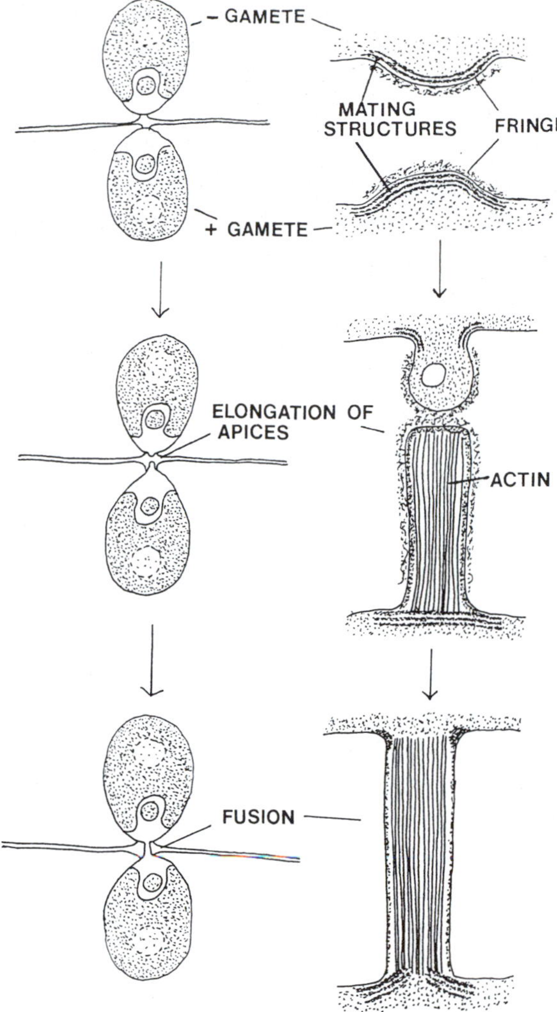

Fig. 5.44 Drawings of fusing cells of *Chlamydomonas reinhardtii* on the left. On the right are enlargements of the apical areas of the fusing cells. (Adapted from Goodenough et al., 1982.)

3 One or more "signals" are transmitted to the cell bodies of the paired cells.

4 As a first response to signaling, cells release an **autolysin** that dissolves the crystalline glycoprotein wall (Imam et al., 1985) surrounding the gametes. The autolysin is an enzyme composed of a single polypeptide with a molecular mass of 62 kilodaltons. Autolysin occurs in an inactive soluble form in vegetative cells. Autolysin is converted to a soluble active form when the cell is transformed into a gamete (Matsuda et al., 1987).

5 As a second response to signaling, the cells activate their **mating structures**. The plus gamete has a mating structure composed of three electron-dense layers (Fig. 5.44) that lie under the plasma membrane in an apical bud, approximately 1 μm in diameter (Goodenough et al., 1982).

The minus gamete has a mating structure composed of two electron-dense plates in a similar apical bud. Covering the plasma membrane of the apical buds of the plus and minus gametes is a **fringe** of fuzzy material that probably contains the recognition apparatus of the plus and minus gametes.

6 The apical bud of the minus gamete elongates slightly into a dome-shaped structure. In the plus gametes, actin microfilaments extend from the mating structure into the bud to form a long narrow fertilization tubule. Cell fusion occurs when the fringe at the tip of the fertilization tubule of the plus gamete strikes the fringe at the tip of the apical bud of the minus gamete. A narrow cytoplasmic bridge is formed, which opens up to allow full cytoplasmic confluence.

7 The adhering flagella of the resulting quadriflagellate cell lose their agglutinative properties, presumably in response to a "signal to dis-adhere," which is transmitted at the time of cell fusion. The remaining steps leading to the formation of a zygote are similar to those of *Chlamydomonas moewusii.*

The attraction of gametes of the opposite strain involves the recognition by each cell of a chemical substance or **agglutinin** on the flagella (Bloodgood, 1991). The agglutinins are formed in the cells and pass out of the flagella tip to coat the flagella. The agglutinins are glycoproteins, male and female cells having different glycoproteins or sexual hormones. The glycoproteins have at least 12 amino acids and 5 sugars, with the female glycoprotein having 36% protein and the male 21% (Foerster et al., 1956). The recognition mechanism has been postulated as follows (Wiese and Metz, 1969; Wiese and Shoemaker, 1970; Wiese and Hayward, 1972): The minus gametes have a protein as the recognition part of their flagellar glycoprotein. This protein has the ability to recognize a particular sugar sequence in the glycoprotein on the flagellum of the plus gametes (possibly a mannose-containing polysaccharide).

Most species of *Chlamydomonas* are found in small ponds and puddles. An unusual habitat is in the snows of some mountains in western North America (Hardy and Curl, 1972). Here *C. nivalis* is responsible for what is known as "red snow." Great numbers of this alga produce the watermelon color associated with these alpine snowfields. In spring and early summer, the algae grow most abundantly near the surface of old, melting snow, although the resting spores can be found to a depth of more than 2 feet. If the snow persists until autumn, then the algal cells remain until heavy snowfalls cover the surface. Upon the arrival of spring, the algal cells migrate upward to the surface of the snow to form another bloom. The algae need melt-water before they will grow, and this need explains their lack of

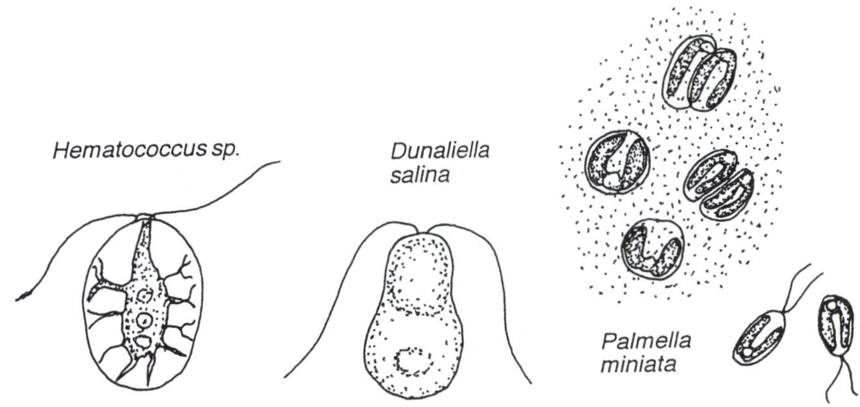

Fig. 5.45 Drawings of motile cells of *Hematococcus* sp. and *Dunaliella salina*, and palmelloid and motile cells of *Palmella miniata*.

growth in the winter. These snow algae support a unique type of fauna, including various species of protozoans, ciliates, rotifers, nematodes, spiders, and springtails. The last organism in this food chain is the segmented snow worm (*Mesenchytraeus*), which like the algae is not prevalent during the winter. During an 1891 expedition to the Malaspina Glacier in Alaska, Dr Israel Russel wrote: "In the early morning before the sunlight touched the snow, its surface was literally covered with small slim black worms, about an inch long and having a remarkable snakelike appearance. These creatures were wriggling over the snow in thousands."

Dunaliella, a green algae that looks like *Chlamydomonas*, has species that have adapted to life in acid waters and species that have adapted to waters high in salt. *Dunaliella acidophila* is an acid-resistant species that exhibits optimal growth at pH 1.0 (Geib et al., 1996). *Dunaliella salina* (Fig. 5.45) has adapted to waters high in salt, where it is common in places such as the Great Salt Lake of Utah. *D. salina* has two varieties. The first variety has small green halotolerant cells that can grow at NaCl concentrations of 0.5M and above. The second variety has large red halophilic cells that can only grow in saline concentrations above 2M.

Species of *Dunaliella* are possibly the most salt-tolerant eukaryotic photosynthetic organisms. *Dunaliella* has two mechanisms that allow it to live in waters of varying salinities (Fisher et al., 1994):

1 *Ion pumps in the plasma membrane* – Plasma membrane proteins are produced when the alga is moved from a low-salinity environment to one of high salinity. These proteins are ion pumps that expel sodium from the protoplasm and control intracellular ion levels.
2 *Production of glycerol* – Lacking a cell wall, *Dunaliella* cells respond to increases or decreases in the external salinity by immediate shrinking or swelling, respectively. Subsequent synthesis or elimination of glycerol

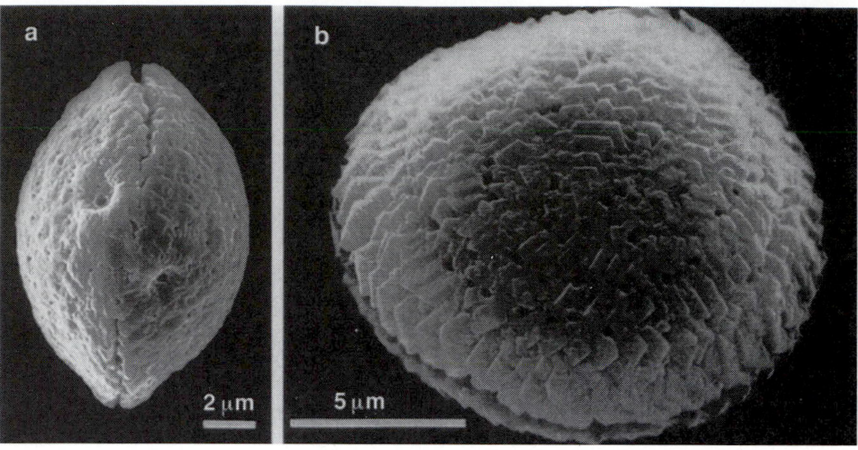

Fig. 5.46 Scanning electron micrographs of zoospores of *Phacotus lenticularis* covered with crystals of calcium carbonate. (From Hepperle and Krienitz, 1996.)

results in an intracellular concentration that balances the external salinity and permits the cells to regain their original volume.

The unicellular *Hematococcus* (Fig. 5.45) accumulates the yellow carotenoid astaxanthin in the cytoplasm in concentrations up to 5% of the dry weight of the cells. The alga can be added to fish and poultry feed to impart a yellow color to the skin. High concentrations of astaxanthin can be induced by growing the cells under high O_2 concentrations, nitrogen limitation or high irradiance (Lee and Ding, 1995; Tan et al., 1995; Ben-Amotz, 1996; Grünewald et al., 1997).

The zoospores of *Phacotus lenticularis* are surrounded by a lorica containing crystals of calcium carbonate (Fig. 5.46). *Phacotus* can occur in high quantities in some lakes (up to 5 million cells per liter) where they have a strong influence on the lake ecosystems (Hepperle and Krienitz, 1996).

Volvocaceae

This family includes those colonial Volvocales in which there is formation of a flat plate (**plakea**) during early development of the colony. The number of cells of a colony is a multiple of two, and each of the cells is surrounded by a gelatinous sheath. Smith (1955) lists three major evolutionary tendencies in the family: (1) an increase in the number of cells in the coenobium; (2) an advance from the condition where all cells are reproductive, to one where only certain cells are reproductive; and (3) an advance from isogamy to oogamy (Nozaki, 1996).

Evolution in the Volvocaceae has resulted in colonies of increasing complexity. The basic cell unit in these organisms is that of *Chlamydomonas*, the cells joining together to form a colony (Fig. 5.47). The individ-

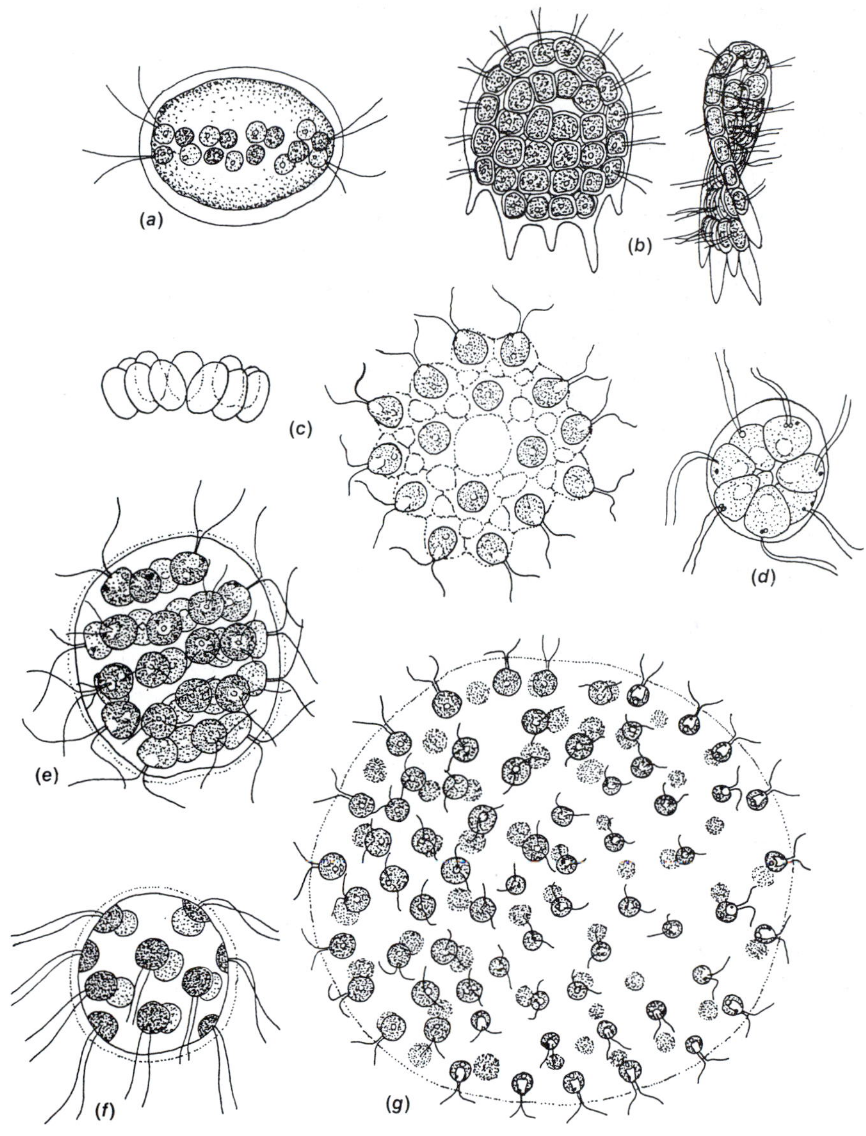

Fig. 5.47 Colonial Volvocales. (*a*) *Stephanoon askenasyi*. (*b*) Front and side views of *Platydorina caudata*. (*c*) Side and front views of *Gonium* sp. (*d*) *Pandorina morum*. (*e*) *Eudorina unicocca*. (*f*) *Volvulina steinii*. (*g*) *Pleodorina* sp. ((*a*)(*d*) after Huber-Pestalozzi, 1961.)

ual cells in the volvocean colony are held together by an extracellular matrix of hydroxyproline-rich glycoproteins that are similar to the extracellular wall of *Chlamydomonas* (Goodenough and Heuser, 1985). Coenobia of most genera exhibit a polarity when swimming, the anterior pole being directed forward. There may also be a morphological difference in the size of the cells or eyespots in the posterior versus the anterior ends. The saucer-shaped colonies of *Gonium*, with 4 to 32 cells, swim with the convex surface forward and have eyespots of a uniform size (Pocock, 1955). In the rest of the Volvocaceae, size differentiation occurs among the eyespots, with larger

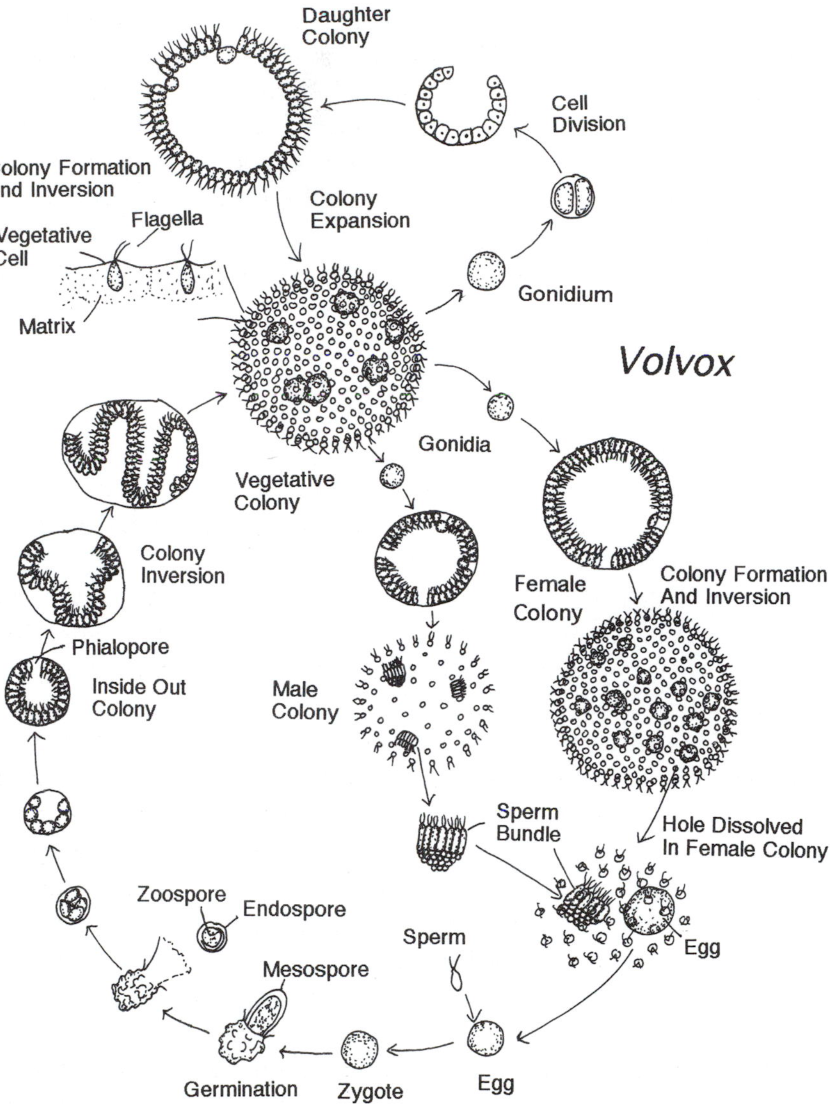

Fig. 5.48 The life history of *Volvox carteri*. (Adapted from Smith, 1944; Kochert, 1968.)

eyespots occurring in cells at the anterior end. This is so in the spherical coenobium of *Pandorina* with 4 to 32 cells, in *Volvulina* with 16 cells in four ranks of four, in *Stephanoon* with 8 or 16 cells in two equatorial ranks, in *Eudorina* with a spherical coenobium of 16 to 128 cells, in *Astrephomene* with a spherical coenobium of 16 to 128 cells, in *Volvox* with spherical colonies of 500 to 40 000 cells and in *Platydorina* with flattened colonies.

The most highly evolved volvocalean genus is *Volvox*, both in its morphological complexity and in its complex oogamous sexual reproduction. Colonies of *Volvox carteri* are oval to spherical with 2000 to 6000 cells arranged in a single layer (Fig. 5.48). Each colony contains a large number of somatic cells and a smaller number of reproductive cells. The protoplast of each somatic cell is typically chlamydomonad and has a single cup-shaped nucleus with a basal pyrenoid, an anterior eyespot, a central nucleus, two contractile vacuoles, and two flagella (Fig. 5.1) (Kochert and Olson, 1970; Pickett-Heaps, 1970). A thick hyaline sheath surrounds each somatic protoplast. The sheath is apparently proteinaceous because pronase dissociates colonies into single cells. Daughter colonies are enclosed in vesicles that expand into the interior of the colony. A watery gelatinous material fills the remaining space inside the colony (Kochert, 1968).

Vegetative, female, and male colonies of *V. carteri* have similar somatic structure, but each contains a different type of reproduction cell. Vegetative colonies contain cells (gonidia) that give rise to daughter colonies by repeated division; female colonies have eggs, which form zygotes after fertilization; and male colonies contain male initial cells, which divide to form sperm bundles. Female and male colonies develop from gonidia as well as the vegetative colonies.

Gonidia are large (80 to 100 μm), spherical, highly vacuolate cells containing numerous pyrenoids. The nucleus is suspended in the center of the cell by thick protoplasmic strands extending from the peripheral layer of cytoplasm. Mature gonidia are not flagellated. Eight to 12 gonidia are usually present in a colony and are arranged in alternating tiers of four.

In asexual reproduction, a gonidium divides to form a daughter vegetative colony that is inverted with the flagellar end of the cells toward the center and the daughter cell gonidia to the outside. A hole (**phialopore**) exists in the daughter colony through which the daughter colony inverts itself. Inversion is initiated by cells throughout the colony changing from pear-shaped to elongate, which causes the colony to contract and the phialopore to open. The cells adjacent to the phialopore become flask-shaped, with long thin stalks at their outer ends. At the same time, the cytoplasmic bridges joining all adjacent cells migrate from the midpoint of the cells to the stalk tips. Together these changes cause the lips of the cells at the phialopore to curl inward. Cells progressively farther from the phialopore become flask-shaped while the closer cells become columnar, causing the lip to curl progressively over the surface of the colony until the colony has turned itself inside out. While still contained with the original gonidial wall, the daughter colony expands, the gelatinous matrix of the daughter colony forms, and the cells separate from one another. Eyespots develop, and the flagella protrude to their full length. Mature daughter colonies rotate slowly in their vesicles before release through pores in the parent colony matrix over each mature daughter colony (Kochert, 1968). It is possible to synchronize asexual division in *Volvox carteri* with alternating periods of light and dark. Under a 32-hour light : 16-hour dark period, gonidia initiate their cleavage division, leading to new colonies at the end of the light period. When the lights come on, new proteins are produced by changes at the translational level (Kirk and Kirk, 1985). Early in the next dark period, the daughter colony is complete and undergoes inversion.

In the formation of female colonies, a gonidium cleaves to form a colony similar to that of a vegetative colony up to the 16- to 32-cell stage. The next division results in cells larger than the somatic cells, the egg initials, which subsequently develop into eggs. Eggs are flagellated at maturity, but the flagella are not of sufficient length to protrude from the matrix of the colony. The maturation and release of the daughter colony are the same as for the vegetative colony.

Usually in *V. carteri*, male colonies are produced from gonidia that are initially the same size as the vegetative and female colonies, but later do not expand to the same size as these other colonies. These are called dwarf male colonies. Dwarf male colonies are formed from gonidia the same way as the vegetative colonies up to the 32- to 64-cell stage. The next division results in larger cells, the male initials, scattered over the surface of the daughter colony. The dwarf male matures similarly to the vegetative cell except that the male colony does not expand as much. Before release of the dwarf male colony from the parent colony, the male initial divides to form a bowl-shaped mass of 64 to 128 cells that undergoes partial inversion to form a sperm bundle which is convex on one side. Each sperm has two flagella and an eyespot. Release of sperm bundles from the dwarf male colonies occurs through individual escape pores after the dwarf male colonies have been released from the parent colonies.

Strains of *V. carteri* are normally heterothallic (male and female colonies are formed in separate parent colonies). In sexual reproduction, the sperm bundles are released from the male colony and swim in the medium. If the sperm bundles come in contact with vegetative colonies, they swim over the surface of the colony with the flagella in contact with the surface. They soon swim off, however. When the sperm bundle comes in contact with a female colony, the flagella of the sperm bundle and the flagella of the female somatic cells bind, causing the sperm bundle to stick to the female colony (Coggin et al., 1979). A fertilization pore is dissolved in the sheath of the female colony, probably by secretion of a proteolytic enzyme (Hutt and Kochert, 1971). A few somatic cells are usually dissolved out of the female colony and swim off. The sperm bundle breaks down into individual sperm, which move in an amoeboid fashion between the somatic cells of the colony or swim with a corkscrew motion through the watery material in the interior of the colony. The sperm fuse with the eggs, forming zygotes that enlarge, develop an orange coloration, and secrete a thick crenulate wall. The parent female colony persists for some time, but eventually it dissociates and releases the zygotes.

After a resting period, the zygote germinates by splitting the outer zygote wall, with the middle layer of the wall (mesospore) protruding through the fissure in the outer layer. The protoplast has two flagella that beat weakly in the watery interior of the mesospore. After the zoospore is released from the outer zygote wall, the mesospore wall breaks down, leaving the protoplast inside the endospore wall. This protoplast then behaves similarly to a gonidial protoplast and divides to form a young colony of approximately 1000 somatic cells with four gonidia in a single tier.

Sexual type is inherited in a 1:1 ratio, and there is no parthenogenetic development of the eggs.

In *Volvox carteri*, the males make and accumulate a sexual inducer that is a 30-kilodalton glycoprotein (Starr and Jaenicke, 1974). The sexual inducer is released when the sperm are released from the sperm packets. The inducer is effective at 6×10^{-17} M. One sexual male releases enough inducer to convert all the related males and females in a volume of 1000 liters from asexual to sexual reproduction. Under the moderate growth conditions present in a large stable body of water, it is normally only the males that produce the sexual inducer. However, it is possible to force asexual males and females to make the inducer by subjugating the cells to a high temperature for a period of time. One hour at 42.5 °C is a sufficient heat shock to induce the formation of sexuality in *V. carteri* asexual males and females. Without the sexual inducer, the asexual males and females will produce asexual gonidia. In the presence of sexual inducer, the males and females produce sperm packets and eggs, respectively. The heat shock response is an adaptation to life in shallow temporary bodies of water where *Volvox* is often found. In the spring, in such bodies of water there is abundant water and the temperature is relatively low. *Volvox* grows asexually under these conditions. As summer progresses, the temperature in these bodies of water rises, and the organisms begin to dry out. The increase in temperature shocks the *Volvox* into producing the sexual inducer and initiating sexual reproduction. This results in the formation of drought-resistant zygospores, which survive the dry conditions and serve as an overwintering spore. The above-described research by Kirk and Kirk (1986) has provided an explanation for Powers's (1908) observation that he had great difficulty finding sexual *Volvox* in large bodies of water. Powers further noted that "in the full blaze of Nebraska sunlight, *Volvox* is able to appear, multiply and riot in sexual reproduction in pools of rainwater of scarcely a fortnight duration." It took another 80 years for Kirk and Kirk to discover the heat-shock phenomenon and to explain Powers's observations.

TETRASPORALES

These algae have immobile vegetative cells that are capable of cell division, unlike those in the Chlorococcales or Volvocales. The colonies are non-filamentous, and flagellated cells are formed by many genera. Asexual reproduction occurs via the formation of zoospores, aplanospores, or akinetes. Sexual reproduction is isogamous, by fusion of biflagellate gametes. Almost all the organisms are freshwater.

The Tetrasporales probably evolved from the Volvocales by the loss of motility in the vegetative condition. Two families will be considered here:

Family 1 Tetrasporaceae: cells with pseudocilia.
Family 2 Palmellaceae: cells without pseudocilia.

Tetrasporaceae

The elongated gelatinous thalli of the Tetrasporaceae have vegetative cells in groups of two to four, with each cell having two pseudocilia. **Pseudocilia** are longer than flagella but are evidently related to them because the pseudocilia have a normal basal body but an abnormal 9 + 0 configuration of microtubules near the base of the pseudocilia (Lembi and Herndon, 1966; Wujek, 1968). The number of microtubules lessens and becomes more irregular as the end of the pseudocilium is approached.

Colonies of *Tetraspora gelatinosa* (Fig. 5.49) are green, amorphous masses with an outer layer of vegetative cells (Klyver, 1929). They are found in quiet freshwater and can be attached or free-floating. Each cell has a large cup-shaped chloroplast with a central pyrenoid and two pseudocilia. Growth of the thallus results from vegetative division of the cells. In the formation of isogametes, a vegetative cell divides two to three times, resulting in four or eight pyriform gametes, each with an eyespot and a cup-shaped chloroplast. The biflagellate gametes break free from the colonial mucilage and fuse with each other at their anterior ends. The quadriflagellate zygote swims for a while before settling, retracting its flagella, and forming a cell wall. The zygote germinates, forming four or eight aplanospores without pseudocilia. These aplanospores enlarge, and when they have reached the size of vegetative cells, they divide to form daughter cells that have pseudocilia. The aplanospores and their daughter cells are held together by mucilage, and the aggregation makes up the typical thallus of *Tetraspora*.

Palmellaceae

Members of the Palmellaceae have their cells united in small gelatinous colonies that are generally amorphous but may be of definite shape.

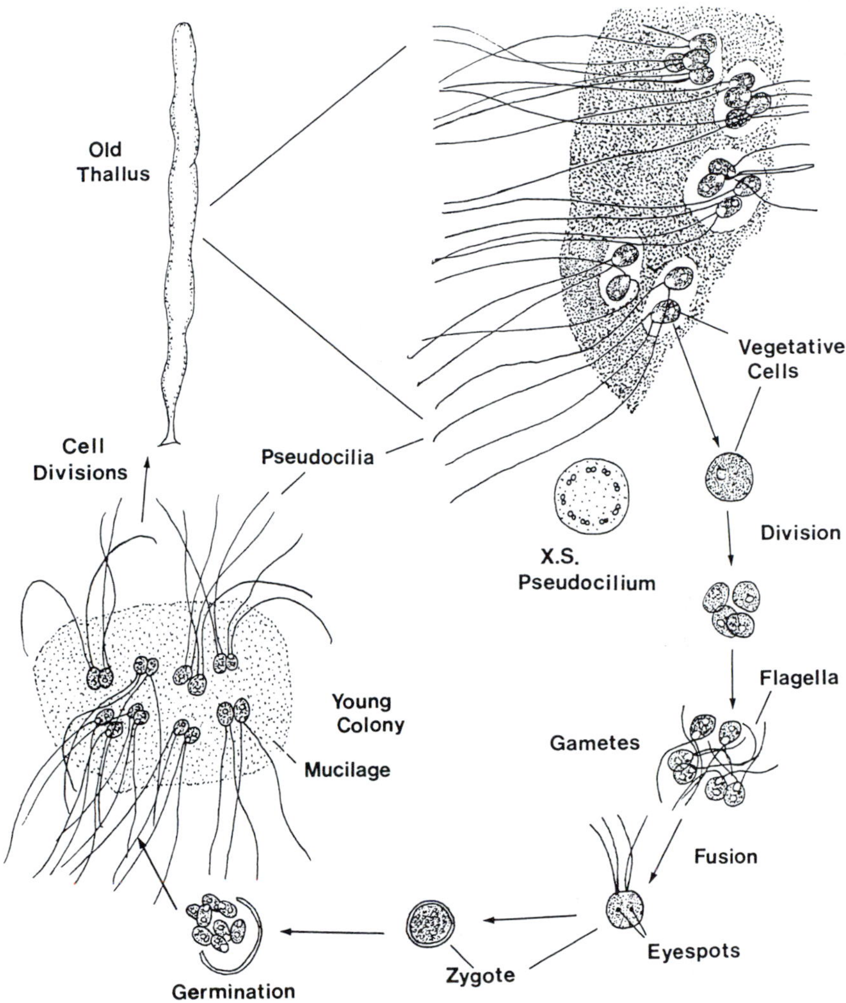

Fig. 5.49 The life cycle of *Tetraspora gelatinosa*. (Adapted from Klyver, 1929.)

Palmella (Fig. 5.45) is a freshwater alga composed of cells united by a gelatinous matrix forming colonies of indefinite shape. Asexual division involves the formation of zoospores, and sexual reproduction occurs by formation of isogametes.

Botryococcus braunii is a free-floating colony of indefinite shape within a hyaline or orange envelope. The colonies are composed of radially arranged cells embedded in a tough mucous envelope. It is common as a former of water blooms, which in the autumn appear reddish owing to the formation of large amounts of oil, which often obscure the cell contents (Fig. 5.50, 5.51). This alga has been postulated as the cause of the boghead coals (e.g., torbonite) and the oil shales of the Tertiary Period (Blackburn and Temperley, 1936; Cane, 1977; Wolf et al., 1985). If these deposits are examined under a microscope, it is possible to see plant remains that are similar

$CH_3\text{-}(CH_2)_7\text{-}CH{=}CH\text{-}(CH_2)_{17}\text{-}CH{=}CH_2$ n-alkadiene C_{20}

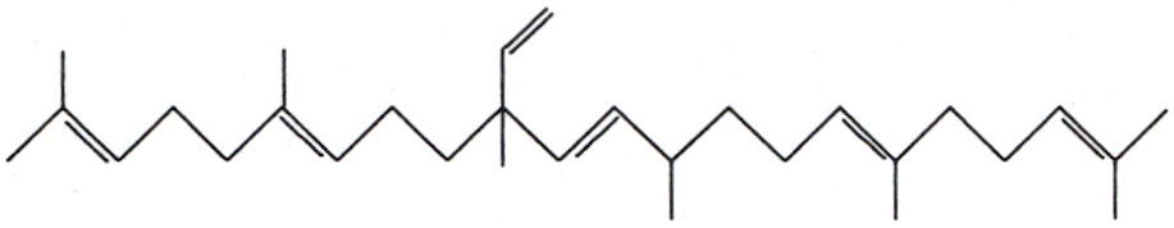

Botryococcene C_{30}

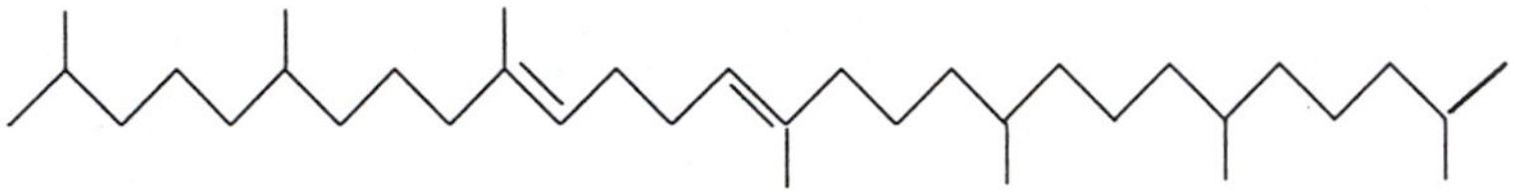

Lycopadiene C_{40}

Fig. 5.50 The structure of hydrocarbons isolated from *Botryococcus braunii*. (Modified from Metzger et al., 1990.)

to extant colonies of *B. braunii*. A hydrocarbon derivative exclusively attributable to *Botryococcus* comprises 1.4% of a Sumatran petroleum (Moldowan and Seifert, 1980). The cultivation of the alga has been proposed as a renewable source of liquid hydrocarbon fuel (Wake and Hillen, 1980; Yamaguchi, 1997). The resting stage of living *B. braunii* contains up to 70% of its dry weight as alkadienes, botryococcenes or lycopadiene (Fig. 5.50) (Metzger et al., 1990). The hydrocarbons are produced primarily during the exponential and linear growth phases. The dense matrix surrounding the cells is impregnated with the hydrocarbons.

SCHIZOGONIALES

The foliose marine algae in this small order have a central stellate chloroplast. The principal genus, *Prasiola*, has a unique type of life history, with meiotic divisions in mature sexual thalli resulting in a haploid apex and a diploid base. There is only one family, the Schizogoniaceae.

Prasiola stipitata consists of small, thin, broadly ovate blades appearing as dirty green patches at or above high-tide level, commonly in the spray zone or areas fouled by bird extrement. The diploid thalli can form either diploid spores or haploid gametes (Fig. 5.52), the spore-forming plants growing higher on the shore than the sexual (gamete-forming) plants (Friedmann, 1959; Friedmann and Manton, 1959).

In the formation of the diploid aplanospore, the vegetative cells in the upper part of the thallus divide, making the upper part of the thallus multilayered. Each cell in this area forms a non-flagellated spore that settles, germinates, and develops into a new diploid plant like the parent (Fig. 5.52).

In the production of gametes, the cells in the upper part of a diploid thallus undergo meiosis, with the subsequent division of the haploid cells

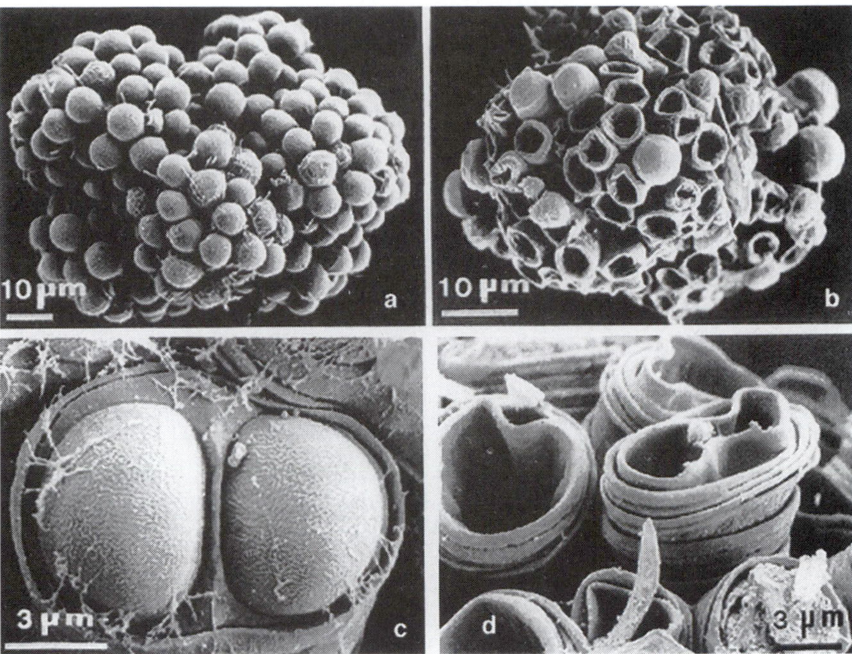

Fig. 5.51 Scanning electron micrographs of *Botryococcus brauni i*. (*a*) Colony from the wild. (*b*) Colony showing the cup-shaped mucilaginous bases. (*c*) Two cells in mucilaginous bases. (*d*) Mucilaginous base with no cells. (From Plain et al., 1993.)

resulting in a multilayered upper part of the thallus. The haploid tissue is divided into a patchwork of rectangular darker and lighter areas containing the male and female cells, respectively. The difference in shading is due to the difference in size of the chloroplasts, the larger female cells having the larger chloroplasts. The gametes are liberated after the thalli are wet by the incoming tide. More male gametes are released because of their smaller size and greater production per unit area. The anteriorly biflagellate male gametes swim around the non-motile egg, and one of the male flagella touches the egg. The flagellum fuses with the egg, a step followed by fusion of the bodies of the gametes. The pear-shaped zygote swims vigorously by means of the posteriorly directed remaining male flagellum. At 5 °C, the zygote swims for several hours, retracts the flagellum, sinks, attaches firmly to the substratum, and develops a cell wall. The zygotes germinate into diploid thalli, completing the life cycle (Friedmann, 1959).

CHLOROCOCCALES

In this order are the algae that have a non-motile vegetative thallus where the thallus comprises only a single cell or a coenobium composed of a definite number of cells arranged in a specific manner. Asexual reproduction occurs by zoospores or aplanospores that are commonly autospores. These autospores are probably no more than daughter cells of the parent

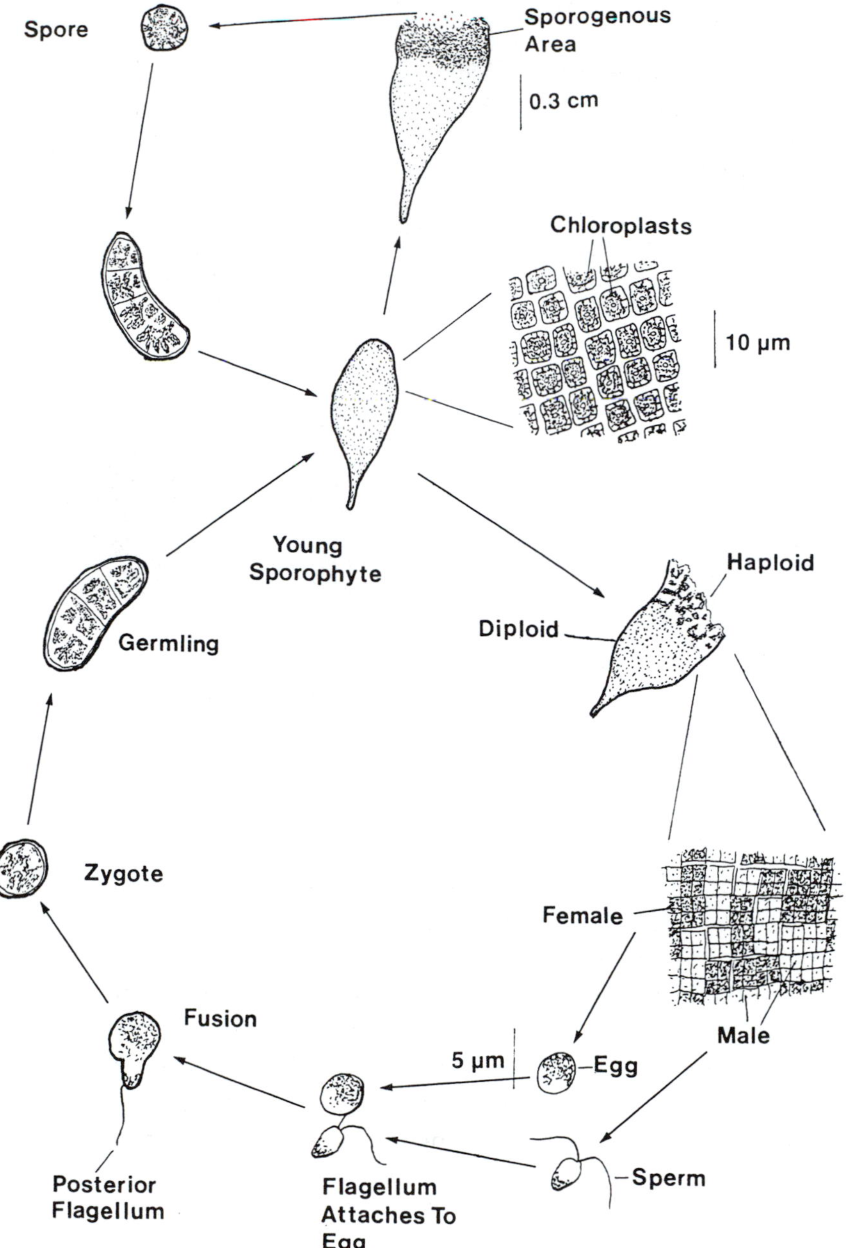

Fig. 5.52 The life cycle of *Prasiola stipitata*. (Adapted from Friedmann, 1959.)

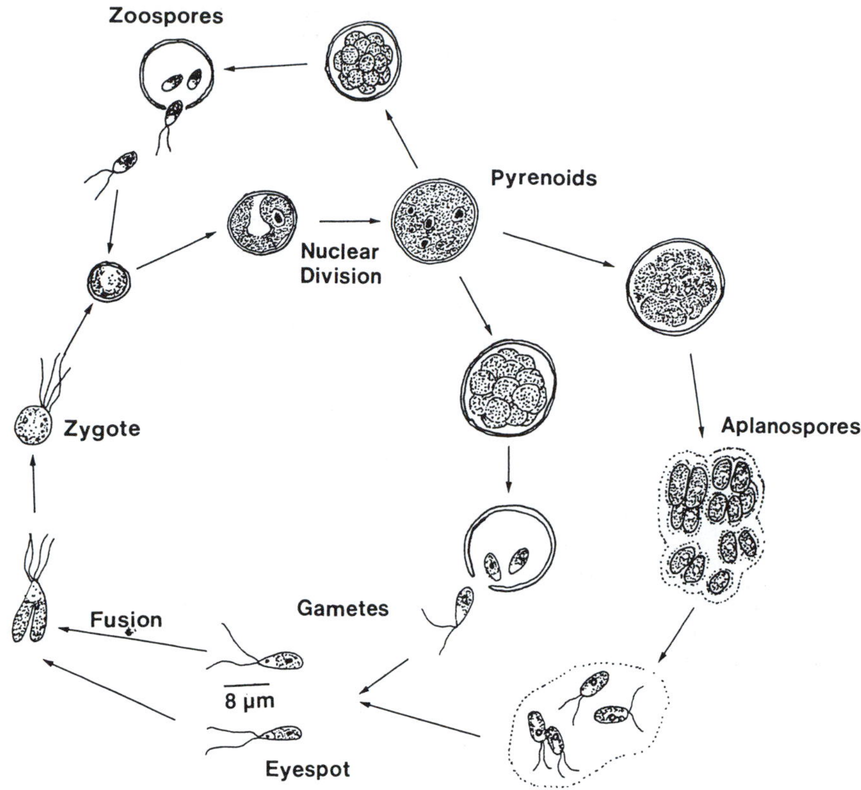

Fig. 5.53 The life cycle of *Chlorococcum humicola*. (Adapted from Bristol, 1919.)

thallus. Sexual reproduction can be isogamous, anisogamous, or oogamous. The order is exclusively freshwater.

Chlorococcum is an alga frequently isolated from beneath the surface of the soil. It sometimes occurs in abundance on damp soil or brickwork. The cells can survive for a long time in the soil. In one study they grew after being in desiccated soil for 59 years (Trainor, 1970). The cells of the same species vary considerably in size, young cells having a thin cell wall and older cells a thick one. The chloroplast in young cells is a massive parietal cup with a single pyrenoid. In older cells the chloroplast becomes diffuse. Asexual reproduction is by zoospores (Fig. 5.53), there never being any vegetative division. The biflagellate zoospores have a cup-shaped chloroplast and an eyespot. Sexual reproduction occurs by the formation of isogametes. Under certain conditions, aplanospores can be produced, which eventually liberate two to four biflagellate gametes. Nearly dormant cells of *Chlorococcum echinozygotum* that have been deprived of nitrogen can be induced to undergo gametogenesis by being placed in fresh medium in darkness without nitrogen (O'Kelley, 1984). If nitrogen is supplied 6 hours after

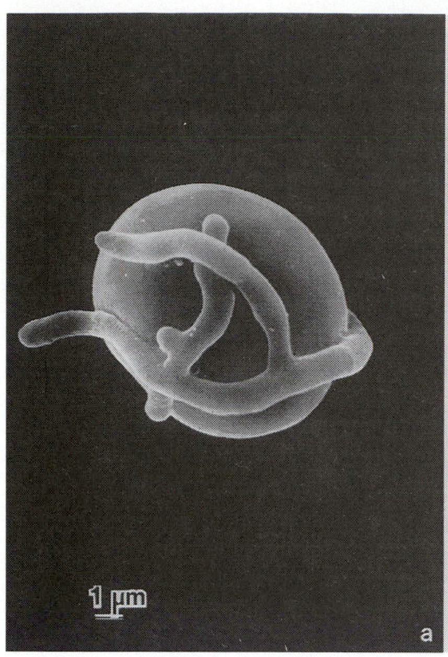

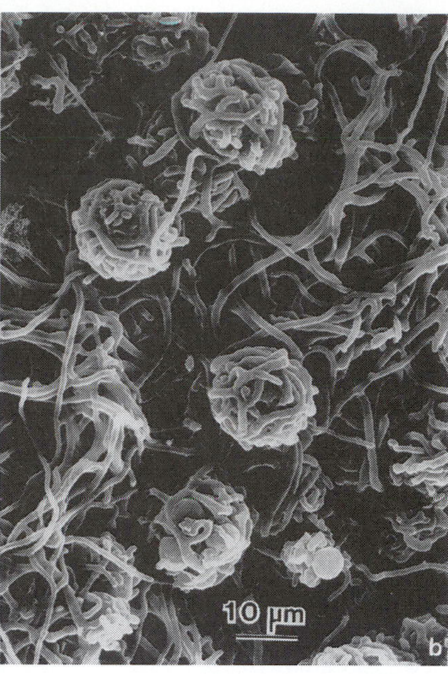

Fig. 5.54 The beginning of a lichen. (*a*) Scanning electron micrograph of the envelopment of a cell of the phycobiont *Trebouxia erici* by the hyphae of the mycobiont *Cladonia cristatella*. (*b*) The mycobiont has completely enveloped the cells of the phycobiont, and the hyphae have formed the thallus of the lichen. (From Ahmadjian and Jacobs, 1981.)

the cells are placed in darkness, zoosporogenesis occurs instead of gametogenesis.

Trebouxia (Figs. 5.54 and 5.55) (sometimes classified in a specialized order, the Pleurastrales; see Mattox and Stewart, 1984) is the most common green alga as a phycobiont (algal partner) in the lichen association. The genus also is found as a free-living alga, and, when free-living, it is usually twice the size of the alga in the lichen symbiosis (see Ahmadjian 1993, for a review). *Trebouxia* has a massive chloroplast with a single pyrenoid. When it is growing in the lichen association, reproduction is normally by autospores, although under wet conditions zoospores may be formed. When the alga is grown in liquid culture, zoospores are formed. Sexual reproduction is isogamous or anisogamous by the fusion of biflagellate gametes. Lichenized *Trebouxia* produces primarily sugar alcohols (80% ribitol) from photosynthetic processes, whereas free-living *Trebouxia* forms much smaller amounts of sugar alcohols (15% ribitol) and greater quantities of other carbohydrates (Green, 1970). Richardson (1973) has summarized the differences between algae growing as phycobionts (algae in the lichen association) and the free-living ones: The free-living algae (1) synthesize less sugar or sugar alcohol, (2) form more polysaccharides, (3) develop cell sheaths not seen in phycobionts, and (4) release less photosynthate into the surrounding medium. *Trebouxia* is able to grow saprophytically in the dark in the absence of light (Ahmadjian, 1960).

Lichen **mycobionts** (the fungal partners in the lichen association) are

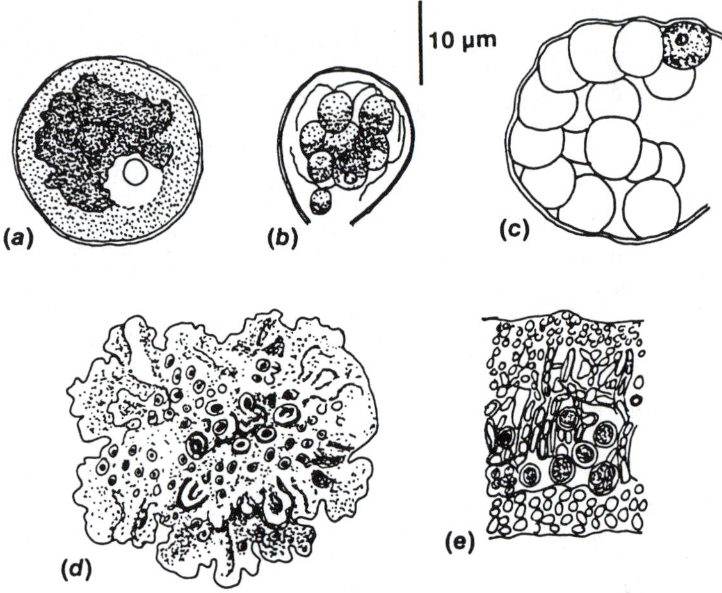

Fig. 5.55 *Trebouxia* sp.: (*a*) vegetative cell; (*b*) zoospores being released; (*c*) aplanospores being released. The *Trebouxia*-containing lichen *Xanthoria parietina*: (*d*) whole thallus; (e) section of thallus showing spherical *Trebouxia* cells. ((*a*)–(*c*) after Ahmadjian, 1960; (*d*),(e) after Fünfstück, 1907.)

able to discriminate between suitable and unsuitable algal partners. The mycobiont *Xanthoria parietina* secretes a protein that will bind only to the cell walls of species of *Trebouxia* and *Pseudotrebouxia*, algal genera that normally make up this lichen symbiosis (Bubrick and Galun, 1980). The cell walls of these algae are characterized by high levels of acidic polysaccharide and a protein coat on the cell wall surface. Members of other algal families do not bind the lichen protein. In the lichen *Cladonia cristatella*, compatible phycobionts have fungal haustoria in most of the phycobiont cells. Fungal haustoria are usually used by the fungus to transfer nutrients from a host to the fungus when the host is a parasitized organism. However, in lichens, transfer of metabolites to the fungus is minimal (Collins and Farrar, 1978; Hessler and Peveling, 1978). This is probably a case of controlled parasitism (Ahmadjian and Jacobs, 1981).

Hydrodictyon reticulatum (Fig. 5.56), or the water net, is a free-floating, relatively rare freshwater alga that forms colonies that are netlike with polygonal or hexagonal meshes, the angles of the net being formed by the union of three of the elongate multinucleate cells. Each cell has a large central vacuole and a reticulate chloroplast with pyrenoids. In asexual reproduction a daughter net is produced inside, and subsequently released from, the parent cell (Pocock, 1960; Marchant and Pickett-Heaps, 1971, 1972a–d). Asexual reproduction begins with the disappearance of pyrenoids, coinciding with the accumulation of starch grains in the fragmenting chloroplasts. Regularly spaced nuclei are surrounded by fragments of chloroplasts. A vesicle forms around and outside of the tonoplast, restrict-

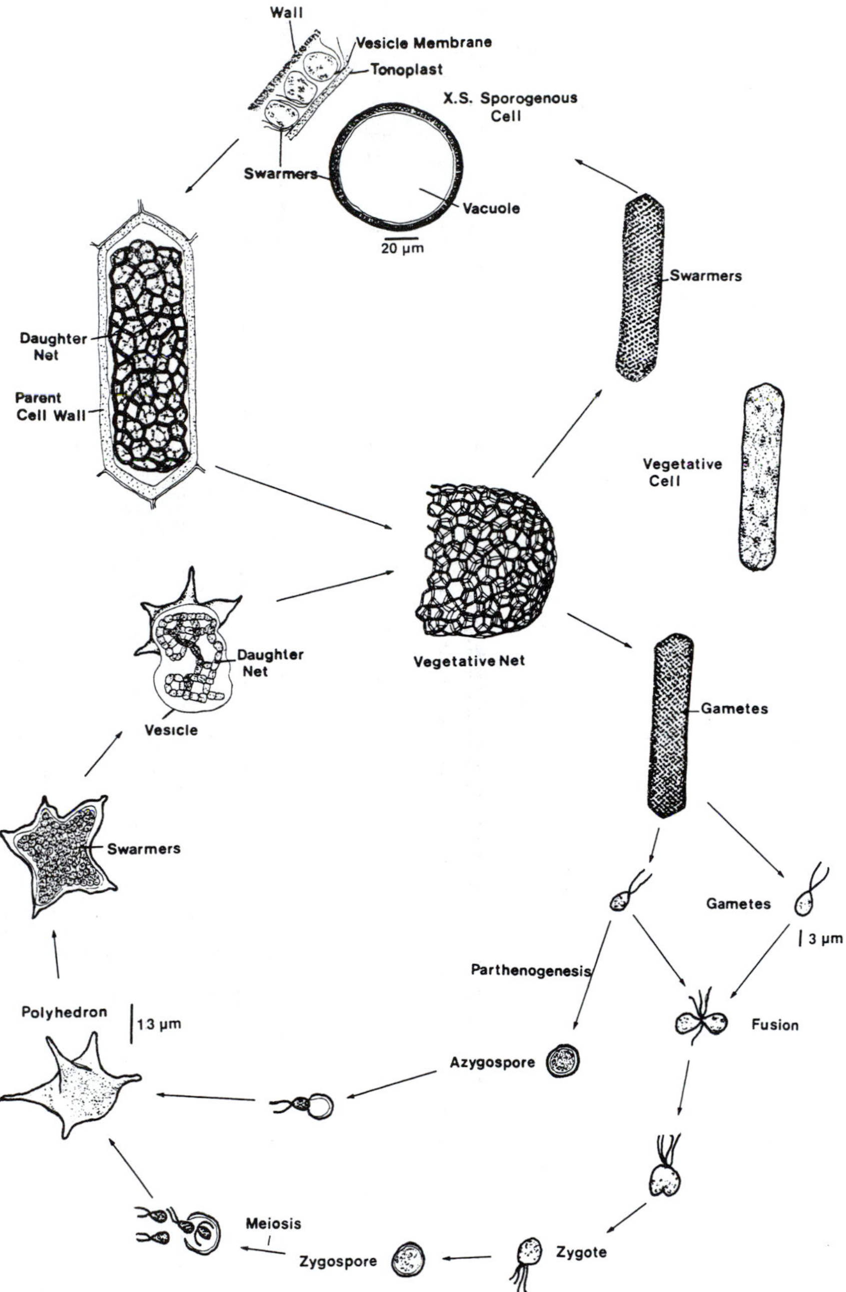

Fig. 5.56 The life cycle of *Hydrodictyon reticulatum*. (Adapted from Pocock, 1960; Marchant and Pickett-Heaps, 1971, 1972a–d.)

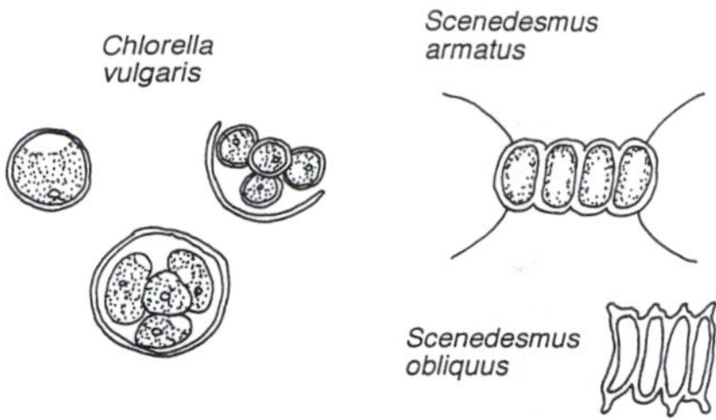

Fig. 5.57 Drawing of *Chlorella vulgaris* (with autospores), *Scenedesmus armatus* and *S. obliquus*.

ing most of the protoplasm to a small area around the periphery of the cell. The protoplasm divides to produce uninucleate, biflagellate zoospores that actively swim about inside the parent cell. The zoospores stop swimming, become joined in certain areas, and retract their flagella. A daughter net has now been formed within the parent cell, which is released to enlarge to a mature colony.

Chlorella cells are spherical with a cup-shaped chloroplast (Fig. 5.57). The only method of reproduction is by daughter cells that resemble the parent cell. *Chlorella* often forms intracellular symbioses with aquatic invertebrates and protozoa (such as *Paramecium*). In these symbioses, *Chlorella* is held in host vacuoles where the *Chlorella* synthesizes and releases maltose into the host vacuole. *Chlorella* is held at a low pH in the host vacuoles. Studies on isolated *Chlorella* cells have shown that maximum synthesis and release of maltose occurs at pH 4–5 (Dorling et al., 1997).

Prototheca lacks chlorophyll and resembles a colorless *Chlorella* although the *Prototheca* cells do have starch-containing amyloplasts (Webster et al., 1968). *Prototheca* causes protothecosis in animals and humans. This disease is more common than is supposed, but because the small round colorless cells resemble yeasts, the disease, at least in animals, is often incorrectly diagnosed. *Prototheca* cells are fairly common in the soil, where they presumably live by digesting organic matter, and it is from here that most infections occur. In animals most of the reported cases have been severe systemic infections, such as massive invasion of the bloodstream, that have resulted in the death of the animal within a short time (van Kruiningen et al., 1969). Such massive infections have been reported in humans but usually only as secondary invaders after the primary invader has seriously lowered the body's defenses. More common in humans is the subcutaneous type of infection that starts initially as a small lesion and spreads slowly through the lymph glands, covering large areas of the body and preventing the sufferer from performing normal work duties (Fig. 5.58). (Mars et al., 1971).

Scenedesmus is a common alga (Fig. 5.57), often occurring as almost a pure culture in plankton. Cells in the colony occur in multiples of 2, with

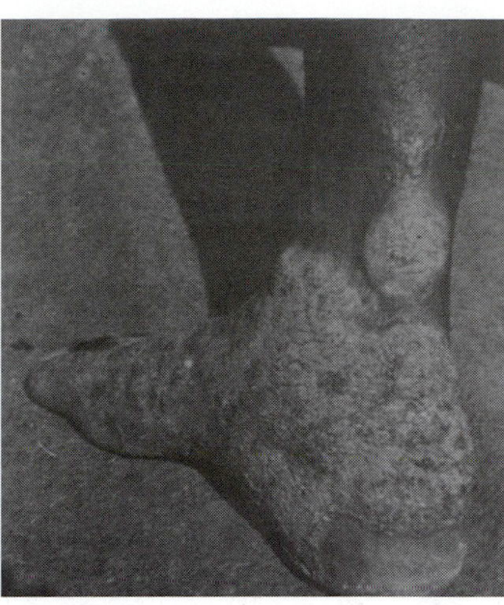

Fig. 5.58 Lesions of protothecosis on the foot and lower leg of a man from Sierra Leone. (From Davies and Wilkinson, 1967.)

four or eight cells being most common. The species differ mostly in the number and type of spines on the cells and the texture of the wall. The uninucleate cells have a single laminate chloroplast. The morphology of the colony can be varied considerably by varying the medium in which the cells are growing (Egan and Trainor, 1989). In a medium with low phosphorus or low salt concentration, the *Scenedesmus* is induced to grow as unicells, resembling the genera *Chodatella* and *Franceia* (Trainor, 1992) and the same species can also be induced to grow with or without spines. Composed of aggregated proteinaceous tubules, the spines probably aid in the flotation of the colonies (Staehelin and Pickett-Heaps, 1975). *Scenedesmus* occasionally forms zoospores when deprived of nitrogen (Trainor, 1963).

Unicellular green algae, such as *Chlorella*, have been extensively investigated as possible new sources of food for an increasing world population (Yamaguchi, 1997). *Chlorella* produces little cellulose or other carbohydrate wall material; thus more of the cell is digestible in *Chlorella* than in an alga with a large amount of cellulose. Exponentially growing *Chlorella* cells contain about 50% protein, 5% chlorophyll, and a large number of vitamins. Much greater quantities of biomass can be obtained per unit area with algae than with higher plants. When the growth of algae is linked to the purification of sewage in oxidation ponds, yields of 112 000 kg hectare^{-1} year^{-1} of dried algae can be obtained. At the same time, the algae are taking up elements such as nitrogen and phosphorus, and reducing the biological oxygen demand (an indication of the organic material in the water) by 85% (McGarry and Tongkasame, 1971). The algae obtained make a very good stock feed, and attempts have been made to utilize it as human food, although the algae in the human digestive tract lead to undesirable intesti-

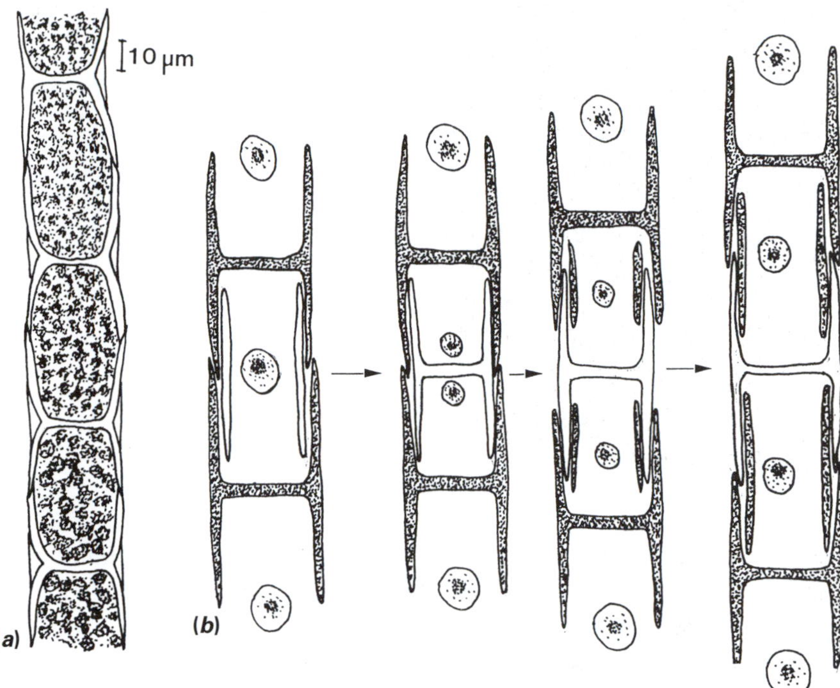

Fig. 5.59 *Microspora crassior.* (*a*) Filament. (*b*) Diagrammatic representation of wall deposition. Initially the interphase cell is enclosed by two H-shaped segments and a central cylinder. At cytokinesis, a cross wall is deposited in the middle of this cylinder. During cell expansion, the two H-shaped segments move apart while a new cylinder is secreted inside the H-shaped segments. ((*b*) after Pickett-Heaps, 1973.)

nal floras that often result in gas in the intestine. A second line of investigation in this area has been to provide food and oxygen for space and submarine travel. In such a setup, carbon dioxide from human respiration is taken up for use in photosynthesis, with the oxygen produced being respired by humans. A man uses about 600 liters of oxygen daily, and a gas exchange unit that would produce this amount of oxygen would also produce almost enough biomass to provide sufficient food for one man as well (Fogg, 1971).

SPHAEROPLEALES

These filamentous Chlorophyceae have a mature lateral wall that consists of segments, rather than having a continuous structure. In *Microspora*, there are H-shaped wall segments that arise from two separate phases of wall secretion (Fig. 5.59) (Pickett-Heaps, 1973). During interphase, cell expansion is accommodated by the interlocking H-segments moving apart. At the same time, a new cylindrical wall is secreted inside these segments. Then during cytokinesis, the newly formed cross wall turns this cylinder into the typical wall segment. A similar wall occurs in the Oedogoniales, but the algae in the Sphaeropleales lack the unusual structure of the Oedogoniales. The wall structure in the Sphaeropleales is similar to that in the Xanthophyceae, and it is probable that the filaments in both algal

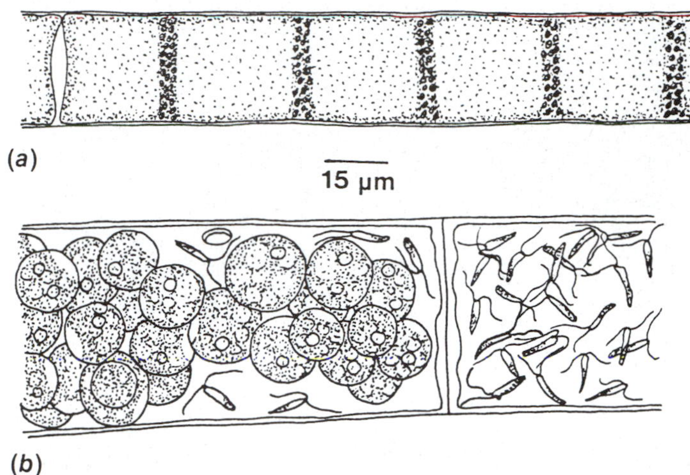

Fig. 5.60 *Sphaeroplea annulina*. (*a*) Part of a vegetative cell. (*b*) Portion of an oogonium in which the eggs are being fertilized, and a portion of an adjoining mature antheridium. (After Smith, 1955.)

groups evolved from a unicellular alga, although independently of each other. The Sphaeropleales probably evolved from an alga in the Chlorococcales (Cáceres et al., 1997).

Algae in the Sphaeropleales (Figs. 5.59, 5.60) typically occur in shallow, freshwater habitats that are inundated only periodically. The ephemeral vegetative stage (often lasting a few weeks or less) occurs during short intervals of flooding, whereas the thick-walled, resistant zygotes persist in the soil through long dry periods, thereby enabling long-term survival of these algae. The zygotes are ornamented and sometimes have **cirri** (Fig. 5.61), curled appendages composed of organic material, on their surface (Hoffman and Buchheim, 1989).

Sphaeroplea has multinucleate cells arranged end to end in unbranched filaments (Fig. 5.60). The alga is freshwater, occurring on periodically wet ground, completing its life cycle within 4 to 5 weeks. Within a cell the cytoplasm is restricted to a number of transverse bands, each band separated by a vacuole. Each cytoplasmic band has several nuclei and a band-shaped chloroplast with several pyrenoids, or numerous discoid chloroplasts. There is also a thin layer of cytoplasm without chloroplasts between the vacuoles and the side walls. Asexual reproduction occurs by fragmentation of the filaments. Sexual reproduction is usually oogamous, with eggs and spermatozoids formed in alternate cells of the same filament or in different filaments. The spindle-shaped biflagellate spermatozoids escape through pores in the side walls of the antheridial cell and swim to the oogonial cells. The eggs, when first formed, are multinucleate, with all the nuclei but one disintegrating. The oogonial cells have pores in their walls through which the spermatozoids enter, swim about the eggs, and eventually fuse with them. Zygotes form a thick ornamented cell with a reddish protoplast, and

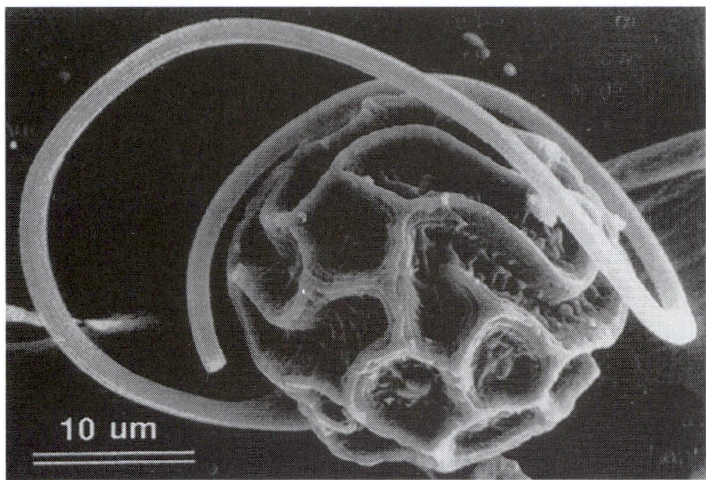

Fig. 5.61 Scanning electron micrograph of a zygote of *Sphaeroplea fragilis* showing the sculptured exterior and a long curved cirrus. (From Hoffman and Buchheim, 1989.)

are released by the decay of the oogonial cell. Zygotes germinate by forming four biflagellate ovoid zoospores, which settle and germinate into a new filament.

CHLOROSARCINALES

The algae in this order have a type of cell division (**desmoschisis**) that results in the formation of a number of cells, each with its own cell wall, within the cell wall of the parent cell. The genera in the Chlorosarcinales lack the plasmodesmata and complexity found in the Chaetophorales and Oedogoniales.

Chlorosarcina occurs free-living in the soil, or it can occur as an endophyte within the epidermis of aquatic vascular plants. Each cell contains a parietal chloroplast and has a cell wall that separates the cell from the other cells inside the old parental cell wall (Fig. 5.62). During **desmoschisis**, the protoplasm of a cell divides at successively perpendicular planes to build up cuboidal packets of cells. Cell walls are formed around the newly divided protoplast, next to the parent cell wall which remains intact. Eventually the colonies fragment to distribute the alga. Asexual reproduction can occur by the formation of four to eight biflagellate zoospores per cell, which are released by softening in one area of the cell wall.

CHAETOPHORALES

The Chaetophorales, along with the Oedogoniales, mark the most complex organization attained within the Chlorophyceae. The Chaetophorales have plasmodesmata, a characteristic of true multicellularity. In many ways, the

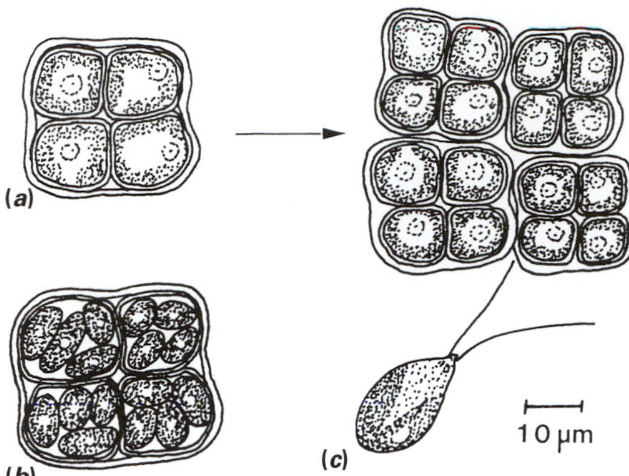

Fig. 5.62 *Chlorosarcina* sp. (*a*) Packet of cells dividing by desmoschisis to produce daughter colonies, each enclosed within the cell wall of the parent cell. (*b*) Colony-forming zoospores. (*c*) Zoospore.

level of organization in the Chaetophorales parallels that attained by the Coleochaetales in the Charophyceae.

The Chaetophorales have branched filaments with uninucleate cells that contain a single parietal chloroplast. Many of the plants exhibit **heterotrichy**; that is, there are *two different types of filaments, one constituting the prostrate system and the other the erect system.* In the past, these organisms were postulated to be the precursors of higher plants because they show the differentiation of the filaments into different systems. An evolutionary progression from *Stigeoclonium* (Fig. 5.63(*b*)) to *Chaetophora* (Fig. 5.63(*c*)) to *Draparnaldia* (Fig. 5.63(*d*)) to *Fritschiella* (Fig. 5.63(*a*)) illustrates increasing differentiation toward what one would expect a primitive land plant to look like. As stated earlier, however, the above algae divide by a phycoplast and are therefore not in the direct line to higher plants.

Stigeoclonium is a common freshwater alga found attached in flowing water. It has a very wide tolerance to organic pollution and can be indicative of heavily polluted water (McLean, 1974; McLean and Benson-Evans, 1974) although it will also grow in unpolluted water. The thallus is differentiated into prostrate and erect portions, with the terminal parts of the branches usually drawn out into colorless hairs (Fig. 5.63(*b*)). Vegetative reproduction can take place by fragmentation, although the erect portion does not grow well after detachment from the prostrate portion. Quadriflagellate zoospores are frequently formed when a thallus is brought into the laboratory, with most of the cells in the smaller branches sporulating a day after being brought in. A cell forms a single zoospore. After settling on its anterior end, the zoospore develops into a new filament. In sexual reproduction, gametes are produced in cells of erect filaments after a cruciate, presumably meiotic, process of division (Simons and van Beem, 1987). The gametes can be biflagellate or quadriflagellate. Fusion of gametes produces a zygote which

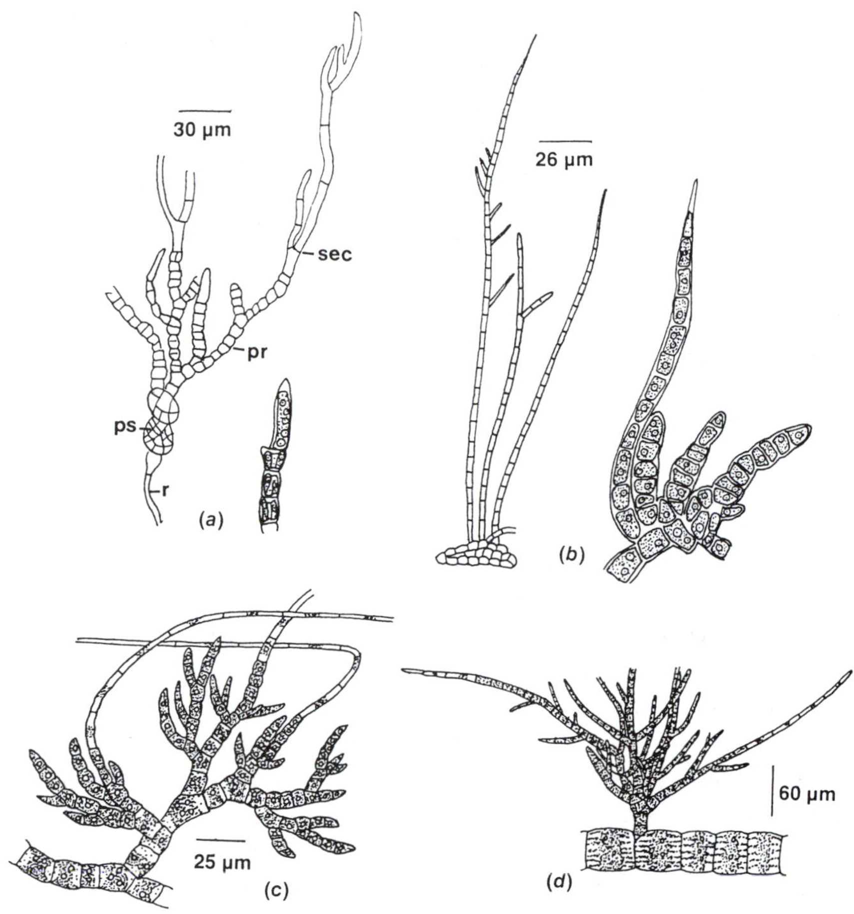

Fig. 5.63 (*a*) *Fritschiella tuberosa* showing prostrate system (ps), protonema (pr), projecting secondary branches (sec), and a rhizoid (r). Also shown is the tip of a branch. (*b*) *Stigeoclonium farctum*, old and young plants with erect and prostrate portions. (*c*) *Chaetophora incrassata*. (*d*) *Draparnaldia glomerata*. ((*a*) after Iyengar, 1932; (*b*) after Butcher, 1932; (*c*),(*d*) after Smith, 1920.)

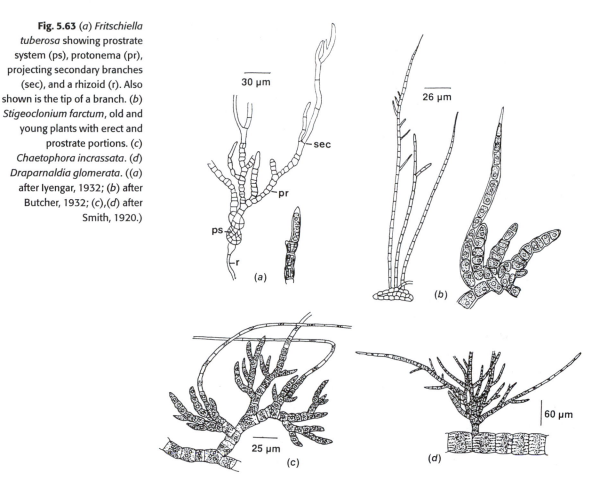

germinates into a juvenile plant in which the original zygote remains visible as an embryonic cell.

OEDOGONIALES

The filamentous, freshwater, uninucleate algae in this order are characterized by a unique type of cell division and the production of motile reproductive cells with a whorl of flagella at one pole (stephanokonts). Sexual reproduction is oogamous, and asexual reproduction can be by zoospores or akinetes.

The Oedogoniales and its single family, the Oedogoniaceae, have only three genera – *Oedogonium, Oedocladium,* and *Bulbochaete. Oedogonium* is unbranched, whereas *Oedocladium* (Fig. 5.64) and *Bulbochaete* are branched. *Bulbochaete* has long colorless hairs (setae) that are usually swollen at the base (Fig. 5.64(*a*)); *Oedocladium* lacks hairs. These algae are

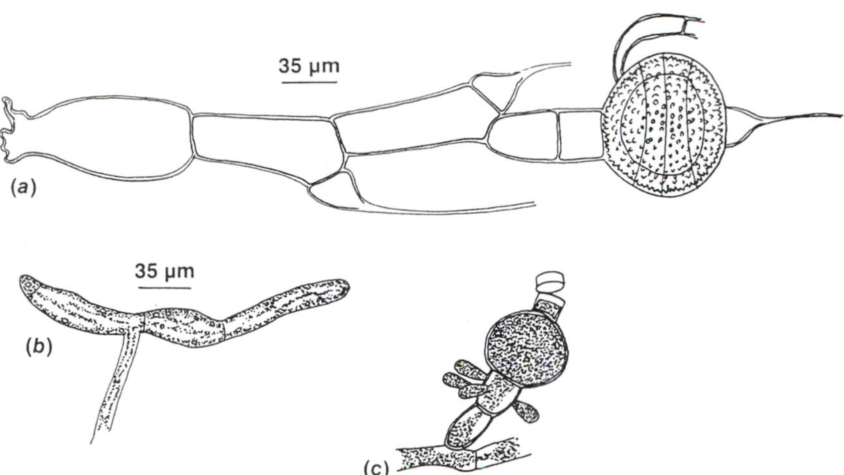

Fig. 5.64 (*a*) *Bulbochaete gigantea*, filament with zygote. (*b*),(*c*) *Oedocladium hazenii*, germling with rhizoid (*b*) and portion of a filament with empty androsporangia, immature dwarf males, and an unfertilized egg (*c*). (After Smith, 1950.)

usually present in permanent bodies of water such as ponds or lakes. If they are growing in moving water, they are seldom in the fruiting condition. Normally fruiting takes places in the summertime. The plants may be epiphytic on aquatic plants and other algae, or they may be free-floating.

Chloroplasts in this order are reticulate, extending from one end of the cell to the other. The many pyrenoids are at the intersections of the reticulum. Plasmodesmata are present between cells, and the ones in *Bulbochaete* are similar to those of higher plants (Fraser and Gunning, 1969).

Cell division involves the breaking of the parent wall and the formation of apical caps. In *Oedogonium*, cell division (Fig. 5.65) is initiated by the formation of a ring under the wall in the upper part of the cell (Hill and Machlis, 1968). The ring enlarges by the coalescence of material produced in the cytoplasm. While the ring is being produced, the nucleus migrates to the center of the cell and divides mitotically. During late telophase, the new cross wall begins to form by means of a phycoplast. The daughter cells elongate, causing a split in the parent wall near the apical ring. This rent in the cell wall is covered by the material in the apical ring, which expands as the cells elongate. Each daughter cell eventually elongates to about the same length as the mother cell, elongation being completed within 15 minutes. The material of the ring becomes the cuticle, and a new cell wall is laid down under it. During the elongation of the daughter cells, the new transverse wall has moved up to the base of the newly formed secondary wall and fused with it. Because cell division is intercalary in *Oedogonium*, division of every cell in the filament and repeated division of the daughter cells result in alternate cells with and without caps of old cell walls. This theoretical condition usually does not occur in nature. Repeated division of the distal daughter

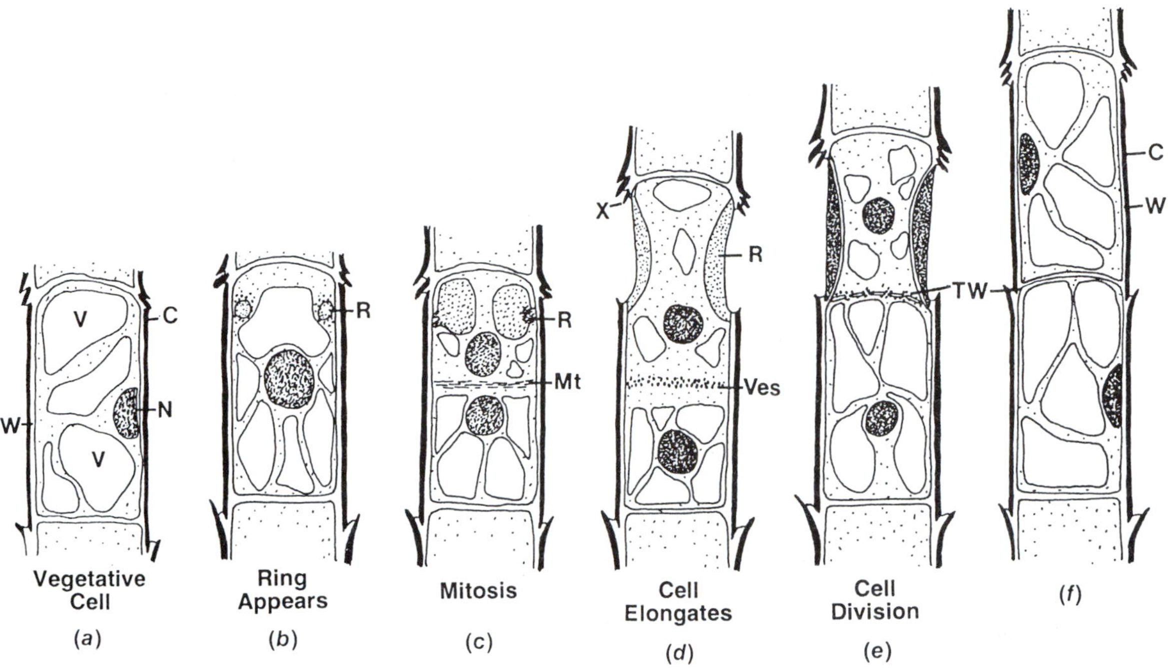

Fig. 5.65 Cell division in *Oedogonium*. (C) Cuticle; (Mt) microtubules; (N) nucleus; (R) ring; (TW) transverse wall; (V) vacuole; (Ves) vesicles; (W) wall; (X) cap. (After Hill and Machlis, 1968.)

cell commonly results in filaments in which a cell with an apical cap is successively followed by several cells without caps.

Asexual reproduction (Fig. 5.66) is by means of zoospores in all three genera. Zoospores are formed singly within a cell and usually only in those cells with apical caps. The earliest sign of zoosporogenesis is the appearance of a small electron-dense mass in an invagination of the nuclear envelope (Pickett-Heaps, 1971). From this the centrioles appear and multiply rapidly, forming two adjacent rows near the nucleus. The nucleus, with its two rows of centrioles, moves to the lateral wall. The two rows of centrioles move apart in the center to form a circle of centrioles under the plasmalemma. The centrioles extrude the flagella and the flagella roots. The Golgi secrete a fibrillar hyaline layer around the zoospore, which makes up the vesicle in which the zoospore is initially encased. The Golgi also secrete mucilage to the base of the zoospore, which probably aids in extrusion of the zoospore. The lateral wall of the parent cell splits at the apical cap, and the zoospore in a vesicle emerges through the aperture. The vesicle has two layers with a ring (Retallack and Butler, 1970); it opens in the area of the ring, releasing the zoospore. The zoospore has about 30 flagella linked together by a striated root at the apical end (Hoffman and Manton, 1962); it swims for about an hour, settles, retracts its flagella, and develops a holdfast that attaches to the substrata. This then develops into a new filament. Aplanospores can also be formed, and resemble oogonia.

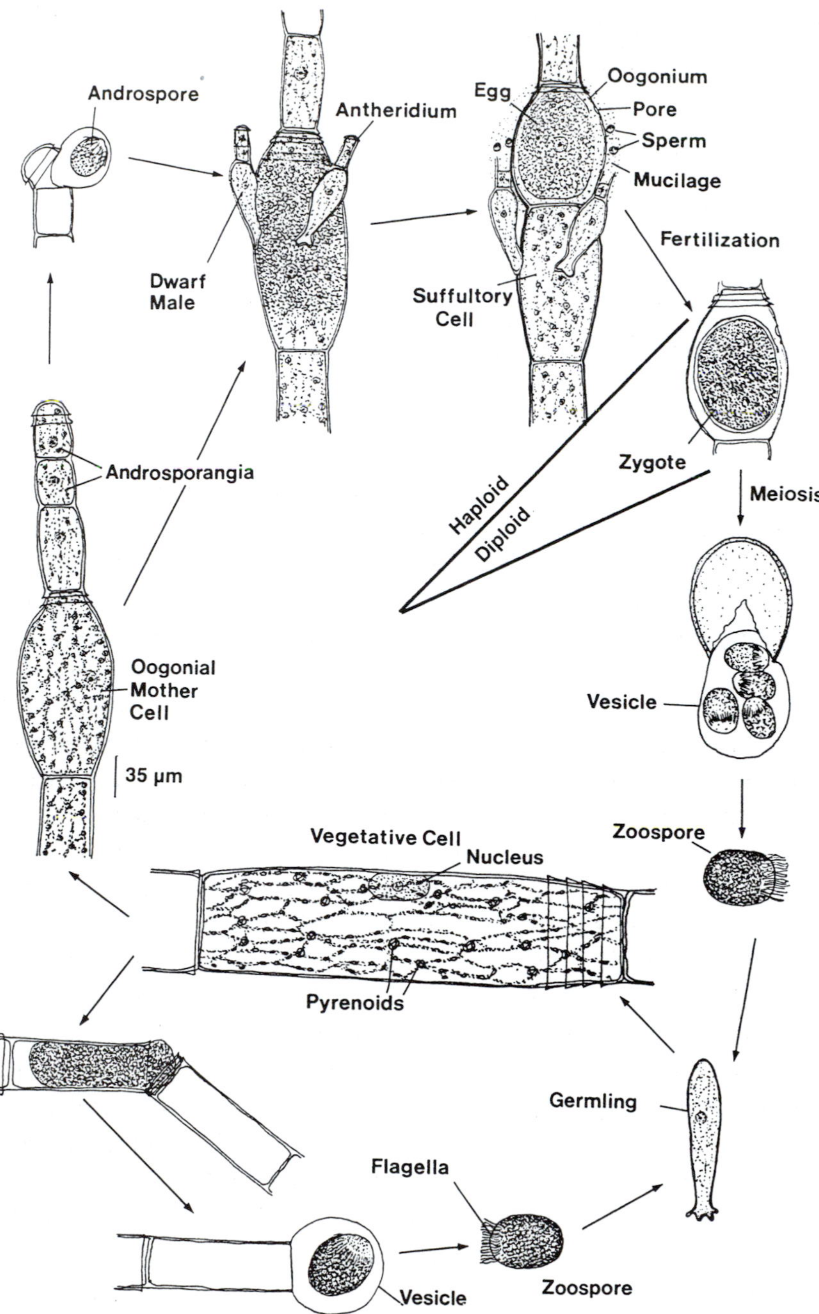

Fig. 5.66 The life cycle of a nannandrous species of *Oedogonium*. (Adapted from Smith, 1955.)

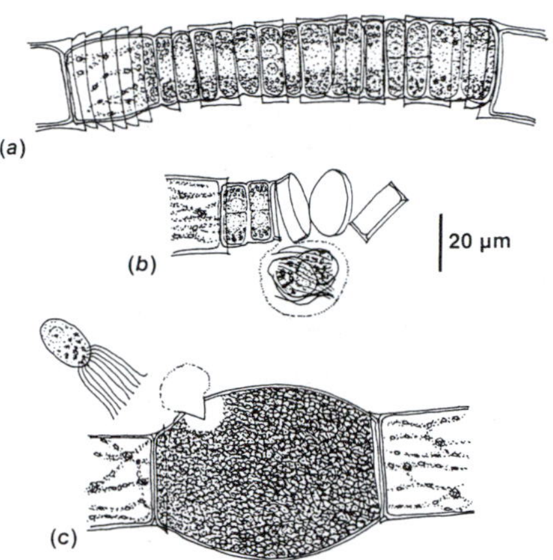

Fig. 5.67 Gametogenesis in the macrandrous *Oedogonium crassum*. (*a*) Antheridia. (*b*) Liberation of sperm. (*c*) Oogonium with a single egg attracting a spermatozoid. (After Smith, 1955.)

Sexual reproduction is oogamous and, depending on the behavior of the male filaments, either macrandrous if the male filament forms the sperm directly, or nannandrous if the sperm are produced in a special dwarf male filament. In many species of *Oedogonium*, it is relatively easy to induce sexual reproduction by placing old filaments in fresh media and saturating the atmosphere with carbon dioxide. Nitrogen limitation has also been reported to induce the formation of oogonia (Singh and Chaudhary, 1990).

Oogonia have a similar developmental pattern in macrandrous (Fig. 5.67) and nannandrous species (Fig. 5.66). The intercalary or terminal oogonial mother cells divide transversely, with one cell becoming the oogonium and the other either a suffultory (supporting) cell or a second oogonial mother cell. If one cell becomes another oogonial mother cell, then the oogonia are produced in series. The oogonia initially are wider than the vegetative cells and have a number of apical caps at their upper end from previous cell divisions. As the oogonium matures, it swells and becomes globose, forming a small pore or crack in the oogonial wall. The protoplast develops into an egg. Just prior to fertilization the central nucleus moves to just under the fertilization pore, and gelatinous material is extruded through the pore.

Macrandrous species form terminal or intercalary antheridia by an unequal division of an antheridial mother cell (Fig. 5.67). In this division, the upper cell is much shorter than the lower. The lower cell then divides repeatedly to produce a series of 2 to 40 antheridia. Each antheridium forms two sperm, which are liberated in a vesicle by transverse splitting of the wall. The sperm then escape from the vesicle. The sperm are similar in structure to the zoospores but smaller and more elongated with a crown of about 30 flagella (Hoffman and Manton, 1963), and are yellowish-green because of their reduced plastids. In a macrandrous species such as *Oedogonium cardiacum* (Hoffman, 1960), the sperm are attracted to the oogonium by a chemotactic substance secreted by the oogonium.

In nannandrous species, male filaments form **androsporangia** in a manner similar to the formation of antheridia by macrandrous species, except only one **androspore** is produced per sporangium. The androspore can develop in one of three ways depending on the environment (Hill et al., 1989).

1 In an environment with low nitrate or ammonium concentration, the androspore produces a cyst that will eventually produce another androspore. The cycle will continue until the nitrate or ammonium concentration in the environment increases.
2 If there is a large amount of nitrate or ammonium in the water, the androspore will produce a vegetative male thallus that is indistinguishable from the parent. This is similar to the situation in nature when a rainstorm washes nutrients into a nutrient-depleted environment.
3 If the environment contains the female pheromone "circein," the androspore is attracted to the female oogonium that is secreting the pheromone. The androspore attaches to the oogonial mother cell and causes the oogonial mother cell to divide into the oogonium and a lower **suffultory cell** to which the androspore is now attached. The production of the substance that attracted the androspore now ceases. If the attachment of the androspores to the oogonial mother cell is prevented by coating the oogonial mother cell with agar, then the oogonial mother cell continues to attract androspores, but the divison of the oogonial mother cell does not occur. Female differentiation beyond the oogonial mother cell stage must be triggered by direct contact of the attached androspore. The main advantage of a nannandrous system is that differentiation of the male is a prerequisite for differentiation of the female, thus ensuring the presence of a male for the receptive female.

The attached androspore on the suffultory cell next elongates and develops into a two-celled dwarf male filament. The growth of the dwarf male is toward the oogonium under the influence of a chemotactic substance secreted by the oogonium. The apical cell of the male filament is an antheridium, which forms two sperm. The oogonium secretes a thick gelatinous sheath that includes the tip of the dwarf male. The sperm are released and trapped in this gel. After about 1 hour of random movement in the gel, the sperm aggregate at a certain spot on the upper third of the oogonium. The sperm extend their anterior end so that this end is flexible (Hoffman, 1973). A protoplasmic papilla suddenly appears through the wall of the oogonium, and all the spermatozoids in the vicinity attach to this papilla by their anterior ends. Within a second the papilla is withdrawn through the pore in the wall, taking with it a single sperm. The remaining sperm hover about the pore for about 5 minutes before they disperse in the gel. Four hormone-controlled steps are thus involved in fertilization: (1) the chemotaxis of the androspores; (2) the directed growth of the dwarf males; (3) the triggering of the division of the oogonial mother cell; and (4) the chemotaxis of the sperm to the prospective opening of the oogonium (Rawitscher-Kunkel and Machlis, 1962).

Nannandrous species are thought to have evolved from macrandrous ones by parthenogenetic germination of sperm to form dwarf male filaments.

The zygote is separated from the oogonial wall by a space, and after fertilization a wall is soon formed around the zygote. As the zygote matures, there is an accumulation of a reddish oil in the protoplasm. Eventually the zygote is liberated by decay of the oogonial wall. It undergoes a rest period during which its nucleus divides by meiosis to form four haploid nuclei. Shortly before germination, the protoplast becomes green and forms four daughter protoplasts, each of which becomes a zoospore. The zoospores are released and develop into filaments in the same manner as in asexual reproduction. The life cycle is thus mostly haploid, the zygote being the only diploid structure.

Fig. 5.68 *Left:* Felix Eugen Fritsch and *right:* Gilbert Morgan Smith.
Felix Eugen Fritsch Born April 26, 1879, in Hampstead, United Kingdom; died May 23, 1954. Dr Fritsch was educated at the University of London and the University of Munich, where he received his PhD. From 1902 to 1911, he was an assistant professor at University College London; from 1905 to 1906, a lecturer at Birkbeck College; and from 1907 to 1911, a lecturer at East London College. In 1911, he became the head of the Department of Botany at Queen Mary College, University of London, where he stayed until retirement in 1948. Dr Fritsch was the author of a number of books, by far the best known being *The Structure and Reproduction of the Algae*, which today is still the most comprehensive treatise on the algae as a whole.
Gilbert Morgan Smith Born January 6, 1885, in Beloit, Wisconsin. Died July 11, 1959. Dr Smith received his BS from Beloit College (1907) and his PhD from the University of Wisconsin (1913). From 1913 to 1925, he was at the University of Wisconsin, ultimately as an associate professor. From 1925 until his retirement, he was a professor at Stanford University. Dr Smith is best known for his books on algae which include *Phytoplankton of the Inland Lakes of Wisconsin*, *The Freshwater Algae of the United States*, *Cryptogamic Botany*, and *Marine Algae of the Monterey Peninsula*.

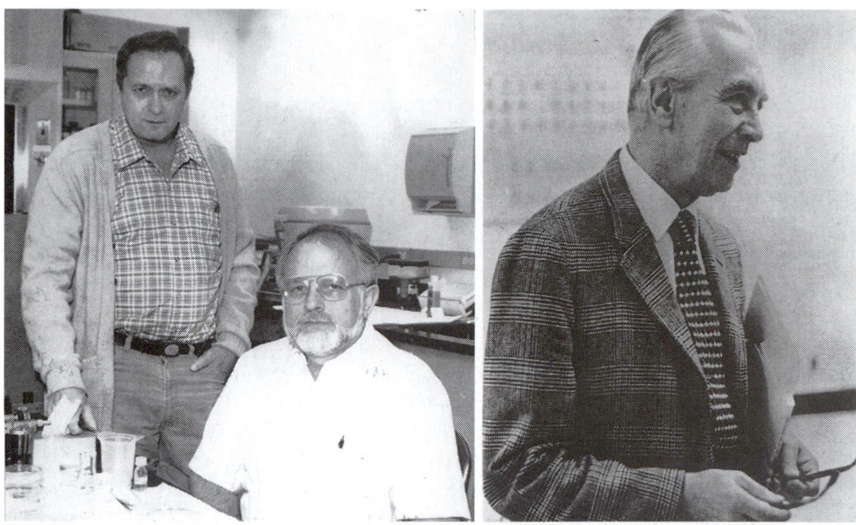

Fig. 5.69 *Left photograph:* Kenneth Stewart (*left*) and Karl Mattox (*right*). *Right photograph:* Luigi Provasoli. (Photograph of Provasoli from *Journal of Phycology* 14:1, 1978.)
Kenneth Stewart Born September 4, 1932, in Moberly, Missouri. Dr Stewart received his BS from Southern Illinois University at Carbondale, Illinois (1954). He served in the Army and worked in civilian life until enrolling at the University of California, Davis, where he received his PhD in 1968. Since 1968, he has been in the Department of Botany, Miami University, Oxford, Ohio.
Karl Mattox Born August 22, 1936, in Cincinnati, Ohio. Dr Mattox received his BS (1958) and MA (1960) from Miami University and his PhD (1962) from the University of Texas. From 1962 to 1966, he was an assistant professor in the Department of Botany at the University of Toronto. In 1966, he moved to Miami University where he is currently a professor in the Department of Botany.
Luigi Provasoli Born February 13, 1908, in Busto Arsizio, Italy. Died 30 October, 1992. Dr Provasoli received his BS (1931) and PhD (1939) from the University of Milan. From 1933 to 1942, he was in the Entomology Department at the University of Milan and, from 1942 to 1945, professor of zoology at the University of Camerina. From 1948 to 1952, he was at St Francis College in Brooklyn where he became head of the department. In 1952, he joined Haskins Laboratories, where he stayed first in New York City and then, when his branch moved, at Yale until his retirement. Dr Provasoli is well known in the field of phycology for his investigations on the cultivation of algae and protozoa. Dr Provasoli was the first editor of the *Journal of Phycology*, the publication of the Phycological Society of America.

References

Ahmadjian, V. (1960). Some new and interesting species of *Trebouxia*, a genus of lichenized algae. *Am. J. Bot.* 47:677–83.

Ahmadjian, V. (1993). *The Lichen Symbiosis*. New York: John Wiley, 250 pp.

Ahmadjian, V., and Jacobs, J. B. (1981). Relationship between fungus and alga in the lichen *Cladonia cristatella* Tuck. *Nature* 289:169–72.

Barlow, S. B., and Cattolico, R. A. (1980). Fine structure of the scale-covered green flagellate *Mantoniella squamata* (Manton et Parke) Desikachary. *Br. Physol. J.* 15:321–33.

Bean, B. (1977). Geotactic behaviour of *Chlamydomonas*. *J. Protozool.* 24:394–401.

Becker, B., Perasso, L., Kammer, A., Salzburg, M., and Melkonian, M. (1996). Scale-associated glycoproteins of *Scherffelia dubia* (Chlorophyta) form high-molecular-weight complexes between the scale layers and the flagellar membrane. *Planta* 199:503–10.

Belcher, J. H., Pennick, N. C., and Clarke, K. J. (1974). On the identity of *Asteromonas propulsum* Butcher. *Br. Physol. J.* 9:101–6.

Ben-Amotz, A. (1996). Effect of low temperature on the stereoisomer composition of β-carotene in the halotolerant alga *Dunaliella bardawil*. *J. Phycol.* 32:272–5.

Bendix, S. W. (1960). Phototaxis. *Bot. Rev.* 26:145–208.

Benson, E. E., Rutter, J. C., and Cobb, A. H. (1983). Seasonal variation in frond morphology and chloroplast physiology of the intertidal alga *Codium fragile* (Suringar) Hariot. *New Phytol.* 95:569–80.

Berger, S., and Kaever, M. J. (1992). *Dascladales: An Illustrated Monograph of a Fascinating Algal Order*. Stuttgart: Thieme. 47 pp.

Berger-Perrot, Y., Thomas, J-Cl., and L'Hardy-Halos, M-Th. (1993). Gametangia, gametes, fertilization and zygote development in *Ulothrix flacca* var. *roscoffensis* (Chlorophyta). *Phycologia* 342:356–66.

Betsche, T., Schaller, D., and Melkonian, M. (1992). Identification and characterization of glycolate oxidase and related enzymes from the endocyanotic algae *Cyanophora paradoxa* and from pea leaves. *Plant Physiol.* 98:887–93.

Blackburn, K., and Temperley, B. N. (1936). *Botryococcus* and the algal coals. *Proc. R. Soc. Edinburgh* 58:841–6.

Bloodgood, R. A. (1991). Transmembrane signaling in cilia and flagella. *Protoplasma* 164:12–22.

Borowitzka, M. A. (1982). Morphological and cytological aspects of algal calcification. *Int. Rev. Cytol.* 74:127–62.

Borowitzka, M.A., and Larkum, A. W. D. (1974). Chloroplast development in the caulerpalean alga *Halimeda*. *Protoplasma* 81:131–44.

Boscov, J. S., and Feinleib, M. E. (1979). Phototactic response of *Chlamydomonas* to flashes of light. II. Response of individual cells. *Photochem. Photobiol.* 30:499–505.

Bouck, G. B., and Morgan, E. (1957). The occurrence of *Codium* in Long Island waters. *Bull. Torrey Bot. Club* 84:384–7.

Bråten, T. (1971). The ultrastructure of fertilization and zygote formation in the green alga. *Ulva mutabilis* Føyn. *J. Cell Sci.* 9:621–35.

Bråten, T. (1973). Autoradiographic evidence for the rapid disintegration of one chloroplast in the zygote of the green alga *Ulva mutabilis*. *J. Cell Sci.* 12:385–9.

Braun, A. (1851). *Betrachtungen über die Erscheinung der Verjungung in der Natur.*, Leipzig.

Bristol, B. M. (1919). On a Malay form of *Chlorococcum humicola* (Naeg.) Rabenh. *J. Linn. Soc. Bot.* 44:473–82.

Brook, A. J. (1989). Barium sulphate crystals and desmids. *Brit Phycol. J.* 24:299–300.

Brown, R. M., Johnson, C., and Bold, H. C. (1968). Electron and phase-contrast microscopy of sexual reproduction in *Chlamydomonas moewusii. J. Phycol.* 4:100–20.

Bubrick, P., and Galun, M. (1980). Proteins from the lichen *Xanthoria parietina* which bind to phycobiont cell walls. Correlation between binding patterns and cell wall cytochemistry. *Protoplasma* 104:167–73.

Butcher, R. N. (1932). Notes on new and little-known algae from the beds of rivers. *New Phytol.* 31:289–309.

Cáceres, E. J., Hoffman, L. R., and Leonardi, P.I. (1997). Fine structure of the male and female gametes of *Atractomorpha porcata* (Sphaeropleaceae, Chlorophyta) with emphasis on the development and absolute configuration of the flagellar apparatus. *J. Phycol.* 33:948–59.

Cane, R. F. (1977). Coorgongite, balkashite and related substances – An annotated bibliography. *Trans. R. Soc. S. Aust.* 101:153–64.

Carothers, Z. B., and Kreitner, G. L. (1967). Studies of spermatogenesis in the Hepaticae. I. Ultrastructure of the *Vierergruppe in Marchantia. J. Cell Biol.* 33:43–51.

Casanova, M. T. (1997). Oospore variation in three species of *Chara* (Charales, Chlorophyta). *Phycologia* 36:274–80.

Chapman, V. J. (1970). *Seaweeds and Their Uses,* 2nd ed. London: Methuen.

Chihara, M. (1963). The life history of *Prasinocladus ascus* as found in Japan, with special reference to the systematic position of the genus. *Phycologia* 3:19–28.

Chihara, M. (1969a). *Ulva arasakii,* a new species of green algae: Its life history and taxonomy. *Bull. Nat. Sci. Mus. (Japan)* 12:849–62.

Chihara, M. (1969b). Culture study of *Chlorochytrium inclusum* from the northeast Pacific. *Phycologia* 8:127–33.

Christie, A. O., and Evans, L. V. (1962). Periodicity in the liberation of gametes and zoospores of *Enteromorpha intestinalis* Link. *Nature* 193:193–4.

Coggin, S. J., Hutt, W., and Kochert, G. (1979). Sperm bundle–female somatic cell interaction in the fertilization process of *Volvox carteri* f. *weismannia* (Chlorophyta). *J. Phycol.* 15:247–51.

Collings, D. A., Wasteneys, G. B., and Williamson, R. E. (1996). Actin microtubule interactions in the alga *Nitella*: analysis of the mechanism by which microtubule depolymerization potentiates cytochalasin's effects in streaming. *Protoplasma* 191:178–90.

Collins, C. R., and Farrar, J. F. (1978). Structural resistances to mass transfer in the lichen *Xanthoria parietina. New Phytol.* 81:71–83.

Daugbjerg, N., Moestrup, Ø., and Arctander, P. (1995). Phylogeny of genera of Prasinophyceae and Pedinophyceae (Chlorophyta) deduced from molecular analysis of the *rbs* gene.*Phycological Research* 43:203–13.

Davies, R. R., and Wilkinson, J. L. (1967). Human protothecosis: Supplementary studies. *Ann. Trop. Med. Parasitol.* 61:112–15.

Deason, T. R. (1983). Cell wall structure and composition as taxonomic characters in the coccoid Chlorophyceae. *J. Phycol.* 19:248–51.

deJesus, M. D., Tabatabai, F., and Chapman, D. J. (1989). Taxonomic distribution of copper–zinc superoxide dismutase in green algae and its phylogenetic importance. *J. Phycol.*25:767–72.

Demets, R., Tomson, A. M., Stegwee, D., and van de Ende, H. (1990). Cell–cell coordination in conjugating *Chlamydomonas* gametes. *Protoplasma* 158:188–99.

Dodds, W. K., and Gudder, D. A. (1992). The ecology of *Cladophora*. *J. Phycol.* 28:415–27.

Domozych, C. R., Plante, K., Blair, P., Paliules, L., and Domozych, D. S. (1993). Mucilage processing and secretion in the green alga *Closterium*. I. Cytology and biochemistry. *J. Phycol.* 29:650–9.

Domozych, D. S., Stewart, K. D., and Mattox, K. R. (1981). Development of the cell wall in *Tetraselmis:* Role of the Golgi apparatus and extracellular wall assembly. *J. Cell Sci.* 52:351–71.

Dorling, M., McAuley, P. J., and Hodge, H. (1997). Effect of pH on growth and carbon metabolism of maltose-releasing *Chlorella* (Chlorophyta). *Eur. J. Phycol.* 32:19–24.

Douglas, A. E. (1985). Growth and reproduction of *Convoluta roscoffensis* containing different naturally occurring algal symbionts. *J. Mar. Biol. Assoc. UK* 65:871–9.

Dreher, T. W., Grant, B. R., and Wetherbee, R. (1978). The wound response in the siphonous alga *Caulerpa simpliciuscula* C. Ag.: fine structure and cytology. *Protoplasma* 96:189–203.

Duncan, T. M., Renzaglia, K. S., and Garbary, D. J. (1997). Ultrastructure and phylogeny of the spermatozoid of *Chara vulgaris* (Charophyceae). *Pl. Syst. Evol.* 204:125–40.

Eckhardt, R., Schnetter, R., and Seibold, G. (1986). Nuclear behaviour during the life cycle of *Derbesia* (Chlorophyceae). *Br. Phycol. J.* 21:287–95.

Egan, P. F., and Trainor, F. R. (1989). The role of unicells in the polymorphic *Scenedesmus armatus* (Chlorophyceae). *J. Phycol.* 25:65–70.

Egerod, L. E. (1952). An analysis of the siphonaceous Chlorophycophyta. *Univ. Calif. Publ. Bot.* 25:325–454.

Ernst, A. (1902). *Dichotomosiphon tuberosus* (A. Br.) Ernst, eine neue oogame Süsswasser-Siphonee. *Beih. Bot. Zentralbl.* 13:115–48.

Feldmann, J. (1950). Sur l'existence d'une alternance de générations entre *l'Halicystis parvula* Schmitz et le *Derbesia tenuissima* (De Not.) Crn. *C. R. Séances Acad. Sci. Paris* 230:322–33.

Feldmann, J., and Feldmann, G. (1947). Quelques algues marines de Roscoff nouvelles pour les côtes de la Manche. *Bull. Soc. Bot. France* 93:234–7.

Fisher, M., Pick, U., and Zamir, A. (1994). A salt-induced 60-kilodalton plasma membrane protein plays a potential role in extreme halotolerence of the alga *Dunaliella*. *Plant Physiol.* 106:1359–65.

Floyd, G. L., Stewart, K. D., and Mattox, K. R. (1972). Cellular organization, mitosis and cytokinesis in the ulotrichalean alga, *Klebsormidium*. *J. Phycol.* 8:176–84.

Foerster, H., Wiese, L., and Braunitzer, G. (1956). Über das agglutinierend wirkende Gynogamon von *Chlamydomonas eugametos*. *Z. Naturforsch.* 11b:315–17.

Fogg, G. E. (1971). Recycling through algae. *Proc. R. Soc. Lond.* [B] 179:201–7.

Fraser, T. W., and Gunning, B. E. S. (1969). The ultrastructure of plasmodesmata in the filamentous green alga *Bulbochaete hiloensis* (Nordst.) Tiffany. *Planta* 88:244–54.

Friedmann, I. (1959). Structure, life history and sex determination of *Prasiola stipitata* Suhr. *Ann. Bot.* 23:571–94.

Friedmann, I., and Manton, I. (1959). Gametes, fertilization and zygote development in *P. stipitata*. *Nova Hedwigia* 1:333–44.

Fukumoto, P., Fuji, T., and Sekimoto, H. (1997). Detection and evaluation of a novel sexual pheromone that induces sexual cell division of *Closterium ehrenbergii* (Chlorophyta). *J. Phycol.* 33:441–5.

Fünfstück, M. (1907). Lichenes. In *Die Naturlichen Pflanzenfamilien*, ed. A. Engler, and K. Prantl. Leipzig: Engelmann.

Geib, K., Golldack, D., and Gimmler, H. (1996). Is there a requirement for an external carbonic anhydrase in the extremely acid-resistant green alga *Dunaliella acidophila*? *Eur. J. Phycol.* 31:273–84.

Gerrath, J. F. (1969). *Penium spinulosum* (Wolle) comb. nov. (Desmidaceae): A taxonomic correction based on cell wall ultrastructure. *Phycologia* 8:109–18.

Gessner, F. (1967). Untersuchungen über das osmotische Verhalten der Grünalge *Valonia ventricosa*. *Helgol. Wiss. Meeresunters.* 15:143–54.

Goddard, R. H., and Dawes, C. J. (1983). An ultrastructural and histochemical study of the wound response in the coenocytic green alga *Caulerpa ashmeadii* (Caulerpales). *Protoplasma* 114:163–72.

Goldstein, M., and Morral, S. (1970). Gametogenesis and fertilization in *Caulerpa*. *Ann. NY Acad. Sci.* 175:660–72.

Goodenough, U. W., and Heuser, J. E. (1985). The *Chlamydomonas* cell wall and its constituent glycoproteins analyzed by the quick-freeze, deep-etch technique. *J. Cell Biol.* 101:1550–68.

Goodenough, U. W., Detmers, P. A., and Hwang, C. (1982). Activation for cell fusion in *Chlamydomonas*: Analysis of wild-type gametes and nonfusing mutants. *J. Cell Biol.* 92:378–86.

Goreau, T. F. (1963). Calcium carbonate deposition by coralline algae and corals in relation to their roles as reef-builders. *Ann. NY Acad. Sci.* 109:127–67.

Graham, J. M., Arancibia-Avile, P., and Graham, L. E. (1996). Physiological ecology of a species of the filamentous green alga *Mougeotia* under acid conditions: light and temperature effects on photosynthesis and respiration. *Limnol. Oceanogr.* 41:253–62.

Graham, J. M., Lembi, C. A., Adrian, H. L., and Spencer, D. F. (1995). Physiological responses to temperature and irradiance in *Spirogyra* (Zygnematales, Charophyceae). *J. Phycol.* 31:334–40.

Graham, L. E. (1984). *Coleochaete* and the origin of land plants. *Am. J. Bot.* 71:603–8.

Graham, L. E., and McBride, G. E. (1979). The occurrence and phylogenetic significance of a multilayered structure in *Coleochaete* spermatozoids. *Am. J. Bot.* 66:887–94.

Graham, L. E., and Kranzfelder, J. A. (1986). Irradiance, daylength and temperature effects on zoosporogenesis in *Coleochaete scutata* (Charophyceae). *J. Phycol.* 22:35–9.

Green, T. G. A. (1970). The biology of lichen symbionts, DPhil. thesis, Oxford.

Griffin, N. J., and Aken, M. E. (1993). Rhythmic settling behavior in *Pyramimonas parkae* (Prasinophyceae). *J. Phycol.* 29:9–15.

Gruber, P. J., Frederick, S. E., and Tolbert, N. E. (1974). Enzymes related to lactate metabolism in green algae and lower land plants. *Plant Physiol.* 53:167–70.

Grünewald, K., Hagen, C., and Braune, W. (1997). Secondary carotenoid accumulation in flagellates of the green alga *Haematococcus lacustris*. *Eur. J. Phycol.* 32:387–92.

Gulliksen, O. M., Hushovd, O. T., Texmon, I., and Nordby, Ø. (1982). Changes in respiration, photosynthesis and protein composition during induced synchronous formation of gametes and zoospores in *Ulva mutabilis* Føyn. *Planta* 156:33–40.

Halldal, P. (1958). Action spectra of phototaxis and related problems in Volvocales, *Ulva*-gametes, and Dinophyceae. *Physiol. Plant.* 11:118–53.

Hanisak, M. D. (1979). Nitrogen limitation of *Codium fragile* ssp. *tomentosoides* as determined by tissue analysis. *Mar. Biol.* 50:333–7.

Hardy, J. T., and Curl, H. C., Jr. (1972). The candy-colored, snow-flaked alpine biome. *Nat. Hist.* 81 (Nov.): 75–8.

Hartshorne, J. N. (1953). The function of the eyespot in *Chlamydomonas*. *New Phytol.* 52:292–7.

Hawes, I., and Smith, R. (1995). Effect of current velocity on the detachment of thalli of *Ulva lactuca* (Chlorophyta) in a New Zealand estuary. *J. Phycol.* 31:875–80.

Hawthorne, D. B., Dreher, T. W., and Grant, B. R. (1981). The wound response in the siphonous alga *Caulerpa simpliciuscula* C. Ag.: II. The effect of wounding on carbon flow. *Protoplasma* 105:195–206.

Head, W. D., and Carpenter, E. J. (1975). Nitrogen fixation associated with the marine macroalga *Codium fragile*. *Limnol. Oceanogr.* 20:815–23.

Hepperle, D., and Krienitz, L. (1996). The extracellular calcification of zoospores of *Phacotus lenticularis* (Chlorophyta, Chlamydomonadales). *Eur. J. Phycol.* 31:11–21.

Hernandez, I., Peralta, G., Perez-Llorens, J. L., Vergara, J. J., and Niell, F. X. (1997). Biomass and dynamics of growth of *Ulva* species in Palmones River estuary. *J. Phycol.* 33:764–72.

Hessler, R., and Peveling, E. (1978). Die Lokalisation von ^{14}C-Assimilaten von Flechtenthalli von *Cladonia incrassata* Floerke und *Hypogymnia physodes* (L.) Ach. *Z. Pflanzenphysiol.* 86:287–302.

Hill, G. J. C., and Machlis, L. (1968). An ultrastructural study of vegetative cell division in *Oedogonium borisianum*. *J. Phycol.* 4:261–71.

Hill, G. J. C., Cunningham, M. R., Byrne, M.M., Ferry, T. P., and Halvorsen, J. S. (1989). Chemical control of androspore morphogenesis in *Oedogonium donnelli* (Chlorophyta, Oedogoniales). *J. Phycol.* 25:368–76.

Hodick, D. (1994). Negative gravitropism in *Chara* protonemata: a model integrating the opposite gravitropic responses of protonemata and rhizoids. *Planta* 195:43–9.

Hoffman, L. (1960). Chemotaxis of *Oedogonium* sperms. *Southwest. Nat.* 5:111–16.

Hoffman, L. (1973). Fertilization in *Oedogonium*. I. Plasmogamy. *J. Phycol.* 9:62–84.

Hoffman, L., and Manton, I. (1962). Observations on the fine structure of the zoospore of *Oedogonium cardiacum* with special reference to the flagellar apparatus. *J. Exp. Bot.* 13:443–9.

Hoffman, L., and Manton, I. (1963). Observations on the fine structure of *Oedogonium* II. The spermatozoid of *O. cardiacum*. *Am. J. Bot.* 50:455–63.

Hoffman, L. R., and Buchheim, M. A. (1989). Zygote appendages (cirri), a new structural feature in the Sphaeropleaceae (Chlorophyceae). *J. Phycol.* 25:149–59.

Hoham, R. W. (1973). Pleiomorphism in the snow alga *Raphidonema nivale* Lagerh. (Chlorophyta), and a revision of the genus *Rhaphidonema* Lagerh. *Syesis* 6:255–63.

Hori, T., Norris, R. E., and Chihara, M. (1982a). Studies on the ultrastructure and taxonomy of the genus *Tetraselmis* (Prasinophyceae). I. Subgenus *Tetraselmis*. *Bot. Mag. (Tokyo)* 95:49–61.

Hori, T., Norris, R. E., and Chihara, M. (1982b). Studies on the ultrastructure and taxonomy of the genus *Tetraselmis* (Prasinophyceae). II. Subgenus *Prasinocladia*. *Bot. Mag. (Tokyo)* 96:385–92.

Hoskin, C. M. (1963). Recent carbonate sedimentation on Alacran Reef, Yucatan, Mexico. *Nat. Acad. Sci. Nat. Res. Council Publ.* 1089, 160 pp.

Huber-Pestalozzi, G. (1961). Volvocales. In *Die Binnengewässer*, Vol. 16, Part 5. Stuttgart, Germany: E. Schweizerbart'sche Verlagsbuchhandlung.

Huizing, H. J., Rietema, H., and Sietsma, J. H. (1979). Cell wall constituents of several siphonaceous green algae in relation to morphology and taxonomy. *Br. Phycol. J.* 14:25–32.

Hutt, W., and Kochert, G. (1971). Effects of some protein and nucleic acid synthesis inhibitors on fertilization in *Volvox carteri*. *J. Phycol.* 7:316–20.

Imam, S. H., Buchanan, M. J., Shin, H. C., and Snell, W. J. (1985). The *Chlamydomonas* cell wall: Characterization of the wall framework. *J. Cell Biol.* 101:1599–1607.

Inouye, I., Hori, T., and Chihara, M. (1990). Absolute configuration analysis of the flagellar apparatus of *Pterosperma cristatum* (Prasinophyceae) and consideration of its phylogenetic position. *J. Phycol.* 26:329–44.

Iyengar, M. O. P. (1932). *Fritschiella*, a new terrestrial member of the Chaetophoraceae. *New Phytol.* 31:329–35.

Johnson, H. J. (1961). *Limestone-Building Algae and Algal Limestones.* Boulder, Colo: Johnson Publ.

Kamitsubo, E. (1980). Cytoplasmic streaming in characean cells: role of subcortical fibrils. *Can J. Bot.* 58:760–5.

Kamiya, R., and Witman, G. B. (1984). Submicromolecular levels of calcium control the balance of beating between the two flagella in demembranated models of *Chlamydomonas. J. Cell Biol.* 98:97–107.

Kantz, T. S., Theriot, E. C., Zimmer, E. A., and Chapman, R. L. (1990). The Pleurophyceae and Micromonadophyceae: a cladistic analysis of nuclear rRNA sequence data. *J. Phycol.* 26:711–21.

Kawaguti, S., and Yamasu, T. (1965). Electron microscopy on the symbiosis between an elysoid gastropod and chloroplasts of a green alga. *Biol. J. Okayama Univ.* 11:57–64.

Kim, Y-S., Oyaizu, H., Matsumoto, S., Watanabe, M-M., and Nozaki, H. (1994). Chloroplast small-subunit ribosomal RNA gene sequence from *Chlamydomonas parkae* (Chlorophyta): Molecular phylogeny of a green alga with a peculiar pigment composition. *Eur. J. Phycol.* 29:213–17.

Kirk, D. L., and Kirk, M. M. (1986). Heat shock elicits production of sexual inducer in *Volvox. Science* 231:51–4.

Kirk, M. M., and Kirk, D. L. (1985). The translational regulation of protein synthesis, in response to light, at a critical stage of *Volvox* development. *Cell* 41:419–28.

Kleinig, H. (1969). Carotenoids of siphonous green algae: A chemotaxonomical study. *J. Phycol.* 5:281–4.

Klintworth, G. K., Fetter, B. F., and Nielsen, H. S. (1968). Protothecosis, an algal infection: Report of a case in man. *J. Med. Microbiol.* 1:211–16.

Klyver, F. D. (1929). Notes on the life history of *Tetraspora gelatinosa* (Vauch.) Desv. *Arch. Protistenk.* 66:290–6.

Kochert, G. (1968). Differentiation of reproductive cells in *Volvox carteri. J. Protozool.* 15:438–52.

Kochert, G., and Olson, L. W. (1970). Ultrastructure of *Volvox carteri.* I. The asexual colony. *Arch. Mikrobiol.* 74:19–30.

Koop, H-U. (1975a). Über den Ort der Meiose bei *Acetabularia mediterranea. Protoplasma* 85:109–14.

Koop. H-U. (1975b). Germination of cysts of *Acetabularia mediterranea. Protoplasma* 84:137–46.

Kornmann, P. (1938). Zur Entwicklungsgeschichte von *Derbesia* und *Halicystis. Planta* 28:464–70.

Kranz, H. D., Mikš, D., Siegler, M-L., Capesius, I., Sensen, C. W., and Huss, V. A. R. (1995). The origin of land plants: Phylogenetic relationships among charophytes, bryophytes, and vascular plants inferred from complete small-subunit ribosomal RNA gene sequences. *J. Mol. Evol.* 41:74–84.

Kriemer, G., and Melkonian, M. (1990). Reflection confocal laser scanning microscopy of eyespots in flagellated green algae. *Eur. J. Cell Biol.* 53:101–11.

Kroger, P., and Hegemann, P. (1994). Photophobic responses and phototaxis in *Chlamydomonas* are triggered by a single rhodopsin photoreceptor. *FEBS Letters* 341:5–9.

Kuroiwa, T. (1985). Mechanisms of maternal inheritance of chloroplast DNA: An active digestion hypothesis. *Microbiol. Sci.* 2:267–70.

Kuroiwa, T., Nakamura, S., Sato, C., and Tsubo, Y. (1985). Epifluorescent microscopic studies on the mechanism of preferential destruction of chloroplast nucleoids of male origin in young zygotes of *Chlamydomonas reinhardtii*. *Protoplasma* 125:43–52.

Kwiatkowski, M. (1996). Changes in ultrastructure of cytoplasm and nucleus during spermiogenesis in *Chara vulgaris*. *Folia Histochem. et Cytobiol.* 34:41–56.

Lee, Y-K., and Ding, S-Y. (1995). Effects of dissolved oxygen partial pressure on the accumulation of astaxanthin in chemostat cultures of *Haematococcus lacustris* (Chlorophyta). *J. Phycol.* 31:922–4.

Lembi, C. A., and Herndon, W. R. (1966). Fine structure of the pseudocilia of *Tetraspora*. *Can. J. Bot.* 44:710–12.

Lewin, R. A. (1957). The zygote of *Chlamydomonas moewusii*. *Can. J. Bot.* 35:795–807.

Lewin, R. A. (1958). The cell walls of *Platymonas*. *J. Gen. Microbiol.* 19:87–90.

Lucas, W. C. (1979). Alkaline band formation in *Chara corallina*. *Plant Physiol.* 63:248–54.

Lynn, R., and Brock, T. D. (1969). Notes on the ecology of a species of *Zygogonium* (Kütz) in Yellowstone National Park. *J. Phycol.* 5:181–5.

McCourt, R. M. (1995). Green algal phylogeny. *Trends in Ecology & Evolution* 10:159–63.

McCourt, R. M., Karol, K. G., Kaplan, S., and Hoshaw, R. W. (1995). Using *rbc*L sequences to test hypotheses of chloroplast and thallus evolution in conjugating green algae (Zygnematales, Charophyceae). *J. Phycol.* 31:989–95.

McCourt, R.M., Meiers, S. T., Karol, K. G., and Chapman, R. L. (1996). Molecular systematics of the Charales. In *Cytology, Genetics and Molecular Biology of Algae*, ed. B. R. Chaudhary, and S. B. Agrawal, pp. 323–6. Amsterdam, Netherlands: SPB Academic Publ.

McGarry, M. G., and Tongkasame, C. (1971). Water reclamation and algae harvesting. *Water Pollut. Control Fed. J.* 43:824–35.

McLean, R. O. (1974). The tolerance of *Stigeoclonium tenue* Kütz. to heavy metals in South Wales. *Br. Phycol. J.* 9:91–5.

McLean, R. O., and Benson-Evans, K. (1974). The distribution of *Stigeoclonium tenue* Kütz. in South Wales in relation to its use as an indicator of organic pollution. *Br. Phycol. J.* 9:83–9.

Maeda, M., Kuroda, K., Iriki, Y., Chihara, M., Nisizawa, K., and Miwa, T. (1966). Chemical nature of major cell wall constituents of *Vaucheria* and *Dichotomosiphon* with special reference to their phylogenetic positions. *Bot. Mag. Tokyo* 79:634–43.

Manton, I., and Parke, M. (1960). Further observations on green flagellates with special reference to possible relatives of *Chromulina pusilla* Butcher. *J. Mar. Biol. Assoc.* 39:275–98.

Marchant, H. J. (1977). Ultrastructure, development and cytoplasmic rotation of seta-bearing cells of *Coleochaete scutata*. *J. Physol* 13:28–36.

Marchant, H. J., and Pickett-Heaps, J. D. (1971). Ultrastructure and differentiation of *Hydrodictyon reticulatum*. II. Formation of zoids within the coenobium. *Aust. J. Biol. Sci.* 24:471–86.

Marchant, H. J., and Pickett-Heaps, J. D. (1972a). Ultrastructure and differentiation of *Hydrodictyon reticulatum*. III. Formation of the vegetative daughter net. *Aust. J. Biol. Sci.* 25:265–78.

Marchant, H. J., and Pickett-Heaps, J. D. (1972b). Ultrastructure and differentiation of *Hydrodictyon reticulatum*. IV. Conjugation of gametes and the development of zoospores and azygospores. *Aust. J. Biol. Sci.* 25:279–91.

Marchant, H. J., and Pickett-Heaps, J. D. (1972c). Ultrastructure and differentiation of *Hydrodictyon reticulatum*. V. Development of polyhedra. *Aust. J. Biol. Sci.* 25:1187–97.

Marchant, H. J., and Pickett-Heaps, J. D. (1972d). Ultrastructure and differentiation of *Hydrodictyon reticulatum*. VI. Formation of the germ net. *Aust. J. Biol. Sci.* 25:1199–213.

Mars, P. W., Rabson, A. R., Rippey, J. J., and Ajello, L. (1971). Cutaneous protothecosis. *Br. J. Dermatol.* 85, *Suppl.* 7:76–84.

Martin, N. C., and Goodenough, U. W. (1975). Gametic differentiation in *Chlamydomonas reinhardtii*. I. Production of gametes and their fine structure. *J. Cell Biol.* 67:587–605.

Maszewski, J., and van Bel, A. J. E. (1996). Different patterns of intercellular transport of lucifer yellow in young and mature antheridia of *Chara vulgaris* L. *Bot. Acta* 109:110–14.

Matsuda, Y., Saito, T., Yamaguchi, T., Koseki, M., and Hayashi, K. (1987). Topography of cell wall lytic enzyme in *Chlamydomonas reinhardtii:* From and location of the stored enzyme in vegetative cell and gamete. *J. Cell Biol.* 104:321–9.

Mattox, K. R., and Stewart, K. D. (1973). Observations on the zoospores of *Pseudodendoclonium basiliense* and *Trichosarcina polymorpha* (Chlorophyceae). *Can. J. Bot.* 51:1425–30.

Mattox, K. R., and Stewart, K. D. (1977). Cell division in the scaly green flagellate *Heteromastix angulata* and its bearing on the origin of the Chlorophyceae. *Am. J. Bot.* 64:931–45.

Mattox, K. R., and Stewart, K. D. (1984). Classification of the green algae; A concept based on comparative cytology. In *Systematics of the Green Algae* ed. D. E. G. Irvine, and D. M. John, pp. 29–72. London and Orlando: Academic Press.

Melkonian, M. (1979). An ultrastructural study of the flagellate *Tetraselmis cordiformis* Stein (Chlorophyceae) with emphasis on the flagellar apparatus. *Protoplasma* 98:139–51.

Melkonian, M. (1980a). Flagellar roots, mating structure and gametic fusion in the green alga *Ulva lactuca* (Ulvales). *J. Cell Sci.* 46:149–69.

Melkonian, M. (1980b). Flagellar apparatus, mating structure and gametic fusion in *Ulva lactuca* (Ulvales, Chlorophyceae). *Br. Phycol. J.* 15:197.

Melkonian, M., and Robenek, H. (1980a). Eyespot membranes in newliy released zoospores of the green alga *Chlorosarcinopsis gelatinosa* (Chlorosarcinales) and their fate during zoospores settlement. *Protoplasma* 104:129–40.

Melkonian, M., and Robenek, H. (1980b). Eyespot membranes of *Chlamydomonas reinhardtii:* A freeze-fracture study. *J. Ultrastruct. Res.* 72:90–102.

Melkonian, M., Schulze, D., McFadden, G. I., and Robenek, H.. (1988). A polyclonal antibody (anticentrin) distinguishes between the two types of fibrous flagellar roots in green algae. *Protoplasma* 144:56–61.

Menzel, D. (1980). Plug formation and peroxidase accumulation in two orders of siphonous green algae (Caulerpales and Dasycladales) in relation to fertilization and injury. *Phycologia* 19:37–48.

Menzel, D. (1986). Visualization of cytoskeletal changes through the life cycle in *Acetabularia. Protoplasma* 134:30–42.

Menzel, D. (1988). How do giant plant cells cope with injury? – The wound response in siphonous green algae. *Protoplasma* 144:73–91.

Mesland, D. A. M., Hoffman, J. L., Caligor, E., and Goodenough, U. W. (1980). Flagellar tip activation stimulated by membrane adhesions in *Chlamydomonas* gametes. *J. Cell Biol.* 84:599–617.

Metzger, P., Allard, B., Casadevall, E., Berkaloff, C., and Couté, A. (1990). Structure and chemistry of a new chemical rare of *Botryococcus braunii* (Chlorophyceae) that produces lycopodiene, a tetraterpenoid hydrocarbon. *J. Phycol.* 26:258–66.

Miller, D. H. (1978). Cell wall chemistry and ultrastructure of *Chlorococcum oleofaciens* (Chlorophyceae). *J. Phycol.* 14:189–94.

Mineyuki, Y., Kataoka, H., Masuda, Y., and Nagai, R. (1995). Dynamic changes in the actin cytoskeleton during the high-fluence rate response of the *Mougeotia* chloroplast. *Protoplasma* 185:222–29.

Moestrup, Ø. (1978). On the phylogenetic validity of the flagellar apparatus in green algae and other chlorophyll *a* and *b* containing plants. *BioSystems* 10:117–44.

Moestrup, Ø., and Hoffman, L. R. (1973). Ultrastructure of the green alga *Dichotomosiphon tuberosus* with special reference to the occurrence of striated tubules in the chloroplast. *J. Phycol.* 9:430–37.

Moestrup, Ø., and Hoffman, L. R. (1975). A study of the spermatozoids of *Dichotomosiphon tuberosus* (Chlorophyceae). *J. Phycol.* 11:225–35.

Moldowan, J. M., and Seifert, W. K. (1980). First discovery of botrycoccane in petroleum. *J.S.C. Chem. Comm.*, 912–14.

Musgrave, A. (1993). Mating in *Chlamydomonas. Prog. Phycol. Res.* 9:193–237.

Nakanishi, K., Nishijima, M., Nishimura, M., Kuwano, K., and Suga, N. (1996). Bacteria that induce morphogenesis in *Ulva pertusa* (Chlorophyta) grown under axenic conditions. *J. Phycol.* 32:479–82.

Nawata, T., Kikuyama, M., and Shihara-Ishikawa, I. (1993). Behavior of protoplasm for survival in injured cells of *Valonia ventricosa*: involvement of turgor pressure. *Protoplasma* 176:116–26.

Neumann, K. (1969). Protonema mit Riesenkern bei der siphonalen Grünalge *Bryopsis hypnoides* und weitere cytologische Befunde. *Helgol. Wiss. Meeresunters.* 19:45–57.

Nitecki, M. H. (1971). *Ischadites abbottae*, a new North American Silurian species (Dasycladales). *Phycologia* 10:263–75.

Nozaki, H. (1996). Morphology and evolution of sexual reproduction in the Volvocaceae (Chlorophyta). *J. Plant. Res.* 109:353–61.

Nossag, J., and Kasprik, W. (1993). The movement of *Micrasterias thomasiana* (Desmidaceae, Zygnematophyceae) in directed blue light. *Phycologia* 32:332–7.

O'Kelley, J. C. (1984). Nitrogen and gamete production in *Chlorococcum echinozygotum. J. Phycol.* 20:220–5.

O'Kelly, C. J., and Floyd, G. L. (1984). Flagellar apparatus absolute orientations and the phylogeny of the green algae. *BioSystems* 16:227–51.

Okuda, K., and Brown, R. M. (1992). A new putative cellulose-synthesizing complex of *Coleochaete scutata.* Protoplasma 168:51–63.

Okuda, K., and Mizura, S. (1993). Diversity and evolution of putative cellulose-synthesizing enzyme complexes in green plants. *Jpn. J. Phycol.* 41:151–73.

Okuda, K., Tsekos, I., and Brown, R. M. (1994). Cellulose microfibril assembly in *Erythrocladia subintegra* Rosenv.: An ideal system for understanding the relationship between synthesizing complexes (TCs) and microfibril crystallization. *Protoplasma* 180:49–58.

Oltmanns, F. (1898). Die Entwickelung der Sexualorgane bei *Coleochaete pulvinata. Flora* 85:1–14.

Oschman, L. J. (1966). Development of the symbiosis of *Convoluta roscoffensis* Graff and *Platymonas* sp. *J. Phycol.* 2:105–11.

Page, J. Z., and Kingsbury, J. M. (1968). Culture studies on the marine green alga *Halicystis parvula – Derbesia tenuissima.* II. Synchrony and periodicity in gamete formation and release. *Am. J. Bot.* 55:1–11.

Parke, M., and Manton, I. (1967). The specific identity of the algal symbiont in *Convoluta roscoffensis. J. Mar. Biol. Assoc. UK* 47:445–64.

Parker, B. C. (1970). Significance of cell wall chemistry to phylogeny in the algae. *Ann. NY Acad. Sci.* 175:417–28.

Pickett-Heaps, J. D. (1970). Some ultrastructural features of *Volvox* with particular reference to the phenomenon of inversion. *Planta* 90:174–90.

Pickett-Heaps, J. D. (1971). Reproduction by zoospores in *Oedogonium*. I. Zoosporogenesis. *Protoplasma* 72:275–314.

Pickett-Heaps, J. D. (1973). Cell division and wall structure in *Microspora. New Phytol.* 72:347–55.

Pickett-Heaps, J. D. (1974). Cell division in *Stichococcus. Br. Phycol. J.* 9:63–73.

Pickett-Heaps, J. D. (1975). *Green Algae, Structure, Reproduction and Evolution in Selected Genera.* Sunderland, MA: Sinauer Assoc.

Pickett-Heaps, J. D. (1976). Cell division in *Raphidonema longiseta. Arch. Protiskenkd.* 118:209–14.

Plain, N., Largeau, C., Derenne, S., and Couté, A. (1993). Variabilité morphologique de *Botryococcus braunii* (Chlorococcales, Chlorophyta) corrélations avec les conditions de croissance et la teneur en lipides. *Phycologia* 32:259–65.

Pocock, M. A. (1955). Studies in North American Volvocales. I. The genus *Gonium. Madrōno* 13:49–64.

Pocock, M. A. (1960). *Hydrodictyon:* A comparative biological study. *J. S. Afr. Bot.* 26:167–319.

Powers, J. H. (1908). Further studies in *Volvox*, with description of three new species. *Trans. Am. Microsc. Soc.* 28:141–75.

Proctor, V. W. (1962). Viability of *Chara* oospores taken from migrating water birds. *Ecology* 43:528–9.

Proskauer, J. (1950). On *Prasinocladus. Am. J. Bot.* 37:59–66.

Provasoli, L, and Pinter, I. J. (1980). Bacteria induced polymorphism in an axenic laboratory strain of *Ulva lactuca* (Chlorophyceae). *J. Phycol.* 16:196–201.

Rawitscher-Kunkel, E., and Machlis, L. (1962). The hormonal integration of sexual reproduction in *Oedogonium. Am. J. Bot.* 49:177–83.

Retallack, B., and Butler, R. D. (1970). The development and structure of the zoospore vesicle in *Bulbochaete hiloensis. Arch. Mikrobiol.* 72:223–37.

Richardson, D. H. S. (1973). Photosynthesis and carbohydrate movement. In *The Lichens*, ed. V. Ahmadjian, and M. E. Hale, pp. 249–88. New York and London: Academic Press.

Rivers, J. S., and Peckol, P. (1995). Summer decline of *Ulva lactuca* (Chlorophyta) in a eutrophic embayment: Interactive effects of temperature and nitrogen availability. *J. Phycol.* 31:223–8.

Roberts, K. (1974). Crystalline glycoprotein cell walls of algae: Their structure, composition and assembly. *Philos. Trans. R. Soc. Lond.* [*B*], 268:129–46.

Roberts, K. R., Sluiman, H. J., Stewart, K. D., and Mattox, K. R. (1981). Comparative cytology and taxonomy of the Ulvaphyceae. III. The flagellar apparatuses of the anisogametes of *Derbesia tenuissima* (Chlorophyta). *J. Phycol.* 17:330–40.

Russell, F. S., and Yonge, C. M. (1963). *The Seas.* 2nd ed. London and New York: Frederick Warne and Co.

Sabnis, D. D., and Jacobs, W. P. (1967). Cytoplasmic streaming and microtubules in the coenocytic marine alga, *Caulerpa prolifera. J. Cell Sci.* 2:465–72.

Sack, F. (1991). Plant gravity sensing. *Int. Rev. Cytol.* 127:193–252.

Sager, R., and Granick, S. (1954). Nutritional control of sexuality in *Chlamydomonas reinhardi. J. Gen. Physiol.* 37:729–42.

Saito, T., and Matsuda, Y. (1984). Sexual agglutinin of mating-type minus gametes in *Chlamydomonas reinhardtii.* II. Purification and characterization of minus agglutinin and comparison with plus agglutinin. *Arch. Microbiol.* 139:95–9.

Saito, t., Tsubo, Y., and Matsuda, Y. (1985). Synthesis and turnover of cell body-agglutinin as a pool of flagellar surface-agglutinin in *Chlamydomonas reinhardtii* gamete. *Arch. Microbiol.* 142:207–10.

Salisbury, J. L., and Floyd, G. L. (1978). Calcium-induced contraction of the rhizoplast of a quadriflagellate green alga. *Science* 202:975–6.

Sawada, T. (1972). Periodic fruiting of *Ulva pertusa* at three localities in Japan. *Proc. 7th Int. Seaweed Symp.*, pp. 229–30. Tokyo: University of Tokyo Press.

Scagel, R. F., Bandoni, R. J., Rouse, G. E., Scofield, W. B., Stein, J. R., and Taylor, T. M. C. (1965). *An Evolutionary Survey of the Plant Kingdom.* Belmont, CA: Wadsworth.

Schlicher, U., Linden, L., Calenberg, M., and Kreimer, G. (1995). G proteins and Ca^{2+}-modulated protein kinases of a plasma membrane-enriched fraction and isolated eyespot apparatuses of *Spermatozopsis similis* (Chlorophyceae). *Eur. J. Phycol.* 30:319–30.

Shimmen, T., and MacRobbie, E. A. C. (1987). Demonstration of two proton translocating systems in tonoplast of permeabilized *Nitella* cell. *Protoplasma* 136:205–7.

Simons, J., and van Beem, A. P. (1987). Observations on asexual and sexual reproduction in *Stigeoclonium helveticum* Vischer (Chlorophyta) with implications for the life history. *Phycologia* 26:356–62.

Simons, J., van Beem, A. P., and de Vries, P. J. R. (1982). Structure and chemical composition of the spore wall in *Spirogyra* (Zygnemataceae, Chlorophyceae). *Acta Bot. Neerl.* 31:359–70.

Singh, H. V., and Chaudhary, B. R. (1990). Nutrient effects on the formation of oogonia in *Oedogonium hatei* (Chlorophyta). *Phycologia* 29:332–7.

Sluiman, H. J. (1983). The flagellar apparatus of the zoospore of the filamentous green alga *Coleochaete pulvinata:* Absolute configuration and phylogenetic significance. *Protoplasma* 115:160–75.

Smith, G. M. (1920). *Phytoplankton of the Inland Lakes of Wisconsin,* Part I. State of Wisconsin Publ.

Smith, G. M. (1944). A comparative study of the species of *Volvox. Trans. Am. Microsc. Soc.* 63:265–310.

Smith, G. M. (1947). On the reproduction of some Pacific Coast species of *Ulva. Am. J. Bot.* 34:80–7.

Smith, G. M. (1950). *The Freshwater Algae of the United States.* New York: McGraw-Hill.

Smith, G. M. (1955). *Cryptogamic Botany,* Vol. 1, 2nd ed. New York: McGraw-Hill.

Smith, G. M. (1969). *Marine Algae of the Monterey Peninsula.* Stanford, Calif: Stanford University Press.

Snell, W. J. (1985). Cell–cell interactions in *Chlamydomonas. Annu. Rev. Plant Physiol.* 36:287–315.

Snell, W. J., and Moore, W. S. (1980). Aggregation-dependent turnover of flagellar adhesion molecules in *Chlamydomonas* gametes. *J. Cell Biol.* 84:203–10.

Sormus, L., and Bicudo, C. E. M. (1974). Polymorphism in the desmid *Micrasterias pinnatifida* and its taxonomical implications. *J. Phycol.* 10:274–9.

Staehelin, L. A., and Pickett-Heaps, J. D. (1975). The ultrastructure of *Scenedesmus* (Chlorophyceae) I. Species with the “reticulate” or “warty” type of ornamental layer. *J. Phycol.* 11:163–85.

Stanley, S. M. (1979). *Macroevolution.* San Francisco: W. H. Freeman. 232 pp.

Stark, L. M., Almodovar, L., and Krauss, R. W. (1969). Factors affecting the rate of calcification in *Halimeda opuntia* (L.) Lamouroux and *Halimeda discoidea* Decaisne. *J. Phycol.* 5:305–12.

Starr, R. C. (1954). Heterothallism in *Cosmarium botrytis* var. *subtumidum*. *Am. J. Bot.* 41:601–7.

Starr, R. C. (1972). A working model for the control of differentiation during development of the embryo of *Volvox carteri* f. *nagariensis*. *Soc. Bot. Fr., Memoires*, pp. 175–82.

Starr, R. C., and Jaenicke, L. (1974). Purification and characterization of the hormone initing sexual morphogenesis in *Volvox carteri* f. *nagariensis Iyengar. Proc. Natl. Acad. Sci. USA* 71:1050–4.

Steinkötter, J., Bhattacharya, D., Semmelroth, I., Bibeau, C., and Melkonian, M. (1994). Prasinophytes form independent lineages within the Chlorophyta: Evidence from ribosomal RNA sequence comparisons. *J. Phycol.* 30:340–5.

Stratmann, J., Paputsoglu, G., and Oertel, W. (1996). Differentiation of *Ulva mutabilis* (Chlorophyta) gametangia and gamete release are controlled by extracellular inhibitors. *J. Phycol.* 32:1009–21.

Suda, S., Watanabe, M. M., and Inouye, I. (1989). Evidence for sexual reproduction in the primitive green alga *Nephroselmis olivacea* (Prasinophyceae). *J. Phycol.* 25:596–600.

Suzuki, K., Iwamoto, K., Yokoyama, S., and Ikawa, T. (1991). Glycolate-oxidizing enzymes in algae. *J. Phycol.* 27:492–8.

Syrett, P. J., and Al-Houty, F. A. A. (1984). The phylogenetic significance of the occurrence of urease/urea amidolyase and glycollate oxidase/glycollate dehydrogenase in green algae. *Br. Phycol. J.* 19:11–21.

Takatori, S., and Imahori, K. (1971). Light reactions in the control of oospore germination of *Chara delicatula. Phycologia* 10:221–8.

Tan, S., Cunningham, F. X., Youmana, M., Grabowski, B., Sun, Z., and Gantt, E. (1995). Cytochrome loss in astaxanthin-accumulating red cells of *Hematococcus pluvialis* (Chlorophyceae): Comparison of photosynthetic activity, photosynthetic enzymes, and thylakoid membrane polypeptides in red and green cells. *J. Phycol.* 31:897–905.

Taylor, W. R. (1960). *Marine Algae of the Eastern Tropical and Subtropical Coasts of the Americas.* Ann Arbor: University of Michigan Press.

Tolbert, N. E. (1976). Glycollate oxidase and glycollate dehydrogenase in marine algae and plants. *Aust. J. Plant Physiol.* 3:129–32.

Trainor, F. R. (1963). Zoospores on *Scenedesmus obliquus. Science* 142:1673–4.

Trainor, F. R. (1970). Survival of algae in a dessicated soil. *Phycologia* 9:111–13.

Trainor, F. R. (1992). Cyclomorphosis in *Scenedesmus armatus* (Chlorophyta): an ordered sequence of ecomorph development. *J. Phycol.* 28:552–8.

Trench, R. K., Greene, R. W., and Bystrom, B. G. (1969). Chloroplasts as functional organelles in animal tissues. *J. Cell Biol.* 42:404–17.

Trench, R. K., Boyle, J. E., and Smith, D. C. (1973a). The association between chloroplasts of *Codium fragile* and the mollusc *Elysia viridis*. I. Characteristics of isolated *Codium* chloroplasts. *Proc. R. Soc. Lond.* [*B*] 184:51–61.

Trench, R. K., Boyle, J. E., and Smith, D. C. (1973b). The association between chloroplasts of *Codium fragile* and the mollusc *Elysia viridis*. II. Chloroplast ultrastructure and photosynthetic carbon fixation in *E. viridis. Proc. R. Soc. Lond.* [*B*] 184:63–81.

van Kruiningen, H. J., Garner, F. M., and Schiefer, B. (1969). Protothecosis in a dog. *Pathol. Vet.* 6:348–54.

van Winkle-Swift, K. P., and Rickoll, W. L. (1997). The zygospore wall of *Chlamydomonas monoica* (Chlorophyceae): morphogenesis and evidence for the presence of sporopollenin. *J. Phycol.* 33:655–65.

Wake, L. V., and Hillen, L. W. (1980). Study of a "bloom" of the oil-rich alga *Botryococcus braunii* in the Darwin River Reservation. *Biotech. Bioeng.* 22:1637–56.

Webster, D. A., Hackett, D. P., and Park, R. B. (1968). The respiratory chain of colorless algae. III. Electron microscopy. *J. Ultrastruct. Res.* 21:514–23.

Wheeler, A. E., and Page, J. Z. (1974). The ultrastructure of *Derbesia tenuissima* (de Notaris) Crouan. I. Organization of the gametophyte protoplast, gametangium, and gametangial pore. *J. Phycol.* 10:336–52.

Wiese, L., and Hayward, P. C. (1972). On sexual agglutination and mating-type substances in isogamous dioecious chlamydomonads. III. The sensitivity of sex cell contact to various enzymes. *Am. J. Bot.* 59:530–6.

Wiese, L., and Metz, C. B. (1969). On the trypsin sensitivity of gamete contact at fertilization as studied with living gametes in *Chlamydomonas. Biol. Bull.* 136:483–93.

Wiese, L., and Shoemaker, D. W. (1970). On sexual agglutination and mating-type substances (gamones) in isogamous heterothallic chlamydomonads. II. The effect of concanavalin A upon the mating-type reaction. *Biol. Bull.* 138:88–95.

Wilbur, K. M., Colinvaux, L. H., and Watabe, N. (1969). Electron microscope study of calcification in the alga *Halimeda* (order Siphonales). *Phycologia* 8:27–35.

Wilcox, L. W., Fuerst, P. A., and Floyd, G. L. (1993). Phylogenetic relationships of four charophycean green algae inferred from complete nuclear-encoded small subunit rRNA gene sequences. *Amer. J. Bot.* 80:1028–33.

Wolf, F. R., Nonomura, A. M., and Bassham, J. A. (1985). Growth and branched hydrocarbon production in a strain of *Botryococcus braunii* (Chlorophyta). *J. Phycol.* 21:388–96.

Woodcock, C. L. F. (1971). The anchoring of nuclei by cytoplasmic microtubules in *Acetabularia. J. Cell Sci.* 8:611–21.

Woodcock, C. L. F., and Miller, G. J. (1973a). Ultrastructural features of the life cycle of *Acetabularia mediterranea.* I. Gametogenesis. *Protoplasma* 77:313–29.

Woodcock, C. L. F., and Miller, G. J. (1973b). Ultrastructural features of the life cycle of *Acetabularia mediterranea.* II. Events associated with the division of the primary nucleus and the formation of cysts. *Protoplasma* 77:331–41.

Wray, J. L. (1977). *Calcareous Algae.* Amsterdam: Elsevier.

Wujek, D. E. (1968). Some observations on the fine structure of three genera in the Tetrasporaceae. *Ohio J. Sci.* 68:187–91.

Yamaguchi, K. (1997). Recent advances in microalgal bioscience in Japan, with special reference to utilization of biomass and metabolites: a review. *J. Appl. Phycol.* 8:487–502.

Zechman, F. W., Theriot, E. C., Zimmer, E. A., and Chapman, R. L. (1990). Phylogeny of the Ulvophyceae (Chlorophyta): cladistic analysis of nuclear-encoded rRNA sequence data. *J. Phycol.* 26:700–10.

Zeigler, J. R., and Kingsbury, J. M. (1964). Cultural studies on the marine green alga *Halicystis parvula–Derbesia tenuissima.* I. Normal and abnormal sexual and asexual reproduction. *Phycologia* 4:105–16.

PART IV · Evolution of one membrane of chloroplast endoplasmic reticulum

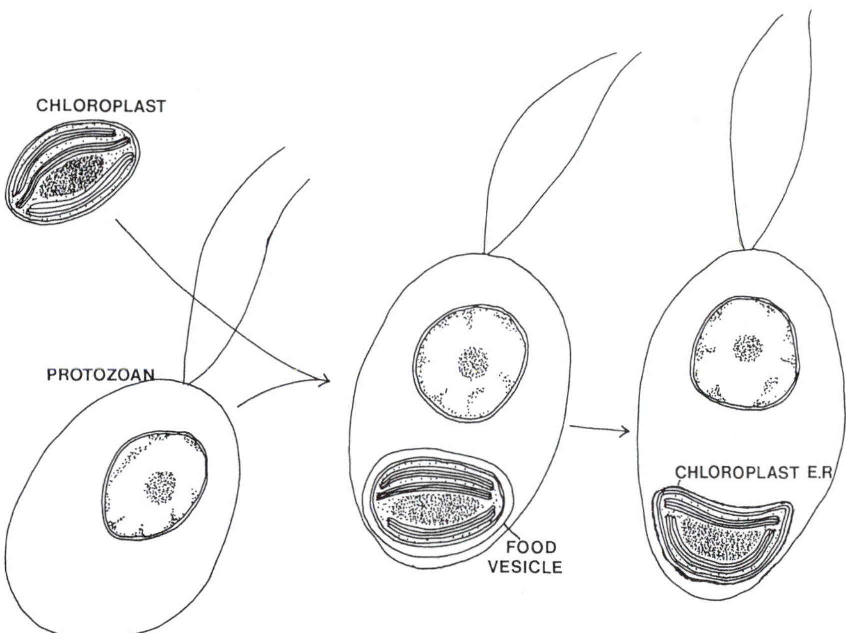

Fig. IV.1 Drawing illustrating the probable sequence of evolutionary events that led to the chloroplast's being surrounded by a single membrane of chloroplast endoplasmic reticulum. Initially a chloroplast was taken up by a phagocytotic protozoan into a food vesicle. An endosymbiosis resulted, with the food vesicle membrane eventually evolving into a single membrane of chloroplast endoplasmic reticulum surrounding the chloroplast.

The Euglenophyta (euglenoids) and Dinophyta (dinoflagellates) are a natural grouping in that they are the only algal groups to have one membrane of chloroplast endoplasmic reticulum. Chloroplast endoplasmic reticulum resulted when a chloroplast from a eukaryotic alga was taken up into a food vesicle by a phagocytotic euglenoid or dinoflagellate (Fig. IV.1) (Lee, 1977; see

Chapter 1, References). Normally the phagocytotic euglenoid or dinoflagellate would have digested the chloroplast as a source of food. However, in this case, the chloroplast was retained in the cytoplasm of the host as an endosymbiont. The host benefited from the association by receiving photosynthate from the endosymbiotic chloroplast. The endosymbiotic chloroplast benefited by the high concentration of carbon dioxide in the acidic environment of the host vesicle (see Chapter 1). Eventually, the food vesicle membrane of the host became the single membrane of chloroplast endoplasmic reticulum surrounding the chloroplast in the euglenoids and dinoflagellates. It is probable that the euglenoids and dinoflagellates evolved through different evolutionary events and that they are not related. The euglenoids probably evolved by the capture of a chloroplast of a green alga, as the chloroplasts of the green algae and euglenoids are very similar. It is not as clear how the chloroplast of the dinoflagellates originated, but it is possible that the chloroplast is derived from an endosymbiotic red algal chloroplast.

6 · Euglenophyta

Euglenophyceae

Euglenoid flagellates occur in most freshwater habitats: puddles, ditches, ponds, streams, lakes, and rivers, particularly waters contaminated by animal pollution or decaying organic matter (Buetow, 1968). Usually larger bodies of purer water, such as rivers, lakes, and reservoirs, have sparser populations of less common euglenoids as planktonic organisms. Marine euglenoids are more common than supposed, with *Eutreptia*, *Eutreptiella*, and *Klebsiella* occurring exclusively in marine or brackish water, and many other genera having one or a few marine species. These occur in the open sea, in tidal zones among seaweeds, and as sand inhabitants on beaches. Brackish species of *Euglena* often color estuarine mud flats green when light intensity is low, the green color disappearing in full sunlight as the euglenoids creep away from the surface. There are also several parasitic euglenoid flagellates, mostly species of *Khawkinea*, *Euglenamorpha*, and *Hegneria*.

Euglenoids are characterized by chlorophylls *a* and *b*, one membrane of chloroplast endoplasmic reticulum, a mesokaryotic nucleus, flagella with fibrillar hairs in one row, no sexual reproduction, and paramylon formed as the storage product in the cytoplasm.

Cell structure

Euglenoid cells have two basal bodies and one or two emergent flagella (Figs. 6.1 and 6.12). Those cells with two emergent flagella are considered to be the more primitive cells. The flagella of the euglenoids are similar to those in dinoflagellates in having a **paraxonemal rod** (**paraxial rod**) that runs the length of the flagellum (Walne and Dawson, 1993) and in having only fibrillar hairs (not composed of microtubules) which are arranged helically along the length of the flagellum (Melkonian et al., 1982). The fibrillar hairs are of two lengths: There is a single helical row of long (3 μm) hairs and two helical rows of short (1.5 μm) hairs in *Euglena* (Bouck et al., 1978). Other genera have flagella similar to *Euglena* (Hilenski and Walne, 1985). There are two basic types of flagellar movement in the class. The first group (including the Eutreptiales and Euglenales) has the flagellum continually motile from

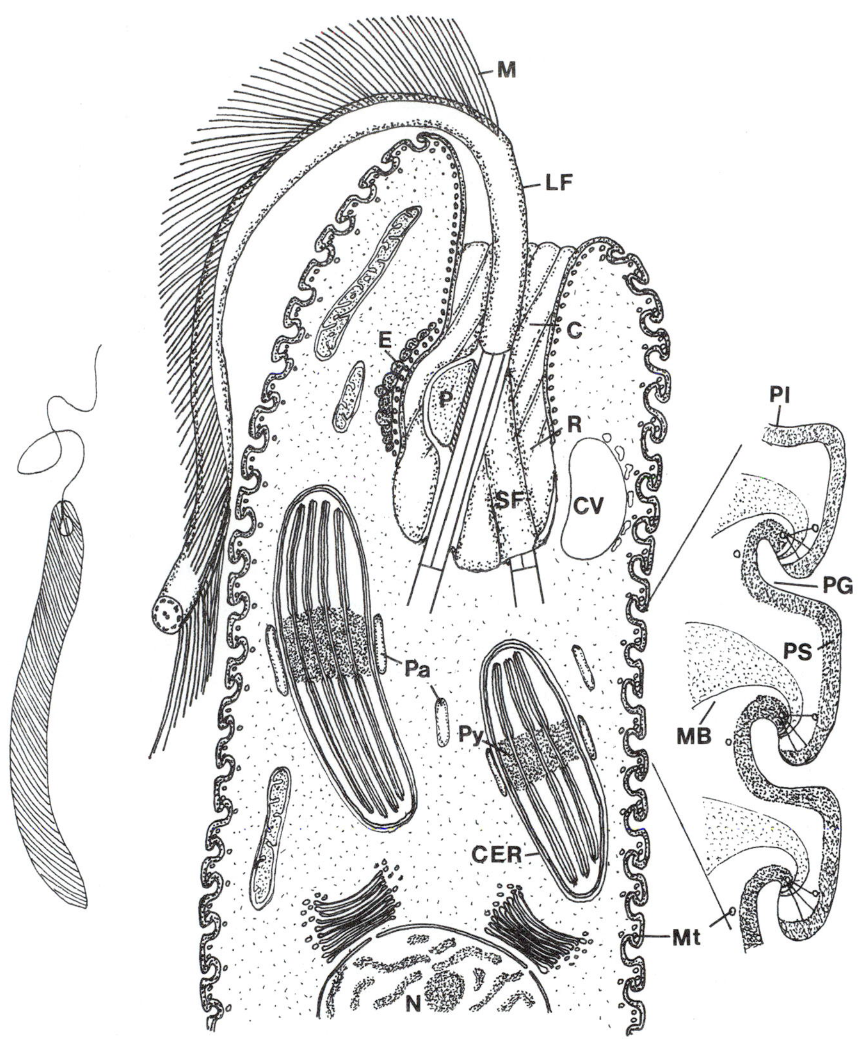

Fig. 6.1 A semidiagrammatic drawing of the fine structure of the anterior part of a *Euglena* cell. (C) Canal; (CER) chloroplast endoplasmic reticulum; (CV) contractile vacuole; (E) eyespot; (LF) long flagellum; (M) mastigonemes; (MB) muciferous body; (Mt) microtubules; (N) nucleus; (P) paraflagellar swelling; (Pa) paramylon; (PG) pellicle groove; (Pl) plasmalemma; (PS) pellicle strip; (Py) pyrenoid; (R) reservoir; (SF) short flagellum. (Adapted from Jahn and Bovee, 1968; Mignot, 1966.)

base to apex, usually resulting in cell gyration with the anterior end of the cell tracing a wide circle. The second group (including *Peranema*, *Entosiphon*, and *Sphenomonas*) has the flagellum held out straight in front of the cell with just the tip mobile, resulting in smooth swimming or gliding locomotion in contact with a substratum or water–air interface (Leedale, 1967). Lee (1954) showed that the forward swimming rate of *Euglena gracilis* is heat-responsive and shows an increase from 15 $\mu m\ s^{-1}$ at 10 °C to a peak of 84 $\mu m\ s^{-1}$ at 30 °C, followed by a drop to 38 $\mu m\ s^{-1}$ at 40 °C.

The **pellicle** is found inside the plasmalemma and is primarily proteinaceous, the chemical composition of the pellicle of *E. gracilis* var. *bacillus* being 80% basic protein and 11.6% lipid, with the remainder being composed of carbohydrate (Barras and Stone, 1965; Bricheux and

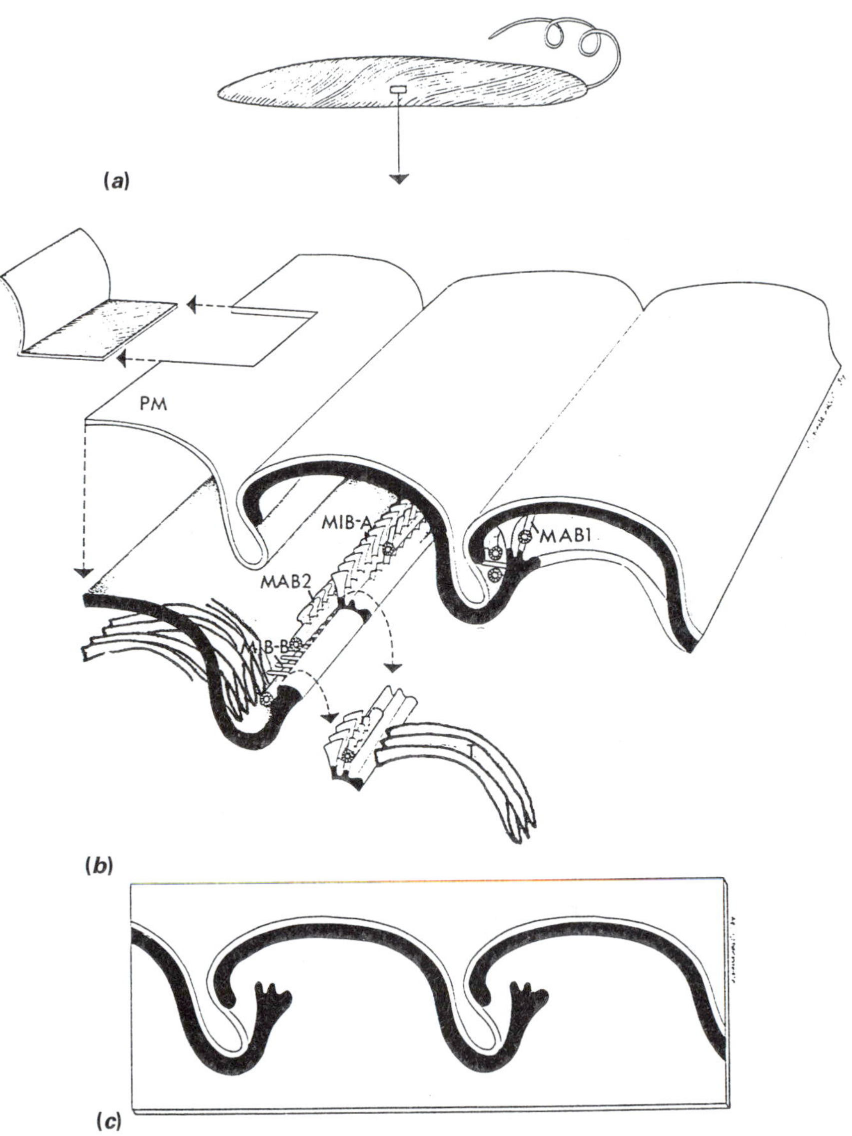

Fig. 6.2 *Euglena gracilis.* Diagrams of a whole swimming cell (*a*) and transverse sections of the cell surface (*b*) and (*c*), illustrating details of the articulating S-shaped strips of the membrane skeleton and the infrastructure associated with strip overlap. The position of the skeleton and bridges seems well suited to mediate the sliding of adjacent strips that is presumed to occur during shape changes. (MAB1, MAB2) Microtubule-associated bridges between the pellicle strips; (MIB-A, MIB-B) microtubule-independent bridges between the pellicle strips; (PM) plasma membrane, (T) traversing fiber. (From Dubreuil and Bouck, 1985.)

Brugerolle, 1987). The pellicle consists of a number of helically wound strips, each strip with a thick edge and a thinner flange end (Figs. 6.1 and 6.2). In the construction of the pellicle, the thick end of one strip fits under the thin flange end of the second strip. This structure gives the pellicle an alternating pattern of ridges and grooves. In *Euglena gracilis,* there are about 40 of the S-shaped strips overlapping at their lateral margins, under the plasma membrane. The region of strip overlap is occupied by a set of microtubule-associated bridges and microtubule-independent bridges (Fig. 6.2) (Dubreuil and Bouck, 1985; Mignot et al., 1987). The form of the

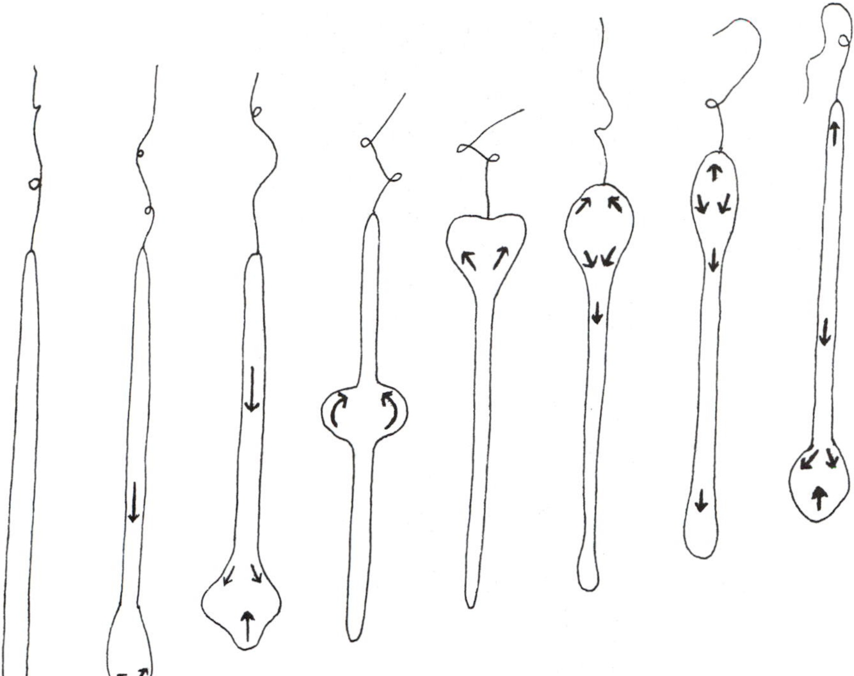

Fig. 6.3 *Astasia klebsii*; successive stages in euglenoid movement, arrows indicating direction of cytoplasmic flow. (After Leedale, 1967.)

cell and the organization of the plasma membrane are determined by this membrane skeleton. Muciferous bodies found under the pellicle strips follow the same helical pattern. Some euglenoids have a flexible pellicle which allows the cells to undergo a flowing movement known as **euglenoid movement** (Figs. 6.3 and 6.4). This type of movement occurs only when the cells are not swimming and results from sideways movement of the pellicle strips (Leedale, 1967; Suzaki and Williamson, 1986).

Euglena gracilis changes its shape two times per day when grown under the synchronizing effect of a daily light–dark cycle. At the beginning of the light period, when photosynthetic capacity is low (as measured by the ability of the cells to evolve oxygen), the population of cells is largely **spherical**. The **mean cell length** of the population **increases to a maximum** in the middle of the light period when photosynthetic capacity is greatest, and then decreases for the remainder of the 24-hour period. The population becomes **spherical** by the end of the 24-hour period when the cycle reinitiates. These changes are also observed under dim light conditions and are therefore controlled by a biological clock and represent a circadian rhythm in cell shape (Lonergan, 1983).

The locomotory flagellum or flagella emerge from an anterior invagination of the cell, which consists of a narrow tubular portion, the **canal**, and a

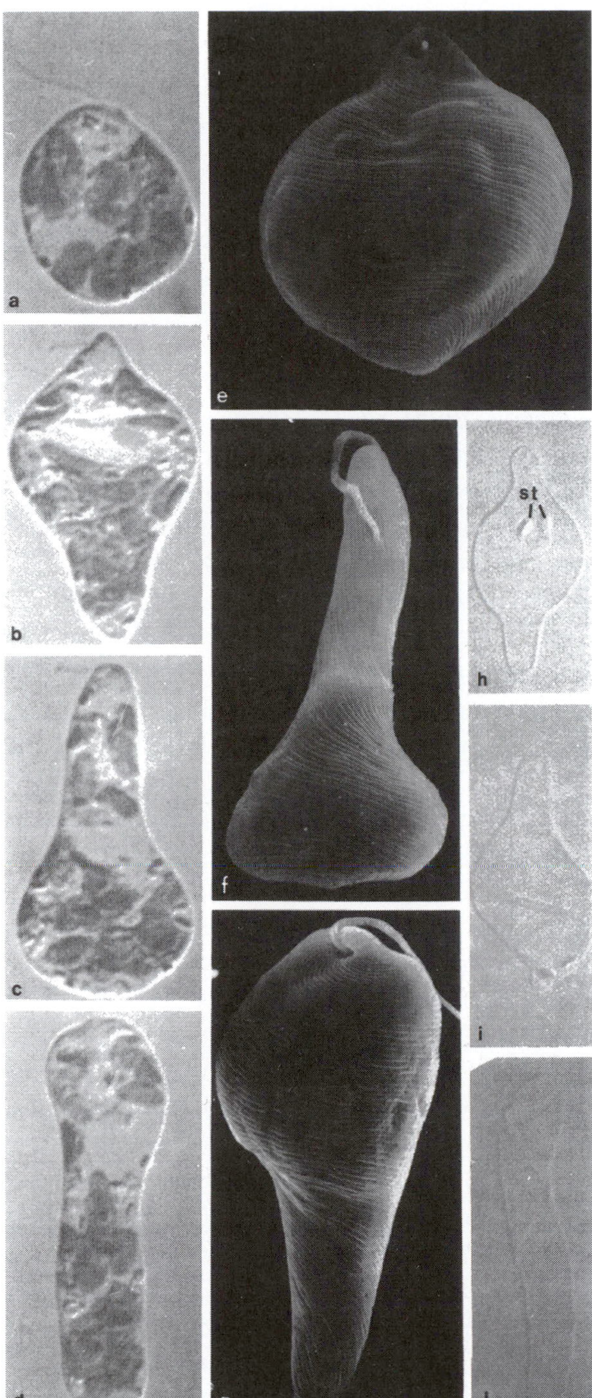

Fig. 6.4 *Euglena gracilis*. (*a*)–(*d*). Sequence of shape changes photographed at 5-second intervals of a cell undergoing euglenoid movements. The nearly spherical cell in (*a*) initiates a forward wave of dilation in (*b*), which reaches the anterior of the cell (*b*), top), and then recovers by an inward flow of cytoplasm to initiate a new wave at (*c*). The new wave progresses forward, and the cell recovers in (*d*). Scanning electron microscope micrographs in (*e*)–(*g*) illustrate the positions of the surface ridges and grooves during selected stages of deformation. Nearly horizontal strips in (*e*) reorient to longitudinal in cells, initiating (*f*) or completing (*g*) the cycle. In (*h*)–(*j*), cell ghosts are shown with the changes in cell shape. (From Dubreuil and Bouck, 1985.)

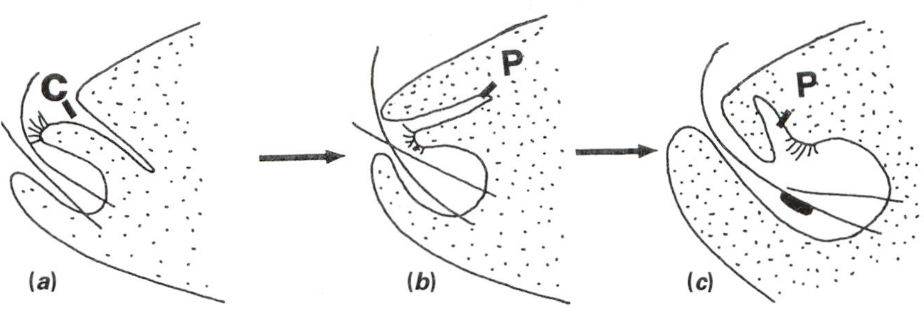

Fig. 6.5 Semi-diagrammatic drawings of the anterior portion of cells of *Bodo* (Kinetoplastida) (*a*), *Peranema* (Euglenophyta) (*b*), and *Colacium* (Euglenophyta) (*c*), illustrating probable evolution of the euglenoids. (C) Cytosome; (P) pocket. (After Willey and Wibel, 1985b.)

spherical or pyriform chamber, the **reservoir** (Figs. 6.1, 6.5 and 6.12). The canal is a rigid structure, whereas the reservoir easily changes shape and is regularly distorted by the discharge of the contractile vacuole. The rigidity of the canal is maintained by microtubules that form a flat helix around the canal, in much the same position as hoops on a barrel. The pellicle lines the canal but not the reservoir, the reservoir being the only part of the cell covered solely by the plasmalemma. A cylindrical pocket arises as an infolding of the plasma membrane of the reservoir in *Euglena, Colacium,* and *Peranema* and is probably common in the euglenoids (Fig. 6.5) (Willey and Wibel, 1985a; Surek and Melkonian, 1986). The reservoir pocket is similar to the cytosome found in protozoa assigned to the Kinetoplastida (Willey and Wibel, 1985b). In the Kinetoplastida (bodonids and trypanosomatids), the cytosome is associated with the uptake of food organisms by phagotrophy, and the reservoir pocket has a similar function in the phagotrophic euglenoids. In the phagotrophic euglenoids such as *Peranema,* the pocket is often called a cytosome. The derivation of the Euglenophyceae from the colorless phagotrophic zooflagellates in the Kinetoplastida has been proposed partly on the basis of the similarities between the cytosome and pocket (Montegut-Felkner and Triemer, 1977).

The contractile vacuole, which is in the anterior part of the cell next to the reservoir, has an osmoregulatory function, expelling excessive water taken into the cell. The contractile vacuole fills and empties at regular intervals of 15 to 60 seconds. It empties into the reservoir, from which the water is carried out through the canal (Leedale, 1967).

Mitochondria are of typical algal type. Colorless euglenoids always have more mitochondria than do equivalent-sized green ones. When green cells are decolorized by heat or streptomycin, they have a sevenfold increase in mitochondrial volume, reflecting a change from autotrophic to heterotrophic nutrition. The formation of two mitochondrial enzymes, fumarase and succinate dehydrogenase, necessary for dark respiration of substrates, is repressed by light (Davis and Merrett, 1974).

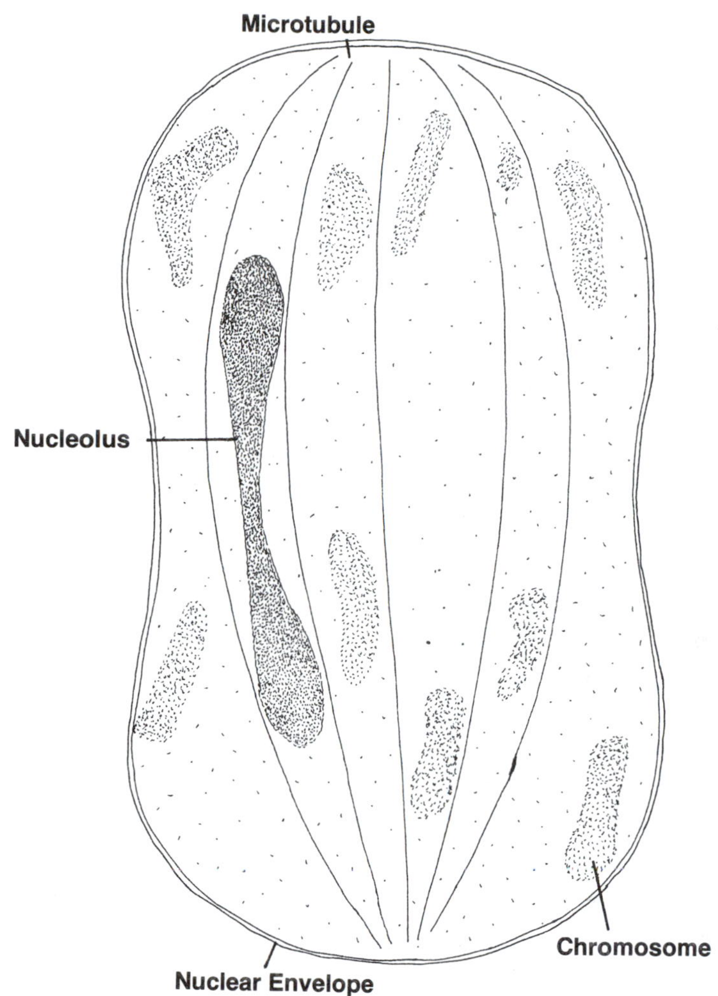

Fig. 6.6 A semi-diagrammatic drawing of an anaphase nucleus of *Euglena* with the nuclear envelope intact, the nucleolus pinching in two, and the chromosomes not attached to the spindle microtubules.

Nucleus and nuclear division

The euglenoid nucleus is of the mesokaryotic type, having chromosomes that are permanently condensed during the mitotic cycle, a nucleolus (endosome) that does not disperse during nuclear division, no microtubules from chromosomes to pole spindles, and a nuclear envelope that is intact during nuclear division (Fig. 6.6). The chromosome number is usually high, and polyploidy probably occurs in some genera (Gravilă, 1996).

Mitosis in euglenoids (Leedale, 1970; Chaly et al., 1977) begins during early prophase with the nucleus migrating from the center of the cell to an anterior position. Microtubules appear in the nucleus, but they do not attach to the chromosomes. At metaphase, bundles of microtubules are among the chromosomes, and the nucleolus has started to elongate along the division axis. In anaphase, the intact nuclear envelope elongates along the division axis, the nucleolus divides, and the daughter chromosomes disperse into the two daughter nuclei.

Eyespot, paraflagellar swelling, and phototaxis

The eyespot (stigma) is a collection of orange-red lipid droplets, independent of the chloroplast (Figs. 6.1 and 6.12). The eyespot is in the anterior part of the cell, curving to ensheath the neck of the reservoir on the dorsal side. In most euglenoids the eyespot consists of a compact group of 20 to 50 droplets, although in *Eutreptia* and *Khawkinea* it may consist of just one or two large droplets. The eyespot has been reported to contain α-carotene and seven xanthophylls (Sperling Pagni et al., 1981), mainly β-carotene (Batra and Tollin, 1964), or a β-carotene derivative, echineone (Krinsky and Goldsmith, 1960). The independence of the eyespot is emphasized by the existence of colorless species with an eyespot but no plastids. One flagellum of all green euglenoids bears a lateral swelling near the transition zone from canal to reservoir; in *Euglena*, the swelling is on the longer flagellum. The swelling is composed of a crystalline body next to the axoneme and inside the flagellar membrane.

All euglenoid species with an eyespot and flagellar swelling exhibit phototaxis, usually swimming away from bright light (negative phototaxis) and away from darkness toward subdued light (positive photoaxis) to accumulate in a region of low light intensity. Upon sudden changes in its environment, the cell responds with a transient sideways turn by swinging out its one emergent flagellum. At low light intensities (less than $1.4\,W\,m^{-2}$), the alga swims toward the light source, whereas at higher light intensities, it moves away from the light source. According to the shading hypothesis (Häder, 1987), positive phototaxis is brought about by repetitive step-down photophobic responses. During forward locomotion, the cells rotate helically with a frequency of 1 to 2 hertz (cycles per second). In lateral light, each time the stigma (eyespot) intercepts the light beam impinging on the paraflagellar body, the flagellum swings out temporarily and turns the front end of the cell toward the light source by a fraction until the cell is aligned in the light direction.

Euglena bleached of its chlorophyll but retaining its eyespot and photoreceptor (paraflagellar swelling) is still positively phototactic, eliminating chlorophyll and chloroplasts in the phototaxis directly. A *Euglena* bleached of all pigments but retaining its photoreceptor is negatively phototactic;

this rules out the possibility that the carotenoids of the eyespot are directly stimulatory in phototaxis. A *Euglena* lacking a photoreceptor and all pigments, like *Astasia,* is no longer phototactic (Jahn and Bovee, 1968). The flagellar swelling is therefore the photoreceptor or light-sensitive organelle. It has a sensitive maximum at 410 nm, explaining the phototactic peak at that wavelength. Positive phototaxis occurs only if the eyespot, with its absorptive range of 400 to 630 nm, periodically shades the photoreceptor.

There is a circadian rhythm in phototaxis in *Euglena,* with phototaxis operative during the light period and not operative during the dark period. Even if light is introduced during the normal dark period, the *Euglena* cell does not respond phototactically. The phototactic response in the direction of the light source plainly involves the photoreceptor, but why this is incapable of reacting during dark periods when the light is reintroduced is not known. Continuous darkness does not eliminate the rhythm unless continued for so long a period that complete bleaching results, and the cells become photonegative. It is possible that the non-operative nature of photaxis during this time is related to mitotic division occurring during this period. Leedale (1959) found that green euglenoids have almost perfectly circadianly synchronized mitotic cycles, mitosis occurring at the beginning of the dark period, requiring one hour of dark to trigger the division. During mitotic division, *Euglena* normally rounds up and loses its flagellum; but even if the flagellum is not shed, the cells are still unable to swim ably during this period. Because phototactic insensitivity coincides with the generation time of *Euglena* during dark hours, it perhaps is of no surprise that motility loss, generation time, and absence of phototaxis should coincide as they do (Jahn and Bovee, 1968).

A phototactic circadian rhythm also exists in the movement of mud flat *Euglena.* The mouth of the River Avon in Bristol, England, is an estuary with very large tidal differences in water level. The mud flats which are exposed at low tide become green in spots owing to an accumulation of *Euglena obtusa* on the surface of the mud (Bracher, 1937; Palmer and Round, 1965). Before the incoming tide floods the mud flats, the *Euglena* cells move back down into the mud, a pattern of behavior that avoids the washing away of the *Euglena* by the tide. *Euglena* cells accumulate at the surface of the mud only when the low tides occur during daylight, no cells being found on the surface during low tides that occur at night (Fig. 6.7). The above observations indicate that phototaxis is the main mechanism involved, with the *Euglena* cells swimming toward the light during the day to reach the surface of the mud. If the *Euglena* cells are brought into the laboratory and placed in constant light (980 lux), the cells display a circadian rhythm in their vertical migration. The phase of this circadian rhythm is determined by the last dark period in nature, be it night or a dark period resulting from the covering of

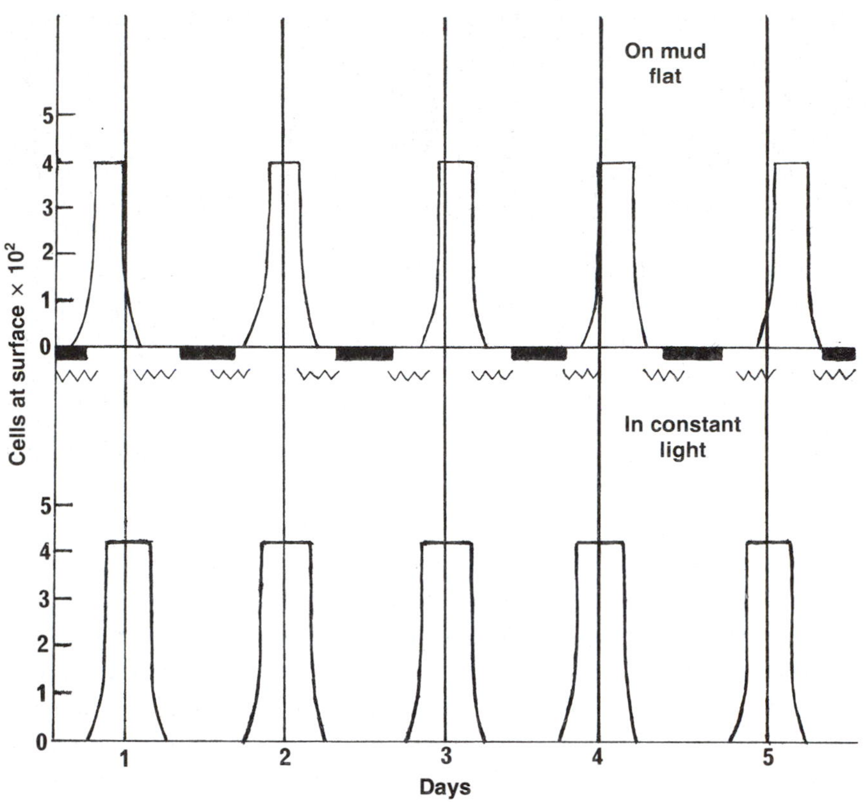

Fig. 6.7 The vertical migration rhythm in *Euglena obtusa* on a mud flat (top curve), and in constant light in the laboratory (bottom curve). The dark bars on the upper abscissa represent night, whereas the wavy lines indicate when the tide covered the mud flats. (After Palmer and Round, 1965.)

the mud flat by murky water at high tide. Thus the rhythm is actually circadian, entrained by the tidally caused light–dark cycle, rather than truly tidal.

Muciferous bodies and extracellular structures

Muciferous bodies occur in helical rows under the pellicle in all species of euglenoid flagellates (Fig. 6.1), straining reactions showing that they contain a water-soluble polysaccharide. Euglenoid cells are permanently coated with a thin slime layer from the muciferous bodies (Rosowski, 1977). It sometimes accumulates at the posterior end of the cell as a trailer of slime, and some species have the habit of sticking to a substratum by their posterior ends. Species with large muciferous bodies eject the contents on irritation and produce a copious slime layer around the cell (Hilenski and Walne, 1983). In *Euglena gracilis*, the slime is composed of glycoproteins and polysaccharides (Cogburn and Schiff, 1984). The envelopes and stalks of *Colacium* (Fig. 6.13 and 6.14) are formed of carbohydrate extruded by mucocysts in the anterior portion of the cell (Willey, 1984). The cylindrical

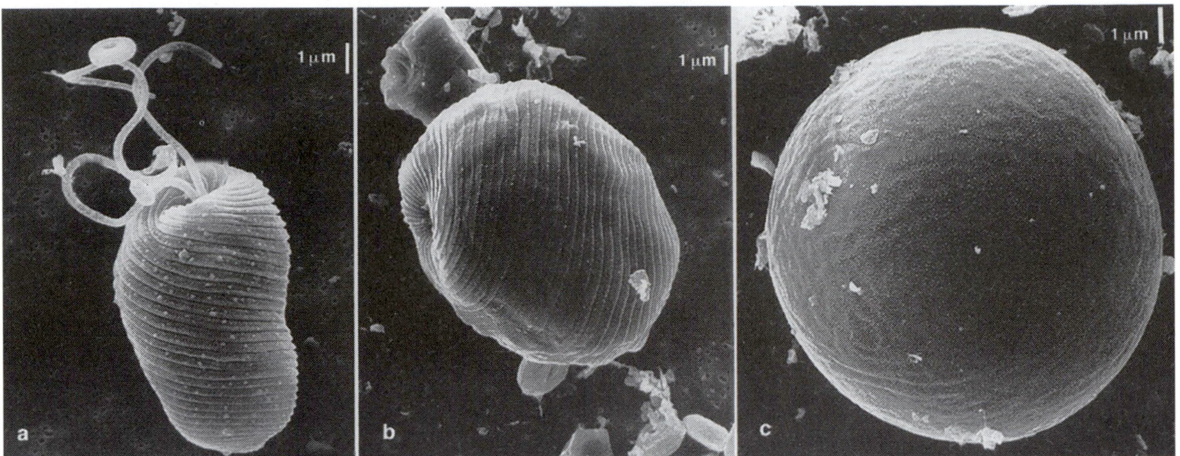

Fig. 6.8 Scanning electron micrographs of the encystment of *Eutreptiella gymnastica*. The vegetative cell (*a*) loses its flagella, forms a large number of paramylon grains and begins to round up (*b*). The cell swells and produces a mucilaginous covering (*c*). (From Olli, 1996.)

stalks are composed of an inner and outer core of mildly acidic carbohydrates. The stalk is continuous with the canal and anterior part of the cell (Willey et al., 1977). The envelopes of such species as *Trachelomonas* are build up by inorganic deposition on a foundation of mucilaginous threads.

Cysts are formed by euglenoids (notably *Euglena* and *Distigma*) as a means of surviving unfavorable periods. The cell rounds off and secretes a thick sheath of mucilage (Fig. 6.8) that survives for months until the cell emerges by cracking the cyst. In conditions of partial desiccation or excessive light, the slime sheath sometimes acts as a temporary cyst, cells emerging from the sheath as soon as conditions improve. In certain genera (*Euglena* and *Eutreptia*), cell division within the slime layer leads to the formation of a palmelloid colony, which may form extensive sheets of cells covering many square feet of mud surface (Leedale, 1967).

Trachelomonas is a large genus of free-swimming green euglenoids, characterized by encasement of the cell in a patterned mineralized envelope with a rimmed apical pore through which emerges the flagellum (Figs. 6.9 and 6.12(*d*)). Most of the species are defined by the form and ornamentation of the envelopes, characteristics that can be changed by varying conditions of growth, especially iron and manganese supply. This has resulted in some described species being growth forms of other species (Pringsheim, 1953). The process of envelope formation involves mineralization of an initial fibrillar envelope which is probably derived as a secretion of the mucilaginous bodies (Pringsheim, 1953; Leedale, 1975). It begins with longitudinal division of the parent protoplast, the two daughter cells rotating within the envelope, and one or both squeezing out through the pore. Each naked daughter cell then secretes a new envelope externally, at first delicate and colorless but already the size and shape of the old one. Under good growth conditions, the envelope slowly becomes thicker and ornamented,

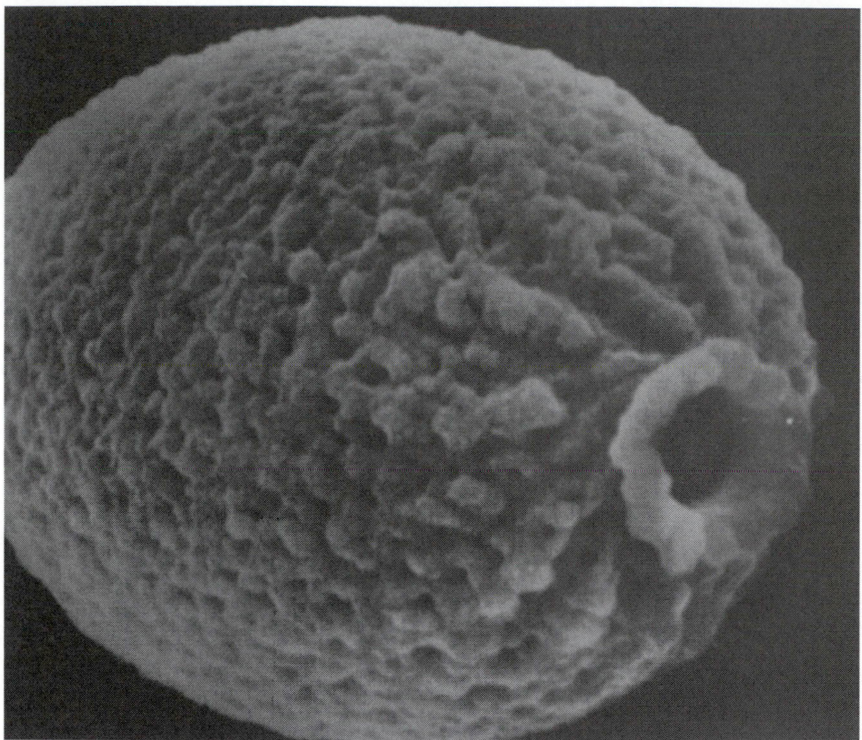

Fig. 6.9 Scanning electron micrograph of a mineralized envelope of *Trachelomonas lefevrei*. (from Dunlap et al., 1983.)

first yellow, then brown. Under conditions of manganese deficiency, the envelope usually remains thin and unornamented. Envelope substructure is species specific, with a given species producing one of two general types of substructure: (1) weft or fibrillar deposits composed primarily of **manganese**, or (2) fine granules composed primarily of **iron** (Dodge, 1975; Leedale, 1975; Walne, 1980; West and Walne, 1980; West et al., 1980).

Pellicular ornamentation occurs in a number of euglenoids, particularly in species of *Phacus* and in the *Euglena spirogyra* complex. The process is related to envelope formation in *Trachelomonas* and stalk formation in *Colacium*. In *E. spirogyra*, ornamentation occurs as warts outside the plasmalemma arranged in rows along pellicular strips. The warts contain a large amount of ferric hydroxide, and the degree of pellicular ornamentation depends on the availability of ferric iron and manganese in the habitat (Leedale, 1967).

Chloroplasts and storage products

Euglenoid chloroplasts are usually discoid or platelike with a central pyrenoid (Figs. 6.1 and 6.12). The chloroplasts are surrounded by two membranes of the chloroplast envelope plus one membrane of chloroplast endoplasmic reticulum; the latter membrane is not continuous with the nuclear membrane. The thylakoids are grouped into bands of three, with two thylakoid bands traversing the finely granular stroma of the pyrenoid. Surrounding the pyrenoid, but outside the chloroplast, is a shield of paramylon grains. Gottlieb isolated the granules in 1850, and showed that they were composed of a carbohydrate that, although isomeric with starch, was not stained with iodine. For this reason, they were termed **paramylon granules**; they have since been shown to be composed of a β-1,3 linked glucan (Barras and Stone, 1968). The paramylon granule is a membrane-bounded crystal composed of two types of segments – rectangular solids and wedges (Kiss et al., 1987). The liquid storage product, chrysolaminarin, can be an alternative storage product in some Euglenophyceae such as *Eutreptiella gymnastica* and *Sphenomonas laevis* where it can occur with solid paramylon grains in the same cell (Leedale, 1967; Throndsen, 1969). Whereas the paramylon usually occurs as a shield of grains, the chrysolaminarin occurs in vacuoles primarily in the anterior part of the cell (Throndsen, 1973).

Cell division, growth, and nutrition

Cell division begins after mitosis and replication of the basal bodies of the flagella has occurred, the original flagella being first lost. The strips of the pellicle also duplicate themselves. Two daughter canals are formed, and there is an inpushing at the anterior end of the cell between the two daughter canals. The cleavage line progresses helically backward between the daughter reservoirs and nuclei, following the helix of the pellicle. Cell division is accompanied by intense euglenoid movement in those genera with an elastic pellicle, involving the movement of the organelles. Partition of organelles is usually equal so that the two daughter cells are identical in size and contents. The mitochondria divide when the cells divide, with the mitochondria being numerous and small during cell division, whereas during the stationary phase they grow and branch to form a network in the peripheral cytoplasm (Calvayrac et al., 1972).

The Euglenophyceae have a number of modes of nutrition, depending on the species involved. No euglenoid has yet been demonstrated to be fully **photoautotrophic** – capable of living on a medium devoid of all organic compounds (including vitamins), with carbon dioxide as a carbon source, nitrates or ammonium salts as a nitrogen source, and light as an energy source. All green euglenoid flagellates so far studied are **photoauxotrophic** – capable of growing in a medium devoid of organic nutrients, with carbon dioxide, ammonium salts, and light, but needing at least one vitamin. *Euglena gracilis* has an absolute requirement for vitamin B_{12} (Hutner and Provasoli, 1955), it having been calculated that between 4900 and 22000 molecules of vitamin B_{12} are necessary for cell division (Carell, 1969). Vitamin B_{12}-starved cells increase in cell volume, sometimes to 10 times the size of control organisms, the cells in the final stage of vitamin B_{12} starvation often being polylobed, polynucleate, and containing more than the normal number of chloroplasts per cell (Bertaux and Valencia, 1971, 1973; Carell, 1969). During vitamin B_{12} starvation, total cellular RNA and protein increase 400% to 500% compared with controls (Carell et al., 1970). The chloroplast number per cell increases during this period, although the ratio of chloroplast protein to total cellular protein remains constant, evidence for the independence of chloroplast division from nuclear division (Bré and Lefort-Tran, 1974). Although the protein increase is 400% to 500% during vitamin B_{12} starvation, the total DNA increases only about 180%, suggesting that a particular step in DNA replication may be preferentially affected by the vitamin (Bré et al., 1975).

As *Euglena* cells age, they become immobile and spherical, with a tendency to form enlarged “giant” cells and to accumulate orange to black pigment bodies. Aging also results in the formation of larger numbers of lysosomes and microbodies with an increase in the degradation of organelles (Gomez et al., 1974). The older cells undergo a shift from carbohydrate to fat oxidation, as evidenced by an increase in malate synthetase, the enzyme involved in the glycolate bypass, important in the oxidation of fats.

The euglenoids belong to the acetate flagellates, having the ability to grow photosynthetically in the light or heterotrophically in the dark. In either state, the fixed carbon is used as a source of energy or as building blocks for cell constituents. The substrates that can be used for heterotrophic growth vary from one species to another, with the permeability of the substrate into the cell probably being the most important factor. As a rule, the most readily utilized substrates are acetic and butryic acids and the corresponding alcohols (e.g., ethanol). The two most commonly used substrates are acetate and ethanol.

Classification

The euglenoids probably arose from the ingestion of a green algal chloroplast by a protozoan in the Kinetoplastida (the bodonids, trypanosomatids, and the closely related protozoan *Isonema*). The Euglenophyta and

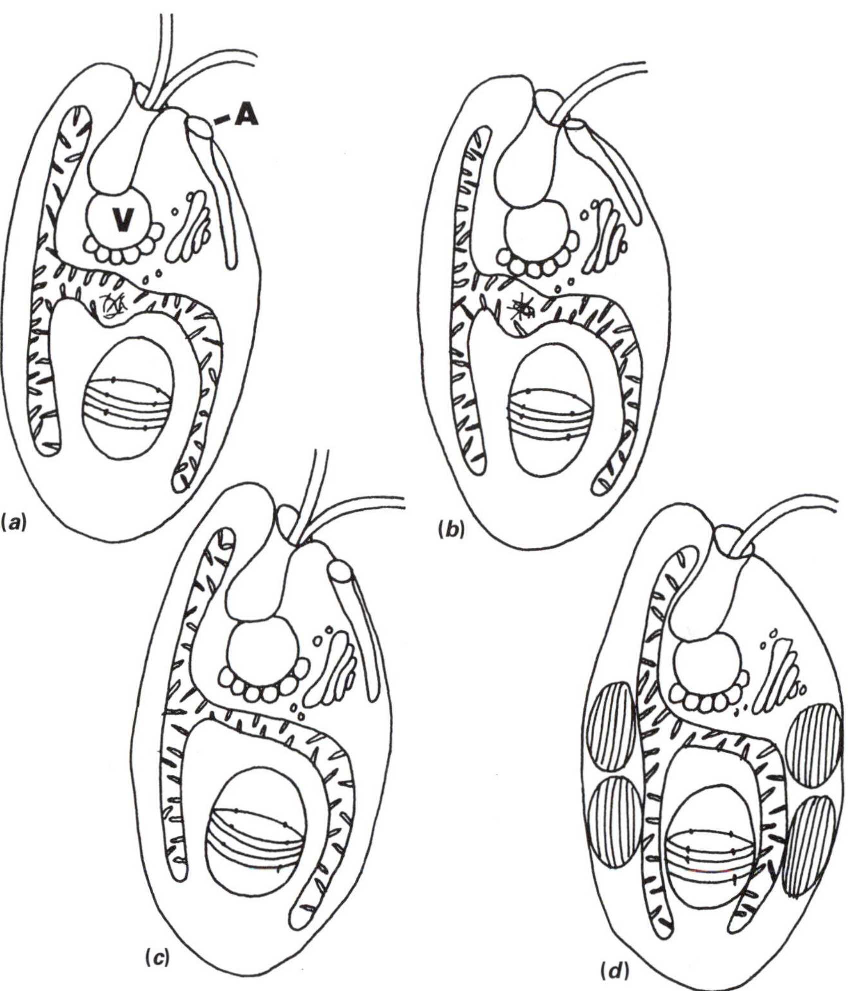

Fig. 6.10 Drawing showing the similarities between bodonids (*a*), trypanosomatids (*b*), phagotrophic euglenoids (*c*), and euglenoids with chloroplasts (*d*). The characteristics the organisms have in common are the external morphology, flagella inserted in a subapical depression, paraflagellar rods, a tubular ingestion apparatus (A) ventral to the flagellar depression, a large contractile vacuole (V) associated with a Golgi apparatus, a single giant mitochondrion with flat cristae, and mitosis occurring without a breakup of the nuclear envelope. (After Kivic and Walne, 1984.)

Kinetoplastida have a number of similarities (Fig. 6.10) (Dawson and Walne, 1994; Montegut-Felkner and Triemer, 1997): Their external morphology is essentially the same, the flagella arise from a subapical depression and have paraxonemal rods, there is a tubular ingestion apparatus (cytosome) ventral to the subapical depression in phagotrophic genera, there is a large contractile vacuole associated with the Golgi apparatus, there is a giant mitochondrion, their sialic acids and glycosphingolipids are unique (Preisfeld and Ruppel, 1995), and mitosis does not involve the breakup of the nuclear envelope. The phagotrophic uptake of a green algal chloroplast by a member of Kinetoplastida followed by the endosymbiotic establishment of a chloroplast would result in an organism similar to the present-day euglenoids. Such an endosymbiotic event would occur when the food vacuole membrane of the protozoan becomes the single membrane of chloroplast endoplasmic reticulum surrounding the two membranes of the chloroplast envelope (see p. 51, 283).

The Euglenophyta appears to be monophyletic with the primitive euglenoid having two flagella, each with a paraxonemal rod, and a special ingestion organelle, similar to the organisms in the Heteronematales (Montegut-Felkner and Triemer, 1997). In subsequent evolution the special ingestion organelle was lost and organisms similar to those in the Eutreptiales evolved (Dawson and Walne, 1994). The reduction in the length of one flagellum resulted in the organisms in the Euglenales (Soloman et al., 1991).

Order 1 Heteronematales: two emergent flagella, the longer flagellum directed anteriorly and the shorter one directed posteriorly during swimming; special ingestion organelle present.

Order 2 Eutreptiales: two emergent flagella, one directed anteriorly and the other laterally or posteriorly during swimming; no special ingestion organelle.

Order 3 Euglenales: two flagella, only one of which emerges from the canal; no special ingestion organelle.

HETERONEMATALES

Here the colorless cells have a special ingestion organelle (Triemer, 1997); and are phagocytic, taking up food particles whole and digesting them in food vesicles. *Peranema trichophorum* is a euglenoid that ingests other cells and detritus (Chen, 1950; Leedale, 1967) (Fig. 6.11). The ingestion apparatus consists of two parallel tapering rods, the hooked anterior ends of which are attached to the stiffened rim of the **cytosome**. The latter is a permanent "mouth" situated in a subapical position independent of the canal opening. There is no permanent "gullet," and food vacuoles are formed at the cytosome only when feeding takes place. *Peranema* normally ingests food particles and living organisms by engulfing them whole into food vacuoles. The ingestion rods are protruded and attached to the surface of the prey, which is then pulled through the cytosome in connection with a wave of euglenoid movement from the *Peranema* cell. With a large prey, such as *Euglena*, the rods are detached, moved, and attached again farther along the prey, so that more of it can be pulled into the predator. By repeated pullings, the whole *Euglena* is engulfed, the process taking up to 15 minutes. A second form of attack, reserved for larger algal cells, consists of cutting and sucking rather than engulfing. Several *Peranema* cells converge on their prey, with their ingestion rods protruded and used

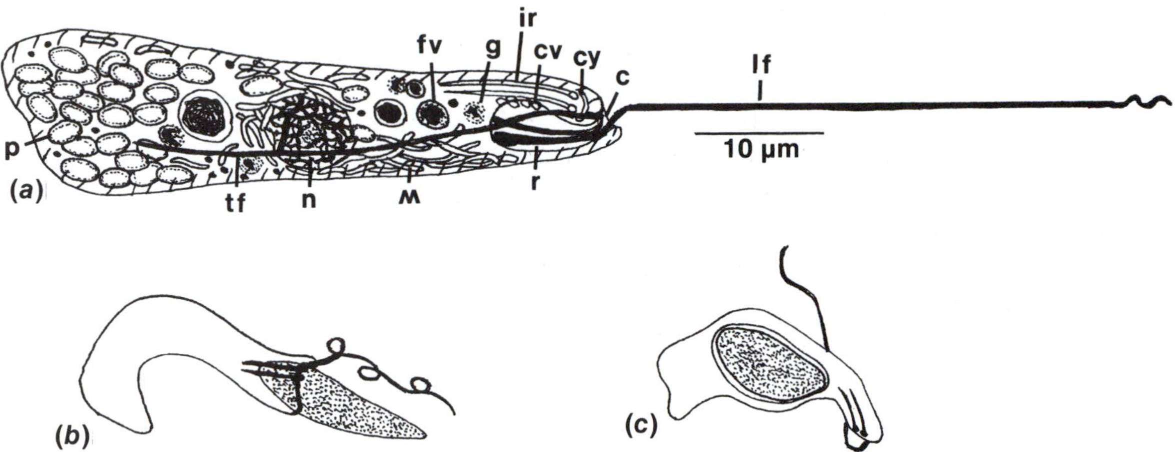

Fig. 6.11 *Peranema trichophorum*. (*a*) General cell structure. (*b*),(*c*) Two stages in the ingestion of a cell of *Euglena* (stippled cell). (c) Canal; (cv) contractile vacuoles; (cy) rim of cytosome; (fv) food vesicle; (g) Golgi; (ir) ingestion rods; (lf) leading flagellum; (m) mitochondrion; (n) nucleus; (p) paramylon; (tf) trailing flagellum. (After Leedale, 1967.)

to rasp a way through the prey's wall or periplast. *Euglena spirogyra* pellicle is cut through in about 10 minutes, with the cell contents sucked out into a temporary food canal below the cytosome. If the prey is large enough, the predators finally enter the cell and engulf what remains of the prey. The food vacuoles decrease in size as digestion proceeds, the indigestible remains being finally ejected through a "defecation area" of constant position at the posterior end of the cell. It is possible to show that chemotaxis is important in directing *Peranema* to its prey by bursting open living algal cells in a suspension of *Peranema*, the peranemas streaming in from all directions for the meal.

EUTREPTIALES

The organisms in the Eutreptiales have two emergent flagella and no special ingestion organelles. *Eutreptia* and *Eutreptiella* (Fig. 6.12(*c*)) are estuarine or marine genera, while *Distigma* is characteristic of acid freshwaters.

EUGLENALES

In this primarily freshwater order, the flagellum without the paraflagellar swelling has been reduced so that it does not emerge from the canal. Common genera in the order are the green photosynthetic *Euglena* (Fig. 6.12(*a*)), *Trachelomonas* (Fig. 6.12(*d*)), and *Phacus* (Fig. 6.12(*e*)), as well as the colorless osmotrophic *Astasia* (Fig. 6.12(*b*)).

Colacium libellae is a member of this order that establishes itself in the rectum of damselfly nymphs during the winter in colder lakes (Fig. 6.13). During the warm summer months the damselfly nymphs and *C. libellae* live

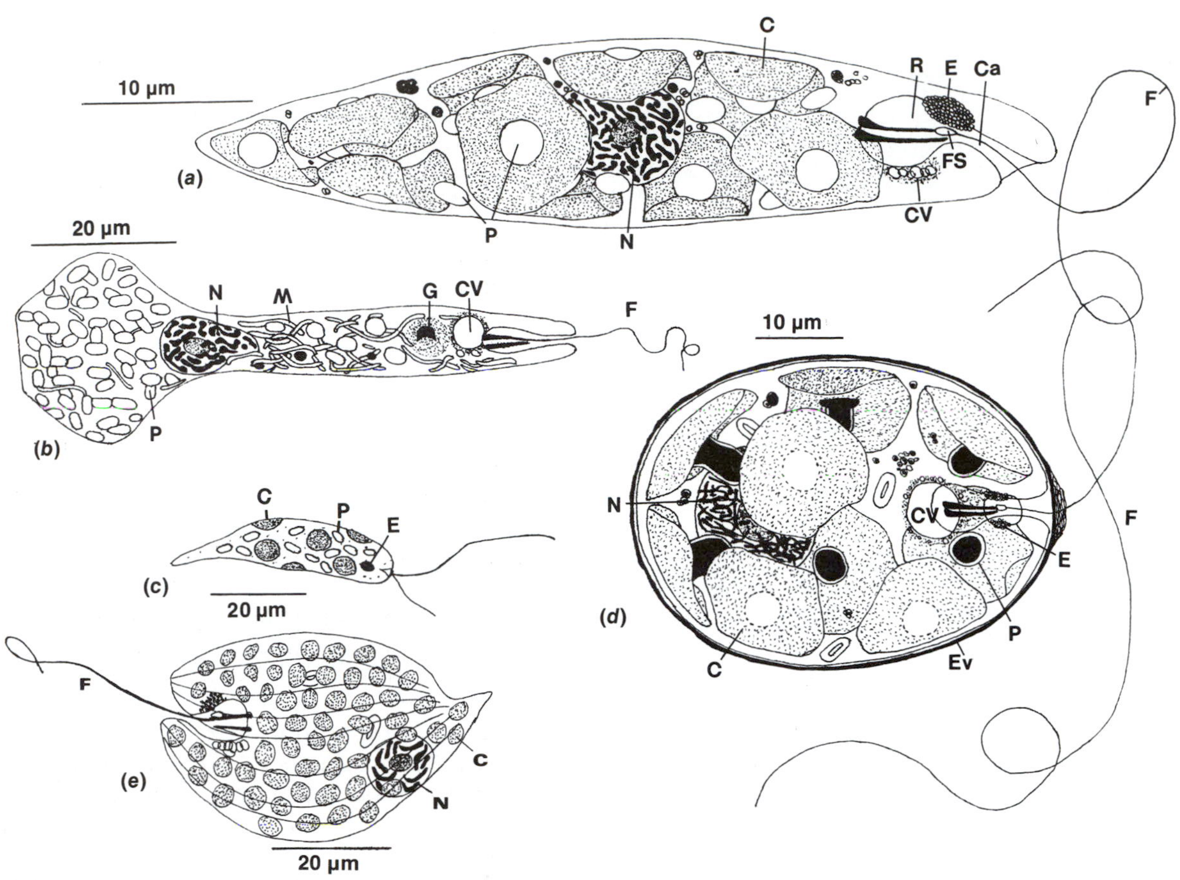

Fig. 6.12 (*a*) *Euglena gracilis*. (*b*) *Astasia klebsii*. (*c*) *Eutreptiella marina*. (*d*) *Trachelomonas grandis*. (*e*) *Phacus triqueter*. (C) Chloroplast; (Ca) canal; (CV) contractile vacuole; (E) eyespot; (Ev) envelope; (F) emergent flagellum; (FS) flagellar swelling; (M) mitochondrion; (N) nucleus; (P) paramylon grains or paramylon sheath around chloroplast; (R) reservoir. (After Leedale, 1967.)

separately. With the onset of winter, the *Colacium* cells attach to the cuticle of the rectum of the damselfly larvae, forming a conspicuous green plug that colors the terminal four segments of the abdomen dark green. In the rectum, the *Colacium* is in a palmelloid state, lacking a flagellar swelling and eyespot. As the peripheral waters of the lake freeze, the damselfly nymphs move to deeper water. Here the damselfly nymphs form a protected, motile, translucent microhabitat for the *Colacium*; the damselfly nymphs probably also provide a source of nutrients for the *Colacium* cells, which helps the alga to survive the unfavorable winter conditions. In spring, the damselfly nymphs swim to warmer water, at which time the *Colacium* cells swim out of the rectum to establish a free-living existence in the summer (Fig. 6.14). If algal-free damselfly nymphs are placed in water with *C. libellae*, the alga will establish itself within 36 hours in the damselfly nymph rectum. Other species are not able to infect the nymphs (Willey et al., 1970, 1973; Willey, 1972; Rosowski and Willey, 1975).

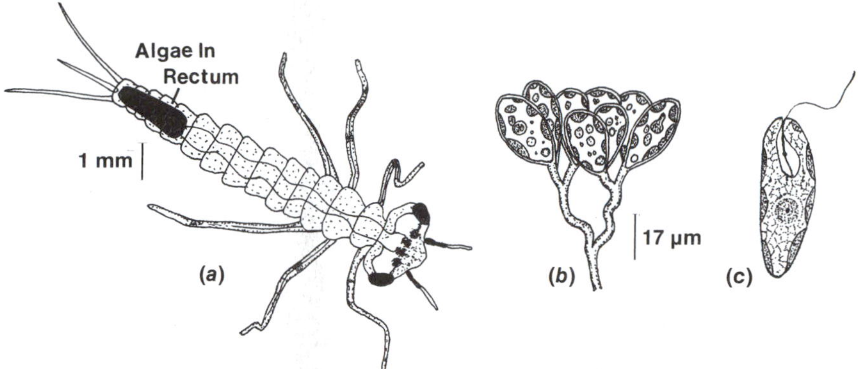

Fig. 6.13 (*a*) Larva of the damselfly *Ischnura verticalis* with a plug of *Colacium libellae* in the rectum. (*b*), (*c*) Colony and single swimming cell of *Colacium vesiculosum*. (*a*) adapted from Rosowski and Willey, 1975; (*b*), (*c*) after Stein and Johnson in Huber-Pestalozzi, 1955.)

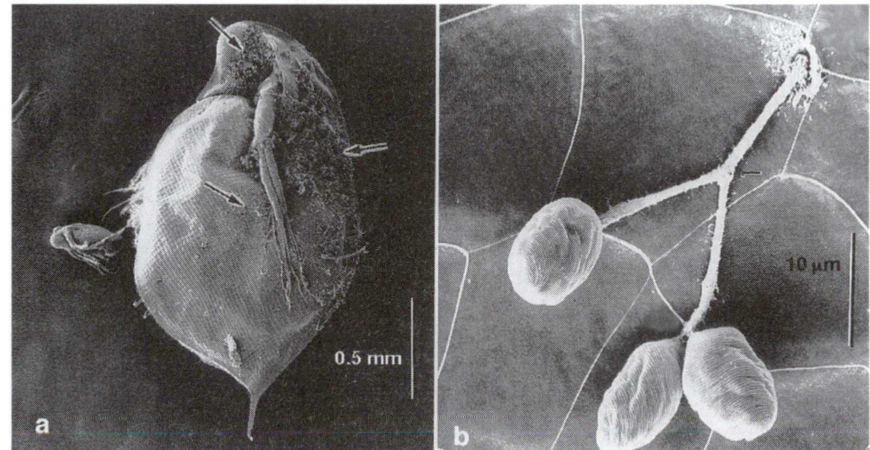

Fig. 6.14 Scanning electron micrographs of the euglenoid *Colacium vesiculosum* on the freshwater arthropod *Daphnia pulex*. (*a*) Arrows point to large concentrations of *Colacium* on *Daphnia*. (*b*) A colony of *Colacium* attached by mucilage stalks to *Daphnia*. (From Al-Dhaheri and Willey, 1996.)

References

Al-Dhaheri, R. S., and Willey, P. L. (1996). Colonization and reproduction of the epibiotic flagellate *Colacium vesiculosum* (Euglenophyceae) in *Daphnia pulex*. *J. Phycol.* 32:770–4.

Barras, D. R., and Stone, B. A. (1965). The chemical composition of the pellicle of *Euglena gracilis* var. *bacillaris*. *Biochem. J.* 97:14P–15P.

Barras, D. R., and Stone, B. A. (1968). Carbohydrate composition and metabolism in *Euglena*. In *The Biology of Euglena*, ed. D. E. Buetow, Vol. 2, pp. 149–91. New York and London: Academic Press.

Batra, P. P., and Tollin, G. (1964). Phototaxis in *Euglena*. I. Isolation of the eyespot granules and identification of the eye-spot pigments. *Biochem. Biophys. Acta* 79:371–8.

Bertaux, O., and Valencia, R. (1971). Effects de la carence B_{12} sur les cellules synchrones de *Euglena gracilis:* Blocage de la division cellulaire, polyploidie et gigantisme. *J. Physiol. (London)* 63:167A.

Bertaux, O., and Valencia, R. (1973). Blocage de la division cellulaire et malformations induites par carence B_{12} chez les cellules synchrones de *Euglena gracilis Z. C. R. Séances Acad. Sci. Paris* 276:753–6.

Bouck, G. B., Rogalski, A., and Valaitis, A. (1978). Surface organization and composition of *Euglena*. II. Flagellar mastigonemes. *J. Cell Biol.* 77:805–26.

Bracher, R. (1937). The light relations of *Euglena limosa* Gard. Part I. The influence of intensity and quality of light on phototaxy. *J. Linn. Soc. Bot.* 51:23–42.

Bré, M. H., and Lefort-Tran, M. (1974). Influence de l'avitaminose B_{12} sur les chloroplastes de l'*Euglena gracilis* Z. en milieu lactate. *C. R. Séances Acad. Sci. Paris* 278:1349–52.

Bré, M. H., Diamond, J., and Jacques, R. (1975). Factors mediating the vitamin B_{12} requirement of *Euglena*. *J. Protozool.* 22:432–4.

Bricheux, G., and Brugerolle, G. (1987). The pellicular complex of euglenoids. II. A biochemical and immunological comparative study of major epiplasmic proteins. *Protoplasma* 140:43–54.

Buetow, D. E. (ed.) (1968). *The Biology of Euglena*, Vols. 1 and 2. New York and London: Academic Press.

Calvayrac, R., Butow, R. A., and Lefort-Tran, M. (1972). Cyclic replication of DNA and changes in mitochondrial morphology during the cell cycle of *Euglena gracilis (Z)*. *Esp. Cell Res.* 71:422–32.

Carell, E. F. (1969). Studies on chloroplast development and replication in *Euglena*. I. Vitamin B_{12} and chloroplast replication. *J. Cell Biol.* 41:431–40.

Carell, E. F., Johnston, P. L., and Christopher, A. R. (1970). Vitamin B_{12} and the macromolecular composition of *Euglena*. *J. Cell Biol.* 47:525–30.

Chaly, N., Lord, A., and Lafontaine, J. G. (1977). A light- and electron-microscope study of nuclear structure throughout the cell cycle in the euglenoid *Astasia longa* (Jahn). *J. Cell Sci.* 27:23–45.

Chen, Y. T. (1950). Investigations of the biology of *Peranema trichophorum* (Euglenineae). *Q. J. Microsc. Sci.* 91:279–308.

Cogburn, J. N., and Schiff, J. A. (1984). Purification and properties of the mucus of *Euglena gracilis* (Euglenophyceae). *J. Phycol.* 20:533–44.

Davis, B., and Merrett, M. J. (1974). The effect of light on the synthesis of mitochondrial enzymes in division-synchronized *Euglena* cultures. *Plant Physiol.* 53:575–80.

Dawson, N. S., and Walne, P. L. (1994). Evolutionary trends in euglenoids. *Arch. Protistendkd.* 144:221–5.

Dodge, J. D. (1975). The fine structure of *Trachelmonas* (Euglenophyceae). *Arch. Protistenk.* 117:65–77.

Dubreuil, R. R., and Bouck, G. B. (1985). The membrane skeleton of a unicellular organism consists of bridged, articulating strips. *J. Cell Biol.* 101:1884–96.

Dunlap, J. R., Walne, P. L., and Bentley, J. (1983). Microarchitecture and elemental spatial segregation of envelopes of *Trachelomonas lefevrei* (Euglenophyceae). *Protoplasma* 117:97–106.

Epstein, H. T., and Schiff, J. A. (1961). Studies of chloroplast development in *Euglena*. IV. Electron and fluorescence microscopy of the proplastid and its development into a mature chloroplast. *J. Protozool.* 8:427–32.

Gomez, M. P., Harris, J. B., and Walne, P. L. (1974). Studies of *Euglena gracilis* in aging cultures. I. Light microscopy and cytochemistry. *Br. Phycol. J.* 9:163–74.

Gravilă, L. (1996). Light and electron microscope studies of euglenoid nuclei. In *Cytology, Genetics and Molecular Biology of Algae*, ed. B. R. Chaudhary, and S. B. Agrawal, pp. 193–213. Amsterdam, The Netherlands: SPB Academic Pub.

Häder, D-P. (1987). Polarotaxis, gravitaxis and vertical phototaxis in the green flagellate, *Euglena gracilis*. *Arch. Microbiol.* 147:179–83.

Hilenski, L. L., and Walne, P. L. (1983). Ultrastructure of mucocysts in *Peranema trichophorum* (Euglenophyceae). *J. Protozool.* 30:491–6.

Hilenski, L. L., and Walne, P. L. (1985). Ultrastructure of the flagella of the colorless phagotroph *Peranema trichophorum* (Euglenophyceae). I. Flagellar mastigonemes. *J. Phycol.* 21:114–25.

Huber-Pestalozzi, G. (1955). Euglenophycean. *Das Phytoplankton des Susswassers*, Vol. 16, Part 4. E. Schweizerbart'sche Verlagsbuchhandlung, Stuttgart.

Hutner, S. H., and Provasoli, L. (1955). Comparative biochemistry of flagellates. In *Biochemistry and Physiology of Protozoa*, ed. S.H. Hutner, and A. Lwoff, Vol. 2, pp. 1–40. New York and London: Academic Press.

Jahn, T. L., and Bovee, E. C. (1968). Locomotive and motile response in *Euglena*. In *The Biology of Euglena* ed. D. E. Búetow I:45–108. New York and London: Academic Press.

Kiss, J. Z., Vasconcelos, A. C., and Triemer, R. E. (1987). Structure of the euglenoid storage carbohydrate, paramylon. *Am. J. Bot.* 74:877–82.

Kivic, P. A., and Walne, P. L. (1984). An evaluation of a possible phylogenetic relationship between the Euglenophyta and Kinetoplastida. *Origins of Life* 13:269–88.

Krinsky, N. I., and Goldsmith, T. H. (1960). The carotenoids of the flagellated alga, *Euglena gracilis*. *Arch. Biochem. Biophys.* 91:271–9.

Lee, J. W. (1954). The effect of temperature on forward swimming in *Euglena* and *Chilomonas*. *Physiol. Zool.* 27:275–83.

Lee, R. E. (1977). Evolution of algal flagellates with chloroplast endoplasmic reticulum from the ciliates. *S. Afr. J. Sci.* 78:179–82.

Leedale, G. F. (1959). Periodicity of mitosis and cell division in the Euglenineae. *Biol. Bull.* 116:162–74.

Leedale, G. F. (1967). *Euglenoid Flagellates*. Englewood Cliffs, NJ: Prentice Hall.

Leedale, G. F. (1970). Phylogenetic aspects of nuclear cytology in the algae. *Ann. NY Acad. Sci.* 175(2):429–53.

Leedale, G. F. (1975). Envelope formation and structure in the euglenoid genus *Trachelomonas*. *Br. Phycol. J.* 10:17–41.

Lonergan, T. A. (1983). Regulation of cell shape in *Euglena gracilis*. I. Involvement of the biological clock, respiration, photosynthesis, and cytoskeleton. *Plant Physiol.* 71:719–730.

Melkonian, M., Robenek, H., and Rassat, J. (1982). Flagellar membrane specializations and their relationship to mastigonemes and microtubules in *Euglena gracilis*. *J. Cell Sci.* 55:115–35.

Mignot, J-P. (1966). Structure et ultrastructure de quelques Euglénomonadines. *Protistologica* 2:51–117.

Mignot, J-P., Brugerolle, G., and Bricheux, G. (1987). Intercalary strip development and dividing cell morphogenesis in the euglenid *Cyclidiopsis acus*. *Protoplasma* 139:51–65.

Montegut-Felkner, A. E., and Triemer, R. E. (1997). Phylogenetic relationships in selected euglenoid genera based on morphological and molecular data. *J. Phycol.* 512–19.

Olli, K. (1996). Resting cyst formation of *Eutreptiella gymnastica* (Euglenophyceae) in the northern coastal Baltic Sea. *J. Phycol.* 32:535–42.

Palmer, J. D., and Round, F. E. (1965). Persistent, vertical-migration rhythms in the benthic microflora. I. The effect of light and temperature on the rhythmic behaviour of *Euglena obtusa*. *J. Mar. Biol. Assoc. UK* 45:567–82.

Preisfeld, A., and Ruppel, H. G. (1995). Detection of sialic acid and glycosphingolipids in *Euglena gracilis* (Euglenozoa). *Arch Protistendk.* 145:251–60.

Pringsheim, E. G. (1953). Observations on some species of *Trachelomonas* grown in culture. *New Phytol.* 52:93–113, 238–66.

Ray, D. S., and Hanawalt, P. C. (1965). Satellite DNA components in *Euglena gracilis* cells lacking chloroplasts. *J. Mol. Biol.* 11:760–8.

Rosowski, J. R. (1977). Development of mucilaginous surfaces in euglenoids. II. Flagellated, creeping and palmelloid cells of *Euglena. J. Phycol.* 13:323–8.

Rosowski, J. R., and Willey, R. L. (1975). *Colacium libellae* sp. nov. (Euglenophyceae), a photosynthetic inhabitant of the larval damselfly rectum. *J. Phycol.* 11:310–15.

Soloman, J. A., Walne, P. L., Dawson, N. S., and Willey, R. I. (1991). Structural characterization of *Eutreptia* (Euglenophyta). II. The flagellar root system and putative vestigial cytopharynx. *Phycologia* 30:402–14.

Sperling Pagni, P. G., Walne, P. L., and Pagni, R. M. (1981). On the occurrence of α-carotene in isolated stigmata of *Euglena gracilis* var. *bacillaris. Phycologia* 20:431–4.

Surek, B., and Melkonian, M. (1986). A cryptic cytosome is present in *Euglena. Protoplasma* 133:39–49.

Suzaki, T., and Williamson, R. E. (1986). Pellicular ultrastructure and euglenoid movement in *Euglena ehrenbergii* Klebs and *Euglena oxyuris* Schmarda. *J. Protozool.* 33:165–71.

Throndsen, J. (1969). Flagellates of Norwegian coastal waters. *Nytt Mag. Bot.* 16:161–216.

Throndsen, J. (1973). Fine structure of *Eutreptiella gymnastica* (Euglenophyceae). *Norw. J. Bot.* 20:271–80.

Triemer, R. E. (1997). Feeding in *Peranema trichophorum* revisited (Euglenophyta). *J. Phycol.* 33:649–54.

Walne, P. L. (1980). Euglenoid flagellates. In *Phytoflagellates,* ed. E. Cox, pp. 165–212. North Holland: Elsevier.

Walne, P. L., and Dawson, N. S. (1993). A comparison of paraxial rods in the flagella of euglenoids and kinetoplastids. *Arch. Protistendk.* 143:177–94.

West, L. K., and Walne, P. L. (1980). *Trachelomonas hispada* var. *coronata* (Euglenophyceae). II. Envelope substructure. *J. Phycol.* 16:498–506.

West, L. K., Walne, P. L., and Bentley, J. (1980). *Trachelomonas hispida* var. *coronata* (Euglenophyceae). III. Envelope elemental composition and mineralization. *J. Phycol.* 16:582–91.

Willey, R. L. (1972). The damselfly (Odonata) hindgut as a host for the euglenoid *Colacium. Trans. Am. Microsc. Soc.* 91:585–93.

Willey, R. L. (1984). Fine structure of the mucocysts of *Colacium calvum* (Euglenophyceae). *J. Phycol.* 20:426–30.

Willey, R. L., and Wibel, R. G. (1985a). The reservoir cytoskeleton and a possible cytosomal homologue in *Colacium* (Euglenophyceae). *J. Phycol.* 21:570–7.

Willey, R. L., and Wibel, R. G. (1985b). A cytosome/cytopharynx in green euglenoid flagellates (Euglenales) and its phylogenetic implications. *BioSystems* 18:369–76.

Willey, R. L., Bowen, W. R., and Durban, E. M. (1970). Symbiosis between *Euglena* and damselfly nymphs is seasonal. *Science* 170:80–1.

Willey, R. L., Durban, E. M., and Bowen, W. R. (1973). Ultrastructural observations of a *Colacium* palmella: The reservoir, eyespot, and flagella. *J. Phycol.* 9:211–15.

Willey, R. L., Ward, K., Russin, W., and Wibel, R. (1977). Histochemical studies of the extracellular carbohydrate of *Colacium mucronatum. J. Phycol.* 13:349–53.

7 · Dinophyta

Dinophyceae

These organisms are important members of the plankton in both fresh and marine waters, although a much greater variety of forms is found in marine members. Generally the Dinophyceae are less important in the colder polar waters than in warmer waters. The highly elaborate Dinophysiales are essentially a tropical group.

A typical motile dinoflagellate (Figs. 7.1 and 7.5) consists of an **epicone** and **hypocone** divided by the transverse **girdle** or **cingulum**. The epicone and hypocone are normally divided into a number of **thecal plates**, the exact number and arrangement of which are characteristic of the particular genus (Figs. 7.1, 7.5, 7.11, 7.12 and 7.28). There is a longitudinal **sulcus** running perpendicular to the girdle. The longitudinal and transverse flagella emerge through the thecal plates in the area where the girdle and sulcus meet. The longitudinal flagellum projects out from the cell, whereas the transverse flagellum is wavelike and is closely appressed to the girdle. The cells can be photosynthetic or colorless and heterotrophic. Photosynthetic organisms have chloroplasts surrounded by one membrane of chloroplast E.R., which is not continuous with the outer membrane of the nuclear envelope. Chlorophylls a and c_2 are present in the chloroplasts, with **peridinin** and **neoperidinin** being the main carotenoids. About half of the Dinophyceae that have been examined by electron microscopy have pyrenoids in the chloroplasts (Dodge and Crawford, 1970). The storage product is **starch**, similar to the starch of higher plants (Vogel and Meeuse, 1968), which is found in the cytoplasm. An eyespot may be present. The nucleus has permanently condensed chromosomes and is called a **dinokaryotic** or **mesokaryotic** nucleus.

Cell structure

THECA

The thecal structure of motile Dinophyceae consists of an outer plasmalemma beneath which lies a single layer of flattened vesicles (Figs 7.1 and 7.2) (Dodge and Crawford, 1970). These vesicles, which usually contain

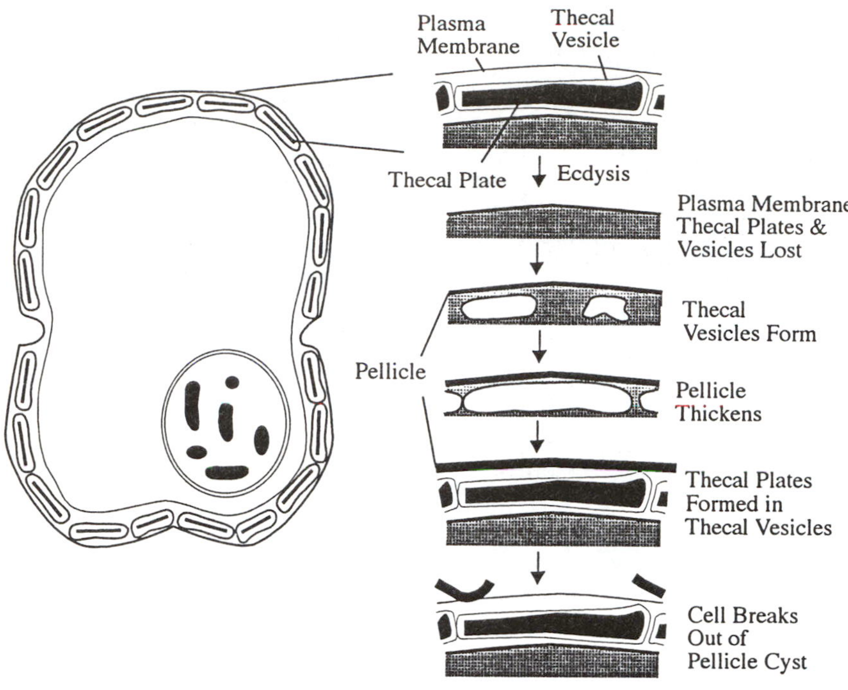

Fig. 7.3 Ecdysis and the formation of new thecal vesicles in *Glenodinium foliaceum*. Ecdysis results in the loss of the plasma membrane, thecal plates and vesicles. A pellicle forms from a layer that existed under the thecal vesicles. New thecal vesicles and plates are formed under the thickening pellicle. When the amphiesma is mature, the wall breaks out of the pellicular cyst. (Drawn from transmission electron micrographs in Bricheux et al., 1992.)

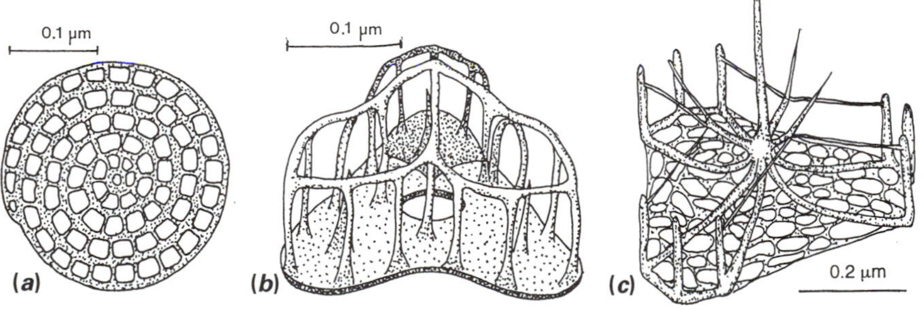

Fig. 7.4 Dinoflagellate scales. (a) *Oxyrrhis marina*. (b) *Heterocapsa niei*. (c) *Katodinium rotundatum*. ((*a*) after Clarke and Pennick, 1976; (*b*) after Morrill and Loeblich, 1983b; (*c*) after Hansen, 1989.)

longitudinal sulcus (Fig. 7.1 and 7.5) (Roberts and Roberts, 1991). The two flagella are inserted into the cell in the area of the intersection of the girdle and sulcus. The longitudinal flagellum usually has a wide basal portion and a thinner apical portion. In the basal portion the flageller sheath surrounds packing material plus the axoneme, whereas at the apical portion the flagellar sheath surrounds only the axoneme, Fibrillar hairs 0.5 μm long and 10 nm wide may cover the entire flagellum (Leadbeater and Dodge, 1967a). Mechanical stimulation of cells of *Ceratium tripos* (Fig. 7.6) causes the longitudinal flagellum to be retracted and folded so that the flagellum lies along the sulcus (Maruyama, 1982). The longitudinal flagellum contains an R-fiber running along the length of the flagellum. Retraction of the

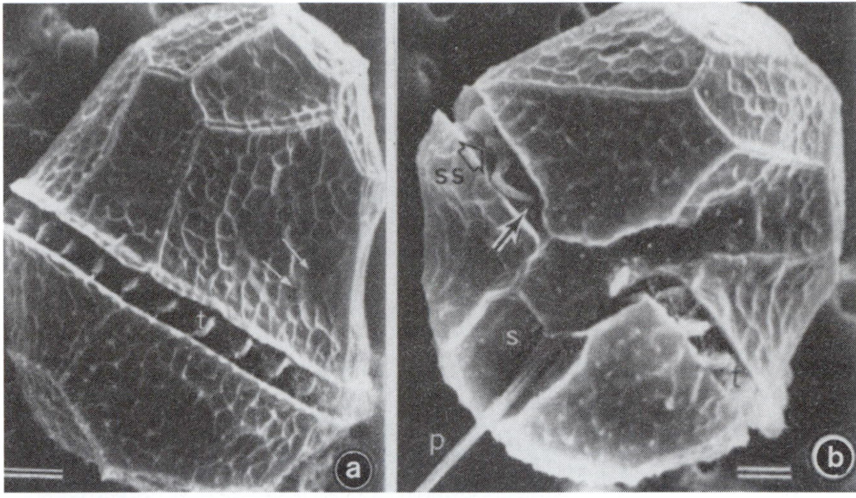

Fig. 7.5 Front and back of cells of *Peridinium cinctum* in the scanning electron microscope. (p) Posterior (longitudinal) flagellum; (s) longitudinal sulcus; (ss) striated strand; (t) transverse flagellum. Bar = 5 μm. (From Berdach, 1977.)

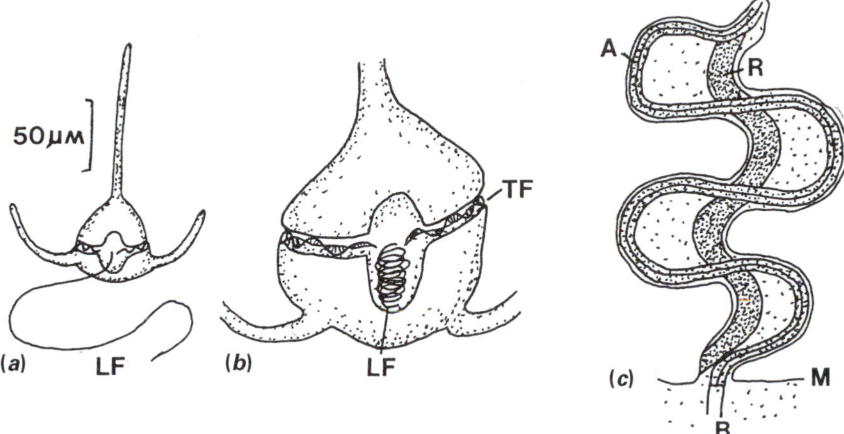

Fig. 7.6 *Ceratium tripos*. (*a*) Vegetative cell with a relaxed longitudinal flagellum (LF) at full length. (*b*) A cell with the longitudinal flagellum (LF) fully contracted and folded in the sulcus. (TF) Transverse flagellum. (*c*) Schematic drawing of a retracted flagellum showing the contracted R-fiber (R) and the axoneme (A). (B) Basal body; (M) plasma membrane. (Adapted from Maruyama, 1982.)

flagellum occurs when the R-fiber contracts to one third of its length. Influx of Ca^{2+} into the longitudinal flagellum causes retraction of the R-fiber (Maruyama, 1985).

The transverse flagellum is about two to three times as long as the longitudinal flagellum and has a helical shape (Figs. 7.7 and 7.8). The transverse flagellum consists of (1) an axoneme whose form approximates a helix, (2) a striated strand that runs parallel to the longitudinal axis of the axoneme but outside the loops of the coil, and (3) a flagellar sheath that encloses both axoneme and striated strand (Berdach, 1977). The striated strand contains **centrin**, a Ca^{2+}-modulated contractile protein (Höhfeld et al., 1988). Contraction of the striated strand leads to supercoiling of the axoneme. On one side of the axoneme, a unilateral row of fibrillar hairs, 2 to 4 μm long and

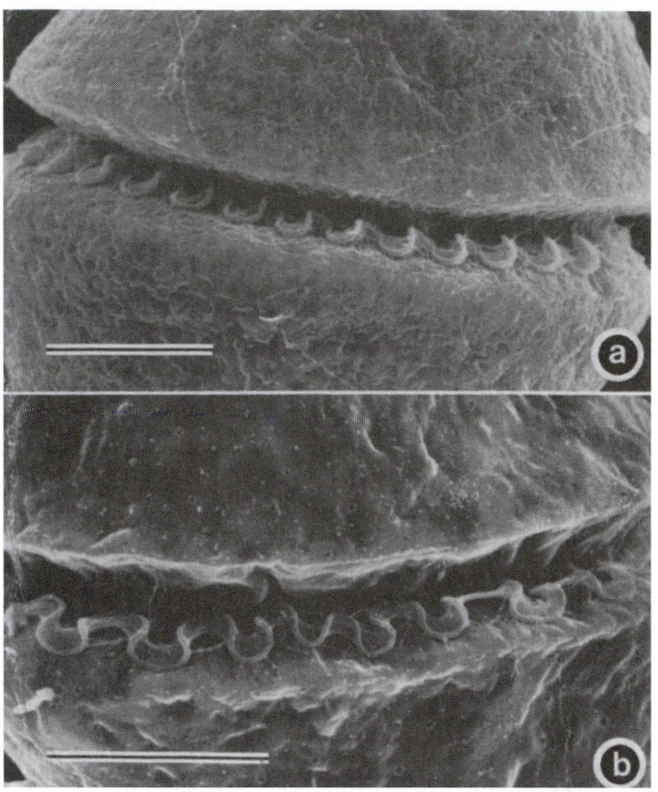

Fig. 7.7 Scanning electron micrograph showing the form of the transverse flagellum in *Gymnodinium sanguineum* (*a*) and *Gyrodinium uncatenum* (*b*). Bar = 10 μm. (From Gaines and Taylor, 1985.)

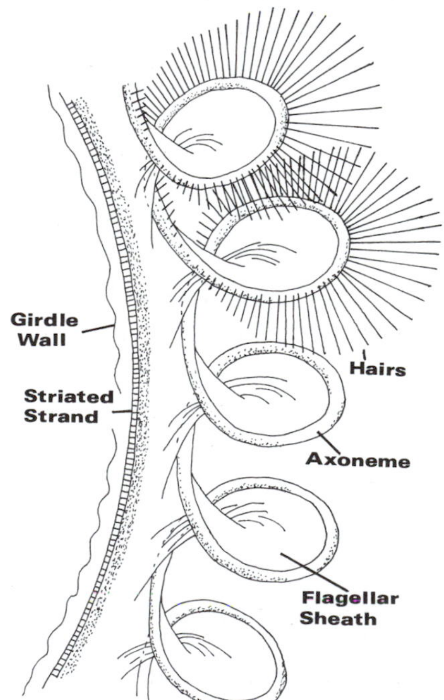

Fig. 7.8 Diagram of part of the transverse flagellum of *Peridinium cinctum*. Only a portion of the fibrillar hairs have been drawn in. (After Berdach, 1977.)

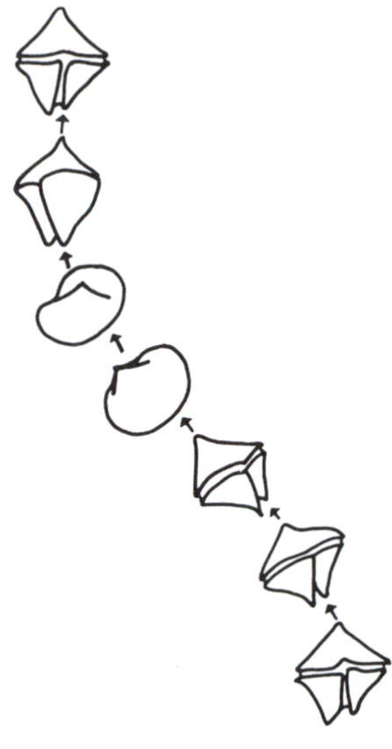

Fig. 7.9 Tracings from successive video frames of *Protoperidinium conicum* swimming, showing the leiotropic rotation of the cell. Time between intervals is 1/5 second. (From Gaines and Taylor, 1985.)

10 nm wide, is attached to the flagellar sheath. The final micrometer of the flagellum lacks the striated strand and hairs, with the axoneme tightly covered by the flagellar sheath. The axoneme is a left-handed screw; waves propagated from the attached end toward the free end create a downward thrust at the outermost edges of the coil. The flagellum causes a direct forward movement while at the same time causing the cell to rotate. The flagellar beat always proceeds counterclockwise when seen from the cell apex (**leiotropic** direction). The cell always rotates in the direction of the flagellar beat (in the leiotropic direction), with the fluid propelled in the opposite direction (in the **dexiotropic** direction) (Fig. 7.9) (Gaines and Taylor, 1985). The longitudinal flagellum swings in a narrow orbit acting as a rudder.

The dinoflagellates are the fastest swimmers among the algae although they are slower than *Mesodinium*, the photosynthetic symbiosis between a ciliate and a cryptophyte (see Chapter 8, Cryptophyta, Symbiotic Associations). Dinoflagellates swim from 200 to 500 $\mu m\ s^{-1}$ (Raven and Richardson, 1984). The cells of *Gonyaulax polyedra* swim at a linear rate of 250 $\mu m\ s^{-1}$ at 20 °C whereas *Gyrodinium* sp. (Fig. 7.36) swim at a mean linear velocity of 319 $\mu m\ s^{-1}$ at 20 °C. Marine dinoflagellates frequently move into deeper, nutrient-rich waters at night and better illuminated waters near the surface during the day. The **diel** (over a 24-hour period) migrations of dinoflagellates are 5 to 10 m in relatively quiet waters. A dinoflagellate swimming at 500 $\mu m\ s^{-1}$ would take about 6 hours of the 24-hour period for 5 m of upward migration around dawn, and a further 6 hours for downward migration around dusk (Raven and Richardson, 1984).

PUSULE

A **pusule** is a saclike structure that opens by means of a pore into the flagellar canal and probably has an osmoregulatory function similar to that of a contractile vacuole. The pusule of *Amphidinium carteri* is representative of those dinoflagellates that have a pusule (Fig. 7.10) (Dodge and Crawford, 1968). In this organism there are two pusules, one associated with each flagellar canal. The pusule consists of about 40 vesicles which open by small pores into the flagellar canal. Whereas the flagellar canal is lined by a single membrane continuous with the plasma membrane, the pusule vesicles are lined by a double membrane.

CHLOROPLASTS AND PIGMENTS

In the Dinophyceae, the chloroplasts are surrounded by the chloroplast envelops and generally one membrane of chloroplast endoplasmic reticu-

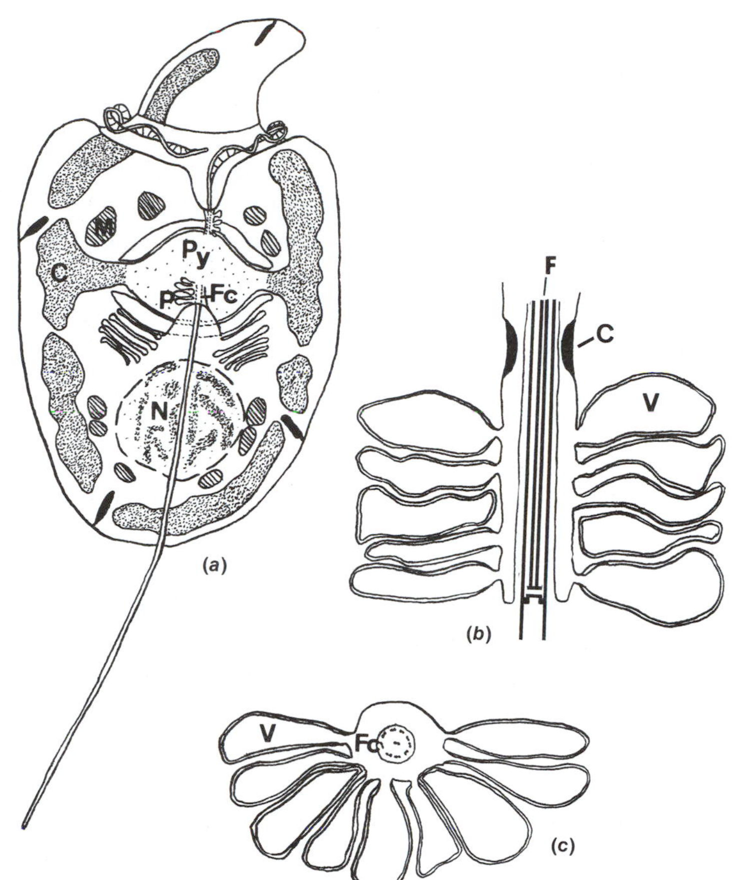

Fig. 7.10 (*a*) Drawing of *Amphidinium carteri*. The small epicone is separated from the hypocone by the encircling girdle, which contains the transverse flagellum. The peripheral chloroplast (C) is shown only at the sides of the cell so that the positions of the nucleus (N), mitochondria (M), pyrenoid (Py), and the pusules (P) associated with the flagellar canals can be seen. (*b*) Longitudinal section through a pusule showing the flagellar pore constriction (C), the pusule vesicles (V), and the flagellum (F). (*c*) Transverse section of a pusule showing the flagellum within the flagellar canal (Fc) and the pusule vesicles (V) opening into the canal. (From Dodge and Crawford, 1968.)

lum, which is not continuous with the nuclear envelope. A few dinoflagellates have chloroplasts not surrounded by chloroplast endoplasmic reticulum (Wilcox et al., 1982). These probably evolved from dinoflagellates with chloroplast endoplasmic reticulum. Pyrenoids may, or may not, be present. Chlorophylls *a* and c_2 are present. Dinoflagellates have a low chlorophyll *a* to carbon ratio and this has been hypothesized as a reason for their slow growth rate when compared to other algal cells of a similar size (Tang, 1996).

The major carotenoids are **peridinin** or occasionally **19′hexanoyloxyfucoxanthin** (a fucoxanthin derivative) (Haxo, 1985). Most of the chlorophyll *a* and peridinin occurs together in a water-soluble protein complex called peridinin–chlorophyll *a*–protein (PCP). Within the chromophore the peridinin and chlorophyll *a* are in a 4:1 ratio (Prézelin and Haxo, 1976). PCP is similar to the phycobiliproteins in cyanobacteria and red algae in that it is on the surface of thylakoids, is water-soluble, and acts as a light-harvesting pigment. If *Glenodinium* (Fig. 7.16) is cultured under decreasing light

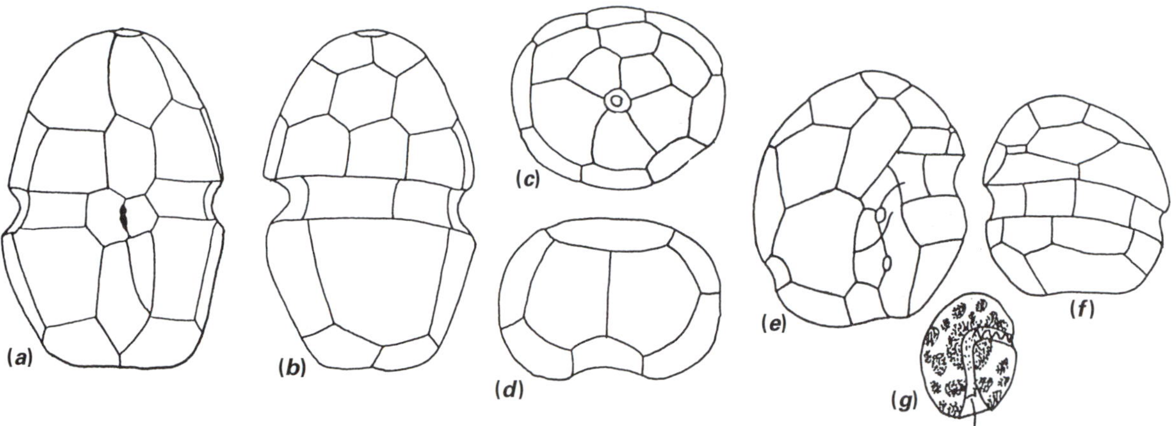

Fig. 7.11 *(a)–(d) Cachonina niei* showing the arrangement of thecal plates: *(a)* ventral, *(b)* dorsal, *(c)* apical, and *(d)* posterior views. *(e)–(g) Cryptothecodinium cohnii*: *(e)* ventral and *(f)* dorsal views; *(g)* living cell. (*(a)–(d)* after Loeblich, 1968; *(e)–(g)* after Chatton, 1952.)

intensity (to 1/12 of the original light intensity) the amount of PCP increases about seven fold, whereas the amount of chlorophyll *a* and peridinin not associated with PCP increases only 1.5 times. Little change occurs in chlorophyll *c* (Prézelin, 1976). This is a type of chromatic adaptation; as the cells receive less light (i.e., grow in deeper water), they produce more PCP, with the peridinin capturing light and passing it to chlorophyll *a*.

NUCLEUS

The organization of the dinophycean nucleus is unlike that of either eukaryotic or prokaryotic cells and has been referred to as **mesokaryotic** or **dinokaryotic** (Rizzo, 1991). A striking feature is the condensed state of the chromosomes, even during interphase (Fig. 7.1). The chemical composition of the chromosomes is also unusual. Eukaryotic chromosomes are for the most part composed of nucleohistone protein and DNA organized into structures called nucleosomes. The DNA of dinophycean chromosomes has very little nucleohistone; and is composed of 2.5-μm-wide fibrils. The histone that is present is not the same as the histone found in eukaryotes, and the chromatin is not organized into nucleosomes (Rizzo and Burghardt, 1982). The pattern of DNA synthesis has also been used as an indication of the level of advancement of the nuclear material in the Dinophyceae (Spector et al., 1981). In prokaryotic cells, a continuous synthesis of DNA occurs over a 24-hour period, whereas in eukaryotic cells the synthesis of DNA is discontinuous (confined to a part of the 24-hour period). In the Dinophyceae, a primitive dinoflagellate (*Prorocentrum*) (Fig. 7.43, 7.44(*b*),(*c*)) shows continuous synthesis of DNA (Filfilan and Sigee, 1977), whereas in the more complex *Cryptothecodinium* (Franker et al., 1974) and *Cachonina* (Fig. 7.11) (Loeblich, 1976) there is discontinuous DNA synthesis. Thus the Dinophyceae appear to be intermediate between

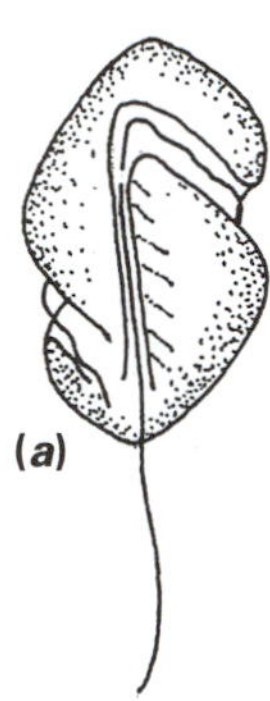

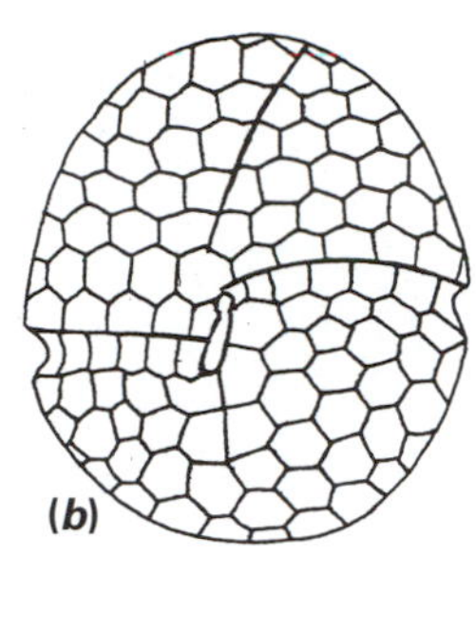

Fig. 7.12 (*a*) *Syndinium turbo*. (*b*) *Gymnodinium neglectum*. ((*a*) after Chatton, 1952; (*b*) after Schiller, 1933.)

eukaryotic and prokaryotic organisms, having some species with continuous, and others with discontinuous, DNA synthesis.

Unicellular eukaryotic organisms usually have between 0.046 and 3 picograms (pg) of DNA per nucleus. Dinoflagellates, however, usually have much more DNA in their nuclei, with values ranging from 3.8 pg per nucleus in *Cryptothecodinium cohnii* (Fig. 7.11(*e*)–(*g*)) to 200 pg per nucleus in *Gonyaulax polyedra* (Sigee, 1984). This implies that a large amount of the DNA is genetically inactive (structural DNA) in dinoflagellates.

Nuclear division shows the following characteristics (Barlow and Triemer, 1988): the nuclear envelope remains intact during division; the nucleolus persists and divides by pinching in two; and the chromosomes are attached to the nuclear envelope. There are several variant forms of mitosis in the Dinophyceae. In the more primitive organisms, such as *Syndinium* (Fig. 7.12(*a*)), tunnels of cytoplasm containing microtubules pass through the dividing nucleus (Fig. 7.13) (Ris and Kubai, 1974). In the more advanced organisms, such as *Oxyrrhis* (Fig. 7.44(*a*)) (Xiao-Ping and Jing-Yan, 1986), microtubules occur inside the dividing nucleus and not in the cytoplasm (Triemer, 1982). In the more primitive type of nuclear division (Fig. 7.13), the onset of division is usually marked by duplication of the flagellar bases from two to four. Throughout this stage the nucleus enlarges, and many Y- and V-shaped chromosomes are found. The nucleus now becomes invaginated, resulting in the formation of from 1 to 15 channels traversing the dividing nucleus (Fig. 7.13). These channels contain tunnels of cytoplasm passing through the nucleus outside of the nuclear membrane. There are bundles of microtubules in these channels, which are not connected to the intact nuclear envelope. The chromosomes are probably attached to the nuclear membrane or a specialized kinetochore in the nuclear membrane. At metaphase there is no formation of a metaphase plate as is common with eukaryotes, and the chromosomes are still scattered. The nucleolus persists throughout the whole nuclear cycle and divides by constricting in the middle. In anaphase the cell and nucleus expand laterally, and the chromosomes move to opposite ends of the

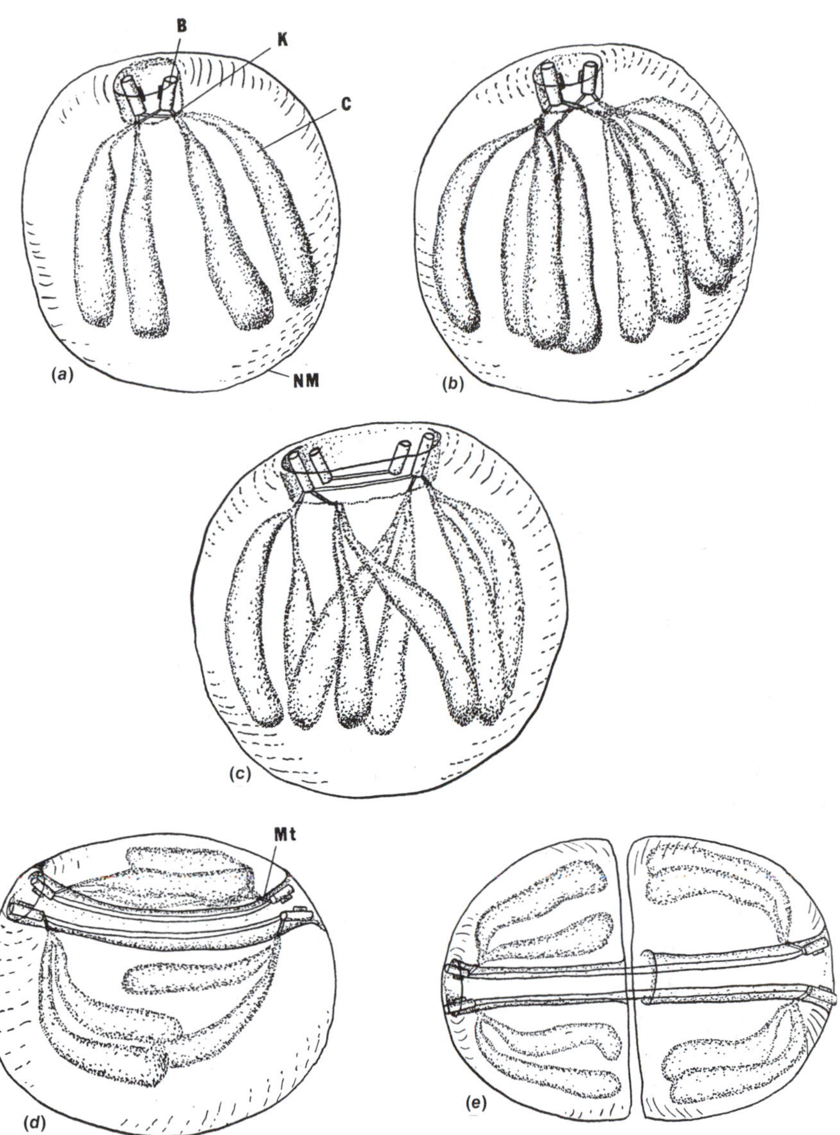

Fig. 7.13 Diagrammatic representation of nuclear division in *Syndinium* sp. (*a*) Interphase. (*b*) Early division; kinetochores and chromosomes have duplicated. (*c*) Early stage of chromosome segregation. Central spindle between separating basal bodies of flagella. (*d*) Late stage of chromosome separation. Central spindle in cytoplasmic channel through nucleus. (*e*) Division of nucleus. (B) Basal body of flagellum; (K) kinetochore; (C) chromosome; (Mt) microtubules; (NM) nuclear membrane. (After Ris and Kubai, 1974.)

nucleus. With continued lateral expansion, the central isthmus eventually severs, and the two daughter nuclei become independent (Leadbeater and Dodge, 1967b; Kubai and Ris, 1969; Ris and Kubai, 1974).

PROJECTILES

The Dinophyceae have a number of different projectiles, which are fired out of the cell when it is irritated, resulting in a sudden movement of the cell in the opposite direction from the discharge.

A **trichocyst** has a membrane-bounded, rod-shaped crystalline core

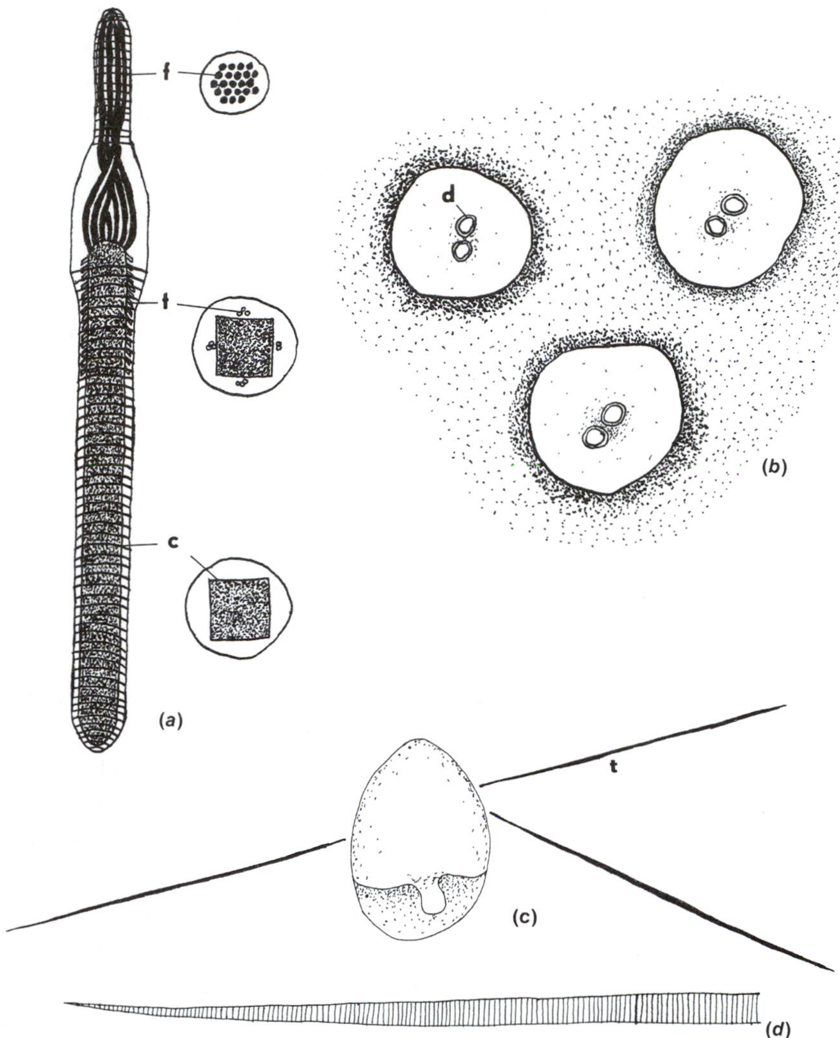

Fig. 7.14 (*a*) Diagram of a longitudinal section and cross sections through a charged trichocyst of *Gonyaulax polyedra*. The single membrane limiting the trichocyst is lined on its inner surface with fine hoops or spirals. Within the membrane is a crystalline core (c) composed of long rods or plates. Along the upper one-third of the core, short tubules (t) protrude downward and outward. At the anterior portion of the core, a series of fibers (f) attach to the core to still finer fibrils, which eventually reach the anterior portion of the enclosing membrane. (*b*) Drawing of a segment of a wall plate of *G. polyedra*. Porelike thin areas in the plate contain two to three slightly ridged discs (d) through which the trichocyst is discharged. (*c*) Cell of *Oxyrrhis* with discharged trichocysts. (*d*) Drawing of the tip of a discharged trichocyst showing striations. ((*a*),(*b*) after Bouck and Sweeney, 1966.)

(Bouck and Sweeney, 1966) (Figs. 7.1 and 7.14). Along the anterior one-third of the core are short, fine, tubular elements that project slightly downward. At the extreme outer end of the core, a group of 20 to 22 fibers extend from the outer part of the core to the enclosing membrane, and still finer fibrils then connect the larger fibrils to the apical portion of the trichocyst membrane. Just within the enclosing membrane are fine, threadlike, opaque hoops. The outer part of the trichocyst membrane is attached to the plasma membrane between the thecal vesicles or to the thecal vesicles beneath round, thin areas of thecal plates that form trichocyst pores. Trichocysts originate in areas rich in Golgi bodies and are probably derived from them initially as spherical vesicles that eventually become spindle-shaped and develop into the trichocysts. On irritation a "charged" trichocyst is converted to a "discharged" trichocyst in a few milliseconds, possibly by the rapid uptake of water. The discharged trichocysts are straight, tapering rods many times longer than the charged trichocyst (up to 200 μm long in *Prorocentrum*). The discharged trichocyst has transverse banding, with a major period of 50 to 80 nm. Although most dinoflagellates have trichocysts, ultrastructural investigations have shown that some do not (i.e., *Gymnodinium neglectum* (Fig. 7.12(*b*)), *Aureodinium pigmentosum, Woloszynskia tylota,* and the symbiotic *Symbiodinium microadriaticum*). The actual benefit of trichocysts to the cell (if any) is still obscure. They could be a mechanism for quick escape as the cells move sharply in the opposite direction of discharge, or they could be able to directly "spear" a naked intruder.

EYESPOTS

Less than 5% of the Dinophyceae contain eyespots, and those that do are mostly freshwater species; yet the eyespots are among the most complex in the algae.

The simplest type of eyespot consists of a collection of lipid globules in the cytoplasm not surrounded by a membrane (e.g., *Woloszynskia coronata*) (Dodge, 1971). A second type of eyespot consists of a row of lipid globules in a plastidlike structure at the cell periphery (e.g., *Peridinium westii, W. tenuissima*) (Messer and Ben-Shaul, 1969; Crawford et al., 1970) (Fig. 7.15).

The eyespot of *Glenodinium foliaceum* (Dodge and Crawford, 1969) is immediately under the anterior portion of the sulcus and is about 6 μm long and 3 μm wide (Fig. 7.16). It is more or less rectangular in outline with a hook-shaped projection at the anterior end. This flattened saclike structure contains two rows of large lipid globules separated by a granular space. Surrounding the eyespot is a triple-membrane envelope identical in

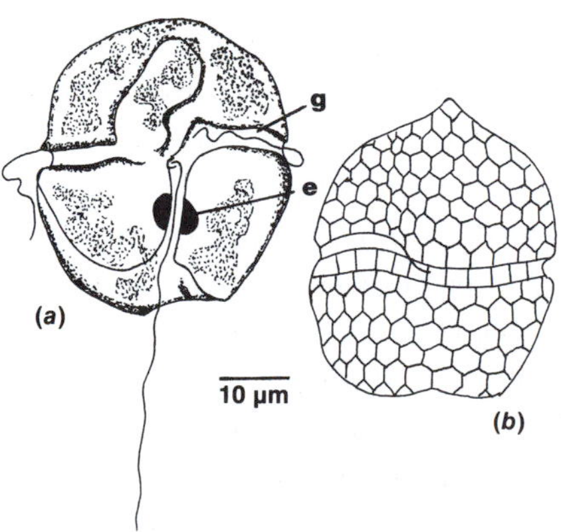

Fig. 7.15 *Woloszynskia tenuissima*. (*a*) Ventral view showing the two flagella, the girdle (g), and the eyespot (e). (b) Side view showing the thecal plates, which were not drawn in (*a*). (After Crawford et al., 1970.)

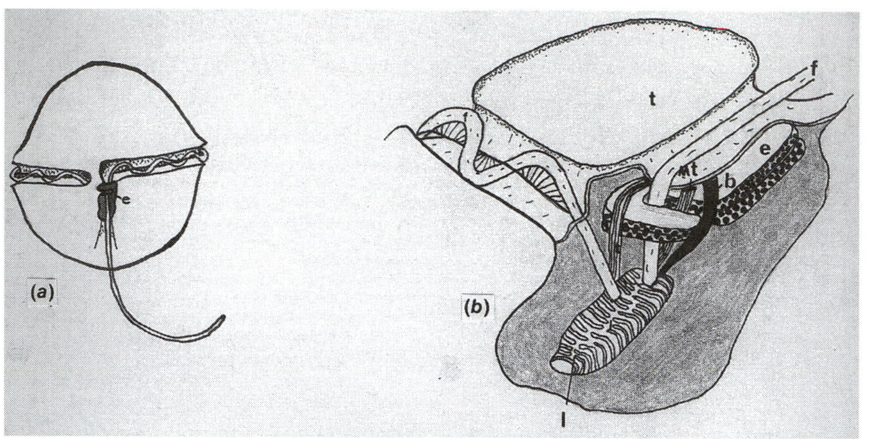

Fig. 7.16 (*a*) A ventral view of a cell of *Glenodinium foliaceum* showing the location of the eyespot (e). (*b*) Three-dimensional diagram of the eyespot area. The two flagella arise just above the lamellar body (l). The eyespot (e) lies beneath the sulcus. (Mt) Microtubular roots; (b) banded root; (t) thecal plate. (After Dodge and Crawford, 1969.)

appearance to that surrounding the chloroplasts. Adjacent to the eyespot is a non-membrane-bounded lamellar body consisting of a stack of flattened vesicles arranged more or less parallel to one another. The lamellar body is about 2 μm long and 0.75 μm wide, and contains up to 50 vesicles. The vesicles are connected at their edges, and also at the ends of the stack, with rough endoplasmic reticulum.

The most complex type of eyespot is found in the Warnowiaceae of the Peridiniales. The eyespot in *Nematodinium armatum* (Moronin and Francis, 1967) and that in *Erythropsis cornuta* (Gruet, 1965) have been studied at the fine-structural level and found to be essentially similar in construction (Figs. 7.17 and 7.18). In *N. armatum*, the eyespot is located toward the rear of the cell alongside the girdle, and consists of a **lens** mounted in front of a **pigment cup**, oriented so that the axis through the center of the lens and pigment cup is nearly perpendicular to the longitudinal axis of the cell body. The lens lies just below the plasmalemma. The pigment cup is made up of three main parts. Most of its wall consists of a single layer of large oblong pigment granules 0.3 μm in diameter. Toward the lip of the cup, the pigment granules are smaller and loosely arranged in several layers. Within the base of the cup is a dense layer of fibrils 33 nm in diameter, parallel to the axis of the eyespot; this layer is covered by a mat of transverse fibrils and has been named the **retinoid** because of its supposed light-receptive function. Above the retinoid is a canal that opens into a furrow of the transverse flagellum; the outermost membrane on the cell surface is continuous with the membrane of the canal. The lens unit is a complex structure completely surrounded by mitochondria. Inside the mitochondria is a granular layer separated by a membrane from the lens. The central core of the lens consists of membrane-bounded domed or concentric layers of dense material; the major bulk of the lens surrounds this

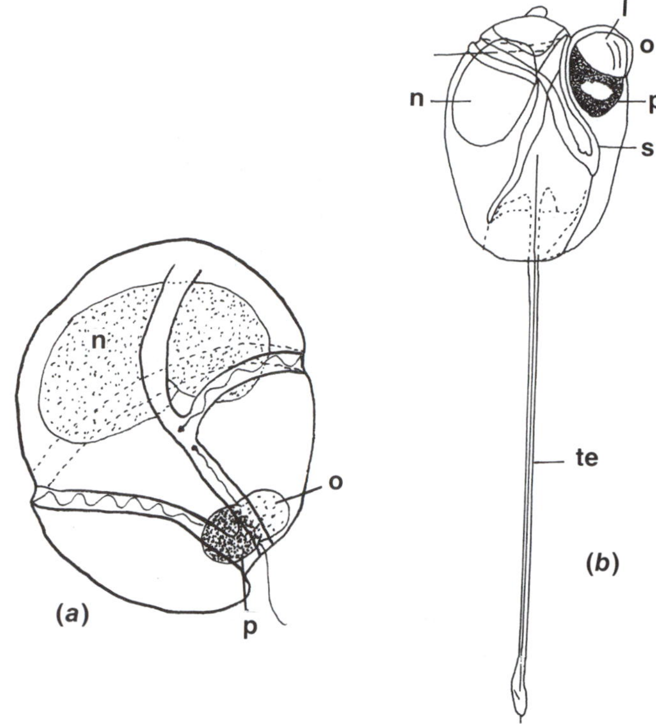

Fig. 7.17 (*a*) *Nematodinium armatum*. (*b*) *Erythropsis cornuta*. (l) Lens; (n) nucleus; (o) ocellus (eyespot); (p) pigment cup; (s) longitudinal sulcus; (te) tentacle.

core and consists of several large, nearly empty lobes. Between the core and the lobes is a network of vesicles of intermediate size.

Many of the Dinophyceae exhibit phototaxis, with the most effective wavelength of light in *Gonyaulax* and *Peridinium* being about 475 nm (blue) and that for *Prorocentrum* about 570 nm (green-yellow) (Halldal, 1958).

ACCUMULATION BODY

This is a large vesicle containing the remains of digested organelles (Fig. 7.41). It is probably similar to the Corps de Maupas of the Cryptophyceae and the digestive vesicles of other flagellates (Zhou and Fritz, 1994). An accumulation body is particularly common in symbiotic Dinophyceae.

Resting spores or cysts or hypnospores and fossil Dinophyceae

The **resting spore** or **cysts** of most dinoflagellates are morphologically distinct when separated from their parent thecae. They are 30 to 70 μm in diameter, smooth or spinose bodies whose cell walls are highly resistant to

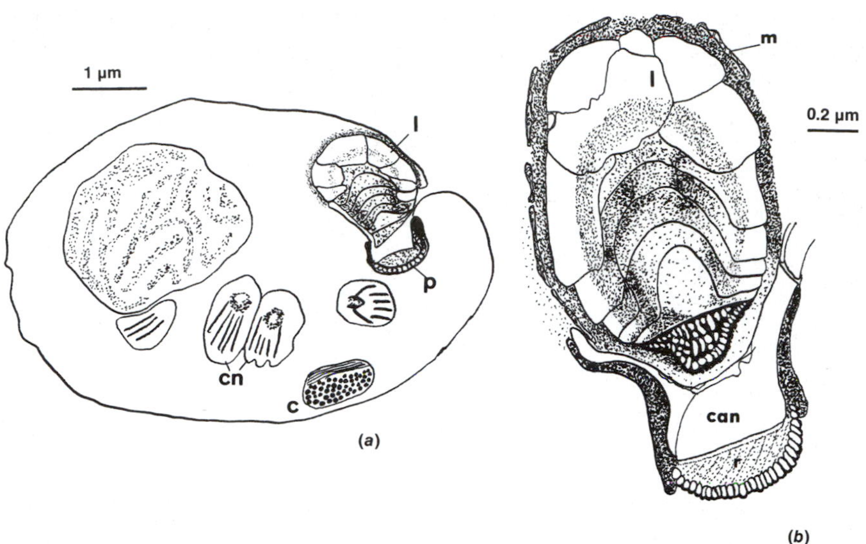

Fig. 7.18 (*a*) A cell of *Nematodinium armatum*, showing the lens (l), pigment cup (p), cnidocysts (cn), and a chloroplast (c). (*b*) A complete eyespot (ocellus) with retinoid (r), canal leading to outside (can), and mitochondria (m) surrounding the lens (l). (After Moronin and Francis, 1967.)

decay and contain chemicals similar to sporopollenin in the pollen of higher plants (Fig. 7.19). The newly formed cysts of *Scrippsiella trochoidea* have ten times more carbohydrate and 1.5%, the respiratory rate of vegetative cells (Brooks and Anderson, 1990).

The process of encystment or resting spore formation, is induced by nutrient deficiency, primarily nitrogen and phosphorus (Blanco, 1995; Chapman and Pfiester, 1995). Melatonin levels increase by several orders of magnitude during encystment and may function in preventing oxidation of the lipids in the cyst (Balzer and Hardeland, 1996). In the freshwater dinoflagellate *Woloszynskia tylota*, encystment involves the following changes (Bibby and Dodge, 1972): (1) replacement of the theca by a thin amorphous outer wall, which gradually thickens by the deposition of material on its inner face; (2) the appearance of a layer of closely packed lipid droplets at the cytoplasmic margin of the mature cyst; (3) the reduction in size or disappearance of cytoplasmic structures such as chloroplasts, Golgi bodies, and pusules; and (4) the enlargement of a central orange-brown body and cytoplasmic vacuoles containing crystals. Cysts are recognizable because they lack chromatophores and have a microgranular brown cytoplasm and a red eyespot (if the organism normally has an eyespot). Calcification of cysts in some genera occurs by the deposition of calcium carbonate crystals in the narrow space between the cell wall and the plasma membrane (Montresor et al., 1997). The cysts of *Ceratium hirundinella* contain an outer silicon layer (Chapman et al., 1982).

Fossilized resting spores, called **hystrichosphaerids** (**hystrichospores**) are included in fossilized cystlike structures of unicellular algae called **acritrachs**. These fossilized dinoflagellates first appeared in the Triassic and

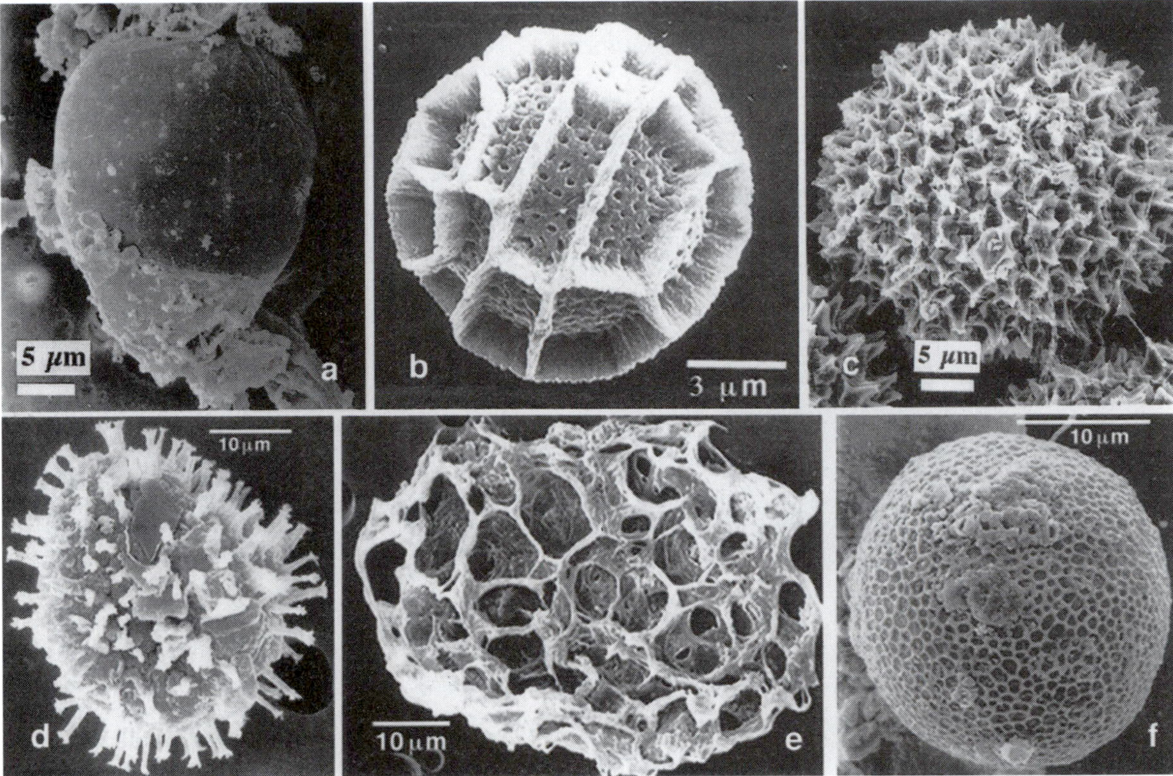

Fig. 7.19 Dinoflagellate cysts. (*a*) *Alexandrium catenella*. (*b*) *Calciodinellum operosum*. (*c*) *Scrippsiella trophoidea*. (*d*) *Gonyaulax grindleyi*. (*e*) *Polykrikos schwartzii*. (*f*) *Gymnodinium catenatum*. ((*d*)–(*f*) from Ellegaard et al., 1994; (*b*) from Montresor et al., 1997; (*a*),(*c*) from Meksumpun et al., 1994.)

reached a peak in the Jurassic and Cretaceous, followed by a decrease in the Tertiary (MacRae et al., 1996). Hystrichosphaerids were discovered independently by paleontologists and classified under a separate taxonomic scheme consisting of only fossil species. Many extant resting spores are identical to hystrichosphaerids of the Tertiary and Quarternary, with the result that there are two names for the same structure.

A small number of dinoflagellates produce siliceous skeletons, with the protoplasm wrapped around the skeleton. The best known species is the heterotrophic non-armored *Actiniscus pentasterias* (Fig. 7.20) (Hansen, 1993).

Toxins

Some Dinophyceae have the ability to produce very potent toxins which cause the death of fish and shellfish during **red tides** when there are dinoflagellate blooms that color the water red. The dinoflagellates become lodged in the gills of the shellfish, and when shellfish are eaten by humans or animals, poisoning results.

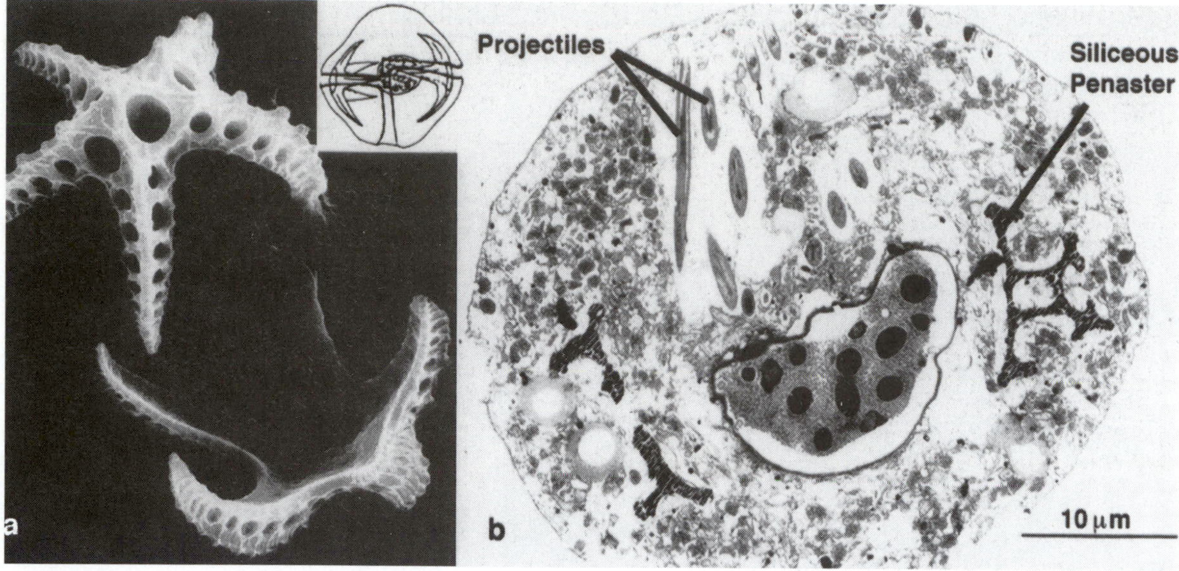

Fig. 7.20 *Actiniscus pentasterias.* (*a*) Scanning electron micrograph of the siliceous internal skeleton (two pentasters). (*b*) Transmission electron micrograph showing two pentasters surrounding the nucleus. (From Preisig, 1994; Hansen, 1993.)

Historically, red tides and paralytic shellfish poisoning have been mentioned many times (Shilo, 1967). One of the plagues that struck Egypt was described in the Bible: "all the waters that were in the river were turned to blood. And the fish that was in the river died; and the river stank, and the Egyptians could not drink the water of the river . . ." (Exodus 7:17). This description is strongly reminiscent of the poisonous red tides. Darwin, in his description of discolored water in 1832 during his voyage on the *Beagle*, graphically described blooms of algae that were dinoflagellates.

Death and illness caused by consumption of poisonous mussels and clams were reported by Captain Cook and Captain George Vancouver during their expeditions to the coast of the Pacific Northwest. An old custom among Indian tribes along the coast of Alaska was to station sentries to watch for the marine luminescence occurring during hot weather, which they understood to be associated with Kal-Ko-O, their name for mussel poisoning.

There are three basic types of poisoning that result from eating contaminated shellfish (Hallegraeff, 1993):

1 **Diarrhetic shellfish poisoning.** This occurs primarily in temperate regions and is caused by species of the plankton dinoflagellates *Dinophysis* and *Prorocentrum* (Fig. 7.21). Diarrhetic shellfish poisoning is caused by okadaic acid, macrolide toxins and yessotoxin (Fig. 1.29) that are produced in the chloroplasts of the dinoflagellates. The toxins are powerful inhibitors of protein phosphatases and induce severe

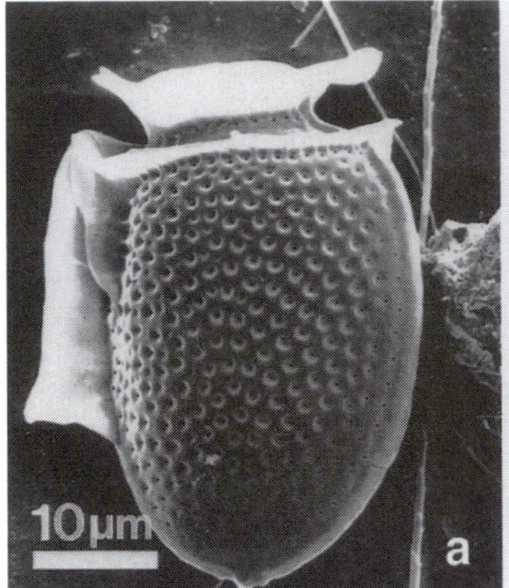

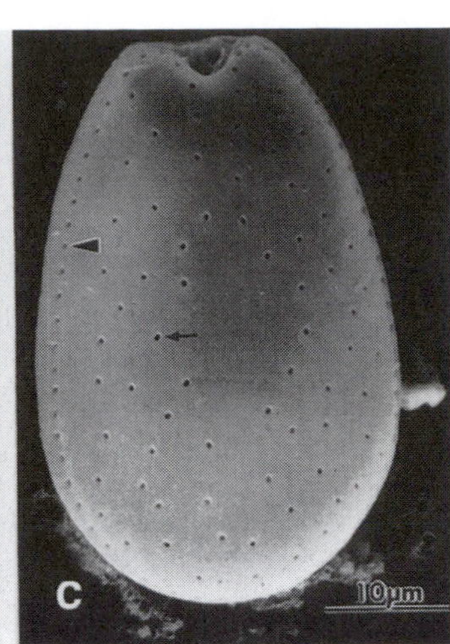

Fig. 7.21 Scanning electron micrographs of dinoflagellates that cause diarrhetic shellfish poisoning. (*a*) *Dinophysis acuminata*. (*b*) *Dinophysis fortii* (*c*) *Prorocentrum lima* with the arrows pointing to pores in the theca. (From Hallegraeff, 1993.)

gastroenteritis (Morton and Tindall, 1995; Suzuki et al., 1997; Zhou and Fritz, 1994).

2 **Ciguatera fish poisoning.** This occurs primarily in tropical regions with the common causative agent being *Gambierdiscus toxicus* (Fig. 7.22), originally described from the Gambiers Islands, South Australia, although it is common circumtropically between 32 °N and 32 °S. *G. toxicus* contains two potent toxins, ciguatoxins and maitotoxins. *Gambierdiscus* is epiphytic on macroalgae that are eaten by herbivorous fish and shellfish, which, in turn, are eaten by humans. In French Polynesia alone, approximately 1000 cases of ciguatera fish poisoning are reported every year (Chinain et al., 1997). The term ciguatera is derived from the Spanish term "cigua" for the turban shell, which was commonly eaten before the illness developed. The typical course of ciguatera fish poisoning is diarrhea for two days, followed by general weakness for one to two days. Occasionally the condition is fatal (Withers, 1982).

3 **Paralytic shellfish poisoning.** This is caused by species of *Alexandrium* (*A. catanella, A. acatenella, A. excavatum, A. tamarense*), *Pyrodinium bahamense* and *Gymnodinum catenatum* (Fig. 7.23). These dinoflagellates produce a group of toxins that are all derivatives of saxitoxin (Fig. 1.29). Saxitoxins have two-positively-charged guanidinium groups that bind to DNA and RNA, interrupting translation and transcription (Andersen and Cheng, 1988).

Alexandrium excavatum is a toxic red-tide dinoflagellate (Fig. 7.24). The vegetative cells divide to produce motile gametes that subsequently fuse. In the early stages of gamete fusion, when superficial contact is apparent, fusing pairs swim poorly and settle in the water. The motile quadriflagellate

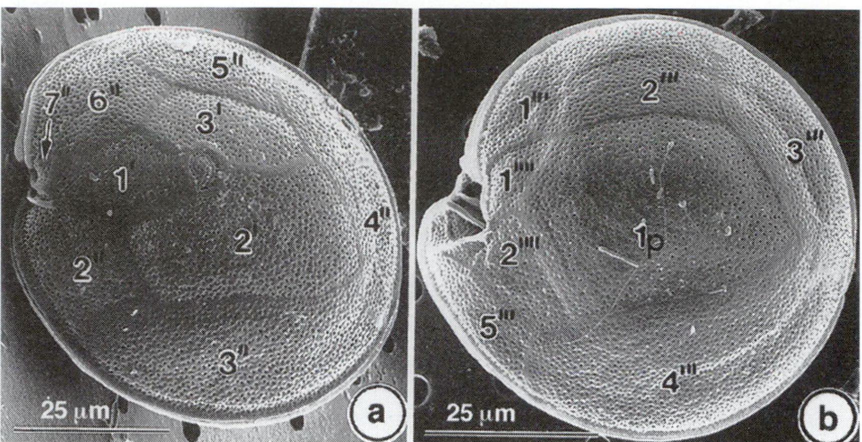

Fig. 7.22 Scanning electron micrographs of *Gambierdiscus toxicus*, the causative organism of ciguatera fish poisoning. (*a*) Epithecal view. (*b*) Hypothecal view. The numbers refer to the arrangements of plates. (From Faust, 1995.)

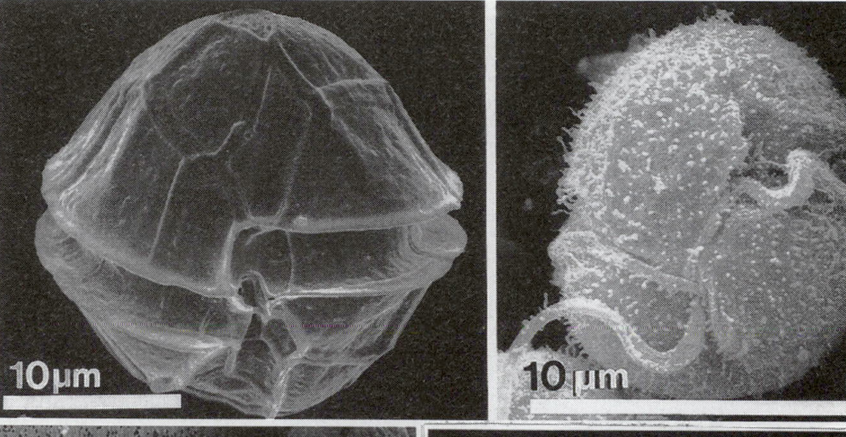

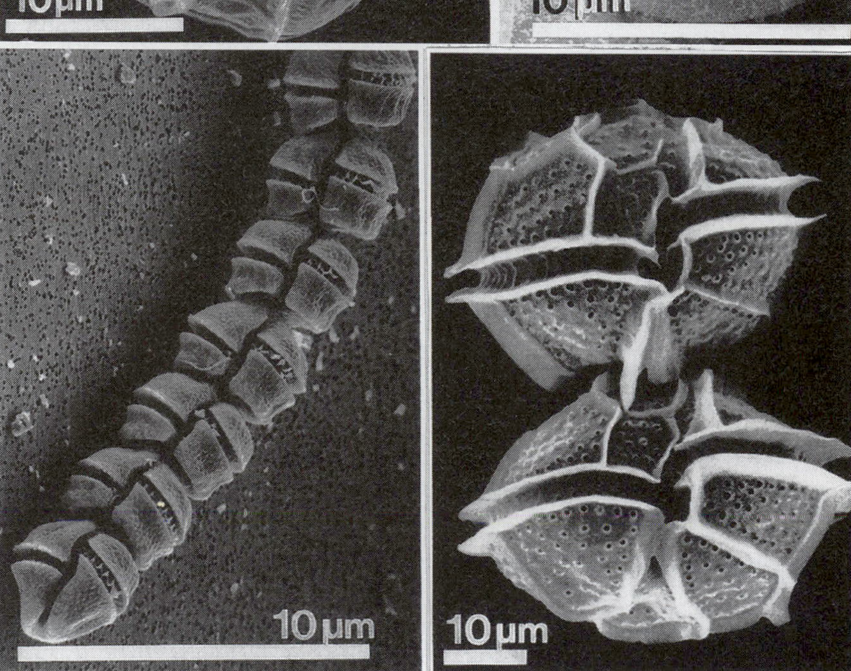

Fig. 7.23 Scanning electron micrographs of poisonous dinoflagellates. *Alexandrium minutum (upper left)*, *Gymnodinium galatheanum (upper right)*, *Gymnodinium catenatum (bottom left)*, *Pyrodinium bahamense* var. *compressum* (bottom right). *Gymnodinium galatheanum* has killed care-reared fish, while the others cause paralytic shellfish poisoning. (From Hallegraeff, 1993.)

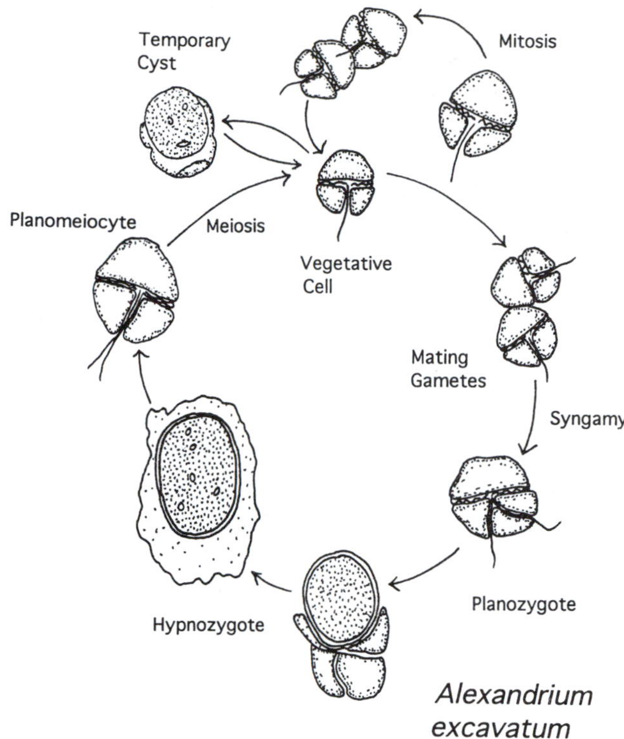

Fig. 7.24 The life cycle of *Alexandrium excavatum*. (Adapted from Destombe and Cembella, 1990.)

planozygotes swim for a few days before losing their flagella and thecal plates, and encyst to form resting cysts (hypnospores) (Fig. 7.19(*a*)) that can survive for 5 to 10 years (Scholin et al., 1995). Following a dormancy period, each hypnozygote undergoes meiosis with two cell divisions to produce haploid vegetative cells, completing the life cycle (Destombe and Cembella, 1990).

Dinoflagellates and oil and coal deposits

Blooms of dinoflagellates have most likely been responsible for some of the oil deposits of the world, including the North Sea oil deposit (Downie, 1956; Gallois, 1976). The best studied oil deposits are the oil shales of the Kimmeridge Clay deposits in England, which vary from less than 100 m in thickness to over 500 m and are composed of a number of different layers: clays and black and brown shales intermixed with occasional thin limestones. The limestones are composed primarily of coccolithophorids, whereas the shales contain yellow-brown organic matter called **kerogen**. Kerogen contains a large amount of amorphous organic matter, from which it is impossible to determine its biological origin. In addition to the amorphous organic matter, there are found large numbers of dinoflagellates and their hystrichospores, as well as some coccoliths. The best known Kimmeridge Clay deposit is the Kimmeridge Coal, a highly bituminous shale, about 80 cm thick. The Kimmeridge Coal has long been used as a solid fuel, and in the latter half of the last century much interest was taken in it as a source of oil. However, the high sulfur content and the relative thinness of the "coal" bed prevented the economic exploitation of the oil.

The Kimmeridge Clay oil shales were formed from algal blooms in seas that were to some degree land-locked, with salinity a little beneath the average salinity of the open ocean. These seas were rich in land-derived nutrients, allowing the water to support blooms of toxic dinoflagellates. These blooms deoxygenated and poisoned the water, providing the temporary anaerobic bottom conditions required for the preservation of organic matter. It is interesting that one of the bivalves commonly poisoned by toxic dinoflagellate blooms today is *Lucinoma borealis*, and the bivalve that forms the most extensive plasters in the Kimmeridge Clay oil shales is the related *Lucina miniscula*.

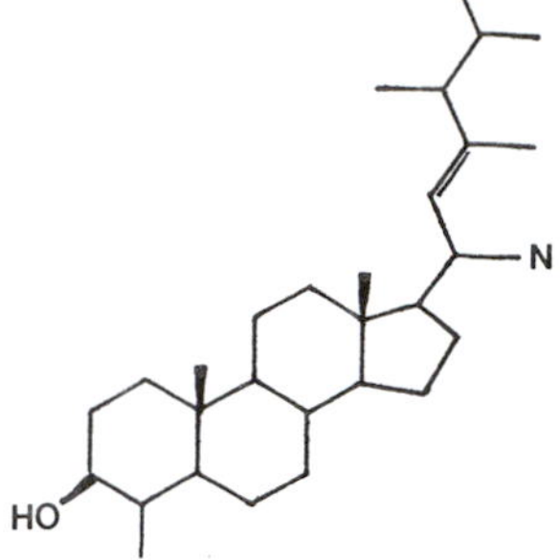

Fig. 7.25 The structure of dinosterol, the major sterol found in dinoflagellates.

Petroleum deposits and ancient sediments contain 4α-methylsteroidal hydrocarbons, which probably originated from 4α-methylsterols in dinoflagellates (Robinson et al., 1984). The dinoflagellates differ from other classes of algae with respect to the dominance of 4α-methylsterols among their sterol components and the uniqueness of certain of these 4α-methylsterol structures. The principal sterol of several marine dinoflagellates, and other organisms with dinoflagellate symbionts, is **dinosterol**) (Fig. 7.25).

A number of factors have been suggested as the cause of red tides:

1 **High surface-water temperatures:** Dinoflagellates favor warm water, and are generally more abundant near the surface. This does not necessarily mean that they occur only in warm seas, because the surface of the sea in normally cool areas may be warmed up during periods of hot, calm weather.
2 **Wind:** A strong, offshore wind aids upwelling, whereas a gentle onshore wind concentrates the bloom near the coast. On the other hand, heavy weather and strong winds disperse the bloom. Storms also result in the death of dinoflagellates and can prevent the development of red tides (Berdalet, 1992).
3 **Light intensity:** There is usually a period of bright, sunny, calm weather before outbreaks.
4 **Nutrients:** Red tides usually occur after an upwelling has stopped, but the nutrients brought to the surface do not, themselves, appear to be the direct cause of these blooms (Grindley and Nel, 1970). It is thought that preceding blooms of diatoms may impoverish the water and reduce one or more of the inorganic nutrients to a level favorable for the growth of

dinoflagellates (but too low for the diatoms), and also allow the production of organic nutrients such as vitamin B_{12}, which are important for their growth.

Red tides usually last longer than do blooms of non-toxic dinoflagellates because red tides are avoided by macrozooplankton that normally feed on dinoflagellates (Fiedler, 1982).

In the Gulf of Maine off the coast of New England (USA), there are different genotypes of the toxic *Alexandrium tamarensis* (Anderson and Keafer, 1987). Here paralytic shellfish poisoning occurs between the months of April and November at a time when the temperatures are favorable for growth of *A. tamarensis.* At the end of the growth period, *A. tamarensis* overwinters as cysts in the bottom sediments. Newly formed cysts have a mandatory 2 to 6 month dormancy period during which germination is not possible. There are two genotypes of *A. tamarensis,* one that occurs in **shallow coastal waters** and one that occurs in **deep coastal waters**. The genotype that lives in **shallow** waters has cysts that will germinate and produce motile, vegetative cells at the end of the dormant period, provided that the temperature is favorable and oxygen is available. The genotype that lives in **deep** coastal waters has germination of the bottom-dwelling cysts controlled by an endogenous clock. The cysts of this genotype are never exposed to favorable temperatures and available oxygen in these deep sediments and, therefore, would never germinate if they relied on the same sensing mechanism that the shallow water genotype uses. The cysts of the deep-water genotype germinate from January to August in these deep sediments, and the motile cells travel upward to the euphotic zone where they are able to grow and multiply. This germination is timed by an endogenous rhythm within the cell that has evolved to enable survival of *A. tamarensis* in deep coastal waters.

Bioluminescence

About, about, in reel and rout
The death-fires danced at night
The water, like a witch's oils
Burnt green, and blue, and white

Samuel Taylor Coleridge
The Rime of the Ancient Mariner

Mariners from early times have marveled at the displays of bioluminescence that accompany large populations of dinoflagellates. The burning seas were at first thought to be of supernatural origin, omens of the pleasure or displeasure of the gods. As science began to usurp the explanation of natural phenomena from religion, the light emitted from friction

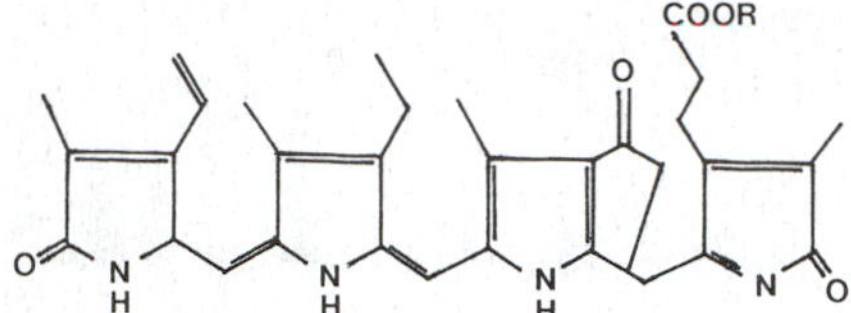

Fig. 7.26 A possible partial structure of dinoflagellate luciferin. (After Dunlap and Hastings 1981; Hastings, 1986.)

between salt molecules or from phosphorus burning in water was invoked; the term **phosphorescence** still survives today from the explanation. By 1800, living cells were suspected, but the last experiments were not settled in favor of a biological origin until 1830 (Sweeney, 1979).

There are two types of light emission in living organisms: (1) **bioluminescence** (**chemiluminescence**), in which *energy from an exergonic chemical reaction is transformed into light energy;* and (2) **photoluminescence**, which is dependent on the *prior absorption of light* (Hastings, 1986). Many marine, but no freshwater dinoflagellates are capable of bioluminescence. The Dinophyceae are the main contributors to marine bioluminescence, emitting a bluish-green (maximum wavelength at 474 nm) flash of light of 0.1-second duration when the cells are stimulated. The luminescent wake of a moving ship or the phosphorescence of tropical bays is usually caused primarily by Dinophyceae.

The compound responsible for bioluminescence is **luciferin** (Fig. 7.26(*b*)), which is oxidized with the aid of the enzyme **luciferase**, resulting in the emissions of light. Luciferin and luciferase are terms for a general class of compounds, and not of a specific chemical structure. Bioluminescence occurs in many organisms in many different phyla, ranging from bacteria to dinoflagellates to jellyfish and brittle stars to worms, fireflies, molluscs, and fish (Hastings, 1986). In bacteria, luciferin is a reduced flavin; in insects it is a (benzo)thiazole nucleus; and in dinoflagellates it is a tetrapyrrole. Likewise, luciferase has different structures, although all luciferases share the feature of being **oxygenases** (enzymes that add oxygen to compounds). The necessity for oxygen in bioluminescence was actually discovered by Boyle in 1667, who showed with his air (vacuum) pump that:

> a piece of shining (bioluminescent) wood . . . gave a vivid light . . . which was manifestly lessened . . . [at] about the seventh suck, losing its light more and more as the air was still further pumped out. . . . Wherefore we let in outward air and had the pleasure to see the seemingly extinguished light revive so fast and perfectly, that it looked to us almost like a little flash of lightening. (Cited in Harvey, 1952, p. 142).

In the basic reaction of bioluminescence (Hastings, 1983), a luciferin is oxidized by a luciferase, resulting in an electronically excited product (P)* which emits a photon (hν) on decomposition:

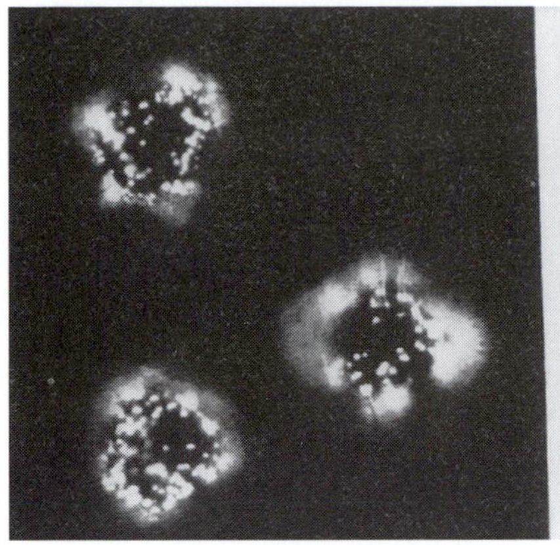
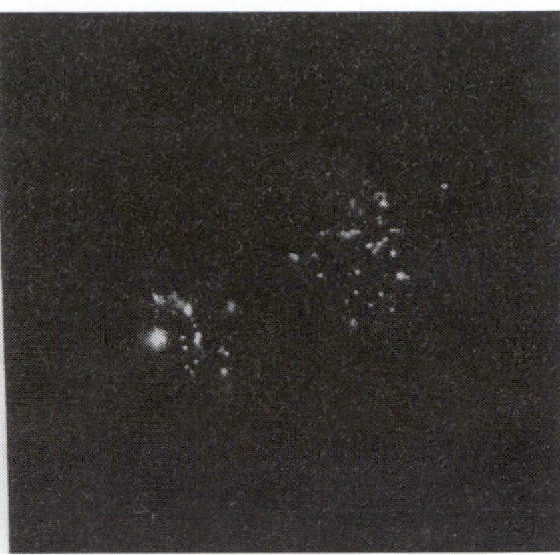

Fig. 7.27 The white spots in these photographs of *Gonyaulax* cells indicate fluorescence and thus localization of luciferin, the substrate that reacts with the enzyme luciferase and oxygen to produce light. The intensity of the luciferin fluorescence is greater in the night phase (*left*) of the circadian cycle of bioluminescence than in the day phase (*right*). The amount of luciferase shows a similar oscillation during the daily cycle. (From Johnson et al., 1985.)

$$\text{luciferin} + O_2 \xrightarrow{\text{luciferase}} (\text{P})^* \rightarrow \text{P} + h\nu$$

Luciferin in the dinoflagellates has been reported to be a linear tetrapyrrole (Fig. 7.26). Associated with dinoflagellate luciferin is a **luciferin-binding protein (LBP)** which sequesters luciferin at alkaline pH and releases it under acidic conditions (Sulzman et al., 1978). It has been postulated that the flash of bioluminescent light may occur simply by a lowering of the pH from 8.0 to 6.5. Agitation of cells depolarizes the vacuolar membrane, allowing a flux of protons (H^+) and acidification of the peripheral cytoplasm (Johnson et al., 1985). Lowering the pH causes two pH-dependent reactions to occur: (1) release of luciferin from its binding protein at acidic pH, and (2) activation of luciferase followed by emission of a photon of blue-green light.

Luciferin, luciferase and luciferase-binding protein occur in particles called **scintillons (flashing units)** that are approximately 0.5 to 1.5 μm in diameter (Sweeney, 1980). Scintillations occur in cytoplasmic invaginations in the vacuolar membrane. Flashes of blue-green light are produced when an action potential proceeds along the vacuolar membrane, causing an outflow of protons from the acidic contents of the vacuole. The resulting drop in the pH in the scintillon causes flashes of light (Nicolas et al., 1991).

In *Gonyaulax polyedra*, most bioluminescence occurs during the night phase of a circadian rhythm (Fig. 7.27) (Fritz et al., 1991). This is due to a tenfold increase in luciferase and luciferin-binding protein during the night phase. Daytime photoinhibition of bioluminescence is considered a mechanism to conserve the energy of the cells when ambient light levels are high enough to render the bioluminescent flash ineffective (Buskey and Swift, 1983). The position of bioluminescence in *Pyrocystis fusiformis* (Fig. 7.28(*a*)) varies over 24 hours (Sweeney, 1980). During the day bioluminescence emanates from a spherical mass of tightly packed vesicles

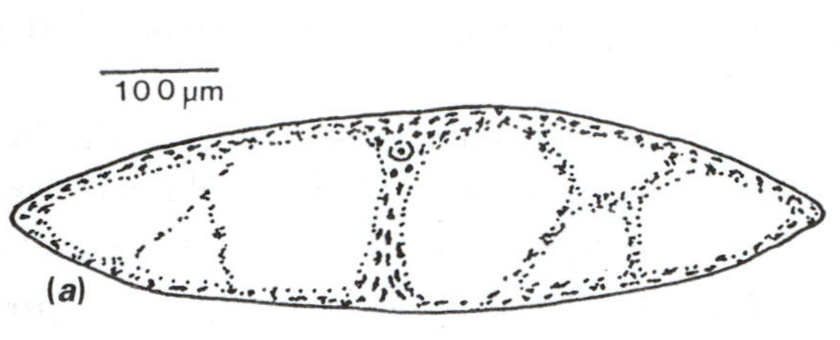

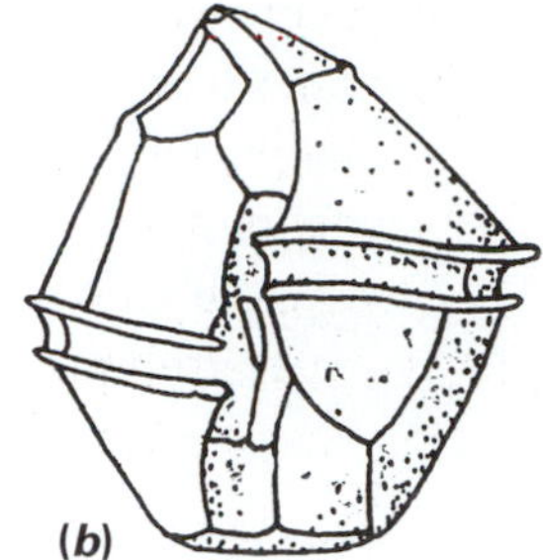

Fig. 7.28 (*a*) *Pyrocystis fusiformis*. (*b*) *Gonyaulax polyedra* ((*b*) after Chatton, 1952.)

near the nucleus, whereas at night bioluminescence occurs in the peripheral cytoplasm. There is a reverse movement of chloroplasts, with the organelles in the cell periphery during the day and the chloroplasts concentrating around the nucleus at night.

Dinoflagellates can emit light in three modes: (1) They can flash when stimulated mechanically, chemically, or electrically; (2) they can flash spontaneously; and (3) late at night they can glow dimly (Sweeney, 1979). The maximum amount of light emitted in a flash differs widely among species, with larger species emitting more light per flash than smaller ones. In a population of dinoflagellates, the cells emit an average of one flash per cell per day (Hastings and Krasnow, 1981). It is not clear whether each cell flashes once and only once during this period, or whether some cells are responsible for a larger share of the flashes whereas others do not emit at all. The nutritional status of a cell influences the brightness of the flash. *Noctiluca* with green algal symbionts (Fig. 7.42) shows an increase in the photons emitted per flash as the intensity of the illumination of the dinoflagellate, and therefore photosynthesis, increases. Different isolates of the same species in genera such as *Dissodinium* and *Pyrocystis* can be bioluminescent or not (Swift et al., 1973; Schmidt et al., 1978).

Pyrocystis (Fig. 7.28(*a*)), one of the most strongly phosphorescent dinoflagellates, is the chief source of diffused phosphorescence in the sea in equatorial regions (Swift et al., 1973). Species of *Pyrocystis* produce at least 1000 times more bioluminescence per cell than members of the genus *Gonyaulax* and about 100 times as much light per cell as *Ceratium fusus, Peridinium pentagonium*, and *Pyrodinium bahamense.*

There are two theories concerning the adaptive value of bioluminescence to dinoflagellates, both of which relate to nighttime grazing of the dinoflagellates:

1 "Burglar alarm" hypothesis. This hypothesis argues that dinoflagellates render themselves dangerous as prey to invertebrate grazers because they generate a signal identifying the location of invertebrate food to individuals two levels up the food chain from the dinoflagellates

(Abrahams and Townsend, 1993). Bioluminescence generated by dinoflagellates serves to attract predators of the grazers of the dinoflagellates.

2 "Startle" hypothesis. In this hypothesis, mechanical stimulation of a bioluminescent dinoflagellate by a grazer produces a flash of light that startles an invertebrate grazer, such as a copepod, and causes the copepod to swim away with its feeding appendage retracted (Buskey and Swift, 1983).

Whichever theory is correct, experiments have shown that copepods consume only half as many dinoflagellates at night, indicating that the bioluminescence is serving as a deterrent to grazing (Buskey et al., 1983).

Rhythms

Many Dinophyceae exhibit rhythmic processes, with the best-known of the rhythms in the algae being those in the dinoflagellate *Gonyaulax polyedra* (Sweeney, 1969). It produces light via bioluminescence by forming its own dinoflagellate-specific luciferin and luciferase. The cells emit a flash of light when the seawater in which they are swimming is given a sharp shake or stirred vigorously. When the luminescence is measured while the culture is being stirred, the amount of light that the cells emit is not always the same and depends upon their recent history (Fig. 7.27). If they have been growing in natural illumination or on a light–dark cycle, the amount of light emitted will be markedly dependent on the time of day when measurements are made. If luminescence is elicited during the day, the amount of light will be low, and very hard stirring will be required to bring about any luminescence at all. But if the cells are stimulated at night, a great deal more light will result, and the most delicate shake will produce a flash. When the amount of luminescence is plotted as a function of the time of day, a graph such as that in Fig. 7.29(*a*), will result. The greatest luminescence will be produced in the middle of the dark period, whereas toward morning, flashes will gradually become smaller and a greater stimulus will be required. The rhythm is circadian [which means literally about (*circa*) a day (*diem*)], as shown by the persistence of changes in brightness of luminescence when the cells are kept in the darkness (Fig. 7.29(*b*)). In continuous darkness the cycles continue for as long as 4 days, but the amplitude becomes successively smaller. If the cells are kept in continuous light, the reduction in amplitude is no longer evident, although all maxima are somewhat lower than in light–dark. Cycles continue for at least 3 weeks in continuous light of appropriate intensity (Fig. 7.30).

When photosynthesis in *Gonyaulax* is measured, either as oxygen production or carbon dioxide fixation, it is also found to be rhythmic. The

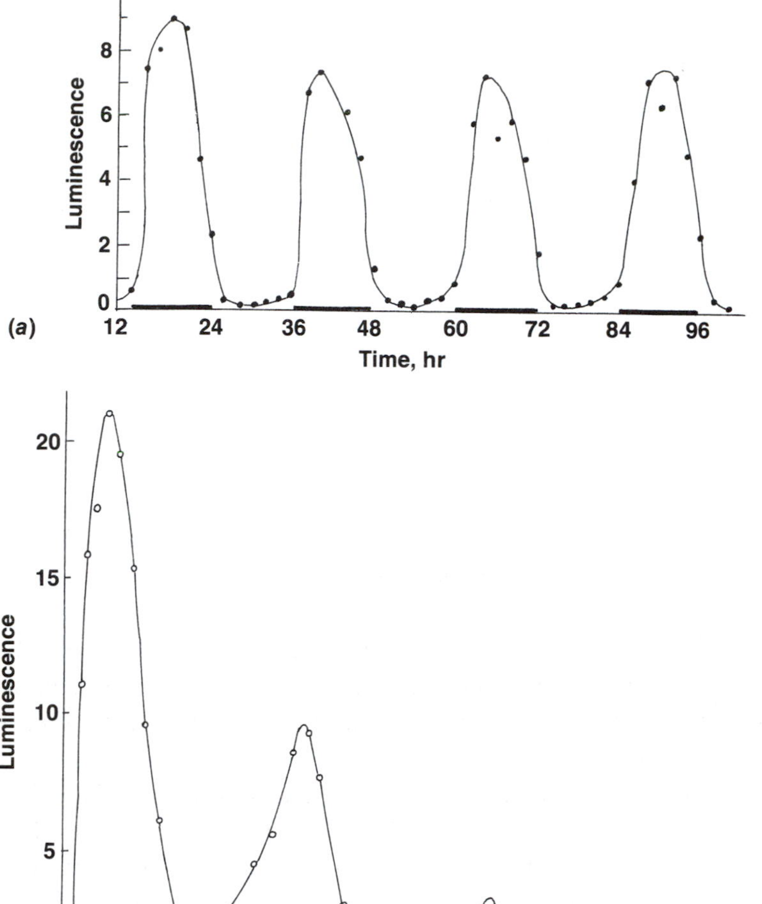

Fig. 7.29 The rhythm of stimulated luminescence in *Gonyaulax polyedra* measured in a light–dark cycle of 12:12 (*a*) and in complete dark (*b*). (After Sweeney, 1969.)

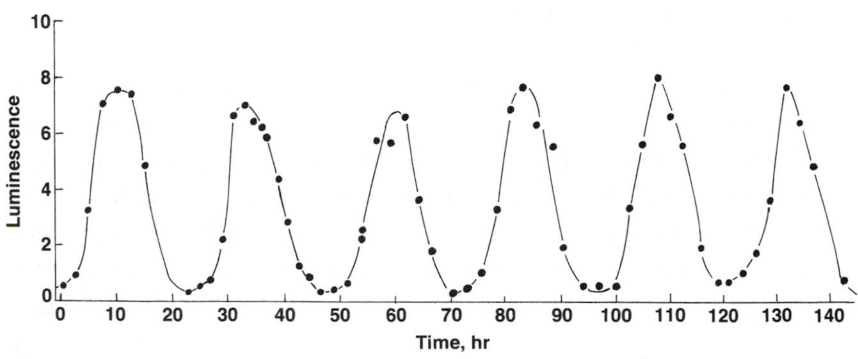

Fig. 7.30 The rhythm of stimulated luminescence in *Gonyaulax polyedra* measured in continuous light (1000 lux). (After Sweeney, 1969.)

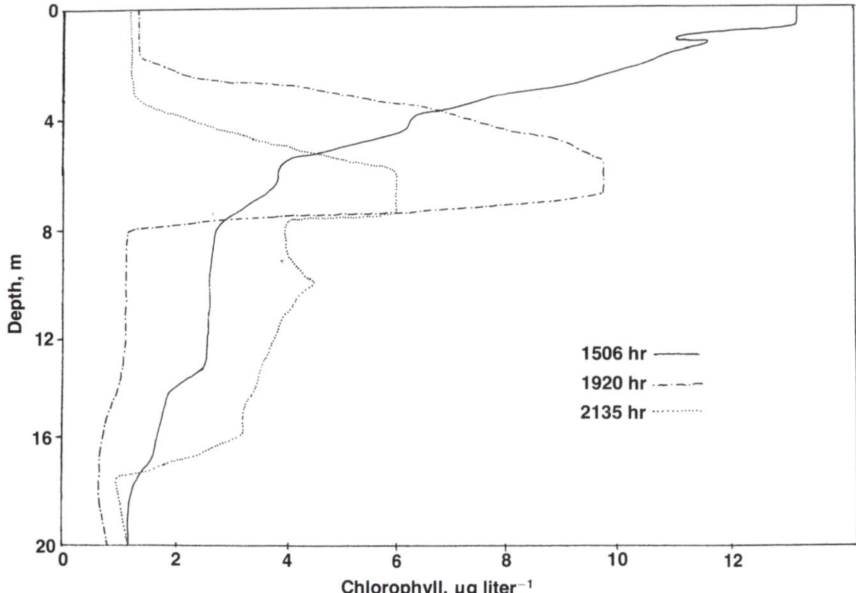

Fig. 7.31 Profiles of chlorophyll *a* during a bloom of dinoflagellates showing the vertical migration of dinoflagellates at different times of the day. (After Eppley et al., 1968.)

rhythm is circadian and continues under conditions of continuous light. The maximum rate of photosynthesis occurs, as one would expect, in the middle of the day. The rhythm in photosynthesis is due to changes in photosystem II (Samuelsson et al., 1983).

A third rhythm with a circadian period in *Gonyaulax* is that of cell division; all cell division occurs during 30 minutes when cultures are in a light–dark cycle. When the light–dark cycle is 12:12, then this 30 minutes spans "dawn." An investigation of other light–dark cycles, such as 7:7, shows that the dark–light transition is not the determining factor because division takes place about 12 hours after the beginning of the dark period, even though this time is considerably after the beginning of the next light period. In continuous light of low intensity where other rhythms of *Gonyaulax* persist, very little cell division takes place, and the average generation time may be as long as 6 days. However, those cells that are ready to divide do so only at the expected time in each 24 hours.

A fourth type of rhythm involves the vertical migration of dinoflagellate cells in the water column (Eppley et al., 1968; Horiguchi and Pienaar, 1988; Lombard and Capon, 1971, Roenneberg and Deng, 1997). Before dawn, the cells rise to the surface, where they form dense clouds (aggregations), and before night fall, they again sink to lower depths (Fig. 7.31). In the marine environment, this vertical migration exposes the cells to several gradients: (1) Nutrients are more concentrated at lower depths (accumulating at the bottom of the ocean or at thermoclines) while surface waters are often prac-

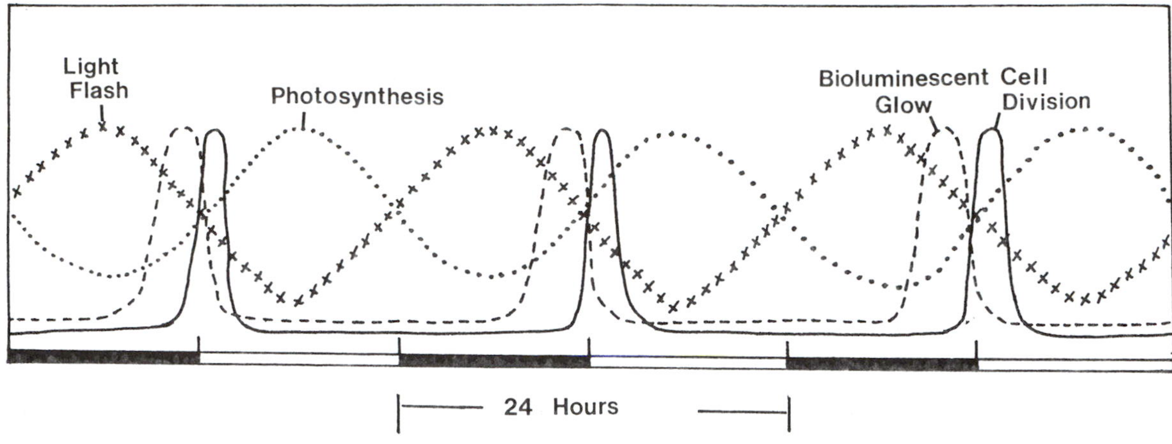

Fig. 7.32 Circadian rhythmicity has been identified in at least four distinct biological processes of the unicellular alga *Gonyaulax*. Although the four rhythms peak at different phases of the 24-hour day, as represented here diagrammatically, they are all synchronized with the circadian clock. Perturbations that reset the clock will phase-shift all four rhythms similarly, suggesting that all of these overt rhythms are driven by a single pacemaker. (After Johnson and Hastings, 1986.)

tically devoid of nutrients. (2) Temperatures at the surface exceed those in deeper waters. (3) Variations in light intensities. (4) In shallow waters, differences in washout by the tidal waters.

In *Gonyaulax,* there exists a control over luminescence, photosynthesis, and cell division, so that each process reaches a maximum and then declines in an orderly fashion during each 24 hours (Fig. 7.32). A single process, a biological clock, may control all of these processes. It appears that the part of the cell that may be the controlling agent is the plasma membrane because there is a rhythmic reorganization of the plasma membrane over a 24-hour period in synchronized cells (Adamich et al., 1976).

Although the phases at which the rhythms peak in *Gonyaulax* are different, it is generally believed that they are all controlled by a single pacemaker (Johnson and Hastings, 1986). The rhythmic processes seem to have no feedback mechanism and thus appear to be driven systems. For example, photosynthesis can be completely inhibited with a specific herbicide, but the bioluminescence rhythm continues and its phase is not altered. Such observations on *Gonyaulax* suggest models of circadian organization like the diagram in Fig. 7.33. The central pacemaker is phased to the solar light cycle by means of a photoreceptor and an entraining pathway, and it controls the expression of the different rhythms, such as cell division, bioluminescence, and photosynthesis.

Two different systems time the circadian rhythm in dinoflagellates, a red-light sensitive system that delays the timing, and a blue-light sensitive system that advances the timing to dawn (Roenneberg and Deng, 1997). These light systems probably stimulate or depress the production of melatonin in the cells. An increase in the concentration of melatonin appears to mark the end of the light phase. There is a circadian rhythm in the production of melatonin, with a rapid increase in melatonin concentration at the

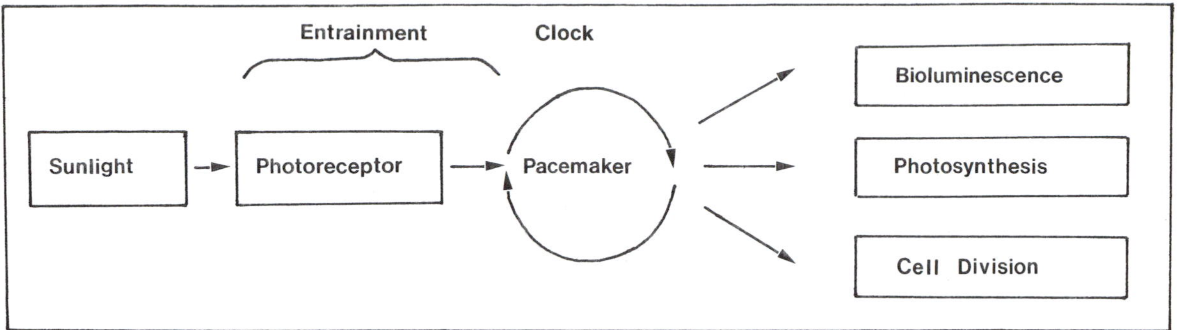

Fig. 7.33 Diagram showing how a single endogenous clock within a *Gonyaulax* cell can control a multiplicity of observed rhythms. (After Johnson and Hastings, 1986.)

end of the light phase and a subsequent decline during the dark phase, reaching a minimal value in the light phase. The concentration of melatonin in dinoflagellate cells is similar to that in the mammalian pineal gland (Balzer and Hardeland, 1996). Melatonin may represent a common basis for photoperiodism in organisms as distant as dinoflagellates and vertebrates, a fact that suggests an ancient mechanism for mediating information about darkness within the circadian cycle (Pöggeler et al., 1991).

Heterotrophic dinoflagellates

Heterotrophic dinoflagellates constitute roughly half of the more than 2000 living dinoflagellate species (Jacobsen and Anderson, 1992). The heterotrophic dinoflagellates are usually phagotrophic with solid particles being ingested into food vesicles, which are broken down into constituent molecules, that are absorbed into the cytoplasm. Phagocytosis involves specialized feeding mechanisms. There are three basic feeding mechanisms (Skovgaard, 1996):

DIRECT ENGULFMENT OF PREY

This occurs in dinoflagellaes that lack a theca.

The large cells of *Noctiluca* have a 300-μm-long food-gathering **tentacle** that is covered with a slimy exudate (Fig. 7.34) (Sweeney, 1978; Nawata and Sibaoka, 1983). Two winglike extensions of the cells form an **oral pouch** at the base of the tentacle. At the bottom of the oral pouch is a **cytosome** that opens like a slit during ingestion of food organisms. Collision of other organisms in the medium with the mucus-covered tentacle tip results in attachment of the other organisms to the tentacle. Continued collisions result in the formation of a clump of cells adhering to the tentacle. The tentacle is primarily in an extended configuration during this time. The tent-

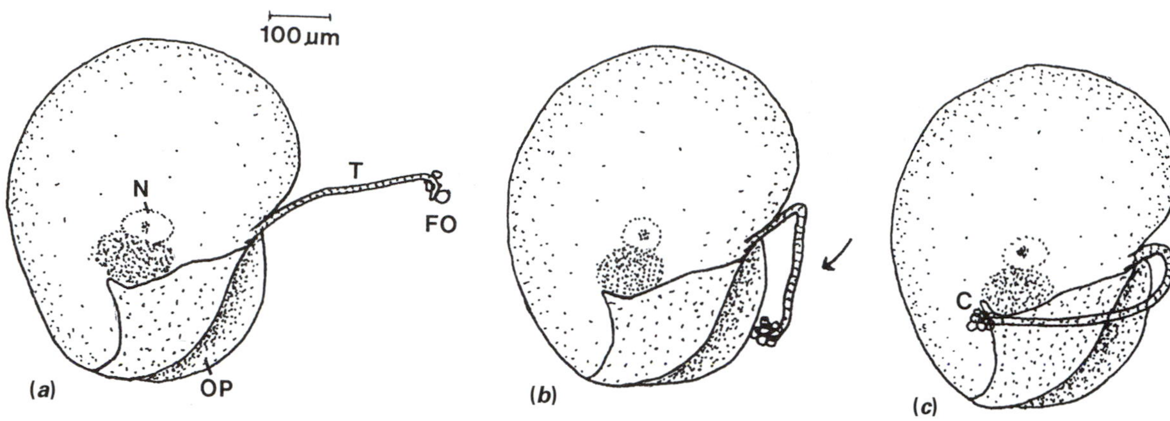

Fig. 7.34 Drawing of the ingestion of food organisms (other algae, bacteria) by *Noctiluca*. (*a*) The tentacle (T) is in an extended configuration. Any food organisms (FO) that collide with the mucus-covered tentacle tip, stick to the tentacle. (N) Nucleus; (OP) oral pouch. (*b*) The tentacle bends back toward the oral pouch. (*c*) The cytosome (C) at the base of the oral pouch opens, the tentacle tip is inserted into the cytosome, and the food organisms are swept into a food vacuole. (After Nawata and Sibaoka, 1983.)

acle now bends so that the tentacle tip moves to the bottom of the oral pouch. The cytoplasm streams vigorously toward the cytosome and then aggregates around the cytosome (Nawata and Sibaoka, 1987). The cytosome opens, the tentacle tip is inserted into it, the food organisms are swept into the cytosome, and the tentacle tip is retracted. The cytosome opening closes, and the food organisms are engulfed in a food vesicle in the cytoplasm. The food organisms are digested in the food vesicle. In the cell under the microscope, long strands of mucus containing food organisms can be seen spirally coiled in the food vacuoles.

PALLIUM FEEDING

This occurs only in thecate species and utilizes a feeding veil, the **pallium**, which emerges from the flagellar pore and encloses the prey. The prey protoplasm is digested by enzymes released into the pallium, and the digestion products are transported into the feeding cell. *Protoperidinium* (Fig. 7.35), *Oblea* and *Zygabikodinium* feed in this manner. These dinoflagellates swim in a straight line until they encounter a prey organism. The dinoflagellate then changes swimming behavior by slowing down and swimming in tight circles around the prey for less than one minute. A thin filament of cytoplasm (about 1 μm in diameter) emerges from the sulcal pore and attaches to the prey (Fig. 7.35). A pseudopod is extended along the filament while the filament is retracted, pulling the prey closer to the dinoflagellate. The pseudopod advances at 2 to 6 $\mu m\ s^{-1}$ until it reaches the prey. The pseudopod conforms to projections, such as spines, and can engulf an organism many times larger than the dinoflagellate. The prey is digested in the pseudopod, with the pseudopod showing active cytoplasmic streaming at a velocity of about 5 $\mu m\ s^{-1}$. A diatom is digested by

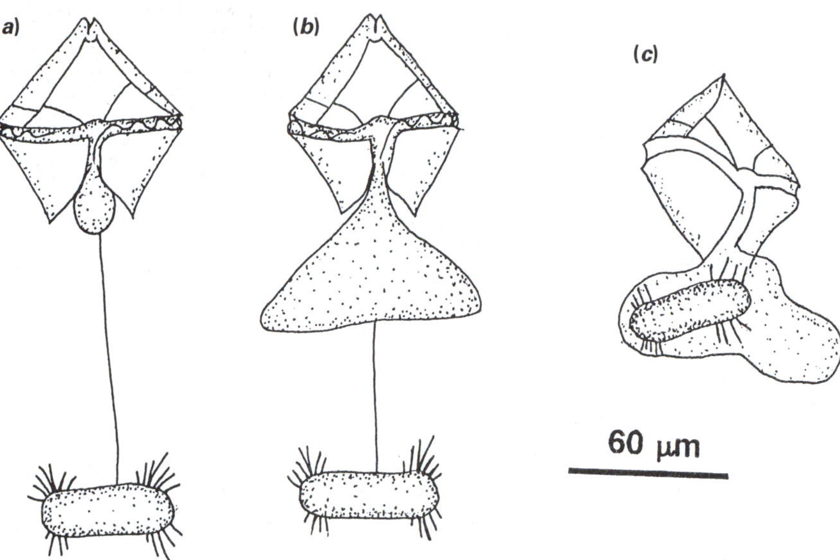

Fig. 7.35 The heterotrophic dinoflagellate *Protoperidinium conicum* feeding on the diatom *Corethron hystria*. Initially the dinoflagellate attaches to the prey by a long thin filament (*a*). Next a pseudopod extends along the filament (*b*) and engulfs the prey (*c*) which is digested. (After Jacobsen and Anderson, 1986.)

the dinoflagellate in about 30 minutes, at which time the pseudopod is rapidly retracted (at a rate of 10 $\mu m\ s^{-1}$) into the dinoflagellate.

PEDUNCLE FEEDING

Gymnodinium fungiforme contains an extensible **peduncle** (Lee, 1977), a projection of cytoplasm full of microtubules, in the epicone just above the intersection of the sulcus and cingulum (Fig. 7.36) (Spero and Moree, 1981). The peduncle can extend from 8 to 12 μm to attach to, and make a hole into the prey (Spero, 1982). The cytoplasm of the prey moves through the peduncle to the dinoflagellate cytoplasm. After feeding is complete, the microtubules of the peduncle, and the peduncle itself, retract back into the dinoflagellate cytoplasm. Feeding is characterized by the aggregation of numbers of the dinoflagellate on the prey organism. A small green alga, such as *Dunaliella*, has five to ten dinoflagellates attached by their peduncles, whereas a ciliate may have 400 to 500 dinoflagellates attached to it. The peduncle of *G. fungiforme* apparently cannot penetrate a cell wall, so the dinoflagellate feeds only on prey that lacks a cell wall, or on injured higher organisms. The marine ciliate *Condylostoma magnum* is not fed on by *G. fungiforme* until it is injured (in the laboratory by piercing it with a micropipette). Within 15 seconds, more than 100 dinoflagellates will congregate around the wound. When a ciliate is taken from a dying culture and placed in a *G. fungiforme* culture, the dinoflagellate cells congregate around the posterior end of the ciliate. Initially small strands of ciliate cytoplasm

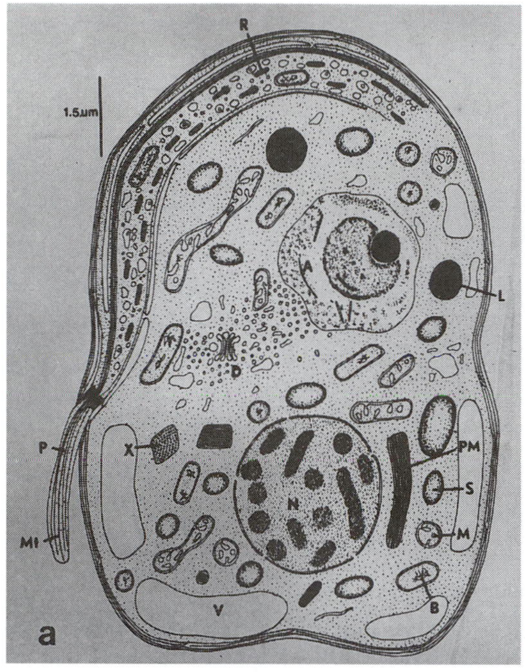

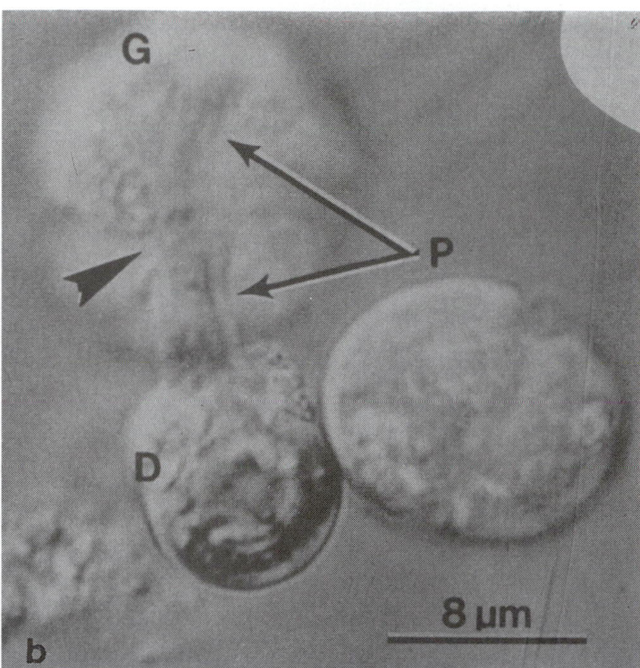

Fig. 7.36 (*a*) A semi-diagrammatic drawing of the fine structure of a cell of the saprophytic isolate of *Gyrodinium lebouriae*. (A) Accumulation body; (B) endosymbiotic bacterium; (D) dictyosome; (L) lipid; (M) mitochondrion; (Mt) microtubules; (N) nucleus; (P) peduncle; (PM) promastigonemes; (R) rod-shaped body; (S) starch; (X) crystalline body. (*b*) Light micrograph of the dinoflagellate *Gymnodinium fungiforme* (G) ingesting the protoplasm of *Dunaliella salina* (D). The peduncle (P) of *G. fungiforme* has attached to *D. salina*, and the protoplasm is passing through the enlarged and extended peduncle into the dinoflagellate. ((*a*) from Lee, 1977; (*b*) from Spero, 1982.)

are pulled off by groups of one to five dinoflagellates. As the ciliate begins to leak cytoplasm through these small wounds, large aggregations of phagotrophic dinoflagellates are formed. Approximately 20 to 30 minutes later, the cilate is completely digested. The attacking cells are small (9 µm long and 6 µm wide), clear cells that swell to 20 times their original size with food vacuoles after ingesting a victim. These large cells barely move or are motionless. *Gymnodinium fungiforme* is attracted to a variety of amino acids and other organic compounds (Spero, 1985). Glycine, taurine, and serine attract the dinoflagellate at threshold detection levels of 10^{-8} M, followed by dextrose at 10^{-7} and alanine, proline, and threonine at 10^{-6} M. Glycine, taurine, and alanine are three of the most abundant free amino acids found in invertebrates and protozoa, which are the major food organisms of *G. fungiforme.*

Pfiesteria piscicida is a heterotrophic dinoflagellate that is responsible for many of the fish kills along the Atlantic coast of the Southeastern United States (Burkholder and Glasgow, 1997). *P. piscicida* belongs to the "ambush–predator" group of dinoflagellates which chemically detect their prey and then swarm in a directed attack on the prey. The prey is immobilized and consumed by the dinoflagellates, which then encyst or disperse. *P. piscicida* produces scaled cysts that reside in sediments. Zoospores (Fig. 7.37) emerge from the cysts when fish are present in the water. The zoospores produce a neurotoxic ichthyotoxin that promotes epidermal slough-

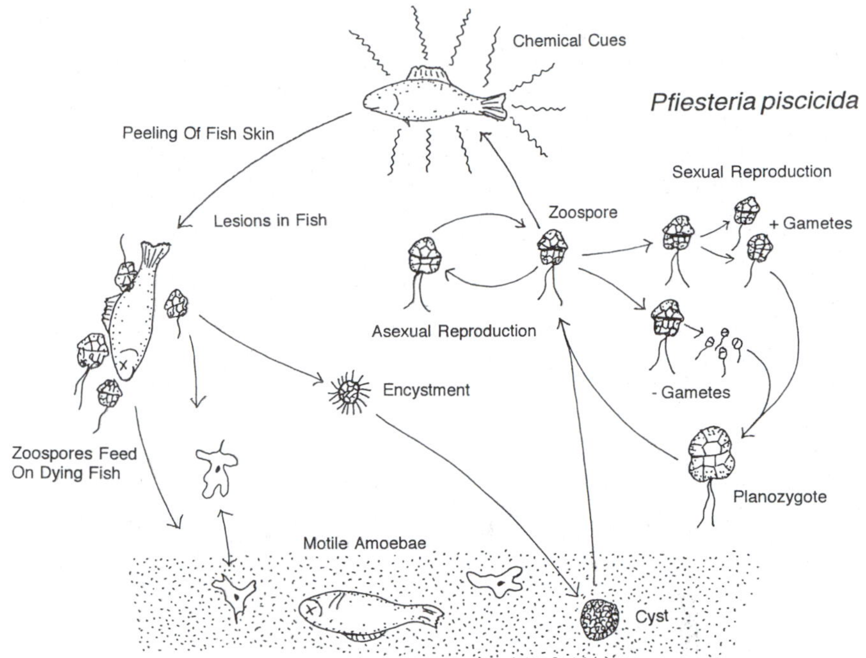

Fig. 7.37 The life cycle of *Pfiesteria piscicida*. (Adapted from Burkholder and Glasgow, 1997.)

ing and the development of deep open lesions in the fish. The zoospores of *P. piscicida* have a feeding organelle (peduncle) which is thin and tapered when the dinoflagellate is not feeding (Fig. 7.38). In the presence of fish, the dinoflagellate zoospores feed on the lesions in the fish by phagocytizing pieces of fish tissue through the peduncle. When the peduncle is engaged in saprophytic feeding on fish tissue, it is swollen with haustoria-like penetrating extensions. The zoospores nearly double in size during feeding, becoming sluggish in movement and engorged with fish-derived materials. The cellulosic plates that cover the zoospore become fused into one continuous coat of hardened armor. The zoospores can undergo vegetative division during this time. Alternatively, they may undergo sexual reproduction by gathering within a gelatinous mass where each zoospore undergoes repeated cytoplasmic divisions to yield more than 100 small plus gametes or two large minus gametes. Gamete fusion occurs if the affected fish are present. The large planozygotes divide to produce four zoospores. The zoospores can transform into amoebae with either filiform or lobose projections (Fig. 7.39) that live in the sediment or in the water column, and which also have the ability to attack fish. If fish are not present, the *P. piscicida* can live by ingesting protozoa, such as the alga *Cryptomonas*. If neither fish or other algae are present as a food source, *P. piscicida* cells decrease in numbers and encyst.

Photosynthetic dinoflagellates can also exhibit phagotrophy. *Amphi-*

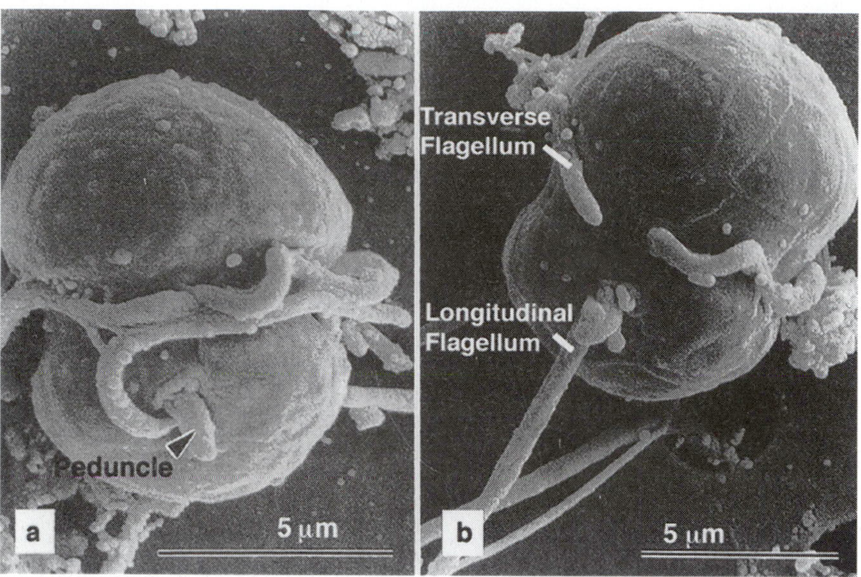

Fig. 7.38 Scanning electron micrographs of zoospores of *Pfiesteria piscicida*. (From Steidinger et al., 1996.)

dinium cryophilum has its own chloroplasts but is phagotrophic with the amount of phagotrophy varying with light intensity (Wilcox and Wedemeyer, 1991). At low light intensities, cells feed phagotrophically and are nearly colorless, whereas at high light levels the cells feed much less frequently, if at all, and are brightly pigmented. *A. cryophilum* inserts a peduncle into the prey organism (Fig. 7.40). Over a period of about 10 minutes the prey protoplasm is drawn into a food vacuole in the dinoflagellate.

Symbiotic dinoflagellates

Symbiotic dinoflagellates, usually assigned to the genus *Symbiodinium*, occur in almost all species of tropical reef-building corals, jellyfish and sea anemones (Cnidaria). *Symbiodinium* cells in the symbiotic state are coccoid. The cells change into typical gymnodinoid form when placed in culture. Cultured and symbiotic *Symbiodinium* differ physiologically. The symbiotic cells grow ten times slower than cultured cells. The host exerts strong control over the translocation of metabolites from the dinoflagellate endosymbiont, resulting in 98% of the carbon fixed by the endosymbiont being released to the host. In contrast, cultured cells release only 10% of their photosynthetically fixed carbon (Trench, 1993). The host animal cells secrete the amino acid taurine which causes the dinoflagellate to release photosynthate outside the cell for absorption by the animal cells (Wang and Douglas, 1997).

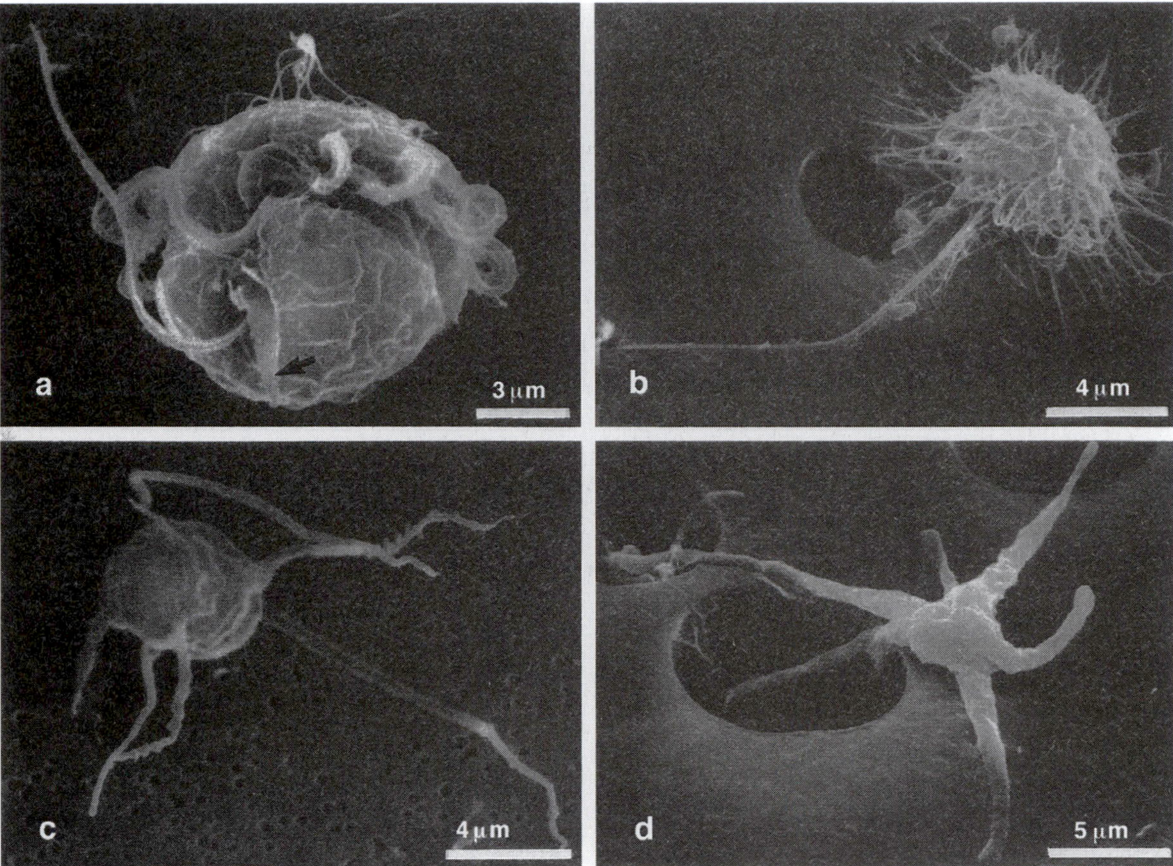

Fig. 7.39 Scanning electron micrographs of stages in the life history of *Pfiesteria piscicida*. (*a*) Zoospore with a retracted peduncle (arrow). (*b*) Encysting zoospore producing chrysophyte-like scales. (*c*) Filiose amoeba. (*d*) Labose amoeba. (From Burkholder and Glasgow, 1997.)

Corals with symbiotic dinoflagellates do not incorporate calcium into their skeletons as fast as corals without symbiotic dinoflagellates. During the daytime, corals with and without symbiotic dinoflagellates calcify at the same rate, but at night calcification is less in corals with symbiotic algae (Marshall, 1996).

In the symbiosis between the marine anemones and marine dinoflagellates, the dry weight ratio of host to alga is about 300 : 1 (Taylor, 1969a). When the anemone is subjected to poor growing conditions, it responds by excreting some of the dinoflagellate cells, which the anemone obviously has difficulty in maintaining. Even under normal growing conditions the algal population divides too rapidly, and the anemone reacts by pruning the population down to a level it can support. In this situation the older dinoflagellate cells in the outer regions of the host (crown and tentacles) respond by thickening the outer secreted layer around the cell wall and subsequently forming cysts. As the thickness of the cyst's outer layer increases, the cells begin to show signs of degeneration. The host then seems to become sensitive to the degenerate condition of the cysts and reacts by

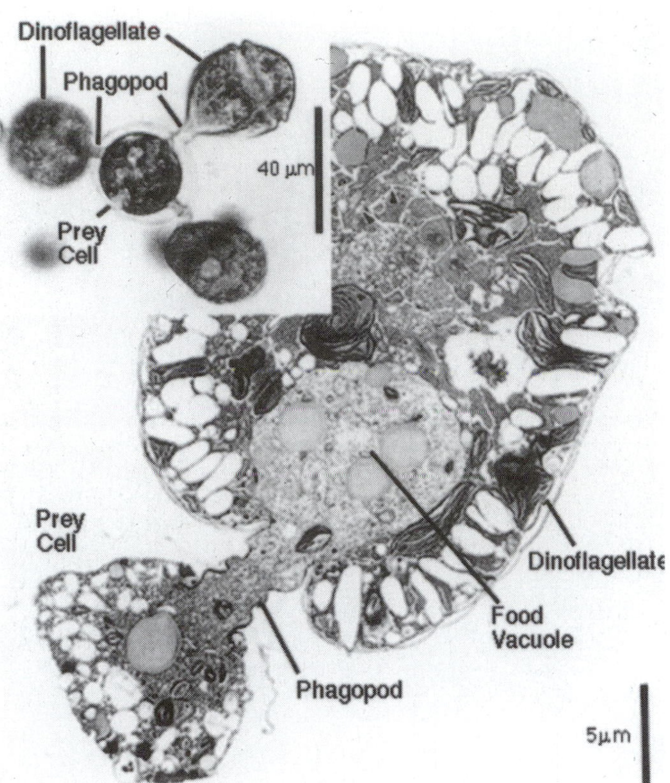

Fig. 7.40 *Amphidinium cryophilum* attached to a prey organism by a peduncle (phagopod). The inset shows a light micrograph of three cells of *A. cryophilum* attached to a prey cell. (From Wilcox and Wedemeyer, 1991.)

removing them from these outer regions and transporting them to the mesenteries as intracellular inclusions of its undifferentiated amoeboid cells. They are stored in the mesenteries in varying states of decomposition until they are finally excreted by the host.

In the flatworm *Amphiscolops langerhansi*, D. L. Taylor (1971) has found that the cells of the dinoflagellate *Amphidinium klebsii* occur exclusively between the cells of the peripheral parenchyma of the host and appear as a conspicuous layer below the composite muscle (Fig. 7.41). Each cell of *Amphidinium* has a specific and uniform orientation within the host. The anterior of the cell is directed toward the central parenchyma, and the posterior with the large nucleus is directed toward the epithelium. By rearing the flatworm from eggs, it is possible to obtain individuals without the dinoflagellate symbiont. When these flatworms are grown in culture with a number of different types of dinoflagellates, only species of *Amphidinium* are able to infect the host. When contact is made, the animal seizes and ingests the alga. Once inside the flatworm, the alga is not retained in the central digestive paranchyma but passes freely between the animal cells to the peripheral parenchyma, where it comes to lie intercellularly below the

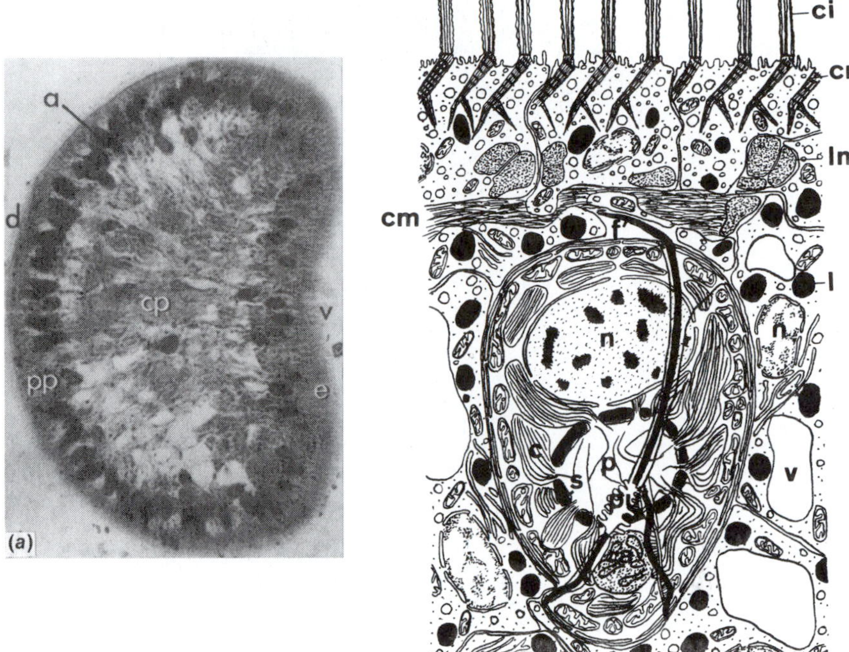

Fig. 7.41 (*a*) Cross-section of the flatworm *Amphiscolops langerhansii* showing the dinoflagellate *Amphidinium klebsii* (a) in the peripheral parenchyma (pp). (cp) Central parenchyma; (d) dorsal surface; (e) epithelium; (v) ventral surface. × 90. (*b*) Diagrammatic reconstruction of the above association. (a) Accumulation body; (c) chloroplast; (ci) cilia; (cm) circular muscle; (cr) ciliary root; (f) flagellum; (l) lipid; (lm) longitudinal muscle; (n) nucleus; (p) pyrenoid; (pu) pusule; (s) starch; (v) vacuole. (After D. Taylor, 1971.)

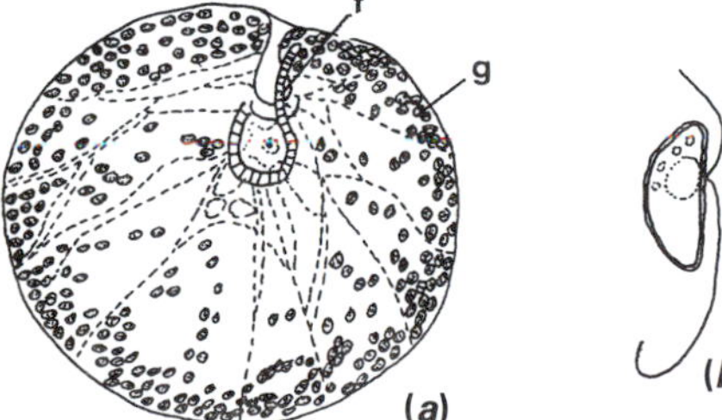

Fig. 7.42 *Noctiluca* containing green algae in its vacuoles. (*a*) Whole cell. (g) Green algae; (t) tentacle. (*b*) Green symbiont. (After Sweeney, 1971.)

composite muscle. The alga remains unchanged within the host and still has its typical two flagella. In nature the association between the flatworm and the dinoflagellate is probably essential for the survival of the animal. It is not known whether the dinoflagellate receives any benefit.

In addition to Dinophyceae living symbiotically inside other organisms, there are other organisms that live inside dinoflagellate cells. *Noctiluca* is normally a colorless dinoflagellate with no chloroplasts. Off the coast of New Guinea there are green *Nociluca* cells with the normal striated tentacle, no chloroplasts, a highly vacuolate protoplasm, and a few food vacuoles (Fig. 7.42) (Sweeney, 1971). The cells, however, have 6000 to 12 000 green flagellates actively swimming in the fluid within the large vacuoles,

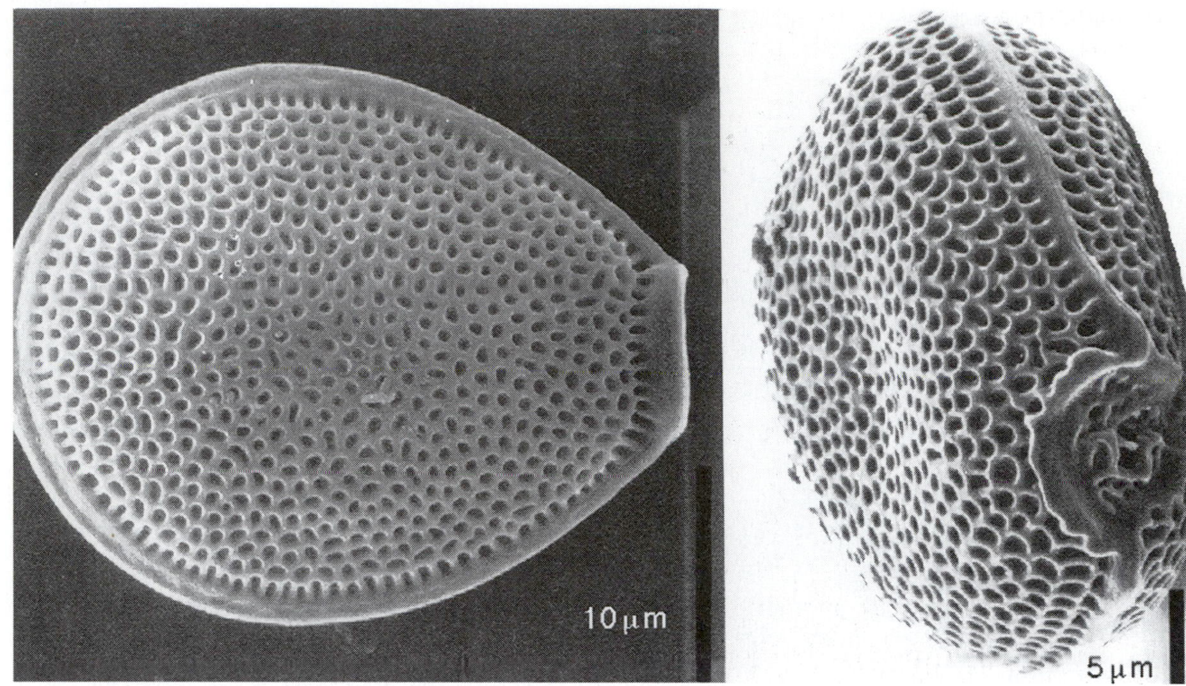

Fig. 7.43 Scanning electron micrographs of *Prorocentrum hoffmanianum*. (From Faust, 1990.)

especially around the periphery of the vacuole. The green flagellate resembles the green alga *Pedinomonas* in being bright green, 2×5 μm in size, with no eyespot, and having a single posterior flagellum. Whether the dinoflagellate gains any advantage from the association is not known.

Classification

There is a single class in the Dinophyta, the Dinophyceae. Four orders are considered here. Molecular studies has shown that the Prorocentrales, Peridinales and Gymnodiniales represent three clear lines of evolution (Zardoya et al., 1995). The Dinophysales are probably related to the Prorocentrales since they are both divided vertically into two halves.

Order 1 Prorocentrales: cell wall divided vertically into two halves; no girdle; two flagella borne at cell apex.

Order 2 Dinophysales: cell wall divided vertically into two halves, cells with elaborate extensions of the theca.

Order 3 Peridiniales: motile cells with an epicone and hypocone separated by a girdle, relatively thick theca.

Order 4 Gymnodiales: motile cells with an epicone and hypocone separated by a girdle, theca thin or reduced to empty vesicles.

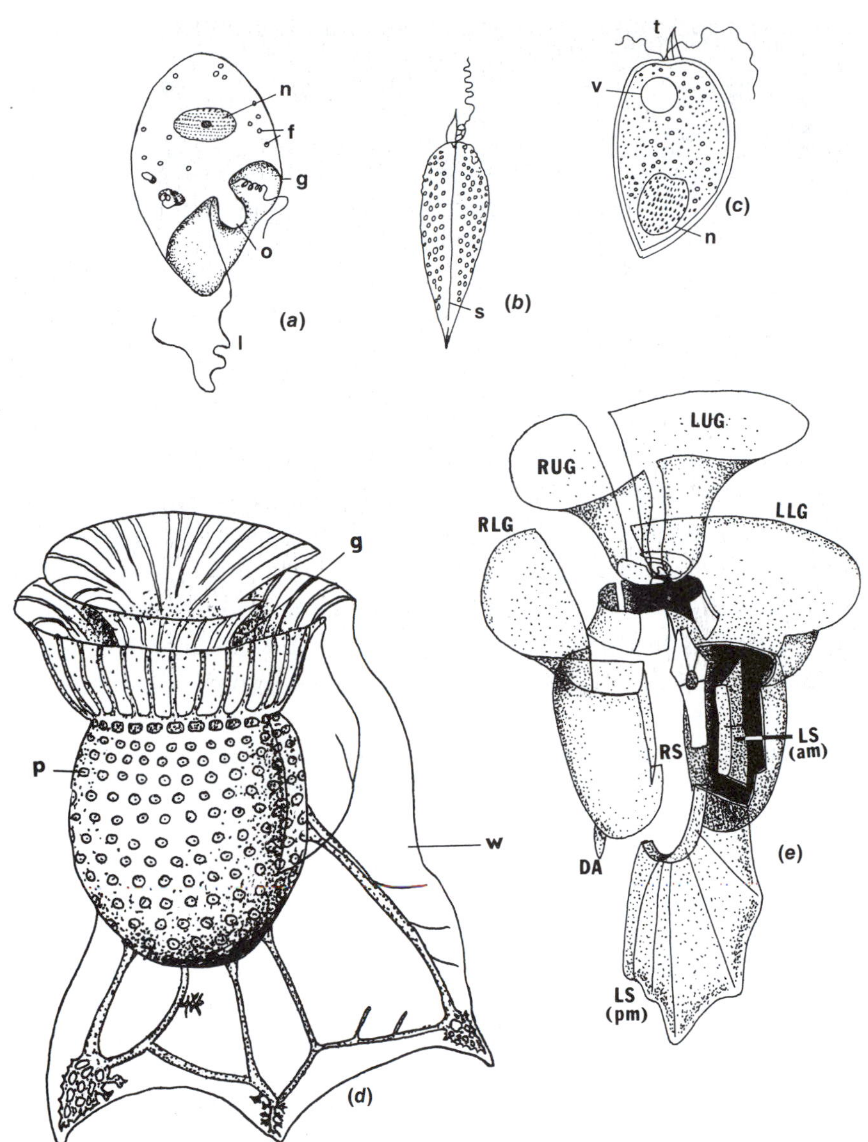

Fig. 7.44 (*a*) *Oxyrrhis marina.* (f) Fat globule; (g) girdle; (l) longitudinal flagellum; (n) nucleus; (o) tentacle. (*b*),(*c*) *Prorocentrum micans*, side (*b*) and front (*c*) views. (n) Nucleus; (s) suture; (v) vacuole. (*d*),(*e*) *Ornithocercus magnificus*, a right-hand view (*d*) and an "exploded" view of the theca (e). (DA) Dorsal accessory moiety of the left sulcal list; (LLG) left lower girdle list; (LS(am)) anterior moiety of the left sulcal list; (LS(pm)) posterior moiety of the left sulcal list; (LUG) left upper girdle list; (RS) right sulcal list; (RLG and RUG) right lower and upper girdle list; (g) girdle; (p) pore; (w) wing. ((*d*),(*e*) after F. Taylor, 1971.)

PROROCENTRALES

These cells have the cell wall divided vertically into two halves, no girdle, and two apically inserted flagella, *Prorocentrum* (Fig. 7.43. 7.44(*b*),(*c*)) can be used as an example of the order. The cell is divided vertically into two halves, each half containing a relatively thick thecal plate, the suture joining them running from the anterior to the posterior end. The cell is flattened parallel to the suture so the two halves are like two watch glasses. The two flagella emerge apically through a pore, and there is a single tooth containing cytoplasm. Usually there are two brownish-yellow chloroplasts, apposed to the two valves. Asexual reproduction takes place by longitudinal division, the daughter cells retaining one valve of the parent and forming a new second valve.

The Prorocentrales are usually considered to be the most primitive of the Dinophyceae because of their thecal structure, the low number of chloroplasts, and their continuous formation of DNA (Taylor, 1980).

DINOPHYSIALES

This order consists of morphologically complex organisms that are mainly in tropical seas, with the cells having adaptations to the floating habit such as elaborate wings (**lists**). One of the more complex organisms in this order is *Ornithocercus* (Fig. 7.44(*d*),(*e*)). The cell is divided vertically into two halves by an anterior posterior suture. The cells have a girdle and sulcus, with the side having the sulcus being the ventral side and opposite side, the dorsal side. The respective flagella lie in the girdle and transverse sulcus. The epicone is usually much smaller than the hypocone, and the edges of the thecal plates next to the girdle and sulcus are expanded into the lists. In asexual reproduction the cell is cleaved vertically, with the two halves separating and the missing half being formed by the protoplasm. The oldest fossil of this group is *Nannoceratopsis* from the Lower Jurassic (Loeblich, 1974, 1976).

PERIDINIALES

The dinoflagellates in this order have relatively thick thecal plates, in contrast to the next order, the Gymnodiniales, which has no, or thin, thecal plates. The algae in this order have the classic dinoflagellate structure (Fig. 7.5) with an epicone and hypocone and two furrows, the transverse girdle and the longitudinal sulcus.

The widely distributed genus *Ceratium* (Fig. 7.45) is markedly asymmetric, with one apical horn and two to three long antapical horns filled with cytoplasm. The girdle is nearly horizontal and divides the body into two approximately equal, but dissimilar halves. In the middle of the ventral surface is a large rhombic hyaline area, which is probably similar to a sulcus. Like other Dinophyceae and Prymnesiophyceae, *Ceratium* is more common in warmer water than in colder polar waters. There are fewer than ten species common in the colder waters of the North Atlantic, whereas more than 20 species are common in the warmer more southerly waters (Graham and Bronikovsky, 1944). *Ceratium* is one of the members of the

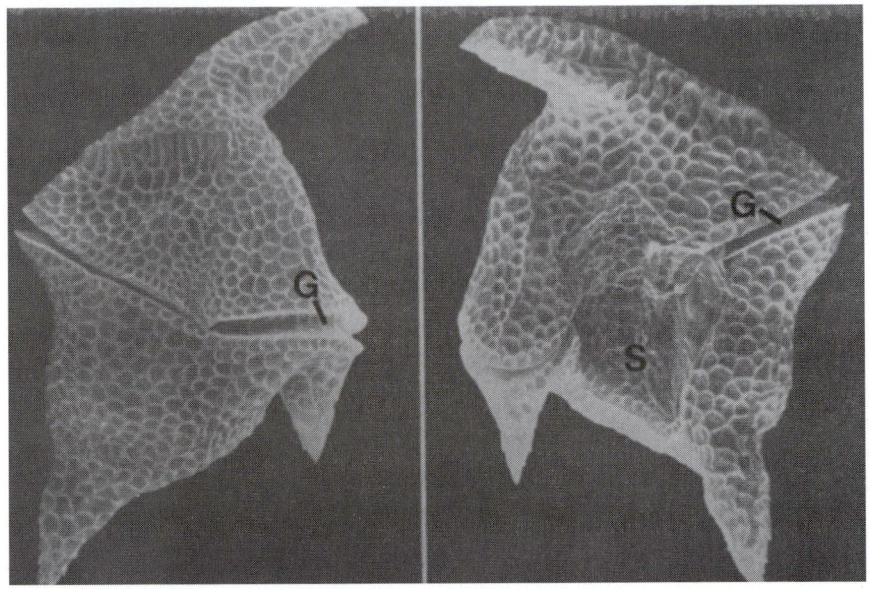

Fig. 7.45 Scanning electron micrographs of both sides of a vegetative cell of *Ceratium cornutum*. (G) Girdle; (S) sulcus. (From Happach-Kasan, 1982.)

phytoplankton that has **shade forms** which show an increase in frequency from the surface to 100 m depth. These shade forms are found only in the relatively sterile warm oceanic waters where the upper layers are usually depleted of nitrogen and phosphorus. The shade forms of *Ceratium* have a survival value in that they are able to take advantage of the higher levels of nitrogen and phosphorus deeper in the water and still are able to receive enough light to keep their photosynthetic rate above their respiratory rate. These shade forms adapt themselves to absorb the maximum amount of light by increasing the surface area of the cell by expansion of the cell body and/or horns and packing these extensions with chloroplasts. Long-horned forms are found among surface forms also, but the shade forms always have their horns crowded with chloroplasts.

Ceratium horridum has a life cycle that is more or less similar to the known life cycle of other Dinophyceae. Von Stosch (1972) has shown that the vegetative cells of *C. horridum* are haploid. The male gametes are smaller than the female gametes; thus fusion in this organism is anisogamous (Fig. 7.46). The male gamete attaches by its ventral side to the ventral side of the female gamete. The cell wall of the male then breaks up into its individual plates, which are taken up into the female cytoplasm. The naked male gamete fuses with the female gamete to produce a zygote. The zygote remains motile (planozygote) through all of the following steps. The planozygote grows for several days, ending up as a large cell with unusually long horns and a thick cell wall. Finally, the nucleus enlarges, and nuclear

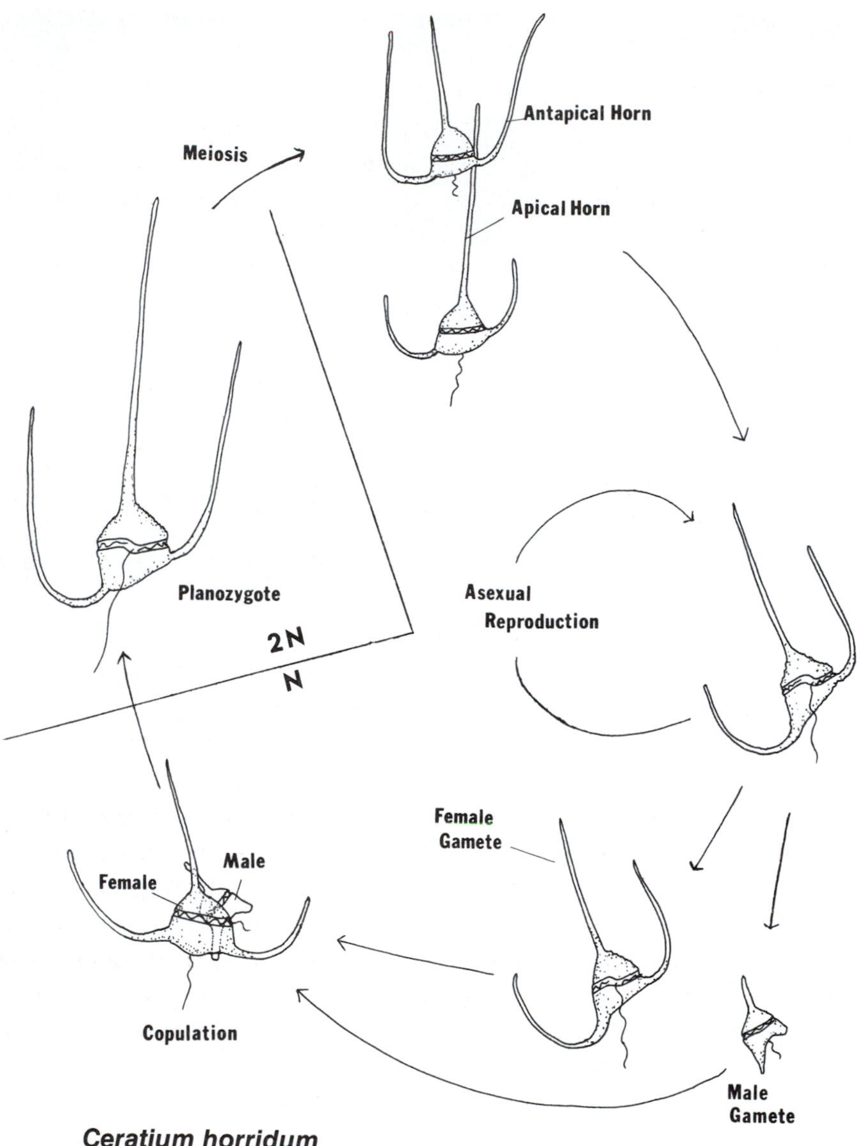

Fig. 7.46 The life cycle of *Ceratium horridum*. (Adapted from von Stosch, 1972.)

cycloses (movement) begins. At its highest speed the chromosomal mass circulates once every 30 seconds. After some hours the motion slows down and stops, and approximately 12 hours later meiosis commences. The first meiotic division halves the chromosome number and gives rise to one flagellate with normally sized antapical, and abnormally long apical horns and a second flagellate with normally sized apical, and abnormally long antapical horns. After 2 to 3 days the second meiotic division takes place to yield haploid vegetative cells.

GYMNODINIALES

The dinoflagellates in this order are similar to those in the previous order, the Peridiniales, except that they have thin or no thecal plates.

The life cycle of *Gymnodinium pseudopalustre* is a representative of the order. Vegetative multiplication results in daughter cells that remain attached to each other for at least 12 hours (Fig. 7.47). Gametes are formed when an actively growing culture at 21 °C is subjected to a short-day treatment (10 hours light) at 15 °C (von Stosch, 1973). Gamete differentiation occurs by divisions that give rise to cells lower in mass and poorer in plastids and pigments than the vegetative cells. Apart from their smaller size and lighter color, the gametes are not obviously different from vegetative cells. They are incapable of living for long as vegetative cells and die if fusion does not occur. The gametes are homothallic; thus gametes from a single strain will fuse.

Initially, small groups of two to ten gametes swarm around each other in a weaving motion. A copulation pair occurs between two gametes, with the pair swimming rapidly while turning slowly on a common axis. Copulation is isogamous although there can be a slight difference in the size of the gametes. The cells are joined together by a hyaline globular bridge slightly below the intersection of the transverse and posterior grooves. After 30 minutes the fusion of protoplasts has begun, with the bridge between the two cells enlarging. At the end of fusion a cell similar to a vegetative cell is attained. The flagella and nuclei of each gamete are still distinct. The zygote grows, at first having the shape of a vegetative cell, and later the epicone elongates. One of the transverse flagella is lost during this development, but both of the posterior flagella persist. The non-motile zygote suddenly rounds off and secretes a preliminary wall. This wall is subsequently inflated, giving rise to a hyaline area between the wall and the protoplast surface. An ornamentation of small separated granules now appears on the protoplast wall, which grow out radially to become spines while the hyaline area increases in width. The preliminary wall then bursts and crumples away to one side of the spore. The preliminary wall is necessary for formation of the hypnospore. The duration of the preliminary wall is only about 9 minutes.

During the next 48 hours the hypnospore matures, with the plastids bleaching and becoming inconspicuous, masses of red oil appearing, the starch becoming indistinct, and the two nuclei now fusing. A thick cellulosic endospore is also secreted under the exospore with its spines. The hypnospores germinate after treatment for 4 weeks in the dark at 3 °C before being returned to light and higher temperature. After the cellulosic endospore has been digested away, approaching release of the swarmer is indicated by a slight contraction of the protoplast so that the transverse groove of the prospective flagellate becomes visible. The space between the spore

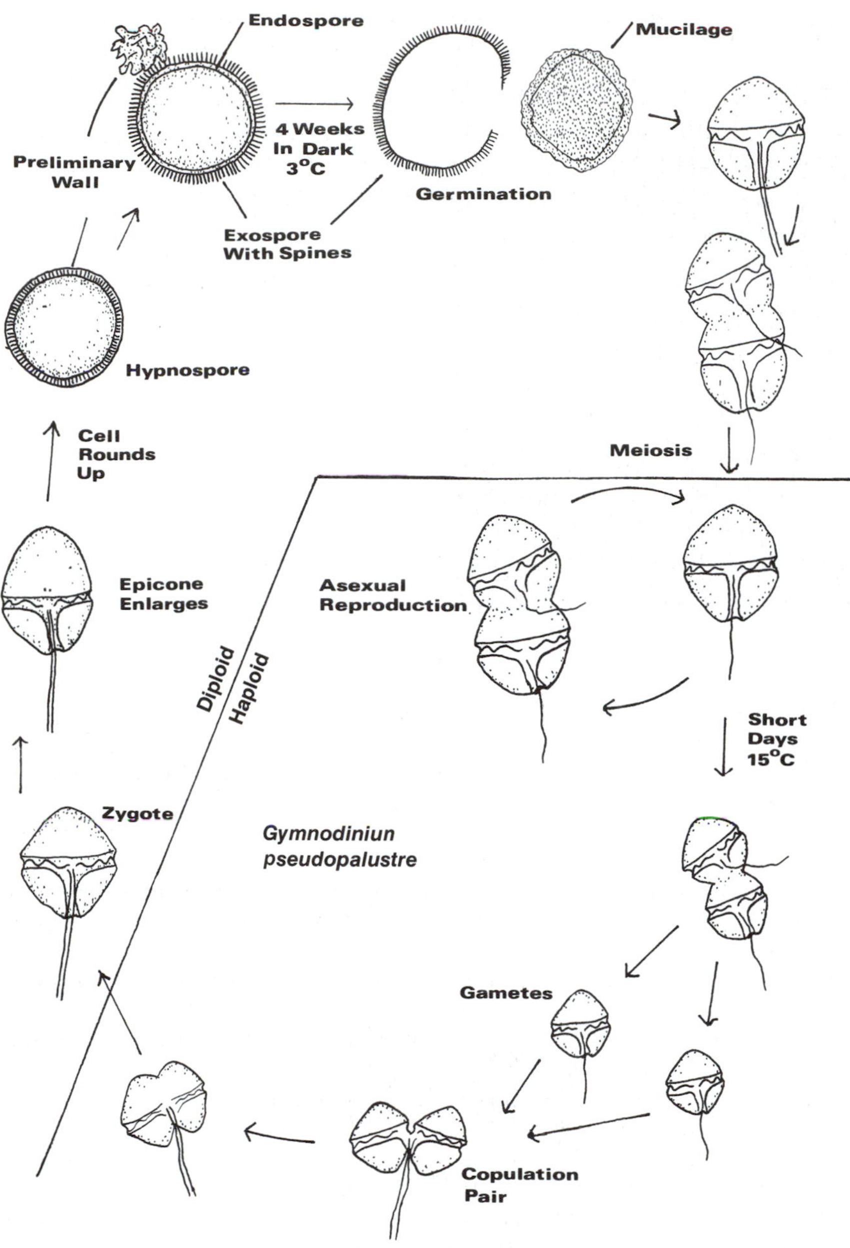

Fig. 7.47 The life cycle of *Gymnodinium pseudopalustre*. (Adapted from von Stosch, 1973.)

wall and the surface of the swarmer is filled with mucilage. Eventually the wall bursts, and the swarmer escapes, enveloped in mucilage. The swarmer frees itself from the mucilage and swims away. The swarmer is rather plump and of oval shape at first, and, apart from red oil globules, nearly colorless; but later it acquires brown pigment, and its form becomes similar to that of the vegetative cell. Two "skiing track" posterior flagella have reappeared. The swarmer then goes through two meiotic divisions, resulting in four haploid flagellates.

Fig. 7.48 Donald M. Anderson. Dr Anderson earned his doctorate in civil engineering at the Massachusetts Institute of Technology in 1977. In 1977, he joined the scientific staff at the Woods Hole Oceanographic Institute, where he is currently a Senior Scientist. Dr Anderson studies the physiology and genetic regulation of toxicity in dinoflagellates, their bloom dynamics and ecology, and the global biogeography of toxic *Alexandrium* species.

References

Abrahams, M. V., and Townsend, L. D. (1993). Bioluminescence in dinoflagellates: a test of the burglar alarm hypothesis. *Ecology* 74:258–60.

Adamich, M., Laris, P. C., and Sweeney, B. M. (1976). In vivo evidence for a circadian rhythm in membranes of *Gonyaulax*. *Nature* 261:583–5.

Anderson, D. M., and Wall, D. (1978). Potential importance of benthic cysts of *Gonyaulax tamarensis* and *G. excavata* in initiating toxic dinoflagellate blooms. *J. Phycol.* 14:224–34.

Anderson, D. M., and Keafer, B. A. (1987). An endogenous annual clock in the toxic marine dinoflagellate *Gonyaulax tamarensis*. *Nature* 325:616–17.

Anderson, D. M., and Cheng, T. P. O. (1988). Intracellular localization of saxitoxins in the dinoflagellate *Gonyaulax tamarensis*. *J. Phycol.* 24:17–22.

Balzer, I., and Hardeland, R. (1996). Melatonin in algae and higher plants – possible new roles as a phytochrome and antioxidant. *Bot. Acta* 109:180–3.

Barlow, S. B., and Triemer, R. E. (1988). The mitotic apparatus of the dinoflagellate *Amphidinium carterae*. *Protoplasma* 145:16–26.

Berdach, J. T. (1977). In situ preservation of the transverse flagellum of *Peridinium cinctum* (Dinophyceae) for scanning electron microscopy. *J. Phycol.* 13:243–51.

Berdalet, E. (1992). Effects of turbulence on the marine dinoflagellate *Gymnodinium nelsonii*. *J. Phycol.* 28:267–72.

Bibby, B. T., and Dodge, J. D. (1972). The encystment of a freshwater dinoflagellate: A light and electron-microscopical study. *Br. Phycol. J.* 7:85–100.

Blanco, J. (1995). Cyst production in four species of neritic dinoflagellates. *J. Plankton Res.* 17:165–82.

Bouck, G. B., and Sweeney, B. M. (1966). The fine structure and ontogeny of trichocysts in marine dinoflagellates. *Protoplasma* 61:205–23.

Bricheux, G., Mahoney, D. G., and Gibbs, S. P. (1992). Development of the pellicle and thecal plates following ecdysis in the dinoflagellate *Glenodinium foliaceaum. Protoplasma* 168:159–71.

Brooks, B. J., and Anderson, D. M. (1990). Biochemical composition and metabolic activity of *Scrippsiella trochoidea* (Dinophyceae) resting cysts. *J. Phycol.* 26:289–98.

Burkholder, J. M., and Glasgow, H. B. (1997). Trophic controls on stage transformation of a toxic ambush-predator dinoflagellate. *J. Euk. Microbiol.* 44:200–5.

Buskey, E. J., and Swift, E. (1983). Behavioral responses of *Acartia hudsonica* to simulated dinoflagellate bioluminescence. *J. Exp. Mar. Biol. Ecol.* 77:43–58.

Buskey, E., Mills, L., and Swift, E. (1983). The effects of dinoflagellate bioluminescence on the swimming behavior of a marine copepod. *Limnol. Oceanogr.* 28:575–9.

Chapman, A. D., and Pfiester, L. A. (1995). The effects of temperature, irradiance, and nitrogen on the encystment and growth of the freshwater dinoflagellates *Peridinium cinctum* and *P. willei* in culture (Dinophyceae). *J. Phycol.* 31:355–9.

Chapman, D. V., Dodge, J. D., Heaney, S. J. (1982). Cyst formation in the freshwater dinoflagellate *Ceratium hirundinella. J. Phycol.* 18:121–9.

Chatton, E. (1952). Classe des dinoflagelles ou peridiniens. In *Traité de Zoologie*, ed. P-P. Grassé, pp. 304–406. Paris: Masson.

Chinain, M., Germain, M., Sako, Y., Pauillac, S., and Legrand, A-M. (1997). Intraspecific variation in the dinoflagellate *Gambierdiscus toxicus* (Dinophyceae). I. Isoenzyme analysis. *J. Phycol.* 33:36–43.

Clarke, K. J., and Pennick, N. C. (1976). The occurrence of body scales in *Oxyrrhis marina* Dujardin. *Br. Phycol. J.* 11:345–8.

Crawford, R. M., Dodge, J. D., and Happey, C. M. (1970). The dinoflagellate genus *Woloszynskia*. I. Fine structure and ecology of *W. tenuissima* from Abbot's Pool, Somerset. *Nova Hedwigia* 19:825–40.

Destombe, C., and Cembella, A. (1990). Mating-type determination, gamete recognition and reproductive success in *Alexandrium excavatum* (Gonyaulacales, Dinophyta), a toxic red-tide dinoflagellate. *Phycologia* 29:315–25.

Dodge, J. D. (1965). Chromosome structure in the dinoflagellates and the problem of the mesocaryotic cell. *Prog. Protozool.* 2:264.

Dodge, J. D. (1971). Fine structure of the Pyrrophyta. *Bot. Rev.* 37:481–508.

Dodge, J. D. (1983). Dinoflagellates: Investigation and phylogenetic speculation. *Br. Phycol. J.* 18:335–56.

Dodge, J. D. (1985). *Atlas of Dinoflagellates.* London: Farrand Press. 184 p.

Dodge, J. D., and Crawford, R. M. (1968). Fine structure of the dinoflagellate *Amphidinium carteri* Hulbert. *Protistologica* 4:231–42.

Dodge, J. D., and Crawford, R. M. (1969). Observations of the fine structure of the eyespot and associated structures in the dinoflagellate *Glenodinium foliaceum. J. Cell Sci.* 5:479–93.

Dodge, J. D., and Crawford, R. M. (1970). A survey of thecal fine structure in the Dinophyceae. *J. Linn. Soc. Bot.* 63:53–67.

Downie, C. (1956). Microplankton from the Kimmeridge Clay. *Q. J. Geol. Soc. Lond.* 112:413–34.

Dunlap, J. C., and Hastings, J. W. (1981). Biochemistry of dinoflagellate bioluminescence: Purification and characterization of dinoflagellate luciferin from *Pyrocystis lunula. Biochemistry* 20:983–9.

Ellegaard, M., Christensen, N. F., and Moestrup, O. (1994). Dinoflagellate cysts from recent Danish marine sediments. *Eur. J. Phycol.* 29:183–94.

Eppley, R. W., Holm-Hansen, O., and Strickland, J. D. H. (1968). Some observations on the vertical migration of dinoflagellates. *J. Phycol.* 4:333–40.

Faust, M. A. (1990). Morphological details of six benthic species of *Prorocentrum* (Pyrrophyta) from a mangrove island, Twin Cays, Belize, including two new species. *J. Phycol.* 26:548–58.

Faust, M. A. (1995). Observation of sand-dwelling toxic dinoflagellates (Dinophyceae) from widely differing sites, including two new species. *J. Phycol.* 31:996–1003.

Fiedler, P. C. (1982). Zooplankton avoidance and reduced grazing responses of *Gymnodinium splendens* (Dinophyceae). *Limnol. Oceanogr.* 27:961–5.

Filfilan, S. A., and Sigee, D. C. (1977). Continuous DNA replication in the nucleus of the dinoflagellate *Prorocentrum micans* (Ehrenberg). *J. Cell Sci.* 27: 81–90.

Franker, C. K., Sakhrani, L. M., Prichard, C. D., and Landen, C. A. (1974). DNA synthesis in *Cryptothecodinium cohnii*. *J. Phycol.* 10:91–4.

Fritz, L., Milos, P., Morse, D., and Hastings, J. W. (1991). *In situ* hybridization of luciferase-binding protein anti-sense RNA to thin sections of the bioluminescent dinoflagellate *Gonyaulax polyedra*. *J. Phycol.* 27:436–41.

Gaines, G., and Taylor, F. J. R. (1984). Extracellular digestion in marine dinoflagellates. *J. Plank. Res.* 6:1057–61.

Gaines, G., and Taylor, F. J. R. (1985). Form and function of the dinoflagellate transverse flagellum. *J. Protozool.* 32:290–6.

Gallois, R. W. (1976). Coccolith blooms in the Kimmeridge Clay and origin of the North Sea oil. *Nature* 259:473–5.

Graham, H. W., and Bronikovsky, N. (1944). The genus *Ceratium* in the Pacific and North Atlantic oceans. *Carnegie Inst. Washington Publ.* 565:1–209.

Grindley, J. R., and Nel, E. A. (1970). Red water and mussel poisoning at Elands Bay, December 1966. *Fish Bull., S. Afr.* 6:36–55.

Grindley, J. R., and Sapeika, N. (1969). The cause of mussel poisoning in South Africa. *S. Afr. Med. J.* 43:275–9.

Gruet, C. (1965). Structure fine de l'ocelle d'*Erythropsis pavillardi* Hetwig, Péridinien Warnowiidae Lindemann. *C. R. Séances Acad. Sci., Paris* 261:1904–7.

Halldal, P. (1958). Action spectra of phototaxis and related problems in Volvocales, *Ulva*-gametes and Dinophyceae. *Physiol. Plant.* 11:118–53.

Hallegraeff, G. M. (1993). A review of harmful algal blooms and their apparent global increase. *Phycologia* 32:79–99.

Hansen, G. (1989). Ultrastructure and morphogenesis of scales in *Katodinium rotundatum* (Lohmann) Loeblich (Dinophyceae). *Phycologia* 28:385–94.

Hansen, G. (1993). Light and electron microscopical observation of the dinoflagellate *Actiniscus pentasterias* (Dinophyceae). *J. Phycol.* 29:486–99.

Happach-Kasan, C. (1982). Beobachtungen zum Bau der Theka von *Ceratium cornutum*, (Ehrenb,) Clap. et Lachm. (Dinophyta). *Arch. Protistenk.* 125:181–207.

Harvey, E. N. (1952). *Bioluminescence.* New York: Academic Press.

Hastings, J. W. (1983). Biological diversity, chemical mechanisms, and the evolutionary origins of bioluminescent systems. *J. Mol. Evol.* 19:309–21.

Hastings, J. W. (1986). Bioluminescence in bacteria and dinoflagellates. In *Light Emission in Plants and Bacteria*, pp. 363–98. New York: Academic Press.

Hastings, J. W., and Krasnow, R. (1981). Temporal regulation in the individual *Gonyaulax* cell. In *International Cell Biology 1980–1981*, Proc. 2nd Int. Cong. on Cell Biology, pp. 815–823. Berlin: Springer-Verlag.

Haxo, F. T. (1985). Photosynthetic action spectrum of the coccolithophorid, *Emiliania huxleyi* (Haptophyceae): 19'-Hexanoyloxyfucoxanthin as antenna pigment. *J. Phycol.* 21:282–7.

Höhfeld, I., and Melkonian, M. (1992). Amphiesmal ultrastructure of dinoflagellates, A reevaluation of pellicle formation. *J. Phycol.* 28:82–9.

Höhfeld, I., Otten, J., and Melkonian, M. (1988). Contractile eukaryotic flagella: Centrin is involved. *Protoplasma* 147:16–24.

Horiguchi, T., and Pienaar, R.N. (1988). Ultrastructure of a new sand-dwelling dinoflagellate *Scrippsiella arenicola* sp. nov. *J. Phycol.* 24:426–38.

Jacobsen, D. M., and Anderson, D. M. (1986). Thecate heterotrophic dinoflagellates: Feeding behavior and mechanisms. *J. Phycol.* 22:249–58.

Jacobsen, D. M., and Anderson, D. M. (1996). Widespread phagocytosis of ciliates and other protists by marine mixotrophic and heterotrophic thecate dinoflagellates. *J. Phycol.* 32:279–85.

Johnson, C. H., and Hastings, J. W. (1986). The elusive mechanism of the circadian clock. *Am. Sci.* 74:29–36.

Johnson, C. H., Inoué, S., Flint, A., and Hastings, J. W. (1985). Compartmentalization of algal bioluminescence: Autofluoresence of bioluminescent particles in the dinoflagellate *Gonyaulax* as studied with image-intensified video microscopy and flow cytometry. *J. Cell Biol.* 100:1435–46.

Klut, M. E., Bisalputra, T., and Antia, N. J. (1985). Some cytochemical studies on the cell surface of *Amphidinium carteri* (Dinophyceae). *Protoplasma* 129:93–99.

Kubai, D. F., and Ris, H. (1969). Division in the dinoflagellate *Gyrodinium cohnii* (Schiller). A new type of nuclear reproduction. *J. Cell Biol.* 40:508–28.

Leadbeater, B. S. C., and Dodge, J. D. (1967a). Fine structure of the dinoflagellate transverse flagellum. *Nature* 213:421–2.

Leadbeater. B. S. C., and Dodge, J. D. (1967b). An electron microscope study of nuclear and cell division in a dinoflagellate. *Arch. Mikrobiol.* 57:239–54.

Lee, R. E. (1977). Saprophytic and phagocytic isolates of the colorless heterotrophic dinoflagellate *Gyrodinium lebouriae* Herdman *J. Mar. Biol. Assoc. UK* 57:303–15.

Loeblich, A. R. (1968). A new marine dinoflagellate genus, *Cachonia*, in axenic culture from the Salton Sea, California with remarks on the genus *Peridinium*. *Proc. Biol. Soc. Wash.* 81:91–96.

Loeblich, A. R. (1974). Protistan phylogeny as indicated by the fossil record. *Taxon* 23:277–90.

Loeblich, A. R. (1976). Dinoflagellate evolution: Speculation and evidence. *J. Protozol.* 23:13–28.

Lombard, E. H., and Capon, B. (1971). Observations on the tide pool behaviour of *Peridinium gregarium*. *J. Phycol.* 7:188–94.

MacRae, R. A., Fensome, R. A., and Williams, G. L. (1996). Fossil dinoflagellate diversity, originations, and extinctions and their significance. *Can. J. Bot.* 74:1687–94.

Marshall, A. T. (1996). Calcification in hermatypic and ahermatypic corals. *Science* 271:637–9.

Maruyama, T. (1982). Fine structure of the longitudinal flagellum in *Ceratium tripos*, a marine dinoflagellate. *J. Cell Sci.* 58:109–23.

Maruyama, T. (1985). Ionic control of the longitudinal flagellum in *Ceratium tripos* (Dinoflagellida). *J. Protozool.* 3:106–10.

Meksumpun, S., Montani, S., and Uematsu, M. (1994). Elemental composition of cell walls of three marine phytoflagellates, *Chattonella antiqua* (Raphidophyceae), *Alexandrium catenella* and *Scrippsiella trochoidea* (Dinophyceae). *Phycologia* 33:275–80.

Messer, G., and Ben-Shaul, Y. (1969). Fine structure of *Peridinium westii* Lemm., a freshwater dinoflagellate. *J. Protozool.* 16:272–80.

Montresor, M., Janofske, D., and Willems, H. (1997). The cyst-theca relationship in *Calciodinellum operosum* emend. (Peridinales, Dinophyceae) and a new approach for the study of calcareous cysts. *J. Phycol.* 33:122–31.

Moronin, L., and Francis, D. (1967). The fine structure of *Nematodinium armatum,* a naked dinoflagellate. *J. Microscopie* 6:759–72.

Morrill, L. C., and Loeblich, A. R. (1983a). Ultrastructure of the dinoflagellate amphisema. *Int. Rev. Cytol.* 82:151–80.

Morrill, L. C., and Loeblich, A. R. (1983b). Formation and release of body scales in the dinoflagellate genus *Heterocapsa. J. Mar. Biol. Assoc. UK* 63:905–13.

Morton, S. L., and Tindall, D. R. (1995). Morphological and biochemical variability of the toxic dinoflagellate *Prorocentrum lima* isolated from three locations at Heron Island, Australia. *J. Phycol.* 31:914–21.

Nawata, T., and Sibaoka, T. (1983). Experimental induction of feeding behavior in *Noctiluca miliaris. Protoplasma* 115:34–42.

Nawata, T., and Sibaoka, T. (1987). Local ion currents controlling the localized cytoplasmic movement associated with feeding initiation of *Noctiluca. Protoplasma* 137:125–33.

Nicolas, M.T., Morse, D., Bassot, J-M. and Hastings, J. W. (1991). Colocalization of luciferin binding protein and luciferase to the scintillons of *Gonyaulax polyedra* revealed by double immunolabeling after fast-freeze fixation. *Protoplasma* 160:159–66.

Pöggeler, B., Balzer, I., Hardeland, R., and Lerchl, A. (1991). Pineal hormone melatonin oscillates also in the dinoflagellate *Gonyaulax polyedra. Naturwissenschaften* 78:268–9.

Preisig, H. R. (1994). Siliceous structures and silicification in flagellated protists. *Protoplasma* 181:1–28.

Preisig, H. R., Anderson, O. R., Corliss, J. O., Moestrup, O., Powell, M. J., Roberson, R. W., and Wetherbee, R. (1994). Terminology and nomenclature of protist cell surface structures. *Protoplasma* 181:1–28.

Prézilin, B. B. (1976). The role of peridinin-chlorophyll *a*-proteins in the photosynthetic light adaptation of the marine dinoflagellate, *Glenodinium* sp. *Planta* 130:225–33.

Prézilin, B. B., and Haxo, F. T. (1976). Purification and characterization of peridinin-chlorophyll α-proteins from the marine dinoflagellates *Glenodinium* sp. and *Gonyaulax polyedra. Planta* 128:133–41.

Raven, J. A., and Richardson, K. (1984). Dinophyte flagella: A cost–benefit analysis. *New Phytol.* 98:259–76.

Ris, H., and Kubai, D. F. (1974). An unusual mitotic mechanism in the parasitic protozoan *Syndinium* sp. *J. Cell Biol.* 60:702–20.

Rizzo, P. J. (1991). The enigma of the dinoflagellate chromosome. *J. Protozool.* 38:246–52.

Rizzo, P. J., and Burghardt, R. C. (1982). Histone-like protein and chromatin structure in the wall-less dinoflagellate *Gyrodinium nelsoni. BioSystems* 15:27–34.

Roberts, K. R., and Roberts, J. E. (1991). The flagellar apparatus and cytoskeleton of the dinoflagellate. *Protoplasma* 164:105–22.

Robinson, N., Eglinton, G., Brassell, S. C., and Cranwell, P. A. (1984). Dinoflagellate origin for sedimentary 4α-methylsteroids and 5α(*H*)-stanols. *Nature* 308:439–42.

Roenneberg, T., and Deng, T-S. (1997). Photobiology of the *Gonyaulax* circadian system. I. Different phase response curves for red and blue light. *Planta* 202:494–501.

Samuelsson, G., Sweeney, B. M., Matlock, H. A., and Prézilin, B. B. (1983). Changes in photosystem II account for the circadian rhythm in photosynthesis in *Gonyaulax polyedra. Plant Physiol.* 73:329–31.

Schiller, J. (1933). Dinoflagellatae. In *Dr L. Rabenhorst's Kryptogamen-Flora,* Vol. 10. pp. 1–617. Leutershausen: Strauss and Cramer.

Schmidt, R. J., Gooch, V. D., Loeblich, A. R., and Hastings, J. W. (1978). Comparative study of luminescent and nonluminescent strains of *Gonyaulax excavata* (Pyrrhophyta). *J. Phycol.* 14:5–9.

Scholin, C. A., Hallegraeff, G. M., and Anderson, D. M. (1995). Molecular evolution of the *Alexandrium tamerense* "species complex" (Dinophyceae): dispersal in the North American and West Pacific region. *Phycologia* 34:472–85.

Schütt, F. (1895). Die Peridineen der Plankton-Expedition. *Ergeb. Plankton Exped. Humboldt-Stiftung.* 4; M.a. A:1–170.

Shilo, M. (1967). Information and mode of action of algal toxins. *Bacteriol. Rev.* 31:180–93.

Sigee, D. C. (1984). Structural DNA and genetically active DNA in dinoflagellate chromosomes. *BioSystems* 16:302–10.

Skovgaard, A., (1996). Engulfment of *Ceratium* spp. (Dinophyceae) by the thecate photosynthetic dinoflagellate *Fragilidium subglobosum. Phycologia* 35:490–9.

Spector, D. L., Vasconcelos, A. C., and Triemer, R. E. (1981). DNA duplication and chromosome structure in the dinoflagellates. *Protoplasma* 105:185–94.

Spero, H. J. (1982). Phagotrophy in *Gymnodinium fungiforme* (Pyrrophyta): The peduncle as an organelle of ingestion. *J. Phycol.* 18:356–60.

Spero, H. J. (1985). Chemosensory capabilities in the phagotrophic dinoflagellate *Gymnodinium fungiforme. J. Phycol.* 21:181–4.

Spero, H. J., and Moree, M. D. (1981). Phagotrophic feeding and its importance to the life cycle of the holozoic dinoflagellate, *Gymnodinium fungiforme. J. Phycol.* 17:43–57.

Steidinger, K. A., Burkholder, J. M., Glasgow, H. B., Hobbs, C. W., Garrett, J. K., Truly, E. W., Noga, E. J., and Smith, S. A. (1996). *Pfiesteria piscicida* gen. et sp. nov. (Pfiesteriaceae fam. nov.), a new toxic dinoflagellate with a complex life cycle and behavior. *J. Phycol.* 32:157–64.

Sulzman, F. N., Krieger, N. R., Gooch, V. D., and Hastings, J. W. (1978). A circadian rhythm of the luciferin binding protein from *Gonyaulax polyedra. J. Comp. Physiol.* 128:251–7.

Suzuki, T., Mitsuya, T., Imai, M., and Yamasaki, M. (1997). DSP toxin contents in *Dinophysis fortii* and scallops collected at Mutsu Bay, Japan. *J. Applied Phycol.* 8:509–15.

Sweeney, B. M. (1969). *Rhythmic Phenomena in Plants.* London and New York: Academic Press.

Sweeney, B. M. (1971). Laboratory studies of a green *Noctiluca* from New Guinea. *J. Phycol.* 7:53–8.

Sweeney, B. M. (1978). Ultrastructure of *Noctiluca miliaris* (Pyrrophyta) with green flagellate symbionts. *J. Phycol.* 14:116–20.

Sweeney, B. M. (1979). The bioluminescence of dinoflagellates. In *Biochemistry and Physiology of Protozoa* (Levandowsky, M., and Hutner, S. H., eds.), Vol. 1, pp. 287–306. New York: Academic Press.

Sweeney, B. M. (1980). Intracellular source of bioluminescence. *Int. Rev. Cytol.* 68:173–95.

Sweeney, B. M. (1982). Microsources of bioluminescence in *Pyrocystis fusiformis* (Pyrrophyta). *J. Phycol.* 18:412–16.

Swift, E., Biggley, W. H., and Seliger, H. H. (1973). Species of oceanic dinoflagellates in the genera *Dissodinium* and *Pyrocystis*: Interclonal and intraspecific comparisons of color and photon yield of bioluminescence. *J. Phycol.* 9:420–6.

Tang, E. P. Y. (1996). Why do dinoflagellates have lower growth rates? *J. Phycol.* 32:80–4.

Taylor, D. L. (1969a). On the regulation and maintenance of algal numbers in zooxanthellae-coelenterate symbiosis, with a note on the nutritional relationship in *Anemonia sulcata. J. Mar. Biol. Assoc. UK* 49:1057–65.

Taylor, D. L. (1971). On the symbiosis between *Amphidinium klebsii* (Dinophyceae) and *Amphiscolops langerhansi* (Turbellaria: Acoela). *J. Mar. Biol. Assoc. UK* 51:301–13.

Taylor, F. J. R. (1971). Scanning electron microscopy of thecae of the dinoflagellate genus *Ornithocercus. J. Phycol.* 7:249–58.

Taylor, F. J. R. (1980). On dinoflagellate evolution. *BioSystems* 13:65–108.

Trench, R.K. (1993). Microlgal–invertebrate symbioses: a review. *Endocytobios Cell Res.* 9:135–75.

Triemer, R. E. (1982). A unique mitotic variation in the marine dinoflagellate *Oxyrrhis marina* (Pyrrophyta). *J. Phycol.* 18:399–411.

Vogel, K., and Meeuse, B. J. D. (1968). Characterization of the reserve granules from the dinoflagellate *Thecadinium inclination* Balech. *J. Phycol.* 4:317–18.

von Stosch, H. A. (1972). La signification cytologique de la "cyclose nucléaire" dans le cycle de vie des Dinoflagellés. *Soc. Bot. Fr., Memoires*, pp. 201–12.

von Stosch, H. A. (1973). Observations on vegetative reproduction and sexual life cycles of two freshwater dinoflagellates, *Gymnodinium pseudopalustre* Schiller and *Woloszynskia apiculata* sp. nov. *Br. Phycol. J.* 8:105–34.

Wang, J-T., and Douglas, A. E. (1997). Nutrients signals and photosynthate release by symbiotic algae. *Plant Physiol.* 114:631–6.

Wilcox, L. W., and Wedemeyer, G. J. (1991). Phagotrophy in the freshwater photosynthetic dinoflagellate *Amphidinium cryophilum*. *J. Phycol.* 27:600–9.

Wilcox, L. W., Wedemeyer, G. J., and Graham, L. E. (1982). *Amphidinium cryophilum* sp. nov. (Dinophyceae), a new freshwater dinoflagellate. II. Ultrastructure. *J. Phycol.* 18:18–30.

Withers, N. (1982). Ciguatera fish poisoning. *Annual Review Medicine* 33:97–111.

Xiao-Ping, G., and Jing-Yan, L. (1986). Nuclear division in the marine dinoflagellate *Oxyrrhis marina*. *J. Cell Sci*. 85:161–75.

Zardoya, R., Costas, E., Lopez-Rodas, V., Garrido-Pertierra, A., and Bautista, J. M. (1995). Revised dinoflagellate phylogeny inferred from molecular analysis of large-subunit ribosomal RNA sequences. *J. Mol. Evol.* 41:637–45.

Zhou, J., and Fritz, L. (1994). Okadaic acid localizes to chloroplasts in the DSP-toxin-producing dinoflagellates *Prorocentrum lima* and *Prorocentrum maculosum*. *Phycologia* 33:455–61.

PART V · Evolution of two membranes of chloroplast endoplasmic reticulum

Algae with two membranes of chloroplast endoplasmic reticulum (chloroplast E.R.) have the inner membrane of chloroplast E.R. surrounding the chloroplast envelope. The outer membrane of chloroplast E.R. is continuous with the outer membrane of the nuclear envelope and has ribosomes on the outer surface (Fig. V.1).

The algae with two membranes of chloroplast E.R. evolved by a secondary endosymbiosis. In the secondary endosymbisosis (Fig. V.1) (Lee, 1977; see Chapter 1 references), a phagocytic protozoan took up a eukaryotic photosynthetic alga into a food vesicle. Instead of being phagocytosed by the protozoan, the photosynthetic alga became established as an endosymbiote within the food vesicle of the protozoan. The endosymbiotic photosynthetic alga benefited from the acidic environment in the food vesicle that kept much of the inorganic carbon in the form of carbon dioxide, the form needed by ribulose bisphosphate/ carboxylase for carbon fixation (see Introduction chapter for further explanation). The host benefited from receiving some of the photosynthate from the endosymbiotic alga. The food vesicle membrane eventually fused with the endoplasmic reticulum of the host protozoan, resulting in ribosomes on the outer surface of this membrane, which became the outer membrane of the chloroplast E.R. Through evolution, ATP production and other functions of the endosymbiont's mitochondrion were taken over by the mitochondria of the protozoan host, and the mitochondria of the endosymbiont were lost. The host nucleus also took over some of the genetic control of the endosymbiont, with a reduction in the size and function of the nucleus of the endosymbiont. The resulting cytology is characteristic of the extant cryptophytes which have a nucleomorph representing the degraded endosymbiont nucleus, as well as starch produced in what remains the endosymbiont cytoplasm.

The type of chloroplast E.R. that exists in the Heterokontophyta and the

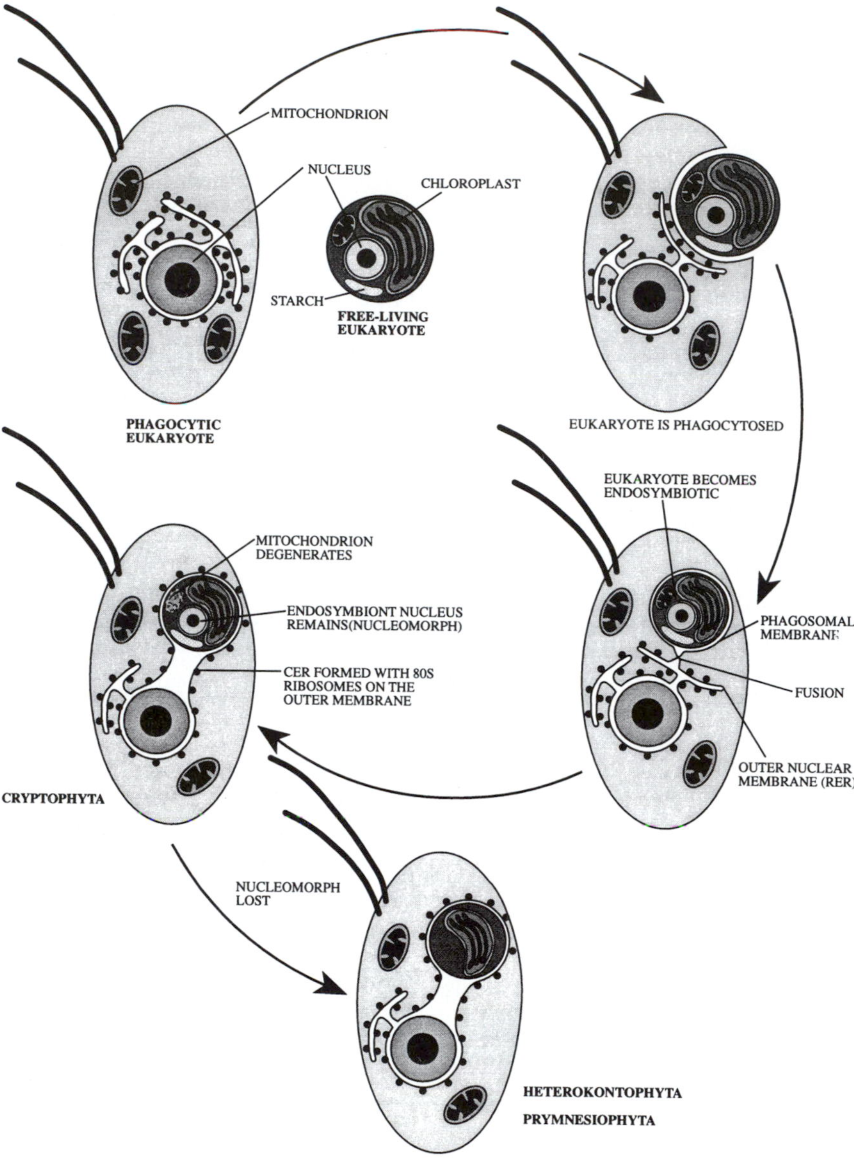

Fig. V.1 The sequence of events that led to the evolution of algae with two membranes of chloroplast endoplasmic reticulum. (Drawing by Brec Clay.)

Prymnesiophyta resulted from further reduction. The nucleomorph was completely lost and storage product formation was taken over by the host. The resulting cell had two membranes of chloroplast envelope surrounding the chloroplast. Outside of this was the inner membrane of chloroplast E.R. that was the remains of the plasma membrane of the endosymbiont. Outside of this was the outer membrane of chloroplast E.R. which was the remains of the food vesicle membrane of the host.

Although the above evolutionary scheme is discussed in one sequence, it is probable that two membranes of chloroplast E.R. evolved at least twice, with one line going to the Cryptophyta, and the second (or more) line leading to the Heterokontophyta and Prymensiophyta.

The organisms with two members of chloroplast E.R. are:

Cryptophyta (cryoptophytes) (Chapter 8): nucleomorph present between inner and outer membrane of chloroplast endoplasmic reticulum; starch is formed in grains between inner membrane of chloroplast endoplasmic reticulum and chloroplast envelope; chlorophylls *a* and *c*; phycobiliproteins; periplast is inside plasma membrane.

Heterokontophyta (heterokonts) (Chapters 9–17): anterior tinsel and posterior whiplash flagellum, chlorophylls *a* and *c*, fucoxanthin, storage product usually chrysolaminarin occurring in vesicles in cytoplasm.

- **Chrysophyceae** (golden-brown algae) (Chapter 9)
- **Synurophyceae** (Chapter 10)
- **Dictyochophyceae** (silicoflagellates) (Chapter 11)
- **Pelagophyceae** (Chapter 12).
- **Bacillariophyceae** (diatoms) (Chapter 13).
- **Raphidophyceae** (chloromonads) (Chapter 14)
- **Xanthophyceae** (yellow-green algae) (Chapter 15)
- **Eustigmatophyceae** (Chapter 16)
- **Phaeophyceae** (brown algae) (Chapter 17)

Prymnesiophyta (haptophytes) (Chapter 18): two whiplash flagella, haptonema present, chlorophylls *a* and *c*, fucoxanthin, scales common outside cell, storage product usually chrysolaminarin occurring in vesicles in cytoplasm.

8 · Cryptophyta

Cryptophyceae

This group is composed primarily of flagellates that occur in both marine and freshwater environments. The cells contain chlorophylls *a* and c_2 and phycobiliproteins that occur inside the thylakoids of the chloroplast. The cell body has a dorsiventral shape, with the cells flattened in one plane. A few members lack chloroplasts and are heterotrophic, but most have a single lobed chloroplast with a central pyrenoid.

Cell structure

There are two apically or laterally attached flagella at the base of a depression. Each flagellum is approximately the same length as the body of the cell (Figs. 8.1 and 8.10). Depending on the species, there are one or two rows of microtubular hairs attached to the flagellum. In *Cryptomonas* sp., the hairs on one flagellum are 2.5 μm long and in two rows whereas the hairs on the other flagellum are only 1 μm long and arranged in a single row (Heath et al., 1970). Small, 150-nm-diameter organic scales (Fig. 8.2) are common on the flagellar surface and sometimes on the cell body (Lee and Kugrens, 1986).

The outer portion of the cell or **periplast** (Gantt, 1971), is composed of the plasma membrane and a plate or series of plates directly under the plasma membrane (Fig. 8.1) (Kugrens and Lee, 1987). The number and shape of these plates are important in the taxonomy of the group. New periplast plates are added in an area adjacent to the vestibulum (Brett and Wetherbee, 1996).

The chloroplast most likely evolved from a symbiosis between an organism similar to the phagocytic cryptomonad *Goniomonas* and a red alga (Kugrens and Lee, 1991; Liaud et al., 1997; McFadden et al., 1994). The chloroplast is surrounded by two membranes of chloroplast endoplasmic reticulum and the two membranes of the chloroplast envelope (Fig. 8.1). Between the outer membrane and the inner membrane of the chloroplast endoplasmic reticulum are starch grains and a nucleomorph (Figs. 8.1 and 8.4). The **nucleomorph** contains nucleic acids that encode rRNAs that are incorporated into ribosomes in the space between the two membranes of chloroplast E.R. (Douglas et al., 1991; Maier et al., 1991; McFadden et al.,

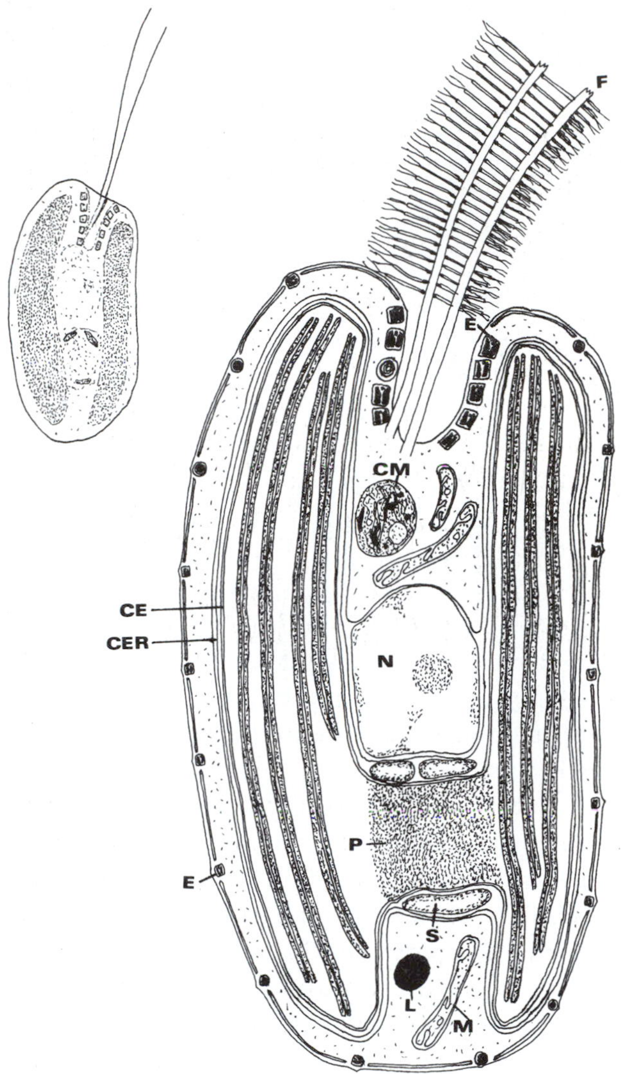

Fig. 8.1 Drawing of a cell of the Cryptophyta as seen in the light and electron microscope. (CE) Chloroplast envelope; (CER) chloroplast endoplasmic reticulum; (CM) Corps de Maupas; (D) dorsal; (E) ejectisome; (L) lipid; (M) mitochondrion; (N) nucleus; (NM) nucleomorph; (P) pyrenoid; (PP) periplast plate; (S) starch; (V) ventral.

1994). The nucleomorph is probably the remnant of the nucleus of the endosymbiont in the event that led to chloroplast E.R. The nucleomorph is surrounded by an envelope that has pores similar to those in a nuclear envelope. The nucleomorph exhibits a rudimentary type of division utilizing microtubules (Morrall and Greenwood, 1982). The nucleomorph divides in preprophase of the main nucleus following basal body replication, but before division of the chloroplast and the chloroplast endoplasmic reticulum (McKerracher and Gibbs, 1982). The only cryptophyte that is known to lack a nucleomorph is *Goniomonas* (Fig. 8.8), a colorless cryptophyte that lacks a plastids. A second colorless alga, *Chilomonas*, is a reduced form of a

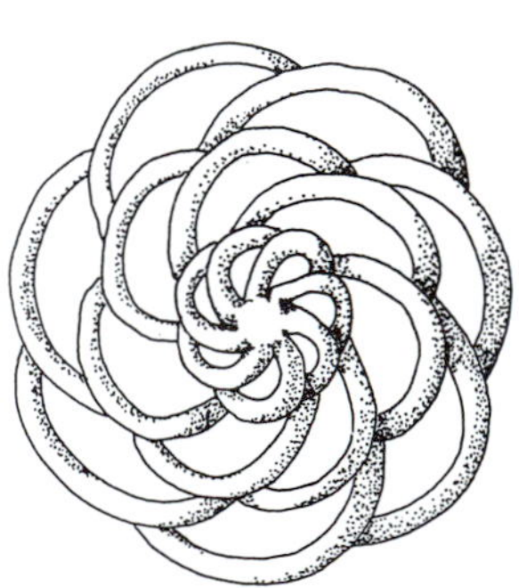

Fig. 8.2 Drawing of the most common type of flagellar scale found in freshwater cryptophytes. (From Lee and Kugrens, 1986.)

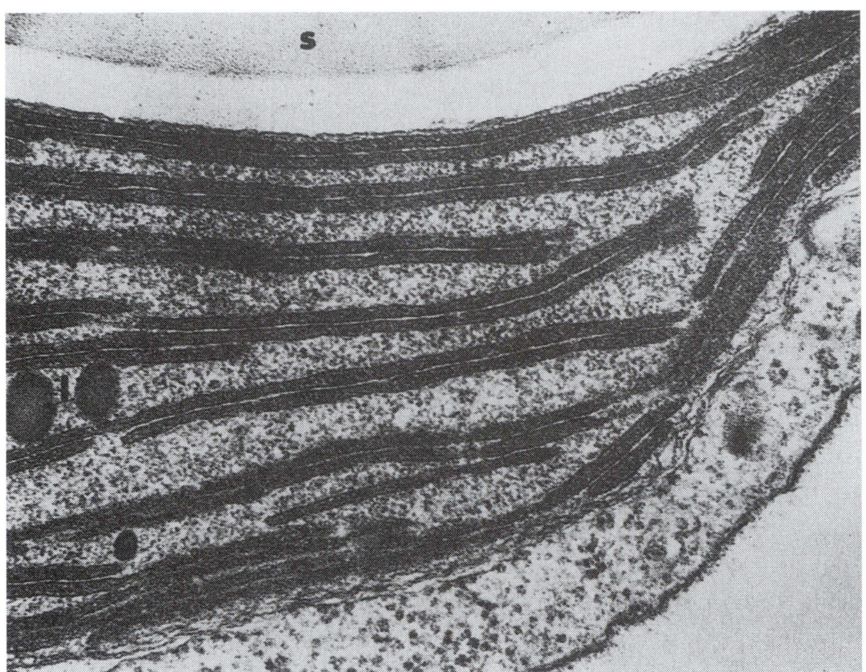

Fig. 8.3 Electron micrograph of part of a chloroplast of *Chroomonas mesostigmatica*. The thylakoids are grouped in pairs, with the dense contents representing the phycobiliproteins. Also present are lipid droplets (l) and a large starch grain (s). × 50 000. (From Dodge, 1969.)

photosynthetic cryptophyte and contains a leucoplast and a nucleomorph (McKerracher and Gibbs, 1982).

In the chloroplast, the thylakoids are grouped in pairs (Fig. 8.3), and there are no connections between adjacent thylakoids. The Cryptophyta is the only group to have this arrangement of thylakoids. Chlorophylls *a* and c_2 are present. The major carotenoid present is α-carotene, and the major xanthophyll, diatoxanthin. There are three spectral types of phycoerythrin and three spectral types of phycocyanin, all of which are different from the phycobiliproteins found in the cyanobacteria and red algae (Hill and Rowan, 1989). The phycobiliproteins are in the intrathylakoid space (inside the thylakoids, (Fig. 8.3) (Gantt et al., 1971; Spear-Bernstein and Miller, 1984), and are not on the stromal side of the thylakoids in phycobilisomes as occurs in the cyanobacteria and red algae. Each photosynthetic cryptophyte has only one species of phycobiliprotein – either a phycoerythrin or a phycocyanin – but never both. No allophycocyanin is present (Gantt, 1979). Allophycocyanin acts as a bridge in the transfer of light energy from phycoerythrin and phycocyanin to chlorophyll *a* of the reaction center in red algae and cyanobacteria. The presence of allophycocyanin may not be necessary in the cryptophytes because of the greater absorption range of cryptophycean phycobiliproteins in conjunction with chlorophyll *c* overlaps the chlorophyll *a* absorption spectrum. There is a variation in the amount of pigments under different light-intensity conditions. Cells of *Cryptomonas*

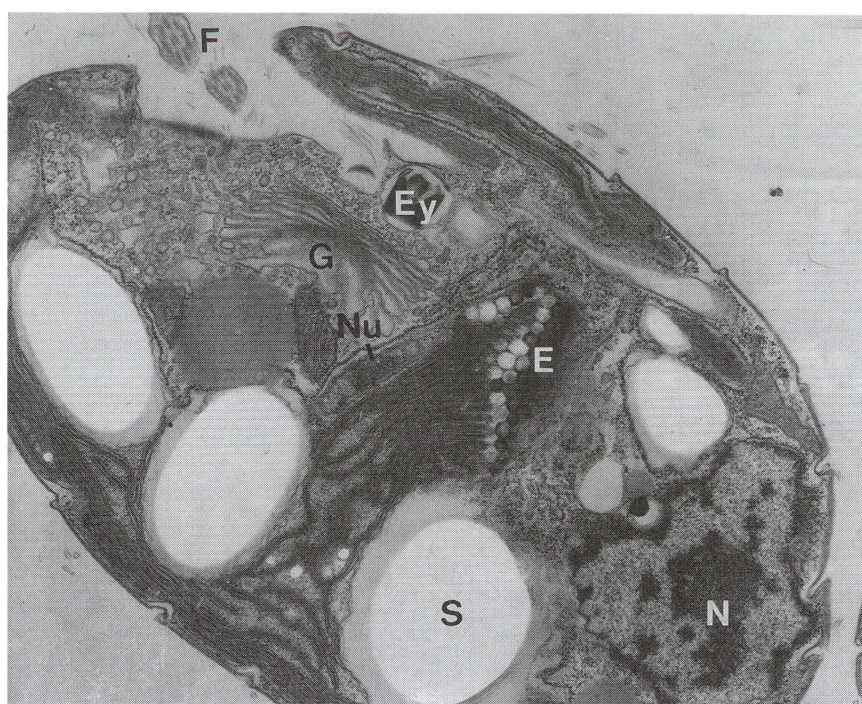

Fig. 8.4 Transmission electron micrograph of a cell of *Chroomonas mesostigmatica* showing the eyespot (E) present in the chloroplast. (Ey) Ejectisome; (F) flagellum; (G) Golgi; (N) nucleus; (Nu) nucleomorph; (S) starch. (Micrograph provided by Paul Kugrens.)

grown under low light-intensity conditions (10 μE m^{-2} s^{-1}) contain twice as much of chlorophylls *a* and c_2, and six times as much phycoerythrin per cell, as those grown under high light-intensity conditions (260 μE m^{-2} s^{-1}) (Thinh, 1983). Under low light-intensity conditions there is a higher concentration of phycoerythrin and the thylakoids are thicker.

The reserve product (similar in appearance to starch grains) is appressed to the pyrenoid area outside of the chloroplast envelope but inside the chloroplast E.R. The cryptophytes are the only algae that form their storage product in this area. The starch is an α-1,4-glucan composed of about 30% amylose and amylopectin. Cryptophycean starch is similar to potato starch and starch found in the green algae and dinoflagellates (Antia et al., 1979).

Some of the Cryptophyceae have eyespots. The eyespots that have been reported consist of lipid granules inside the chloroplast envelope. In *Chroomonas mesostigmatica*, the red eyespot is in the center of the cell (Fig. 8.4) and is an extension of the chloroplast beyond the pyrenoid (Dodge, 1969). In *Cryptomonas rostella*, the eyespot is beneath the chloroplast membrane near the depression. Some of the Cryptophyceae exhibit positive phototaxis, with maximum sensitivity of the colorless *Chilomonas* being in the blue at 366 nm (Halldal, 1958).

The Cryptophyceae have projectiles called **ejectisomes**, which are of different structure from the trichocysts of the Dinophyceae and which are

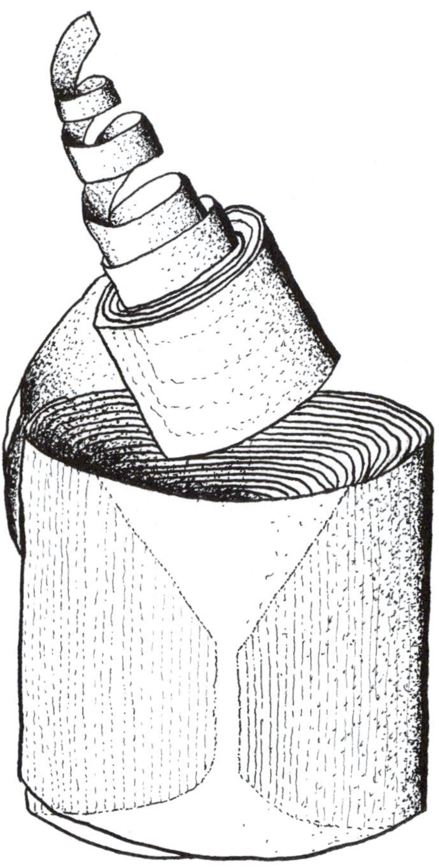

Fig. 8.5 General organization of an ejectisome showing the two subparts. (After Hovasse et al., 1967.)

probably closely related to the R-bodies of the kappa particles of the ciliates (Hovasse et al., 1967; Kugrens et al., 1994). Within a cell there are usually large ejectisomes near the anterior depression and smaller ejectisomes around the cell periphery (Figs. 8.1, 8.4 and 8.8). Both sizes of ejectisomes have the same structure; they are made up of two unequal-sized bodies enclosed within a single membrane (Fig. 8.5). Each of these bodies is a long tape curled up on a very tight spiral. The tape is tapered, with the greatest width being on the outside of the ejectisome. The smaller body is joined to the first and sits at an angle within the V-shaped portion of the larger body. The two bodies actually constitute one long tape with two spirals. The bodies are always arranged so that the smaller body is near the surface. The ejectisomes discharge when the organism is irritated (Fig. 8.6), the discharged ejectisome being a long tubular structure with a short portion at an angle to the long portion. The discharged small ejectisome from the cell periphery is 4 μm long, whereas that of a larger ejectisome from under the anterior depression is about 20 μm long. The discharge of the ejectisome results in a movement of the organism in the opposite direction. The discharge of the ejectisome could function as an escape mechanism, or it could be a direct defense mechanism causing damage to an offending organism. Ejectisomes originate in vesicles in the area of Golgi bodies.

The **Corps de Maupas** is a large vesicular structure in the anterior portion of the cell (Fig. 8.1). Its main function is probably that of disposing of unwanted protoplasmic structures by digestion (Lucas, 1970a,b).

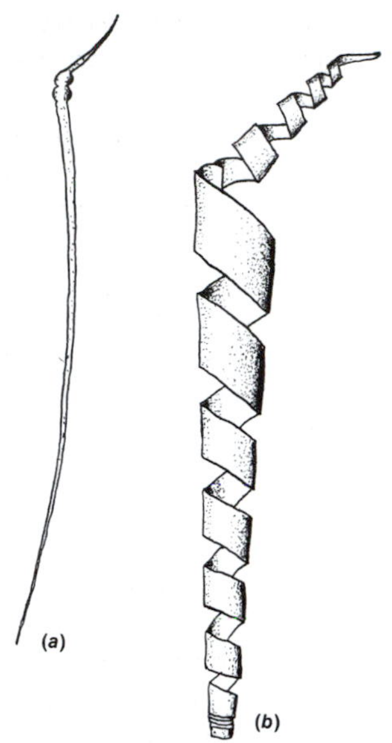

Fig. 8.6 (*a*) A drawing of a discharged ejectisome. (*b*) A model of an ejectisome being fired to the outside of the cell. (After Hovasse et al., 1967.)

Ecology

In comparison with other algal groups, the Cryptophyta appear to be especially light sensitive, often forming the deepest living populations in clear oligotrophic lakes (Nauwerk, 1968). In higher mountain and north temperate lakes, cryptomonads and other flagellates are present in the water column throughout the winter. Because of the low light intensity under snow and ice cover, these algae concentrate in surface waters to receive sufficient light from net photosynthesis (Wright, 1964; Pechlauer, 1971). Survival at these extremely low light levels depends not only on a highly efficient photosynthetic system, but also on slow rates of cell respiration at low water temperatures and reduced winter zooplankton grazing. In spring, with the disappearance of snow and resulting sudden increase in light in Arctic and mountain lakes, cryptomonads suffer considerable light stress, such that the biomass maximum moves to deeper waters (Kalff and Welch, 1974).

Cryptophytes will often undergo diel vertical migrations with an amplitude less than 5 meters. In small humic forest lakes, species of *Cryptomonas* are positively phototactic in the morning, moving into the phosphorus-depleted upper layer. Later in the day the cells move away from the uppermost water layer, avoiding high levels of irradiance, and move into the phosphorus-rich hypolimnion (Arvola et al., 1991). A further advantage of this cycle is the reduction of grazing pressure by zooplankton which often migrate in the reverse direction.

Symbiotic associations

Mesodinium rubrum is a marine planktonic holotrich ciliate of extremely wide geographical distribution that colors the water in which it is growing red. It has been recorded from neritic locations such as bays and fjords; away from the coast it is usually associated with regions of upwelling and in such conditions the blooms have been recorded as extending over areas as large as 100 square miles. The color of the ciliate (Fig. 8.7) is due to numerous reddish-brown chloroplasts, which belong to a single cryptophycean alga that lives endosymbiotically inside the ciliate (Hibberd, 1977). The

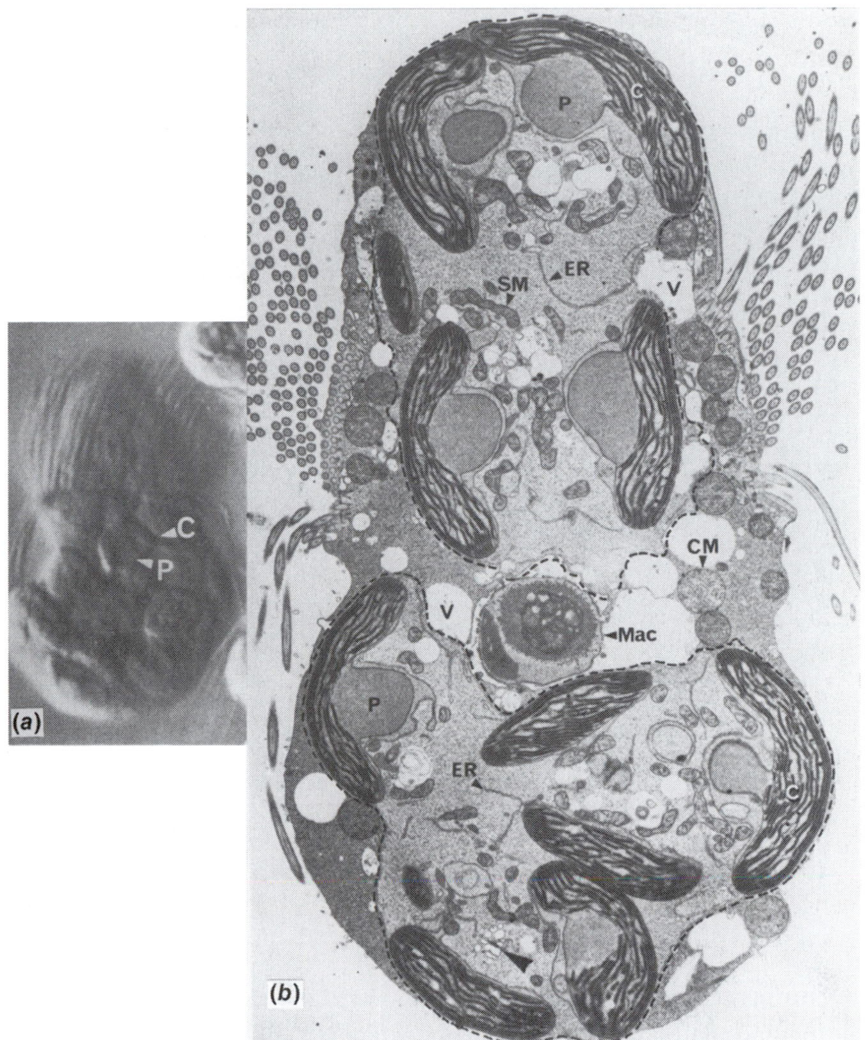

Fig. 8.7 *Mesodinium rubrum* with its cryptomonad symbiont. (*a*) Light microscopical section of the ciliate showing the chloroplast (C) and pyrenoid (P) of the cryptomonad endosymbiont. (*b*) Electron microscopical section. The dotted lines indicate the boundary between the cytoplasm of the ciliate and that of the cryptomonad symbiont; the difference in density of the two cells is particularly clear. The symbiont nucleus, one of the macronuclei, and the micronucleus of the ciliate are out of plane of the section. (CM) Ciliate mitochondrion; (ER) endoplasmic reticulum; (Mac) macronucleus; (P) pyrenoid; (SM) symbiont mitochondrion; (V) vacuole. The large arrowhead indicates a possible region of Golgi activity. ×4500. (From Hibberd, 1977.)

cryptophyte is surrounded by a single membrane, and has a nucleus and the normal cytology and pigments of the Cryptophyceae. The endosymbiotic cryptophyte is able to fix ^{14}C in the light, evolve oxygen in photosynthesis, and assimilate ^{32}P, indicating that it is a functioning autotroph. The association is probably similar to that of symbiotes in other classes, with the endosymbiont providing the host with photosynthate and the host providing the endosymbiont with a protected environment. Blooms of *Mesodinium rubrum* are a regular feature of upwelling ecosystems. The organism has three characteristics that enable it to compete effectively with other autotrophic plankton (Smith and Barber, 1979). (1) It is motile, swimming at rates of 2.0 to 7.2 m h^{-1}, an order of magnitude greater than the maximum swimming speeds attained by dinoflagellates. (2) It has strong phototropisms, being being positively phototactic in an increasing light

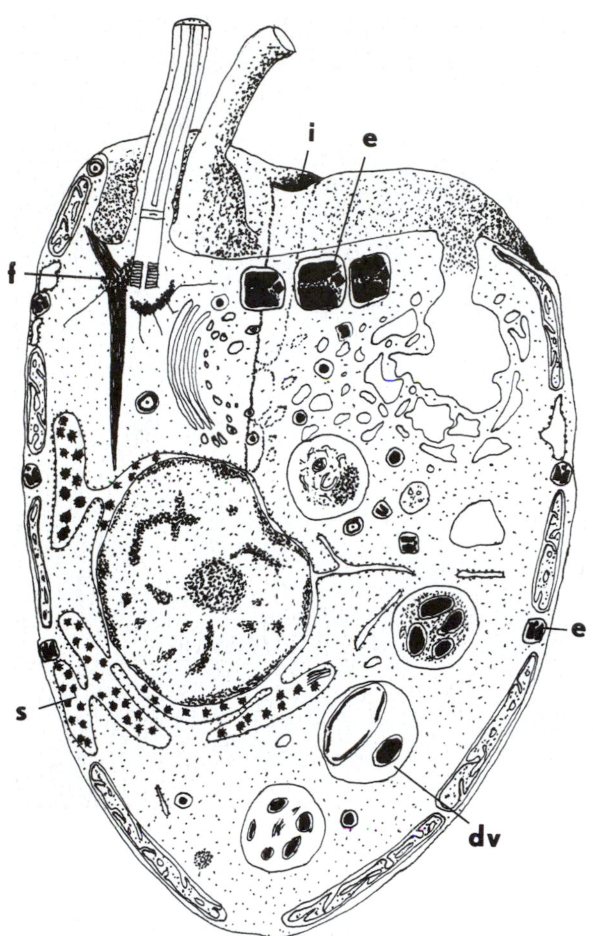

Fig. 8.8 Reconstruction of a cell of *Geniomonas truncata*. (f) Flagellar roots; (s) storage granules; (e) ejectisomes; (dv) digestive vacuole; (i) infundibulum. (After Mignot, 1965.)

regime in the morning and negatively phototactic in decreasing light and in nutrient-depleted waters. (3) It has extremely high photosynthetic rates (1000 to 2000 mg C m^{-3} h^{-1}), equaling the highest ever observed for oceanic plankton. Conventional dinoflagellate or diatom blooms typically have only 60 to 70 mg C m^{-3} h^{-1}.

Classification

The Cryptophyceae can be divided into three orders (Clay et al., 1999):

Order 1 Goniomonadales: colorless cells with no plastids.

Order 2 Cryptomonadales: cells usually reddish in color with chloroplasts containing the phycobiliprotein Cr-phycoerythrin.

Order 3 Chroomonadales: cells usually blue-green in color with the chloroplasts containing the phycobiliprotein Cr-phycocyanin.

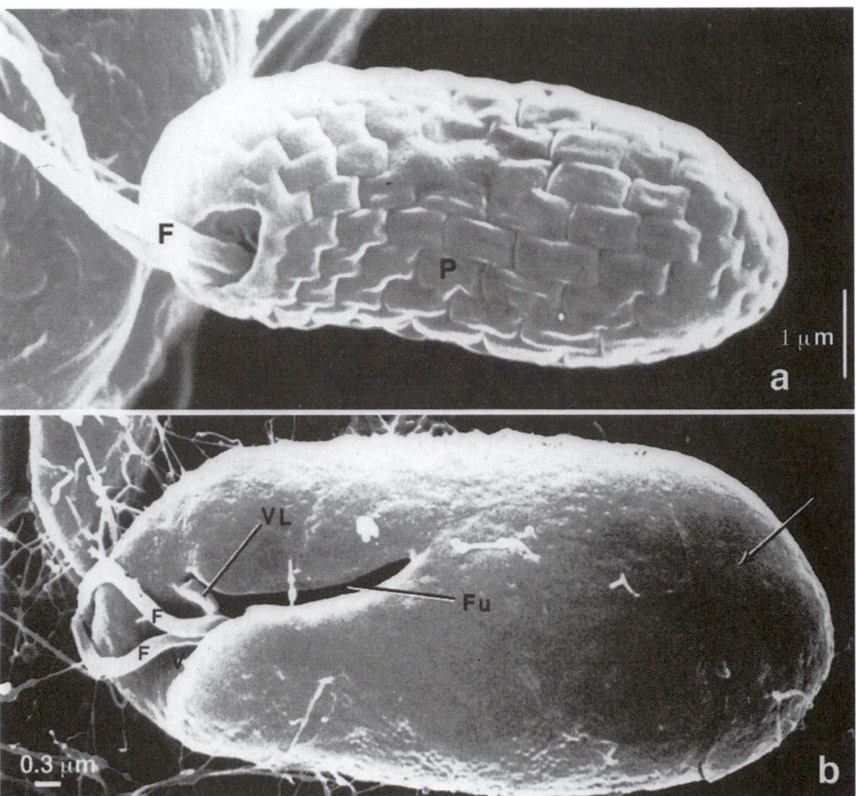

Fig. 8.9 Scanning electron micrographs of *Chroomonas oblonga* (*a*) and *Cryptomonas* sp (*b*). *Chroomonas oblonga* has multiple periplast plates (P) under the plasma membrane, no furrow is present and the flagella (F) arise from an anterior vestibular depression. *Cryptomonas* sp. has a smooth surface that is produced by a single periplast plate under the plasma membrane. The furrow (Fu) is an extension of the anterior vestibulum. A vestibular ligule (VL) overlaps the vestibulum. (From Kugrens et al., 1986.)

GONIOMONADALES

Goniomonas (Fig. 8.8) is a freshwater flagellate that is the sole alga in the order. *Goniomonas* is colorless and does not contain a plastid. Food organisms are taken up by an anterior tubular invagination, the infundibulum, and digested in food vacuoles in the cytoplasm. Storage granules occur inside an extension of the outer membrane of the nuclear envelope. Large ejectisomes occur under the anterior plasma membrane and small ejectisomes occur between the periplast plates.

CRYPTOMONADALES

Cells in this order are usually reddish in color because of the presence of the phycobiliprotein Cr-phycoerthyrin in the typically bilobed chloroplast, joined in the center by the pyrenoid. These algae have an asymmetric shape which can be attributed, in part, to a subapical depression called the vestibulum which may extend internally to form a gullet or progress along the ventral surface into a furrow (Fig. 8.9(*b*) and 8.10(*b*)) (Kugrens and Lees, 1991). Large ejectisomes occur in rows under the furrow. Sexual reproduc-

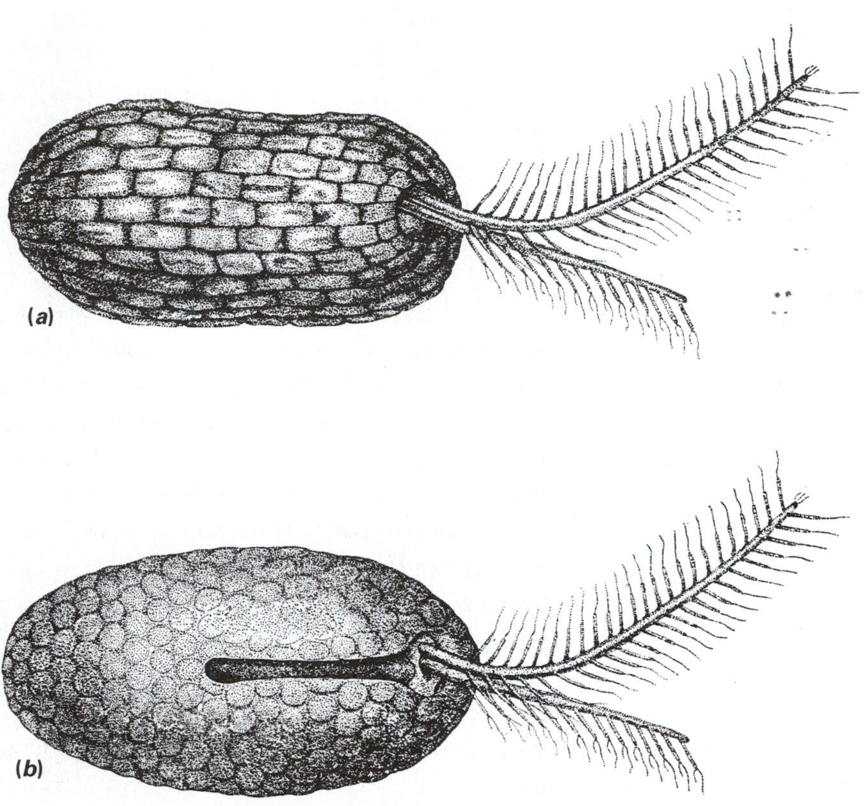

Fig. 8.10 Drawings of *Chroomonas* sp. (*a*) and *Cryptomonas* sp. (*b*). In both cells, the flagella have two rows of hairs on one flagellum and one row of hairs on the other flagellum. *Cryptomonas* has a ventral furrow running from the anterior vestibulum to which the flagella are attached. *Chroomonas* has only the anterior vestibulum. Both genera have a periplast composed of a number of plates. (Drawn by Georgia Heisterkamp.)

tion occurs in *Cryptomonas* sp. (Fig. 8.11) (Kugrens and Lee, 1988). Vegetative cells act as gametes. Initially, two cells line up with the ventral side of one cell next to the ventral side of the second cell. The posterior end of one cell attaches to the midventral area of the second cell, and cell fusion begins. Continued plasmogamy produces an *r*-shaped configuration in the fusing gametes. When the cells have fused completely, a quadriflagellate, cone-shaped cell is formed, which eventually becomes spherical. The quadriflagellate cell divides, presumably by meiosis, to form the haploid vegetative cells.

CHROOMONADALES

Cells in this order are usually blue-green in color with the chloroplasts containing the phycobiliprotein Cr-phycocyanin. *Chroomonas* (Figs. 8.9(*a*) and 8.10(*a*)) has a bilobed chloroplast joined in the center by a pyrenoid. The anterior end has a vestibulum, in which the two flagella are attached (Kugrens et al., 1987). Large ejectisomes occur under the plasma membrane of the vestibulum.

Fig. 8.11 The life cycle of Cryptomonas sp. (Adapted from Kugrens and Lee, 1988.)

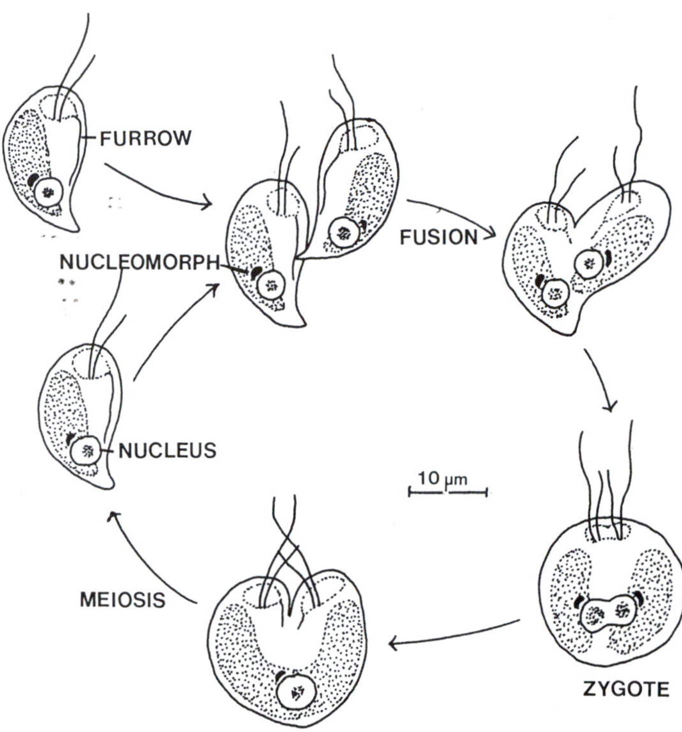

Fig. 8.12 *Left:* Øjvind Moestrup and *right:* Michael Melkonian.

Fig. 8.12 *cont.* **Michael Melkonian** Born October 4, 1948. Dr Melkonian received a Diploma in Biology (1974) and his PhD (1978) from the University of Hamburg. From 1978 to 1983, he was a Wissenschaftlicher Assistent at the Botanisches Institut, University of Munster and in 1983 Habilitation (Venia Legendi) in Botany and Privatdozent at the Botanisches Institut. Dr Melkonian is well known for his many investigations on the fine structure of the algae, particularly the green algae.
Øjvind Moestrup Born December 15, 1941, in Marvede, Denmark. Dr Moestrup received his Cand. scient. from the University of Copenhagen in 1969 and his Dr scient. from the University of Copenhagen in 1983. Since 1970, Dr Moestrup has been employed by the University of Copenhagen where he is now chairman of the Institut of Sporeplanter. Dr Moestrup worked with, and was influenced by, two prominent phycologists, Dr Tyge Christensen at Copenhagen and Professor Irene Manton at the University of Leeds, Dr Moestrup's main research interests are the systematics and cytology of the algae.

References

Antia, N. J., Cheng, J. Y., Foyle, R. A. J., and Percival, E. (1979). Marine cryptomonad starch from autolysis of glycerol-grown *Chroomonas salina*. *J. Phycol.* 15:57–62.

Arvola, L., Ojala, A., Barbosa, F., and Heaney, S.I. (1991). Migration behaviour of three cryptophytes in relation to environmental gradients: an experimental approach. *Br. Phycol. J.* 26:361–73.

Brett, S. J., and Wetherbee, R. (1996). Periplast development in Cryptophyceae. II. Development of the inner periplast component in *Rhinomonas pauea*, *Proteomonas sulcata* [haplomorph], *Rhodomonas baltica*, and *Cryptomonas ovata*. *Protoplasma* 192:40–8.

Clay, B. R., Kugrens, P., and Lee, R. E. (1999). The classification of the Cryptophyta. *Encycl. Life Sci.* (in press).

Dodge, J. D. (1969). The ultrastructure of *Chroomonas mesostigmatica* Butcher (Cryptophyceae). *Arch. Mikrobiol.* 69:266–80.

Douglas, S. E., Murphy, C. A., Spencer, D. F., and Gray, M. W. (1991). Molecular evidence that cryptomonad algae are evolutionary chimeras of two phylogenetically distinct unicellular eukaryotes. *Nature* 350:148–51.

Gantt, E. (1971). Micromorphology of the periplast of *Chroomonas* sp. (Cryptophyceae). *J. Phycol.* 7:177–84.

Gantt, E. (1979). Phycobiliproteins of Cryptophyceae. In *Biochemistry of Protozoa*, ed. N. Levandowsky, and S. A. Hutner, pp. 121–37. New York: Academic Press.

Gantt, E., Edwards, M. R., and Provasoli, L. (1971). Chloroplast structure of the Cryptophyceae. Evidence for phycobiliproteins within the intrathylakoidal spaces. *J. Cell Biol.* 48:280–90.

Halldal, P. (1958). Action spectra of phototaxis and related problems in Volvocales: *Ulva*-gametes and Dinophyceae. *Physiol. Plant.* 11:118–53.

Heath, I. B., Greenwood, A. D., and Griffiths, H. B. (1970). The origin of flimmer in *Saprolegnia*, *Dictyuchus*, *Synura* and *Cryptomonas*. *J. Cell Sci.* 7:445–61.

Hibberd, D. J. (1977). Observations on the ultrastructure of the cryptomonad endosymbiont of the red water ciliate *Mesodinium rubrum*. *J. Mar. Biol. Assoc. UK* 57:45–61.

Hill, D. R. A., and Rowan, K. S. (1989). The biliproteins of the Cryptophyceae. *Phycologia* 28:455–3.

Hovasse, R., Mignot, J. P., and Joyon, L. (1967). Nouvelles observations sur les trichocystes des Cryptomonadines et les "R bodies" des particules kappa de *Paramecium aurelia* Killer. *Prostistologica* 3:241–55.

Kalff, J., and Welch, H. E. (1974). Phytoplankton production in Char Lake, a natural polar lake, Cornwallis Is., Northwest Territories. *J. Fish. Res. Board Can.* 31:621–36.

Kugrens, P., and Lee, R. E. (1987). An ultrastructural survey of cryptomonad periplasts using quick-freezing freeze-fracture techniques. *J. Phycol.* 23:365–76.

Kugrens, P., and Lee, R. E. (1988). Ultrastructure of fertilization in a cryptomonad. *J. Phycol.* 24:385–93.

Kugrens, P., and Lee, R. E. (1991). Organization of cryptomonads. In *The Biology of Free-living Heterotrophic Flagellates*, ed. P. J. Patterson, and J. Larsen, pp. 219–33. Oxford: Clarendon Press.

Kugrens, P., Lee, R. E., and Andersen, R. A. (1986). Cell form and surface patterns in *Chroomonas* and *Cryptomonas* cells (Cryptophyta) as revealed by scanning electron microscopy. *J. Phycol.* 22:512–22.

Kugrens, P., Lee, R. E., and Andersen, R. A. (1987). Ultrastructural variations in cryptomonad flagella. *J. Phycol.* 23:511–18.

Kugrens, P., Lee, R. E., and Kugrens, P. (1986). The occurrence and structure of flagellar scales in some freshwater cryptophytes. *J. Phycol.* 22:549–52.

Lee, R. E., and Corliss, J. O. (1994). Ultrastructure, biogenesis, and functions of extrusive organelles in selected non-ciliate protists. *Protoplasma* 181:164–90.

Liaud, M-F., Brandt, U., Scherzinger, M., and Cerff, R. (1997). Evolutionary origin of cryptomonad microalgae. Two novel chloroplast/cytosol-specific GAPDH genes as potential markers of ancestral endosymbiont and host cell components. *J. Mol. Evol.* 44 (Suppl. 1): 528–37.

Lucas, I. A. N. (1970a). Observations on the fine structure of the Cryptophyceae. I. The genus *Cryptomonas*. *J. Phycol.* 6:30–8.

Lucas, I. A. N. (1970b). Observations on the ultrastructure of representatives of the genera *Hemiselmis* and *Chroomonas* (Cryptophyceae) *Br. Phycol. J.* 5:29–37.

McFadden, G. I., Gilson, P. R., and Hill, D. R. A. (1994). *Goniomonas*: rRNA sequences indicate that this phagotrophic flagellate is a close relative of the host component of cryptomonads. *Eur. J. Phycol.* 29:29–32.

McKerracher, L., and Gibbs, S. P. (1982). Cell of nucleomorph division in the alga *Cryptomonas*. *Can. J. Bot.* 60:2440–52.

Maier, U-G., Hofmann, C. J. B., Eschbach, S., Wolters, J., and Igloi, G. (1991). Demonstration of nucleomorph-encoded small subunit ribosomal RNA in cryptomonads. *Mol. Gen. Genet.* 230:155–60.

Mignot, J. P. (1965). Etrude ultrastructurale de *Cyathomonas truncata* From. (Flagellé Cryptomonadine). *J. Microscopie* 4:239–52.

Morrall, S., and Greenwood, A. D. (1982). Ultrastructure of nucleomorph division in species of Cryptophyceae and its evolutionary implications. *J. Cell Sci.* 54:311–28.

Nauwerk, A. (1968). Das Phytoplankton des Latnjajaure 1954–55. *Schweiz. Z. Hydrol.* 30:188–216.

Pechlauer, R. (1971). Factors that control the production rate and biomass of phytoplankton in high-mountain lakes. *Mitt. Int. Ver. Theor. Angew. Limnol.* 19:124–5.

Smith; W. O., and Barber, R. T. (1979). A carbon budget for the autotrophic ciliate *Mesodinium rubrum*. *J. Phycol.* 15:27–33.

Spear-Bernstein, L., and Miller, K. R. (1984). Unique localization of the phycobiliprotein light-harvesting pigment in the Cryptophyceae. *J. Phycol.* 25:412–19.

Thinh, L-V. (1983). Effect of irradiance on the physiology and ultrastructure of the marine cryptomonad, *Cryptomonas* strain Lis (Cryptophyceae). *Phycologia* 22:7–11.

Wright, R. T. (1964). Dynamics of a phytoplankton community in an ice-covered lake. *Limnol. Oceanogr.* 9:163–78.

9 · Heterokontophyta

Chrysophyceae

The Chrysophyceae are distinguished chemically by having chlorophylls *a*, c_1, and c_2 (Andersen and Mulkey, 1983) and structurally by two flagella inserted into the cell perpendicular to each other, one photoreceptor on the short flagellum that is usually shaded by an eyespot in the anterior portion of the chloroplast, contractile vacuoles in the anterior portion of the cell, chloroplast endoplasmic reticulum, and radially or biradially symmetrical silica scales (if they are present). The storage product is chrysolaminarin. Many members of the class produce statospores enclosed in a silicified wall with a terminal pore.

Most of the species in the Chrysophyceae are freshwater and occur in soft waters (low in calcium). Many of the freshwater species are in the plankton of lakes where they are present in abundance. The coccoid and filamentous genera are found mostly in cold springs and brooks, where they occur as gelatinous or crustous growths on stones and woodwork. Most of the Chrysophyceae are sensitive to changes in the environment and survive the unfavorable periods as statospores.

Cell structure

FLAGELLA AND EYESPOT

Many of the Chrysophyceae have a tinsel flagellum that is inserted at the anterior end of the cell parallel to the cell axis (Fig. 9.1) and a whiplash flagellum that is inserted approximately perpendicular to the tinsel flagellum. The whiplash flagellum is often reduced to a short stub. The hairs on the tinsel flagellum are usually tripartite microtubular hairs (Hill and Outka, 1974), although tripartite and fibrillar hairs have been reported in *Ochromonas* (Bouck, 1971) (Fig. 9.2). Flagellar scales have been reported in a couple of genera (Andersen, 1982). The posterior whiplash flagellum is usually the shorter flagellum and has a swelling at its base on the side toward the cell (Fig. 9.1). This flagellar swelling contains an electron-dense area referred to as the photoreceptor. The flagellar swelling contains **retinal**,

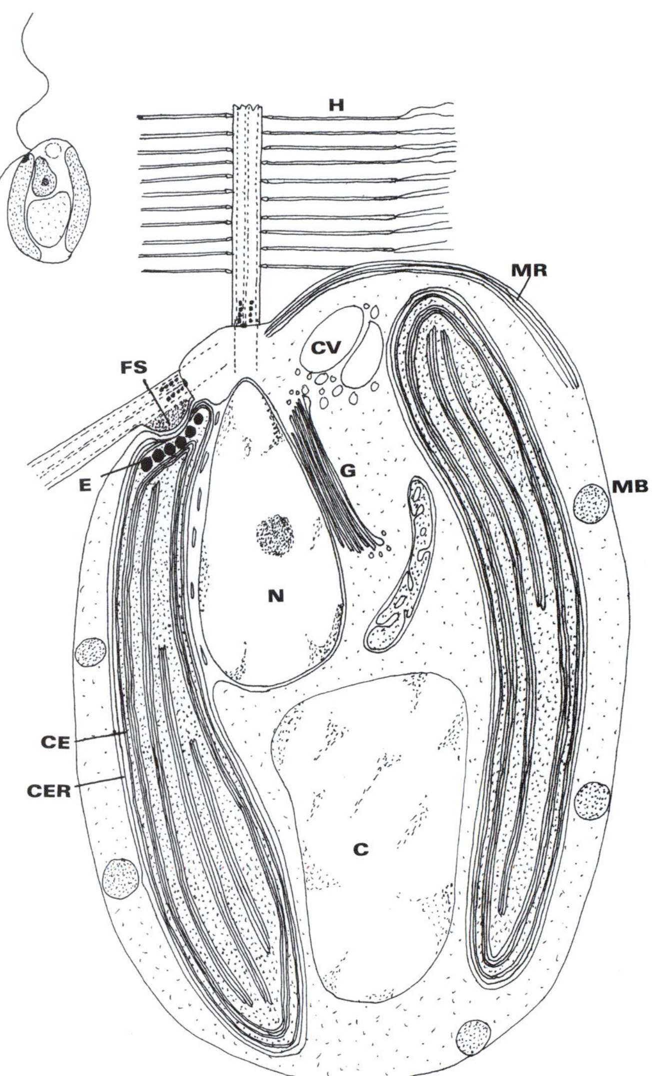

Fig. 9.1 Semi-diagrammatic drawing of a light and electron microscopical view of the basic organization of a cell of the Chrysophyceae. (C) Chrysolaminarin vesicle; (CE) chloroplast envelope; (CER) chloroplast endoplasmic reticulum; (CV) contractile vacuole; (E) eyespot; (FS) flagellar swelling; (G) Golgi body; (H) hair of the anterior flagellum; (MB) muciferous body; (MR) microtubular root of flagellum; (N) nucleus. (Adapted from Hibberd, 1976.)

the chromophore of rhodopsin-like proteins, suggesting that a rhodopsin-like protein is the photoreceptor in the Chrysophyceae (Walne et al., 1995). The flagellar swelling fits into a depression of the cell immediately beneath which, inside the chloroplast, is the eyespot. The eyespot consists of lipid globules inside the anterior portion of the chloroplast, between the chloroplast envelope and the first band of thylakoids.

In *Ochromonas*, the long tinsel flagellum beats in one plane and pulls the cell forward, whereas the shorter flagellum is flexed over the anterior eyespot and appears to play little role. During forward movement the cell rotates because of the shape of the body.

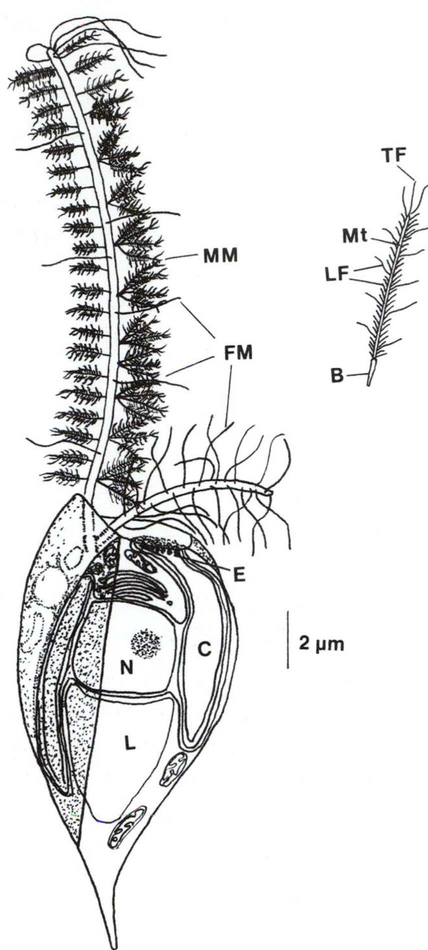

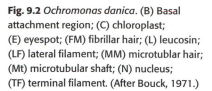

Fig. 9.2 *Ochromonas danica*. (B) Basal attachment region; (C) chloroplast; (E) eyespot; (FM) fibrillar hair; (L) leucosin; (LF) lateral filament; (MM) microtublar hair; (Mt) microtubular shaft; (N) nucleus; (TF) terminal filament. (After Bouck, 1971.)

INTERNAL ORGANELLES

The chloroplasts are parietal and usually only a few in number, often only one or two. Chlorophylls *a*, c_1, and c_2 are present, with the main carotenoid being fucoxanthin. The chloroplasts are surrounded by two membranes of chloroplast E.R., the outer membrane of which is usually continuous with the outer membrane of the nuclear envelope (Fig. 9.1). The thylakoids are usually grouped three to a band.

The chloroplast of *Ochromonas danica* will form a small proplastid if the cells are grown in the dark (Gibbs, 1962). The proplastid contains a single thylakoid, a few small vesicles and a large number of dense granules. The chlorophyll *a* content of the proplastid is about 1.2% of the mature chloroplast. On exposure of the proplastid to light, vesicles appear that fuse to form the thylakoids. After 2 days in the light, the chloroplasts have reached their mature form, although it takes 8 days for them to acquire their full complement of chlorophyll.

Pyrenoids are common in chloroplasts of the Chrysophyceae. They consist of a granular area that is different in appearance from the stroma. Few, if any, thylakoids traverse the pyrenoid area.

The storage product is **chrysolaminarin** (**leucosin**), a β-1,3 linked glucan, supposedly found in a posterior vesicle (Fig. 9.1). The actual function of the so-called chrysolaminarin vesicle may be more complex than supposed with the discovery of microorganisms in the chrysolaminarin vesicle of *Ochromonas* (Daley et al., 1973; Dubowsky, 1974). The chrysolaminarin vesicle is much larger in organisms grown in the dark on a synthetic medium than in cells grown in the light. The opposite would be expected if the vesicle stored chrysolaminarin, the accumulation product of photosynthesis in the light. It may be that the structure may also function as a digestive vesicle, breaking down material taken up by the cell into building blocks for metabolism and growth.

The single nucleus is pear-shaped (pyriform), with its narrow anterior end extended in the direction of the basal bodies (Fig. 9.1). There is a single, large Golgi body which lies against the nucleus in the anterior part of the cell, often in a concavity in the nuclear envelope. Contractile vacuoles are common, usually occurring in the anterior part of the cell next to the Golgi apparatus. There is often a complex system of vesicles associated with the contractile vacuoles similar to the pusule system in the Dinophyceae. Lipid bodies can also be found in the protoplasm. In young cells there are usually few lipid bodies; however, as the cell ages, the lipid bodies become larger and more numerous until they fill the protoplasm.

Two different types of projectiles occur in the Chrysophyceae, muciferous bodies and discobolocysts, the former like the muciferous bodies in the Prymnesiophyta, Raphidophyta, and Dinophyta. The muciferous bodies (Fig. 9.1) contain granular material and are bounded by a single membrane. On discharge the contents of the vesicle often form a fibrous network outside the cell. The **discobolocysts** are similar to the muciferous bodies and have been described in the most detail by Hibberd (1970), in *Ochromonas tuberculatus* (Fig. 9.3). The discobolocysts are in the outer layer of cytoplasm and consist of a single membrane-bounded vesicle with a hollow disc in the outward-facing part of the vesicles. The discharge of the discobolocyst is explosive, taking place by the expansion of the projectile into a thin thread 6 to 11 μm long, the disc being at the tip of the mucilage. As the discharge occurs, the cell jerks violently under the recoil to a distance of 5 μm. After the discharge of either muciferous bodies or discobolocysts, the protoplast recovers without any deleterious effects. Both of the projectiles originate in the area of the Golgi apparatus.

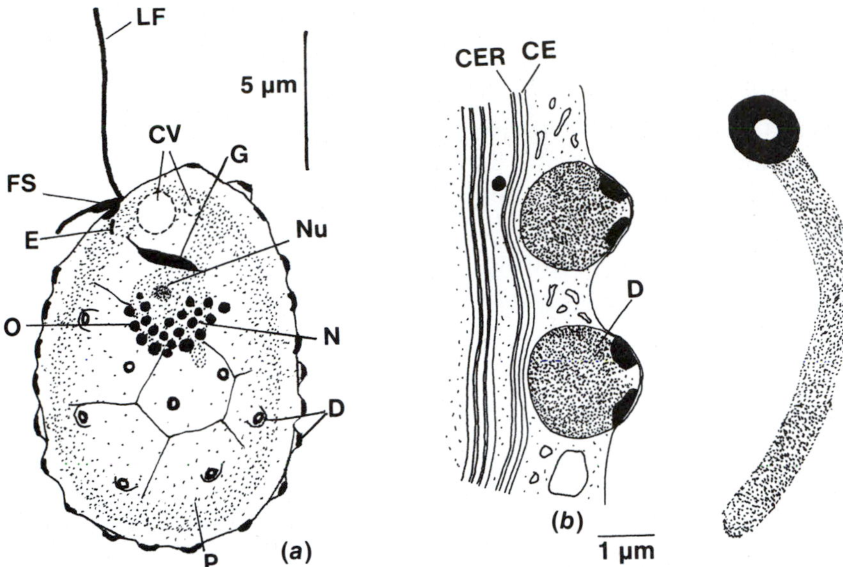

Fig. 9.3 (*a*) *Ochromonas tuberculatus*. (*b*) Charged and discharged discobolocysts. (CE) Chloroplast envelope; (CER) chloroplast endoplasmic reticulum; (CV) contractile vacuole; (D) discobolocyst; (E) eyespot; (FS) flagellar swelling; (G) Golgi body; (LF) long flagellum; (N) nucleus; (Nu) nucleolus; (O) oil; (P) plastid. (After Hibberd, 1970.)

EXTRACELLULAR DEPOSITS

Cell walls composed of cellulose (Herth and Zugenmaier, 1979), loricas, and silicified scales and walls, occur in some of the Chrysophyceae (Preisig, 1994). Silica scales, such as those in *Paraphysomonas* (Figs. 9.4 and 9.5), are radially or biradially symmetrical. The scales are arranged loosely outside the plasma membrane without any clearly defined pattern. Like the Bacillariophyceae and Synurophyceae, the scales of the Chrysophyceae are formed inside a silica deposition vesicle that is derived from endoplasmic reticulum (Schnepf and Deichgraber, 1969). This arrangement differs from that of the calcified scales of the Prymnesiophyceae and the Chlorophyta, which are formed by the Golgi apparatus. Scale formation is also similar to frustule formation in the diatoms in that the addition of germanium (a competitive inhibitor of silicon utilization) to the growth medium results in inhibition of growth (Lee, 1978). Organic scales without mineralization also occur in the Chrysophyceae. In *Chromulina placentula*, there is a single layer of organic scales covering the posterior portion of the cell, the scales being very similar to those found in the Prymnesiophyceae (Throndsen, 1971).

In one of the orders (Parmales), silicified walls occur which are composed of five or eight parts (Figs. 9.15 and 9.16) (Booth and Marchant, 1987). Loricas, such as those in *Pseudokephyrion pseudospirale* and *Kephyrion rubri claustri*, can also be mineralized. Manganese can occur in loricas as needle-like structures, whereas iron occurs as granular deposits (Dunlap et al., 1987).

Some of the species have a **lorica** (an envelope around the protoplast, but

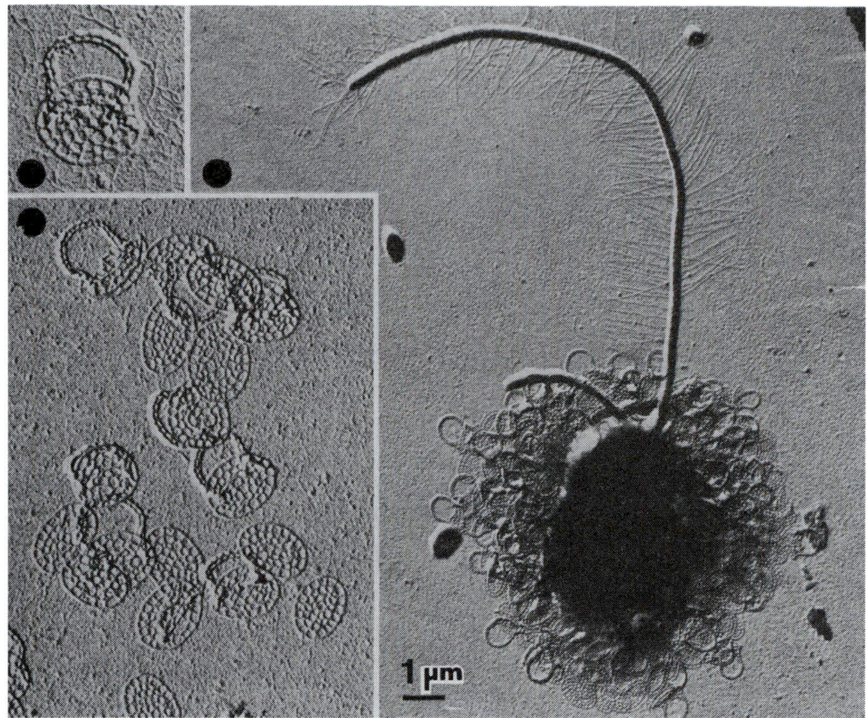

Fig. 9.4 Transmission electron micrographs of *Paraphysomonas sigillifera* showing a whole cell with a long tinsel flagellum and a short whiplash flagellum. Also included are higher-magnification micrographs of the scales. (From Thomsen et al., 1981.)

not generally attached to the protoplast as a wall is). In *Dinobryon*, the lorica is composed of an interwoven system of microfibrils (Kristiansen, 1969). The formation of the lorica begins when a small funnel-shaped piece arises from the cell. The protoplast then rotates on its axis following a spiral course and secreting the remainder of the lorica. When the lorica is complete, the protoplast withdraws to its base (Fig. 9.6). Several strains of *Ochromonas malhamensis* and *O. sociabilis* produce a delicate lorica consisting of a 10- to 20-μm hollow stalk and a cup-shaped envelope that encloses a long protoplasmic filament and the basal half of the cell, respectively. The lorica of *O. malhamensis* is composed of the polysaccharide chitin (Herth et al., 1977). The lorica can be mineralized as is the case with *Chrysococcus rufescens* (Fig. 9.7). This alga has a basically mucilaginous lorica surrounding the cell (Belcher, 1969). In young cells the lorica is very transparent, probably consisting just of mucilage. At a later stage, dark needle-shaped crystals appear, which are primarily oriented parallel to the cell surface. The crystals ultimately unite to form a felty mass outside the cell.

Anthophysa vegetans (Fig. 9.11(*b*),(*c*)) is a freshwater colonial alga that produces a mineralized stalk that can contain salts of calcium, iron or manganese. The minerals in the environment determine color and the mineral composition of the stalk (Lee and Kugrens, 1989).

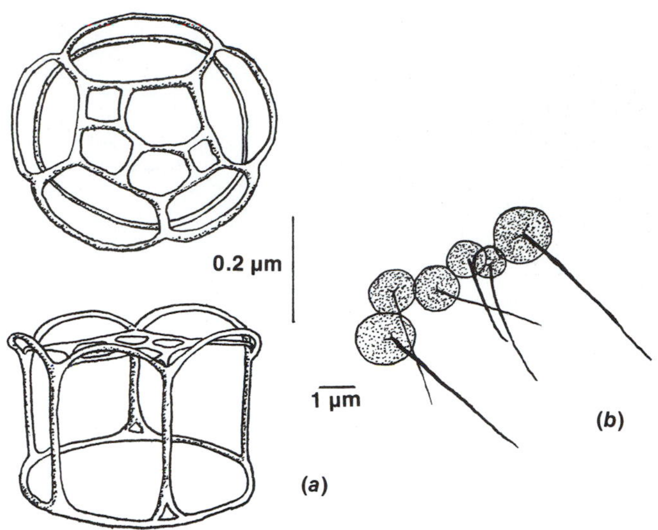

Fig. 9.5 (*a*) Typical crown scales of *Paraphysomonas butcheri*. (*b*) Pointed scales of *P. vestita*. ((*a*) after Pennick and Clarke, 1972; (*b*) after Manton and Leedale, 1961.)

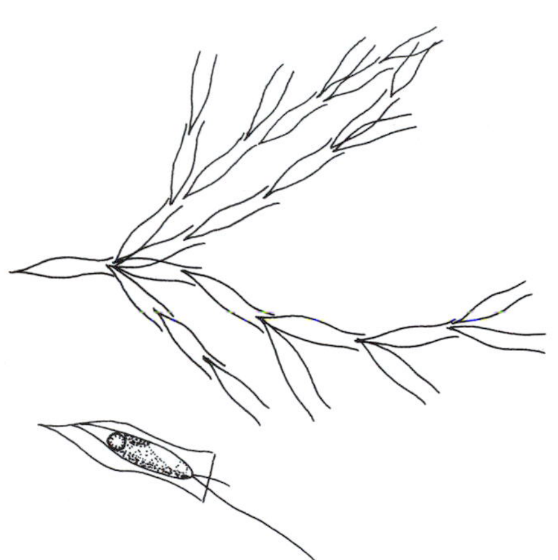

Fig. 9.6 *Dinobryon pediforme*.

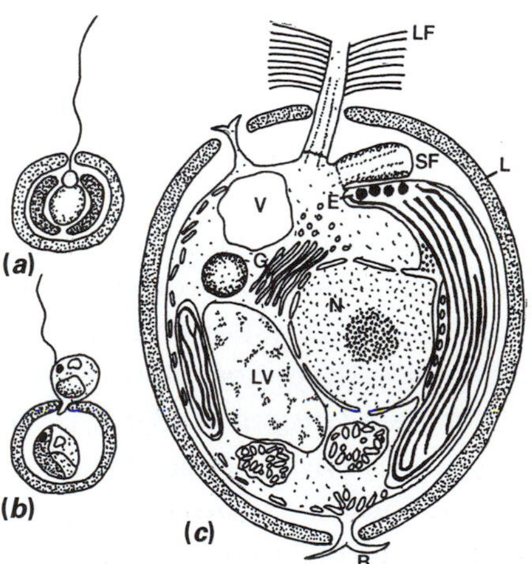

Fig. 9.7 *Chrysococcus rufescens*. (*a*) Whole cell. (*b*) Cell undergoing reproduction. (*c*) Ultrastructure of vegetative cell. (B) Branched cytoplasmic process; (E) eyespot; (G) Golgi; (L) lorica; (LV) leucosin (chrysolaminarin) vesicle; (LF) long flagellum; (N) nucleus; (V) contractile vacuole. ((*c*) after Belcher, 1969)

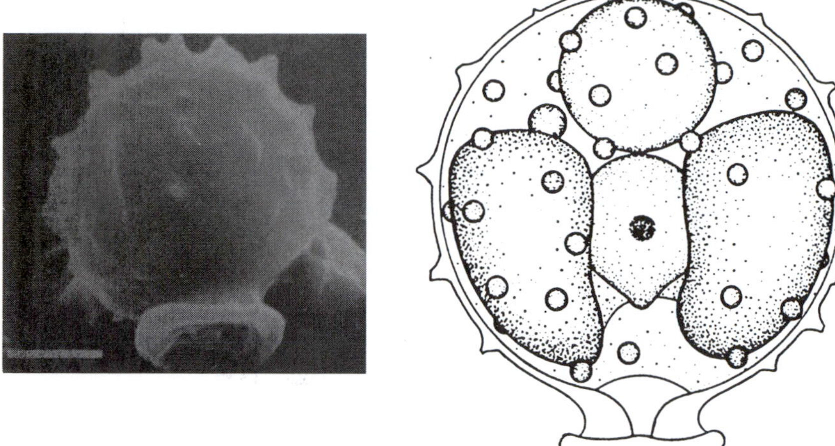

Fig. 9.8 Statospore of *Ochromonas sphaerocystis*. Bar = 5 μm. (From Andersen, 1982.)

Statospores

The formation of a **cyst** or **statospore** or **resting spore** is one character by which a member of the Chrysophyceae or Synurophyceae may be unequivocally recognized. **Statospores** are mostly spherical, ellipsoidal, or ovate in shape, and the outer surface may be smooth or variously ornamented with warts, spines, or arms (Fig. 9.8). Wall ornamentation is species specific. The statospore has a pore with a collar that is closed by a plug. A vegetative cell forms a statospore internally. In the formation of a statospore the cells become non-motile, any projectiles are discharged, and there is considerable contractile vacuole activity (Sheath et al., 1975; Hibberd, 1977). A nearly spherical vesicle called the silica deposition vesicle is formed in the cytoplasm, and silica is deposited inside the vesicle (Fig. 9.9) (Preisig, 1994). A complete sphere of silica is formed, interrupted only by the developing pore and collar. The nucleus, chloroplast, flagellar basal bodies, mitochondria, Golgi body, chrysolaminarin vesicle, and ribosomes are segregated to the inside of the silica deposition vesicle, whereas outside there are mitochondria, ribosomes, contractile vacuoles, and small vesicles. After the spines, pore, and collar have been formed in the silica deposition vesicle, a plug is formed by the cytoplasm in the pore area. With the formation of the unsilicified plug, contact is lost between the protoplasm inside the statospore and that outside, with the inner membrane of the silica deposition vesicle becoming the new plasmalemma. When a statospore germinates, there is a dissolution of the plug or separation of it from the spore wall. The protoplast then moves out of the statospore by amoeboid motion, forming flagella as it moves out.

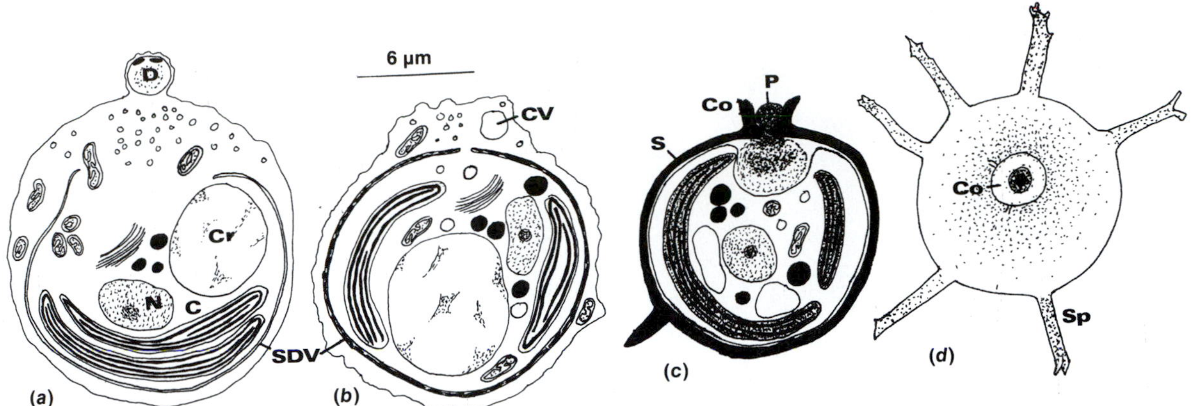

Fig. 9.9 Formation of a statospore or cyst in *Ochromonas tuberculata*. (*a*)–(*c*) Fine structural drawings of the formation of the statospore. (*d*) A mature statospore as seen from the collar end. (C) Chloroplast; (Co) collar of statospore; (Cr) chrysolaminarin vesicle; (CV) contractile vacuole; (D) discobolocyst; (N) nucleus; (P) plug in pore of statospore; (S) statospore wall; (SDV) silica deposition vesicle; (Sp) spine. (Adapted from Hibberd, 1977.)

Nutrition

Nutrition in the Chrysophyceae can be either phototrophic, phagotrophic or mixotrophic (photosynthetic organism capable of taking up particles and molecules from the medium). *Dinobryon* (Fig. 9.6 and 9.12) is a mixotrophic chrysophyte that can have half of the carbon in the cells derived from ingestion of microorganisms (Caron et al., 1993). *Dinobryon* has the ability to compete with crustaceans, rotifers and ciliates in capturing microorganisms. Under low-light conditions, *Dinobryon* can ingest an average of three bacterial cells every five minutes (Bird and Kalff, 1986). Chrysophytes have the ability to ingest prey up to 30 times larger than themselves. The ability to take up microorganisms appears to be about the same for pigmented and non-pigmented chrysophytes (Zhang et al., 1996).

Like other mixotrophic chrysophytes, *Epipyxis pulchra* phagocytizes food particles that are both living (bacteria, small algae or even cells of its own kind) and non-living (detritus, fecal material) (Wetherbee and Andersen, 1992). During prey gathering, the long flagellum, which is adorned with stiff hairs, beats rapidly to direct a strong current towards the cell while the short, smooth flagellum moves very little. When a potential food particle is drawn by current to contact the flagellar surface, the long flagellum stops beating and positions itself, in concert with the short flagellum, to seize the prey between the two flagella (Fig. 9.10). Both flagella briefly rotate the prey before selecting or rejecting it. If selected as food, the prey is held in place until a complex collecting cup emanates out and engulfs the prey. The cup plus the prey, now a food vacuole, is then retracted back into the cell proper.

Ochromonas (Fig. 9.2 and 9.3) is a photosynthetic mixotrophic chrysophyte that has been used in many nutrition studies. *Ochromonas* requires other organisms or organic compounds in the medium in order to survive.

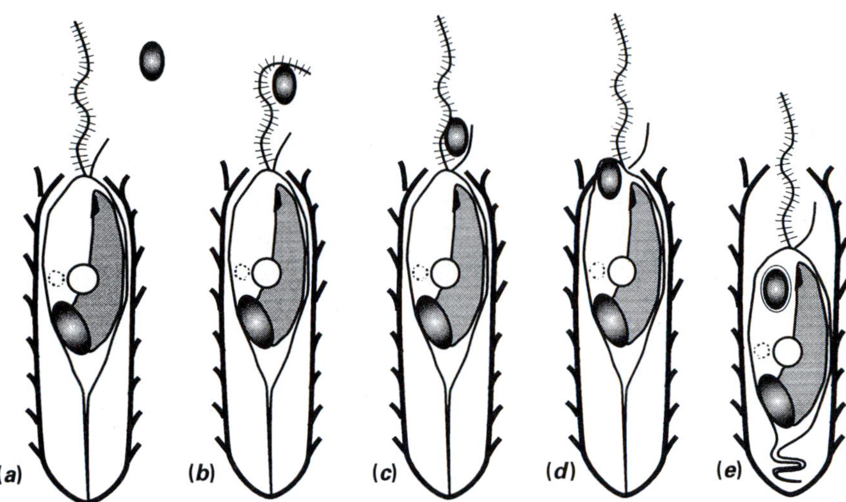

Fig. 9.10 Phagotrophy in *Epipyxia pulchra*. The cell has a posterior stalk by which it is attached to a lorica. (*a*) The long tinsel flagellum beats in such a way that water and suspended particles are drawn to the cell. (*b*) A particle is seized by the long tinsel flagellum. (*c*) The particle is maneuvered between the two flagella. (*d*) A feeding cap from the cell envelopes the particle. (*e*) The particle is enclosed within a food vacuole within the cytoplasm. The stalk has pulled the cell into the lorica. (Modified from Wetherbee and Andersen, 1992.)

Ochromonas danica can use bacteria as a sole source of carbon in the light, but not in the dark. Normally, phagotrophy supplements food sources by providing vitamins and other growth substances. *Ochromonas* will even exhibit cannibalism by ingesting other *Ochromonas* cells (Daley et al., 1973). Experiments growing *Ochromonas* on a chemically-defined medium has shown that the cells need vitamin B_{12} in the medium. Also, *Ochromonas* is not able to utilize inorganic nitrogen and needs peptones or some other similar compounds as a source of nitrogen.

Isofloridoside, a product of photosynthesis in *Ochromonas malhamensis*, is used to adapt the cells to changes in the dissolved substances (osmotic pressure) of the medium. As the osmotic pressure of the medium increases, *O. malhamensis* responds by increasing the concentration of isofloridoside inside the cell, preventing water loss to the medium (Kauss, 1967).

Ecology

Fauré-Fremiet (1950) has shown that *Chromulina psammobia* moves up to the surface of mud flats at low tide, and as the tide comes in, the flagellate moves down in the mud so as not to be washed away. The basis for the migration is a reversal in the phototactic response, the cells being positively phototactic at low tide and negatively phototactic at high tide. The presence of mud is not necessary for the changes in phototaxis, which will continue for 6 to 7 days in the laboratory, more or less synchronized with the tide. After this the phototaxis changes gradually until the cells are always positively phototactic.

Dinobryon is a freshwater alga that is almost never found in waters with a high concentration of phosphorus. This observation led some investigators to conclude that high concentrations of phosphorus are inhibitory to growth of the alga. More detailed work (Lehman, 1976) showed this not to be true, that *Dinobryon* in culture grows well at high concentrations of phosphorus. What happens in nature is that other algae grow faster and outcompete *Dinobryon* at higher concentrations of phosphorus; only when **vernal** (spring) **blooms** of diatoms or other phytoplankton have reduced phosphorus to a level that inhibits their own growth will *Dinobryon* compete effectively. This demonstrates a basic rule among algae (Eppley et al., 1969; Fuhs et al., 1972), which is *algae that have efficient uptake of nutrients (are able to utilize low levels of a nutrient) usually have lower maximum intrinsic growth rates than algae that are less efficient in taking up nutrients.* These differences represent two variations of an adaptive scheme for nutrient utilization. Some species exploit a resource-laden environment rapidly, whereas others display measured efficiency in utilizing an energy source. Energy trade-offs among competing intracellular processes might preclude organisms from excelling at both simultaneously.

Dinobryon uses phosphate not only as inorganic phosphate but also as organically bound phosphate in moderate-sized molecules (glycerophosphate, uridylic acid, adenylic acid) (Lehman, 1976). It can also use nitrogen either from inorganic sources or from organic molecules (urea, glycine, uridylic acid, adenylic acid). Its capacity to utilize these organic sources and its phagocytosis of whole particles imply that *Dinobryon* is suited to occupy the water column when death and decline of previous bloom-forming organisms release products of cell breakdown to solution. It is also at these times, when ambient levels of dissolved inorganic nutrients have become somewhat depleted, that the potentials of the cells for effective uptake of nutrients show their advantage.

Although high concentrations of phosphorus do not inhibit the growth of *Dinobryon*, high concentrations of potassium do (Lehman, 1976). The concentrations necessary for inhibition are commonly found in many lakes. As an example, in one Swedish lake, *D. sertularia* appears in June a few weeks after the ice breaks, with the alga persisting in the cool water until early autumn (Willén, 1961). Its period of abundance matches exactly with a summertime decrease in potassium caused partly by the precipitation of potassium by clay colloids and partly by the uptake by littoral vegetation (Ahl, 1966). The disappearance of *Dinobryon* corresponds to increased concentration of potassium as the element is released from dying vegetation.

Classification

Three of the orders of the Chrysophyceae (Preisig, 1995) will be considered here:

Order 1 Chromulinales: cells with the flagella inserted into the anterior portion of the cell.

Order 2 Parmales: cells with siliceous walls composed of five or eight plates.

Order 3 Chrysomeridales: motile naked cells with laterally inserted flagella and an eyespot, motile cells similar to those in the brown algae.

CHROMULINALES

All of the organisms in this order have a unicell with two apically inserted flagella somewhere in their life history. One of the flagella is tinsel with mastigonemes and directed forward, while the second flagellum is whiplash (lacking mastigonemes) and is inserted at approximately 90° to the tinsel flagellum (e.g., *Ochromonas*, Figs. 9.2 and 9.3). In some of the genera (e.g., *Chromulina*, Fig. 9.11(*d*)) the whiplash flagellum is reduced to a stub. There are usually two parietal chloroplasts, a central nucleus, and a large posterior chrysolaminarin vesicle.

Within the order there is a progression from the unicellular to the colonial form (Fig. 9.11) as exemplified by *Uroglena* (Fig. 9.11(*a*)) and *Anthophysa* (Fig. 9.11(*b*),(*c*)). Some of the genera, such as *Dinobryon* (Fig. 9.6 and 9.12), have cells surrounded by a lorica. In *Chrysococcus* (Fig. 9.7), the cell is surrounded by a lorica that has pores in it.

In the Chromulinales there is a frequent tendency toward loss of photosynthetic activity and adaptation to various forms of phagotrophy and chemo-organotrophy. Associated with this is a diminution of the chloroplast as in *Anthophysa* (Fig. 9.11(*b*),(*c*)), which has a leukoplast with a pigmented eyespot. The end of this reduction is represented by *Paraphysomonas* (Fig. 9.4), which has no trace of an eyespot or plastid. Along with the reduction in the chloroplast have arisen adaptations for feeding, such as in *Anthophysa*, which beats its long flagellum, creating a water current toward the cells, which brings with it bacteria from as far away as 200 μm. The bacteria strike the anterior end of the cells and are ingested at the point where they touch. The bacteria sink into the cytoplasm, which closes behind them, and are completely in the cell within 2 to 3 seconds of contact (Belcher and Swale, 1972b).

In *Dinobryon cylindricum*, sexual reproduction is heterothallic and dioeceous, and morphologically and physiologically anisogamous (Fig. 9.12) (Sandgren, 1981). Cells of the second or third tier of the colony (the basal cell being the oldest) are the best potential gamete-producing cells. Gametes are produced mainly in exponentially growing populations. Female cells release a chemical **erogen** that causes male cells to divide once. One of the male cells remains in the lorica, whereas the other swims away as a naked

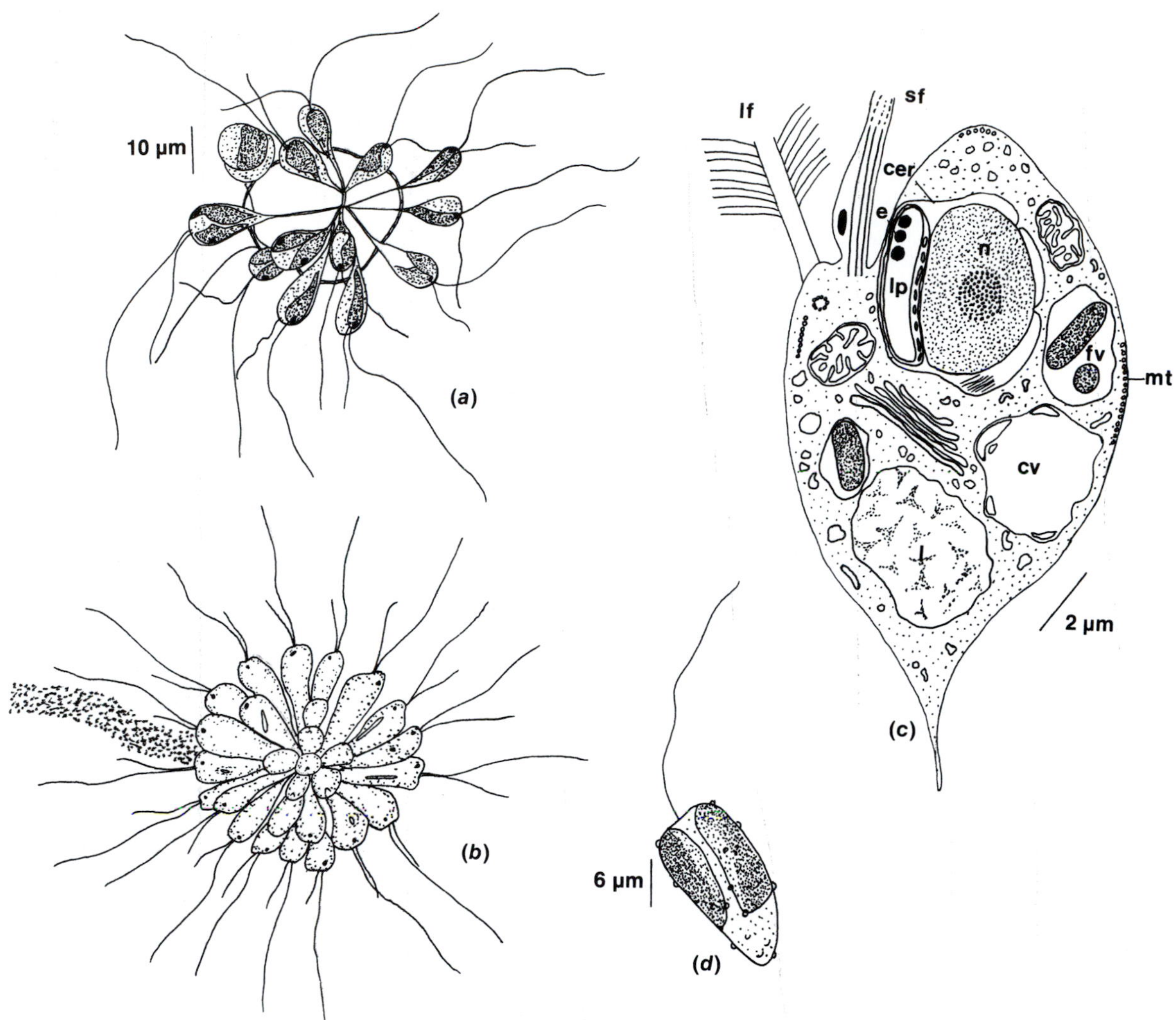

Fig. 9.11 (*a*) *Uroglena conradii.* (*b*),(*c*) *Anthophysa vegetans.* (*d*) *Chromulina conica.* (cer) Chloroplast endoplasmic reticulum; (cv) contractile vacuole; (e) eyespot; (fv) food vacuole; (l) leucosin; (lf) long flagellum; (lp) leucoplast; (mt) microtubules; (n) nucleus; (sf) short flagellum. ((*a*),(*d*) after Schiller, 1929; (*b*) after Pringsheim, 1946; (*c*) after Belcher and Swale, 1972b.)

male gamete, instead of attaching to the lorica mouth and building a new lorica as a vegetative cell would. The non-loricate male gamete (which is structurally similar to an *Ochromonas* cell) swims to the female cell in its lorica, the flagella become oriented parallel to each other, and fusion occurs at the anterior end of each cell. Plasmogamy results in a quadriflagellate planozygote that fills the lorica. Nuclear fusion does not occur. After 30 minutes, the flagella have been lost and the zygote creeps to the lorica mouth by amoeboid movement. A thin cellulose encystment vesicle is formed as a lorica extension, the zygote rounds up, and a binucleate statospore (Fig. 9.13) is formed. In *D. divergens,* most statospores settle to the lake bottom and germinate early the next year (Sheath el al., 1975).

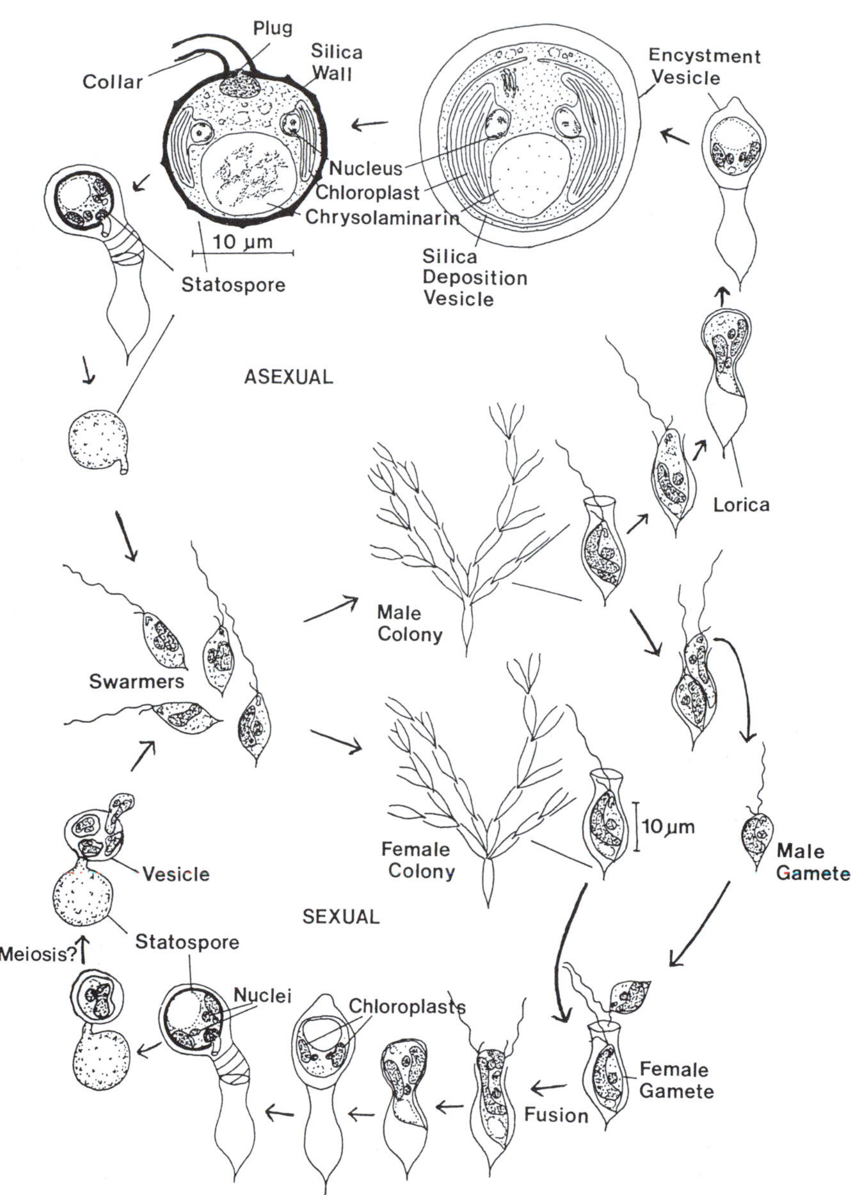

Fig. 9.12 The life cycle of *Dinobryon*. (Adapted from Sheath et al., 1975; Sandgren, 1980, 1981.)

Statospore germination occurs by the formation of a cellulosic vesicle at the statospore pore, followed by a presumably meiotic cleavage into four daughter cells that migrate into the vesicle. The *Ochromonas*-like swarmers escape to form new vegetative cells.

Asexually produced statospores of *Dinobryon cylindricum* may occur in exponential or stationary-phase populations, depending upon the clones involved (Sandgren, 1980, 1981). Asexual statospores (cysts) are formed at a

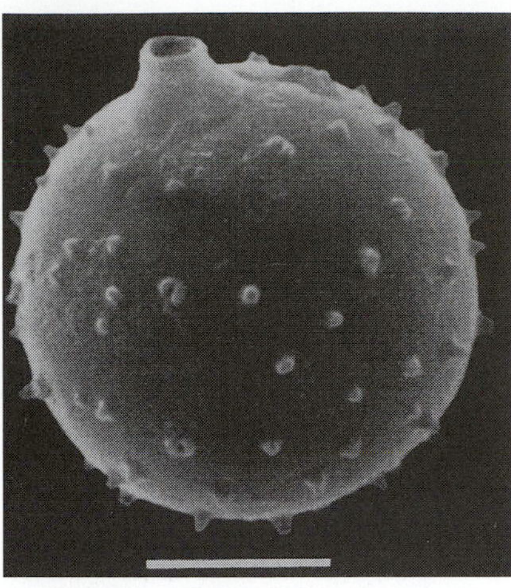

Fig. 9.13 Scanning electron micrograph of a cyst of *Dinobryon cylindricum*. Bar = 5 μm. (From Sandgren, 1983b.)

much lower frequency (0.05% or less) than sexual statospores unless the population is placed in a nitrogen-depleted environment where asexual encystment can reach 4%. The continual production of a low number of the resistant statospores by a *Dinobryon* population acts as a hedge against a rapid unfavorable change in environmental conditions that would kill the vegetative cells.

The first event in statospore formation is vegetative cell enlargement so that the cell begins to bulge out of the lorica. The Golgi form vesicles that fuse to form a continuous silica deposition vesicle in the peripheral cytoplasm. The silica deposition vesicle has a pore that eventually will be the statospore pore. Precipitation of silica proceeds in the silica deposition vesicle to form the silicified statospore wall. The portion of the cytoplasm that is exterior to the wall retracts through the pore in the cyst wall into the statospore interior. The small amount of cytoplasm outside the wall forms the collar around the pore. A plug is formed in the cyst wall, sealing the statospore protoplasm. The mature statospore has two nuclei, two plastids, and a rich supply of energy reserves in the form of oil and chrysolaminarin.

Chrysophaera (Fig. 9.14) is an example of a colonial alga in the order. This organism has a dominant spherical non-motile stage with cells adhering to each other and the substrate in irregular clusters. The smallest individuals are unicellular, but individuals with up to 256 cells within a common mucilaginous envelope are not uncommon. All the cells within the envelope have a long tinsel flagellum and a short stub of the second flagellum (Belcher, 1974). When a colony is placed in distilled water, it begins to swell within 15 minutes, and the mucilage gelatinizes on one side. The zoospores are released a few minutes later and swim for about 2 minutes before they shed their flagella, settle down, form a mucilaginous envelope, and become identical in appearance to the youngest coccoid stages.

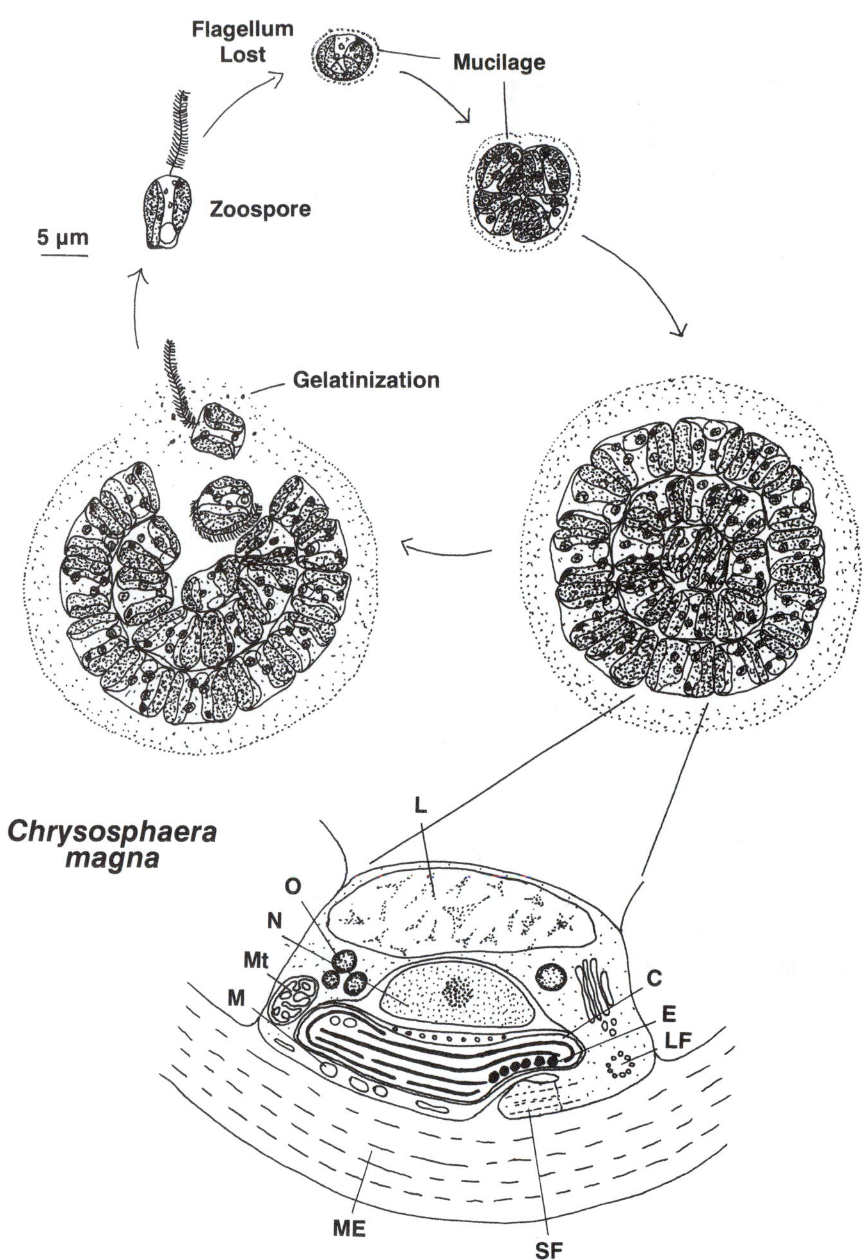

Fig. 9.14 The life cycle of *Chrysosphaera magna*. (C) Chloroplast; (E) eyespot; (L) leucosin vesicle; (LF) basal body of long flagellum; (M) muciferous body; (Mt) mitochondria; (ME) mucilaginous envelope; (N) nucleus; (O) oil; (SF) short flagellum. (After Belcher, 1974.)

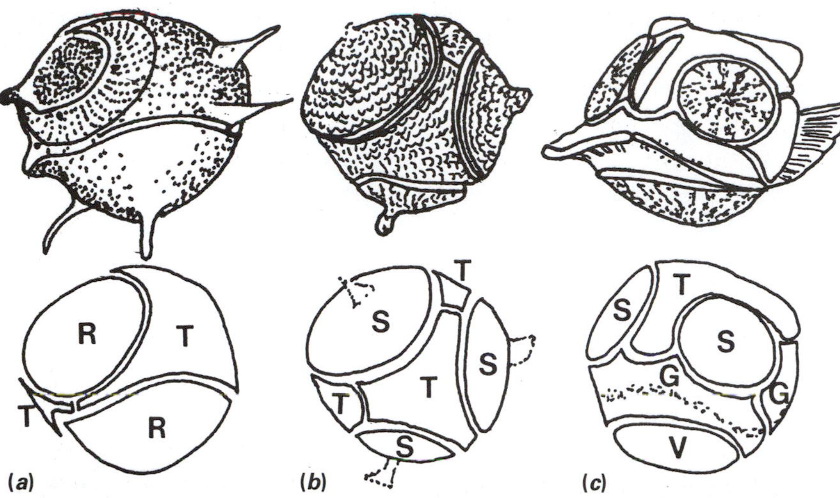

Fig. 9.15 Diagram of three genera in the Parmales. (*a*) *Pentalamina*. (*b*) *Tetraparma*. (*c*) *Triparma*. (G) Girdle plate; (R) round plate; (S) shield plate; (T) triradiate plate; (V) ventral plate, (After Booth and Marchant, 1987.)

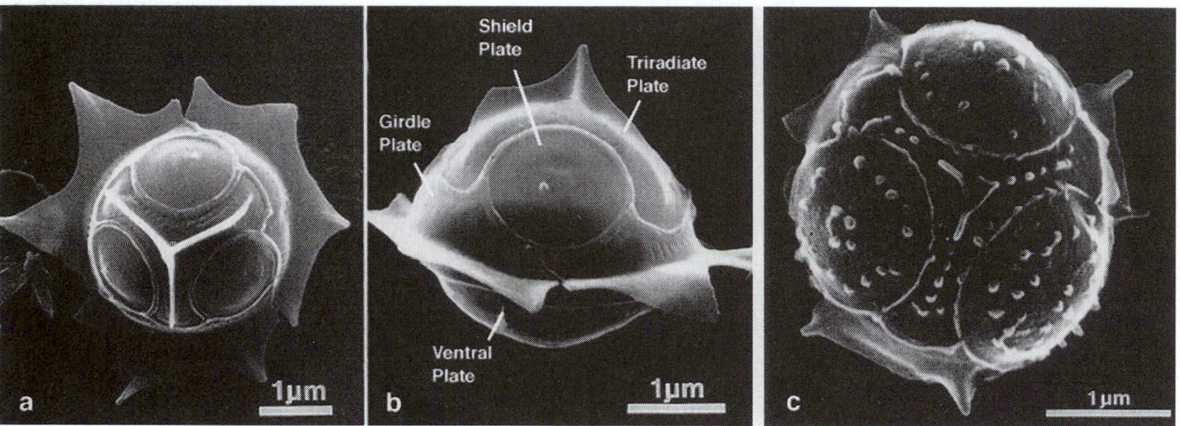

Fig. 9.16 (*a*), (*b*) *Triparma laevis*. (*c*) *Triparma strigata*. (From Kosman et al., 1993.)

PARMALES

The Parmales consist of small cells, generally 2 to 5 μm in diameter, each with a chloroplast (Marchant and McEldowney, 1986) and a silicified cell wall composed of five to eight plates (Figs. 9.15 and 9.16). In the Pentalaminaceae there are five wall plates whereas in the Octolaminaceae there are eight wall plates (Booth and Marchant, 1987). Members of the Parmales occur at concentrations of 10^5 cells per liter in polar and subpolar marine waters, making them one of the more abundant groups in these waters.

Four different shapes of plates occur around the cells (Figs. 9.15 and 9.16) (Booth and Marchant, 1987). A **shield plate** or **round plate** is a circular plate with or without a central knob or process. A **triradiate plate** has three arms equally spaced, each arm fitting between two shield plates. A **ventral plate** is

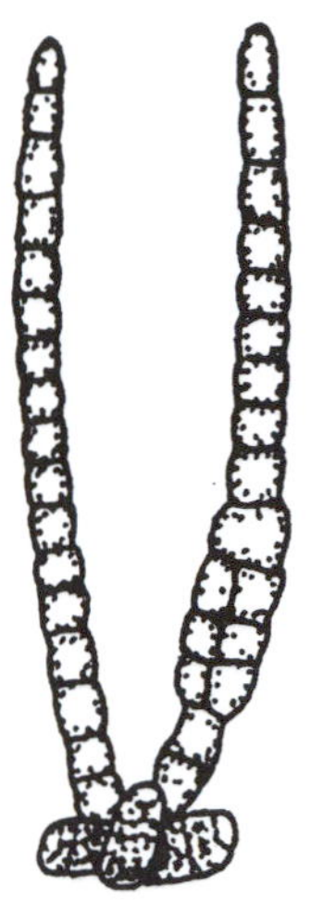

Fig. 9.17 *Giraudyopsis stelliger*. (After Dangeard, 1966.)

a round plate of greater diameter than that of shield plates; it is found in *Triparma* on the opposite side of the sphere from the triradiate plate. **Girdle plates** are three oblong plates in *Triparma*, juxtaposed end to end to form a ring around the ventral plate between it and the other four plates. Ornamentation of various types (papillae, wings, spines, keels) occurs on the plates.

CHRYSOMERIDALES

These algae have zoospores with the flagella more or less laterally inserted into the cell body. There is an eyespot in the chloroplast and the accessory pigment violaxanthin is present. The similarities in the zoospores of the algae in this order has led to the speculation that the Phaeophyceae (brown algae) probably evolved from an alga similar to *Giraudyopsis* in this order (Fig. 9.17) (O'Kelly, 1989; Saunders et al., 1997). The algae in the Chrysomeridales, however, lack the unilocular and plurilocular sporangia, as well as the plasmodesmata and alginates characteristic of the Phaecophyceae.

Fig. 9.18 *Left:* G. Benjamin Bouck and *right:* Adolph Pascher. (Photograph of Pascher from *Archiv fur Protistenkunde*, vol. 98 1952.)

G. Benjamin Bouck Born October 25, 1933, in New York City. Dr Bouck received his BS from Hofstra University (1956) and his M.A. (1958) and PhD (1961) from Columbia University. In 1962, he was a postdoctoral fellow with Keith Porter at Harvard University. From 1962 to 1971, he was at Yale University, rising to associate professor; from 1968 to 1969, he was a visiting associate professor at the University of California, Berkeley; and from 1971 to the present, he has been a professor at the University of Illinois at Chicago. Dr Bouck is known for his thorough cytological investigations on crysophytes, euglenoids, and brown algae.

Adolph Pascher Born May 31, 1881, in Tusset, Bohmerwalde, Province of Krummau, Austria-Hungary (an area today near the West German border); he died in May 1945. Dr Pascher was brought up in the Bohmerwalde, a snowy mountainous area, where he attended small schools until he went to the University of Prague. Here he was able to use his first microscope, a tool he was to use for the rest of his life. He graduated from the University of Prague in 1909. From 1909 to 1912, at a time when he suffered from a chronic illness, he studied pharmaceutical botany. From 1912 to 1927, he worked as a lecturer at different universities including the German Technical University in Prague. During this time he did phycological research at a hydrobiological station in Hirschberg, Germany, and at another hydrobiological station that he founded in Upper Bohemia in Czechoslovakia. In 1927, he became director of the Botanical Institute and Gardens at the University of Prague. In the early 1930s, Dr Pascher became involved in the Nazi Brown Shirt movement which resulted in a hiatus of his numerous publicatins from 1933 to 1938. In 1939, he started publishing again while remaining active in the Nazi administration. With the fall of the Third Reich, he shot himself in May 1945. Dr Pascher could be considered as the most prominent phycologist of modern time. As stated by Prescott (History of Phycology, in *Manual of Phycology*) "His work exerted a greater influence than that of any other phycologist in clarifying modern concepts of algal taxonomy and phylogeny . . . it was his erudite interpretation of facts, nevertheless, which placed the studies of Pascher in a corner stone position."

References

Ahl, T. (1966). Chemical conditions in Ösbysjön, Djursholm. 2. The major constituents. *Oikos* 17:175–6.

Andersen, R. A. (1982). A light and electron microscopical investigation of *Ochromonas sphaerocystis* Matvienko (Chrysophyceae): The statospore, vegetative cell and its peripheral vesicles. *Phycologia* 21:390–8.

Anderson, R. A., and Mulkey, T. J. (1983). The occurrence of chlorophylls c_1 and c_2 in the Chrysophyceae. *J. Phycol.* 19:289–94.

Belcher, J. H. (1969). A morphological study of the phytoflagellate *Chrysococcus rufescens* Klebs in culture. *Br. Phycol. J.* 4:105–17.

Belcher, J. H. (1974). *Chrysophaera magna* sp. nov., a new coccoid member of the Chrysophyceae. *Br. Phycol. J.* 9:139–44.

Belcher, J. H., and Swale, E. M. F. (1971). The microanatomy of *Phaeaster pascheri* Scherffel (Chrysophyceae), *Br. Phycol. J.* 6:157–69.

Belcher, J. H., and Swale, E. M. F. (1972a). Some features of the microanatomy of *Chrysococcus cordiformis* Naumann, *Br. Phycol. J.* 7:53–9.

Belcher, J. H., and Swale, E. M. F. (1972b). The morphology and fine structure of the colourless colonial flagellate *Anthophysa vegetans* (O. F. Müller) Stein. *Br. Phycol. J.* 7:335–46.

Bird, D. F., and Kalff, J. (1986). Bacterial grazing by planktonic lake algae. *Science* 231:493–4.

Booth, B. C., and Marchant, H. J. (1987). Parmales, a new order of marine chrysophytes, with descriptions of three new genera and seven new species. *J. Phycol.* 23:245–60.

Bouck, G. B. (1971). The structure, origin, isolation, and composition of the tubular mastigonemes of the *Ochromonas* flagellum. *J. Cell Biol.* 50:362–84.

Caron, D. A., Sanders, R. W., Lim, E-L., Marrasé, C., Amaral, L. A., Whitney, S., Aoki, R. B., and Porter, K. G. (1993). Light-dependent phagotrophy in the freshwater mixotrophic chrysophyte *Dinobryon cylindricum. Microbial Ecology* 25:93–111.

Daley, R. J., Morris, G. P., and Brown, S. R. (1973). Phagotrophic ingestion of a blue-green alga by *Ochromonas. J. Protozool.* 20:58–61.

Dangeard, P. (1966). Sur le nouveau genre *Giraudyopsis* P.D. *Le Botantiste* 49:99–108.

Drebes, G. (1977). Sexuality. In *The Biology of the Diatoms*, ed. D. Werner, *Botanical Monographs*, Vol. 13, pp. 250–83. Berkeley: Univ. Calif. Press.

Dubowsky, N. (1974). Selectivity of ingestion and digestion in the chrysomonad flagellate *Ochromonas malhamensis. J. Protozool.* 21:295–8.

Dunlap, J. R., Walne, P. L., and Preisig, H. R. (1987). Manganese mineralization in chrysophycean loricas. *Phycologia* 26:394–6.

Eppley, R. W., Rogers, J. N., and McCarthy, J. J. (1969). Half-saturation constants for uptake of nitrate and ammonium by marine phytoplankton. *Limnol. Oceanogr.* 14:912–20.

Fauré-Fremiet, E. (1950). Rythme de marée d'une *Chromulina psammophile. Bull. Biol. Fr. Belg.* 84:207–14.

Fuhs, G. W., Demmerle, S. D., Canelli, E., and Chen, M. (1972). Characterization of phosphorus-limited algae (with reflections on the limiting nutrient concept). *Am. Soc. Limnol. Oceanogr. Spec. Symp.* 1:113–32.

Geitler, L. (1930). Ein grunes Filarplasmodium und andere neue Protisten. *Arch. Protistenk.* 69:515–36.

Gibbs, S. P. (1962). Nuclear envelope–chloroplast relationships in algae. *J. Cell Biol.* 14:433–44.

Herth, W., and Zugenmaier, P. (1979). The lorica of *Dinobryon. J. Ultrastruct. Res.* 69:262–72.

Herth, W., Kuppel, A., and Schnepf, E. (1977). Chitinous fibrils in the lorica of the flagellate chrysophyte *Poteriochromonas stipitata* (Syn. *Ochromonas malhamensis*). *J. Cell Biol.* 73:311–21.

Hibberd, D. J. (1970). Observations on the cytology and ultrastructure of *Ochromonas tuberculatus* sp. nov. (Chrysophyceae), with special reference to the discobolocysts. *Br. Phycol. J.* 5:119–43.

Hibberd, D. J. (1971). Observations on the cytology and ultrastructure of *Chrysoamoeba radians* Klebs (Chrysophyceae). *Br. Phycol. J.* 6:207–23.

Hibberd, D. J. (1976). The ultrastructure and taxonomy of the Chrysophyceae and Prymnesiophyceae (Haptophyceae): A survey with some new observations on the ultrastructure of the Chrysophyceae. *Bot. J. Linn. Soc.* 72:55–80.

Hibberd, D. J. (1977). Ultrastructure of cyst formation in *Ochromonas tuberculata* (Chrysophyceae). *J. Phycol.* 13:309–20.

Hibberd, D. J. (1978). The fine structure of *Synura sphagnicola* (Korsh.) Korsh. (Chrysophyceae). *Br. Phycol. J.* 13:403–12.

Hill, F. G., and Outka, D. E. (1974). The structure and origin of mastigonemes in *Ochromonas minute* and *Monas* sp. *J. Protozool.* 21:299–312.

Kauss, H. (1967). Metabolism of isofluoridoside (O-α-D-galactopyranosyl-$[1 \rightarrow 1]$-glycerol) and osmotic balance in the freshwater alga *Ochromonas. Nature* 214:1129–30.

Kosman, C. A., Thomsen, H. A., and Ostergaarad, J. B. (1993). Parmales (Chrysophyceae) from Mexican, Californian, Baltic, Arctic and Antarctic waters with a description of a new subspecies and several new forms. *Phycologia* 32:116–28.

Kristiansen, J. (1969). Lorica structure in *Chrysolykos* (Chrysophyceae). *Bot. Tidsskr.* 64:162–8.

Lee, R. E. (1978). Formation of scales in *Paraphysomonas vestita* and the inhibition of growth by germanium dioxide. *J. Protozool.* 25:163–6.

Lee, R. E., and Kugrens, P. (1989). Biomineralization in *Anthophysa vegetans* (Chrysophyceae). *J. Phycol.* 25:591–6.

Lehman, J. T. (1976). Ecological and nutritional studies on *Dinobryon* Ehrenb: Seasonal periodicity and the phosphate toxicity problem. *Limnol. Oceanogr.* 21:646–64.

Manton, I., and Leedale, G. F. (1961). Observations on the fine structure of *Paraphysomonas vestita* with special reference to the Golgi apparatus and the origin of scales. *Phycologia* 1:37–57.

Marchant, H. J., and McEldowney, A. (1986). Nanoplankton siliceous cysts from Antarctica are algae. *Mar. Biol. (Berl.)* 92:53–7.

O'Kelly, C. J. (1989). The evolutionary origin of the brown algae: information from the studies of motile cell ulltrastructure. In *The Chrysophyte Algae: Problems and Perspectives*, ed. J. C. Green, B. S. C. Leadbeater, and W. L. Diver, pp. 255–78. Oxford: Clarendon Press.

Pennick, N. C., and Clarke, K. J. (1972). *Paraphysomonas butcheri* sp. nov. a marine colourless, scale-bearing member of the Chrysophyceae. *Br. Phycol. J.* 7:45–8.

Preisig, H. R. (1994). Siliceous structures and silicification in flagellated protists, *Protoplasma* 181:29–42.

Preisig, H. R. (1995). A modern concept of chrysophyte classification. In *Chrysophyte Algae*, ed. C. D. Sandgren, J. R. Smol, and J. Kristiansen, pp. 46–74. Cambridge: Cambridge University Press.

Pringsheim, E. G. (1946). On iron flagellates. *Philos. Trans. R. Soc.* [*B*] 232:311–42.

Sandgren, C. D. (1980). An ultrastructural investigation of resting cyst formation in *Dinobryon cylindricum* Imhof (Chrysophyceae, Chrysophycophyta). *Protistologica* 16:259–76.

Sandgren, C. D. (1981). Characteristics of sexual and asexual resting cyst (statospore) formation in *Dinobryon cylindricum* Imhof (Chrysophyta). *J. Phycol.* 17:199–210.

Sandgren, C. D. (1983a). Survival strategies in chrysophycean flagellates: Reproduction and formation of resistant resting cysts. In *Survival Strategies in the Algae*, pp. 23–48. Cambridge: Cambridge University Press.

Sandgren, C. D. (1983b). Morphological variability in populations of chrysophycean resting cysts. I. Genetic (interclonal) and encystment temperature effects on morphology. *J. Phycol.* 19:64–70.

Sandgren, C. D., and Flanagin, J. (1986). Heterothallic sexuality and density dependent encystment in the chrysophycean alga *Synura petersenii* Korsch. *J. Phycol.* 22:206–16.

Saunders, G. W., Potter, D., and Andersen, R. A. (1997). Phylogenetic affinities of the Sarcinochrysidales and Chrysomeridales (Heterokonta) based on analyses of molecular and combined data. *J. Phycol.* 33:310–18.

Schiller, J. (1929). Neue Chryso- and Cryptomonaden aus Altwässern der Donau bei Wien. *Arch. Protistenk.* 66:436–58.

Schnepf, E., and Deichgraber, G. (1969). Uber die Feinstruktur von *Synura petersenii* unter besonderer Berücksuchtigung der Morphogenesis ihrer Kieselschuppen. *Protoplasma* 68:85–106.

Sheath, R. G., Hellebust, J. A., and Sawa, T. (1975). The statospore of *Dinobryon divergens* Imhoff: Formation and germination in a subarctic lake. *J. Phycol.* 11:131–8.

Thomsen, H. A., Zimmerman, B. Moestrup. Ø., and Kristiansen, J. (1981). Some new freshwater species of *Paraphysomonas* (Chrysophyceae). *Nord. J. Bot.* 1:559–81.

Throndsen, J. (1971). *Apedinella* gen. nov. and the fine structure of *A. spinifera* (Throndsen) comb. nov. *Norw. J. Bot.* 18:47–64.

Walne, P. L., Passarelli, V., Lenzi, P., Barsanti, L., and Gualtieri, P. (1995). Isolation of the flagellar swelling and identification of retinal in the phototactic flagellate, *Ochromonas danica* (Chrysophyceae). *J. Euk. Microbiol.* 42:7–11.

Wetherbee, R., and Andersen, R. A. (1992). Flagella of a chrysophycean alga play an active role in prey capture and selection. Direction observations on *Epipyxis pulchra* using image enhanced video microscopy. *Protoplasma* 166:1–7.

Willén, T. (1961). The phytoplankton of Ösbysjön Djursholm. 1. Seasonal and vertical distribution of the species. *Oikos* 12:36–79.

Zhang, X., Watanabe, M. M., and Inouye, I. (1996). Light and electron microscopy of grazing by *Poterioochromonas malhamensis* (Chrysophyceae) on a range of phytoplankton taxa. *J. Phycol.* 32:37–46.

10 · Heterokontophyta

Synurophyceae

The Synurophyceae are closely related to the Chrysophyceae (Ariztia et al., 1991). The Synurophyceae differ, however, from the Chrysophyceae in the following: the Synurophyceae have chlorophylls *a* and c_1, the flagella are inserted into the cell approximately parallel to one another (Fig. 10.1), there is a photoreceptor near the base of each flagellum, there is no eyespot and the contractile vacuole is in the posterior portion of the cell (Andersen, 1987; Andersen and Mulkey, 1983; Lavau et al., 1997). Chloroplast endoplasmic reticulum is present in a few species, but absent in most. Usually cells are covered by bilaterally symmetrical scales.

In the Synurophyceae, silica scales commonly occur outside the cell (Figs. 10.1 and 10.2). The scales are bilaterally symmetrical and are formed

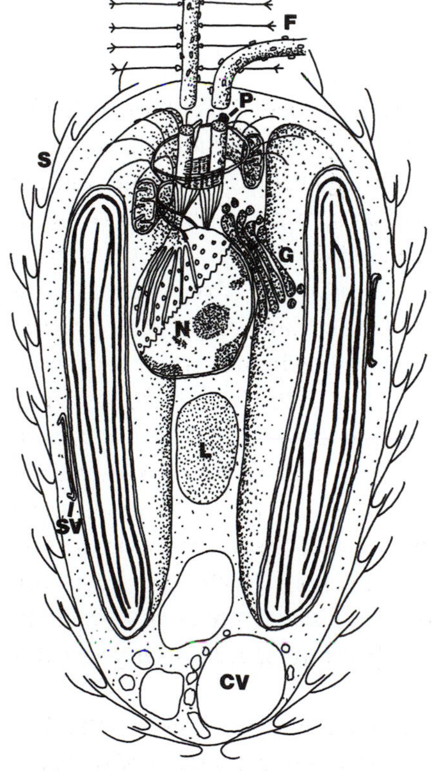

Fig. 10.1 Semi-diagrammatic drawing of the cytology of *Synura*, showing the characteristic cytology of the Synurophyceae. (CV) Contractile vacuole; (F) flagella; (G) Golgi; (L) chrysolaminarin vesicle; (N) nucleus; (P) photoreceptor; (S) scale; (SV) scale vesicles. (Adapted from Andersen, 1985.)

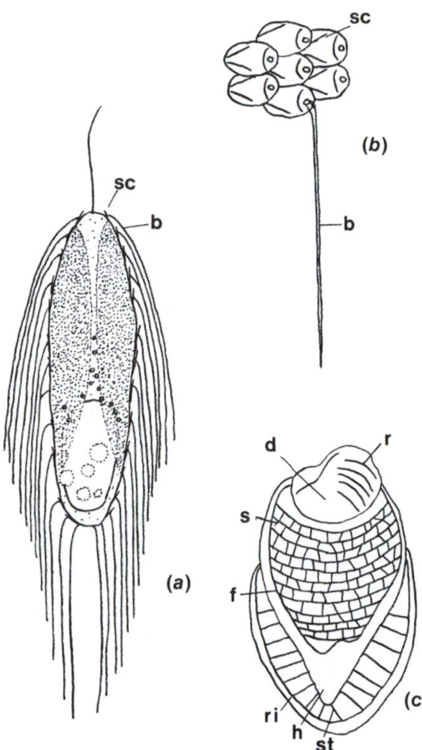

Fig. 10.2 *Mallomonas zellensis*. (*a*) Whole cell. (*b*) Scales. (*c*) A scale with the bristle removed. (b) Bristle; (d) dome; (f) flange; (h) hood; (r) ribs; (ri) rim; (s) shield; (sc) scale; (st) strut. (After Fott, 1962.)

in endoplasmic reticulum that is flattened against the chloroplast. The form of the scale vesicle is controlled by actin microfilaments in the cytoplasm (Brugerolle and Bricheux, 1984). The presence of germanium in the medium results in inhibition of scale formation (Klaveness and Guillard, 1975). The scales are carried in the scale vesicle to the plasma membrane where the plasma membrane and the scale vesicle fuse, releasing the scales outside the cell (Beech et al., 1990). The scales are held next to the cell in a organic envelope (Ludwig et al., 1996), which is either hyaline or yellow-brown, the latter appearance being due to the impregnation of iron salts. The scales of the Synurophyceae are commonly composed of a number of parts, such as the dome, shield, and bristle of *Mallomonas* (Lavau and Wetherbee, 1994) (Fig. 10.2). The scales of the Synurophyceae are overlapped precisely so that the anterior end of one scale overlaps the right margin of the scale to its left (Leadbeater, 1990). The scales are cemented together to form a scale case by the organic envelope (Fig. 10.4). This precise arrangement of scales differs from the loosely arranged scales of the Chrysophyceae.

Tessellaria volvocina (Fig. 10.3) is a colonial member of the class that has a large number of cells encased within a multilayer covering of siliceous

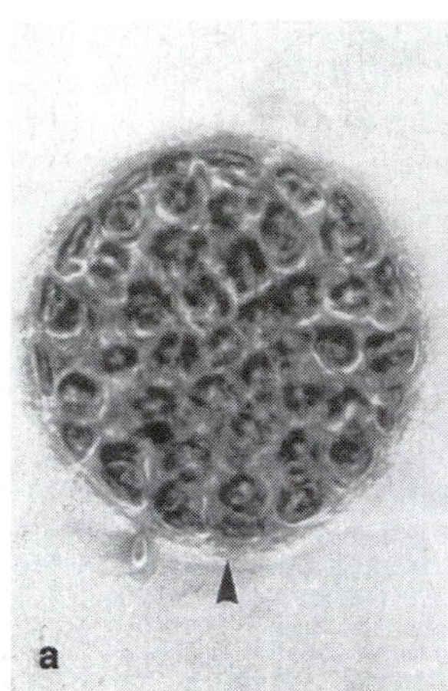

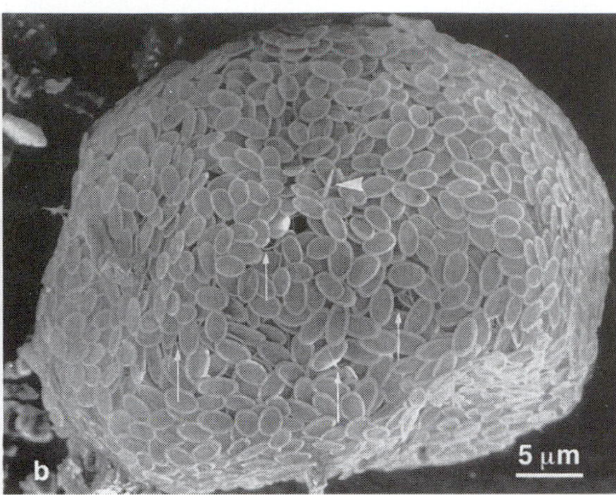

Fig. 10.3 *Tessellaria volvocina.* (*a*) Light micrograph. Arrow head points to scale case. (*b*) Scanning electron micrograph showing overlapping of plate scales and rare spine scales (arrow head). (From Tyler et al., 1989.)

Fig. 10.4 *Synura uvella.* (*a*) Light micrograph. (*b*) Scanning electron micrograph of a colony of cells. (*c*) Scanning electron micrograph of a single cell showing the two flagella emerging from an apical pore in the scale case. Bar = 10 μm. (From Andersen, 1985.)

scales. The individual cells of *T. volvocina* do not have a covering of scales (Tyler et al., 1989; Pipes and Leedale, 1992).

Analysis of lake sediments often reveals the presence of the silicified scales of the Synurophyceae as well as the silicified frustules of diatoms (Smol et al., 1984). Analysis of the species that produced these silicified deposits can often result in a history of environmental conditions in the lake. Diatoms usually do not grow in waters below a pH of 5.8 to 6, whereas several members of the Synurophyceae thrive in acidic lakes (Saxby-Rouen et al., 1997). As environmental concerns over the acidification of lakes by acid rains increase, these species will probably be more widely used as indicators of lake acidification.

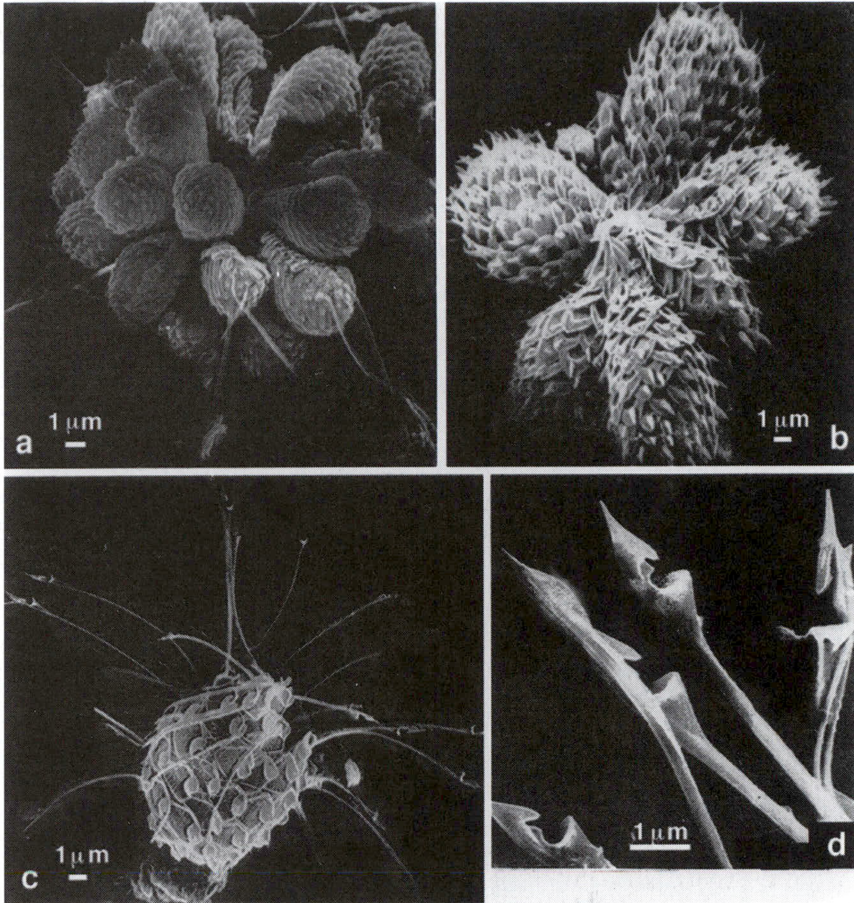

Fig. 10.5 Scanning electron micrographs: (*a*) *Synura petersenii*, whole colony. (*b*) *S. echinulata*. (*c*) *Mallomonas acaroides*, whole cell. (*d*) *M. acaroides*, distal end of helmet bristles. Bar = 1 μm. (From Dürrschmidt, 1984.)

Unlike the Chrysophyceae, the Synurophyceae are not phagocytotic, probably because the fixed scale-case is solidly formed and bacteria and food particles cannot easily reach the cell membrane. The cysts (statospores or stromatocysts) of the Synurophyceae are usually more elaborate than those of the Chrysophyceae.

Classification

The Synurophyceae share affinities with the Bacillariophyceae, Chrysophyceae, and Phaeophyceae, and it is possible to include them in any of these groups (Andersen, 1987; Lavau et al., 1997). Because of their silicified outer covering, they have been called "flagellated diatoms"; yet they have many of the characteristics of the Chrysophyceae such as statospores.

There is a single class, the Synurales. All of the organisms are flagellates.

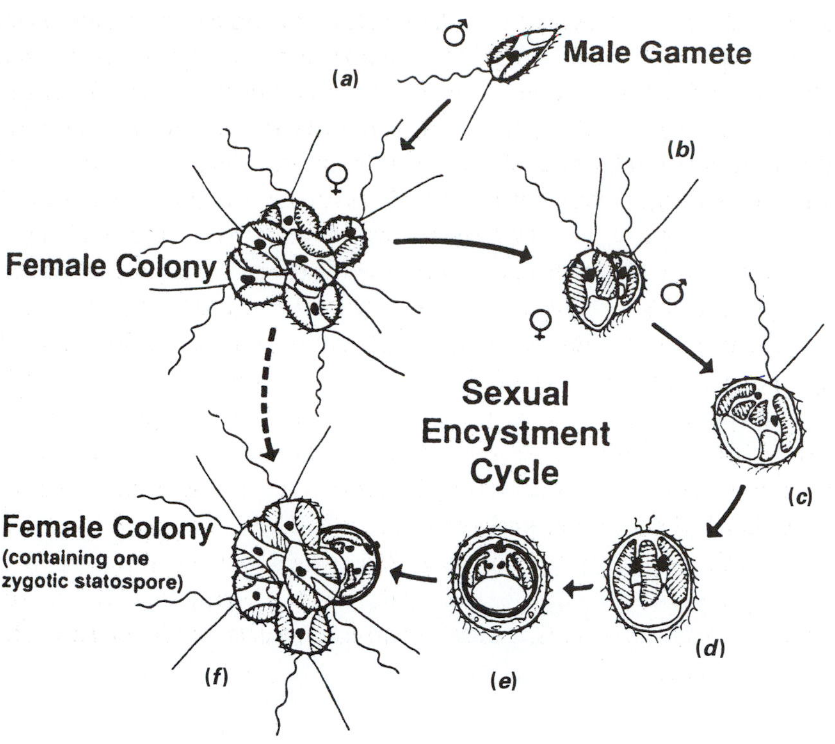

Fig. 10.6 Process of sexual copulation and zygotic encystment in *Synura petersenii*. (*a*) Initial contact of a solitary male gamete with a female gamete cell within a colony. (*b*) An early stage in plasmogamy characterized by the lateral fusion of cells, each with active flagellar pairs. (*c*) A biflagellate planozygote stage with the fusion cell containing four plastids, two nuclei, and two chrysolaminarin starch vesicles. Note that the planozygote is completely covered with scales. (*d*) An early stage in zygotic encystment during which the remaining flagella are lost and significant rearrangement of cytoplasmic organelles occurs to establish a symmetrical orientation. Fusion of chrysolaminarin vesicles occurs, and the nuclei become associated with the inner margins of the plastids. (*e*) A mature zygotic statospore containing four reduced plastids and two nuclei. Some cytoplasm and the vegetative scale layer remain outside the cyst wall. (*f*) A completely mature zygotic statospore as it ultimately appears, taking the place of the female gamete in the colony. (From Sandgren and Flanagin, 1986.)

Siphonaceous forms do not exist, as true filaments cannot be formed by organisms with a silica scale-case, and coccoid forms would be similar to diatoms. The diatoms *Corethron*, *Rhizosolenia*, and *Attheya* are similar in appearance to coccoid Synurophyceae (Andersen, 1987).

Synura (Fig. 10.4) is a widely distributed colonial member of the class. The life cycle of *Synura petersenii* (Figs. 10.5(*a*) and 10.6) has been described by Sandgren and Flanagin (1986). Vegetative colonies of *S. petersenii* normally consist of a number of biflagellate cells joined at their posterior end. The cells are covered with silicified scales (Leadbeater, 1990). Each cell has two parietal chloroplasts, a central nucleus, and a large posterior chrysolamarin vacuole. Sexual reproduction in *S. petersenii* is isogamous and heterothallic. Gametes appear when two compatible clones are mixed. No special culture conditions are required to induce sexual behavior. Actively growing cell populations are continually receptive to mating when mixed with a sufficient number of cells from a compatible clone. Male gametes are solitary biflagellate cells that appear similar to vegetative cells. The solitary male gametes are probably released from the colonies of the male clone. The female gametes occur in colonies or as solitary cells and are also similar in appearance to vegetative cells. Male and female gametes initially contact each other anteriorly; this is followed by lateral fusion. The silica scale layers mingle so that the resulting fusion cell maintains a complete scale layer. The biflagellate planozygote undergoes minor shape changes and either oscillates in place or swims sluggishly for at least 4 to 8 hours. During this time, the four chloroplasts assume a symmetrical orientation around the flagellar bases as is typical of vegetative *Synura* cells, and the two large chrysolaminarin vesicles fuse to occupy the cell posterior. The flagella are lost, and the zygote becomes spherical as cyst wall deposition begins. After 6 to 8 hours, a morphologically mature zygote statospore is produced. Mature zygospores are observed as solitary objects if the females gametes are solitary. If the female gametes are in a colony, then the zygospores take the place of the female gametes in the female colonies. Female colonies can have as many as eight statospores, apparently as the result of numerous cells serving as female gametes. The cyst is an unornamented sphere, 13 to 16 μm in diameter and has a single, small recessed pore that lacks a surrounding collar. Presumably, statospores germinate meiotically to produce haploid cells. The number of statospores produced is in the range of 1% to 20% of the vegetative cell density. In addition, some statospores are produced asexually and can be seen in the clones before they are mixed.

References

Andersen, R. A. (1985). The flagellar apparatus of the golden alga *Synura uvella:* Four absolute orientations. *Protoplasma* 128:94–106.

Anderson, R. A. (1987). Synurophyceae classis nov., a new class of algae. *Am. J. Bot.* 74:337–53.

Anderson, R. A., and Mulkey, T.J. (1983). The occurrence of chlorophylls c_1 and c_2 in the Chrysophyceae. *J. Phycol.* 19:289–94.

Ariztia, E. V., Andersen, R. A., and Sogin, M. L. (1991). A new phylogeny for chromophyte algae using 16-S-like rRNA sequences from *Mallomonas papillosa* (Synurophyceae) and *Tribonema aequale* (Xanthophyceae). *J. Phycol.* 27:428–36.

Beech, P. L., Wetherbee, R., and Pickett-Heaps, J. D. (1990). Secretion and development of bristles in *Mallomonas splendens* (Synurophyceae). *J. Phycol.* 26:112–22.

Brugerolle, G., and Bricheux, G. (1984). Actin microfilaments are involved in scale formation of the chrysomonad cell *Synura. Protoplasma* 133:203–12.

Dürrschmidt, M. (1984). Studies on scale-bearing Chrysophyceae from the Giessen area, Federal Republic of Germany. *Nord. J. Bot.* 4:123–43.

Fott, B. (1962). Taxonomy of *Mallomonas* based on electron micrographs of scales. *Preslia* 34:69–84.

Friedman, A. L., and Alberte, R. S. (1984). A diatom light-harvesting pigment–protein complex. Purification and characterization. *Plant Physiol.* 76:483–9.

Hibberd, D. J. (1978). The fine structure of *Synura sphagnicola* (Korsh.) Korsh. (Chrysophyceae). *Br. Phycol. J.* 13:403–12.

Klaveness, D., and Guillard, R. R. L. (1975). The requirement for silicon in *Synura petersenii* (Chrysophyceae). *J. Phycol.* 11:349–55.

Lavau, S., and Wetherbee, R. (1994). Structure and development of the scale case of *Mallomonas adamas* (Synurophyceae). *Protoplasma* 181:259–68.

Lavau, S., Saunders, G. W., and Wetherbee, R. (1997). A phylogenetic analysis of the Synurophyceae using molecular data and scale case morphology. *J. Phycol.* 33:135–51.

Leadbeater, B. S. C. (1990). Ultrastructure and assembly of the scale case in *Synura* (Synurophyceae Andersen). *Br. Phycol. J.* 25:117–32.

Ludwig, M., Lind, J. L., Miller, E. A., and Wetherbee, R. (1996). High molecular mass glycoprotein associated with the siliceous cell scales and bristles of *Mallomonas splendens* (Synurophyceae) may be involved in cell surface development and maintenance. *Planta* 199:219–28.

Pipes, L. D., and Leedale, G. F. (1992). Scale formation in *Tessellaria volvocina* (Synurophyceae). *Br. Phycol. J.* 27:11–19.

Sandgren, C. D., and Flanagin, J. (1986). Heterothallic sexuality and density dependent encystment in the chrysophycean alga *Synura petersenii* Korsch. *J. Phycol.* 22:206–16.

Saxby-Rouen, K. J., Leadbeater, B. S. C., and Reynolds, C. S. (1997). The growth response of *Synura petersenii* (Synurophyceae) to photon flux density, temperature, and pH. *Phycologia* 36:233–43.

Smol, J. P., Charles, D. F., and Whitehead, D. R. (1984). Mallomonadacean microfossils provide evidence of recent lake acidification. *Nature* 307:628–30.

Tyler, P. A., Pipes, L. D., Croome, R. L., and Leedale, G. F. (1989). *Tessellaria volvocina* rediscovered. *Br. Phycol. J.* 24:329–37.

11 · Heterokontophyta

Dictyochophyceae

These golden-brown algae are characterized by tentacles or rhizopodia on basically amoeboid vegetative cells (Moestrup, 1995; Preisig, 1995). Amoeboid cells are relatively rare among the algae, being mostly restricted to the Dictyochophyceae and the Xanthophyceae (Hibberd and Chretiennot-Dinet, 1979). The algae in the Dictyochophyceae have been previously classified in the Chrysophyceae, although molecular evidence shows them to be most closely related to the Pelagophyceae (Cavalier-Smith et al., 1995) or Eustigmatophyceae (Daugbjerg and Andersen, 1997).

Classification

The Dictyochophyceae can be divided into three orders (Preisig, 1995):

Order 1 Rhizochromulinales: Marine and freshwater unicells with tentacles.
Order 2 Pedinellales: Unicells with a long anterior flagellum and a second flagellum reduced to a basal body, usually three to six chloroplasts (if chloroplasts are present), marine and freshwater.
Order 3 Dictyocales: Marine unicells with an external silicified skeleton.

RHIZOCHROMULINALES

This order contains the more primitive organisms in the order (O'Kelly and Wujek, 1995). *Rhizochromulina* (Fig. 11.1(*a*),(*b*)) has amoeboid non-flagellated vegetative cells with many fine beaded filipodia and a single golden-brown chloroplast (Hibberd and Chretiennot-Dinet, 1979). The fusiform zoospore has a single tinsle flagellum with a second basal body in the protoplasm (Fig. 11.1(*b*)). *Chrysoamoeba* (Fig. 11.1(*d*)) lives as a solitary amoeba for the greater part of its life cycle, transforming into swimming cells with a single long flagellum only for short periods. In *Phaeaster* (Fig. 11.1(*c*)), the anterior portion of the cell is drawn out into rhizopodia.

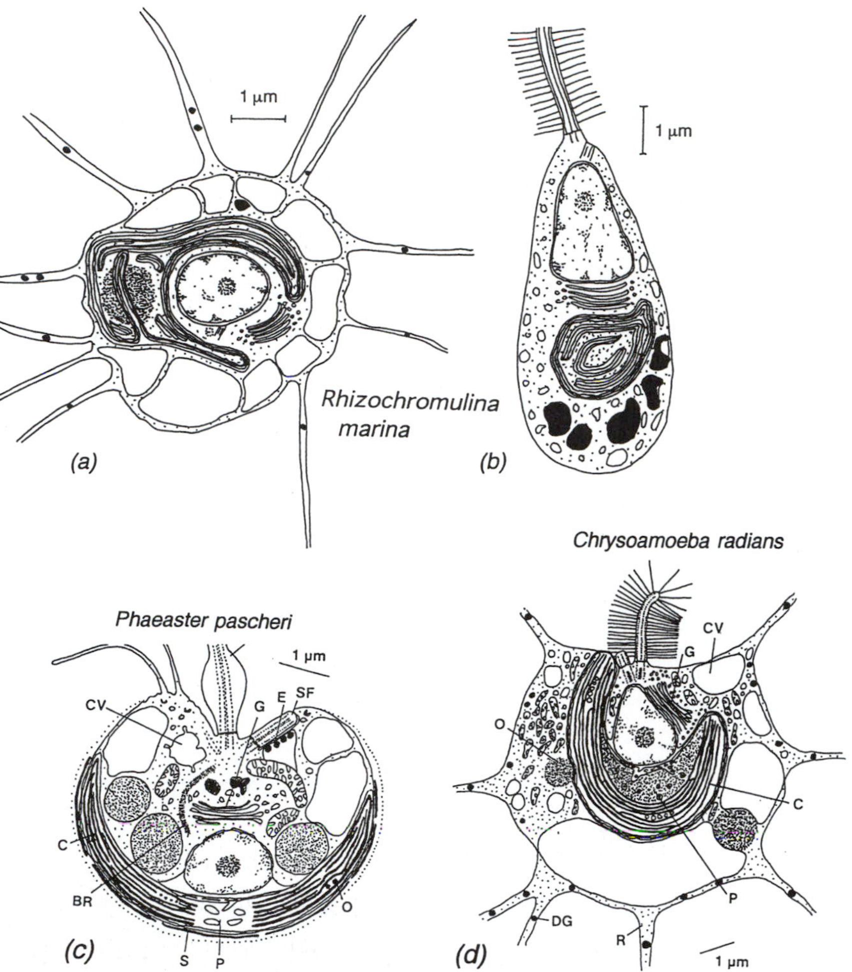

Fig. 11.1 (*a*) *Rhizochromulina marina*. Vegetative cell (*a*) and zoospore (*b*). (*c*) *Phaeaster pascheri*. (*d*) *Chrysoamoeba radians*. (BR) Basal root; (C) chloroplast; (CV) contractile vacuole; (DG) dense granule; (E) eyespot; (G) Golgi body; (LF) long flagellum; (O) oil; (P) pyrenoid; (R) rhizopodium; (S) scale; (SF) short flagellum. ((*a*),(*b*) redrawn from Hibberd and Chretiennot-Dinet, 1979, (*c*) redrawn from Belcher and Swale, 1971; (*d*) redrawn from Hibberd, 1971.)

PEDINELLALES

The pedinellids are unique in containing genera that are phototrophic (*Apedinella* (Fig. 11.2(*c*)), *Pseudopedinella*), mixotrophic (able to photosynthesize and take up organic compounds) (*Pedinella* (Fig. 11.2(*a*),(*b*)) and phagotrophic (*Actinomonas, Ciliophrys*). The organisms in this order have three interconnected microtubules (triads) that course from the nuclear envelope through tentacles (if they are present) to the plasma membrane (Fig. 11.2(*a*)) (Daugbjerg, 1996). A long apical flagellum is extended into a lateral wing by a paraxonemal rod (Fig. 11.2). The apical flagellum is inserted in a pit and there is a second flagellum that is reduced to a basal body. The basal bodies are at a slight angle to each other. The cells are radially symmetrical with a large central nucleus and a posterior Golgi apparatus. There are usually three to six chloroplasts present if the organism is phototrophic

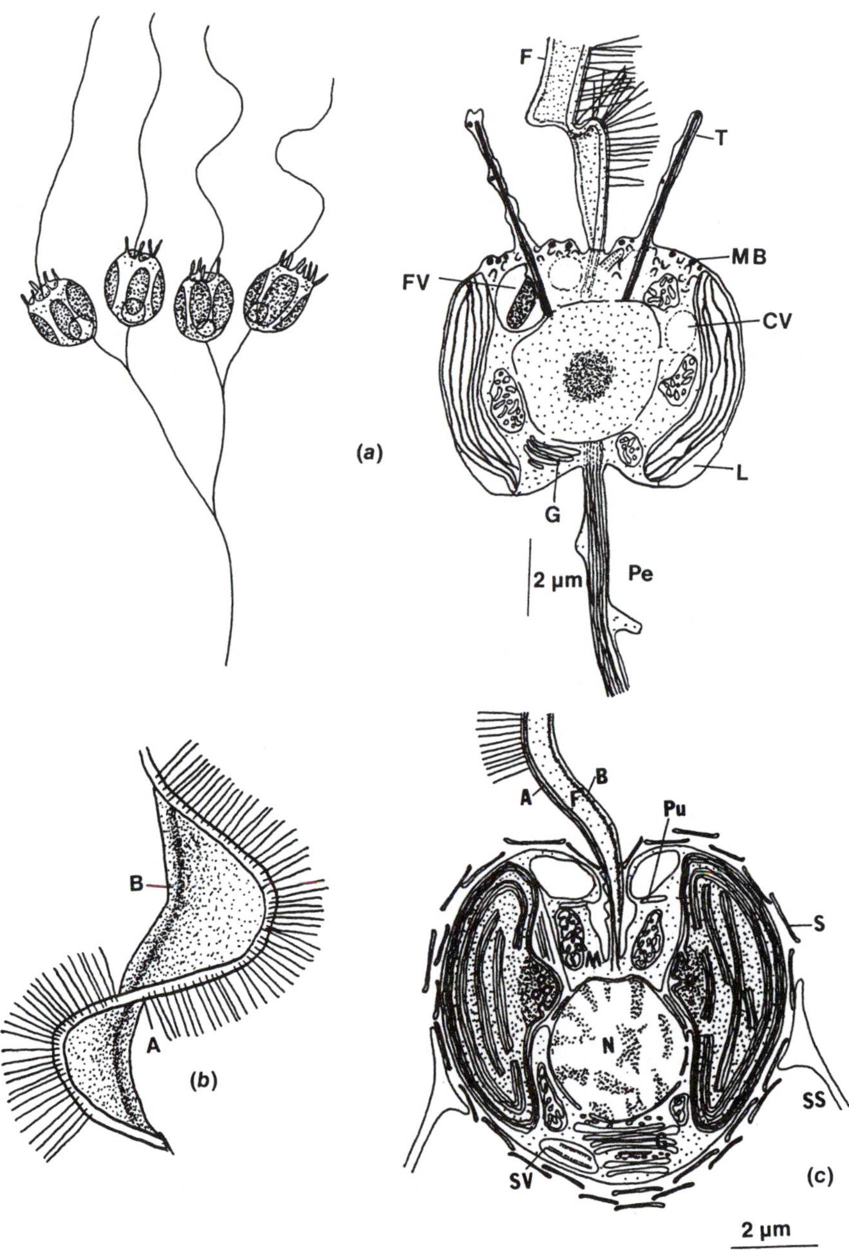

Fig. 11.2 (*a*) *Pedinella hexacostata* in the light and electron microscope. (*b*) Structure of the winged flagellum of *P. hexacostata*. (*c*) *Apedinella spinifera*. (A) Axoneme of flagellum; (B) paraxonema rod of flagellum; (CV) contractile vacuole; (F) flagellum with hairs; (FV) food vacuole; (G) Golgi body; (L) leucosin; (M) mitochondrion; (MB) muciferous body; (N) nucleus; (Pe) peduncle; (Pu) pusule; (S) scale; (SS) spined scale; (SV) scale vesicle; (T) tentacle. ((*a*),(*b*) after Swale, 1969; (*c*) after Throndsen, 1971.)

(whereas in the Chrysophyceae there are usually only one or two chloroplasts). Some genera, such as *Apedinella* (Fig. 11.2(*c*)), have scales attached to the plasma membrane by microligaments (Koutoulis et al., 1988).

A posterior trailing stalk, associated with a system of vacuoles, can be present. *Pedinella* (Fig. 11.2(*a*)) cells rotate while swimming, trailing the stalk behind. The stalk appears to be sticky, and swimming cells will often adhere to a substrate by means of the stalk. *Pedinella* can undergo phagotro-

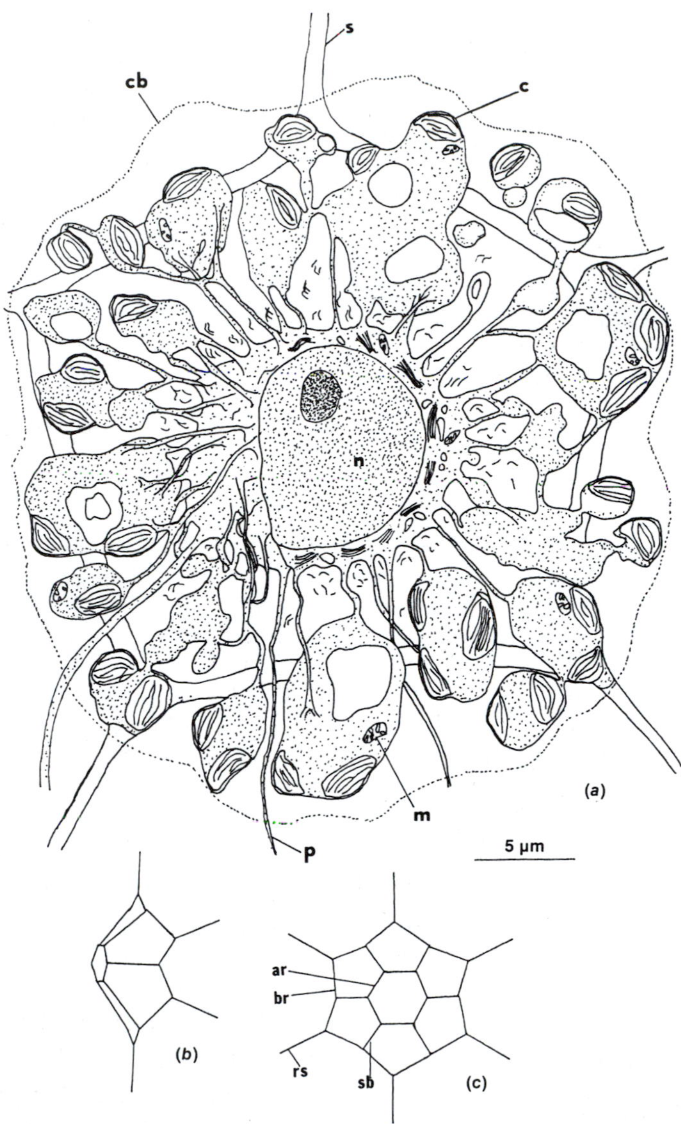

Fig. 11.3 (*a*) A drawing of the fine structure of *Dictyocha fibula*. (*b*), (*c*) Side and front views of the skeleton of *Dictyocha*. (ar) Apical ring; (br) basal ring; (c) chloroplast; (cb) cell boundary; (m) mitochondrion; (n) nucleus; (p) pseudopodium; (rs) radial spine; (s) silica skeleton; (sb) supporting bar. (After van Valkenburg, 1971a,b.)

phy and ingest other small cells. Bacteria are passed down the flagellum and adhere to the plasma membrane, just outside the row of tentacles. Secretions of the muciferous bodies in this region provide an adhesive that sticks the cells to the plasmalemma. Usually within a minute, the bacteria become enveloped in a sheet of cytoplasm that is extruded from the cell.

DICTYOCALES

The Dictyocales or silicoflagellates are a group of cosmopolitan marine flagellates, presently represented by only one extant genus *Dictyocha*, although an abundance of taxa have been described from fossil siliceous

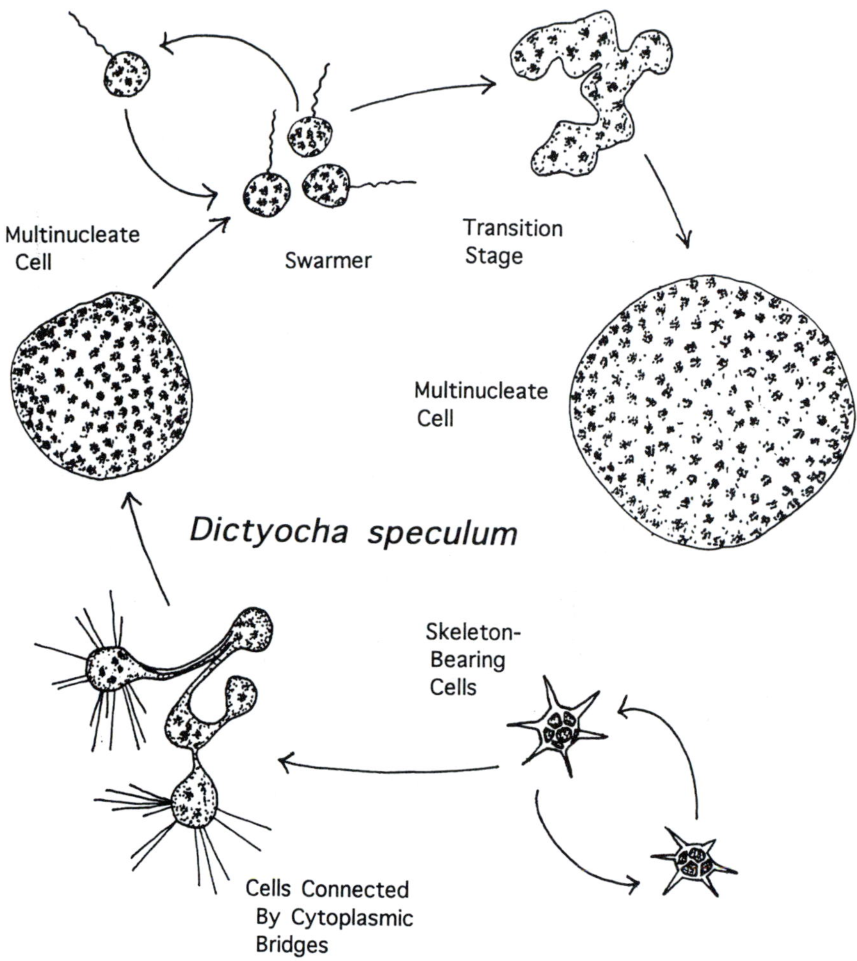

Fig. 11.4 Growth stages of the silicoflagellate *Dictyocha speculum*. (Adapted from Henriksen et al., 1993.)

skeletons (Henriksen et al., 1993). The silicoflagellates have one emergent flagellum and skeleton of hollow siliceous rods outside of the protoplasm. The skeleton can be a simple ring, an ellipse, or a triangle, but it is often much more complex. In *Dictyocha* (Fig. 11.3), the skeleton is composed of a series of peripheral polygons surrounding a central hexagon. The nucleus is in the center of the protoplasm with a number of cytoplasmic processes extending from the central mass. The chloroplasts are usually in the cytoplasmic processes (van Valkenburg, 1971a,b).

In *Dictyocha speculum*, the skeleton-bearing cells multiply vegetatively by mitotic division (Fig. 11.4) (Henriksen et al., 1993). Cells connected by bridges develop and give rise to large spherical cells without skeletons that become multinucleate. Uninucleate swarmers with a single flagellum develop in the large spherical cells. The swarmers are released and grow into large multinucleate cells, which are probably a form of resting cell. All of the cells are of the same ploidy level and sexual reproduction is not known.

The silicoflagellates originated in the Cretaceous, with the earlier forms having a simpler skeleton than those existing today. Fossils of silicoflagellates are common in calcareous chalks along with members of the Prymnesiophyceae. Silicoflagellates constitute a prominent part of the phytoplankton in the colder seas of today.

References

Belcher, J. H., and Swale, E. M. F. (1971). The microanatomy of *Phaeaster pascheri* Scherffel (Chrysophyceae). *Br. Phycol. J.* 6:157–69.

Cavalier-Smith, T., Chao, E. E., and Allsopp, T. E. P. (1995). Ribosomal RNA evidence for chloroplast loss within Heterokonta: pedinellid relationships and a revised classification of ochristan algae. *Arch. Protistenkd.* 145:209–20.

Daugbjerg, N. (1996). *Mesopedinella arctica* gen. et sp. nov. (Pedinelles, Dictyochophyceae) I. fine structure of a new marine phytoflagellate from Arctic Canada. *Phycologia* 35:435–45.

Daugbjerg, N., and Andersen, R. A. (1997). A molecular phylogeny of heterokont algae based on analyses of chloroplast-encoded *rbc*L sequence data. *J. Phycol.* 33:1031–41.

Henriksen, P., Knipschildt, F., Moestrup, Ø., and Thomsen, H. A. (1993). Autecology, life history and toxicology of the silicoflagellate *Dictyocha speculum* (Silicoflagellata, Dictyochophyceae). *Phycologia* 32:29–39.

Hibberd, D. J. (1971). Observations on the cytology and ultrastructure of *Chrysoamoeba radians* Klebs (Chrysophyceae). *Br. Phycol. J.* 6:207–23.

Hibberd, D. J., and Chretiennot-Dinet, M-J. (1979). The ultrastructure and taxonomy of *Rhizochromulina marina* gen. et sp. nov., an amoeboid marine chrysophyte. *J. Mar. Biol. Assn., UK* 59:179–93.

Koutoulis, A., McFadden, G. I., and Wetherbee, R. (1988). Spine-scale reorientation in *Apedinella radians* (Pedinelles, Chrysophyceae): the microarchitecture and immunocytochemistry of the associated cytoskeleton. *Protoplasma* 147:25–41.

Moestrup, Ø. (1995). Current status of chrysophyte 'splinter groups': synurophytes, pedinellids, silicoflagellates. In *Chrysophyte Algae*, ed. C. D. Sandgren, J. R. Smol, and J. Kristiansen, pp. 75–91. Cambridge: Cambridge University Press.

O'Kelly, C. J., and Wujek, D. E. (1995). Status of the Chrysamoebales (Chrysophyceae): observations on *Chrysamoeba pyrenoidefera, Rhizochromulina marina* and *Lagynion delicatulum.* In *Chrysophyte Algae,* ed. C. D. Sandgren, J. R. Smol and J. Kristiansen, pp. 361–72. Cambridge: Cambridge University Press.

Preisig, H. R. (1995). A modern concept of chrysophyte classification. In *Chrysophyte Algae,* ed. C. D. Sandgren, J. R. Smol, and J. Kristiansen, pp. 46–74. Cambridge: Cambridge University Press.

Swale, E. M. F. (1969). A study of the nanoplankton flagellate *Pedinella hexacostata* Vysotshiĭ by light and electron microscopy. *Br. Phycol. J.* 4:65–86.

Throndsen, J. (1971). *Apedinella* gen. nov. and the fine structure of *A. spinifera* (Throndsen) comb. nov. *Norw. J. Bot.* 18:47–64.

van Valkenburg, S. D. (1971a). Observations on the fine structure of *Dictyocha fibula* Ehrenberg. I. The skeleton. *J. Phycol.* 7:113–18.

van Valkenburg (1971b). Observations on the fine structure of *Dictyocha fibula* Ehrenberg. II. The protoplast. *J. Phycol.* 7:118–32.

12 · Heterokontophyta

Pelagophyceae

The Pelagophyceae are a group of basically unicellular algae that are cytologically similar to the Chrysophyceae, in which they were previously classified (Andersen et al., 1993). The cells are very small (3–5 μm) members of the ultraplankton and appear as small spheres with indistinct protoplasm under the light microscope. Recent studies on the sequences of small-subunit RNA nucleotides in these algae have shown them to be closely related to each other and distinct from other members of the Heterokontophyta (Saunders et al., 1997). While these algae have been shown to be distinct from other members of the Heterokontophyta based on molecular data, they do not have cytological or morphological characters which are very different from other members of the phylum.

Pelagomonas calceolata is a very small (1.5 μm × 3 μm) ultraplanktonic marine alga with a single tinsel flagellum and basal body, and a single chloroplast and mitochondrion (Fig. 12.1(*c*)) (Andersen et al., 1993).

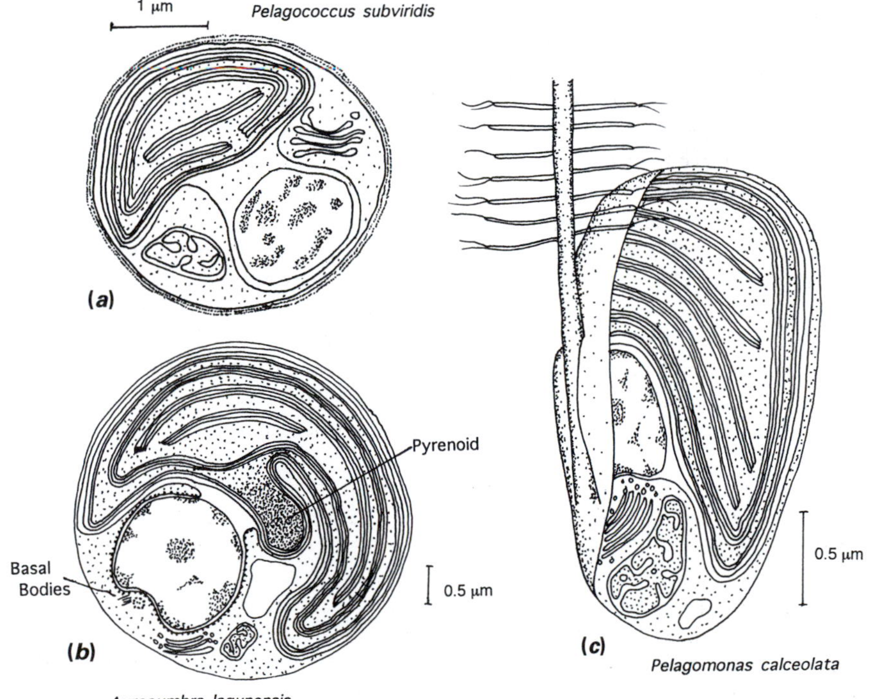

Fig. 12.1 (*a*) *Pelagococcus subviridis* showing the single chloroplast, nucleus and mitochondrion. (*b*) *Aureoumbra lagunensis* with the characteristic stalked pyrenoid and two basal bodies near the nucleus. (*c*) *Pelagomonas calceolata* showing the single flagellum fitting into a groove in the cell. ((*a*) adapted from Vesk and Jeffrey, 1987; (*b*) adapted from DeYoe et al., 1997; (c) adapted from Andersen et al., 1993.)

Another member of the marine ultraplankton is *Pelagococcus subviridis* a green-gold spherical non-motile cell (2.5–3.0 μm) with a single chloroplast, mitochondrion and nucleus (Fig. 12.1(*a*)) (Vesk and Jeffrey, 1987).

Members of the class are economically important because some of the algae produce "brown tides." *Aureoumbra Iagunensis* (Fig. 12.1(*b*)) is the causative agent of brown tides in Texas (DeYoe et al., 1997), while *Aureococcus anophagefferens* forms brown tides along the coasts of New Jersey, New York and Rhode Island. The numbers of cells in brown tides can be so large that they can exclude light from the benthic eelgrass (*Zostera marine*), resulting in elimination of the eelgrass. The larvae of the bay scallop feed off eelgrass and the bay scallop industry was virtually wiped out for a number of years after a brown tide in the waters off the northeast United States (Nicholls, 1995).

References

Andersen, R. A., Saunders, G. W., Paskind, M. P., and Sexton, J. P. (1993). Ultrastructure and 18S rRNA gene sequence for *Pelagomonas calceolata* gen. et sp. nov. and the description of a new algal class, the Pelagophyceae classis nov. *J. Phycol.* 29:701–15.

DeYoe, H. R., Stockwell, D. A., Bidigare, R. R., Latasa, M., Johson, P. W., Hargraves, P. E., and Suttle, C. A. (1997). Description and characterization of the algal species *Aureoumbra lagunensis* gen. at sp. nov. and referral of *Aureoumbra* and *Aureococcus* to the Pelagophyceae. *J. Phycol.* 33:1042–8.

Nicholls, K. H. (1995). Chrysophyte blooms in the plankton and neuston of marine and freshwater systems. In *Chrysophyte Algae, Ecology, Phylogeny and Development*, ed. C. D. Sandgren, J. P. Smol, and J. Kristiansen, pp. 181–213. Cambridge: Cambridge University Press.

Saunders, G. W., Potter, D., and Andersen, R. A. (1997). Phylogenetic affinities of the Sarcinochrysidales and Chrysomeridales (Heterokonta) based on analysis of molecular and combined data. *J. Phycol.* 33:310–18.

Vesk, M., and Jeffrey, S. W. (1987). Ultrastructure and pigments of two strains of the picoplanktonic alga *Pelagococcus subviridis* (Chrysophyceae). *J. Phycol.* 23:322–36.

13 · Heterokontophyta

Bacillariophyceae

The Bacillariophyceae or the diatoms probably evolved from a scaly member of the Chrysophyceae (similar to the organisms in the Parmales) or Synurophyceae (Sorhannus et al., 1995; Medlin et al., 1993). The diatoms are unicellular, sometimes colonial algae found in almost every aquatic habitat as free-living photosynthetic autotrophs, colorless heterotrophs, or photosynthetic symbiotes (Schmaljohann and Röttger, 1978). They may occur as plankton or periphyton, with most brownish-green films on substrates such as rocks or aquatic plants being composed of attached diatoms. The cells are surrounded by a rigid two-part boxlike cell wall composed of silica, called the **frustule**. The chloroplasts contain chlorophylls *a*, c_1, and c_2 with the major carotenoid being the golden-brown fucoxanthin, which gives the cells their characteristic color.

In discussing diatoms and silica, there is often confusion over terminology in regard to silicon. **Silicon** is the element. **Silica** is a short convenient designation for **silicon dioxide** (SiO_2) in all of its crystalline, amorphous, and hydrated or hydroxylated forms. **Silicate** is any of the ionized forms of monosilicic acid [$Si(OH)_4$] (Iler, 1979).

Cell structure

The two-part frustule surrounds protoplasm that has a more or less central nucleus suspended in a system of protoplasmic threads. The chloroplasts occupy most of the cell (Fig. 13.1) usually as two parietal plastids although sometimes as numerous discoid plastids. The storage product, chrysolaminarin, occurs in vesicles in the protoplasm.

CELL WALL

The characteristic feature of the Bacillariophyceae is their ability to secrete an external wall composed of silica, the frustule. It is constructed of two almost equal halves, the smaller fitting into the larger like a Petri dish (Figs. 13.1, 13.13 and 13.32). The outer of the two half-walls is the **epitheca**, and the inner the **hypotheca**. Each theca is composed of two parts: the **valve**, a

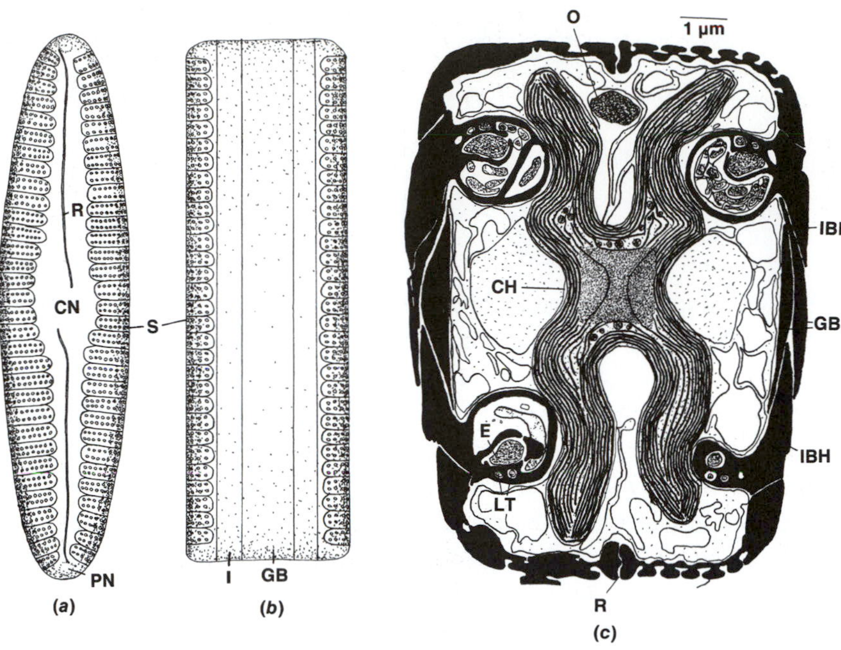

Fig. 13.1 Light microscopical drawing of valve (*a*) and girdle (*b*) views of the diatom *Mastogloia*. (*c*) A drawing of a transverse section of *M. grevillei* in the electron microscope. (Ch) Chloroplast; (CN) central nodule; (E) elongate chamber of a septum; (GB) girdle band; (I) intercalary band; (IBE) intercalary band of the epitheca; (IBH) intercalary band of the hypotheca; (LT) locule tubule; (O) oil body; (PN) polar nodule; (R) raphe; (S) stria. ((*c*) Adapted from Stoermer et al., 1965a.)

more or less flattened plate, and the **connecting band**, attached to the edge of the valve. The two connecting bands, one attached to each valve, are called the **girdle** (von Stosch, 1975). Sometimes the connecting bands themselves are called girdle bands. Occasionally there are one or more additional bands between the valve and the girdle, which are called **intercalary bands.** When an appreciable part of the edge of the valve is bent inward, this portion is called the **mantle** or **valve-jacket**. The girdle bands, often furnished with minute teeth, hold the valves together by their edges. The value margin thus butts onto the end of the girdle band and is usually connected to it by a pectinaceous film. If this film is destroyed, the valve and girdle bands separate.

The siliceous material of the frustule is laid down in certain regular patterns that leave the wall ornamented. According to Hendey (1964), the ornamentation of diatoms can be divided into four basic types: (1) **centric** and radial, where the structure is arranged according to a central point, for example, *Coscinodiscus* (Fig. 13.2(*a*)); (2) **trellisoid**, where the structure is arranged uniformly over the surface without reference to a point or a line, for example, *Eunotia* (Fig. 13.2(*b*)); (3) **gonoid**, where the structure is dominated by angles, for example, *Triceratium* (Fig. 13.2(*c*)); (4) **pennate**, where the structure is symmetrically arranged upon either side of a central line – for example, *Navicula* (Fig. 13.2(*d*)).

Some pennate diatoms have a raphe system composed of the **raphe** (a

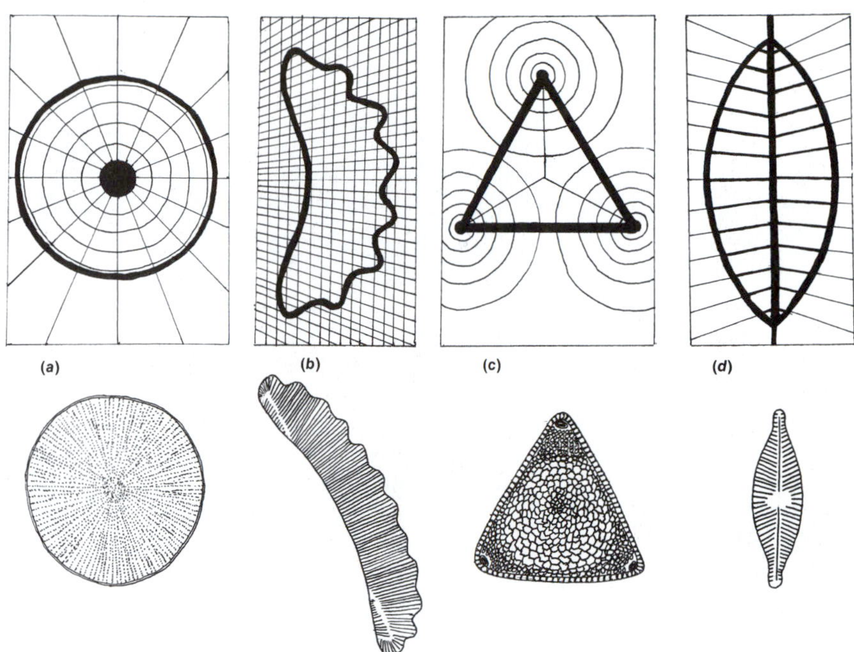

Fig. 13.2 The basic patterns of ornamentation in the Bacillariophyceae: (*a*) centric and radial (example: *Coscinodiscus*). (*b*) Trellisoid, with structure arranged margin to margin (example: *Eunotia*). (*c*) Gonoid, with structure supported by angles (example: *Triceratium*). (*d*) Pennate, symmetrical about an apical line (example: *Navicula*). (After Hendey, 1964.)

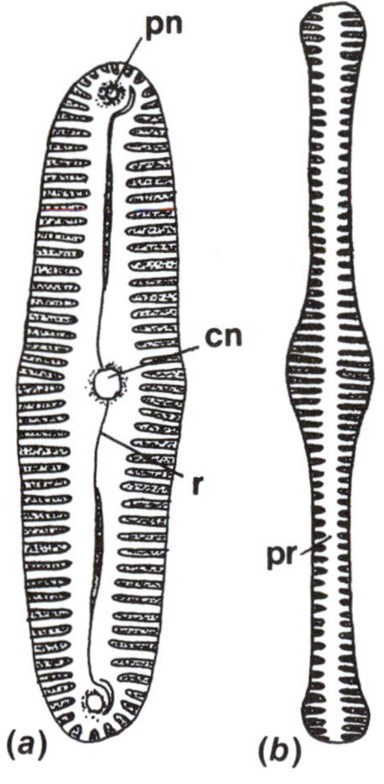

Fig. 13.3 (*a*) A cell with a raphe system (*Pinnularia viridis*). (cn) Central Nodule; (pn) polar nodule; (r) raphe. (*b*) A cell with a pseudoraphe (pr) (*Tabellaria fenestrata*).

longitudinal slot in the theca), divided into two parts by the **central nodule** (Figs. 13.3, 13.5 and 13.6). Each half of the raphe terminates in a swelling of the wall called the **polar nodule**. The ornamentation in the pennate diatoms is **bilaterally symmetrical** around the raphe. In those pennate diatom valves that do not have a raphe system, there is instead an unornamented area running down the center of the valve, which is called the **pseudoraphe** (Fig. 13.3). The raphe is not a simple cleft in the wall but is instead an S-shaped slit that is wider at the outer and inner fissue and thinner in the middle partition region (Fig. 13.1).

Besides the raphe there are basically two types of wall perforations within the Bacillariophyceae: the simple **pore** or hole, and the more complex **loculus** or **areola** (Fig. 13.4) (Hendey, 1964; Ross and Sims, 1972). The pore consists of a simple hole within a usually homogeneous silicified wall that is frequently strengthened by **ribs** and **costae** (Fig. 13.7(*a*) and 13.39). If the pore is occluded by a plate, then it is called a poroid. The loculus consists of a usually hexagonal chamber in the wall that is separated from other loculi by vertical spacers, which often have pores in them to allow for communication between loculi (Fig. 13.7(*b*),(*c*)). At one end of the loculus is a **sieve membrane** (**pore membrane, velum, cribrum**) (Figs. 13.8 and 13.9). The sieve membrane can be on the outside (an inward-opening loculus) or on the inside (an outward-opening loculus) (Fig. 13.7(*b*),(*c*)). The structure of the valve wall with loculi thus resembles a honeycomb. Pores or

Fig. 13.4 Scanning electron micrographs of centric diatoms. (*a*) *Thalassiosira lacustris*. Undulated hypovalve with strutted processes and part of epitheca showing distal pattern of the band openings. (*b*) *Cyclotella striata*. A frustule showing the epitheca, intercalary and girdle bands. (*c*),(*d*) *Thalassiosira gessneri*. Valve showing undulation, aerolation, and subcentral and marginal rings of strutted processes (*c*). Higher magnification of labiate and strutted processes (*d*). (*e*) *Skeletonema costatum* showing the marginal ring of strutted processes used in chain formation. ((*a*),(*c*),(*d*) from Hasle and Lange, 1989; (*b*) from Prasad et al., 1990; (*e*) from Medlin et al., 1993.)

loculi (**punctae**) in a single row are referred to as a **stria** (plural **striae**) (Figs. 13.1, 13.5, 13.6 and 13.28). The girdle bands in some diatoms have the same sculpturing as the value, but in many others sculpturing on the girdle and valve is different. In some cases, the girdle bands have no sculpturing.

Special pores (mucilage or slime pores) through which mucilage is secreted are known in many diatoms. In the pennate diatoms, these pores usually occur singly near one or both poles of the valve and generally occupy thickenings in the walls.

The valve surface can have extensions, called **processes**, whose main function appears to be to maintain contact between contiguous cells and to assist colony formation. These processes are given different names: **Cornutate** processes are hornlike; **strutted** processes are ones that have been reduced to a boss at the apex of a valve (Fig. 13.4); **spinulae** are very small processes; **awns** or **setae** are hollow and elongated (Fig. 13.24).

The frustule is composed of **quartzite** or hydrated amorphous silica that may also have small amounts of aluminum, magnesium, iron, and titanium mixed with it (Mehta et al., 1961; Lewin, 1962). Diatom frustules from marine plankton contain 96.5% SiO_2 and 1.5% Al_2O_3 or Fe_2O_3 (Rogall, 1939). The inorganic component of the frustule is enveloped by an organic com-

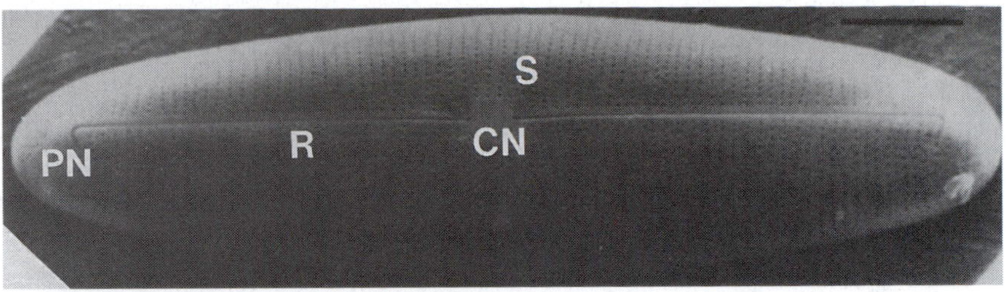

Fig. 13.5 Scanning electron micrograph of the whole valve of *Navicula delognei*. (CN) Central nodule; (R) raphe; (PN) polar nodule; (S) stria. (From Lobban, 1983.)

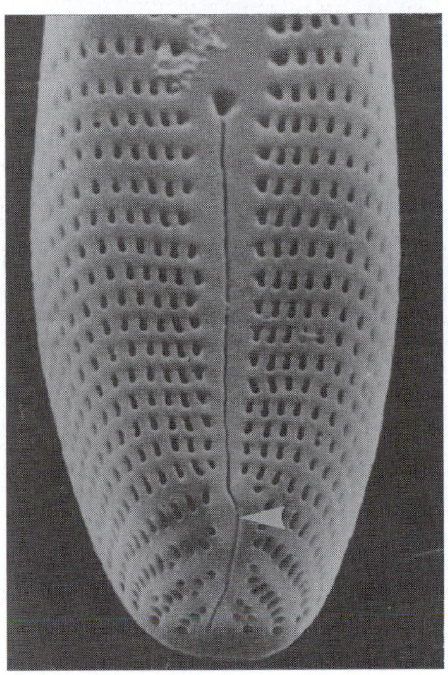

Fig. 13.6 Scanning electron micrograph of a frustule of *Rhoicosphenia curvata* showing the raphe (arrowhead). Bar = 2 μm. (From Mann, 1982.)

ponent or "skin" (Reimann et al., 1965), the latter composed of amino acids and sugars (Coombs and Volcani, 1968; Hecky et al., 1973) with the amino acid hydroxyproline and collagen present (Nakajima and Volcani, 1969).

Diatoms produce five types of mucilaginous aggregations: (1) tubes, (2) pads, (3) stalks, (4) fibrils, and (5) adhering films (Fig. 13.10, 13.11 and 13.12) (Hoagland et al., 1993). Diatoms are ubiquitous fouling microorganisms. Diatoms foul by attaching to submerged structures by secreting insoluble mucilages combined with glycoprotein in the form of the adhering films, pads, stalks, fibrils and tubes. *Achnanthes longipes* is a common marine fouling diatom that is highly resistant to toxic antifouling coatings (Johnson et al., 1995). It produces a stalk that elevates it above the toxic coatings on the ship bottom (Fig. 13.12). Fouling of ship bottoms increases the frictional drag, leading to excess fuel consumption. Cleaning ship hulls costs millions of dollars each year, leading to an additional loss of revenue.

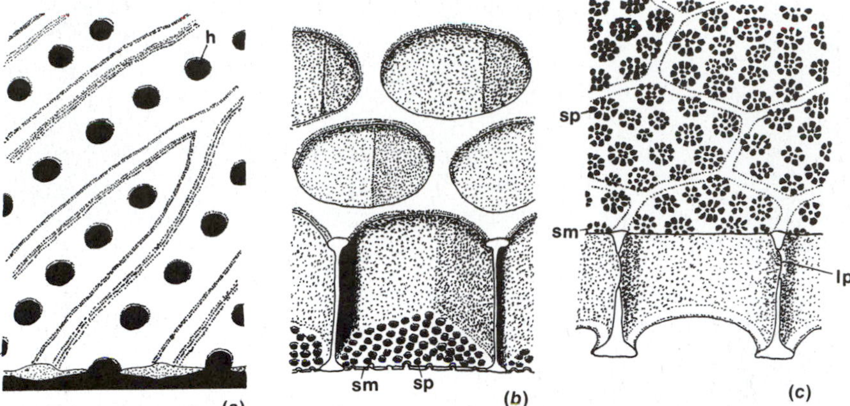

Fig. 13.7 The types of openings in frustule walls. (*a*) Hole or pore (*Chaetoceros didymos* var. *anglica*). (*b*) Loculus opening outward (*Coscinodiscus lineatus*). (*c*) Loculus opening inward (*Thalassiosira wailesii*). (h) Hole; (lp) lateral pore or pass pore; (sm) sieve membrane; (sp) sieve pore. (After Hendey, 1971.)

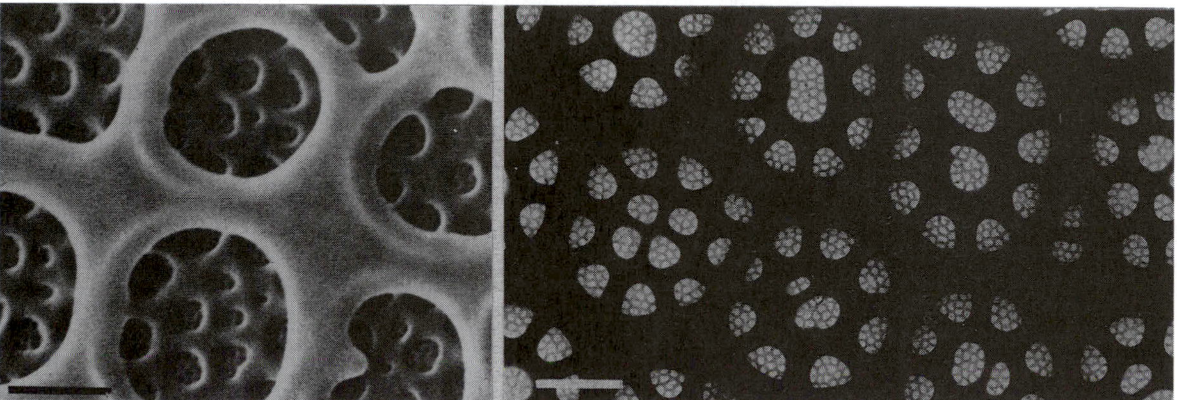

Fig. 13.8 *Left:* Scanning electron micrograph of an acid-cleaned valve of *Coscinodiscus wailesii* showing the three-dimensional structure of the areolae. Each areola has a large hole (foramen) on the side of the valve next to the protoplasm. On the outside of the valve, each areola is covered with a cribrum. The pores of the cribrum are covered by delicate sieves, the cribella. *Right:* Transmission electron micrograph of an acid-cleaned valve, showing the structure of the sieve membrane (cribrum) and the sievelike cribrella occluding the pores. Bar = 1 μm. (From Schmid and Volcani, 1983.)

CELL DIVISION AND THE FORMATION OF THE NEW WALL

The normal asexual method of reproduction is by division of one cell into two, the valves of the parent cell becoming the epithecas of the daughter cells with each daughter cell producing a new hypotheca (Fig. 13.13). As a result of cell division, one of the daughter cells is of the same size as the parent cell, and the other is smaller. As the size of the cell decreases, so does the relative width to height and the morphology of the cell; in other words, the smaller cells are not geometrically proportional to the larger ones from which they arise.

Diatoms have an absolute requirement for silicon if cell division is to take place. The predominant form of soluble silicon available to diatoms in most natural waters is monomeric silicic acid (monosilicic acid, orthosilicic acid). Silicon is taken up by the diatom cells as the ionized form of monosilicic acid, $Si(OH)_3O^-$ (Riedel and Nelson, 1985). Monosilicic acid is actively transported into diatom cells. *Navicula pelliculosa* (Fig. 13.20(*a*)) has been shown to have a carrier-mediated transport of monosilicic ions,

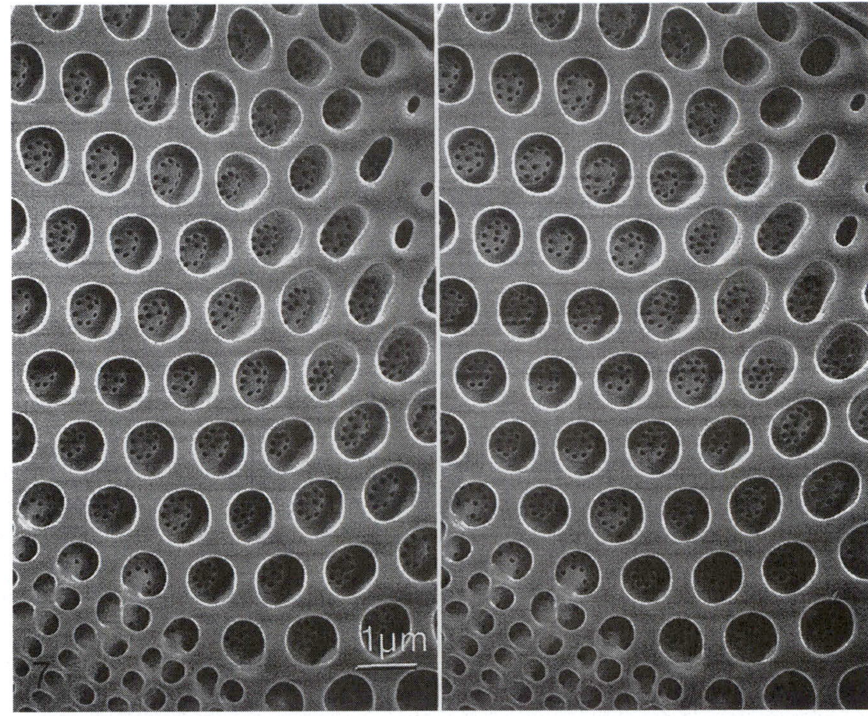

Fig. 13.9 Stereopair illustrating details of the external areolar pattern and structure in *Mastogloia angulata*. If a stereo viewer is not available, bring the photograph about 6 inches from your eyes and cross your eyes to view in three dimensions the loculi, each with a sieve membrane with sieve pores, at the bottom of the loculi. (From Navarro, 1993.)

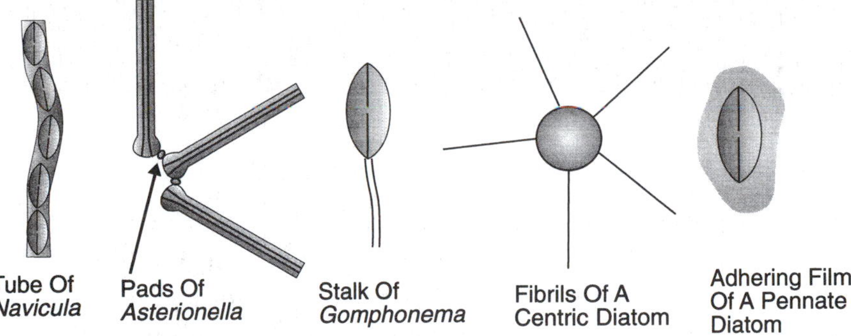

Fig. 13.10 Forms of extracellular mucilage in diatoms. (Modified from Hoagland et al., 1993.)

with changes occurring in transport activity during the cell cycle (Sullivan, 1976, 1977). Uptake of ions of monosilicic acid is confined to a period of the cell cycle following cytokinesis and prior to the separation of the two daughter cells (Sullivan, 1977). Thus uptake of monosilicic acid and cell division are tightly coupled in time. The uptake of monosilicic acid requires substantial amounts of energy in the carrier-mediated uptake mechanisms (Sullivan, 1976, 1977). Photosynthesis is the process that provides the energy for the uptake of monosilicic acid. Because monosilicic acid is not stored in diatom cells to any great extent (Mehard et al., 1974), the diatoms

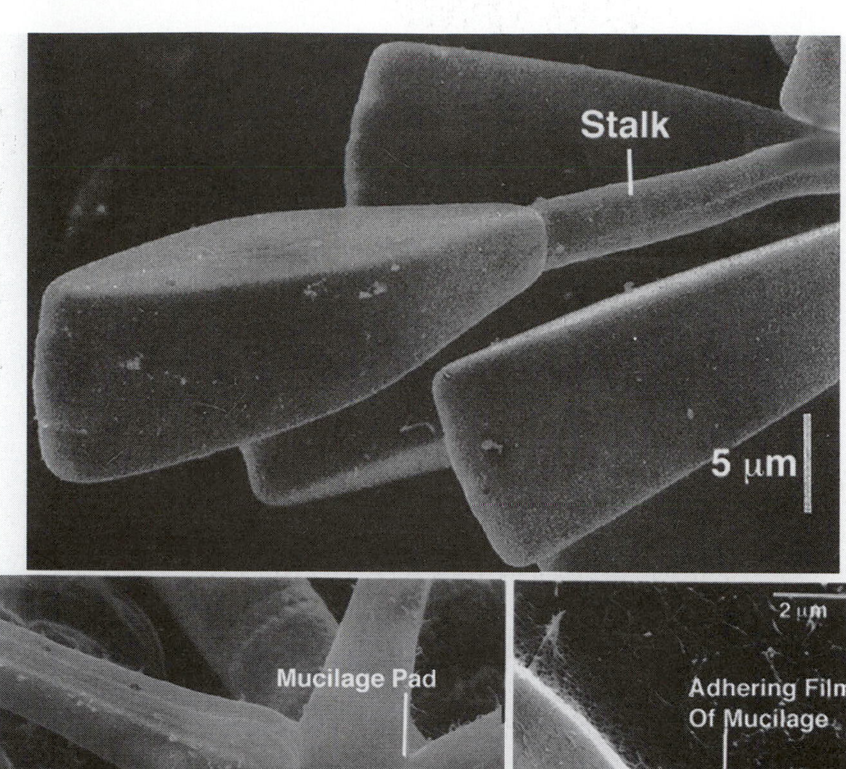

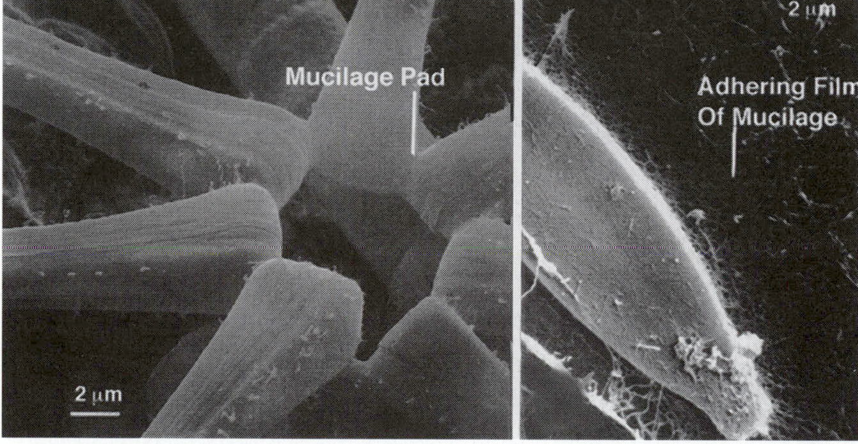

Fig. 13.11 Forms of extracellular mucilage in diatoms. *Top*. Mucilage stalk of *Gomphonema*. *Bottom left*. Colony of *Asterionella formosa* formed by mucilage pads attached to both valve faces. *Bottom right*. Fibrous adhering films of *Cymbella microcephala*. (From Hoagland et al., 1993.)

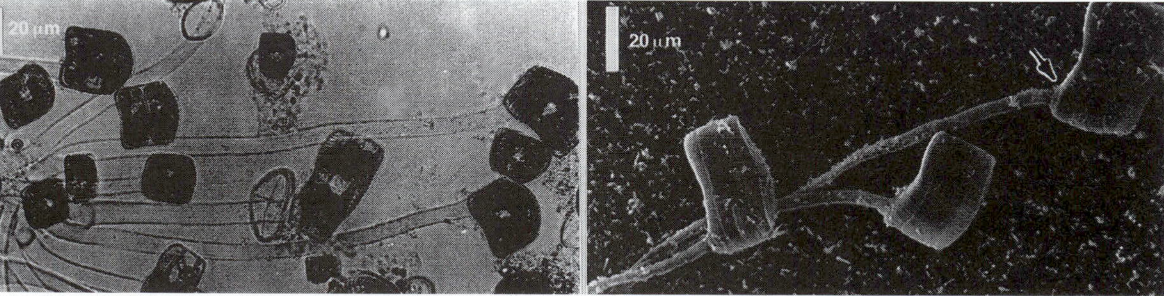

Fig. 13.12 Light micrograph and scanning electron micrograph of the common biofouling diatom *Achnanthes longipes* showing the extracellular stalks (arrow). (From Johnson et al., 1995.)

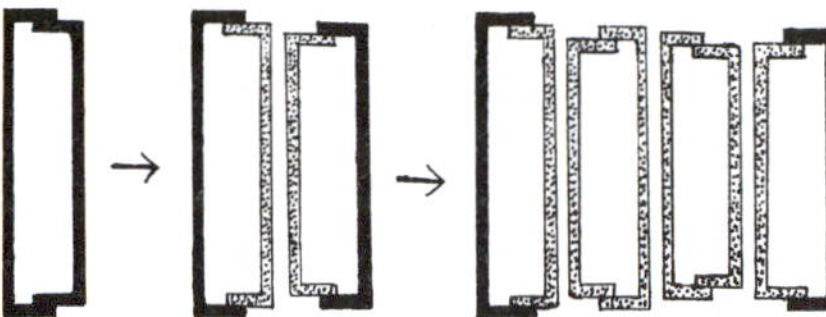

Fig. 13.13 Diagrammatic representation of a diatom cell in girdle view showing the reduction in size through two divisions. (After Hendey, 1964.)

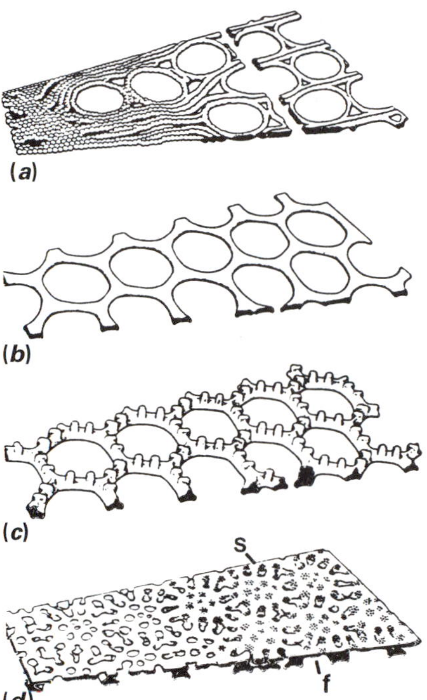

Fig. 13.14 Drawings of the formation of the frustule of the centric diatom *Coscinodiscus wailesii*. Initially silica is deposited in the part of the frustule that will be next to the plasma membrane in the mature frustule (*a*) and (*b*). Subsequently the exterior portion of the frustule is deposited (*c*) and (*d*). The mature frustule has a sieve membrane (s) on the exterior and a foramen (f) on the interior of the areolate. (From Schmid and Volcani, 1983.)

are usually forced to divide during the day in order to use the energy from photosynthesis to take up monosilicic acid. Nelson and Brand (1979) examined 14 marine diatom species in a light–dark cycle and found that ten of the species divided during the light phase, whereas four divided continuously during the light and dark period. This finding contrasts with planktonic marine algae from other classes, where division occurs only during the dark period. The division of the diatoms during the light period reflects the necessity for large amounts of energy from photosynthesis for the uptake and utilization of silicon.

Prior to cell division the cell elongates, pushing the epitheca away from the hypotheca, and the nucleus divides. After the protoplasm has divided (Fig. 13.15) into two by the invagination of the plasmalemma, the Golgi bodies produce translucent vesicles which collect beneath the plasmalemma (Stoermer et al., 1965b). These vesicles then fuse to form the **silicalemma** or membrane of the **silicon deposition vesicle** (Li et al., 1989). The vesicle gradually expands and assumes the shape of a new valve (Schmid and Volcani, 1983). Two silicon deposition vesicles are formed per cell, with

soleirolii require dormancy to be broken by a dark period at –2 °C to 15 °C for a minimum of 4 to 5 weeks. Thus *E. soleirolii* thrives in colder seasons in northern Europe and is induced to form resting spores by rising temperatures and perhaps concomitant nutrient deficiencies. The resting spores are viable for up to 3 years, but would normally last the summer before dormancy would be broken by the cold temperature of autumn, resulting in germination of the resting spores.

Resting cells have the same morphology as vegetative cells and do not form a protective layer, thereby differing from resting spores. Growing vegetative cells of *Amphora coffaeformis* require 4 weeks in the dark to form resting cells (Anderson, 1975). Formation of resting cells initially involves autophagic activity, with the breakdown of existing structures. Large vacuoles decrease in size, and many small ones develop; the mitochondria become fewer; and large lipid bodies are formed (Fig. 13.16). The resting cells contain as much chlorophyll as the vegetative cells, and the whole cell appears to be a parsimonious assemblage of organelles prepared to resume metabolism and growth upon return to favorable conditions. During the summer, diatom cells sink into deep water below the euphotic zone, where they form resting cells and become dormant. Such cells remain viable for at least 2 months in such an environment (Anderson, 1976). Their respiratory rate is 20% that of normal cells, and their photosynthetic capability is very low. Vertical mixing brings these cells and nutrient-rich water to the surface, and within 2 days the cells become active and begin to reproduce. Viable resting cells of diatoms have been collected at 6150-m water depth in the North Atlantic.

Auxospores

Auxospore formation is a second mechanism (in addition to resting spores) for reestablishing the original size of the cell. The auxospores are formed by the fusion of two gametes. In the centric and gonoid diatoms, the male gamete is motile, whereas the female gamete (egg) is non-motile (Figs. 13.35 and 13.36). In the pennate and trellisoid diatoms, both gametes are nonflagellated (Figs. 13.38 and 13.40). There are reports of auxospores arising without any fusion of gametes, but most of these reports need verification. The actual formation of auxospores will be dealt with later when we discuss life histories; but, generally speaking, the protoplast escapes from the parent frustule, enlarges, and secretes an organic wall. In *Navicula cuspidata*, the auxospore undergoes bipolar elongation, rupturing the polysaccharide wall (Fig. 13.17). The auxospore secretes a specialized siliceous wall, the **perizonium**, which consists of multiple overlapping

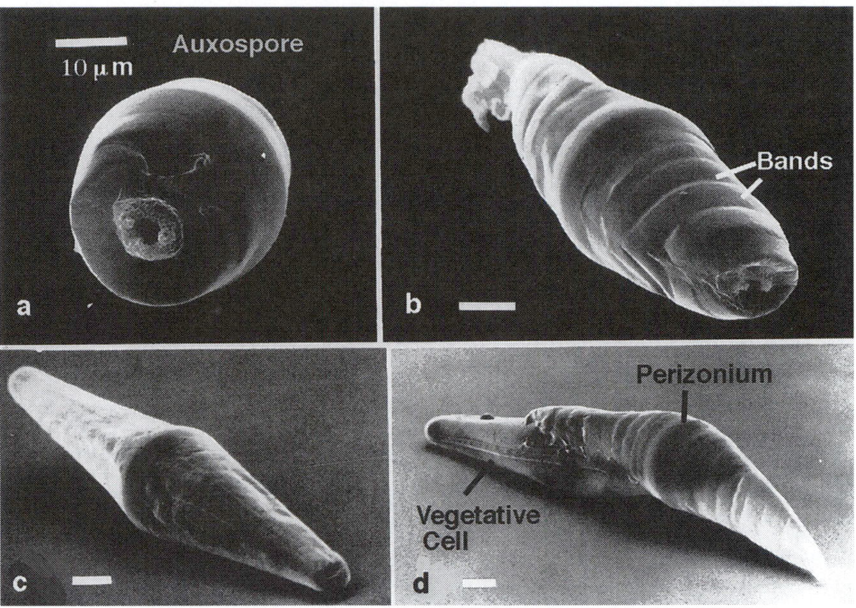

Fig. 13.17 Scanning electron micrographs of sexual stages of *Navicula cuspidata*. (*a*) Immature auxospore. (*b*) Auxospore midway through elongation. The underlying perizonal bands are evident. (*c*) Mature auxospore showing biconical shape of the perizonium. (*d*) Vegetative cell emerging from the perizonium. (From Cohn et al., 1989.)

bands. A vegetative cell with normal valve morphology is formed within the perizonium. Eventually, the vegetative cell ruptures the perizonium and emerges (Cohn et al., 1989). Mass sedimentation of empty siliceous frustules and perizonia can occur at the end of a bloom of diatoms that has terminated in sexual reproduction (Crawford, 1995).

Sexual reproduction in diatoms can occur only after two general conditions have been met (Edlund and Stoermer, 1997). First, cells must reach a minimum size range, typically 30–40% of their maximum size. Second, there must be the presence of correct environmental conditions. These include combinations of temperature, light, nutrients, trace metals, organic growth factors, and osmolarity (Potapova and Snoeijs, 1997). Contrary to most other algal groups, sexuality is primarily a means of size restoration, and is not normally a factor in dormancy or dispersal (Edlund and Stoermer, 1997).

Motility

Some diatoms are able to glide over the surface of a substrate, leaving a mucilaginous trail in their wake. **Gliding** *is restricted to those pennate diatoms with a raphe and those centric diatoms with labiate processes.* The gliding movement is characterized by large fluctuations in the velocity of movement, with great changes in speed occurring within tenths of seconds (Edgar, 1979). In pennate diatoms, the path of the diatom seems to be essen-

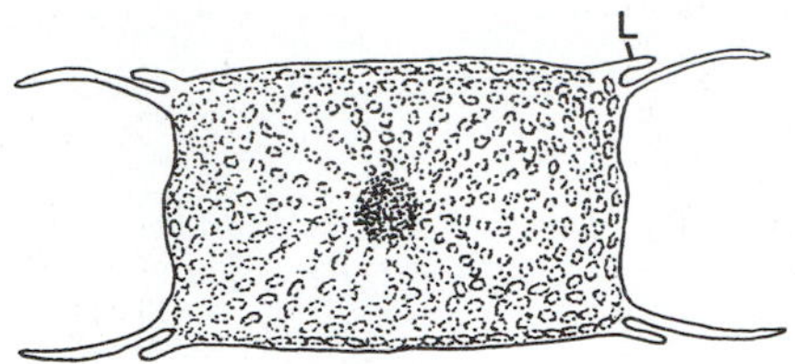

Fig. 13.18 Drawing of a side view of *Odontella sinensis* showing the location of the labiate processes (L). (After Pickett-Heaps et al., 1986.)

tially dependent on the shape of the raphe. Nultsch (1956) distinguished at least three types: (1) the *Navicula* type, with a straight movement; (2) the *Amphora* type, in which the path is usually curved; and (3) the *Nitzschia* type, which always exhibits curved pathways with two different radii. The observed rates of gliding in diatoms vary from 2 to 14 μm s^{-1} at room temperature (Cohn and Weitzell, 1996). *Nitzschia palea* is able to penetrate into 2% agar and to move within the medium. The less solid the substrate is, the slower the movement of the diatom; the rate of movement of *Nitzschia putrida* was found to be 2.7 μm s^{-1} on glass but only 0.8 μm s^{-1} on agar (Wagner, 1934). Many of the diatoms exhibit backward and forward movements in which the direction may alternate at intervals of a minute. According to von Denffer (1949), the motility of *N. palea* is dependent on light. In liquid cultures the cells tend to agglutinate into spherical clumps, loosely held together by mucilage. When they are transferred to a glass slide, the cells move away from one another. These movements do not take place in darkness.

Those pennate diatoms that glide have bundles of actin microfilaments running parallel to the raphe (Pickett-Heaps et al., 1979a,b). The microfilament bundles may serve to orient crystalloid bodies containing mucilaginous material in the cytoplasm immediately below the raphe. On an appropriate stimulus, the mucilaginous material is released into the raphe system from the area of the central or terminal pore (Drum and Hopkins, 1966). The mucilaginous material then streams in the raphe in one direction until it strikes an object to which it adheres. If the object is fixed, then the streaming in the raphe forces the diatom to move in the opposite direction. Nearly all motile diatoms must adhere to the substratum in the area of their raphe in order for movement to occur.

Some of the benthic centric diatoms with labiate processes (Figs. 13.4(*d*), 13.18 and 13.19) are capable of gliding when attached to a substrate. The labiate processes have a pore in the center, and the mucilage is secreted through the pore. In *Actinocyclus subtilis* (Fig. 13.19), secretion of the mucilage through the labiate process causes the diatom to move forward while rotating at the same time (Medlin et al., 1986).

Diatoms that are attached to a substrate and are motile on the substrate

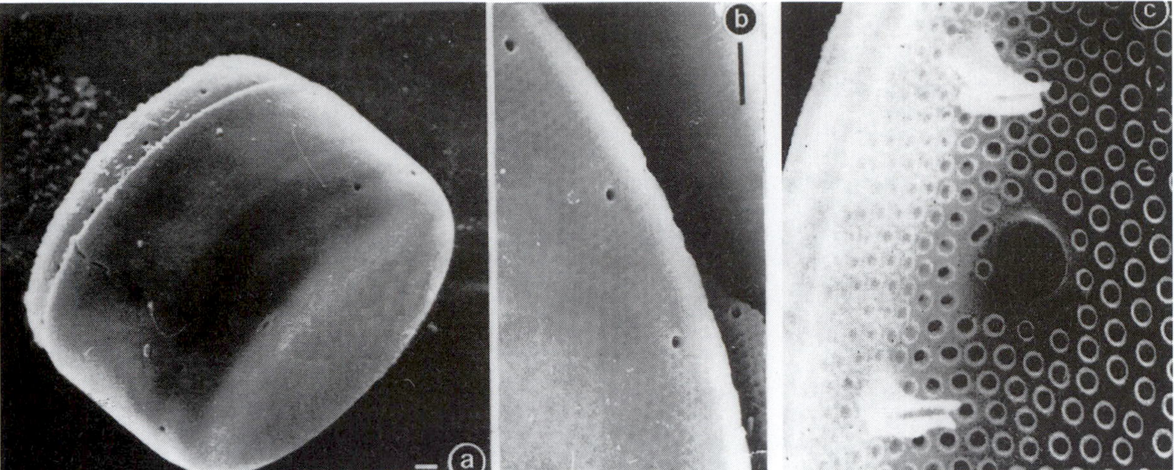

Fig. 13.19 Scanning electron micrographs of *Actinocyclus subtilis*. (*a*) Whole cell. (*b*) Detail of the labiate processes, showing how they open to the outside of the cell by a simple hole. (*c*) Internal view of an acid-cleaned valve, showing the internal structure of two labiate processes. Internally, each labiate process consists of a short tube that projects from the valve mantle to a slitlike opening. Bar = 1 μm. (From Andersen et al., 1986.)

have the advantages of (1) being held in position in moving water; (2) avoiding burial by moving up and over sediments; (3) moving to colonize vacant areas; and (4) moving to areas with more light and/or nutrients (Medlin et al., 1986).

Rhythmic phenomena

It is possible to synchronize the division of diatom cells in a culture in a couple of different ways. Lewin (1966) has shown that removal of silicon from cultures of *Navicula pelliculosa* (Figs. 13.20(*a*)) stops growth of the cells at a stage prior to cytokinesis. When silica is added to the culture, all of the cells then divide synchronously. Another way of obtaining syncronized cell divisions is by keeping the diatoms in the dark for a long period followed by exposure to light. In *Nitzschia palea* (Fig. 13.20(*b*)) (von Denffer, 1949), the shortest time that can be obtained between cell divisions is 16 hours. If the cells are grown on an 8 hours light : 8 hours dark cycle, synchronously dividing cells are obtained. If the cycle is shortened to 6 hours light : 6 hours dark, then cell division occurs every second dark period because there is insufficient time for the diatom to prepare itself for the next division.

Fauré-Fremiet (1951) was the first to observe that the diatom *Hantzschia amphioxys* (Fig. 13.20(*c*)) accumulates on mud flats in certain areas during low tides, causing brown spots on the mud. The cells can glide rapidly at a rate of half their length in a second and can reverse their phototactic response. During low tide the cells are positively phototactic, so that they come to the surface of the mud, and at high tide they are negatively photo-

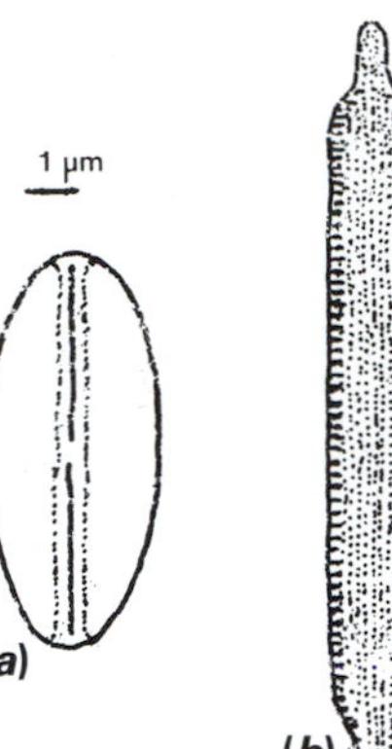

Fig. 13.20 (*a*) *Navicula pelliculosa*.
(*b*) *Nitzschia palea*.
(*c*) *Hantzschia amphioxys*.

tactic, so that they descend into the mud. At high tide they also exude a mucilaginous material, which causes them to stick to one another and to sand grains. This rhythmic phenomenon prevents the cells from being washed away by the tide.

Physiology

For growth of *Navicula pelliculosa* (Fig. 13.20(*a*)), silicon cannot be replaced by any of the elements similar to it in physical and chemical properties or in atomic radius, such as Ge, C, Sn, Pb, As, P, B, Al, Mg, or Fe (Lewin, 1962). Lewin (1966) has shown that with a mixture of algae and other organisms, concentrations of germanium dioxide (GeO_2) above 1.5 mg liter^{-1} will specifically suppress the growth of diatoms. In cultures of many types of algae, diatom contaminants frequently grow so fast that they obscure the growth of the algae the investigator is working with. The finding that GeO_2 specifically inhibits diatom growth was a welcome one for phycologists working on algal cultures. In experiments with 14 pure cultures of diatoms, 1 mg liter^{-1} of GeO_2 reduced the growth rate significantly; 10 mg liter^{-1} was even more inhibitory to growth and in a few cases killed the cells. *Phaeodactylum tricornutum*, a diatom with little or no silicified wall, was the least sensitive to GeO_2 of the diatoms tested. By increasing the amount of SiO_2 in solution, it is possible to reverse the inhibitory effect of GeO_2 on growth. The results indicate that GeO_2 is a specific inhibitor of silicate utilization because concentrations of GeO_2 as high as 400 mg liter^{-1} have no effect on respiration. The minimum inhibitory concentrations of GeO_2 depends to a certain extent on the pH and concentration of the nutrient medium. Such inhibitory concentrations are rarely, if ever, reached in natural waters, the highest concentration reported being that of a mineral spring in Niigata Prefecture, Japan, which contained only 0.03 mg $liter^{-1}$ of Ge, with normal concentrations in seawater being about 5×10^{-5} mg $liter^{-1}$.

In addition to responding adversely to germanium in solution, diatoms are sensitive to copper. Erickson (1972) has shown that *Thalassiosira pseudonana* has its growth inhibited at concentrations of copper as low as 5 μg $liter^{-1}$. The presence of detritus affects the toxicity to a certain extent. Concentrations of 0.25 ppm copper as $CuSO_4 \cdot 5H_2O$ are normally used to control algal blooms without affecting fish in freshwater lakes. Copper toxicity and silicon metabolism are linked in several marine diatoms. *Skeletonema costatum* (Fig. 13.4(*e*) and 13.23(*a*)) and *Thalassiosira pseudonana* exhibit slower growth in the presence of copper. Inhibition of growth by copper is alleviated by increasing the silicic acid concentration in the media (Morel et al., 1978; Rueter et al., 1981). It is probable that copper interferes with the silicic acid transport site across the plasma membrane and slows uptake of silicic acid (Rueter et al., 1981; Rueter, 1983).

The effects of heavy metals on diatoms can be divided into three groups (Thomas et al., 1980): (1) Cu, Zi, and Ge affect the biochemical pathway of silicon metabolism; (2)Hg, Cd, and Pb interfere with cell division and cause morphologically distorted cells to be produced; and (3) Cr, Ni, Se, and Sb have no effects up to a concentration of 1 μM, well above the concentrations that show effects with other toxic metals.

Some normally photosynthetic diatoms are able to grow under heterotrophic conditions. In two studies (Lewin, 1953; Lewin and Lewin, 1960), a number of cultures of photosynthetic diatoms were found that would grow heterotrophically, usually only with glucose as a carbon source. One of these diatoms, *Cyclotella cryptica* (Hellebust, 1971), can grow in the dark in an organic medium with glucose (but not lactate or tryptone) as the sole carbon source. When the organism is growing in the light, it does not have the mechanism for the utilization of glucose in the medium. It requires about 24 hours in the dark in a glucose medium before it is able to use the glucose as a carbon source. This lag period indicates that the lack of light induces an uptake and/or assimilation system for the glucose. White (1974) suggested that such facultative heterotrophy enables these diatoms to settle into bottom deposits, live heterotrophically for long periods, then rise and begin photosynthesis again. Although the above diatoms still have functional chloroplasts, there are some apochlorotic diatoms lacking functional chloroplasts (Lewin and Lewin, 1967). The latter, which are species of *Nitzschia*, are able to grow with lactate or succinate as the sole organic carbon source.

Most Bacillariophyceae require vitamin B_{12} (cyanocobalamin) for growth, although a few estuarine forms do not. The amount of vitamin B_{12}

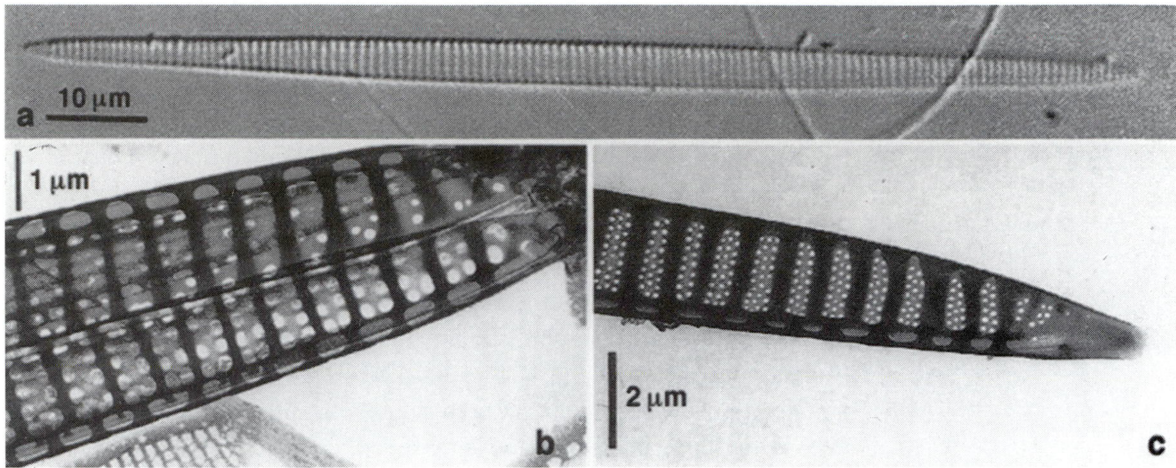

Fig. 13.21 *Pseudo-nitzschia*, a diatom responsible for amnesic shellfish poisoning. (*a*) Light micrograph of one valve. (*b*), (*c*) Transmission electron micrographs of the ends of cleaned valves showing the ribs and poroids. (From Hasle, 1995.)

needed is very small, having been estimated by Guillard and Cassie (1963) to be from 5 to 13.8 molecules μm^{-3} for *Skeletonema costatum* (Fig. 13.4(*e*) and 13.23(*a*)).

Toxins

Amnesic shellfish poisoning occurs when shellfish filter the diatom *Pseudo-nitzschia multiseries* (Fig. 13.21) from marine waters. Subsequent ingestion of the shellfish by man and birds results in memory loss (amnesia), abdominal cramps, vomiting, disorientation and even death. Amnesic shellfish poisoning was first recognized in 1987 in Prince Edward Island, Canada, where it caused 3 deaths and 105 other affected individuals after blue mussels were eaten (Subba Rao et al., 1988). The diatom produces domoic acid, a derivative of the neuroexcitatory amino acid L-glutamic acid (Fig. 13.22). Domoic acid is especially prevalent in moribund cells of the diatom and can be induced by depriving the cells of nutrients, particularly silicate and phosphate (Pan et al., 1996).

Ecology

Diatoms comprise the main component of the open-water marine flora and a significant part of the freshwater flora. Attached diatoms can be characterized by the brown scums found on various kinds of substrata, as well as the fluffy brown growths caused by abundant epiphytic diatoms. The pennate diatoms are represented in about equal numbers in the freshwater

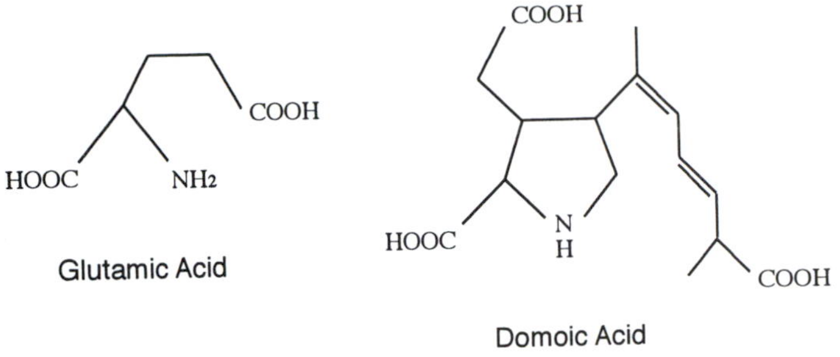

Fig. 13.22 The structure of domoic acid, the causative chemical of amnesic shellfish poisoning, and its analog, L-glutamic acid.

and marine habitats, whereas the centric and gonoid diatoms are present predominantly in the marine environment. Generally speaking, in the marine environment the colder the water is, the greater the diatom population. The populations of diatoms in the open oceans usually have a large number of species with the total number of organisms being low, in contrast to the diatoms living close to the shore, where the total number of diatoms is very high, but the number of different species within the population is low. This situation is actually fairly typical of oligotrophic, unenriched waters (the open ocean) versus eutrophic, enriched waters (the coastal waters receiving enrichment from the land).

MARINE ENVIRONMENT

The maintenance of oceanic diatoms in the water column involves some adaptation of the cells to make them buoyant. The silicified wall of diatoms has a density of about 2.600, comprising up to 50% of the dry weight of the cell (see Smayda, 1970, for a review). The density of seawater varies from 1.021 to 1.028; therefore, planktonic diatoms must have cellular components that are lighter than seawater to achieve neutral or positive buoyancy. The density of the cytoplasm in marine organisms is slightly heavier than seawater, varying from 1.030 to 1.100. This leaves only the vacuole as a likely source of positive buoyancy in diatoms. In fact, most planktonic diatoms have a large vacuole whose contents, while being isotonic, are lighter than seawater. For example, the vacuolar sap of the planktonic diatom *Ditylum brightwelli* (Fig. 13.23(*b*),(*c*)) has a density of 1.0202, less than that of seawater. The vacuole of these diatoms contains lighter ions than the surrounding seawater: The concentration of Na^+ is great relative to K^+ (the latter element is heavier by about 40%), there is a relatively high concentration of light NH_4^+ ions, and divalent relatively heavy ions, especially SO_4^{2-}, are excluded. This ionic mechanism of buoyancy is probably important in planktonic diatoms as long as they are above a certain size (20 μm diameter in *Ditylum*), below which the size of the vacuole is too small to provide the

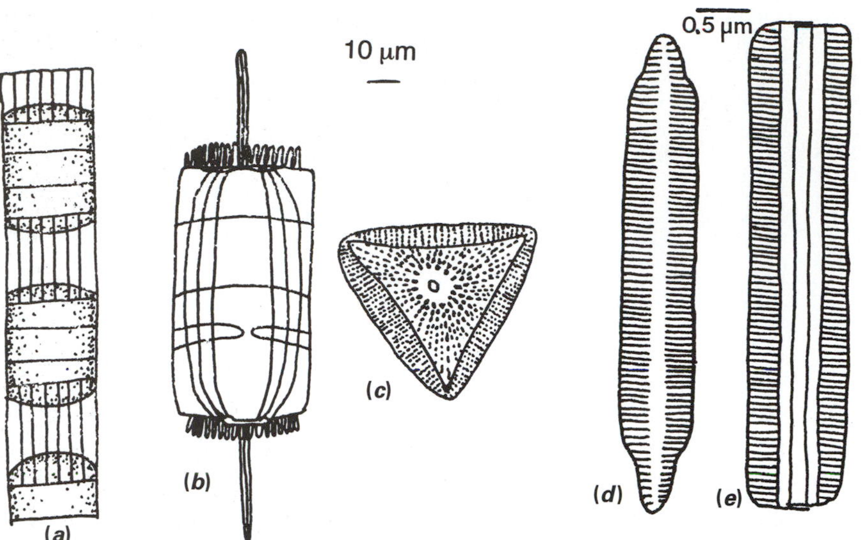

Fig. 13.23 (*a*) *Skeletonema costatum*. (*b*), (*c*) *Ditylum brightwelli.* (*b*) Girdle view. (*c*) Valve view. (*d*), (*e*) *Fragilaria virescens*, valve and girdle view.

necessary lift to the cell. The ionic mechanism of buoyancy cannot be applied to the flotation of freshwater phytoplankton because of the low quantity of salts in the water. In freshwater, most colonial diatoms are encased in a gelatinous sac of low density which aids the flotation of the organisms. Settling in the water column is slowed by increasing the surface area relative to the volume, thereby increasing the drag of the cell in the water. In diatoms this can be accomplished by setae (*Chaetoceros*, Fig. 13.24) or by cells shaped as discs (*Coscinodiscus*, Fig. 13.29(*a*) and 13.32), ribbons (*Fragilaria*), or elongate forms (*Rhizosolenia*, Fig. 13.25(*c*)). The aggregation of diatoms into chains increases the settling rate of the cells because it decreases the surface area. Increasing the size of diatom cells results in an increase in the ascension rate. The largest diatom cells can ascend up to 8 meters per hour (Moore and Villareal, 1996).

Planktonic diatoms can vary in density during the day, moving up and down in the water column. This causes a constant flow of water over the surface of the diatom, enabling a better absorption of nutrients. *Ditylum brightwelli* (Fig. 13.23(*b*),(*c*)) is a diatom with a variation in sedimentation rates, the greatest rate of settling occurring during the latter half of the light period and the least settling occurring at the end of the dark period (Fisher and Harrison, 1996). The insignificance of fat deposits in buoying up cells can be seen here, where the cells contain the maximum amount of fat during the period of maximum settling (Andersen and Sweeney, 1977). Sedimentation of diatoms in the ocean continues in many cases until an increased density of the water, due to either a thermocline or greater water depth, causes them to remain steady in the water column. In the central

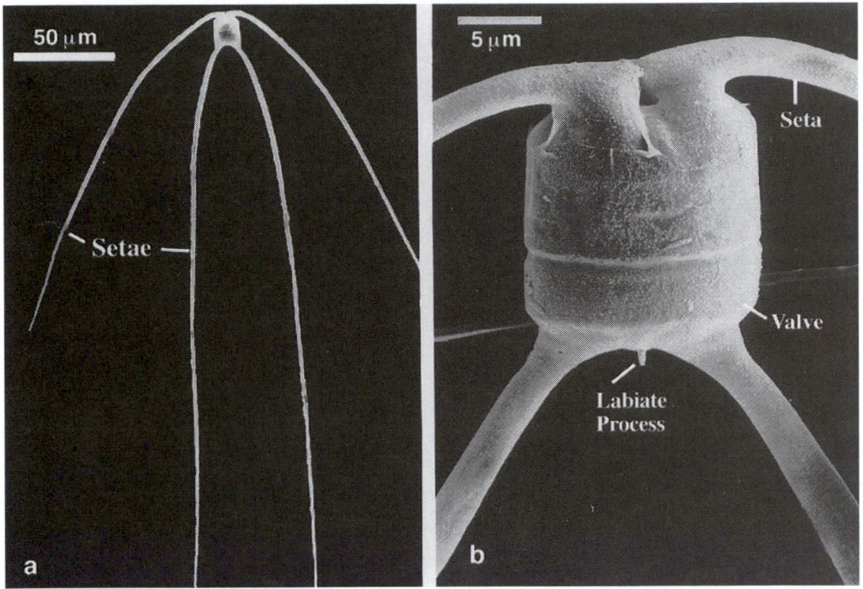

Fig. 13.24 Scanning electron micrograph of *Chaetoceros peruvianus* showing the setae and a labiate process. (From Pickett-Heaps et al., 1994.)

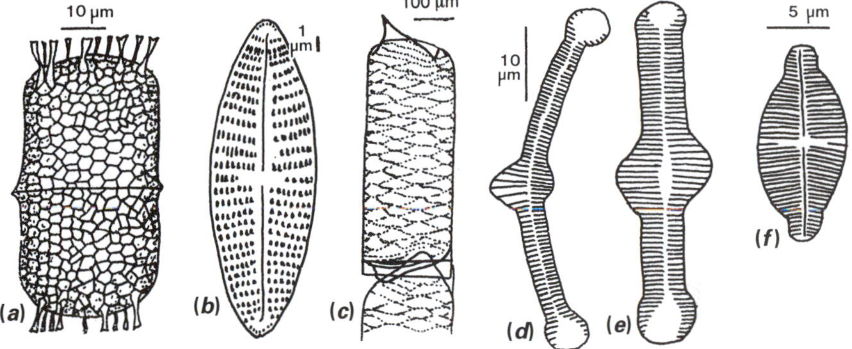

Fig. 13.25 (*a*) *Stephanopyxis turris*. (*b*) *Navicula glaciei*. (*c*) *Rhizosolenia castracanei*. (*d*) *Taballaria fenestrata*. (*e*) *Tabellaria flocculosa*. (*f*) *Achnanthes exigua*.

north Pacific Ocean, there is widespread deep (110- to 130-m depth) chlorophyll layer containing pigments of living algae, not detritus. This chlorophyll maximum contains growing organisms and is not an accumulation of moribund cells (Venrick et al., 1973; Jeffrey, 1976). Diatoms make up a good proportion of these cells and are able to grow there by virtue of an adaptation of the cells to the decreased light (Jeffrey and Vesk, 1977). The light that penetrates down to this depth is composed principally of low-intensity blue-green light with a maximum at about 480 nm. It has been shown with the marine diatom *Stephanopyxis turris* (Fig. 13.25(*a*) and 13.29(*c*)) that blue-green light increases the amount of chlorophyll up to 100% when compared to white light, with an appropriate increase in chloroplasts but with no change in the chlorophyll : carotenoid ratio. At the same time, blue-green light enhances the photosynthetic fixation of carbon dioxide. Thus the diatoms living at great water depth receiving only blue-green light "switch on" a mechanism for more efficient photon capture of light and increased fixation of carbon dioxide. They are therefore able to remain viable and grow at a water depth that has a low amount of light.

Diatoms form a large proportion of the algae associated with sea ice in the Arctic and Antarctic. Sea ice covers a large proportion of each area, reaching a maximum of 22 000 000 km^2 in the Antarctic and 15 600 000 km^2 in the Arctic. Three distinct microalgal communities are associated with sea ice (Palmisano and Sullivan, 1982):

1 Microalgae in surface melt pools. A special type of surface melt pool occurs in the Antarctic, where the sea ice joins the land. Here vertical tidal movements cause a series of cracks in the sea ice through which seawater passes (Whitaker and Richardson, 1980). This seawater mixes with snow on top of the ice, forming the basis of a dynamic environment of slush or semifrozen infiltration ice, termed the coastal tide-crack overflow area. This area contains moderate amounts of chlorophyll derived from almost pure growth of *Navicula glaciei* (Fig. 13.25(*b*)).
2 Microalgae within the sea ice that form in minute brine channels, and pockets produced within the sea ice due to salt exclusion during ice formation.
3 **Epontic** or bottom-ice algae living on the bottom of the ice. Bottom ice algae are dominated by pennate diatoms that produce a spring bloom before the ice breaks up. They form dense populations, often attaining standing crops of 100 mg m^{-2}, or more, of chlorophyll *a* despite chronically low temperatures (roughly 1.0 °C to –1.8 °C) and irradiance (a few percent or less of incident surface irradiance) in their growth environment. These algae thrive in this environment because of their shade adaptation to low irradiance (Smith et al., 1987). This bloom is distinct from the phytoplankton bloom which is dominated by the prymnesiophyte *Phaeocystis* in the Antarctic and by centric diatoms in the Arctic.

Diatoms associated with sea ice suffer a decline in growth and photosynthetic rate in water. As light and temperature decrease, diatoms accumulate storage products which are used up in the winter months. In addition, the sea ice diatoms increase their ability to take up sugars (such as glucose) up to 60 times. The increased heterotrophic uptake or organic compounds is probably enough to sustain the cells during the winter months, at which time the diatoms are subjected to little grazing pressure.

Nitrogen-limited diatoms will display a several-fold increase in uptake for ammonia when exposed to a saturating concentration of the nutrient (Suttle and Harrison, 1988). The enhanced uptake consists of two phases, a short-lived period of very-high uptake ("surge uptake") and a longer sustainable phase. The magnitude of the response is species specific and may be important in dictating competitive advantage in oligotrophic areas. This may be a mechanism by which phytoplankton in nutrient-depleted areas sequester nitrogen that is distributed in ephemeral micropatches.

Symbioses between nitrogen-fixing (diazotrophic) bacteria and cyanobacteria, and the diatom genera *Rhizosolenia* and *Hemiaulus* are common in warm oligotrophic seas (Villareal, 1989). *Rhizosolenia castracanei* (Fig. 13.25(*c*)) and *R. imbricata* var. *shrubsolei* form free-floating diatom mats in oligotrophic central oceanic regions, such as the Sargasso Sea and the North Pacific Gyre. In these areas, the fixation of nitrogen by the endosymbionts contribute a significant amount of nitrogen to the ecosystem (Martinez et al., 1983). The *Rhizosolenia* cells migrate vertically in the column at a rate of several meters per hour and can be found to a depth of 150 meters (Richardson et al., 1996; Villareal and Carpenter, 1994). Cells that are deprived of nitrogen are negatively buoyant and sink below the euphotic zone into waters that are relatively high in nutrients. After taking up nutrients, the cells become positively buoyant and move up into the euphotic zone and resume photosynthesis.

Diatom blooms often occur in the marine habitat, and an important factor in these blooms is probably the concentration of vitamin B_{12}, an essential vitamin for the growth of many diatoms, in the water. This vitamin is produced primarily by bacteria. There most likely exists an oscillating balance between the decay of phytoplankton due to lack of vitamin B_{12}, and the growth of bacteria due to the nutrients released by the decaying phytoplankton. The bacteria release vitamin B_{12}, and diatom growth is stimulated again (Haines and Guillard, 1974).

FRESHWATER ENVIRONMENT

Many planktonic diatoms have regular annual fluctuations in growth that can be attributed to environmental conditions. *Asterionella formosa* is one of these diatoms, a common freshwater planktonic diatom that forms large spring growths and smaller autumn ones (Fig. 13.26) (Lund, 1949, 1950). During the winter months, light and temperature limit the growth of the diatoms; lower temperature affects respiration more than photosynthesis so that the compensation point is lowered, and the possibility that photosynthesis will exceed respiration is increased. With the beginning of spring and consequent increase in temperature and light, there is a rapid growth of cells until the middle of spring, when another factor becomes limiting – this time the concentration of dissolved silica in the water; the growth of diatoms then results in much of the dissolved silica being incorporated into

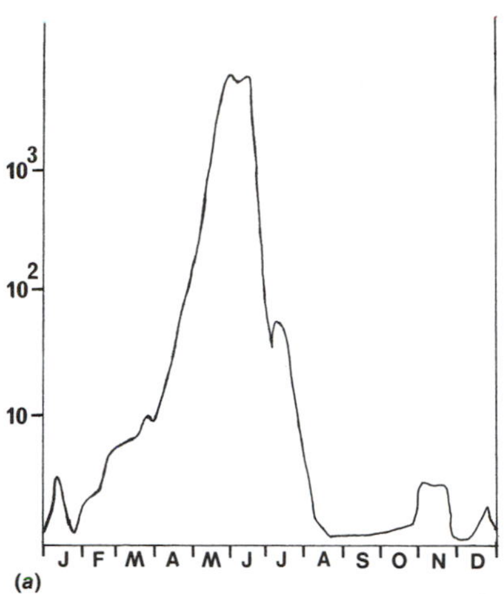

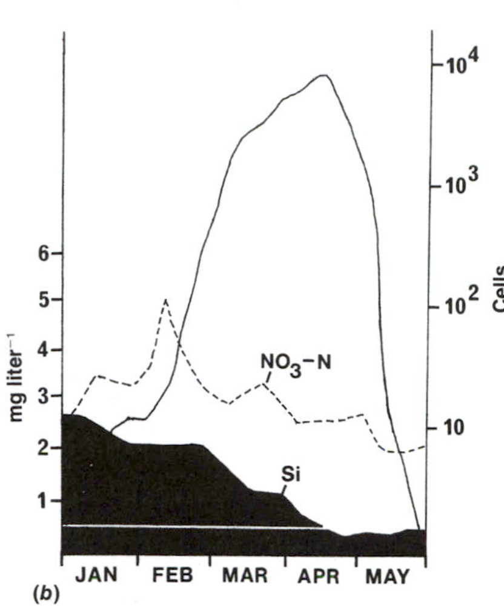

Fig. 13.26 (*a*) Seasonal distribution of live cells of *Asterionella* in Lake Windermere, England, plotted on a logarithmic scale in the 0- to 5-m water column. The horizontal axis represents the months of 1947. (*b*) Numbers of *Asterionella formosa* (solid line) and the concentrations of dissolved silica and nitrate-nitrogen in mg liter^{-1} in the 0- to 5-m water column in Esthwaite Water, England. The white line represents the 0.5-mg liter^{-1} concentrations of silica necessary for growth of diatoms. ((*a*) after Lund, 1949; (*b*) after Lund, 1950.)

the siliceous frustules. Once the concentration of dissolved silica reaches 0.5 mg liter^{-1}, most diatoms cease growth. Because the silica is now limiting, the number of diatoms falls off rapidly and increase only to a lesser degree in the fall when more dissolved silica is available from the breakdown of the frustules of the spring population and the inflow of water containing silica from streams. There are other factors, such as grazing by invertebrates and fungal parasitism, that also check the growth of *Asterionella,* but they are only of secondary significance.

The diatoms also make up a large part of the periphyton or the attached algae in freshwaters. These attached diatoms in streams have two opposing factors acting on their growth (Reisen and Spencer, 1970): (1) any increase in current that retards the attachment of the diatoms to the substrate, and (2) the increase in current that results in an increase in the growth of the diatoms. This leads to the observation that the diatoms growing in fast-flowing streams usually have a higher total standing crop in the long term than those in slow-flowing streams, although initially the fast-flowing water results in a lower number of diatoms being attached to the substrate. The diatoms in the fast-flowing water are, however, able to grow much faster and soon grow so rapidly that they overcome the initial disadvantage caused by low initial attachment numbers.

Attached diatoms in standing waters have good growth in spring, as do the planktonic diatoms, but do not show as marked a decrease in growth when the concentration of silica in water reaches 0.5 mg liter^{-1} or less. This can be seen by the growth of the diatom *Tabellaria flocculosa* (Fig. 13.25(*e*)) on stems of the reeds *Phragmites communis* and *Schoenoplectus lacustris* (Fig. 13.27) (Knudson, 1957). In the May–June period, when the dissolved

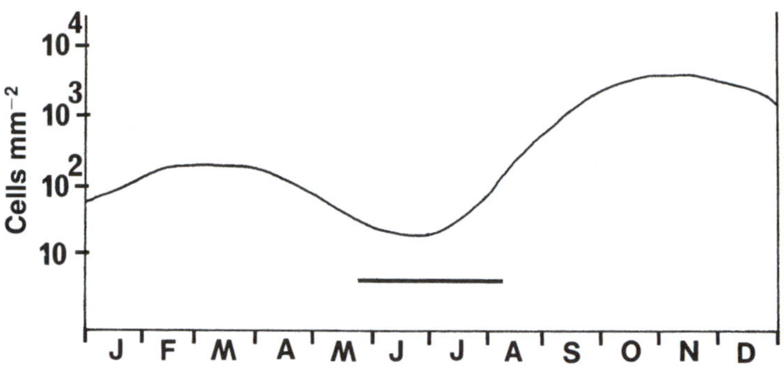

Fig. 13.27 Seasonal fluctuations of epiphytic *Tabellaria flocculosa* on *Phragmites communis* and *Schoenoplectus lacustris* stems. The straight line represents the period when dissolved silica is less than 0.5 mg liter^{-1}. (After Knudson, 1957.)

silica in the water is usually less than 0.5 mg liter^{-1}, there is a decrease in the number of diatoms, but not as great as that of planktonic diatoms, probably because of leaching of silica from the stems of the host plants. A second difference between planktonic and epiphytic diatom seasonal growth is that in the epiphytic diatoms, maximum growth normally occurs in the winter, possibly owing to the secretion by the host of some organic products that are used by epiphytic diatoms.

In freshwater habitats, diatoms often comprise the dominant algal flora in thermal waters between 30 and 40 °C. Fairchild and Sheriden (1974) showed that *Achnanthes exigua* (Fig. 13.25(*f*)) isolated from a hot spring showed optimum photosynthesis at 42 °C, with maximum and minimum temperatures for growth at 44 and 10 °C, respectively – characteristics of a thermophilic organism. This particular diatom is usually associated with an algal mat covered by a shallow layer of water in these hot springs.

Fossil diatoms

After the death of diatom cells the frustules usually dissolve, but under certain circumstances they remain intact and accumulate at the bottom of any water where the diatoms occur (Figs. 13.28 and 13.29) (Smith, 1955). Where conditions are exceptionally favorable and long continued, such accumulations may reach considerable thickness. Deposits of fossil diatoms, known as diatomaceous earth or *kieselguhr*, are found in various parts of the world. None of these deposits originated earlier than the Cretaceous; some diatomaceous earths originated in freshwaters, others in the ocean. However, known deposits of marine species are found inland and above the ocean as a result of geological changes. The best-known and most extensive deposits of marine species are those at Lompoc, California, where the beds are miles in extent and 200 m in thickness (Fig. 13.30). Most of the commercially available diatomaceous earth comes from California, where

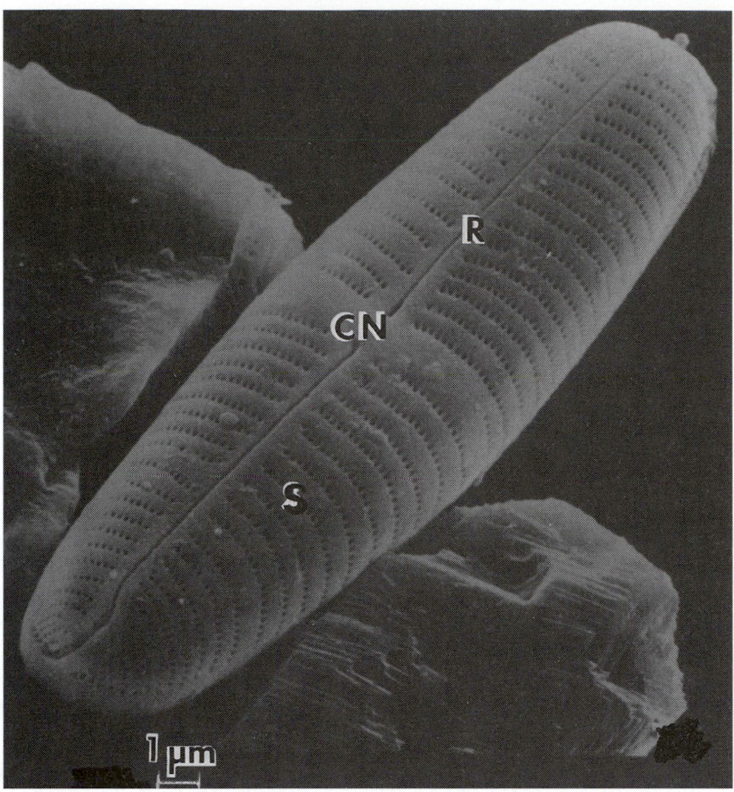

Fig. 13.28 Scanning electron micrograph of an epitheca of the fossil diatom *Navicula subfossilis*. (CN) Central nodule; (R) raphe; (S) striae. (From Lichti-Federovich, 1980.)

the deposits are worked as open quarries. In quarrying, the overburden of soil is removed, and the diatomaceous earth is mined. Diatomaceous earth is also obtained from lakes in Florida by dredging with a suction pump and carrying the material through sluiceways to settling tanks. The material from some deposits can be used directly; that from other deposits must be incinerated to remove organic substances.

The industrial uses of diatomaceous earth are varied. One of the first uses was as a mild abrasive in toothpaste and metal polishes. Diatomaceous earth was also used as an absorbent for liquid nitroglycerin to make dynamite that could be transported with comparative safety. The inert medium used in the present-day manufacture of dynamite is wood meal. Probably the most extensive industrial use of diatomaceous earth is in the filtration of liquids, especially those of sugar refineries. Another major use is in the insulation of boilers, blast furnaces, and other places where a high temperature is maintained.

Analysis of sediments containing remains of diatoms can provide information on past environmental conditions of lakes. This technique has been applied to late glacial and postglacial diatom deposits in the English

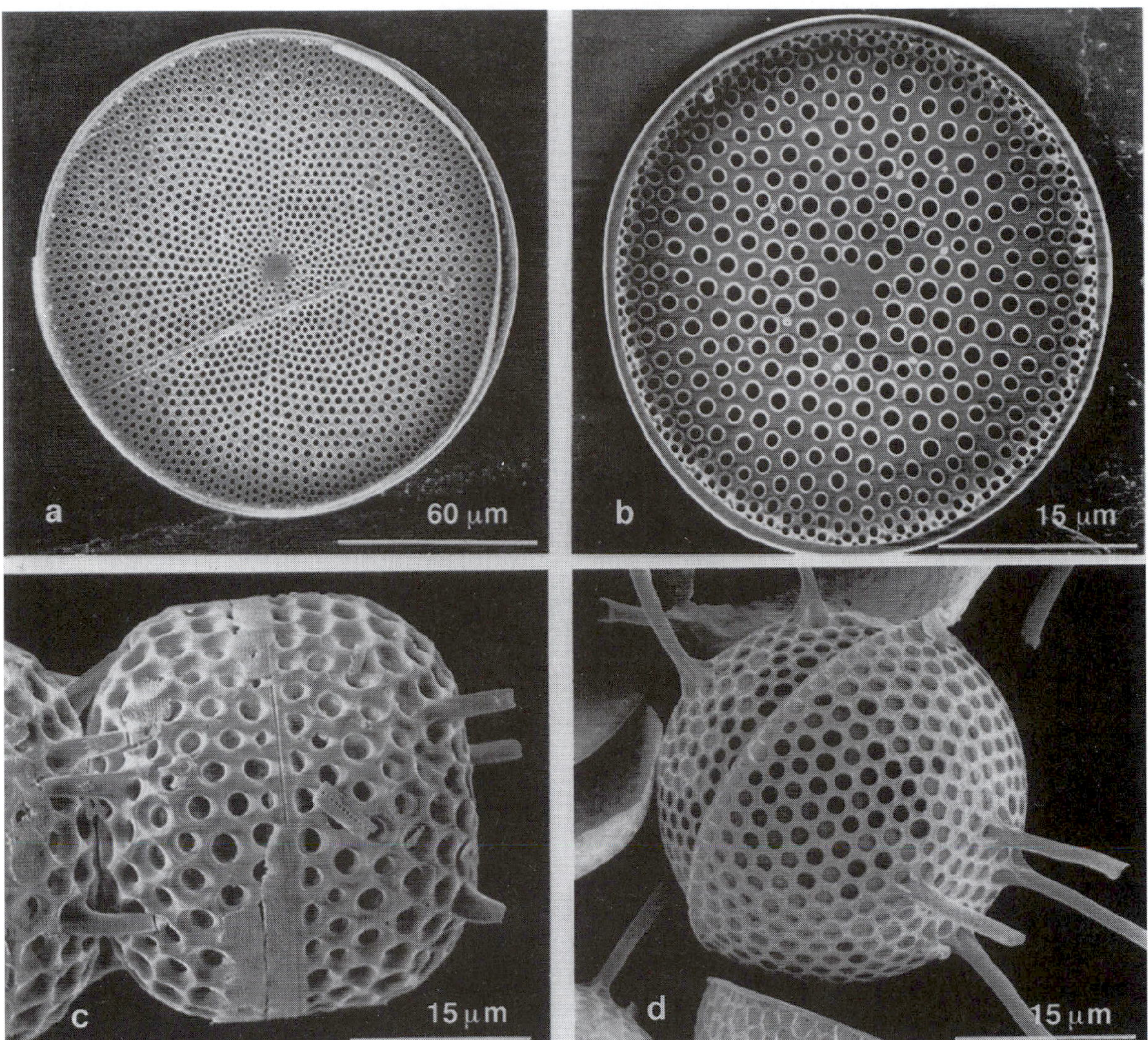

Fig. 13.29 Scanning electron micrographs of fossil diatoms compared to frustules of living diatoms from the same genera. (*a*) *Coscinodiscus radiatus* from Cretaceous sediments. (*b*) *Coscinodiscus radiatus* from recent plankton. (*c*) *Stephanopyxis turris* from Cretaceous sediments. (*d*) *Stephanopyxis broschi* from recent plankton. (From Medlin et al., 1993.)

lake district (Round, 1957, 1961). During the end of the last glacial period, there were few diatoms present, with the sediments being mostly of inorganic origin, indicating that the lakes initially were very oligotrophic and lacked planktonic diatoms. This period was followed by diatoms characteristic of melt waters under semi-arctic conditions (*Melosira arenaria*, Fig. 13.31(*a*). *Cyclotella antique*, Fig. 13.31(*b*)) or characteristic of base-rich (high alkalinity) lakes (*Rhopalodia*, Fig. 13.31(*c*); *Gyrosigma*, Fig. 13.31(*d*); *Cymatopleura*, Fig. 13.31(*e*); *Campylodiscus*, Fig. 13.31(*f*)) indicating leaching of base-rich components from surrounding rocks into the lakes. The next area of sediments showed a decrease in alkaline-water species and an increase in acid-water species (*Eunotia*, Fig. 13.2(*b*); *Anomoeoneis*;

Fig. 13.30 Diatomaceous earth mine at Lompoc, California, with the deposits of diatomaceous earth in the background. (Courtesy of Johns-Manville Corp.)

Frustulia, Fig. 13.31(*g*); *Tabellaria*, Fig. 13.25(*d*),(*e*)), a trend that has continued up to recent times. This trend is related to a decrease in bases being leached from the rocks because no new rock surfaces have been exposed, and to a decline in the stands of birch/pine on the land. At the top of the sediments are remains of *Asterionella formosa*, indicating a recent change of waters to less acidic, more eutrophic conditions, a state related to man's activities. *A. formosa* is a diatom that grows readily in eutrophic waters and shows changes in colony morphology, depending on nutrients in the water. Under optimum growth conditions, the colonies average eight cells, whereas under phosphate limitation the number of cells per colony drops to two, and under silicate limitation the number of cells per colony increases to 20. Under both P and Si limitations, there is a decrease in growth rate (Tilman et al., 1976).

Classification

The diatoms probably originated about 200 million years ago in the late Permian (Medlin et al., 1997) from a scaly member of the Chrysophyceae or Synurophyceae (Medlin et al., 1993; Sorhannus et al., 1995). Two scales evolved into valves, while other scales formed the girdle bands. The first diatom had a centric organization with two dome-shaped valves and

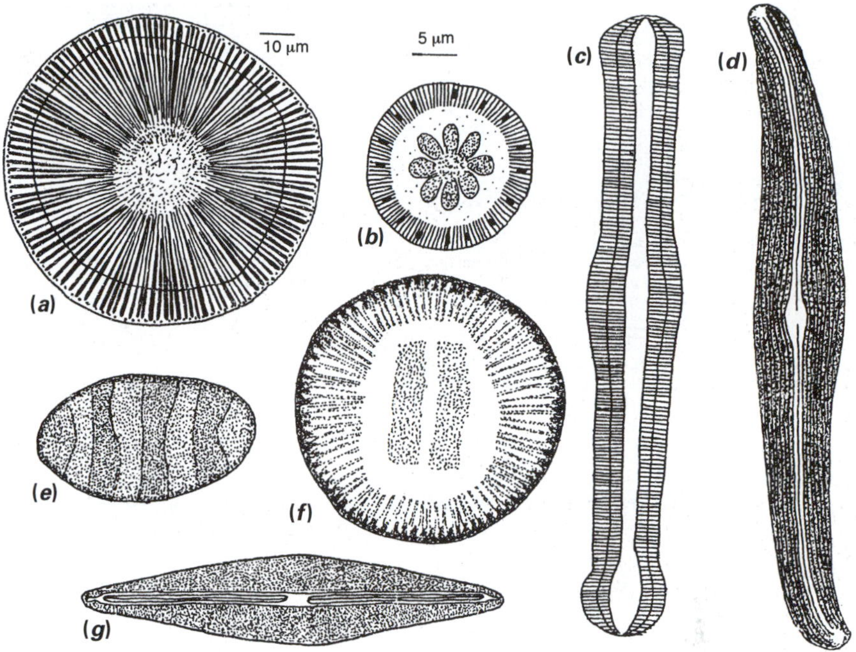

Fig. 13.31 (*a*) *Melosira arenaria*. (*b*) *Cyclotella antique*. (*c*) *Rhopalodia gibba*. (*d*) *Gyrosigma attenuatum*. (*e*) *Cymatopleura elliptica*. (*f*) *Campylodiscus clypeus*. (*g*) *Frustulia rhomboides*.

many scale-like bands (Round et al., 1990). The first recognizable fossils are centric diatoms from the early Cretaceous with pennate diatoms being recorded from the late Cretaceous. The first pennate diatom fossils were araphid (no raphe) with raphid diatoms appearing in the middle Eocene.

The Bacillariophyceae can be divided into two orders as follows:

Order 1 Biddulphiales: radial or gonoid ornamentation; many chloroplasts; no raphe; resting spores formed; motile spermatozoids with a single tinsel flagellum; oogamous sexual reproduction.

Order 2 Bacillariales: pennate or trellisoid ornamentation; one or two chloroplasts; raphes possibly present with gliding; no flagellated spermatozoids; sexual reproduction by conjugation.

BIDDULPHIALES

The Biddulphiales or Centrales is primarily a marine planktonic group having cells with gonoid or centric ornamentation (Fig. 13.32). *Melosira*, a common golden-brown diatom found in marine and freshwater environments, consists of cylindrical cells with a greater length than breadth (Figs. 13.33 and 13.35). There are a number of girdle bands which form incomplete rings around the cell without their ends meeting. Both the valve and the girdle bands are ornamented with small spines. The diatom is usually filamentous, with the cells joined end to end by their valves so that to the

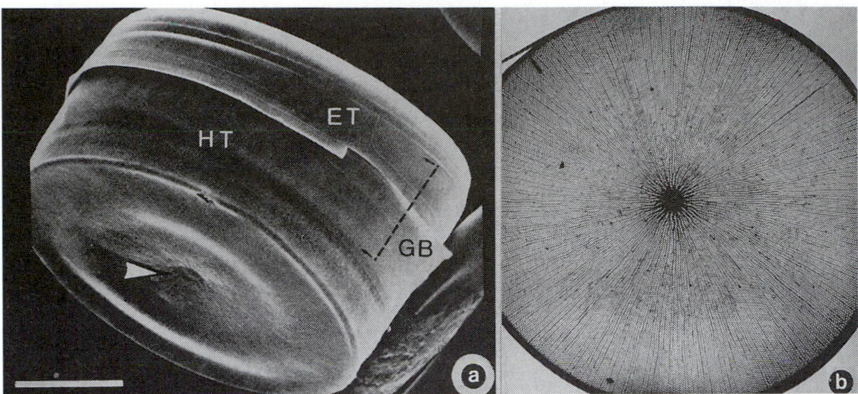

Fig. 13.32 Frustule architecture of the centric diatom *Coscinodiscus wailesii*. (*a*) Scanning electron micrograph of a whole cell. (ET) Epitheca; (HT) hypotheca; (GB) girdle bands. The rows of areolae of the valve radiate from a depressed central area (arrowhead). Bar = 100 μm. (*b*) Transmission electron micrograph of an isolated valve showing the radiating rows of areolae. Bar = 50 μm. (From Schmid and Volcani, 1983.)

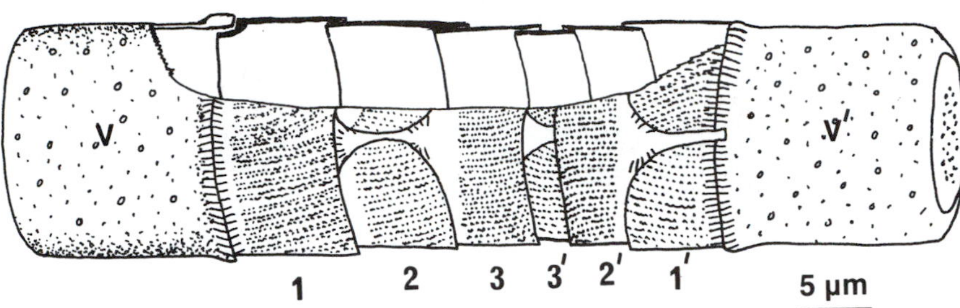

Fig. 13.33 Semi-diagrammatic representation of a cell of *Melosira varians* composed of two valves, *v* and *v*′, two girdle band series: 1,2,3 and, underlapping these, the younger series 1′,2′,3′. (After Crawford, 1971.)

inexperienced eye it resembles a filamentous chrysophycean alga with the cells always in girdle view. Gametogenesis is initiated after the cells have undergone a number of mitotic cell divisions that have reduced the size of the frustule below a certain critical limit. Von Stosch (1951) has elucidated the following events in the life cycle of *Melosira varians* (Fig. 13.35). In the male cells, gametogenesis starts with the nucleus undergoing two meiotic divisions to produce four haploid nuclei. A plasma membrane is formed around each nucleus and a small amount of parent cell protoplasm. The remaining protoplasm of the mother cell now lacks a nucleus and begins to degenerate. A single flagellum is produced by each male gamete, the epitheca separates from the hypotheca, and the male gametes are released into the medium. In the female cells, while the nucleus is undergoing the first meiotic division, one of the girdle bands separates from the edge of the valve, leaving an area of protoplasm open to the medium. The male gamete swims to the area of the female cell where the girdle band has separated from the valve edge. During telophase of the first meiotic division of the female cell, the cytoplasm of the male gamete fuses with that of the female cell. One of the nuclei of the first meiotic division in the female cell degenerates while the other nucleus undergoes a second meiotic division. From the

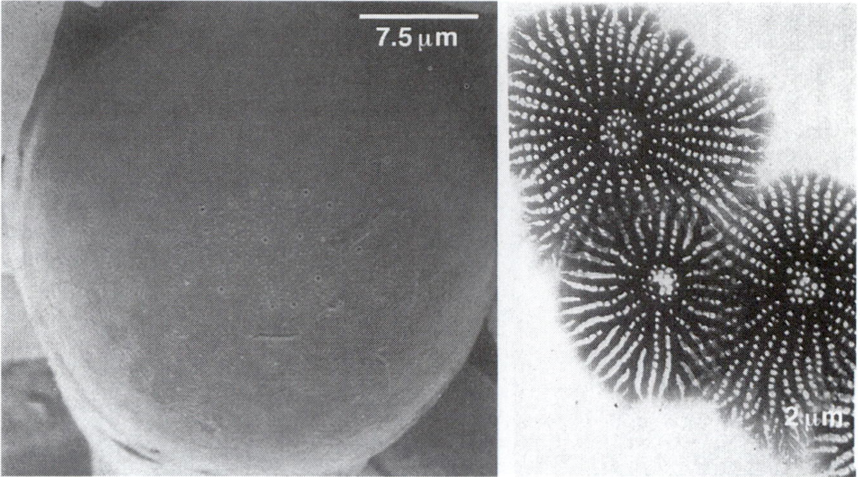

Fig. 13.34 Electron micrographs of an auxospore of *Melosira nummuloides* (*left*) showing the scales covering the auxospore (*right*). (From Medlin et al., 1993.)

second meiotic division, one of the female nuclei again degenerates, and the other becomes the egg nucleus. The male nucleus fuses with the egg nucleus to form the zygote or auxospore nucleus. The resulting auxospore (Fig. 13.34) then swells, pushing apart the hypotheca and epitheca until the auxospore is attached to both or one theca by a small protuberance only. The auxospore has an outer organic wall and an inner wall made of siliceous scales (Crawford, 1974). A new cell wall of maximum dimensions is now produced within the auxospore. This cell divides to give rise to a new filament.

Melosira is capable of producing dormant cells in freshwater environments. The dormant cells have the protoplasm condensed to a dark-brown protoplasmic mass in one part of the frustule (Sicko-Goad et al., 1986). The dormant cells settle into the sediment, where they are capable of surviving for up to 20 years. Normally, however, the dormant cells are swept up by normal recirculation of lake waters during lake overturns (Lund, 1959). Within 24 hours of being swept out of the sediments, the dormant cells of *Melosira* differentiate into vegetative cells.

Chaetoceros (Fig. 13.24 and 13.36) has more than 160 species, the largest number of any planktonic diatom. The genus is more or less widespread in warm and cold waters. The life cycle of *Chaetoceros diadema* has been elucidated by Hargraves (1972) (Fig. 13.36). *Chaetoceros diadema* is a colonial marine diatom, with cells united by their long setae to form filaments. The different phases of the life cycle are characterized by different sizes and shapes of cells. Gametogenesis is initiated at temperatures between 2 and 5 °C (French and Hargraves, 1985). The cells that produce the male gametes are about 1.5 to 2 times as wide as high. These cells divide to produce 32 male gametes, presumably by meiosis, which are released as uniflagellate

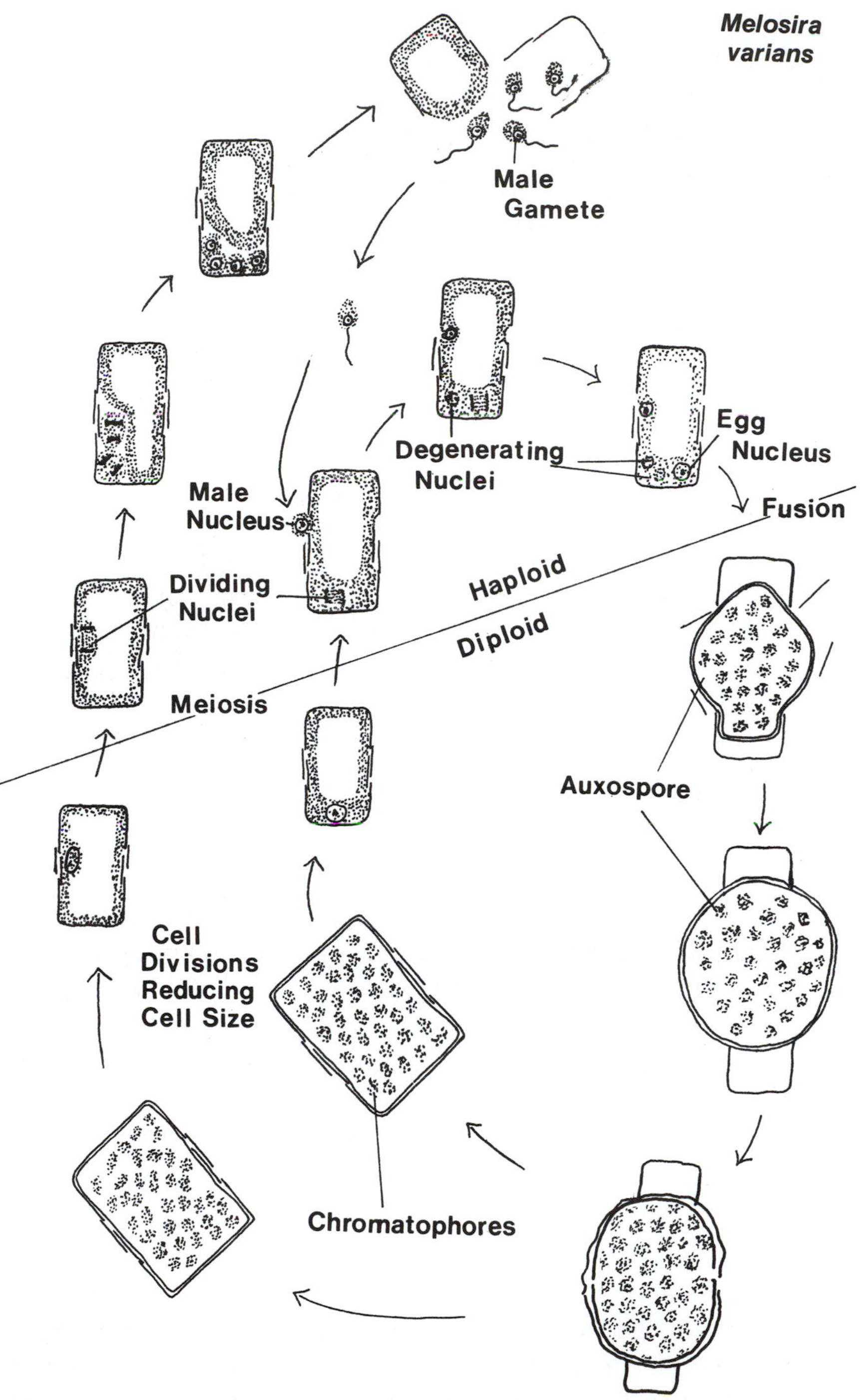

Fig. 13.35 The life cycle of *Melosira varians*. (Adapted from von Stosch, 1951.)

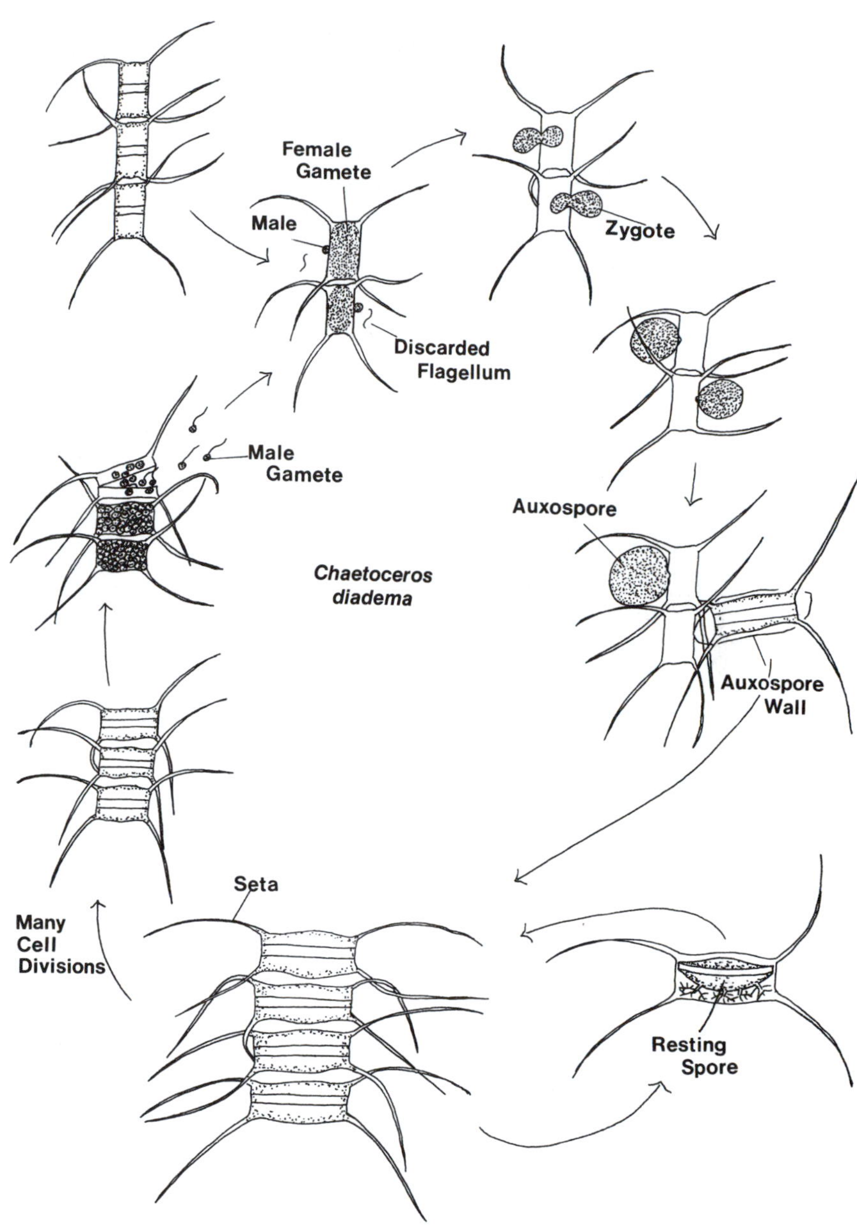

Fig. 13.36 The life cycle of *Chaetoceros diadema*. (Adapted from Hargraves, 1972.)

swarmers. The cells that produce the female gametes are about twice as high as wide. The exact nuclear events that lead to the production of the egg are not known, but the male gamete swims to the girdle area of the female cell, casts off its flagellum, and fuses with the female gamete inside the frustule. The cell contents of the auxospore-forming cell are then extruded through the girdle region into a thin hyaline envelope. This cell then produces the auxospore, which, in turn, produces new cells with frustules of

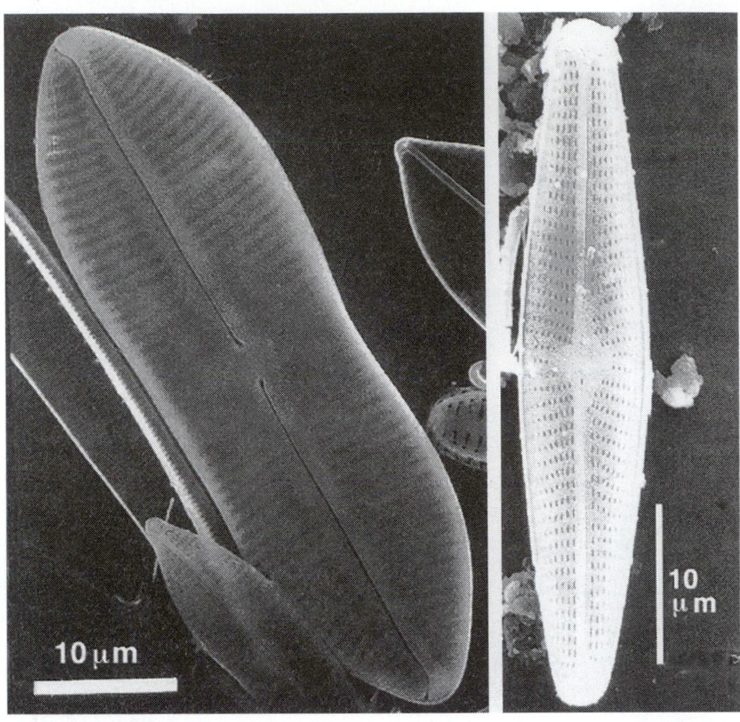

Fig. 13.37 Scanning electron micrographs of the pennate diatoms *Pinnularia quadratarea* (*left*) and *Navicula radiosa* (*right*). (From Poulin, 1993.)

maximum size. These large daughter cells remain attached to the auxospore parent wall through two divisions of the daughter cells. The hyaline envelope of the auxospore is sloughed off after the first division. Any of the vegetative cells have the ability to produce resting spores after nitrogen depletion of the culture medium at temperatures that allow vegetative growth (French and Hargraves, 1985). The resting spores germinate to produce cells larger than the original cells. The resting spores occur within the parent cell as heavily siliceous bodies, having usually convex, often dissimilar, upper and lower surfaces bearing spines.

BACILLARIALES

The Bacillariales or Pennales order contains cells that occur both in freshwater and in marine environments. These cells have either pennate or trellisoid ornamentation. The life cycle of these organisms involves fusion of two gametes by conjugation; and common genera such as *Nitzschia, Navicula* (Fig. 13.37), *Amphora, Cymbella,* and *Pinnularia* (Fig. 13.37) have essentially the same life cycle. In *Pinnularia,* the cells have rounded poles with more or less parallel sides (Fig. 13.38). Auxospore formation is a sexual process initiated after cell divisions have reduced the cell to a certain critical size. Two cells come together and invest themselves in a

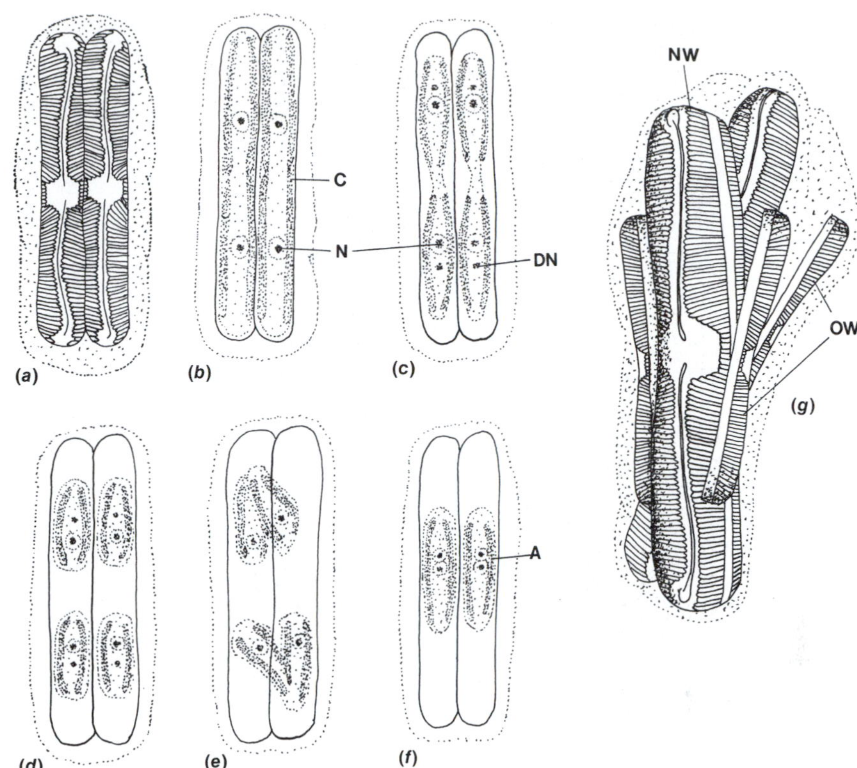

Fig. 13.38 Auxospore formation in *Pinnularia*. (*a*) Two cells come to lie next to each other. (*b*)–(*d*) Meiosis occurs in each mother cell, resulting in two gametes per mother cell, each with one functional nucleus (N) and one degenerate nucleus (DN). (*e*) One gamete from each mother cell passes to the other mother cell. (*f*) There is one auxospore (A) per mother cell frustule. (*g*) Each auxospore has produced a new large daughter cell (NW), with the old cell walls (OW) of the mother cells still embedded in the mucilage. (C) Chloroplast. (After Hendey, 1964.)

mucous envelope (Hendey, 1964). The nucleus in each cell undergoes two meiotic divisions followed by cytokinesis into two cells. Two gametes are thus formed per mother cell, each protoplast receiving one functional and one degenerating nucleus. One gamete from each mother cell moves to the contiguous mother cell to fuse with the passive gamete. In this manner, one zygote (auxospore) is formed within each mother cell. These auxospores become greatly enlarged, and eventually new siliceous frustules of maximum size are produced within each auxospore. The four discarded valves of the mother cells are often seen attached to the newly formed frustules.

Cocconeis is a genus of pennate diatoms with a pseudoraphe and rows of pores covered with a velum (Fig. 13.39). A slightly different type of sexual reproduction occurs in *Cocconeis placentula* var. *klinoraphis* (Geitler, 1927) (Fig. 13.40). Two mother cells come to lie next to each other, each cell undergoing two meiotic divisions with one nucleus degenerating after each division to yield one gamete per cell with one functional nucleus and two degenerating nuclei. The frustules of both mother cells are now forced open, and the gametes emerge and fuse. A spherical auxospore raised on a stalk is formed from the fusion.

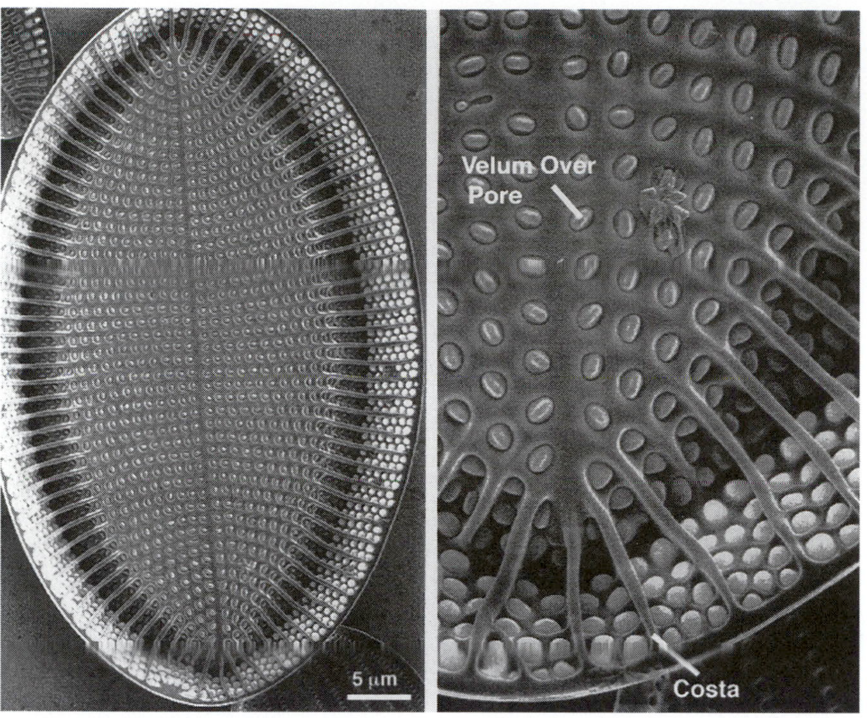

Fig. 13.39 Scanning electron micrographs of *Cocconeis scutellum* showing a pseudoraphe with rows of pores, each covered with a velum. (From Navarro, 1993.)

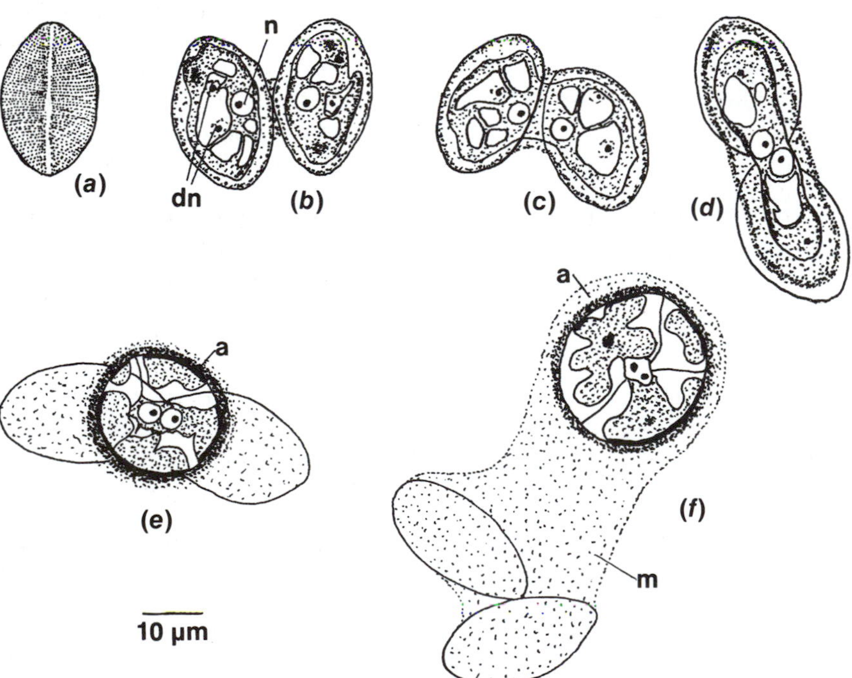

Fig. 13.40 Auxospore formation in *Cocconeis placentula* var. *klinoraphis*. (*a*) Valve view of the frustule. (*b*)–(*d*) Two cells attached by a mucilage pad: The nuclei undergo meiosis and the protoplasts fuse. (*e*) Young auxospore. (*f*) Mature auxospore on mucilage stalk. (a) Auxospore; (dn) degenerate nucleus; (m) mucilage stalk; (n) functional nucleus. (After Geitler, 1927.)

Fig. 13.41 Frank E. Round Dr Round graduated from the University of Birmingham. His initial research was carried out with Dr J. W. Lund at the Freshwater Biological Association, Windermere, Cumbria, where he studied the ecology of the epipelic algal populations in English Lake District. He was a lecturer at the University of Liverpool and the University of Birmingham before moving, in 1955, to the University of Bristol where he is a professor of botany. Dr Round's publications have included *The Biology of the Algae* and *Ecology of the Algae*. His main research interests have been the diatoms, particularly their structure in the scanning electron microscope, and the systematics and evolution of the group.

References

Andersen, L. W. J., and Sweeney, B.M. (1977). Dial changes in sedimentary characteristics of *Ditylum brightwelli*: Changes in cellular lipid and effects of respiratory inhibitors and ion transport modifiers. *Limnol. Oceanogr.* 22:539–52.

Andersen, R. A., Medlin, L. K., and Crawford, R. M. (1986). An investigation of the cell wall components of *Actinocyclus subtilis* (Bacillariophyceae). *J. Phycol.* 22:466–79.

Anderson, O.R. (1975). The ultrastructure and cytochemistry of resting cell formation in *Amphora coffaeformis* (Bacillariophyceae). *J. Phycol.* 11:272–81.

Anderson, O.R. (1976). Respiration and photosynthesis during resting cell formation in *Amphora coffaeformis* (Ag.) Kütz. *Limnol. Oceanogr.* 21:452–6.

Beattie, A., Hirst, E. L., and Percival, E. (1961). Studies on the metabolism of the Chrysophyceae. Comparative structural investigations on leucosin (chrysolaminarin) separated from diatoms and laminarin from brown algae. *Biochem. J.* 79:531–6.

Branton, D., Cohen, C. M., and Tyler, T. (1981). Interaction of cytoskeletal proteins on the human erythrocyte membrane. *Cell* 24:24–32.

Brzezinski, M. A., and Conley, D. J. (1994). Silicon deposition during the cell cycle of *Thalassiosira weissflogii* (Bacillariophyceae) determined using dual rhodamine 123 and propidium iodide staining. *J. Phycol.* 30:45–55.

Cohn, S. A., and Weitzell, R. E. (1996). Ecological consideration of diatom cell motility. I. Characterization of motility and adhesion in four diatom species. *J. Phycol.* 32:918–39.

Cohn, S. A., Spurck, T. P., and Pickett-Heaps, J. D. (1989). Perizonium and initial valve formation in the diatom *Navicula cuspidata* (Bacillariophyceae). *J. Phycol.* 25:15–26.

Cooksey, K. E., and Cooksey, B. (1974). Calcium deficiency can induce the transition from oval to fusiform cells in cultures of *Phaeodactylum tricornutum* Bohlin. *J. Phycol.* 10:89–90.

Coombs, J., and Volcani, B. E. (1968). Studies on the biochemistry and fine structure of silica shell formation in diatoms. Chemical changes in the wall of *Navicula pelliculosa* during its formation. *Planta* 82:280–92.

Crawford, R. M. (1971). The fine structure of the frustule of *Melosira varians* C. A. Agardh, *Br. Phycol. J.* 6:175–86.

Crawford, R. M. (1974). The auxospore wall of the marine diatom *Melosira nummuloides* (Dillw.) C. Ag. and related species. *Br. Phycol. J.* 9:9–20.

Crawford, R.M. (1995). The role of sex in the sedimentation of a marine diatom bloom. *Limnol. Oceanogr.* 40:200–4.

Davis, C. O., Hollibaugh, J. T., Seibert, D. L. R., Thomas, W. H., and Harrison, P. J. (1980). Formation of resting spores by *Leptocylindrus danicus* (Bacillariophyceae) in a controlled experimental ecosystem. *J. Phycol.* 16:296–302.

Dawson, P. A. (1973). Observations on the structure of some forms of *Gomphonema parvulum* Kütz. III. Frustule formation. *J. Phycol.* 9:353–64.

Drum, R. W., and Hopkins, J. T. (1966). Diatom locomotion: An explanation. *Protoplasma* 62:1–33.

Edgar, L. A. (1979). Diatom locomotion: Computer assisted analysis of cine film. *Br. Phycol. J.* 14:83–101.

Edlund, M. B., and Stoermer, E. F. (1997). Ecological, evolutionary, and systematic significance of diatom life histories. *J. Phycol.* 33:897–918.

Erickson, S. J. (1972). Toxicity of copper to *Thalassiosira pseudonana* in unenriched inshore seawater. *J. Phycol.* 8:318–23.

Fairchild, E., and Sheriden, R. P. (1974). A physiological investigation of the hot spring diatom *Achnanthes exigua*. Grün. *J. Phycol.* 10:1–4.

Fauré-Fremiet, E. (1951). The tidal rhythm of the diatom *Hantzschia amphioxys*. *Biol. Bull.* 100:173–7.

Fisher, A. E., and Harrison, P. J. (1996). Does carbohydrate content effect the sinking rate of marine diatoms. *J. Phycol.* 32:360–5.

French, F. W., and Hargraves, P. E. (1985). Spore formation in the life cycles of the diatoms *Chaetoceros diadema* and *Leptocylindrus danicus*. *J. Phycol.* 21:477–83.

Geitler, L. (1927). Somatische Teilung, Reduktionsteilung, Copulation und Parthenogenese bei *Cocconeis placentula*. *Arch. Protistenk.* 59:506–49.

Gross, F. (1940). The development of isolated resting spores into auxospores in *Ditylum brightwelli* (West). *J. Mar. Biol. Assoc. UK* 24:375–80.

Guillard, R. R. L., and Cassi, V. (1963). Minimum cyanocobalamin requirements of some marine centric diatoms. *Limnol. Oceanogr.* 8:161–5.

Haines, K. C., and Guillard, R. R. L. (1974). Growth of vitamin B_{12}-requiring marine diatoms in mixed laboratory cultures with vitamin B_{12}-producing marine bacteria. *J. Phycol.* 10:245–52.

Hargraves, P. E. (1972). Studies on marine planktonic diatoms. I. *Chaetoceros diadema* (Ehr.) Gran.: Life cycle, structural morphology, and regional distribution. *Phycologia* 11:247–57.

Hasle, G. R. (1995). *Pseudo-nitzschia punges* and *P. multiseries* (Bacillariophyceae): nomenclature history, morphology and distribution. *J. Phycol.* 31:428–35.

Hasle, G. R., and Lange, C. B. (1989). Freshwater and brackish water *Thalassiosira* (Bacillariophyceae): taxa with tangentially undulated valves. *Phycologia* 28:120–35.

Hecky, R. F., Mopper, K., Kilham, P., and Degens, E. T. (1973). The amino acid and sugar composition of diatom cell walls. *Mar. Biol.* 19:323–31.

Hellebust, J. A. (1971). Glucose uptake by *Cyclotella cryptica:* Dark induction and light inactivation of transport system. *J. Phycol.* 7:345–49.

Hendey, N. I. (1964). An introductory account of the smaller algae of British coastal waters. Part V. Bacillariophyceae (diatoms). *Fish. Invest. Ser. IV.* H.M.S.O., London.

Hendey, N. I. (1971). Electron microscope studies and the classification of diatoms. In *The Micropaleontology of Oceans,* ed. B. M. Funnell, and W. R. Riedel, pp. 625–31. Cambridge: Cambridge University Press.

Hoagland, K. D., Rosowski, J. R., Gretz, M. R., and Romer, S. C. (1993). Diatom extracellular polymeric substances: Function, fine structure, chemistry, and physiology. *J. Phycol.* 29:537–66.

Iler, R. K. (1979). *The Chemistry of Silica.* New York: John Wiley.

Jeffrey, S. W. (1976). A report of green algal pigments in the central north Pacific Ocean. *Mar. Biol.* 37:33–7.

Jeffrey, S. W., and Vesk, M. (1977). Effect of blue-green light on photosynthetic pigments and chloroplast structure in the marine diatom *Stephanopyxis turris. J. Phycol.* 13:271–9.

Johnson, L. M., Hoagland, K. D., and Gretz, M. R. (1995). Effects of bromide and iodide on stalk secretion in the biofouling diatom *Achnanthes longipes* (Bacillariophyceae). *J. Phycol.* 31:401–12.

Knudson, B.M. (1957). Ecology of the epiphytic diatom *Tabellaria flocculosa* (Roth) Kütz. var. *flocculosa* in three English lakes. *J. Ecol.* 45:93–112.

Lauritis, J. A., Coombs, J., and Volcani, B. E. (1968). Studies on the biochemistry and fine structure of silica shell formation in diatoms. IV. Fine structure of the apochlorotic diatom *Nitzschia alba* Lewin and Lewin. *Arch. Mikrobiol.* 62:1–16.

Lewin, J. C. (1953). Heterotrophy in diatoms. *J. Gen. Microbiol.* 9:305–13.

Lewin, J. C. (1962). Silicification. In *Physiology* and *Biochemistry of the Algae,* ed. R. A. Lewin, pp. 445–55. New York: Academic Press.

Lewin, J. C. (1966). Silicon metabolism in diatoms. V. Germanium dioxide, a specific inhibitor of diatom growth. *Phycologia* 6:1–12.

Lewin, J. C., and Lewin, R. A. (1960). Auxotrophy and heterotrophy in marine littoral diatoms. *Can. J. Microbiol.* 6:127–34.

Lewin, J. C., and Lewin, R. A. (1967). Culture and nutrition of some apochlorotic diatoms of the genus *Nitzschia. J. Gen. Microbiol.* 46:361–7.

Li, C-W., Chu, S., and Lee, M. (1989). Characterizing the silica deposition vesicle of diatoms. *Protoplasma* 151:156–63.

Lichti-Federovich, S. (1980). *Navicula subfossilis* sp. nov., the dominant taxonomic entity of a Pleistocene assemblage from Ellesmere Island, N.W.T. *Can. J. Bot.* 58:1334–40.

Lobban, C. S. (1983). Colony and frustule morphology of three tube-dwelling diatoms from eastern Canada. *J. Phycol.* 19:281–9.

Lowe, R. L. (1975). Comparative ultrastructure of the valves of some *Cyclotella* species (Bacillariophyceae). *J. Phycol.* 11:415–24.

Lund, J. W. G. (1949). Studies on *Asterionella.* I. The origin and nature of the cells producing seasonal maxima. *J. Ecol.* 37:389–419.

Lund, J. W. G. (1950). Studies on *Asterionella formosa* Hass. II. Nutrient depletion and the spring maximum *J. Ecol.* 38:1–14, 15–35.

Lund, J. W. G. (1959). Buoyancy in relation to the ecology of freshwater phytoplankton, *Br. Phycol. Bull.* 7:17.

McQuoid, M. R., and Hobson, L. A. (1996). Diatom resting stages. *J. Phycol.* 32:889–902.

Mann, D. G. (1982). Structure, life history and systematics of *Rhoicosphenia* (Bacillariophyta). I. The vegetative cell of *Rh. curvata. J. Phycol.* 18:162–76.

Mann, J. E., and Myers, J. (1968). On pigments, growth and photosynthesis of *Phaeodactylum tricornutum*. *J. Phycol.* 4:349–55.

Martinez, L. A., Silver, M. W., King, J. M., and Alldredge, A. L. (1983). Nitrogen fixation by floating diatom mats: A source of new nitrogen to oligotrophic ocean waters. *Science* 221:152–154.

Medlin, L. K., Crawford, R. M., and Andersen, R. A. (1986). Histochemical and ultrastructural evidence for the function of the labiate process in the movement of centric diatoms. *Br. Phycol. J.* 21:297–301.

Medlin, L. K., Williams, D. M., and Sims, P. A. (1993). The evolution of the diatoms (Bacillariophyta). I. Origin of the group and assessment of the monophyly of its major divisions. *Eur. J. Phycol.* 28:261–75.

Medlin, L. K., Kooistra, W. H. C. F., Gersonde, R., Sims, P. A., and Wellbrock, U. (1997). Is the origin of the diatoms related to the end-Permian mass extinction? *Nova Hedwigia* 65:1–11.

Mehard, C. W., Sullivan, C. W., Azam, F., and Volcani, B. F. (1974). Role of silicon in diatom metabolism. IV. Subcellular localization of silicon and germanium in *Nitzschia alba* and *Cylindrotheca fusiformis*. *Physiol. Plant.* 30:265–72.

Mehta, S. C., Venkataraman, G. S., and Das, S. C. (1961). The fine structure and the cell wall nature of *Diatoma hiemale* var. *mesodan* (Her.) Grun. *Rev. Algol., N.S.* 6:49–52.

Moore, J. K., and Villareal, T. A. (1996). Size-ascent rate relationships in positively buoyant marine diatoms. *Limnol. Oceanog.* 41:1514–20.

Morel, N. M. L., Rueter, J. G., and Morel, F. M. M. (1978). Copper toxicity to *Skeletonema costatum*. *J. Phycol.* 14:43–8.

Nagai, S., Hori, Y., Manabe, T., and Imai, I. (1995). Restoration of cell size by vegetative cell enlargement in *Coscinodiscus wailesii* (Bacillariophyceae). *Phycologia* 34:533–55.

Nakajima, T., and Volcani, B. E. (1969). 3,4-Dihydroxyproline: a new amino acid from diatom cell walls. *Science* 164:1400–1.

Navarro, J. N. (1993). Three-dimensional imaging of diatom ultrastructure with high resolution low-voltage SEM. *Phycologia* 32:151–6.

Nelson, D. M., and Brand, L. E. (1979). Cell division periodicity in 13 species of marine phytoplankton in a light : dark cycle. *J. Phycol.* 15:67–75.

Nultsch, W. (1956). Studien über die Phototaxis der Diatomeen. *Arch. Protistenk.* 101:1–68.

Palmisano, A. C., and Sullivan, C. W. (1982). Physiology of sea ice diatoms. I. Response of three polar diatoms to a simulated summer–winter transition. *J. Phycol.* 18:489–98.

Pan, Y., Subba Rao, D. V., and Mann, K. H. (1996). Changes in domoic acid production and cellular chemical composition of the toxigenic diatom *Pseudo-nitzschia multiseries* under phosphate limitation. *J. Phycol.* 32:371–81.

Paul, J. S. (1979). Osmoregulation in the marine diatom *Cylindrotheca fusiformis*. *J. Phycol.* 15:280–4.

Pickett-Heaps, J. D., Tippit, D. H., and Andreozzi, J. A. (1979a). Cell division in the pennate diatom *Pinnularia*. III. The valve and associated organelles. *Biol. Cellulaire* 35:195–8.

Pickett-Heaps, J. D., Tippit, D. H., and Andreozzi, J. A. (1979b). Cell division in the pennate diatom *Pinnularia*. IV. Valve morphogenesis. *Biol. Cellulaire* 35:199–206.

Pickett-Heaps, J. D., Hill, D. R. A., and Wetherbee, R. (1986). Cellular movement in the centric diatom *Odontella sinensis*. *J. Phycol.* 22:334–9.

Pickett-Heaps, J. D., Carpenter, J., and Koutoulis, A. (1994). Valve and seta (spine) morphogenesis in the centric diatom *Chaetoceros peruvianus* Brightwell. *Protoplasma* 181:269–82.

Potapova, M., and Snoeijs, P. (1997). The natural life cycle in the wild populations of *Diatoma moniliformis* (Bacillariophyceae) and its disruption in an aberrant environment. *J. Phycol.* 33:924–37.

Poulin, M. (1993). *Craspedopleura* (Bacillariophyta), a new diatom genus of arctic sea ice assemblages. *Phycologia* 32:223–33.

Prasad, A. K. S. K., Nienow, J. A., and Livingstone, R. J. (1990). The genus *Cyclotella* (Bacillariophyta) in Choctawhatchee Bay Florida, with special reference to *C. striata* and *C. choctawhatcheeana* sp. nov. *Phycologia* 29:418–36.

Reimann, B. E. F., Lewin, J. C., and Volcani, B. E. (1965). Studies on the biochemistry and fine structure of silica shell formation in diatoms. I. The structure of the cell wall of *Cylindrotheca fusiformis* Reimann and Lewin. *J. Cell Biol.* 24:39–55.

Reisen, W. K., and Spencer, D. J. (1970). Succession and current demand relationships of diatoms on artifical substrates in Prater's Creek, South Carolina, *J. Phycol.* 6:117–21.

Richardson, T. L., Ciotti, A. M., Cullen, J. J., and Villareal, T. A. (1996). Physiological and optical properties of *Rhizosolenia formosa* (Bacillariophyceae) in the context of open-ocean vertical migration. *J. Phycol.* 32:741–57.

Riedel, G. F., and Nelson, D. M. (1985). Silicon uptake by algae with no known Si requirement. II. Strong pH dependence of uptake kinetic parameters in *Phaeodactylum tricornutum* (Bacillariophyceae). *J. Phycol.* 21:168–71.

Robinson, D. H., and Sullivan, C. W. (1987). How do diatoms make silicon biominerals? *Trends Biochem. Sci.* 12:151–4.

Rogall, E. (1939). Über den Feinbau der Lieselmembran der Diatomeen. *Planta* 29:279–91.

Ross, R., and Sims, P. A. (1972). The fine structure of the frustule in centric diatoms: A suggested terminology. *Br. Phycol. J.* 7:139–63.

Round, F. E. (1957). The late-glacial and post-glacial diatom succession in the Kentmere valley deposit. I. Introduction, methods and flora. *New Phytol.* 56:98–126.

Round, F. E. (1961). The diatoms of a core from Esthwaite water. *New Phytol.* 60:43–59.

Round, F. E., Crawford, R. M., and Mann, D. G. (1990). *The Diatoms.* Cambridge: Cambridge University Press. 747 pp.

Rueter, J. G. (1983). Effect of copper on growth, silicic acid uptake and soluble pools of silicic acid in the marine diatom, *Thalassiosira weisflogii* (Bacillariophyceae). *J. Phycol.* 19:101–4.

Rueter, J. G., Chisholm, S. W., and Morel, F. M. M. (1981). Effects of copper toxicity on silicic acid uptake and growth in *Thalassiosira pseudonana. J. Phycol.* 17:270–8.

Schmaljohann, R., and Röttger, R. (1978). The ultrastructure and taxonomic identity of the symbiotic algae of *Heterostegina depressa* (Foraminifera: Nummulitidae). *J. Mar. Biol. Assoc. UK* 58:227–37.

Schmid, A-M. M. (1994). Aspects of morphogenesis and function of diatom cell walls with implication for taxonomy. *Protoplasma* 181:43–60.

Schmid, A-M. M., and Volcani, B. E. (1983). Wall morphogenesis in *Coscinodiscus wailesii* Gran and Angst. I. Valve morphology and development of its architecture. *J. Phycol.* 19:387–402.

Schütt, F. (1896). Bacillariales, In *Naturlichen Pflanzenfamilien 1*(1b):31–150.

Sicko-Goad, L., Stoermer, E. F., and Fahnensteil, G. (1986). Rejuvenation of *Melosira granulata* (Bacillariophyceae) resting cells from anoxic sediments of Douglas Lake, Michigan. I. Light microscopy and ^{14}C uptake. *J. Phycol.* 22:22–8.

Smayda, T. J. (1970). The suspension and sinking of phytoplankton in the sea. *Oceanogr. Mar. Biol. Annu. Rev.* 8:353–414.

Smith, G. M. (1955). *Cryptogamic Botany*, Vol. 1: *Algae and Fungi*. New York: McGraw-Hill.

Smith, R. E. H., Clement, P., Cota, G. F., and Li, W. K. W. (1987). Intracellular photosynthetic allocation and the control of Arctic marine ice algal production. *J. Phycol.* 23:124–32.

Sorhannus, U., Gasse, F., Perasso, R., and Baroin Tourancheau, A. (1995). A preliminary phylogeny of diatoms based on 28S ribosomal RNA sequence data. *Phycologia* 34:65–73.

Stoermer, E. F., Pankratz, H. S., and Drum, R. W. (1965a). The fine structure of *Mastogloia grevillei* Wm. Smith, *Protoplasma* 59:1–13.

Stoermer, E. F., Pankratz, H. S., and Bowen, C. C. (1965b). Fine structure of the diatom *Amphipleura pellucida*. II. Cytoplasmic fine structure and frustule formation. *Am. J. Bot.* 52:1067–78.

Subba Rao, D. V., Quilliam, M. A., and Pocklinton, R. (1988). Domoic acid – a neurotoxic amino acid produced by the marine diatom *Nitzschia pungens* in culture. *Can. J. Fish. Aquat. Sci.* 45:2076–9.

Sullivan, C. W. (1976). Diatom mineralization of silicic acid. I. $Si(OH)_4$ transport characteristics in *Navicula pelliculosa*. *J. Phycol.* 12:390–6.

Sullivan, C. W. (1977). Diatom mineralization of silicic acid. II. Regulation of $Si(OH)_4$ transport rates during the cell cycle of *Navicula pelliculosa*. *J. Phycol.* 13:86–91.

Suttle, C. A., and Harrison, P. J. (1988). Rapid ammonium uptake by freshwater phytoplankton. *J. Phycol.* 24:13–16.

Thomas, W. H., Hollibaugh, J. T., and Seibert, D. L. R. (1980). Effects of heavy metals on the morphology of some marine phytoplankton. *Phycologia* 19:202–9.

Tilman, D., Kilham, S. S., and Kilham, P. (1976). Morphometric changes in *Asterionella formosa* colonies phosphate and silicate limitation. *Limnol. Oceanogr.* 21:883–6.

Venrich, E. L., McGowan, J. A., and Mantyla, A. W. (1973). Deep maxima of photosynthetic chlorophyll in the Pacific Ocean. *Fish. Bull.* 71:41–52.

Villareal, T. A. (1989). Division cycles in the nitrogen-fixing *Rhizosolenia* (Bacillariophyceae)- *Richelia* (Nostocaceae) symbiosis. *Br. Phycol. J.* 24:357–65.

Villareal, T. A., and Carpenter, E. J. (1994). Chemical composition and photosynthetic characteristics of *Ethmodiscus rex* (Bacillariophyceae): evidence for vertical migration. *J. Phycol.* 30:1–8.

von Denffer, D. (1949). Die planktonische Massenkultur pennatur Grunddiatomeen. *Arch. Mikrobiol.* 14:159–202.

von Stosch, H. A. (1951). Entwicklungsgeschichtliche Untersuchungen an zentrischen Diatomeen. I. Die Auxosporenbildung von *Melosira varians*. *Arch. Mikrobiol.* 16:101–35.

von Stosch, H. A. (1975). An amended terminology of the diatom girdle. *Nova Hedwigia, Beih.* 53:1–28.

von Stosch, H. A., and Fecher, K. (1979). "Internal thecae" of *Eunotia soleirolii* (Bacillariophyceae): Development, structure and function as resting spores. *J. Phycol.* 15:233–43.

Wagner, J. (1934). Beuträge zur Kenntnis der *Nitzschia putrida* Benecke, insbesonders ihrer Bewegung. *Arch. Protistenk.* 82:86–113.

Whitaker, T. M., and Richardson, M. G. (1980). Morphology and chemical composition of a natural population of an ice-associated Antarctic diatom *Navicula glaciei*. *J. Phycol.* 16:250–57.

White, A. W. (1974). Growth of two facultatively heterotrophic marine centric diatoms. *J. Phycol.* 10:292–300.

14 · Heterokontophyta

Raphidophyceae

The Raphidophyceae, or chloromonads, have chlorophyll *a* and *c*, and two membranes of chloroplast endoplasmic reticulum. The anterior flagellum is commonly tinsel, whereas the posterior flagellum is naked (Figs. 14.1, 14.2, 14.3). The freshwater species of the Raphidophyceae are green, whereas the marine forms are yellowish and contain the carotenoid fucoxanthin (Vesk and Moestrup, 1987). The closest relatives of the Raphidophyceae are the Eustigmatophyceae and the Chrysophyceae (Cavalier-Smith and Chao, 1996).

Most marine species are assigned to the genus *Chattonella* (Fig. 14.3), *Fibrocapsa* (Fig. 14.1(*b*)) or *Heterosigma* (Fig. 14.1(*a*)). Toxic red-tide blooms of the marine *Chattonella antigua* and *Heterosigma carterae* (Taylor, 1992) have occurred in the Seto Inland Sea in Japan (Watanabe et al., 1988). These red tides occurred in the summer when a salinity and temperature stratification occurred at a depth of 5–10 meters. There was little mixing of waters above and below the stratified layer resulting in the upper layer being deficient in nutrients while the bottom layer was anaerobic. *Heterosigma carterae* flourishes under these conditions because it has a daily vertical migration of 10 to 15 meters and this allows the alga to move between the stratified layers. This migration enables *H. carterae* to use the nutrients in the lower layers, and light and oxygen in the upper layer, resulting in the red-tide blooms of the organism. The migration is correlated with the production and degradation of cytoplasmic fat particles (Wada et al., 1987). *Heterosigma carterae* has a wide salinity tolerance in culture (3 to 50‰) and loses its motility at temperatures below 10 °C. At 5 to 10 °C, the alga forms non-motile masses capable of surviving in continuous darkness for up to 15 weeks. This is a far longer dark period than the motile cells can tolerate and still live. In Narragansett Bay, Rhode Island, blooms of *H. carterae* occur during a period of low nitrogen concentration, but at a time when phosphate levels are near a yearly maximum (Tomas, 1979).

Chattonella antiqua is a marine raphidophyte that produces blooms in coastal Japan. In 1972, a bloom killed 500 million dollars worth of caged yellow-tail fish in the Seto Inland Sea (Okaichi, 1989). *C. antiqua* has the highest toxicity during the early to mid-logarithmic phase. After reaching the stationary phase, the toxicity decreases markedly (Khan et al., 1996).

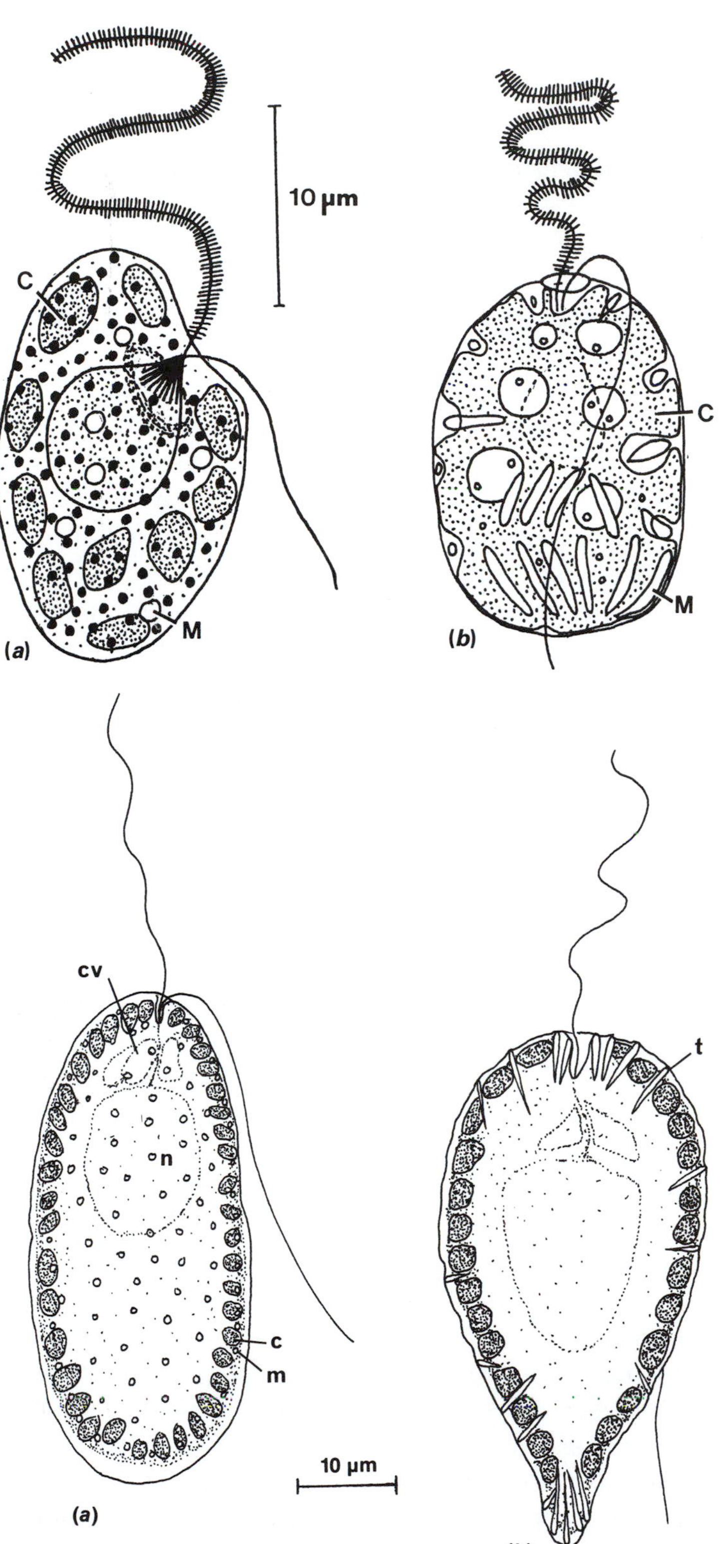

Fig. 14.1 (*a*) *Heterosigma carterae*. (*b*) *Fibrocapsa japonica*. (C) Chloroplast; (M) mucocyst. ((*a*) after Leadbeater, 1969; (*b*) after Hara and Chihara, 1985.)

Fig. 14.2 (*a*) *Vacuolaria virescens*, showing chloroplasts (c), muciferous bodies (m), flagella, contractile vacuoles (cv), and the nucleus (n); (*b*) *Gonyostomum semen* with chloroplasts, trichocysts (t), contractile vacuoles, and the nucleus. (After Mignot, 1967.)

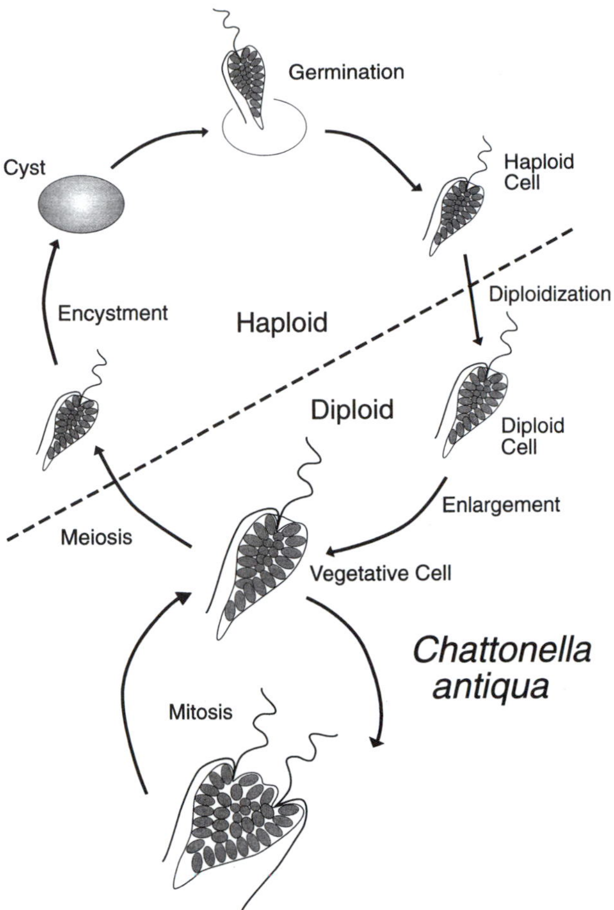

Fig. 14.3 The life cycle of *Chattonella antiqua*. (Adapted from Yamaguchi and Imai, 1994.)

The life cycle of *Chattonella antiqua* involves a vegetative propagation phase and a non-motile dormant phase (Fig. 14.3) (Yamaguchi and Imai, 1994). The vegetative diploid cells grow by binary fission under normal growth conditions. Small haploid cells are produced when the nutrients are depleted in the medium. These haploid cells change into cysts under low-light conditions and spend several months dormant in bottom sediments. The period of dormancy usually lasts from the end of summer to the following spring and is enforced by low temperatures (Imai and Itoh, 1987). Swarmers germinate from the cysts and somehow become diploid, although how diploidization occurs is not known. The resulting diploid vegetative cells complete the life history.

In freshwater, Raphidophyceae are associated with mud bottoms or ponds with abundant macrophytes. *Vacuolaria* contains chlorophylls *a* and *c* as well as the carotenoid β-carotene and the xanthophylls lutein epoxide and antheraxanthin (Chapman and Haxo, 1966; Guillard and Lorenzen, 1972). The numerous chloroplasts have three thylakoids per band, and

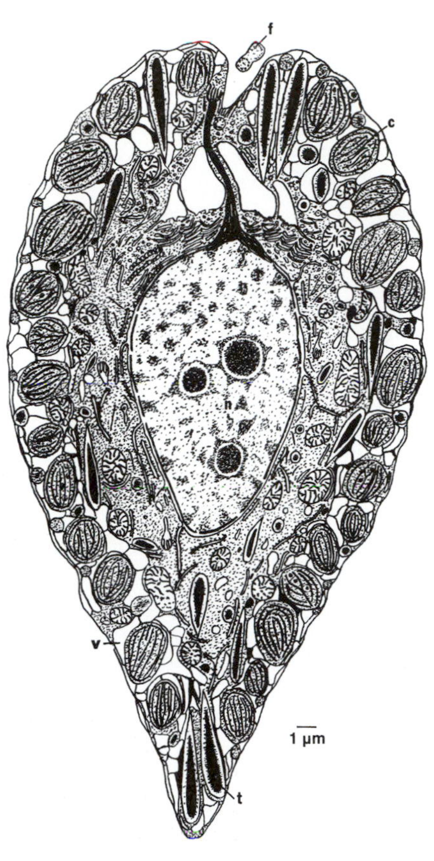

Fig. 14.4 *Gonyostomum semen*, general fine-structural organization, showing chromatophore (c), flagella (f), nucleus (n), trichocyst (t), and vacuole (v). (After Mignot, 1967.)

there are two membranes of chloroplast E.R. around the chloroplasts (Koch and Schnepf, 1967; Mignot, 1967). *Vacuolaria* (Fig. 14.2(*a*)) can be free-swimming or in a palmelloid state. In the palmelloid condition, the cells are more or less spherical, 35 to 50 μm in diameter, yellow-green with each cell surrounded by a thick gelatinous matrix. In the free-swimming state, the organisms vary considerably in shape from broadly ovate to elongate. The flagella emerge from a simple notch slightly to one side of the anterior ventral region. The anteriorly directed flagellum has microtubular hairs, whereas the posteriorly directed flagellum is whiplash. There is no stigma, but the organism is positively phototactic. *Vacuolaria* divides only in the palmelloid state. Mitosis is typically eukaryotic, with the nuclear membrane remaining intact (Heywood and Godward, 1972). No sexual reproduction has been reported in *Vacuolaria.* Occasionally the cells form cysts with the resistant sheaths surrounding the cells (Spencer, 1971).

Vacuolaria has muciferous bodies (mucocysts) in the peripheral cytoplasm (Fig. 14.2(*a*)), which are similar in structure to those of the Prymnesiophyceae and Chrysophyceae. The genus *Gonyostomum* (Figs. 14.2(*b*) and 14.4) has trichocysts, which appear to be the same as those in the Dinophyceae (Mignot, 1967).

References

Cavalier-Smith, T., and Chao, E. E. (1996). 18S rRNA sequence of *Heterosigma carterae* (Raphidophyceae), and the phylogeny of heterokont algae (Ochrophyta). *Phycologia* 35:500–10.

Chapman, D. J., and Haxo, F. T. (1966). Chloroplast pigments of the Chloromonadophyceae. *J. Phycol.* 2:89–91.

Guillard, R. R. L., and Lorenzen, C. J. (1972). Yellow-green algae with chlorophyllide *c*. *J. Phycol.* 8:10–14.

Hara, Y., and Chihara, M. (1985). Ultrastructure and taxonomy of *Fibrocapsa japonica* (Class Raphidophyceae). *Arch. Protistenk.* 130:133–41.

Heywood, P., and Godward, M. B. E. (1972). Centromeric organization in the chloromonadophycean alga *Vacuolaria virescens*. *Chromosoma* 39:333–9.

Imai, I., and Itoh, K. (1987). Annual life cycle of *Chattonella* spp., causative flagellates of noxious red tides in the Inland Sea of Japan. *Marine Biology* 94:287–92.

Khan, S., Arakawa, O., and Onoue, Y. (1996). A toxicological study of the marine phytoflagellate, *Chattonella antiqua* (Raphidophyceae). *Phycologia* 35:239–44.

Koch, W., and Schnepf, E. (1967). Einige elektronen-mikroskopische Beobachtungen an *Vacuolaria virescens* Ciénk. *Arch. Mikrobiol.* 57:196–8.

Leadbeater, B. S. C. (1969). A fine structural study of *Olisthodiscus luteus* Carter. *Br. Phycol. J.* 4:3–17.

Mignot, J-P. (1967). Structure et ultrastructure de quelques Chloromonadines. *Protistologia* 3:5–24.

Okaichi, T. (1989). Red tide problems in the Seto Inland Sea, Japan. In *Red Tides: Biology, Environmental Science and Toxicology*, ed. T. Okaichi, D. M. Anderson, and T. Nemoto, pp. 137–42. New York: Elsevier Science Publishing.

Spencer, L. B. (1971). A study of *Vacuolaria virescens* Cienkowski. *J. Phycol.* 7:274–9.

Taylor, F. J. R. (1992). The taxonomy of harmful marine phytoplankton. *G. Bot. Ital.* 126:209–19.

Tomas, C. R. (1979). *Olisthodiscus luteus* (Chrysophyceae). III. Uptake and utilization of nitrogen and phosphorus. *J. Phycol.* 15:5–12.

Tomas, C. R. (1980). *Olisthodiscus luteus* (Chrysophyceae). V. Its occurrence, abundance and dynamics in Narragansett Bay, Rhode Island, *J. Phycol.* 16:157–66.

Vesk, M., and Moestrup, Ø. (1987). The flagellar root system in *Heterosigma akashiwo* (Raphidophyceae). *Protoplasma* 137:15–28.

Wada, M., Hara, Y., Kato, M., Yamada, M., and Fujii, T. (1987). Diurnal appearance, fine structure and chemical composition of fatty particles in *Heterosigma akashiwo* (Raphidophyceae). *Protoplasma* 137:134–9.

Watanabe, M., Kohata, K., and Kunugi, M. (1988). Phosphate accumulation and metabolism by *Heterosigma akashiwo* (Raphidophyceae) during diel vertical migration in a stratified microsm. *J. Phycol.* 24:22–8.

Yamaguchi, M., and Imai, I. (1994). A microfluorometric analysis of nuclear DNA at different stages in the life history of *Chattonella antiqua* and *Chattonella marina* (Raphidophyceae). *Phycologia* 33:163–70.

15 · Heterokontophyta

Xanthophyceae

The Xanthophyceae contain primarily freshwater algae with a few marine representatives. The class is characterized by motile cells with a forwardly directed tinsel flagellum and a posterior directed whiplash flagellum (Fig. 15.1). The chloroplasts contain chlorophyll *a* and *c* (Sullivan et al., 1990), lack fucoxanthin, and are colored yellowish-green. The eyespot in motile cells is always in the chloroplast, and the chloroplasts are surrounded by two membranes of chloroplast endoplasmic reticulum. The outer membrane of the chloroplast E.R. is usually continuous with the outer membrane of the nucleus. In most cells the wall is composed of two overlapping halves. Molecular data has shown that the Xanthophyceae is most closely related to the Phaeophyceae (Ariztia et al., 1991; Potter et al., 1997). Although the class is commonly called the Xanthophyceae, the proper name is the Tribophyceae since there is no genus in the class that can lend its name to Xanthophyceae (Hibberd, 1981).

Cell structure

CELL WALL

The cell walls of two Xanthophyceae, *Tribonema aequale* (Cleare and Percival, 1972) and *Vaucheria* spp. (Parker et al., 1963), are composed of cellulose. In *Vaucheria* spp., cellulose comprises 90% of the wall, with the remaining portion being amorphous polysaccharides composed primarily of glucose and uronic acids.

Many of the organisms in the class have walls that are composed of two overlapping halves that fit together as do the two parts of the bacteriologist's Petri dish. The two-part nature of the wall cannot be delineated with the light microscope unless the cells have been treated with certain reagents such as concentrated potassium hydroxide. A typical example is the wall of *Ophiocytium majus* (one study places *Ophiocytium* in the closely related Eustigmatophyceae based on the sequence of mitochondrial cytochrome oxidase (Ehara et al., 1997)), which is tubular in shape (Fig. 15.2). The wall is composed of two parts, a cap of constant size fitted over a tubular basal portion. As the cell grows and increases in length, the tubular basal portion elongates, but the cap remains the same size. The rims of both the cap and

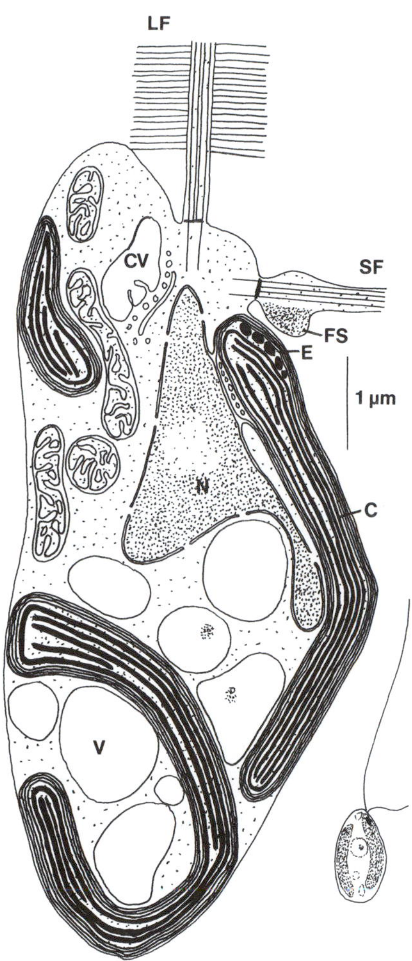

Fig. 15.1 Light and electron microscopical drawing of a zoospore of a typical member of the Xanthophyceae, *Mischococcus sphaerocephalus*. (C) Chloroplast; (CV) contractile vacuole; (E) eyespot; (FS) flagellar swelling; (LF) long flagellum with hairs; (N) nucleus; (SF) short flagellum; (V) vacuole. (Adapted from Hibberd and Leedale, 1971.)

the tube are finely tapered and overlap for some distance so that the bipartite condition is not normally apparent in living cells. A layer of intercalary material, presumably functioning as a cementing substance, separates the cap from the tubular part of the wall. Filamentous genera, such as *Tribonema* (Fig. 15.3), have a wall composed of H-shaped pieces. These alternately overlap each other so that each protoplast is enclosed by halves of two successive H-pieces (Lokhorst and Star, 1988).

CHLOROPLASTS AND FOOD RESERVES

Two membranes of chloroplast E.R. surround the chloroplasts, the outer membrane of chloroplast E.R. being continuous with the outer membrane of the nuclear envelope (Hibberd and Leedale, 1971). The thylakoids are grouped into bands of three, and in many genera there is a pyrenoid in the chloroplast (Marchant, 1972). The eyespot consists of a layer of globules

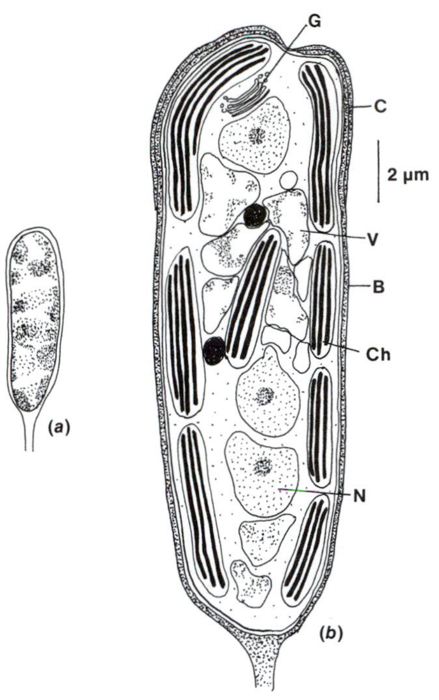

Fig. 15.2 *Ophiocytium majus*. (*a*) Vegetative cell. (*b*) Semidiagrammatic drawing of the fine structure of a vegetative cell. (B) Basal tubular portion of wall; (C) cap of wall; (Ch) chloroplast; (G) Golgi; (N) nucleus; (V) vesicle. (Adapted from Hibberd and Leedale, 1971.)

beneath the chloroplast envelope at the anterior end of the chloroplast (Fig. 15.1). Where the short flagellum passes over the eyespot, the flagellar sheath is dilated into the flagellar swelling, which is closely applied to the plasmalemma in the area of the eyespot.

Chlorophyll *a* and *c* are present in the chloroplasts (Sullivan et al., 1990), with the major carotenoids being diadinoxanthin, heteroxanthin and vaucheriaxanthin ester.

Mannitol and glucose accumulate during photosynthesis in the plastids (Cleare and Percival, 1973). The principal storage product is probably a β-1,3 linked glucan similar to paramylon, although lipids have been suggested as also being important.

Asexual reproduction

Xanthophycean organisms multiply asexually by fragmentation, zoospores, and aplanospores. In addition, they have the ability to form specialized resting spores. Fragmentation is limited to the tetrasporine and filamentous colonies, and is due to the breaking of the colony into parts.

Zoospores are formed by a majority of the genera. The zoospores are biflagellate, with the forward tinsel flagellum usually being four to six times longer than the shorter, posterior whiplash flagellum (Fig. 15.1). The zoospores are naked and usually pyriform (pear-shaped). Zoospore production

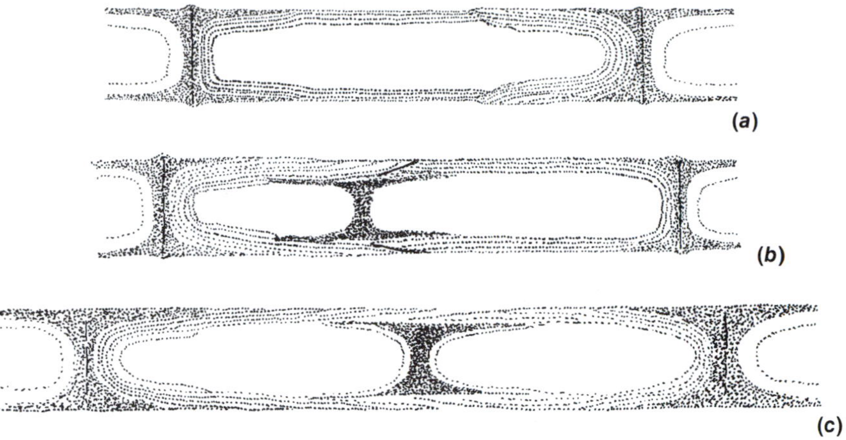

Fig. 15.3 Wall structure of *Tribonema bombycinum* after treatment with potassium hydroxide. (*a*) Two H-pieces articulated to enclose a single protoplast. (*b*), (*c*) Recently divided cell showing the intercalation of a new H-piece. (After Smith, 1938.)

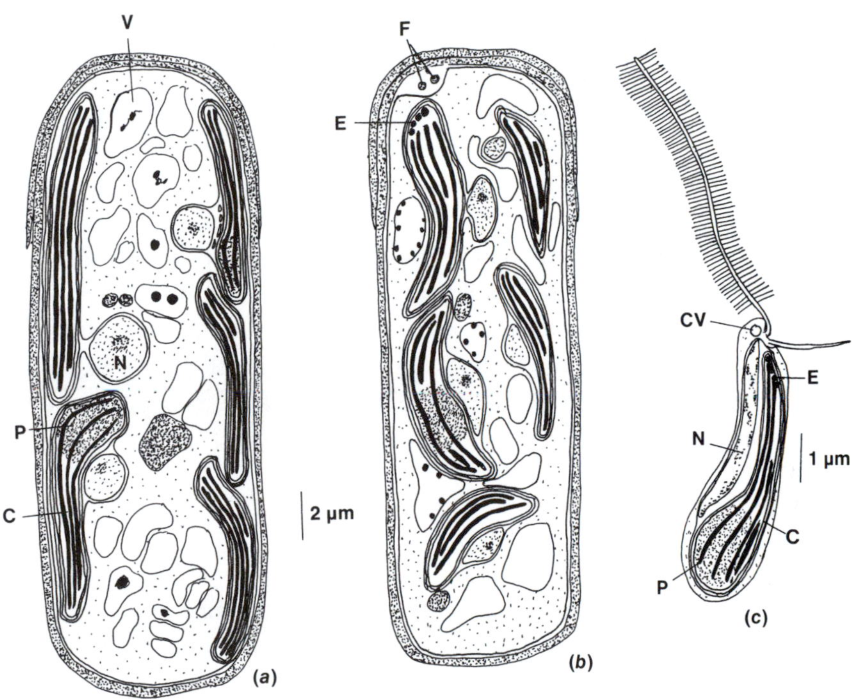

Fig. 15.4 *Pseudobumilleriopsis pyrenoidosa*. (*a*) Vegetative cell. (*b*) Cell undergoing zoosporogenesis. (*c*) Zoospore; (C) chloroplast; (CV) contractile vacuole; (E) eyespot; (F) flagella; (N) nucleus; (P) pyrenoid; (V) vacuole. (Adapted from Deason, 1971.)

has been studied at the fine-structural level in *Pseudobumilleriopsis pyrenoidosa* by Deason (1971) (Fig. 15.4). This alga has rod-shaped cells with several nuclei and laminate chloroplasts. Vegetative cells prior to zoosporogenesis have the nuclei and vacuoles in the center of the cell, whereas the chloroplasts are flattened against the plasmalemma. The first indication of cleavage in zoosporogenesis is the appearance of vacuoles between the ends of adjacent chloroplasts. The chloroplasts move away from the plasmalemma, and each becomes associated with a nucleus. The vacuoles then coalesce and separate the nucleus–chloroplast pairs, each of which becomes a zoospore. Basal bodies are present near the nuclei of the vegetative cells; the basal bodies migrate to one end of the chloroplast as cleavage begins and produce flagella early in zoosporogenesis. One to 16 zoospores are produced, which are released by dissolution and/or separation of the sporangial walls where they overlap. In the zoospore, the chloroplast is massive and has a pyrenoid. The nucleus is elongate, and there are two or more contractile vacuoles present.

Instead of producing zoospores, the entire protoplast may produce a single aplanospore or divide into a number of parts, each of which becomes an aplanospore. In some cases, environmental conditions determine whether the alga reproduces by zoospores or aplanospores. Submerged thalli of *Botrydium* produce zoospores; those living on damp soil produce aplanospores (Rostafiński and Woronin, 1877) (Fig. 15.5). An aplanospore liberated from a parent cell can grow directly into a new plant, or it may give rise to zoospores, which develop into new plants.

Some flagellates and rhizopodial cells produce cysts or statospores endogenously, similar to those in the Chrysophyceae. In their formation, there is an internal delimitation of a spherical protoplast that is separated from the peripheral portion of the mother cell's protoplast by a membrane. The endogenously differentiated protoplast then secretes a wall with two overlapping halves. Vegetative cells can also change directly into sporelike resting stages, with much thicker walls and more abundant food reserves than the vegetative cells. These sporelike cells, in which the spore wall is not distinct from the parent wall, are called akinetes and are usually found in filamentous genera.

Sexual reproduction

There are few reliable reports of sexual reproduction in the Xanthophyceae. Sexual reproduction has only been established in three genera: *Botrydium* (Fig. 15.5), *Tribonema*, and *Vaucheria* (Fig. 15.7). In the first two genera, both gametes are flagellated, whereas in *Vaucheria* reproduction is oogamous.

Classification

Three of the orders will be considered here.

Order 1 Tribonematales: filamentous organisms, not coenocytic.
Order 2 Botrydiales: Globose multinucleate thallus with colorless rhizoids.
Order 3 Vaucheriales: Filamentous coenocyte.

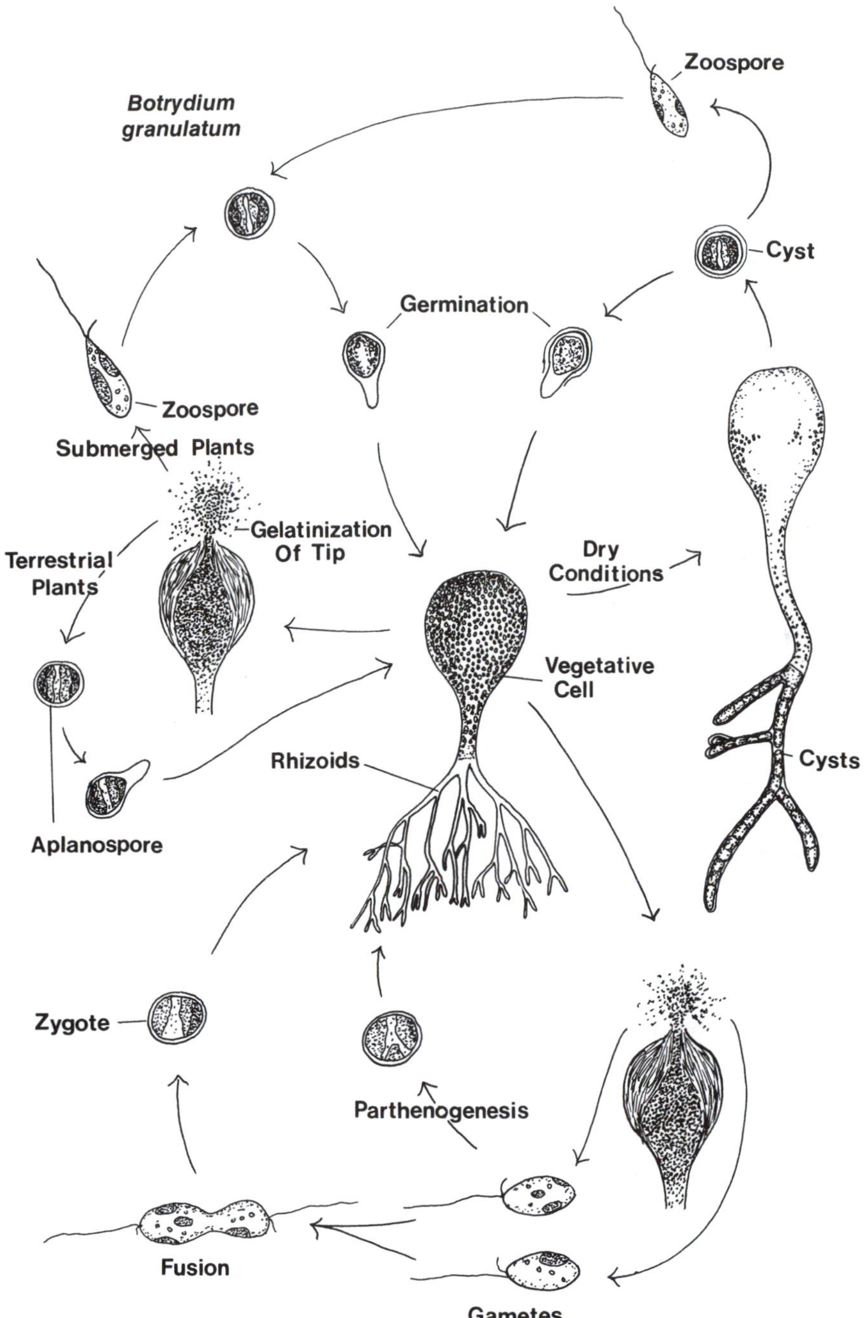

Fig. 15.5 The life cycle of *Botrydium granulatum*. (Adapted from Rostafiński and Woronin, 1877; Kolkwitz, 1926; Rosenberg, 1930.)

TRIBONEMATALES

The algae in this order have cylindrical cells uniseriately united end to end in branched or unbranched filaments. *Tribonema* is composed of barrel-shaped cells that are two to five times longer than they are wide (Fig. 15.3). The wall is composed of two H-shaped pieces overlapping in the middle of the cell. The protoplast is uninucleate and contains a number of discoid chloroplasts. Asexual reproduction is by fragmentation of the filaments, by zoospores, or by aplanospores. Aplanospores are produced more frequently than zoospores and are released by the pulling apart of the two portions of the cell wall. Akinetes can also be formed by the filaments (Smith, 1950). Sexual reproduction is isogamous, one of a uniting pair of gametes coming to rest and withdrawing its flagella just before the other swims up to it and unites with it (Scherffel, 1901).

BOTRYDIALES

Botrydium is a unicellular multinucleate alga consisting of a usually globose aerial portion with chloroplasts and a colorless rhizoidal portion that penetrates the soil (Fig. 15.5). The shape of the aerial part is influenced by environmental conditions. It is usually elongate when growing in shaded habitats and spherical when growing in brightly illuminated places (Moore and Carter, 1926). The aerial portion has a relatively tough wall within which is a delicate layer of cytoplasm containing many nuclei and chloroplasts. The branched rhizoidal system has no chloroplasts but does have many nuclei. The cells are incapable of vegetative division, and the only method by which new plants may be formed is by production of zoospores or aplanospores. According to Rakován and Fridvalsky (1970), the formation of either aplanospores or zoospores begins at night, and the cells must be illuminated 8 to 9 hours after the beginning of the process for flagella to develop. If there is no illumination, aplanospores develop. In the formation of these spores, three to five chloroplasts become associated with a nucleus in the mother cell, and cleavage occurs with each zoospore containing the above organelles. If the spore is an aplanospore, then a wall is secreted; if it is a zoospore, no wall is formed. Motile gametes are apparently formed in a similar manner (Iyengar, 1925; Miller, 1927), and sexual reproduction can be isogamous or anisogamous, with the cells being either homothallic or heterothallic. The gametes become apposed by their anterior ends when uniting in pairs to form a zygote. Gametes that have not fused develop parthenogenetically into thalli. A germinating zygote develops directly into a new vegetative thallus.

Botrydium also produces cysts or resting spores during periods of dry

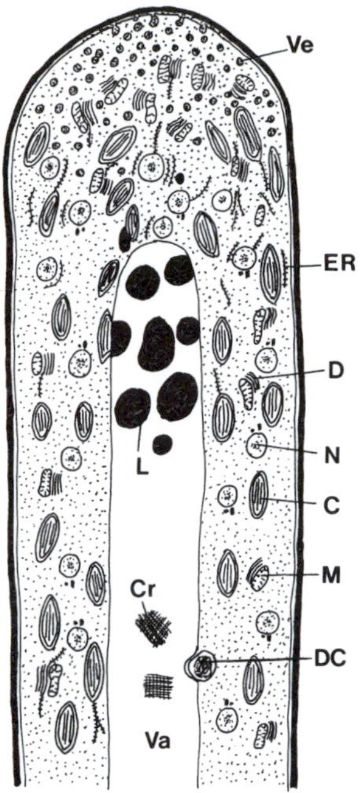

Fig. 15.6 Schematic representation of the tip of a vegatative filament of *Vaucheria dillwynii*. (C) Chloroplast; (Cr) crystal; (D) dictyosome; (DC) degenerate chloroplast; (ER) endoplasmic reticulum; (L) lipid body; (M) mitochordrion; (N) nucleus; (Va) vacuole; (Ve) vesicle. (After Ott and Brown, 1974a.)

conditions (Miller, 1927). In *B. granulatum*, the protoplast migrates into the rhizoids where division occurs to produce a large number of thick-walled cysts. These cysts can either germinate directly to form a new thallus or give rise to zoospores.

VAUCHERIALES

There is only one genus, *Vaucheria*, in this order. *Vaucheria* is a sparsely branched coenocytic alga that can be either terrestrial or aquatic. *Vaucheria* has a relatively thin cell wall within which the cytoplasm is restricted to the periphery of the coenocyte, with the center being occupied by a large central vacuole (Fig. 15.6). In the cytoplasm the numerous elliptical chloroplasts with pyrenoids are to the outside, whereas the nuclei are toward the center. Growth of the filaments is restricted to the apex which has a large number of vesicles, mitochondria, and dictyosomes. Chloroplasts, nuclei, and the large central vacuole are not found at the growing tip (Ott and Brown, 1974a). The large central vacuole contains lipids, degenerated chloroplasts, and crystals and extends the entire length of the filament except for the area immediately behind the growing tip. Cytoplasmic streaming takes place in the area of the large central vacuole and directly involves the nuclei, mitochondria, and their associated dictysomes. The cytoplasmic streaming involves two separate systems, the first based on microtubules that move the nuclei, and the second based on microfilaments that move the mitochondria and their associated dictyosomes. The chloroplasts do not migrate in patterns of definite streaming but have a more or less random movement, not associated with either microtubules or microfilaments.

Although *Vaucheria* can develop transverse septa that block off injured portions of the coenocyte, there is little reproduction by accidental breaking of filaments. Asexual reproduction of aquatic individuals is usually by means of multiflagellate, multinucleate zoospores (Birckner, 1912) (Fig. 15.7), which are produced singly in club-shaped sporangia at the swollen ends of filaments. In their production large numbers of chloroplasts and nuclei stream into the tip of the filament, the central vacuole decreases in size, and the tips appear dark green. A band of colorless protoplasm now appears at the base of the developing sporangium, which breaks in the middle to form two protoplasts. The two protoplasts approach each other, and a septum is formed. Within the sporangium, vesicles are produced (Ott and Brown, 1974b), around which nuclei become oriented with a pair of basal bodies between each nucleus and the vesicle membrane. Flagella are produced through the vesicle membrane, and the vesicles migrate to the plasmalemma. The nuclei with their flagella pairs thus come to lie in the

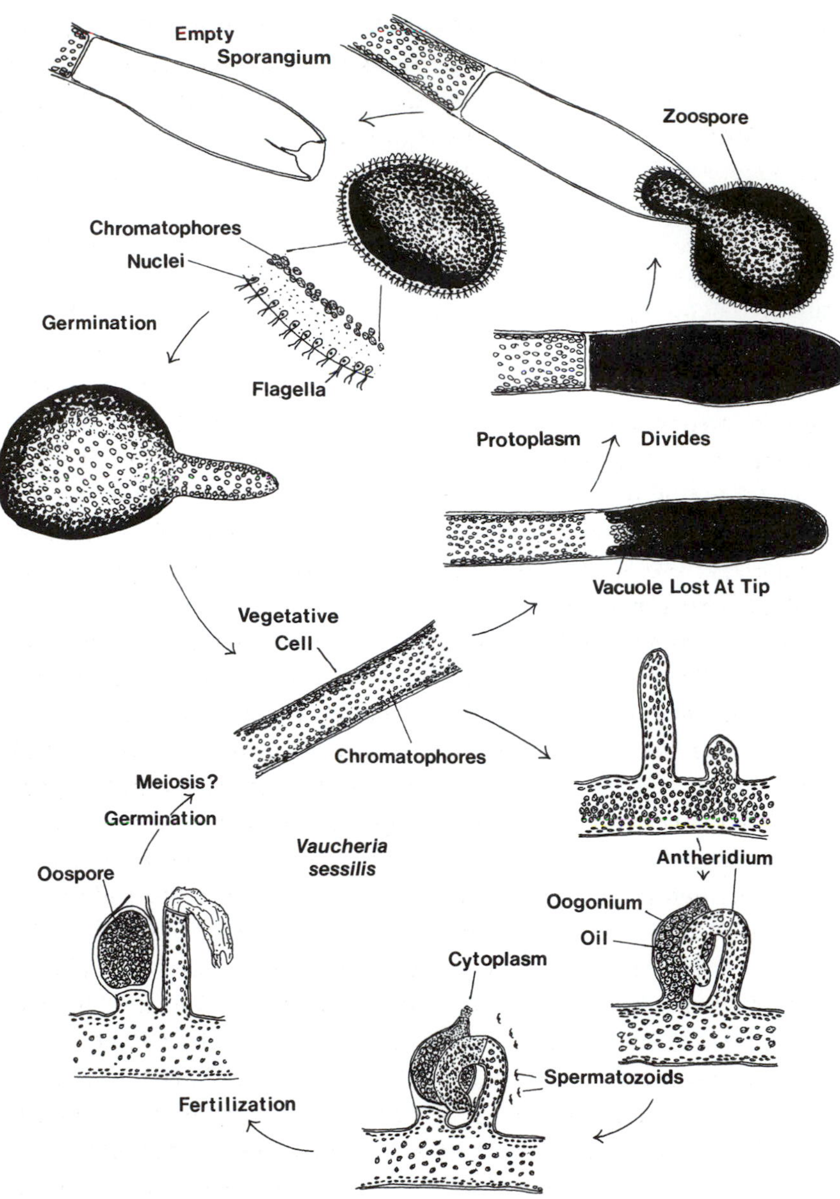

Fig. 15.7 The life cycle of *Vaucheria sessilis*.

peripheral area of the cell. The wall at the apex of the sporangium gelatinizes, forming a narrow aperture; the zoospore pushes its way through the aperture and swims in the medium. The nuclei in the sporangium are separated from each other by a number of vacuoles, and one flagellum of each pair is slightly longer than the other (Greenwood et al., 1957). There is no eyespot or pyrenoid in the zoospore (Greenwood, 1959). The zoospores are sluggish in their movements, swimming for only about 15 minutes. On coming to rest, the flagella are withdrawn, and a thin wall is secreted.

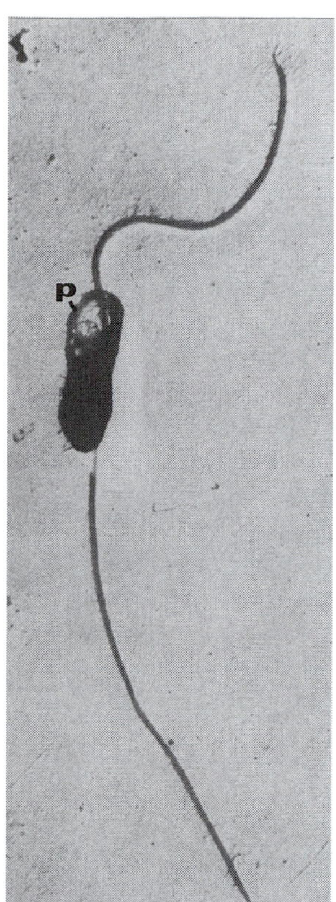

Fig. 15.8 Shadowcast whole mount of a sperm of *Vaucheria synandra*. A proboscis (p) is present on the anterior part of the cell. (From Moestrup, 1970.)

Germination follows almost immediately by the protrusion of one or two tubular outgrowths, one of which attaches itself to the substratum by means of a colorless lobed holdfast. Instead of the production of zoospores, terrestrial individuals may have the entire contents of the sporangium develop into an aplanospore. Zoospores can be obtained if vegetative filaments kept moist for some days are soaked in water, or transferred from a nutritive solution into distilled water, or removed from running water to still water (Klebs, 1896; Starr, 1964).

Sexual reproduction is oogamous (Fig. 15.7), the cells usually being homothallic. Sex organs are common on filaments growing in damp soil or in quiet water, but are infrequent if they are growing in flowing water. The antheridia and oogonia are borne adjacent to each other and on a common lateral branch or on adjacent lateral branches. The sex organs are cut off by a septum. The oogonium has a single egg and is filled with oil and chloroplasts. The mature oogonium produces a beak, the tip of which gelatinizes, forming a aperture. A portion of colorless cytoplasm of the egg projects through the aperture, and the egg contracts.

The antheridia usually develop as strongly curved cylindrical tubes that become cut off by a septum, usually fairly high up in the tube. The mature antheridium has the spermatozoids produced in a specific area between the central and peripheral cytoplasm (Moestrup, 1970). The central and peripheral cytoplasm contain those parts of the cytoplasm that are not included in the spermatozoids: the chloroplasts, vacuoles, and many mitochondria (Ott and Brown, 1978). An aperture appears in the antheridium, and the spermatozoids are released. The spermatozoids are cylindrical posteriorly but have a flattened proboscis in the anterior portion (Fig. 15.8). There is a forward-projecting tinsel flagellum with two lateral rows of hairs, and a slightly longer trailing smooth flagellum. The nucleus is elongated and wormlike, as are the three or four mitochondria. There is neither a chloroplast nor an eyespot, but there is a Golgi body near the basal bodies of the flagella. The proboscis consists of eight or nine microtubules running beneath the plasmalemma with vesicles in between the microtubules.

Fertilization is accomplished by the spermatozoids fusing with the egg protoplasm through the aperture in the oogonium. The zygote secretes a wall, and the oil droplets fuse to form a small number of central droplets. The oospore is colored by the oil and the degeneration products of chlorophyll. It remains in the oogonium until it is liberated by the decay of the oogonial wall. The oospore then remains dormant for a few months before germinating, probably by meiosis, into a new filament.

Vaucheria is one of the algae that exhibit chloroplast orientation movements in the light. In the dark, the chloroplasts are uniformly distributed in the peripheral cytoplasm (Fig. 15.9); in low-intensity light, they are oriented

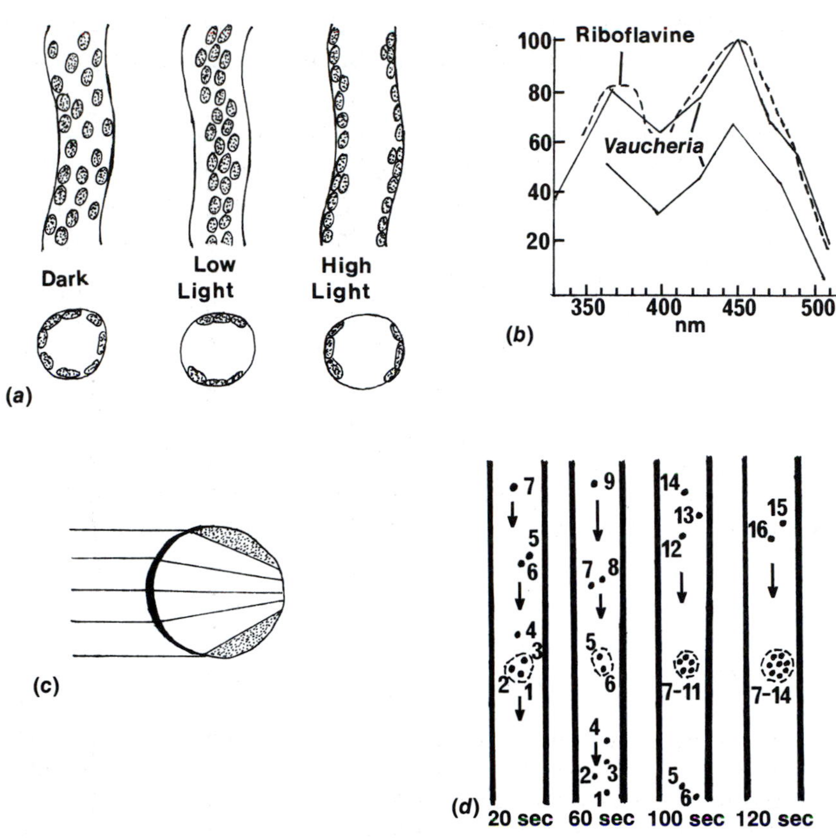

Fig. 15.9 (*a*) Position of chloroplasts of *Vaucheria* in the dark, and in low light and high light intensities. (*b*) Action spectrum of chloroplast orientation movement in *Vaucheria* under low light (lower curve) and high light (upper curve) intensities. Vertical axis is relative quantum efficiency. (*c*) Diagrammatic representation of the lens effect of light passing through a *Vaucheria* coenocyte. (*d*) Movements of single chloroplasts in *Vaucheria* under low light intensity. A beam of light (dashed area) causes accumulation of chloroplasts in the illuminated area. (After Haupt and Schönbohm, 1970.)

to the top and bottom of the coenocyte to trap the maximum light; in high light intensities, they are along the sides of the filament, thereby receiving less light (Fischer-Arnold, 1963; for a review, see Haupt and Schönbohm, 1970). The action spectrum for chloroplast movement in *Vaucheria* is similar to that of higher plants and closely resembles that of the action spectrum of flavins, from which it has been concluded that flavins located in the plasma membrane act as the photoreceptive pigments (Blatt, 1983). That the photoreceptor is not the chloroplast can be demonstrated with a microbeam of light. Irradiation with a microbeam results in chloroplast movement, whether or not the chloroplasts receive light. In addition to being able to perceive light, the cell also has the ability to determine the direction of the light. The theoretical basis for this perception is derived from a lens effect by the cell on the light that it receives (Fig. 15.9). The cell acts as a collecting lens, focusing light to the rear of the cell. This results in the light's bypassing some parts of the cell along the flanks, thus establishing a front-to-rear gradient. The movement of the chloroplasts relies on energy from two metabolic processes, respiration and photosynthesis. Inhibitors of respiration and photosynthesis abolish the chloroplast movement, whereas inhibitors of photosynthesis just slow down the chloroplast movement. If ATP is added along with photosynthetic inhibitors, chloroplast movement is normal.

There are two possible mechanisms to explain chloroplast movement. In the first or "active movement" the chloroplast moves relative to the rest of the protoplasm, whereas in the second or "passive movement" the protoplasm moves, carrying with it the chloroplast and other organelles. In *Vaucheria*, passive movement occurs. If chloroplast movement is followed at high or low intensity, it can be seen that not only chloroplasts, but also other organelles and inclusions are rearranged by light. Furthermore, if only a small area of the filament is irradiated with a spot of light, a strong accumulation of cytoplasm plus inclusions can be observed at this place (Fig. 15.9). The organelles are actually trapped in a portion of the cytoplasm as they stream through the cell. Illumination of an area of *Vaucheria* results in the formation of an actin fiber network that acts as a trapping mechanism (Blatt, 1983).

References

Ariztia, E. V., Andersen, R. A., and Sogin, M. L. (1991). A new phylogeny for chromophyte algae using 16S-like rRNA sequences from *Mallomonas papillosa* (Synurophyceae) and *Tribonema aequales* (Xanthophyceae). *J. Phycol.* 27:428–36.

Birckner, V. (1912). Die Beobachtung von Zoosporenbildung bei *Vaucheria aversa* Hass. *Flora* 104:167–71.

Blatt, M. R. (1983). The action spectrum for chloroplast movements and evidence for blue-light–photoreceptor cycling in the *Vaucheria*. *Planta* 159:267–76.

Cleare, M., and Percival, E. (1972). Carbohydrates of the freshwater alga *Tribonema aequale*. I. Low molecular weight and polysaccharides. *Br. Phycol. J.* 7:185–93.

Cleare, M., and Percival, E. (1973). Carbohydrates of the freshwater alga *Tribonema aequale*. II. Preliminary photosynthetic studies with ^{14}C. *Br. Phycol. J.* 8:181–4.

Deason, T. R. (1971). The fine structure of sporogenesis in the Xanthophycean alga *Pseudobumilleriopsis pyrenoidosa*. *J. Phycol.* 7:101–7.

Descomps, S. (1963). Observations sur l'infrastructure de l'enveloppe des chloroplastes de *Vaucheria* (Xanthophycees). *C.R. Séances Acad. Sci., Paris* 257:727–9.

Ehara, M., Hayashi-Ishimura, Y., Inagaki, Y., and Ohama, T. (1997). Use of a deviant mitochondrial genetic code in yellow-green algae as a landmark for segregating members within the phylum. *J. Mol. Evol.* 45:119–24.

Falk, H. (1967). Zum Feinbau von *Botrydium granulatum* Grev. (Xanthophyceae). *Arch. Mikrobiol.* 58:212–27.

Falk, H., and Kleining, H. (1968). Feinbau und Carotenoide von *Tribonema* (Xanthophyceae). *Arch. Mikrobiol.* 61:347–62.

Fischer-Arnold, G. (1963). Untersuchungen über die Chloroplastenbewegung bei *Vaucheria sessilis*. *Protoplasma* 56:495–506.

Greenwood, A. D. (1959). Observations on the structure of the zoospores of *Vaucheria*. II. *J. Exp. Bot.* 10:55–68.

Greenwood, A.D., Manton, I., and Clarke, B. (1957). Observations on the structure of the zoospores of *Vaucheria. J. Exp. Bot.* 8:71–86.

Haupt, W., and Schönbohm, E. (1970). Light-oriented chloroplast movements. In *Photobiology of Microorganisms*; ed. P. Halldal, pp. 283–307. London: Wiley-Interscience.

Hibberd, D. J. (1981). Notes on the taxonomy and nomenclature of the algal classes Eustigmatophyceae and Tribophyceae (synonym Xanthophyceae). *Bot. J. Linn. Soc.* 82:93–119.

Hibberd, D. J., and Leedale, G. F. (1971). Cytology and ultrastructure of the Xanthophyceae. II. The zoospore and vegetative cell of coccoid forms, with special reference to *Ophiocytium majus* Naegeli. *Br. Phycol. J.* 6:1–23.

Iyengar, M. O. P. (1925). Note on two species of *Botrydium* from India. *J. Indian Bot. Soc.* 4:193–201.

Klebs, G. (1896). *Die Bedingungen de Fortpflanzung bei einigen Algen und Pilzen.* Jena.

Kolkwitz, R. (1926). Zur Ökologie und Systematic von *Botrydium granulatum* (L) Grev. *Ber. Dtsch. Bot. Ges.* 44:533–40.

Lokhorst, G. M., and Star, W. (1988). Mitosis and cytokinesis in *Tribonema regulare* (Tribophyceae, Chrysophyta). *Protoplasma* 145:7–15.

Luther, A. (1899). Ueber *Chlorosaccus*, eine neue Gattung der Süsswasseralgen, nebst einigen Bemerkungen zur Systematic verwandter Algen. *Bih. Kgl. Svensk. Vetersk.-Ak. Handl.* 24, Afd. 3, No. 13:1–22.

Maekawa, F. (1953). [Title in Japanese.] *J. Jpn. Bot.* 28:105–10.

Marchant, H. J. (1972). Pyrenoids of *Vaucheria woroniniana.* Heering. *Br. Phycol. J.* 7:81–4.

Miller, V. (1927). Untersuchungen über die Gattung *Botrydium* Wallroth. II. Spezieller Teil. *Ber Dtsch. Bot. Ges.* 45:161–70.

Moestrup, Ø. (1970). On the fine structure of the spermatozoids of *Vaucheria sescuplicaria* and on later stages in spermatogenesis. *J. Mar. Biol. Assoc. UK* 50:513–23.

Moore, G. T., and Carter, N. (1926). Further studies on the subterranean algal flora of the Missouri Botanical Garden. *Ann. Mo. Bot. Gard.* 13:101–40.

Ott, D. W., and Brown, R. M., Jr. (1974a). Developmental cytology of the genus *Vaucheria.* I. Organization of the vegetative filament. *Br. Phycol. J.* 9:111–26.

Ott, D. W., and Brown, R. M., Jr. (1974b). Developmental cytology of the genus *Vaucheria.* II. Sporogenesis in *V. fontinalis* (L) Christensen. *Br. Phycol. J.* 9:333–51.

Ott, D. W., and Brown, R. M. (1978). Developmental cytology of the genus *Vaucheria.* IV. Spermatogenesis. *Br. Phycol. J.* 13:69–85.

Parker, B. C., Preston, R. D., and Fogg, G. E. (1963). Studies of the structure and chemical composition of the cell walls of Vaucheriaceae and Saprolegniaceae. *Proc. R. Soc. Lond.* [*B*] 158:435–45.

Potter, D., Saunders, G. W., and Andersen, R. A. (1997). Phylogenetic relationships of the Raphidophyceae and Xanthophyceae as inferred from nucleotide sequences of the 18S ribosomal RNA gene. *Am. J. Bot.* 84:966–72.

Rakován, J. N., and Fridvalsky, L. (1970). Electron microscope studies on the gonidiogenesis of *Botrydium granulatum* (L) Grev. (Xanthophyceae). *Ann. Univ. Sci. Budap. Sect. Biol.* 12:209–12.

Rosenberg, M. (1930). Die geschlechtliche Fortpflanzung von *Botrydium granulatum* Grev. *Oesterr. Bot. Z.* 79:289–96.

Rostafiński, J., and Woronin, M. (1877). Ueber *Botrydium granulatum. Bot. Zig.* 35:649–71.

Scherffel, A. (1901). Kleiner Beitrag zur Phylogenie einiger Gruppen neiderer Organismen. *Bot. Ztg.* 59:143–58.

Smith, G. M. (1938). *Cryptogamic Botany*, Vol. 1. *Algae and Fungi*. New York and London: McGraw-Hill.

Smith, G. M. (1950). *The Fresh-Water Algae of the United States*, 2nd ed. New York and London: McGraw Hill.

Starr, R. C. (1964). The culture collection of algae at Indiana University. *Am. J. Bot.* 51:1013–44.

Sullivan, C. M., Entwisle, T. J., and Rowan, K. S. (1990). The identification of chlorophyll *c* in the Tribophyceae (=Xanthophyceae) using spectrophotofluorometry. *Phycologia* 29:285–91.

16 · Heterokontophyta

Eustigmatophyceae

The Eustigmatophyceae is a class that is closely related to the Xanthophyceae, but differs in having an eyespot outside of the chloroplast instead of inside the chloroplast as occurs in the Xanthophyceae (Fig. 16.1) (Hibberd and Leedale, 1970). Other characteristics of the class include a basal swelling of the tinsel flagellum adjacent to the eyespot, only chlorophyll *a*, chloroplasts without girdle lamellae and no peripheral ring of DNA, and chloroplast endoplasmic reticulum not connected to the nuclear envelope (Schnepf et al., 1996).

The name Eustigmatophyceae was chosen (Hibberd and Leedale, 1971)

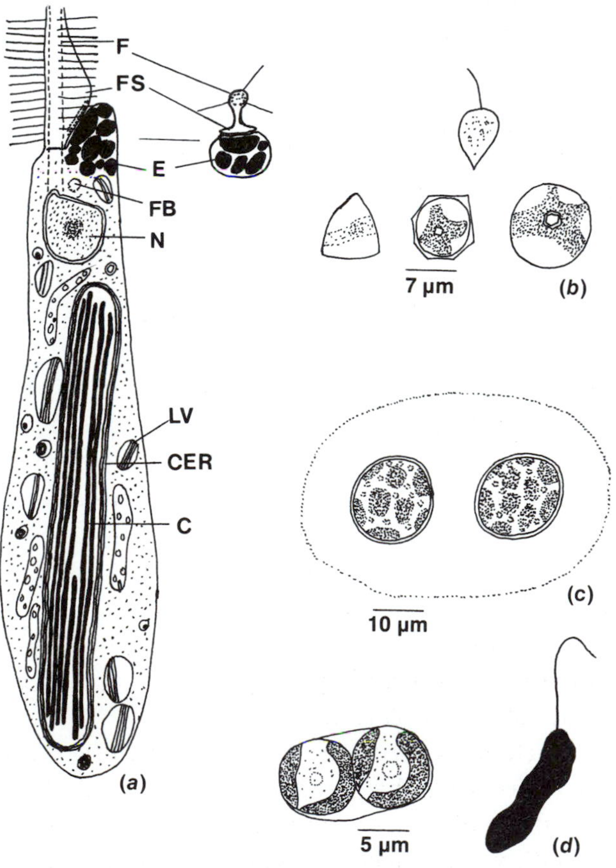

Fig. 16.1 (*a*) Diagrammatic representation of the basic morphology of a zoospore of the Eustigmatophyceae. (C) Chloroplast; (CER) chloroplast endoplasmic reticulum; (E) eyespot; (F) long flagellum; (FB) basal body of short flagellum; (FS) flagellar swelling; (LV) lamellate vesicles; (N) nucleus. (*b*) *Polyhedriella helvetica*, zoospore and vegetative cells. (*c*) *Chlorobotrys regularis*. (*d*) *Pleurochloris magna*, vegetative cell and zoospore. ((*a*) after Hibberd and Leedale, 1972; (*b*) after Fritsch and John, 1942; (*c*) after Smith, 1950; (*d*) after Boye Petersen, 1932.)

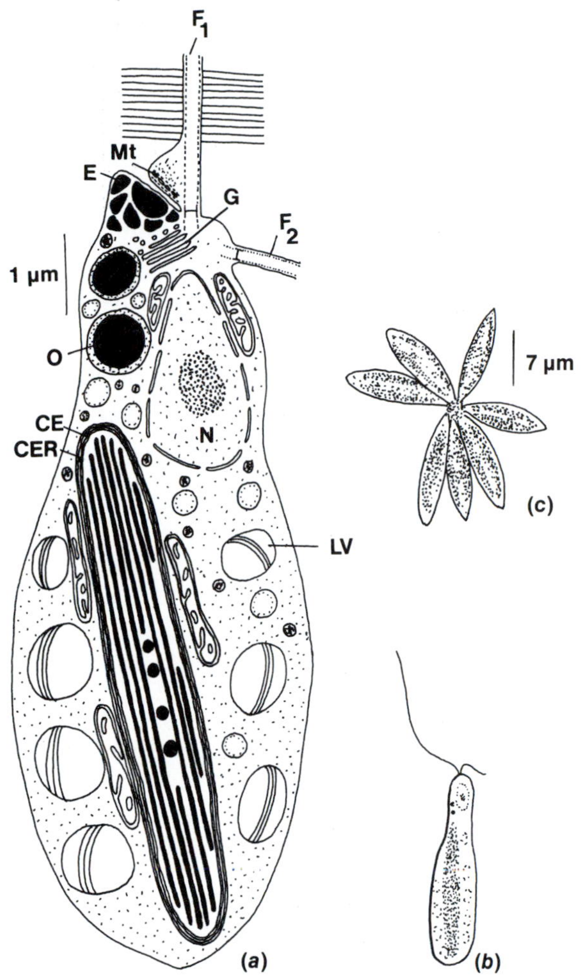

Fig. 16.2 *Pseudocharaciopsis texensis*: (*a*), (*b*) zoospores; (*c*) vegetative cells. (CE) Chloroplast envelope; (CER) chloroplast endoplasmic reticulum; (E) eyespot; (F_1) long flagellum; (F_2) short flagellum; (G) Golgi; (LV) lamellate vacuoles; (Mt) microtubules; (N) nucleus; (O) oil body. (After Lee and Bold, 1973.)

because of the large size of the eyespot in the zoospore (Figs. 16.1 and 16.2). This is the primary difference compared to cells of the Xanthophyceae. The eyespot is a large orange-red body at the anterior end of the motile cell, and is completely independent of the chloroplast. It consists of an irregular group of droplets with no membrane around the whole complex of droplets. At the base of the tinsel flagellum the flagellar sheath is extended to form a T-shaped flagellar swelling (Figs. 16.1 and 16.2), which is always closely appressed to the plasmalemma in the region of the eyespot. In turn, in the eyespot there is a large droplet closely applied to the plasmalemma in the area of the flagellar swelling.

The chloroplasts of the Eustigmatophyceae have chlorophyll *a* and β-carotene, with the two major xanthophylls being violaxanthin and vaucheriaxanthin (Whittle and Casselton, 1969; Antia and Cheng, 1982), the only difference in pigments compared to the Xanthophyceae being the presence of violaxanthin and the absence of antheraxanthin. Violaxanthin is the major light-harvesting pigment in the Eustigmatophyceae (Owens et al., 1987).

Most of the species produce zoospores with only a single emergent flagellum (*Pleurochloris commutata; P. magna*, Fig. 16.1(*d*); *Polyedriella helvetica*, Fig. 16.1(*b*); *Vischeria punctata*; and *V. stellata*, Hibberd and Leedale, 1972), but there is a second basal body present, indicating that the cells had a biflagellate ancestor. The emergent flagellum is tinsel with microtubular hairs, and the flagellum is inserted subapically. Two of the algae in the class, *Ellipsoidion acuminatum* and *Pseudocharaciopsis taxensis* (Lee and Bold, 1973), have zoospores with a long forward tinsel flagellum and a short posteriorly directed smooth flagellum. One of the organisms, *Chlorobotrys regularis* (Fig. 16.1(*c*)), does not form zoospores (Hibberd, 1974).

References

Antia, N. J., and Cheng, J. Y. (1982). The keto-carotenoids of two marine coccoid members of the Eustigmatophyceae. *Br. Phycol. J.* 17:39–50.

Boye Petersen, J. (1932). Einge neue Erdalgen. *Arch. Protistenk.* 76:395–408.

Fritsch, F. E., and John, R. P. (1942). An ecological and taxonomic study of the algae of British soils. II. Consideration of the species observed. I. Chlorophyceae. *Ann. Bot. N.S.* 6:371–95.

Hibberd, D. J. (1974). Observations on the cytology and ultrastructure of *Chlorobotrys regularis* (West) Bohlin with special reference to its taxonomic position in the Eustigmatophyceae. *Br. Phycol. J.* 9:37–46.

Hibberd, D. J., and Leedale, G. F. (1970). Eustigmatophyceae – a new algal class with unique organization of the motile cell. *Nature* 225:758–60.

Hibberd, D. J., and Leedale, G. F. (1971). A new algal class, the Eustigmatophyceae. *Taxon* 20:523–25.

Hibberd, D. J., and Leedale, G. F. (1972). Observations on the cytology and ultrastructure of the new algal class Eustigmatophyceae. *Ann. Bot. N.S.* 36:49–71.

Lee, K. W., and Bold, H. C. (1973). *Pseudocharaciopsis texensis* gen. et sp. nov., a new member of the Eustigmatophyceae. *Br. Phycol. J.* 8:31–7.

Owens, T. G., Gallagher, J. C., and Alberte, R. S. (1987). Photosynthetic light-harvesting function of violaxanthin in *Nannochloropsis* spp. (Eustigmatophyceae). *J. Phycol.* 23:79–85.

Schnepf, E., Niemann, A., and Wilhelm, C. (1996). *Pseudostaurastrum limneticum*, a Eustigmatophycean alga with astigmatic zoospores: morphogenesis, fine structure, pigment composition and taxonomy. *Arch. Protistenk.* 146:237–49.

Smith, G. M. (1950). *The Freshwater Algae of the United States*, 2nd ed. New York and London: McGraw-Hill.

Whittle, S. J., and Casselton, P. J. (1969). The chloroplast pigments of some green and yellow-green algae. *Br. Phycol. J.* 4:55–64.

17 · Heterokontophyta

Phaeophyceae

The Phaeophyceae, or brown algae, derive their characteristic color from the large amounts of the carotenoid fucoxanthin in their chloroplasts as well as from any phaeophycean tannins that might be present. The chloroplasts also have chlorophylls *a*, c_1, and c_2. There are two membranes of chloroplast E.R., which are usually continuous with the outer membrane of the nuclear envelope. The storage product is laminarin. There are no unicellular or colonial organisms in the order, and the algae are basically filamentous, pseudoparenchymatous, or parenchymatous. They are found almost exclusively in the marine habitat, there being only four genera containing freshwater species, that is, *Heribaudiella*, *Pleurocladia*, *Bodanella*, and *Sphacelaria* (Fig. 17.1) (Schloesser and Blum, 1980). A number of marine forms penetrate into brackish water, where they often form an important part of the salt marsh flora. These brackish water plants have almost totally lost the ability to reproduce sexually, and propagate by vegetative means only. Most of the Phaeophyceae grow in the intertidal belt and the upper littoral region. They dominate these regions in colder waters, particularly in the Northern Hemisphere, where the number of phaeophycean species is less than that of the Rhodophyceae, but the number of phaeophycean plants is much greater. In the tropics, the only place where large numbers of Phaeophyceae are found is the Sargasso Sea of the Atlantic.

The Phaeophyceae probably evolved from an organism in the Chrysomeridales of the Chrysophyceae, which have motile cells similar to those in the Phaeophyceae (O'Kelley, 1989). The algae in the Chrysomeridales, however, lack unilocular and plurilocular sporangia and plasmodesmata, and therefore cannot be classified in the Phaeophyceae.

Cell structure

The cell structure is in many ways similar to that of the Chrysophyceae, Prymnesiophyceae, Bacillariophyceae, and Xanthophyceae, which are closely related to the Phaeophyceae. The main difference lies in the large amounts of extracellular polysaccharides surrounding the protoplast.

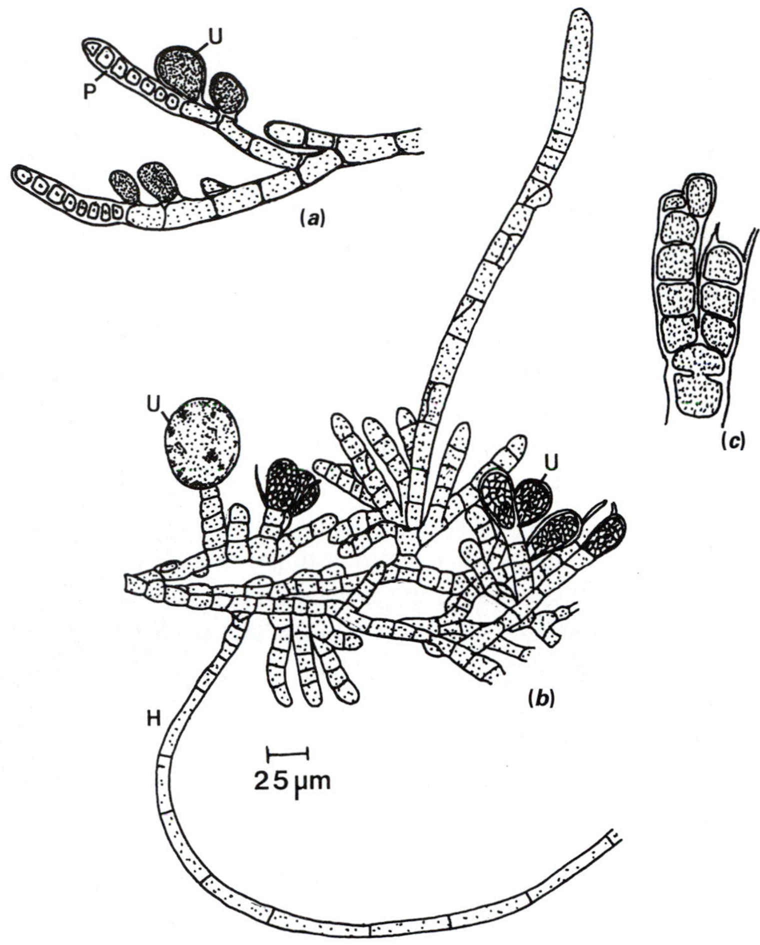

Fig. 17.1 Some freshwater brown algae. (*a*) *Pleurocladia lacustris*. (*b*) *Sphacelaria lacustris*. (*c*) *Heribaudiella fluviatilis*. (H) Hair; (P) pleurilocular sporangia; (U) unilocular sporangium. ((*b*) after Schloesser and Blum, 1980.)

CELL WALLS

Phaeophycean cell walls are generally composed of at least two layers, with **cellulose** making up the main structural skeleton (Kloareg and Quatrano, 1988). The amorphous component of the cell wall is made up of **alginic acid** and **fucoidin**, whereas the mucilage and cuticle are composed primarily of alginic acid (Evans and Holligan, 1972a; Vreeland, 1972). Alginic acid is basically made up of β-1,4 linked mannuronic acid units that have a variable amount of guluronic acid units attached through C-1 and C-4 linkages (Fig. 1.7). Fucoidin is composed primarily of α-1,2 linked sulfated fucose units, with a lesser amount of α-1,4 linked sulfated fucose units (Fig. 1.7). The relative quantities of alginic acid and fucoidin vary between different species, different parts of the plant, and different environments.

Calcification of the wall occurs only in some species of *Padina* where

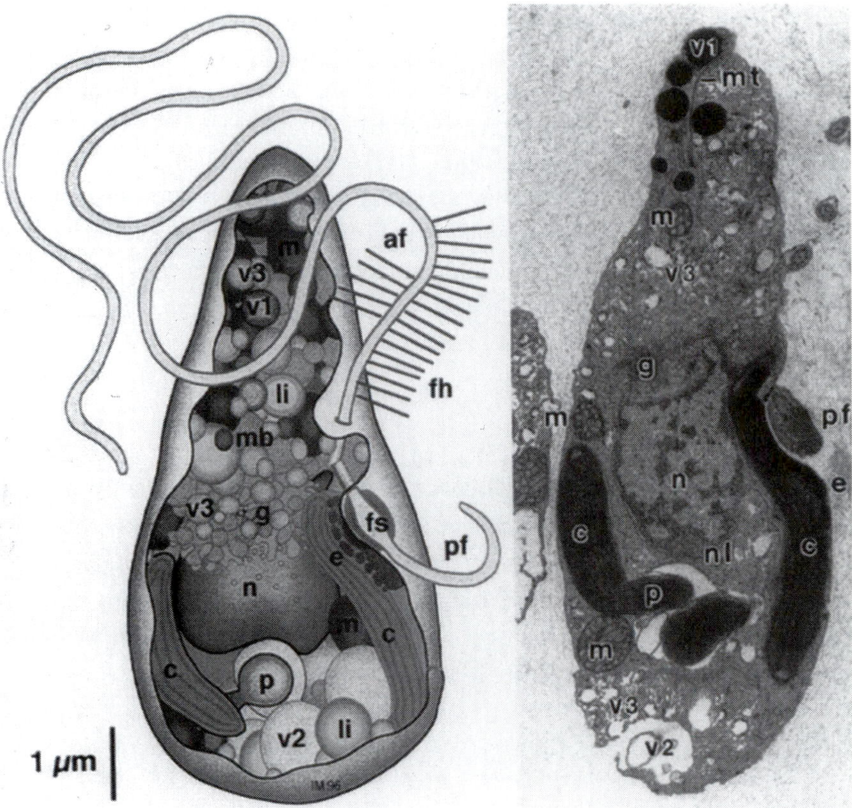

Fig. 17.2 *Left:* Diagrammatic representation of a male gamete of *Ectocarpus siliculosus* showing the distribution of cellular organelles. *Right:* Transmission electron micrograph of a thin section of a male gamete of *E. siliculosus*. (af) Anterior flagellum; (c) chloroplast; (e) eyespot; (fh) flagellar hairs (present along entire length, for clarity only shown on part of the flagellum); (fs) proximal swelling of the posterior flagellum; (g) Golgi apparatus; (li) lipid body; (m) mitochondrion; (mb) microbody; (n) nucleus; (p) pyrenoid; (pf) posterior flagellum; (v1) physode; (v2) storage granule; (v3) vesicles with cell wall or adhesive material. (From Maier, 1997a.)

calcium carbonate is deposited as needle-shaped crystals of aragonite in concentric bands on the surface of the fanlike thallus (Borowitzka et al., 1974).

The parenchymatous Phaeophyceae have plasmodesmata or pores between most of the cells. These pores are bounded by the plasmalemma, and protoplasm is continuous from one cell to the next through them. In the Laminariales, Fucales, and Dictyotales, the pores are grouped in primary pit areas, whereas in the more primitive parenchymatous Phaeophyceae the plasmodesmata are scattered in the cell wall (Bisalputra, 1966; Bourne and Cole, 1968; Cole, 1970).

FLAGELLA AND EYESPOT

Generally the motile cells of the Phaeophyceae (always zoospores or gametes, as there are no motile vegetative cells) have a long anterior tinsel flagellum with tripartite hairs and a shorter posteriorly directed whiplash flagellum (Fig. 17.2) (Bouck, 1969; Loiseaux and West, 1970). The Fucales are an exception to this, with the posterior flagellum of the spermatozoid being longer than the anterior flagellum. The posterior flagellum usually has a swelling near the base, and this swelling fits into a depression of the cell immediately above the eyespot. The eyespot consists of 40 to 80 lipid globules arranged in a single layer between the outermost band of the thylakoids, and the chloroplast envelope. The **eyespot** (**stigma**) acts as a concave mirror focusing light onto the flagellar swelling, which is the photoreceptor site for phototaxis in brown-algal flagellate cells (Kawai et al., 1996). Light at 420 and 460 nm is most effective in phototaxis in the brown algae, and is probably detected by a flavin-like substance in the flagellar swelling of the posterior flagellum (Kawai et al., 1991).

CHLOROPLASTS AND PHOTOSYNTHESIS

The chloroplasts of the Phaeophyceae have three thylakoids per band and are surrounded by the chloroplast envelope and two membranes of chloroplast E.R. (Fig. 17.3). The outer membrane of the chloroplast E.R. is generally continuous with the outer membrane of the nuclear envelope in the Ectocarpales but appears to be discontinuous in the Dictyotales, Laminariales, and Fucales. Membrane-bounded tubules are common in the area between the chloroplast E.R. and chloroplast envelope where the latter two are not closely appressed (Bouck, 1965; Evans, 1968). Microfibrils of DNA occur in the plastids, and in *Sphacelaria* sp. there is a ring-shaped genophore inside the outermost band of thylakoids (Bisalputra and Burton, 1969). The DNA microfibrils are both linear and circular and are attached to the thylakoid membranes. The plastids contain chlorophylls a, c_1, and c_2, with the major carotenoid being fucoxanthin.

All the phaeophycean orders have representatives with pyrenoids (Chi, 1971), but their presence, even in one species, can vary according to the stage of the plant. If the species is one that has pyrenoids only in some stages, then a pyrenoid is usually present in the eggs and/or sporelings but absent in the macroscopic phase, spermatozoids, and/or zoospores (Bourne and Cole, 1968; Evans, 1968; Bisalputra et al., 1971). A pyrenoid in the Phaeophyceae is usually a stalklike structure set off from the main body of the chloroplast and containing a granular substance not traversed by thylakoids. Surrounding the pyrenoid but outside the chloroplast E.R. is a membrane-bounded sac that presumably contains the reserve product. The long-term storage product is **laminarin**, a β-1,3 linked glucan. The sugar alcohol, D-**mannitol** is, however, the accumulation product (up to about 25% of the dry weight of some *Laminaria* species in the autumn) of photosynthesis.

In a number of brown algae, the mannitol concentration in the cell increases or decreases as the salinity of the surrounding medium increases

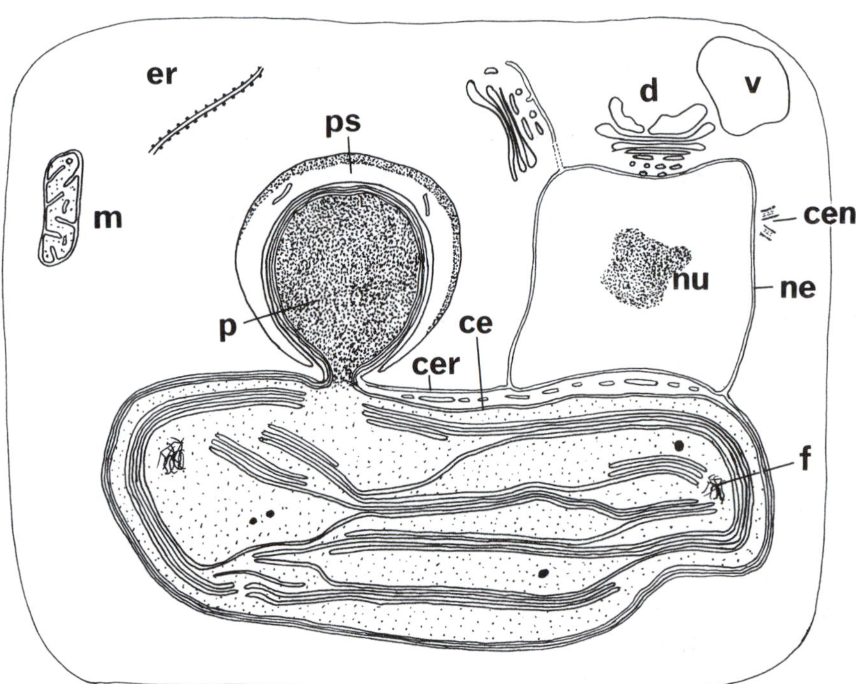

Fig. 17.3 Diagram of a hypothetical brown algal cell. (ce) Chloroplast envelope; (cen) centrioles; (cer) chloroplast endoplasmic reticulum; (d) dictyosome; (er) endoplasmic reticulum; (f) DNA fibrils; (m) mitochondrion; (ne) nuclear envelope; (nu) nucleolus; (p) pyrenoid; (ps) pyrenoid sac; (v) vacuole. (From Bouck, 1965.)

or decreases (Reed et al., 1985). This osmoregulatory mechanism prevents the cells from bursting in hypotonic media or shrinking in hypertonic media. The increase in mannitol concentration occurs in the dark as well as in the light, showing that photosynthesis is not involved in the process.

The brown algae are unique among the algae in having uptake of inorganic carbon, and therefore photosynthetic carbon fixation, stimulated by blue light (Forster and Dring, 1994). Most of the Phaeophyceae live in the littoral zone where they receive a generous amount of light. Photosynthesis is usually limited by the supply of inorganic carbon in this environment. The Phaeophyceae have evolved a mechanism to increase the amount of inorganic carbon uptake, but only when the cells are illuminated by blue light, thereby conserving the energy required for the process when the cells are in the dark. The only brown algae that do not have this mechanism are the fucoids, which appear to have evolved a separated carbon-concentrating mechanism.

PHYSODES AND PHAEOPHYCEAN TANNINS

Phaeophycean tannins (**fucosan**) are stored in **physodes** (**fucosan vesicles**) in the cytoplasm of many brown algae. These vesicles contain a colorless, highly refractive acidic fluid that stains red with vanillin and hydrochloric acid. The tannins are non-glycosidic (do not contain sugars), have strong reducing action, produce precipitates with lead acetate, and in aqueous solutions are astringent to the taste. Also they are readily oxidized in the air, resulting in the formation of a brown or black pigment, **phycophaein**, giving dried brown algae their characteristic black color. Experiments have shown that the yellow solutions of tannins produced by *Fucus vesiculosus* will inhibit the growth of a number of unicellular algae at tannin concentrations of 25 to 150 $\mu g\ ml^{-1}$ (McLachlan and Craigie, 1964). Toxicity of phenols apparently depends on a denaturation of proteins, resulting in lysis of cells. The function of the tannins and phenols produced by algal cells is probably to regulate the occurrence and abundance of endo- and epiphytes on the host plants (McLachlan and Craigie, 1966; Kiirikki, 1996).

The physode vesicles originate in the plastids where the tannins are evidently produced, and then extruded from the plastid area into the cytoplasm in a vesicle. Once in the cytoplasm the vesicles have the ability to fuse and form larger vesicles (Evans and Holligan, 1972b).

Eggs of the Phaeophyceae contain phenolic vesicles just under the plasma membrane that are discharged outside the cells by exocytosis after fertilization. It has been postulated that the discharge of these phenolic vesicles has a toxic effect on spermatozoids and acts as a polyspermy block before the primary wall is secreted (Clayton and Ashburner, 1994). These peripheral phenolic vesicles are distinguishable from physodes, which also contain phenolic compounds, but which are significantly larger and tend to be localized around the egg nucleus.

Life history

The unilocular sporangium is generally considered to be the site of meiosis, the haploid zoospores that are released forming the gametophyte generation. The gametophyte then produces the gametes, which fuse to form the zygote (Bell, 1997). Although meiotic divisions have been considered the rule in the unilocular sporangium, a disturbingly large number of investigations have not found this to be the case. In these investigations there is a "direct" type of life history, with no meiosis or fusion occurring. More research is needed in this area to clarify the situation. In the phaeophycean life cycle, a **plethysmothallus** is a filamentous stage (or one composed of compacted filaments) that can multiply itself by spores (usually zoospores from plurilocular sporangia) (Papenfuss, 1951).

The thallus of many Phaeophyceae is relatively large and complex with a number of different types of growth that include the following: (1) **diffuse**, with most of the cells of the plant capable of cell division; (2) **apical**, with a single cell at the apex giving rise to the cells beneath it; (3) **trichothallic**, in

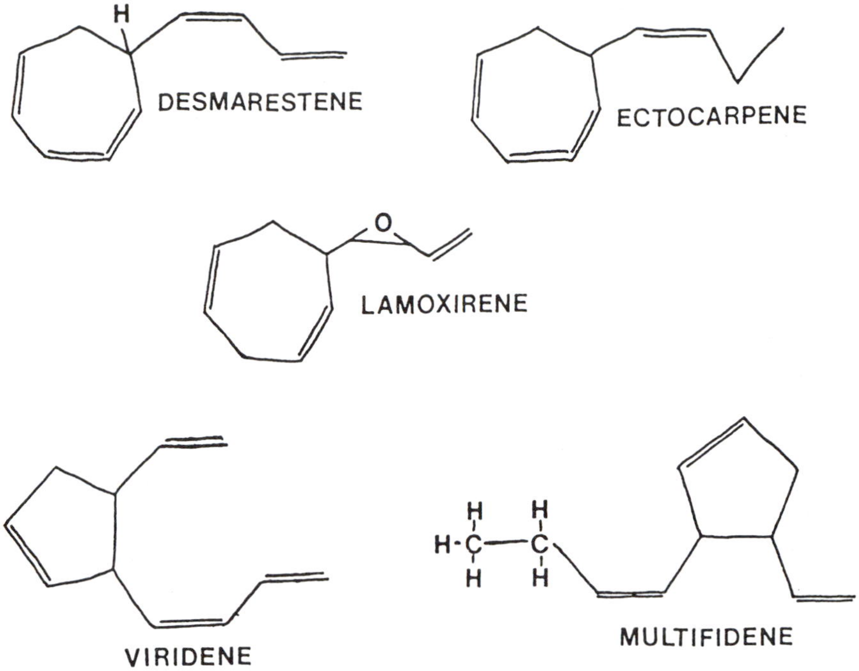

Fig. 17.4 The structure of some of the sexual hormones of the brown algae. (After Jaenicke et al., 1974; Müller et al., 1982; Müller et al., 1985b.)

which a cell divides to form a hair above and the thallus below; (4) **promeristem**, with a non-dividing apical cell controlling a large number of smaller, dividing promeristematic cells beneath it; (5) **intercalary**, with a zone of meristematic cells forming tissue above and below the meristem; (6) **meristoderm**, with a layer of usually peripheral cells dividing **periclinally** (parallel to the surface of the thallus) to form a tissue below the meristoderm (usually cortex) and occasionally dividing **anticlinally** (perpendicular to the surface of the thallus) to add more cells to the meristoderm. Some Phaeophyceae have division of daughter cells without subsequent enlargement to give a **polysiphonous** structure, or one with parallel vertically elongate cells in rows. Daughter cells can form laterals or branches that branch in a **distichous** manner (branches disposed in two equal rows). **Monopodial** growth is a type of growth with one main axis.

Investigations on the sexual hormones of the brown algae in the 1970s and 1980s by Müller and his associates represent the major recent advancement in the knowledge of this group (Müller, 1982; Maier and Müller, 1986). A **sexual hormone** (**sirenine**, **pheromone**) is a diffusible substance that coordinates cellular activities during sexual reproduction. Two types of biological effects mediated by sexual hormones occur in the brown algae: (1) the explosive discharge of spermatozoids from antheridia, and (2) the attraction of male gametes by female gametes or eggs. All sexual hormones in the brown algae are unsaturated hydrocarbons (have at least one double

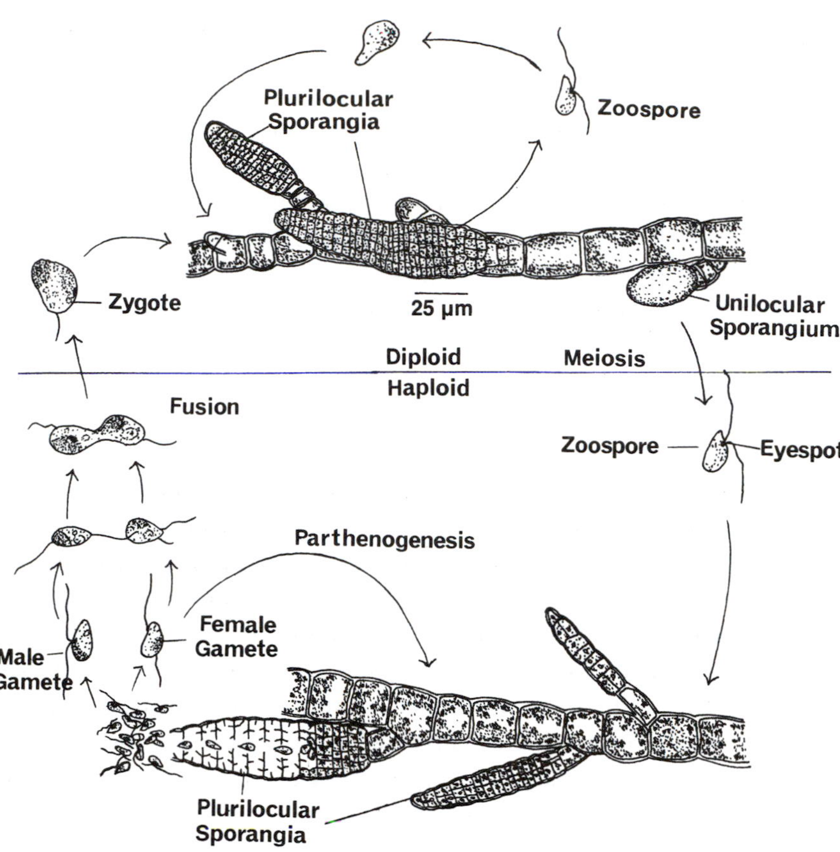

Fig. 17.5 The life cycle of *Ectocarpus siliculosus*.

with intercalary cell divisions confined to certain areas of the filaments. Some workers divide up the family into different genera on the basis of cytology and morphology, whereas others consider that the family contains the single genus *Ectocarpus* (Russell and Garbary, 1978).

The life cycle of *E. siliculosus* (Fig. 17.5) can be taken as representative of the family (Papenfuss, 1935). The haploid and diploid phases are both filamentous, but the diploid filaments have longer cells than the haploid filaments. The diploid plants produce unilocular and plurilocular sporangia either on the same plant or on separate plants. These sporangia discharge their zoospores between 0600 and 1200 hours. The mother cell of a unilocular sporangium can be distinguished from a branch initial by the spherical shape and large nucleus of the mother cell. The cell is initially vacuolate, but the physodes and vacuoles are soon extruded from the cell and become lodged in the wall (Loiseaux, 1973). The chloroplasts and nuclei of the unilocular sporangium divide in regular sequence, with the chloroplasts next to the wall and the nuclei in the center of the cell (Knight, 1929). The nuclei divide meiotically. A chloroplast then becomes associated with a nucleus, and a zoospore is delimited around it. A small perforation occurs at the apex of the unilocular sporangium, and up to 32 haploid zoospores ooze out of the sporangium in a gelatinous matrix. The perforation is small, and the zoospores are constricted to a dumbbell shape as they pass through the pore. The zoospores remain in a spherical mass at the apex of the sporangium for a short time, then become free and swim off individually. The zoospores seldom remain motile for more than 30 minutes and are relatively large, being twice the size of the gametes and zoospores from plurilocular sporangia. The zoospores germinate within 2 to 3 hours to produce haploid filaments.

The plurilocular organs are modified lateral branches that are divided into as many as 660 cubical cells, each containing a motile cell. The plurilocular sporangia on the diploid filaments produce zoospores that remain motile for 3 to 5 hours, settle, and within 2 to 5 hours germinate to produce diploid filaments like the parent. The germ tube of the sporeling arises from the narrow, anterior flagellated end of the zoospore, which is always oriented toward the light. The plurilocular organs on the haploid filaments are smaller than those on the diploid filaments, and produce either zoospores or gametes. The motile gametes are all of the same size but differ physiologically. The female gametes settle down about 5 minutes after liberation and secrete a sexual hormone called **ectocarpene** [all-*cis*-1-(cycloheptadien-2′,5′-yl)-1-butene] (Fig. 17.4) (Müller et al., 1971). Male gametes (Fig. 17.2) (Maier, 1997a,b) move very rapidly (269 μm per second) in a straight line in open seawater when no female gametes are around (Müller, 1978). The motile male gametes (which can remain motile for up to 8 hours) swim in circular paths on encountering ectocarpene, the diameter of the circular path decreasing in response to increasing ectocarpene concentration (Müller, 1982). As soon as the female gamete is reached, a firm contact is established between the apical part of the front flagellum of the male gamete and the plasma membrane of the female gamete. The posterior ends of the two gametes fuse to form the zygote. The process of fusion takes about 20 seconds, and after fusion the zygote loses its attraction for male gametes as indicated by the dispersion of the male gametes near the zygote. The zygotes take 2 to 3 days to germinate, and the sporelings develop more slowly than those from diploid zoospores. Some of the unfused gametes have the ability to germinate parthenogenetically to give rise to haploid filaments again. This germination is slow, requiring 36 to 48 hours. Clonal populations of *E. siliculosus* from different parts of the world are interfertile, indicating that there has been little genetic isolation of the species (Müller, 1979). Ectocarpene isolated from *E. siliculosus* attracts male gametes of two other species of *Ectocarpus*, indicating that ectocarpene is not species specific (Müller and Gassmann, 1980).

Ectocarpene can be artificially synthesized and is effective in attracting male gametes.

Ectocarpus has a fairly wide tolerance to changes in temperatures and salinity. Several studies (Boalch, 1961; Edwards, 1969) have shown that *E. siliculosus* will grow and produce sporangia at temperatures between 10 and 29 °C. Müller (1962) found that at 13 °C this species produces unilocular sporangia, at 19 °C plurilocular sporangia, and at 16 °C both types of sporangia. It has the ability to grow in salinities from 0.5 to 1.5 times that of seawater at 20 °C, and 0.25 to 1.75 times that of seawater at 15 °C. The alga is an obligate photoautotroph and will not grow on any supplied carbon source in the dark. Although it will not grow in the dark, it has the ability to survive up to 150 days of darkness and still remain viable.

Ralfsiaceae

These algae have a basal layer of branched, radiating, laterally coalesced filaments attached to the substratum by the cell wall or by rhizoids. From this basal layer, chloroplast-bearing filaments arise that are compacted together to give a firm tissue. *Ralfsia* (Fig. 17.6) probably has an isomorphic alternation of generations. The brown crustlike diploid plants produce unilocular sporangia at the base of loosely associated multicellular paraphyses. Zoospores are most likely produced by meiosis in the unilocular sporangium, and the zoospores give rise to haploid crusts. These gametophytes form plurilocular sporangia that are terminal on erect filaments and probably produce motile cells that can act as gametes or zoospores (Kylin, 1934; Edelstein et al., 1968; Hollenberg, 1969).

Some of the algae that have been placed in this family are the alternate phase of the life cycle of other higher algae and as such have been removed from the family.

Scytosiphonaceae

The Scytosiphonaceae, Dictyosiphonaceae, and Punctariaceae are sometimes grouped together in a separate order, the Dictyosiphonales. The members of these three families all have parenchymatous thalli resulting from diffuse or apical growth, and both gametes are motile. These algae, though, have many structural similarities to the more complex members of the Ectocarpales such as the Splachnidiaceae, and will be considered as members of the Ectocarpales, as suggested by Russell and Fletcher (1975).

In the Scytosiphonaceae, growth is diffuse although in some older plants growth can be suprabasal. The macroscopic phase of the plant produces plurilocular sporangia, but not unilocular sporangia. *Scytosiphon lomen-*

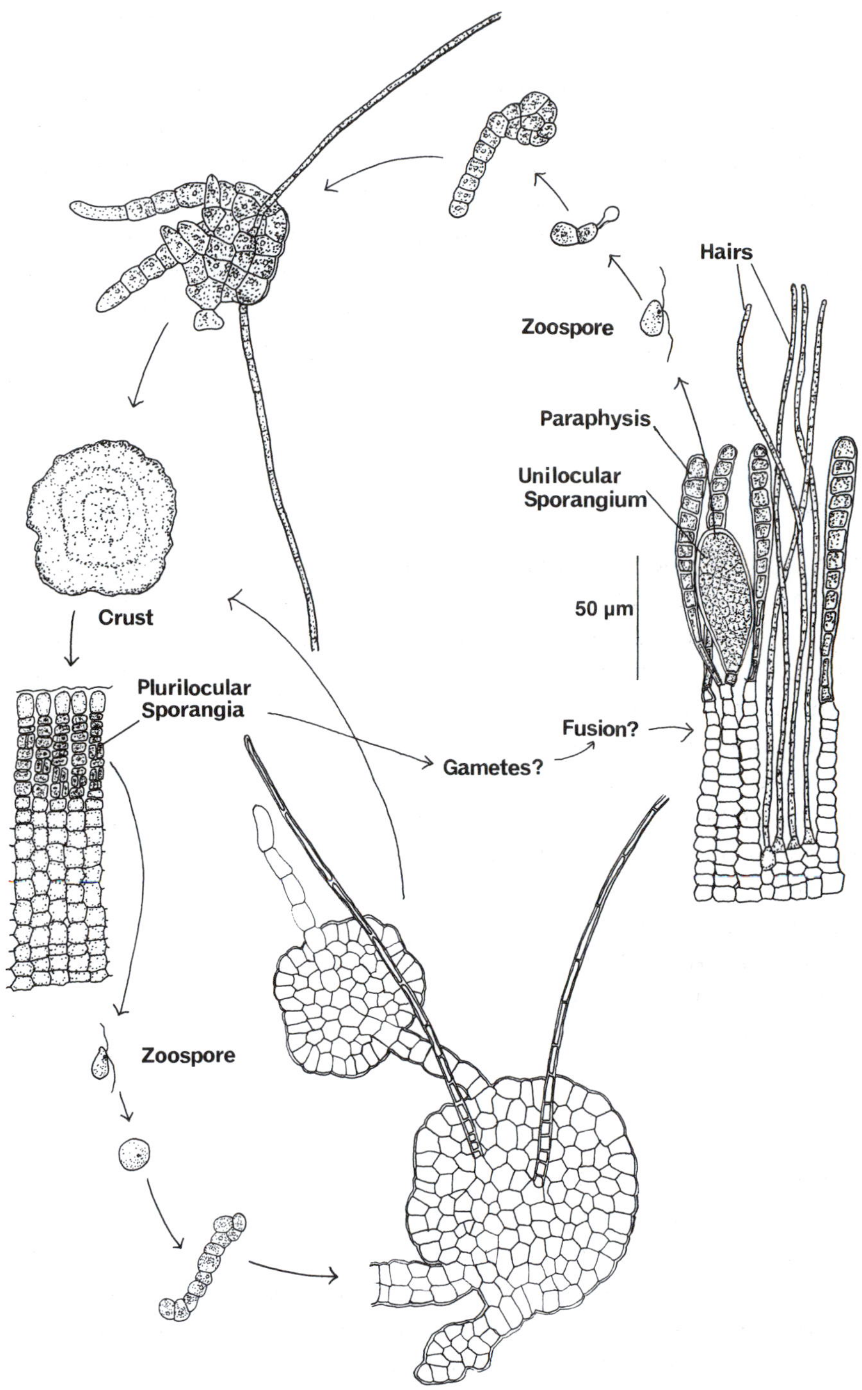

Fig. 17.6 The life cycle of *Ralfsia confusa*. (Adapted from Hollenberg, 1969.)

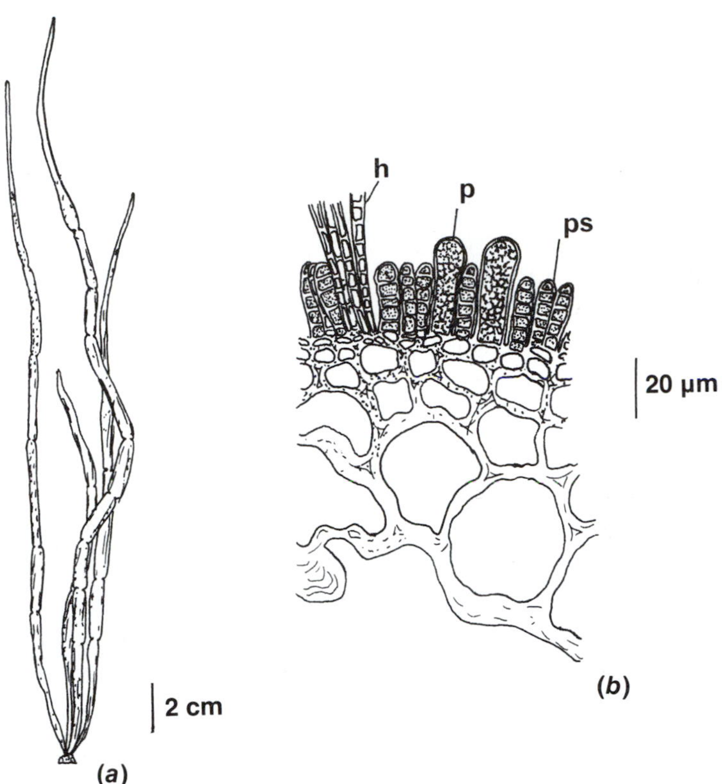

Fig. 17.7 *Scytosiphon lomentaria*. (*a*) Whole plant. (*b*) Portion of a section of the hollow plant showing hairs (h), paraphyses (p), and plurilocular sporangia (ps). (After Taylor, 1957).

taria (Fig. 17.7) is a common intertidal rock-pool alga. It has a narrow, cylindrical thallus up to 50 cm long, gradually tapering from apex to base. The thallus has occasional constrictions, and the plants grow in tufts with smaller individuals showing no or few constrictions. *Petalonia* (Fig. 17.8) has flat leafy fronds consisting of larger medullary cells covered with smaller cortical cells. In the North Atlantic, *Petalonia fascia* and *Scytosiphon lomentaria* occur in the same area. This area is limited on the north by the 0 °C summer isotherm and on the south by the 17 °C winter isotherm (Fig. 17.9) (van den Hoek, 1982).

The life cycle of such organisms as *Petalonia* and *Scytosiphon* has precipitated some controversy. The workers are more or less divided into the North American School (Wynne, 1969; Edelstein et al., 1970; Loiseaux, 1970; Kapraun and Boone, 1987), who claim that meiosis and fusion of gametes do not occur, and the Japanese and Australian school (Nakamura, 1965, 1972; Tatewaki, 1966; Clayton, 1980), who believe that meiosis occurs in the unilocular sporangia and that the motile cells from the plurilocular sporangia function as gametes. The difference in results may be due to different local strains of the same alga. Both groups agree, however, that there are two different morphological phases, a macroscopic phase that

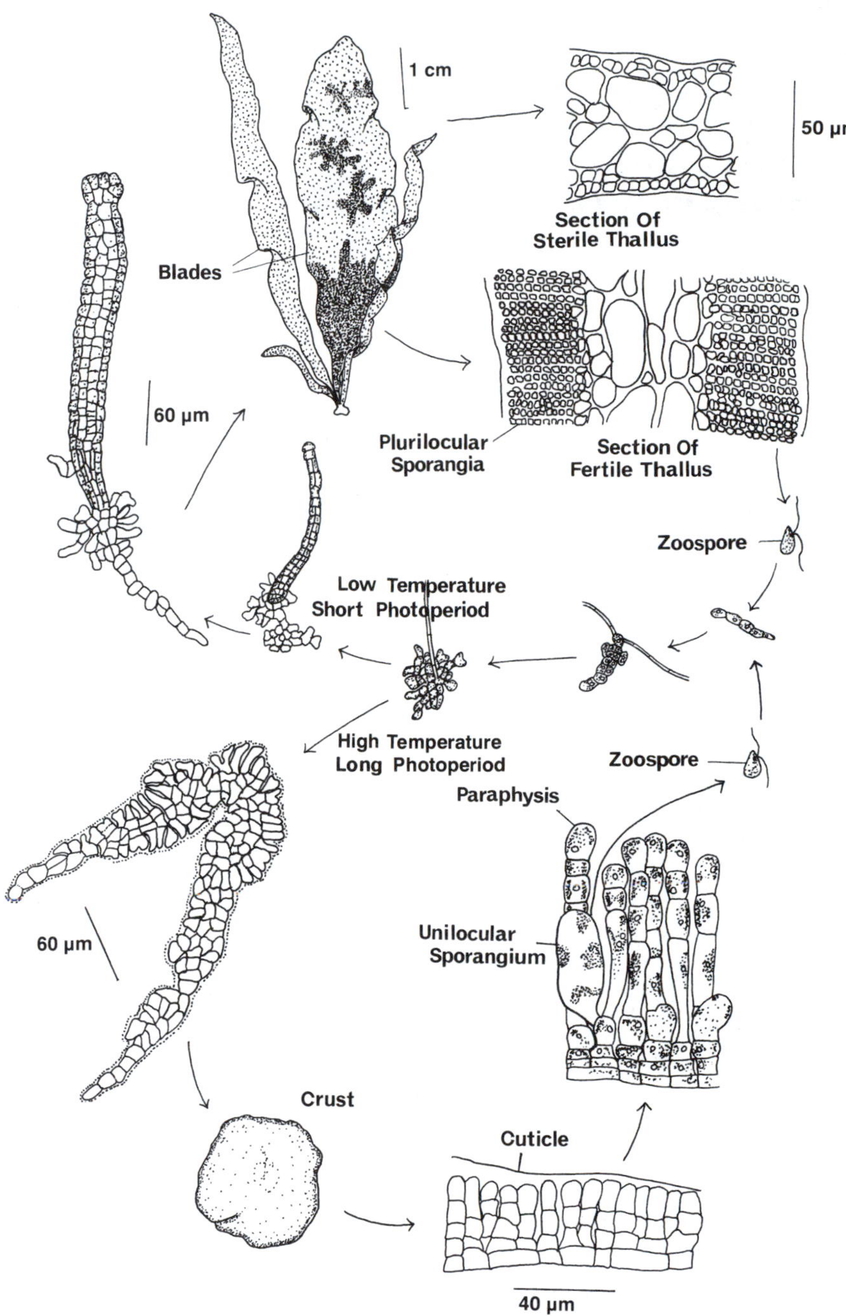

Fig. 17.8 The life cycle of *Petalonia fascia*. (After Smith, 1969; Wynne, 1969.)

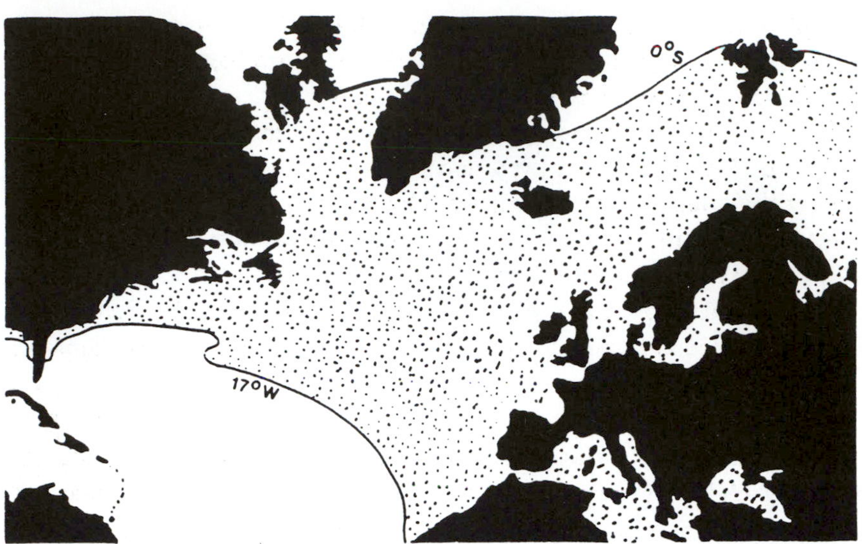

Fig. 17.9 The distribution of *Petalonia fascia* and *Scytosiphon lomentaria* is limited on the north by the 0 °C summer water isotherm and on the south by the 17 °C winter isotherm. Both species occur in a similar region, although there are ecotypes over the whole range. (after van den Hoek, 1982.)

produces plurilocular organs and a crustose phase that forms unilocular sporangia.

According to Wynne (1969), *P. fascia* (Fig. 17.8) has the following life cycle. The foliose thalli are annual and usually have erect lanceolate blades on a small discoid holdfast. The blades produce plurilocular sporangia that release negatively phototactic zoospores, with a period of motility lasting up to 24 hours. These zoospores settle and form filamentous germlings with the protoplast not being evacuated from the spore cell. The filaments branch, with the branches spreading laterally to form discs, with each cell having a single parietal plastid and a large pyrenoid. These discs can develop along two different lines, depending on the environmental conditions. Under short days and low temperature, the discs produce erect, uniseriate processes, the blade initials; these then become parenchymatous by subsequent longitudinal divisions, yielding the upright flattened blade. The surface cell of the blade undergoes numerous anti- and periclinal divisions to form dense lateral files that are the plurilocular organs. The sori cover most of the thallus except for the holdfast and the margin of the blades. The zoospores of the plurilocular organs settle and germinate to form new discs like the original ones. Under long days and high temperature, the discs develop into polystromatic crusts that resemble species of *Ralfsia*. After 4 weeks in culture these crusts reach maturity and are composed of a basal layer of cuboidal cells seldom exceeding six to ten cells in thickness and a layer of paraphyses supported by the basal layer. The paraphyses consist of four to six cells with a conspicuous cuticle covering the paraphyses. This cuticle is evidently secreted by the terminal cells of the paraphyses. A unilocular sporangium is formed by the basal cell of a paraphysis undergoing an unequal division to produce a cell that protrudes from one side. This cell enlarges laterally and upward, a basal cell is laid down, and a unilocular sporangium is formed. The sporangium cleaves up into 128, 256, or more zoospores, which are released by dissolution of the apex of the sporangium. The zoospores free themselves from the enveloping mucilage, swim away exhibiting negative phototaxis, and, after several hours, settle down. The germlings give rise to the original discs again. Therefore, in the above life cycle the motile cells produced by both the unilocular and plurilocular organs behave similarly, germinating to form discs.

From the above discussion it can be seen that the morphology of the plant is dependent on environmental conditions. In addition, Lüning and Dring (1973) have shown that the type of light will affect the morphology of *Petalonia* and *Scytosiphon*. Among other effects, in red light the prostrate system consists of sparsely branched, uniseriate filaments, whereas under blue or white light it consists of profusely branched filaments. Hsiao (1969, 1970) showed that there are certain minimal concentrations of iodine in the water necessary for the different forms of *P. fascia*. In order to have the formation of the crustlike *Ralfsia* stage, 4.0×10^{-5} M KI was necessary, and, for the formation of blades, 4.0×10^{-6} M KI. Filamentous stages of the alga will grow in an iodine-free medium. He also found that the crusts and blades grow at moderate temperatures, whereas the filamentous stages grow at the more extreme temperatures.

Splachnidiaceae

This family contains one genus, *Splachnidium*, a perennial alga of the Southern Hemisphere. The plant consists of a gelatinous, monopodially branched, hollow thallus attached by a basal disc (Fig. 17.10). The alga is interesting because, along with *Notheia* (Nizamuddin and Womersley, 1960; Gibson and Clayton, 1987), it has a number of similarities with the Fucales and may represent a link with this order (Saunders and Kraft, 1995). Like members of the Fucales, *Splachnidium* forms its reproductive bodies in conceptacles, the thallus is divided into a medulla with hyphae and a cortex, and there is a type of trichothallic growth (found in young fucalian sporophytes). The conceptacles originate in the immediate origin of the apex by localized cell division, leading to overarching of certain parts of the surface. Hairs are produced from the inner surface of the conceptacles and project through the ostiole, a situation similar to that in the Fucales. The large unilocular sporangia containing the zoospores arise successively from cells of the inner lining. The zoospores germinate to form filamentous thalli that

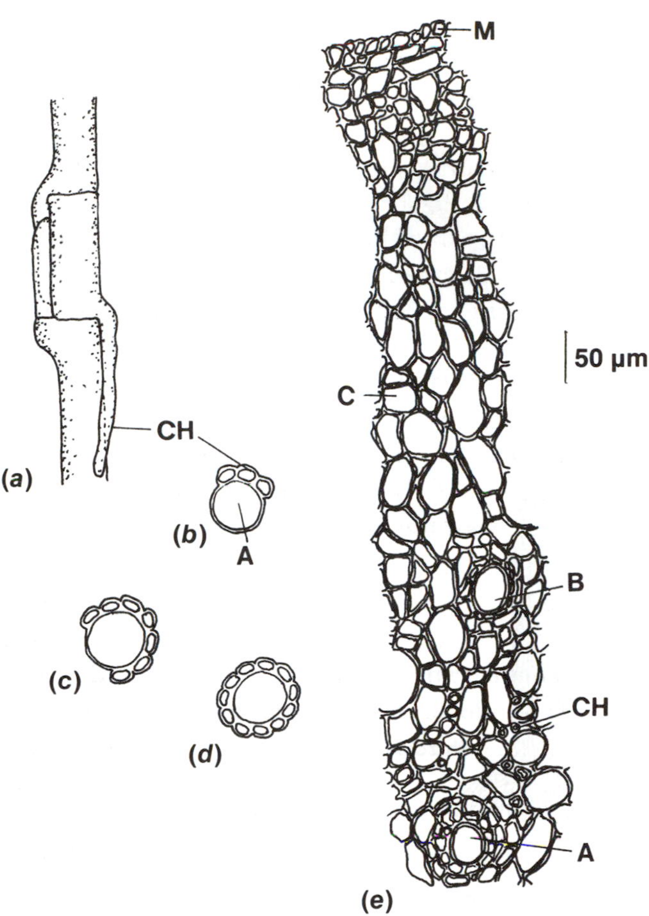

Fig. 17.12 *Desmarestia*. (*a*)–(*d*) Cortication of an axial cell (A) by corticating hyphae (CH). (*e*) Partial section of a mature thallus. (B,C) Laterals; (M) meristoderm.

with the outer layer that is produced behaving as a meristem. This meristem then produces the cells of the cortex to the inside, with the result that the axial cells and laterals are progressively buried. In addition to this primary growth, the cells of the inner cortex are capable of secondary growth, enlarging to form hyphae that push their way downward between the cells of the cortex. Trumpet hyphal cells occur in the medulla. These cells have perforate end walls with callose and probably function in conduction of nutrients as do the sieve filaments in the Laminariales (Moe and Silva, 1981).

The life cycle of *D. aculeata* (Fig. 17.11) has been elucidated by Schreiber (1932) and Chapman and Burrows (1971). The unilocular sporangia are produced on the sporophytes in the winter, and are formed by the tangential division of a surface cell of the cortex. Meiosis apparently occurs in the production of a few biflagellate zoospores formed in each small unilocular sporangium. The zoospores have an eyespot and a single chloroplast.

Released zoospores settle, lose their flagella, and round up within 24 hours. The cell contents move out into a germ tube immediately after settling. The sporelings produce female and male gametophytes in roughly similar numbers, the male gametophytes having small cells and being less densely pigmented than the female gametophytes. Gametophytes grow vegetatively only in red light, with differentiation of oogonia and antheridia occurring under blue or white light (Müller and Lüthe, 1981). About 11 days to 3 weeks after germination of spores, conical antheridia and lateral oogonia appear on the gametophytes. The tubular oogonia dehisce apically to liberate the egg, which usually adheres to the gelatinized aperture. A single spermatozoid is released from each antheridium through a narrow apical aperture. The freshly released eggs secrete three volatile chemicals that cause the antheridia to burst and attract free spermatozoids to the eggs. The three sexual hormones are **desmarestene**, **ectocarpene**, and **viridene** (Fig. 17.4). Desmarestene is the most potent of the three sexual hormones (Müller et al., 1982). Sporophyte development begins with the production of a tube from one side of the zygote and a lightly pigmented rhizoid from the opposite pole. The initial tube goes on to produce an oppositely branched uniseriate filament, which forms a trichothallic meristem below most of the lateral branches. Cortication of the thallus begins, as has already been described, to produce a mature sporophyte and complete the life cycle. Gametophytes of *Desmarestia* (Andersen, 1982) and the Laminariales have several features in common, including clusters of unicellular antheridia, vertically elongated intercalary oogonia, attachment of the extruded egg to the oogonial apex, and growth of the young sporophyte upon the oogonial apex.

Some of the species of *Desmarestia* accumulate large amounts of malic acid, lowering the pH of the vacuolar sap to as low as 2. In collecting seaweeds, the species of *Desmarestia* should be kept separate because some of their cells will rupture, releasing acid and killing the other seaweeds.

In the Antarctic waters, members of the Desmarestiales provide the bulk of the biomass of benthic seaweeds. They are perennial, covering large areas of bottom to depths of about 40 m. The largest and most abundant species (*D. anceps* and *D. menziesii*) form thickets, but not the protective canopy characteristic of many kelps. The Antarctic possesses the only cold-water flora without Laminariales, although in sub-Antarctic waters there are vast stands of kelps (*Macrocystis* and *Lessonia*) (Moe and Silva, 1977).

The Desmarestiales and Laminariales probably had a common phylogenetic origin (Müller and Lüthe, 1981; Müller et al., 1985b; Tan and Druehl, 1996). This conclusion is based on the similarities between the two orders, similarities that include (1) vegetative development of the gametophytes in red light; (2) requirement of white or blue light for development of antheridia and oogonia; (3) existence of spermatozoid-releasing and -attracting factors secreted by eggs; (4) unusually long and flexible hind flagella; (5) lack of eyespots in spermatozoids; and (6) formation of sexual organs by the gametophyte, representing an exhaustive and almost lethal effort for the gametophytes.

CUTLERIALES

This order contains only two genera, *Cutleria* and *Zanardinia.* The genera show an alternation of generations that is heteromorphic in *Cutleria* and isomorphic in *Zanardinia.* The thallus is flattened, bladelike, or disclike, with entirely or partially trichothallic growth. The sporophytes produce only unilocular sporangia, whereas the gametophytes are heterothallic and markedly anisogamous.

Cutleria is a warm-water plant of the Northern Hemisphere that may be closely related to *Saccorhiza* in the Laminariales (Rousseau et al., 1997). The gametophyte is an erect, flattened blade with numerous dichotomies (Fig. 17.13). Growth is trichothallic at the base of many erect uniseriate hairs at the upper margin of the blade. The cells that are cut off below the hairs contribute to the thallus. The innermost of these cells gradually enlarge to form the medulla, whereas the outer ones undergo divisions to form the cortex. The gametophytes are heterothallic, with the sex organs developing in clusters of the surface of the thallus. A superficial epidermal cell may develop directly into a male gametangium, or it may develop into a branched hair that bears several gametangia. The male gametangium consists of a stalk cell on which there are 20 or more tiers of cells, each tier composed of eight cells. The protoplast of each cell forms a biflagellate male gamete, which escapes through a pore in the gametangial wall to the outside. Female gametangia develop similarly to the male, but with a smaller number of larger cells. Female gametangia are four to seven tiers high, with only four cells in each tier. Free-swimming male gametes are pyriform, with a single reddish chloroplast at the place of flagella insertion. Free-swimming female gametes are also pyriform, but are much larger and have a dozen or so chloroplasts. The female gametes release a highly volatile, low-molecular-weight compound, **multifidene** (Fig. 17.4) that attracts the male gametes (Müller, 1974). When the gametes fuse, the male gametes are actively swimming while the female are sluggish or immobile. Fusion of the two nuclei follows within a few hours, and the zygote begins to develop into the sporophyte within a day. Unfertilized female gametes develop parthenogenetically into gametophytes.

The zygote germinates to produce the sporophyte, which was first described as a separate genus, *Aglaozonia.* At first, growth is trichothallic

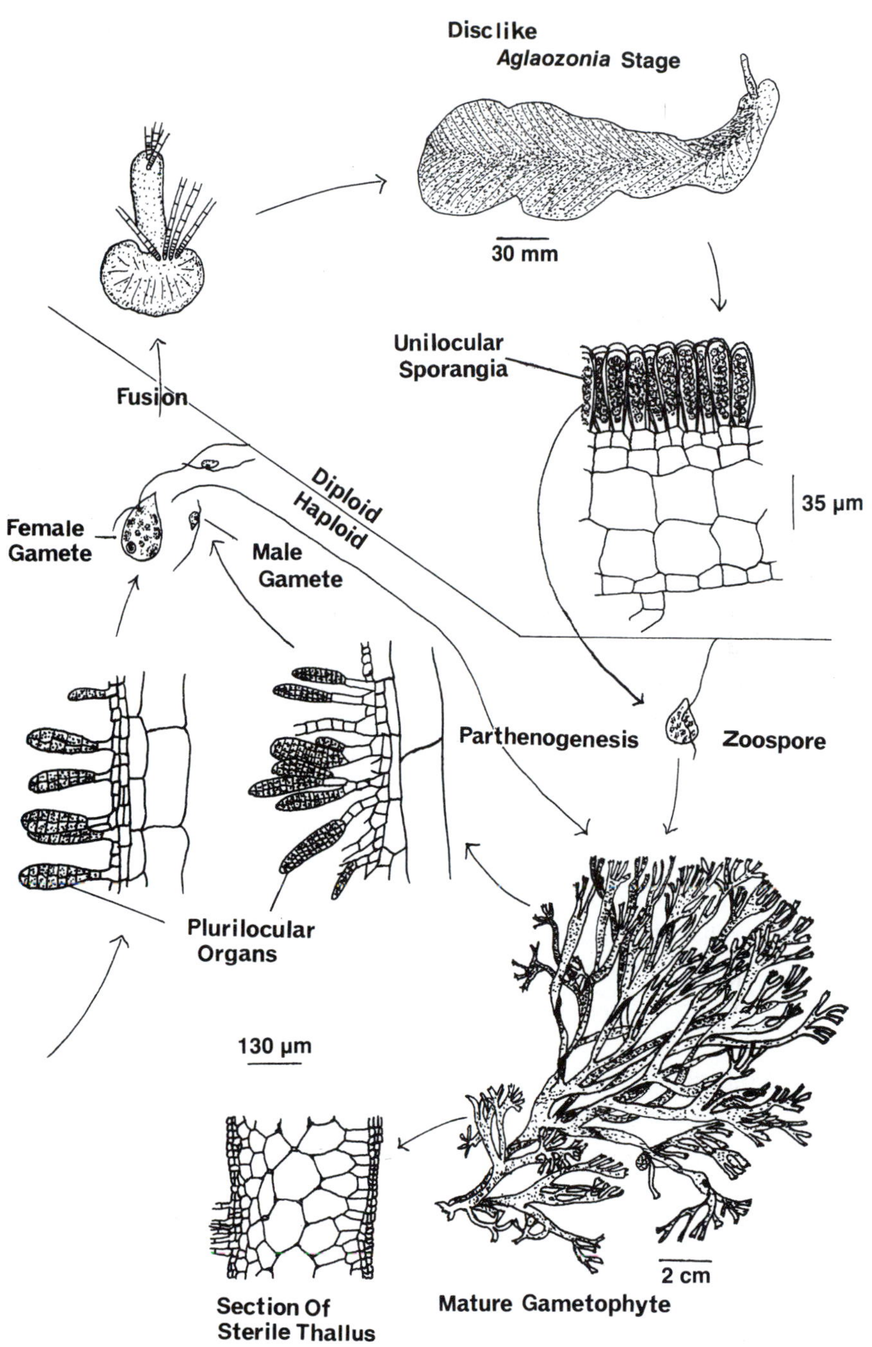

Fig. 17.13 The life cycle of *Cutleria multifida*. (Adapted from Kuckuck, 1899; Savaugeau, 1899.)

and vertically upward into a columnar structure. Upward growth ceases when the plant is about 10 days old, and all further growth is laterally outward from the base of the column. Repeated cell division at the base of the column forms a flat, disclike tissue that expands laterally as a result of division and redivision of the marginal cells. The sporophyte is homologous to a minute erect thallus subtended by an enlarged fertile holdfast. The disclike portion of the thallus is several cells thick, and the outer cells are differentiated into an epidermis-like layer. The holdfast is attached to the substratum by numerous multicellular rhizoids growing from the ventral epidermal cells. The unilocular sporangia are formed in sori on the dorsal surface of the sporpohyte. A single epidermal cell divides into one to six stalk cells and a single terminal unilocular sporangium. In the unilocular sporangium 8, 16, or 32 large, pyriform, haploid zoospores are formed, each with several chloroplasts. The zoospores escape through a large apical pore in the sporangial wall, swarm for 10 to 90 minutes, then settle down, round up, and secrete a wall. The zoospores then divide to form the gametophyte.

Although germlings from zygotes and from zoospores of the *Aglaozonia* sporophytes have not been grown to maturity in culture, they have been grown to a sufficiently advanced stage to show that the two stages are alternate generations of each other. In Europe, the sporophyte is perennial and fruits in winter or spring, whereas the gametophyte is a spring annual that disappears during the summer.

The affinities of the Cutleriales appear to be with those Ectocarpalean genera with trichothallic growth. Fossils of a plant similar in structure to *Cutleria*, called *Limnophycus paradoxa*, have been described from Miocene (25 million year old) deposits in Germany.

LAMINARIALES

The members of this order are parenchymatous with growth from an intercalary meristem between the stipe and blade. The plants have an alternation of a large sporophyte with a microscopic gametophyte. Sexual reproduction is oogamous. With the exception of *Chorda* (the most primitive member of the order) and *Saccorhiza*, the Laminariales lack an eyespot and an associated flagellar swelling in the motile cells (Henry and Cole, 1982; Henry, 1987).

The Laminariales are very large plants that are usually distributed in the colder waters of the world. Many of the genera have sporophytes that are able to grow vegetatively in warmer waters, but their microscopic gametophytes fail to produce gametes above 10 to 15 °C, thereby preventing the distribution of plants in waters warmer than this (Fig. 17.23). The Laminariales probably evolved relatively recently, between 16 and 20 million years ago, in

Fig. 17.14 A small plant of *Pterygophora californica*. (After Smith, 1969.)

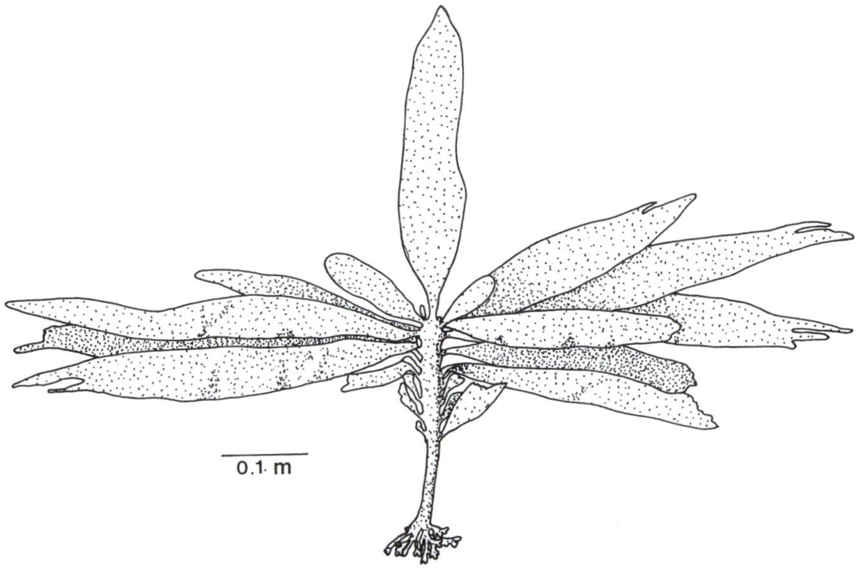

the North Pacific during a strong polar cooling trend (Estes and Steinberg, 1988; Saunders and Druehl, 1992).

Morphology and anatomy

With the exception of the genus *Chorda*, the sporophytes are differentiated into a holdfast, stipe, and blade (Figs. 17.14 and 17.19). An intercalary meristem between the stipe and blade adds tissue to both. The blade length often remains about the same because the increase in length at the base often equals the loss by abrasion at the apex. The blades of most genera last for one year, but in many cases the stipe and base of the blade are perennial. Examples of recorded life-spans for kelps are *Pterygophora californica* (13 years) (Fig. 17.14), *Laminaria hyperborea* (11 to 18 years) (Fig. 17.22(*d*)), and *Macrocystis pyrifera* (4 to 8 years) (Fig. 17.32) (Lobban, 1978). The most frequent cause of death appears to be the tearing of the alga from the rocks by storms. The blades usually stop growing in late summer and begin to disintegrate in the autumn after the plant has discharged its zoospores. The sporophytes are very durable and will withstand a great deal of stress without breaking. When members of the order are washed up on the beach, the plant is usually intact, with the haptera firmly attached to a stone that has come loose from the bottom. The haptera accomplish their firm attachment to the substratum by the growth of rhizoidal cells from the outer meristoderm cells (Tovey and Moss, 1978). These rhizoidal cells fill every microscopic crevice of the substratum until an exact profile of the sub-

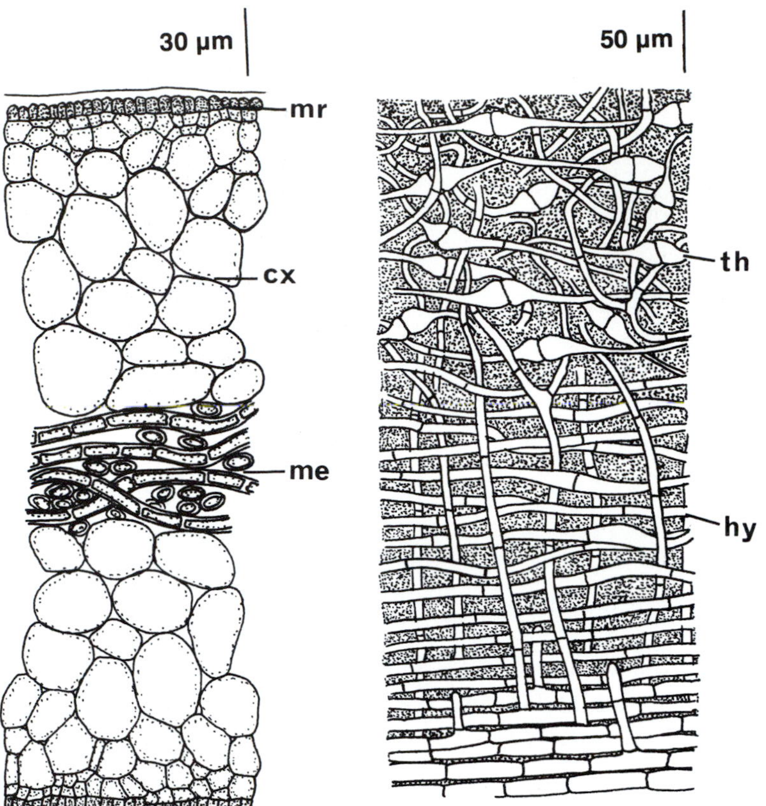

Fig. 17.15 Sections of lamina (*left*) and the central portion of a stipe (*right*) of a member of the Laminariales. (cx) Cortex; (hy) hyphae; (me) medulla; (mr) meristoderm; (th) trumpet hyphae.

stratum is built up. Mucilage is also secreted by the cells. The haptera grow downward, not in response to gravity but because they are negatively phototropic, growing away from light (Buggeln, 1974).

There are three different tissues in the sporophyte: the central medulla, the cortex, and the epidermis (Figs. 17.15 and 17.17). The haptera lacks a medulla, but all three tissues are present in the stipe and blade. The stipe and blade have the same anatomy, the only difference being the cylindrical to elliptical shape of the stipe and the flattened shape of the blade. At the thallus surface are the photosynthetic, meristematic cells of the meristoderm, which add to the girth of the organ (Fig. 17.15). The meristoderm is composed of small cells that cut off daughter cells to the inside, which in turn form the cells of the outer cortex. The meristoderm is usually covered with a layer of mucilage. The meristoderm in the blade is active throughout the life of the blade, dividing primarily periclinically. In the stipe of some genera (e.g., *Laminaria*), the meristematic activity is transferred to a cortical layer at a depth of four to eight cells beneath the surface, with the result that the tissue to the outside is shed as the stipe increases in width.

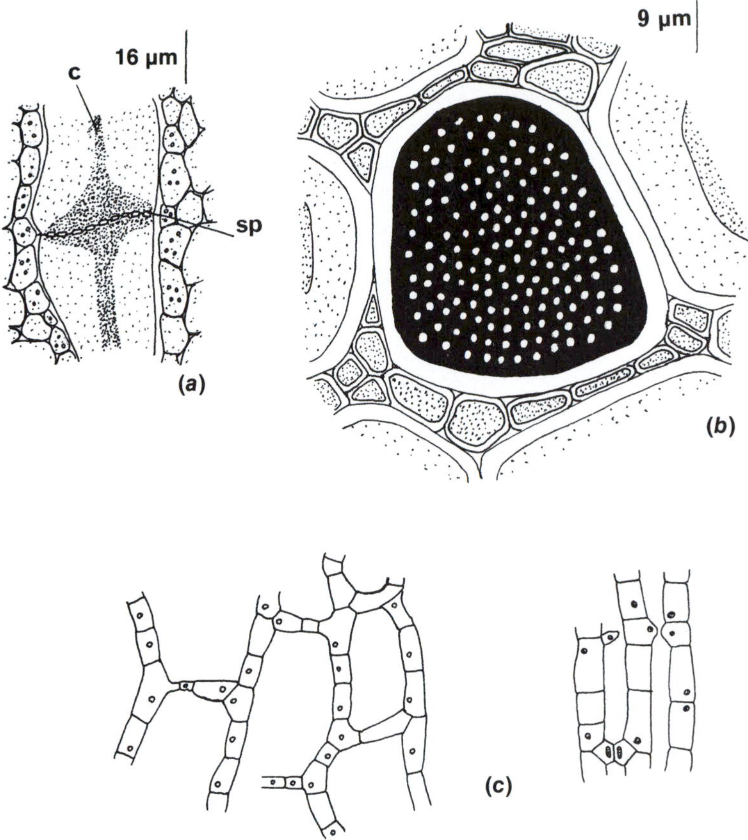

Fig. 17.16 (*a*),(*b*) Sections of a sieve cell from the medulla of an organism in the Laminariales. The sieve plate (sp) has pores with associated callose (c). (*c*) Cells of the inner cortex and medulla with cross connections. ((*a*),(*b*) after Scagel 1971.)

Inside the meristoderm are the larger cells of the inner and outer cortex followed by the tangled elongate cells that make up the medulla. The outer cortex differentiates cells to the inner cortex, whereas the inner cortex differentiates cells to the medulla. The cells of the inner cortex and medulla often form cross connections between adjacent cells (Fig. 17.16(*c*)). In their formation, two adjacent parent cells cut off small cells that elongate toward each other. When they meet, the end walls dissolve, and the cells are continuous. Another type of cells are the **hyphae**, which originate as outgrowth of cells of the cortex and develop into slender, often branched cells of considerable length that grow into the mucilage of the medulla. The medullary cells are in longitudinal rows; and, because they do not have the ability to divide after they are formed, they are drawn out into long cells by the elongation of the thallus from expansion of cells and the meristematic activity of the meristoderm. These medullary elements are often called **trumpet hyphae** because as they are drawn out, the centers become constricted whereas the septal areas maintain their original diameter (Figs. 17.15 and 17.16). Another name for the trumpet hyphae is **sieve cells**

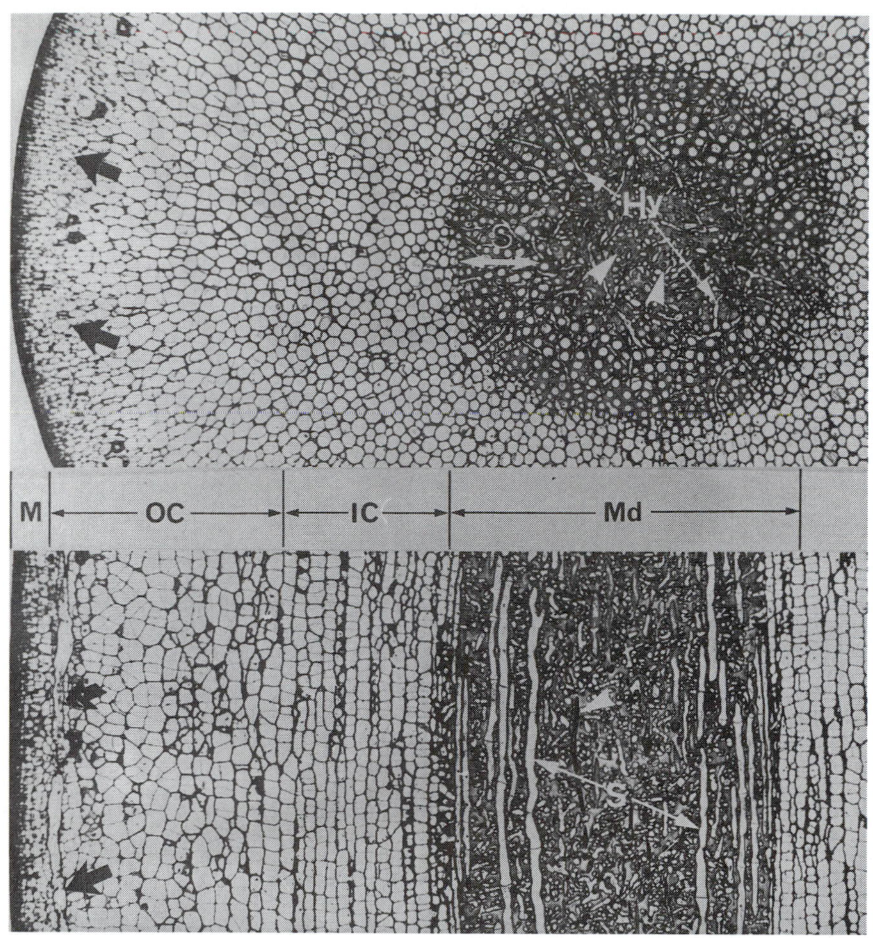

Fig. 17.17 Cross (*upper*) and nearly radial longitudinal (*lower*) sections through a stipe of *Macrocystis integrifolia*, showing the anatomy of a member of the Laminariales. The mucilage ducts (black arrows) are arranged in a ring at the fringe of the outer cortex (OC). Note the sieve elements in radial rows (S) in the outer region of the medulla and the obliterated sieve elements (white arrowheads) in the center of the medulla. Hyphal cells (Hy) anastomose through the medulla. (IC) Inner cortex; (M) meristoderm; (Md) medulla; (OC) outer cortex. (From Shih et al., 1983.)

because there are sieve plates with pores separating the cells. The pores of the sieve plates appear to have evolved from plasmodesmata, which are common in other Phaeophyceae.

Within the Laminariales there is an evolutionary development of sieve cells (Sideman and Scheirer, 1977). In *Laminaria*, the sieve plates have pores ranging from 0.06 to 0.09 μm in diameter; no **callose** associated with the pores; and nuclei, vacuoles, and mitochondria in the sieve cells. As evolution proceeded through *Alaria* (Fig. 17.33) and *Nereocystis* (Fig. 17.31) to *Macrocystis* (Fig. 17.32), the pores in the sieve plates became larger, callose became associated with the pores, and the cells lost organelles. *Macrocystis* has sieve cells with 2.4- to 6.0-μm pores, callose associated with the pores, and only mitochondria in the cells. The sieve cells of *Macrocystis* differ from sieve elements of the angiosperms in that there are no companion cells present, and there is no large central vacuole. The presence of large

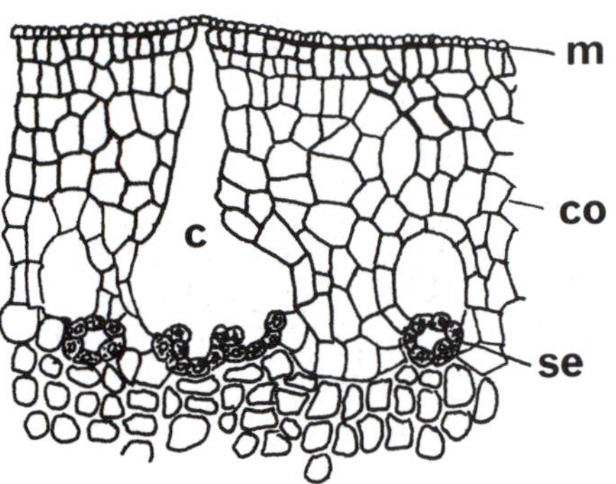

Fig. 17.18 *Laminaria cloustoni*, transverse section of stripe showing mucilage canals (c) and secretory cells (se). (m) Meristoderm; (co) cortex. (After Guignard, 1892.)

numbers of mitochondria in sieve cells of *Macrocystis* probably reflects the lack of a companion cell, with the mitochondria producing the energy necessary for cell processes. In the Laminariales, there is active transport of the products of photosynthesis through the sieve cells, mostly as mannitol (Parker, 1965). The rate at which the mannitol moves depends on the type of sieve cell present (Lüning et al., 1972). In the well-developed sieve cells of *Macrocystis*, the organic products move as fast as 65 to 78 cm hour^{-1}, both toward the base for storage in the basal region and toward the apex to provide the rapidly growing apex with substrates (Sargent and Lantrip, 1952; Parker, 1963, 1965). In *Laminaria* and *Saccorhiza*, with their less developed sieve walls, the rate of transport is about five times slower than in *Macrocystis* (Emerson et al., 1982).

In the stipes and blades of some of the Laminariales, there is an interconnected system of mucilage canals in the cortex (Grenville et al., 1982) (Figs. 17.17 and 17.18). In *Laminaria saccharina* and *Laminaria hyperborea*, these mucilage canals are lined by secretory cells which are linked by plasmodesmata (Evans et al., 1973). The secretory cells produce and secrete fucoidin into the canal, from which it goes to the outside of the thallus. These are the only cells that secrete fucoidin in the thallus, and, like secretory cells in other organisms, they have numerous large Golgi bodies surrounding the nucleus.

Life cycle

The life cycle of a member of the Laminariales involves the alternation of a large sporophyte with a microscopic gametophyte (Figs. 17.19 and 17.20). The sporophyte produces unilocular sporangia intermingled with paraphyses in sori that are induced to form by short-day conditions (tom Dieck, 1991). First, a superficial cell divides to form a basal cell and a paraphysis. The basal cell widens, and the paraphysis elongates. The upper end of the

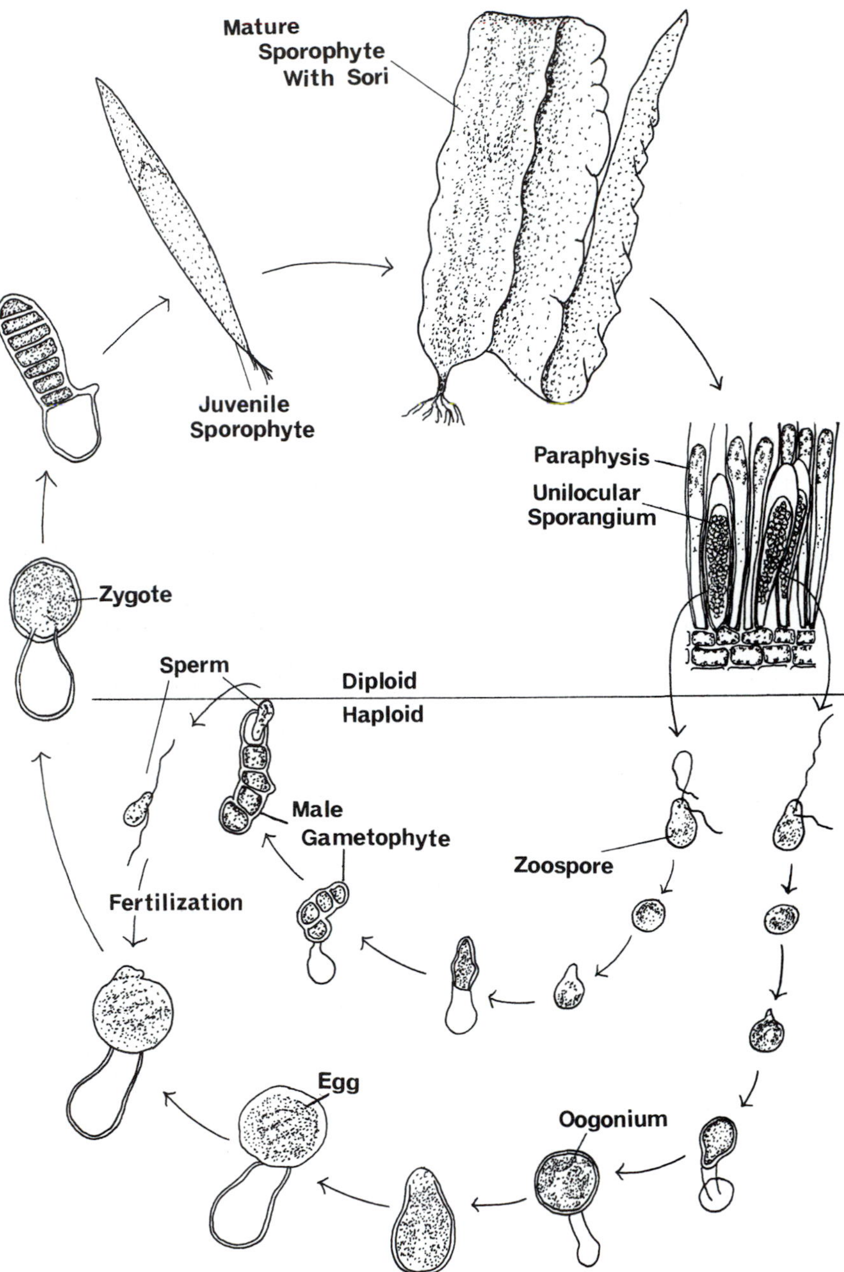

Fig. 17.19 The life cycle of *Laminaria japonica*. (After Cheng, 1969.)

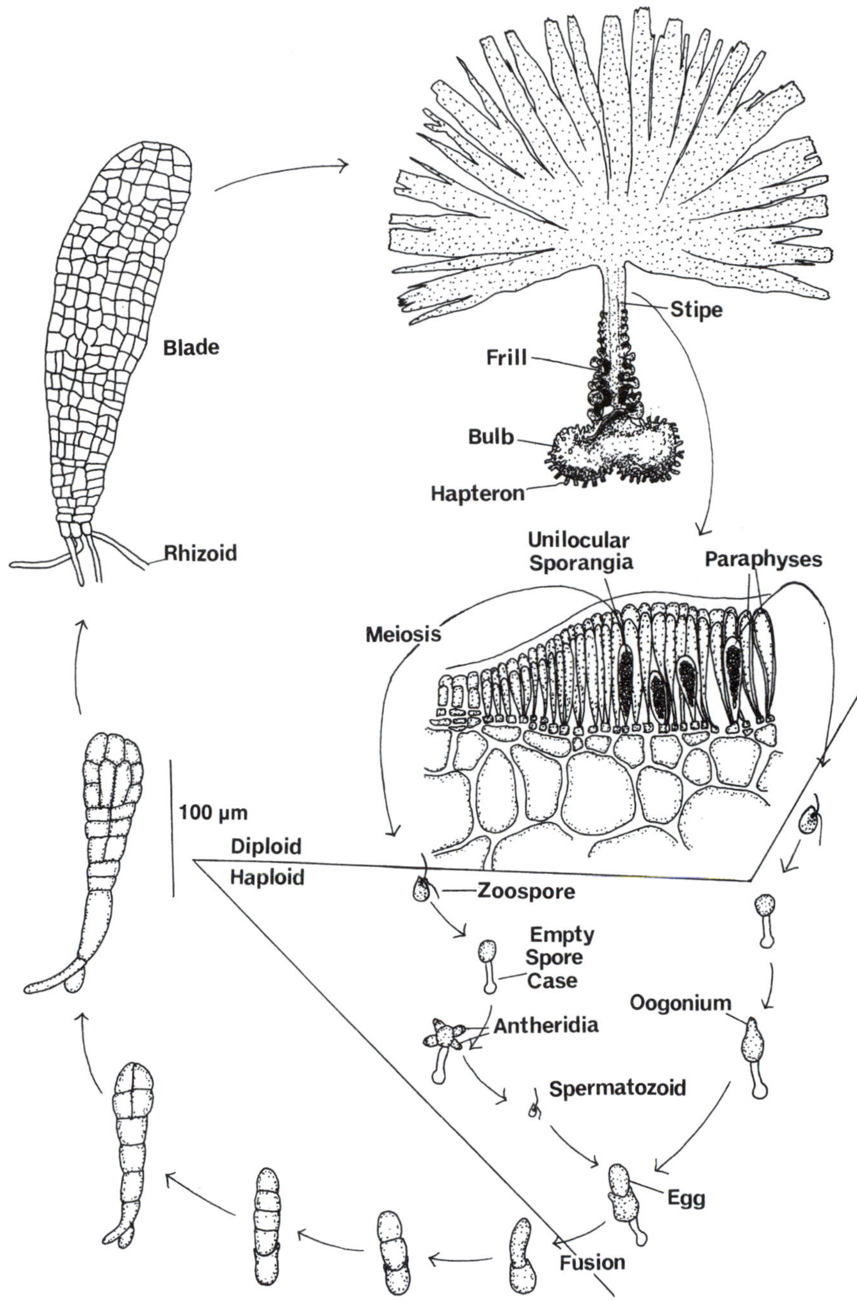

Fig. 17.20 The life cycle of *Saccorhiza* sp. (Adapted from Norton and Burrows, 1969.)

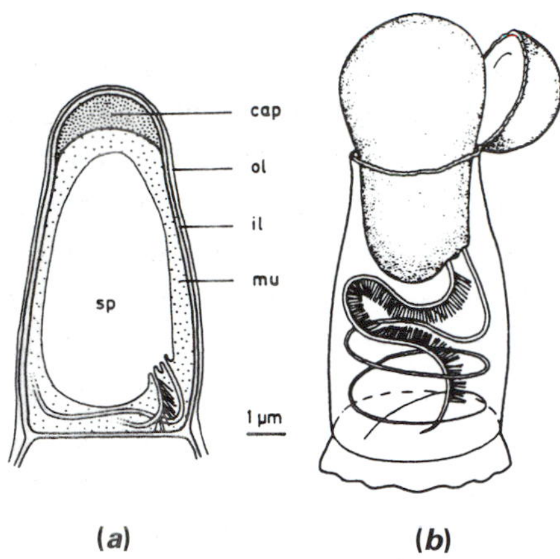

Fig. 17.21 Drawings of the antheridium of *Laminaria digitata*. (*a*) Mature antheridium. (il) Inner layer of cell wall; (ol) outer layer of cell wall; (mu) mucilage, (sp) spermatozoid. (*b*) Release of spermatozoid. The cap is pushed away and the spermatozoid is forced out of the antheridium. (From Maier, 1982.)

paraphysis becomes swollen and mucilaginous, forming a covering over the basal cells. The basal cell now produces a unilocular sporangium next to the paraphysis. Thirty-two (*Laminaria*) to 128 (*Saccorhiza*) haploid zoospores are formed in the unilocular sporangium (Motomura et al., 1997), and the zoospores are released through the thickened apex of the sporangium. The zoospores have a single chloroplast (in *Chorda*, they have a number of chloroplasts) and may or may not have an eyespot (Evans, 1966). The zoospores are positively chemotactic towards nutrients (Amsler and Neushel, 1989) and can be transported for several kilometers (Reed et al., 1988) during the 48 hours that they swim about. After settling, the zoospores produce the gametophytes. The gametophytes in most of the Laminariales are dioecious, with separate male and female plants. However, the primitive genus *Chorda* is monoecious, and there are conflicting reports as to whether the gametophytes of *Saccorhiza* are monoecious or dioecious (Norton, 1972; Henry, 1987). In *Laminaria*, Evans (1965) showed that there is probably an X/Y sex-determining mechanism, with segregation taking place at the meiotic division in the unilocular sporangium. The zoospores contain glycoproteins in small vesicles in the peripheral cytoplasm that are released when the zoospores settle (Oliveira et al., 1980). These glycoproteins adhere the cells to the substratum. The settled zoospore secretes a thin wall around itself, with a slender germ tube emerging; the protoplasm of the zoospore moves out of the original spore and into a swelling at the tip of the germ tube; a wall is formed between the swelling and the original spore; the cell in the swelling now divides to form the gametophyte. It is only at this stage that the female gametophyte looks different from the male. The male gametophyte has smaller cells and is more branched than the female. The male gametophytes produce small colorless antheridia (Fig. 17.21). In the female gametophyte, elongated oogonia are formed that produce a single egg. Under long-day conditions (16 hours light:8 hours dark), eggs are released during the dark cycle, mostly during the first 30 minutes of darkness (Lüning, 1981). The release is apparently controlled by a circadian rhythm. After the female cell has emerged, the thick plastic edges of the wall contract and form a platform on which the egg remains for some time. The sexual hormone **lamoxirene** is secreted by the eggs as they are released in at least 21 species in the families Laminariaceae, Alariaceae, and Lessoniaceae (Lüning and Müller, 1978; Müller et al., 1979, 1985b). Spermatozoids are ejected from the antheridia within a few seconds of exposure to lamoxirene. The spermatozoids are attracted to the eggs where fertilization takes place (Motomura, 1991). The chemical formula of lamoxirene is 1-(1′,2′-*cis*-epoxibut-3′-enyl)-cyclohepta-2,5-diene; it has a molecular weight of 162 and the empirical formula $C_{11}H_{14}O$ (see Fig. 17.4) (Marner et al., 1984). The sexual hormones ectocarpene and desmarestene (Fig. 17.4) are also present, but they do not have hormonal activity in *Laminaria* or *Macrocystis*. Ectocarpene is the male attractant in *Ectocarpus* (see Ectocarpales, Ectocarpaceae, above), whereas desmarestene is the sperm-releasing and -attracting factor in *Desmarestia* (see Desmarestiales, above). In the Laminariales, the zygote germinates to form a flat proembryo that subsequently germinates into a mature sporophyte.

Environmental conditions, particularly light and temperature, usually control the life cycle in the Laminariales. Sporophytes will normally not grow at temperatures above 18 to 20°C (Cheng, 1969; Nakahara and Nakamura, 1973), and sori will not be produced at these temperatures. If a mature sorus of *Laminaria* is placed in water at 20 °C, the sporangia cease to discharge zoospores and disintegrate. The gametophytes are also subject to environmental control. Gametophytes will grow for varying periods of time before forming gametangia. In some cases oogonia are produced by the settled zoospore, whereas at other times the gametophyte grows indefinitely without forming gametes. Lüning and Dring (1972, 1975) showed that if gametophytes are grown in red light at 15 °C, they will grow indefinitely without ever becoming fertile. If these gametophytes are subjected to 6 to 12 hours of blue light, they will produce gametes. Also the gametophytes will not produce gametes if the temperature is above 10 to 12 °C (Sundene, 1963; Vadas, 1972; Nakahara and Nakamura, 1973); or if the water has less than 5 μg liter^{-1} of NO_3-N (Hsiao and Druehl, 1973). The gametophytes have the ability to withstand long periods of darkness and then resume growth when light again becomes available (Kain, 1966). Thus in nature the size of the gametophyte and the time of gametogenesis are probably controlled by environmental conditions.

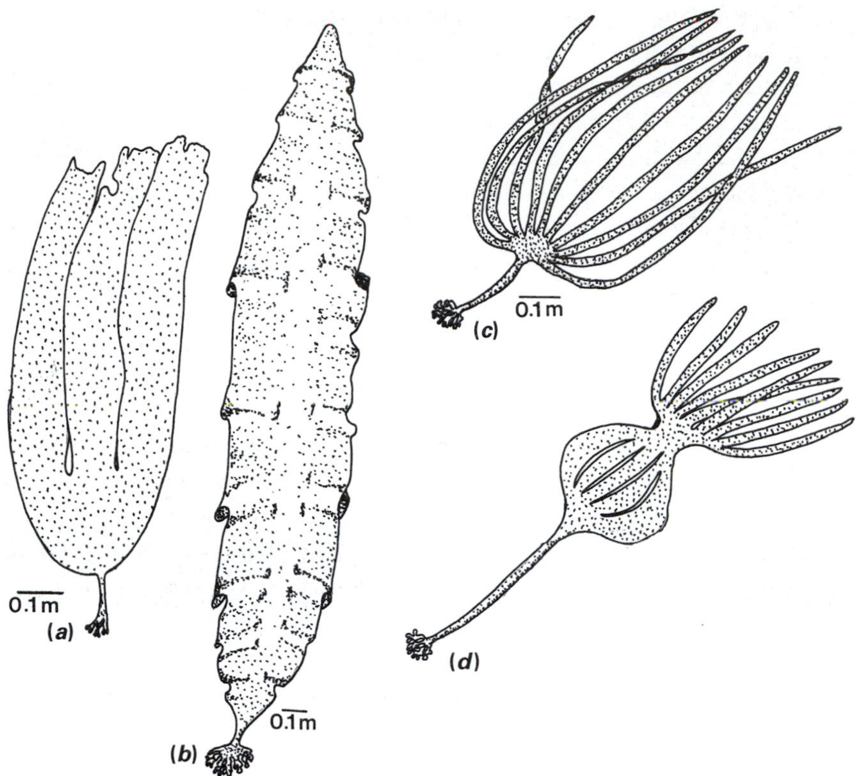

Fig. 17.22 Morphology of some species of *Laminaria*. (*a*) *L. groenlandica*. (*b*) *L. saccharina*. (*c*) *L. digitata*. (*d*) *L. hyperborea*. ((*a*) after Scagel, 1971.)

Ecology

Laminaria plants from northern waters are generally much larger with longer stipes and have a greater blade area than those from more southerly waters, a phenomenon that is probably due to their being older plants rather than their having greater growth rates (Larkum, 1972). Different species of *Laminaria* have different life-spans. *Laminaria hyperborea* (Fig. 17.22(*d*)) is a fairly long-lived species (Kain, 1976), with some plants living to at least 8 years in most populations, and a recorded life of 18 years. *Laminaria saccharina* (Fig. 17.22(*b*)) growing subtidally has a normal life-span of 3 years, during which maximum growth in length and width occurs during the second growing season. However, intertidal populations of *L. saccharina* are only annuals (Druehl and Hsiao, 1977). Breakage of the frond is most easily repaired in the first growing season of subtidal plants. The distribution range of sporophytes in the Laminariales is usually greater than that of the gametophytes, indicating a greater tolerance of unfavorable conditions by the sporophytes. *Laminaria saccharina* produces reproductive tissue at all times of the year, but the greater number of fruiting

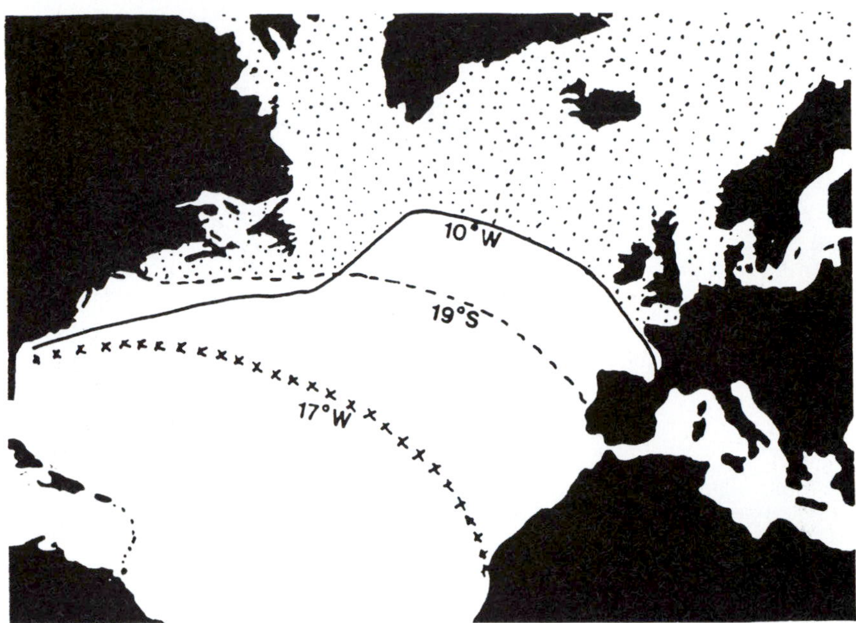

Fig. 17.23 A map of the distribution of *Laminaria digitata* (dotted area). The southern reproductive boundary equals the 10 °C winter (W) water isotherm. The southern lethal boundary equals the 19 °C summer (S) water isotherm. The southern growth boundary would equal the 17 °C winter (W) water isotherm. (After van den Hoek, 1982.)

plants occur during the summer and winter months. Reproductive sori are never found on blade areas less than 6 months old, a time that corresponds with the final expansion period of the tissue.

The southern lethal boundary of *Laminaria digitata* (Fig. 17.22(*c*)) corresponds to an isotherm that represents an August mean temperature of 19 °C (Fig. 17.23) (van den Hoek, 1982). Above this temperature, both the gametophyte and sporophyte die. The southern growth boundary corresponds to an isotherm that represents a February mean temperature below 17 °C. Below this temperature, both the gametophyte and sporophyte are able to grow. The southern reproductive boundary of the female gametophyte (which determines production of sporophytes) is at the isotherm representing the February mean temperature of 10 °C. It is not necessary that the temperature reach this low point every winter at the southern limit, as the species can still survive in the form of perennial sporophytes. Thus the lethal boundary of *Laminaria digitata* is further north than the potential growth boundary. Although in the winter plants can drift southward and grow, they are killed by high temperature in the following summer. Only where the winter temperature reaches 10 °C or colder does the alga actually reproduce. In the eastern North Atlantic, the distribution of *L. digitata* is determined by the 10 °C winter isotherm, which is further north than the 19 °C summer isotherm representing the lethal boundary. However, in the western North Atlantic, the 19 °C summer isotherm is north of the 10 °C

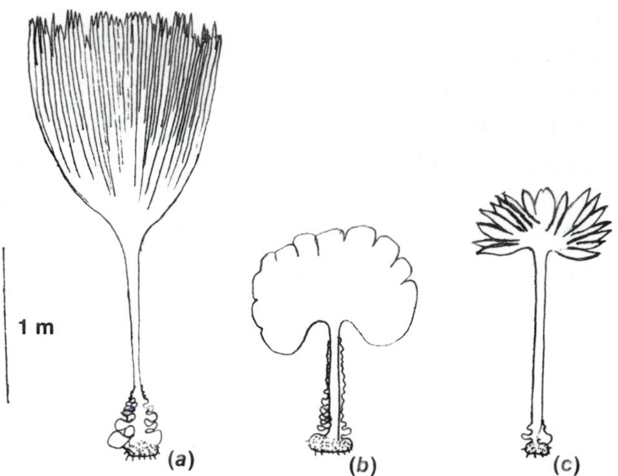

Fig. 17.24 *Saccorhiza polyschides* sporophytes from (*a*) strong current, (*b*) weak current, and (*c*) area with wave action. (After Norton and Burrows, 1969.)

winter isotherm. The actual distribution is thus limited by the 19 °C summer isotherm in the western North Atlantic and the 10 °C winter isotherm in the eastern North Atlantic.

Laminaria hyperborea (Fig. 17.22(*d*)) and *L. digitata* (Fig. 17.22(*c*)) form dense underwater forests in the eastern cold temperate northern Atlantic. Similarly, *Laminaria solidungula* produces dense growths in the Alaskan Beaufort Sea. These kelps exhibit a seasonal growth pattern with a phase of fast growth prevailing from early winter to early summer, and a summer and autumn phase with a reduced or complete cessation of growth. These growth patterns are an ecological strategy, whereby storage of photosynthate during the well-lit summer and autumn enables the algae to start growth by remobilization and translocation of stored carbohydrates early in the dark winter when nutrient supply is optimal due to plankton remineralization. This **circannual** (approximately annual) **rhythm** is under endogenous photoperiodic control (Henley and Dunton, 1997; Schaffelke and Lüning, 1994).

Another factor affecting the morphology of the sporophyte is the environment in which the plant is growing. *Saccorhiza polyschides* sporophytes growing in weak current produce curved blades that are heart-shaped (cordate) at their base (Fig. 17.24). These blades lack subdivisions (digits) and are so fragile that they tear under their own weight when removed from the water. In contrast, plants growing in strong current develop very long, flat, tough blades, narrowly triangular (cuneate) at the base and divided into as many as 30 digits. In habitats without current but exposed to wave action, the sporophytes produce short, flat, extremely tough blades with only three to ten digits. Anatomically, the greater toughness of blades is the result of a larger number of cortical cells increasing the thickness of the thallus (Norton, 1969; Norton and Burrows, 1969). Somewhat similar results with *Laminaria digitata* (Sundene, 1962a) and *Alaria esculenta* (Sundene, 1962b) have been reported.

Many species of *Laminaria*, such as *L. hyperborea*, produce very thick stands or forests of sporophytes beneath the low-tide mark. As the depth increases, the plants become sparser, forming an open "park" (Larkum, 1972). *Macrocystis* and *Laminaria* have primary production rates that rank among the highest in the world, reaching an annual net production in the range of 1000 to 2000 g m^{-1} of carbon (Mann and Chapman, 1975). The forests of the giant kelp *Macrocystis pyrifera* form continuous beds up to 8 km long and 1 km wide along the Pacific Coast of North and South America (Gaines and Roughgarden, 1987). The extent of these forests and the density of plants vary greatly in space and time because of storms (Dayton et al., 1984), herbivores, and predators (Duggins, 1980), and also major current features such as El Niño. The decline in canopy area due to large winter storms of the 1982–83 El Niño was particularly dramatic (Dayton and Tegner, 1984).

These stands of *Macrocystis* and *Laminaria* are often depopulated by storms which dislodge the plants and cast them on the beach. A small mollusc, *Patina pellucida*, frequently browses the holdfast system, thereby weakening the attachment of the plants to the substratum and making them more susceptible to storm damage. Sea urchins, such as *Paracentrotus lividus* will also eat members of the Laminariales, and one study (Norton, 1978) showed that, in their appetites, they exhibit a preference for *Saccorhiza polyschides* to *Laminaria saccharina*. In the last two centuries there has been a marked decrease in the shallow-water kelp forests in the North Pacific, due to the near extinction of sea otter. Prior to this, the sea otters ate large quantities of sea urchins that ate the kelp. With their normal predators gone, the sea urchins were free to devastate the kelp forests (Estes and Steinberg, 1988).

The species of Laminariales present in a particular area is determined by the environment of the area. Around Vancouver Island, British Columbia, Druehl (1967) noticed that there were three prevalent forms of *Laminaria: L. saccharina* and a long- and a short-stipe form of *L. groenlandica* (Fig. 17.22(*a*)). The two forms of *L. groenlandica* were found in surf, the long-stipe form in heavy surf, and the short-stipe form in moderate surf. *Laminaria saccharina* was found only in areas that had no surf. In the United Kingdom, Boney (1966) found that *Laminaria hyperborea* dominates in regions of moderate or severe wave action, whereas *L. saccharina* is in sheltered areas.

Even in the long term, populations of *Laminaria* can vary. Walker (1956) reported that a *Laminaria* population off the coast of Scotland varied in density over a 10-year period, and that this variation showed a strong correlation with sunspot activity and weather changes.

Concentric rings of dense tissue at the base of the stipe can be used to indicate the age of perennial laminarian algae such as *Laminaria* and *Ecklonia*. Dark rings are produced by cortical meristem during slow growth in autumn and winter, and pale rings are produced in winter and spring (Novaczek, 1981; Klinger and DeWeede, 1988).

Metabolism and composition

In the common British Laminariales (*L. hyperborea, L. digitata,* and *L. saccharina*) (Fig. 17.22), the proportion of laminarin and mannitol in the dry matter increases steeply during the active photosynthetic period from April to September. On the other hand, the proportion of alginic acid and cellulose in the dry matter decreases during this period. The opposite occurs from October to April, with the relative amounts of alginic acid and cellulose increasing. These variations are much greater in the frond, which has a high growth rate, than in the stipe, where the growth is slower.

Black (1954a) showed that in a mature frond of *L. saccharina*, where the part near the tip was 7 months old, there was a marked variation in composition along the length. Near the stipe (i.e., the actively growing region and, therefore, the youngest) there was, on a fresh-weight basis, about 3% mannitol and little or no alginic acid or laminarin. About a third of the way along the frond mannitol was at a maximum of about 6%, with laminarin 2% and alginic acid 2.5%, whereas two-thirds of the way up the frond the mannitol content was only 2%, with laminarin at 6% and alginic acid 4%. Variations in the composition of whole fronds can, therefore, be due largely to changes in proportions of old and new tissues.

According to Percival and McDowell (1967), seasonal variations in the Laminariaceae are consistent with the following observations:

1 Mannitol is the first product of photosynthesis to accumulate in appreciable quantities and is the main carbohydrate in tissues that are increasing by active cell division.
2 During photosynthesis in tissues that are largely growing by cell enlargement, there is an increase in the proportion of dry solids, made up of mannitol and salts of alginic acid and laminarin. Protein and cellulose are also being synthesized during this period. Formation of mannitol and laminarin continues after the other constituents have built up to a constant level in each unit of tissue, thereby increasing the dry solids content and reducing the proportion of alginate, cellulose, and protein on a dry-weight basis. Thus, in late summer, mannitol and laminarin are at high

levels and alginic acid, cellulose, and protein are at a minimum on this basis.

3 Laminarin may be formed from mannitol, and during active growth mannitol can be formed faster than its rate of conversion into laminarin so that both substances increase in amount. When growth slows down or stops owing to lack of nutrients, shortage of light, or low temperatures, laminarin increases with loss of mannitol. In late summer, there may be a temporary reduction in mannitol content owing to depletion of phosphate in the water, whereas the laminarin content does not drop until later.

4 During spore formation and periods when respiration is greater than photosynthesis, both laminarin and mannitol are used up. As there is little change in the amounts of other constituents, the proportion of alginate, cellulose, and protein, calculated on a dry-weight basis, increases.

Economic uses

The main group of Phaeophyceae with economic uses are the Laminariales or **kelps**. The word "kelp" in Europe refers primarily to the burnt ash of seaweeds. In America, however, the large brown seaweeds are known as kelps, as well as the ash that is prepared from them.

The first use of kelps was for kelp ash. The brown seaweeds were collected and dried on the shore, and the dried seaweeds were burned in a kiln with the product after burning being a hard cake. The cake constituted the kelp ash. The first use of kelp ash began sometime in the seventeenth century when French peasants used it for glazing pottery and making low-quality glass. This use lasted for about 200 years until the discovery of Barilla soda, made from certain coastal salt-rich plants. This substance produced a better-quality glass, and kelp ash ceased to be used for glass. In 1811, it was discovered that kelp ash contained large amounts of iodine. Using the best seaweeds properly burned, the kelp contained 1.4% to 1.8% iodine, or about 15 kg per ton. At that time iodine was in demand as a cure for goiter, an enlargement of the thyroid gland caused by lack of iodine (even today much of the table salt consumed is iodized, although not from kelp ash). In 1846, there were 20 manufacturers of iodine in Glasgow alone. Subsequently the discovery of mineral deposits of iodine, particularly in Chile, caused a decline in the kelp ash industry.

The current industrial use of kelps is for the alginate that they contain, which has a variety of uses. Algin was first discovered by Stanford in the early 1880s, although it was not obtained in a purified state till 1896 by Krefting. Algin comprises about 10% of the dry weight of the kelps (Smith, 1955), and is mostly the salt of alginic acid. The main area of algin production is the

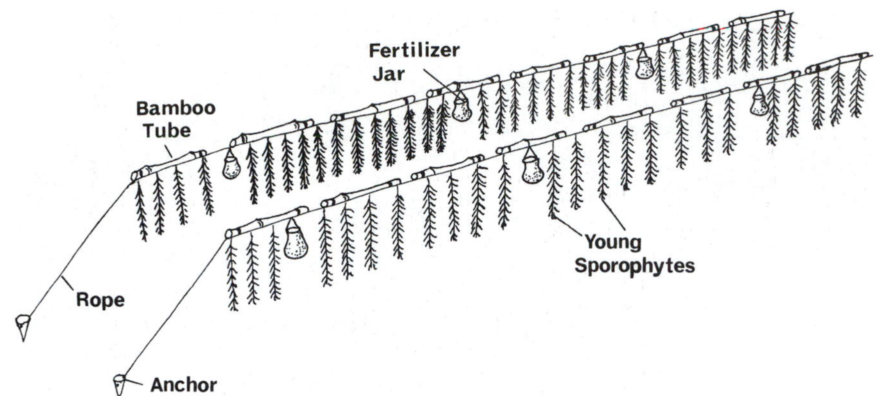

Fig. 17.26 Single-line bamboo rafts used for kelp cultivation in mainland China. The porous earthenware jars are filled with fertilizer that seeps out into the ocean. (After Cheng, 1969.)

water attached to the bamboo tubes. With proper care, sporophytes may reach 3 m or more in length within 4 to 5 months, and are then harvested (Cheng, 1969) (Fig. 17.27).

Classification

The Laminariales are divided into four families:

Family 1 Chordaceae: sporophyte hollow, whiplike, and not differentiated into a stipe and blade.

Family 2 Laminariaceae: transition zone with intercalary meristem not subdivided so that there is a simple primary stipe; sori not on special organs.

Family 3 Lessoniaceae: transition zone with intercalary meristem subdivided so that there are a number of secondary stipes in addition to the primary stipe.

Family 4 Alariaceae: sori borne on special sporophylls.

Chordaceae

This family has a single genus, *Chorda* (Fig. 17.28). The plant is an annual, hollow, whiplike alga that grows in the sublittoral regions of the Northern Hemisphere. The cylindrical sporophyte is long (up to 2.6 m) but seldom wider than 1 cm. The genus differs from other Laminariales in two characteristics: The sporophyte has a meristematic zone beneath the apex, and the young sporophyte has one or more apical hairs. In both of these characteristics, the genus resembles some members of the Dictyosiphonaceae of the Ectocarpales, such as *Dictyosiphon*. *Chorda* has characteristics that might be considered intermediate between the Ectocarpales and the Laminariales.

Fig. 17.27 Harvesting *Laminaria* grown on rafts in mainland China. (From Cheng, 1969.)

Chorda has a life cycle similar to *Laminaria* except that the gametophyte is monoecious (oogonia and antheridia on the same thallus) (Maier, 1984). Freshly released eggs give off a sexual hormone that causes explosive discharge of spermatozoids from the antheridia and subsequent chemotaxis toward the egg (Maier et al., 1984; Müller et al., 1985b).

Laminariaceae

In this family the sporophyte is divided into a holdfast, stipe, and blade. The blade is produced whole by the intercalary meristem even though it may later fragment into a number of digits. *Laminaria* has already been discussed. *Saccorhiza bulbosa* (Fig. 17.20) is an annual found on the Atlantic shores of Europe and North Africa. The mature sporophyte has a divided

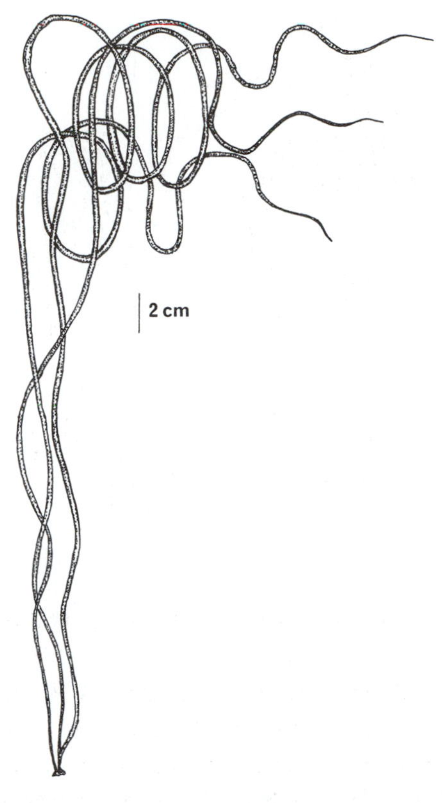

Fig. 17.28 *Chorda filum*. (After Taylor, 1957.)

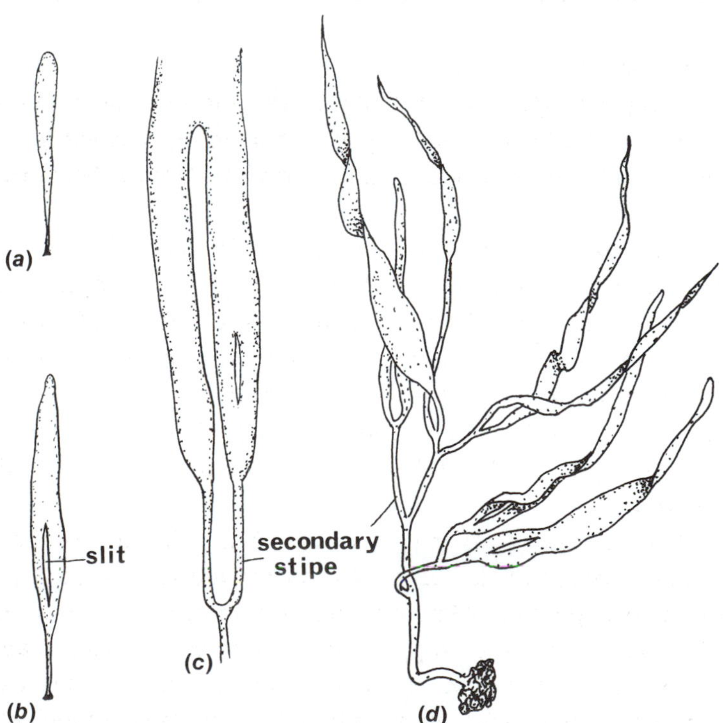

Fig. 17.29 *Lessonia nigrescens*. ((*a*)–(*c*)) Successive stages in the splitting of the blade to produce secondary stripes and secondary blades. (*d*) Mature plant. ((*a*)–(*c*) after Reinke, 1903; (*d*) after Postels and Ruprecht, 1840.)

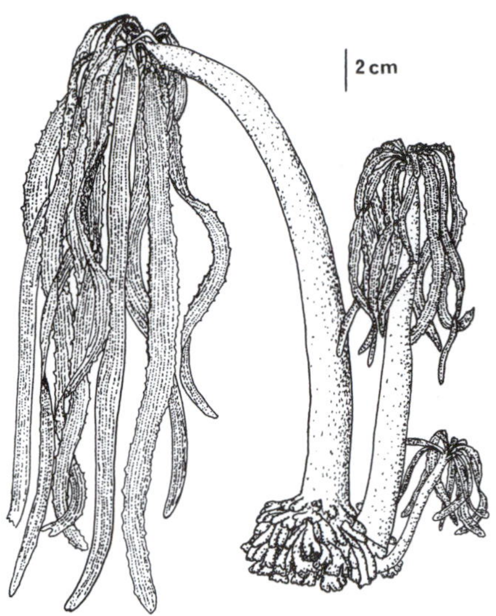

Fig. 17.30 *Postelsia palmaeformis*. (After Smith, 1969.)

(digitate) blade, sometimes exceeding 2 m in length, borne at the end of a flattened stipe. The stipe is spirally twisted in its lower portion, and below the twist there is a large inverted bell-shaped outgrowth that covers the holdfast and the basal part of the stipe. The bell develops in the young plant from the intercalary basal meristem, and the stipe subsequently develops undulating wings. The blades have cryptostomata with hairs.

Lessoniaceae

Although any splitting of the blade in the Laminariaceae does not extend down into the basal intercalary meristem, in the Lessoniaceae it does. This means that the secondary blades produced have their own secondary stipes.

Lessonia nigrescens, a species from the Southern Hemisphere, has entire young blades that soon develop a median split, resulting in a blade with two segments, each with its own secondary stipe (Fig. 17.29). The blades continue to split until the mature, much-divided sporophyte is formed. *Postelsia palmaeformis* (sea palm) has a short thick primary stipe that supports the secondary blades and stipes (Fig. 17.30). The plants are found on the Pacific Coast of North America in habitats exposed to the full violence of the waves.

The largest of all the algae are found in the genera *Nereocystis*, *Pelagophycus*, and *Macrocystis*. *Nereocystis luetkeana* (bull kelp) has a tough whiplike stipe up to 25 m long that terminates in a large air bladder (Fig. 17.31). This air bladder supports the secondary stipes and blades that hang down from the surface of the sea. The remarkable aspect of this large seaweed is that it is an annual plant that can grow as much as 6 cm a day.

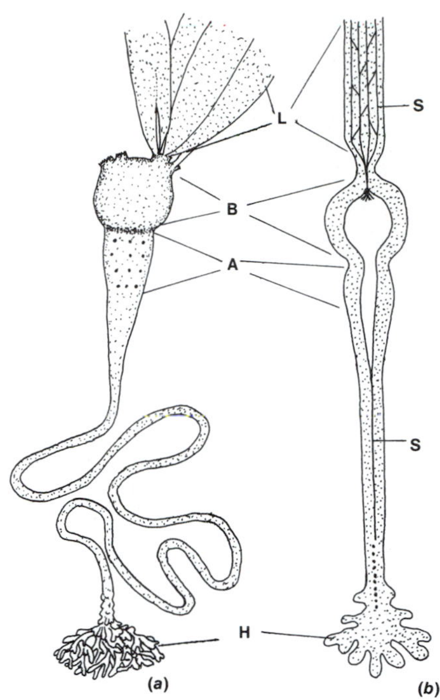

Fig. 17.31 Sporophyte of *Nereocystis*. (*a*) Whole plant. (*b*) Semidiagrammatic longitudinal section. (A) Apophysis of stipe; (B) gas-filled bulb of stipe; (H) holdfast; (L) lamina; (S) sieve filaments. (After Nicholson, 1970.)

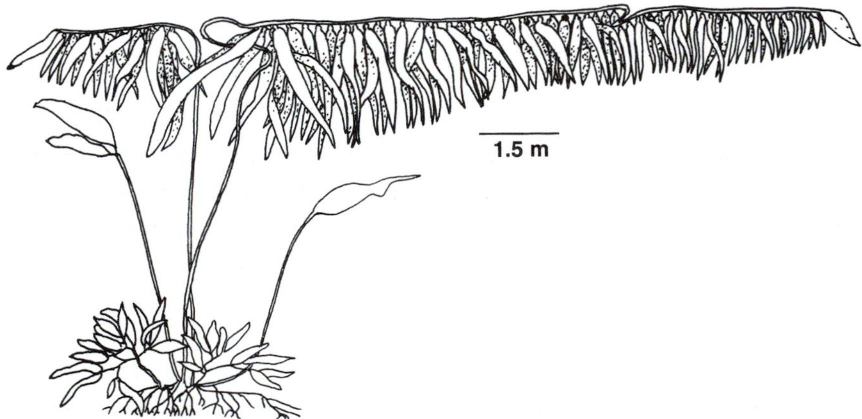

Fig. 17.32 *Macrocystis pyrifera*. (After Chapman, 1970.)

Julescraneia grandicornis is a fossil alga from Miocene diatomite of California that resembles *Nereocystis* (Parker and Dawson, 1966). *Macrocystis* (Fig. 17.32) may grow up to 50 m in length and has a life of 5 years, although the individual secondary blades have a life of only 6 months. As in *Lessonia* there is successive splitting of the primary blade, but in *Macrocystis* the growth of one of the two segments is arrested. This leads to a long curtain type of thallus. Each of the segments has an air bladder at the base of the secondary stipe, allowing the secondary blade to hang down from the surface of the water.

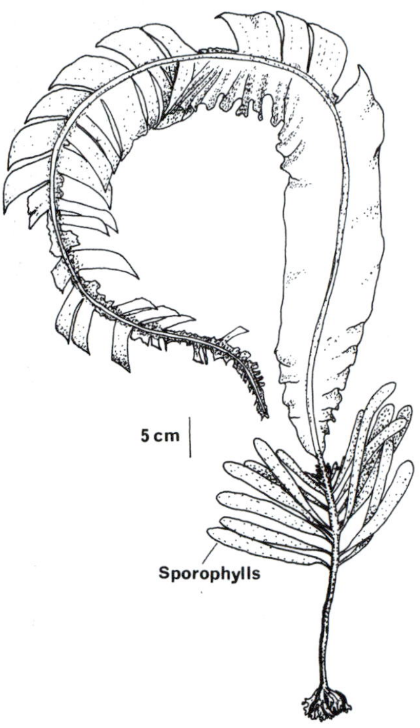

Fig. 17.33 *Alaria esculenta*. The old distal part of the blade is partly eroded away, and the sporophylls are fully developed. (After Taylor, 1957.)

Alariaceae

The sporophytes in this family have the sori formed on special sporophylls. *Alaria* (Fig. 17.33) has a lamina with a wavy margin and a midrib. The short stipe produces thick tongue-shaped sporophylls in the summer, which, after maturation of the sori, are shed during autumn and winter, leaving scars on the stipe. During the winter the blade wears down to the basal meristematic zone, and a new blade is produced the following season.

SPHACELARIALES

This order is characterized by an apical meristematic cell that divides transversely to produce the daughter cells. *Sphacelaria* grows attached to rocks or other algae and has one or more freely branched shoots arising from a discoid holdfast. The apical cell undergoes transverse divisions, with subsequent longitudinal septation of the daughter cells to produce a polysiphonous structure. Although the maturing axes and branches undergo septation into smaller and smaller cells, they do not enlarge; thus the diameter of the filament is essentially the same from the base to the apex (Fig. 17.34). Older axes, though, may become corticated by downward-growing filaments. The erect axes are usually abundantly branched, usually in a regular, distichous manner.

A characteristic method of asexual reproduction is by means of **propagula** (Fig. 17.34), which are small specialized branchlets of distinctive form that are produced throughout the vegetative parts of the plants. They are formed much more frequently than sporangia or gametangia. Each propagule has an apical cell and usually two to three protuberances. After falling from the parent plant and contacting a suitable substrate, the propagule develops into a new plant. Propagula are formed only at temperatures above 12 °C and under daylight conditions longer than 12 hours (Colijn and van den Hoek, 1971).

In *S. bipinnata*, the sporophyte forms both unilocular and plurilocular sporangia terminally on branches (Fig. 17.34). The plurilocular sporangia produce zoospores that re-form the parent sporophyte. Meiosis occurs in the production of zoospores in the unilocular sporangia (Clint, 1927). Over 200 zoospores are released through an apical pore in the unilocular sporangium (Papenfuss, 1934). The zoospores germinate to presumably form gametophytes that are similar to the sporophytes. The gametophytes produce plurilocular gametangia of one type, which release isogamous gametes. The fusion of gametes takes place while they are motile and produces a quadriflagellate zygote that may continue moving for several hours. The life cycle of another species, *S. furcigera*, involves anisogamy and unisexual gametophytes, which are somewhat smaller than the sporophytes (van den Hoek and Flinterman, 1968). The life cycle is controlled by temperature and photoperiod.

The family is related to the Ectocarpaceae through the sphacelarian genus *Choristocarpus*, a uniseriate filamentous alga.

DICTYOTALES

This order has organisms that grow by means of an apical cell or by a marginal row of apical cells. There is an isomorphic alternation of erect, flattened, parenchymatous thalli. A distinctive character of this order is the modification of the unilocular sporangia to produce four to eight large aplanospores. Sexual reproduction is oogamous. The Dictyotales are common in warmer waters throughout the world.

Dictyota dichotoma has a single apical cell that forms the flattened annual thallus (Fig. 17.35). The mature thallus consists of three layers: a middle layer composed of large cells with few or no chloroplasts surrounded on both sides by a layer of small cells densely packed with chloroplasts. Gametophytes form sex organs in projecting sori. Gametogenesis can be artificially induced by exposure of the gametophytes to blue light (Kumke, 1973). An oogonium develops from a surface cell that divides into a stalk cell and the oogonium proper. Each oogonium produces a single egg,

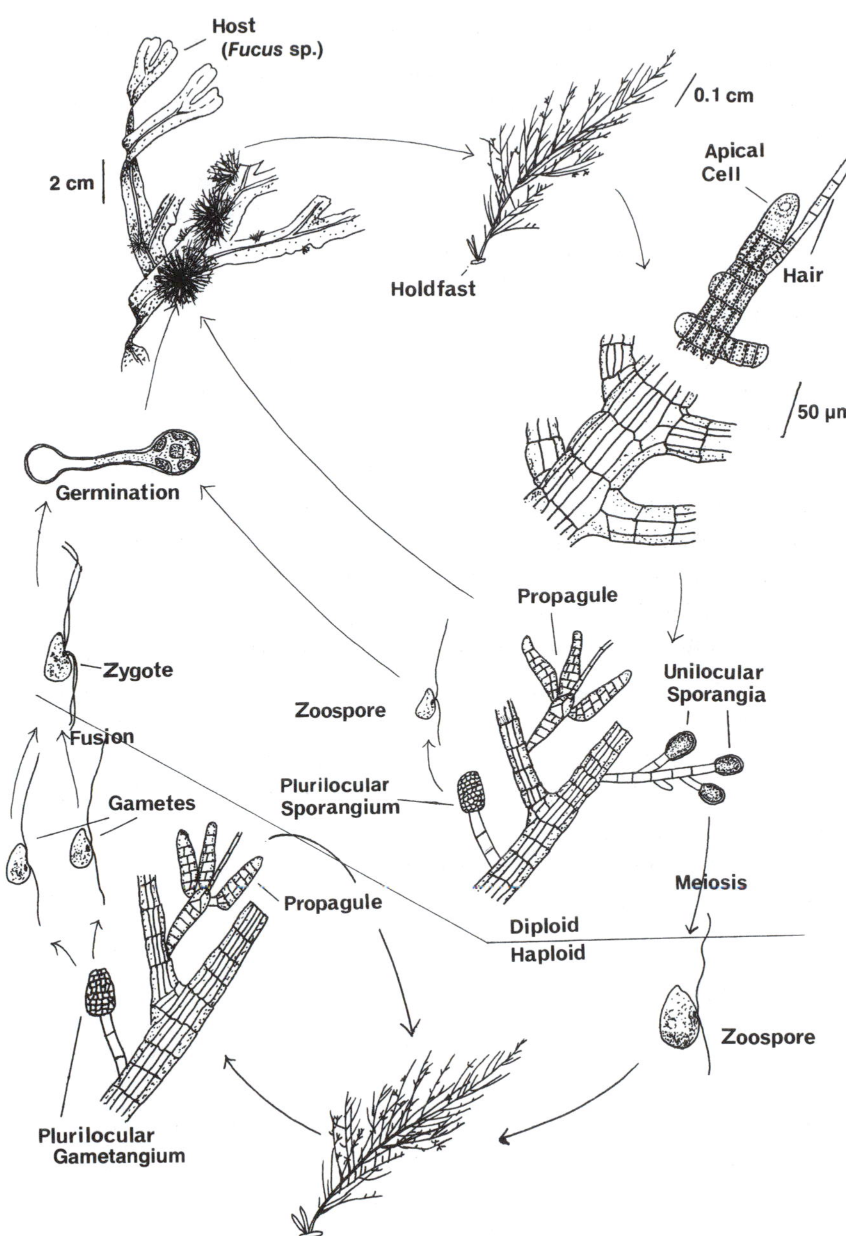

Fig. 17.34 The life cycle of *Sphacelaria* (*S. cirrhosa and S. bipinnata*). (Adapted from Savaugeau, 1900–14; Papenfuss, 1934; Taylor, 1957.)

which is liberated through the gelatinized apex of the wall. There are usually 25 to 50 oogonia in a sorus with sterile oogonia at the margin. The deep-brown color of the female sori contrasts with the white glistening spots that comprise the male sori. The male sori can be recognized early in their development by the disintegration of the chloroplasts in the cells. Like the oogonia, the antheridia develop from surface cells. These cells enlarge and divide horizontally into a stalk cell and a primary spermatogenous cell. This

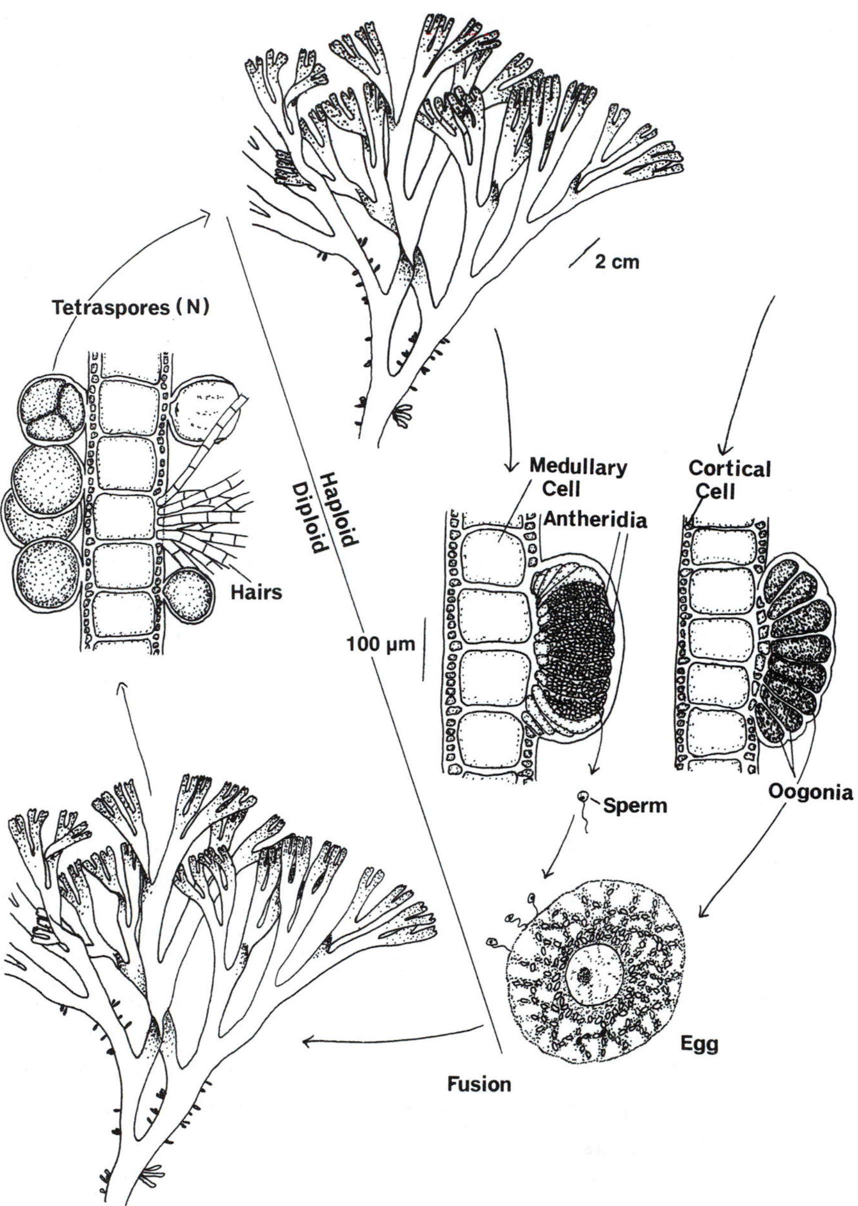

Fig. 17.35 The life cycle of *Dictyota dichotoma*. (Adapted from Thuret and Bornet, 1878; Williams, 1898; Taylor, 1960.)

cell divides and redivides in vertical and horizontal planes into between 650 and 1500 compartments (Williams, 1904). The content of each locule becomes a pear-shaped sperm with a single, laterally inserted, tinsel flagellum and an anterior eyespot (Phillips and Clayton, 1993). Although there is only one emergent flagellum, a second basal body is present (Manton, 1959), indicating a derivation from a biflagellate ancestor. The mature sperms are set free by dissolution of the walls of the antheridium.

The male sorus is surrounded by elongated sterile cells that are regarded as undeveloped antheridia. The sperm fertilizes the egg to produce the zygote that germinates into the sporophyte; unfertilized eggs can germinate parthenogenetically but seldom develop normally and soon abort. The sporophytes produce haploid aplanospores (tetraspores) on the surface of the thallus. The tetrasporangia occur singly or in small groups. The naked tetraspores are released by gelatinization of the apex of the sporangium, and soon after liberation the large motionless spores secrete a cellulose wall and develop into the gametophytes.

In *D. dichotoma*, the gametes are released at regular intervals. This was first noticed by Williams (1905) in Great Britain, where the gametes are released fortnightly. Müller (1962) showed that moonlight is the synchronizing factor for the release. When he grew the alga in natural light, gametes were released every 14 to 15 days. If the alga was grown under artificial conditions with a 14 hours light: 10 hours dark cycle, then few gametes were released, and there was no synchronony. If the artificially lighted cultures had the lights left on all night, then 10 days later a burst of gametes was released. The lights being left on all night simulated moonlight.

The only calcified genus in the Phaeophyceae, *Padina*, is in the Dictyotales.

FUCALES

The organisms in this order are parenchymatous with growth from an apical cell. The haploid generation is reduced to the egg and sperm, with the remainder of the life cycle being diploid. The gametes are borne in special cavities, the **conceptacles**, and gametic union is always oogamous. Conceptacles may be scattered over the surface of the thallus, but more frequently they are limited to the inflated tips of special branches, the **receptacles.** The Fucales are worldwide in distribution, but those of the Arctic and north temperate seas differ considerably from those of the Antarctic and south temperate waters. *Fucus* (Fig. 17.38) is a common genus in northern waters, whereas in tropical and subtropical waters *Sargassum* (Figs. 17.43 and 17.44) is present. In Australian waters, *Cystophora* is a predominant member of the flora, with the large *Durvillea* being common in sub-Antarctic waters.

Morphology and anatomy

The genus *Fucus* will be used as the representative genus in this order (Fig. 17.38). The thallus is much branched and is supported by a short narrow stalk that is attached to a discoid holdfast. The branching is dichotomous, with each flattened segment having a prominent central midrib surrounded on both sides by a narrower wing. The wings usually bear scattered **cryptoblasts**, which are basically sterile conceptacles with large numbers of hairs, that facilitate the uptake of nutrients from the seawater (Hurd et al., 1993). At certain times of the year, the tips of the branches are swollen into receptacles that contain the fertile conceptacles. The inflation of the receptacles is due to the production of a large amount of mucilage.

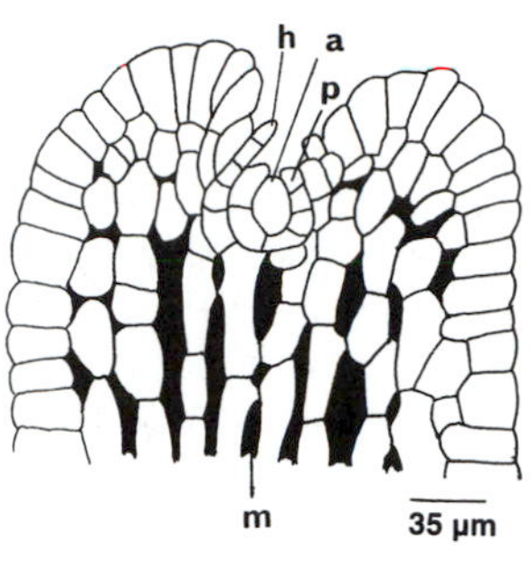

Fig. 17.36 Upper portion of a mature juvenile thallus of *Fucus*, showing the apical depression, apical cell (a), remnants of terminal hairs (h), the promeristem (p), and the medulla (m). (After Oltmanns, 1889.)

Each branch has an apical cell at its apex (Fig. 17.36). The apical cell divides several times a year, resulting in the formation of a dichotomy or fork, with one arm of the fork being longer than the other. The apical cell in mature Fucaceae is a four-sided pyramid with a flattened base. In the other families and in young Fucaceae, the apical cell is a three-sided pyramid. According to Moss (1967), the apical cell itself does not divide (except in the formation of forks), and instead stimulates the cells around it to divide. Thus a meristematic zone extends beneath and around the sides of the apical cell and is called the promeristem. The cells of the promeristem, similar to meristematic cells in other organisms, are very small. At the sides of the promeristem the cells enlarge and divide only transversely, yielding the flattened wings. Mucilage is deposited between the derivatives of the promeristem, causing the files of cells to separate, remaining in contact only where there are pits. The apical cell shows apical dominance, inhibiting the development of laterals beneath it (Moss, 1965, 1970). If the apical cell is destroyed, then the underlying lateral will develop into a new apical cell.

The second meristem in the apical area of the thallus is the meristoderm or outer row of cells derived from the promeristem. This is a closely packed layer of brick-shaped cells that divides anticlinally at first and then periclinally to produce new tissues to the inside.

The anatomy of the Fucales is similar to that of the Laminariales, with a mucilaginous cuticle covering the epidermal layer of cells (Figs. 17.36 and 17.37). Inside this is the cortex with the medulla in the center. Hyphae are produced by the inner cortical cells, but there are no trumpet hyphae present. There is an orientation of organelles within the epidermal cells. They have an outer layer of alginic acid vesicles with a basal nucleus and chloroplasts (Fulcher and McCully, 1969; Rawlence, 1973). The cap of alginic acid vesicles may shield the chloroplasts and nucleus from intense illumination, especially at low tide when the plants are usually exposed. The organelles of the cortical cells are arranged just the reverse of the epidermal cells, having an outer layer of chloroplasts. The chloroplasts of the medullary and hyphal cells are much reduced.

Like the Laminariales, the Fucales are able to translocate organic materials (Floc'h and Penot, 1972). Mannitol is the form of photosynthate translocated. The growing apex acts as a sink, with the mannitol translocated to

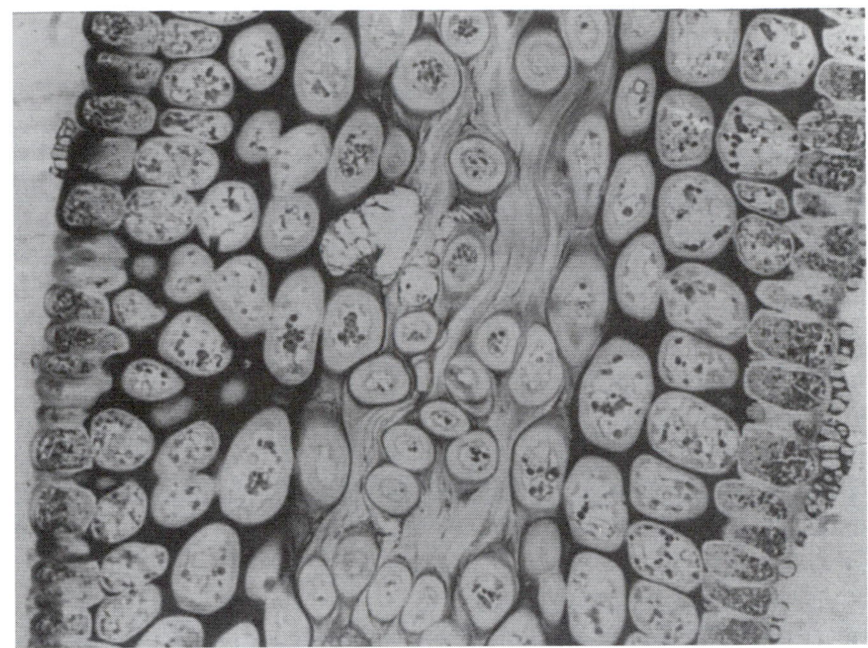

Fig. 17.37 *Fucus virsoides*. Transverse section of a mature thallus. A mucilaginous cuticle covers the outer epidermal layer of cells. Inside the epidermal layer is the cortex with the medulla in the center. (From Mariani et al., 1985.)

the growing apex from the blades of the alga (Diouris, 1989). Structurally, the Fucales have a system of conducting elements very similar to the Laminariales. In the Fucales, the sieve elements run from the apical meristem to the base of the plant. The medullary elements have sieve plates (1 μm thick) at their ends. The pores in the sieve plates enable a continuous system of cytoplasm for the translocation of materials both longitudinally and transversely through the cross connections. The pores in the sieve plates are 0.1 μm^2 or less, which makes them smaller than the pores reported in the Laminariales (*Alaria* has pores ranging in size from 0.1 to 0.3 μm^2) (Moss, 1983).

Air bladders (gas vesicles) originate not far from the apex as a result of growth of the surface layers of cells accompanied by an increase in the thickness of the cortex. This leads to the rupture of the medulla, the remnants of which are commonly around the edge of the hollow. The air bladders are apparently filled with gases similar to those in the atmosphere.

Fucus is subjected to the rigors of the intertidal zone, and the plants are constantly being damaged by ice, browsing animals, and heavy wave action. When vegetative tissues are damaged, regeneration of new tissue in the form of adventitious branches occurs readily from wound surfaces in the midrib region of the thallus. After being damaged the cells of the exposed tissue divide to form a callouslike layer that differentiates the adventitious branches (Fulcher and McCully, 1969).

Life cycle

Fucus will again be used as an example of a typical member of the Fucales (Fig. 17.38). The gametes are borne in conceptacles that are similar to the cryptoblasts except that the colorless hairs are restricted to a small area near the aperture. The wall of the conceptacle is lined with flat cells that bear branched paraphyses with few chloroplasts. The conceptacles originate from an initial that is a superficial cell of the thallus. The initial divides into an outer tongue cell and an inner basal cell. The tongue cell either degenerates or does not contribute to the development of the conceptacle. The basal cell then divides to form the floor of the conceptacle. At the same time, the cells surrounding the original initial grow and divide so that the derivatives of the initial cell become open to the outside. The mature conceptacle consequently has cells lining the floor derived from the conceptular initial and the cells lining the walls derived from the cells surrounding the original initial.

The plants are either monoecious or dioecious, and in the monoecious forms the antheridia and oogonia can be in the same or different conceptacles. The antheridia are usually formed on paraphyses. Antheridial parent cells are distinguished by dense cytoplasm with few vacuoles, a large central vacuole, and chloroplasts with only a few thylakoids (Berkaloff and Rousseau, 1979). Following meiosis, which occurs during the first two divisions of the primary nucleus, the nuclei undergo four mitoses. Mature antheridia thus contain 64 nuclei, each of which becomes incorporated into a spermatozoid. The wall of the mature antheridium is composed of two layers. At liberation, the outer wall ruptures, releasing the inner wall containing the spermatozoids and mucilage. This packet passes out of the conceptacle and into the sea, where the inner wall gelatinizes at one or both ends, releasing the spermatozoids.

The spermatozoids are spherical at first and then uncoil to give the elongated biflagellate form (Fig. 17.39). There is an eyespot consisting of a single layer of pigment globules inside a reduced chloroplast. The basal portion of the posterior flagellum is closely applied to the plasmalemma in the area of the eyespot. The anterior portion of the cell contains 13 microtubules that make up the **proboscis**, a structure that may function in the detection of the female sex attractant. The proboscis microtubules pass from the area of the basal bodies, extending themselves in one plane in front of the spermatozoid, and then pass beneath the plasmalemma to the posterior portion of the cell (Manton and Clarke, 1950, 1951, 1956).

The oogonia are usually borne on a stalk cell that is embedded in the wall of the conceptacle. The oogonial cell undergoes three nuclear divisions, yielding eight haploid nuclei. The cytoplasm then cleaves into eight eggs.

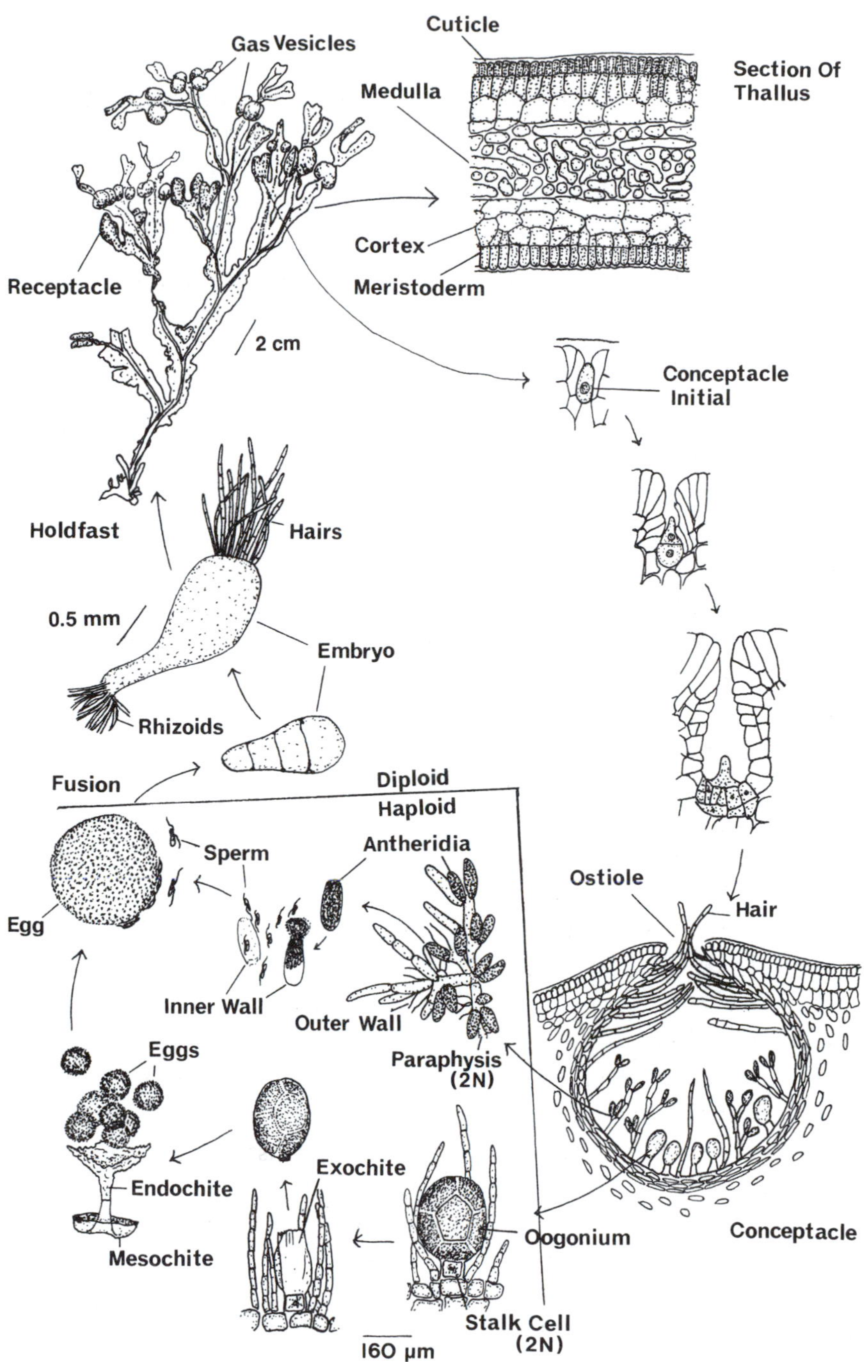

Fig. 17.38 The life cycle of *Fucus* sp. (*F. vesiculosus* and *F. serratus*). (Adapted from Thuret, 1854; Oltmanns, 1889; Nienburg, 1931; Taylor, 1957.)

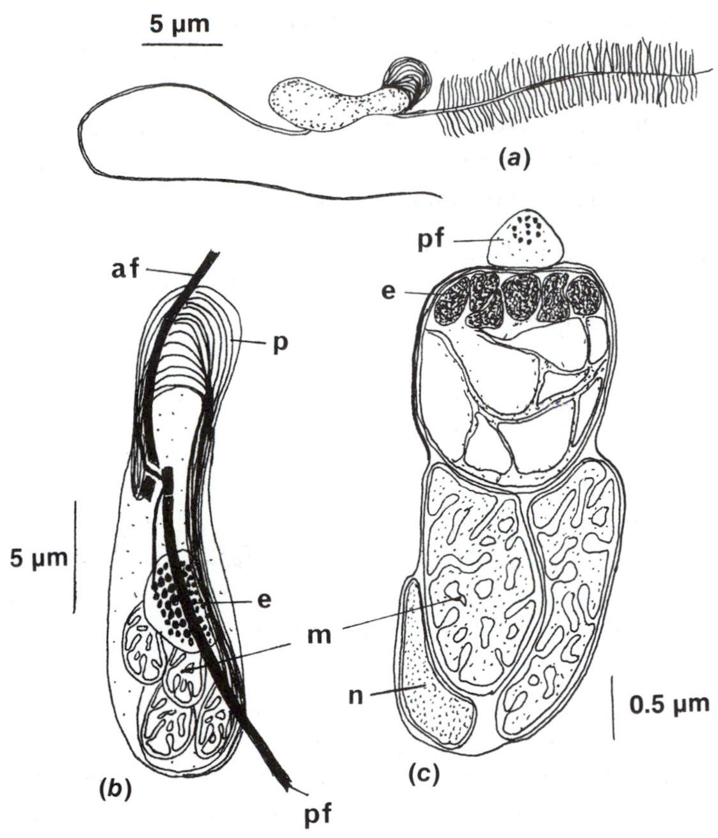

Fig. 17.39 Spermatozoid of *Fucus*. (*a*) Whole spermatozoid. (*b*) Semidiagrammatic drawing showing the anterior flagellum (af), eyespot (e), mitochondria (m), proboscis (p), and the posterior flagellum (pf). (*c*) Section through a spermatozoid illustrating the close appression of the posterior flagellum (pf) to the eyespot area (e) of the reduced chloroplast. (m) Mitochondrion; (n) nucleus. (After Manton and Clarke, 1956.)

The wall of the oogonium is composed of three layers, the thin outer layer or **exochite**, the thick middle layer or **mesochite**, and the thin inner layer or **endochite**. When the oogonium is mature, the exochite ruptures, releasing the packet of eggs, still surrounded by the other two wall layers, into the sea. In the sea, the mesochite ruptures apically, slips backward, and exposes the eggs within the endochite. The endochite rapidly dissolves, releasing the eggs.

The sperm are attracted to the eggs by a species-non-specific pheromone, **fucoserraten**, released by the eggs (Müller and Jaenicke, 1973). The species-specific recognition between eggs and sperm is based on specific oligosaccharides on the eggs and sperm. The oligosaccharide side chains of the egg-surface glycoproteins contain fucosyl, mannosyl, and/or glucosyl residues (Wright et al., 1995a). The surface of the egg is not homogeneous, but instead is organized into different domains, each containing different glycoproteins (Stafford et al., 1992). Likewise, the sperm contain glycoproteins organized into domains on the anterior flagellum plasma membrane, the mastigonemes of the anterior flagellum and the sperm body (Jones et al., 1988). On reaching the egg surface, sperm exhibit a characteristic behavior involving movement over the plasma membrane of the egg and a probing or "searching" of the egg membrane with the anterior flagellum (Brawley, 1991). The glycoproteins on the sperm eventually bind to the complementary glycoproteins on the egg. This results in two "blocks" to further sperm penetration (Wright et al., 1995b):

1 A "fast block" within seconds caused by depolarization of the plasma membrane due to electrogenic Na^+ and Ca^{2+} influx. Excess sperm detach from the egg surface following depolarization.
2 A "slow block" results from the formation of a cell wall around the zygote by the release of cellulose and alginates from cortical vesicles to form the **glycocalyx** (Motomura, 1994).

The male nucleus migrates to the female nucleus along associated microtubules. As it migrates, the nuclear envelope of the male nucleus breaks up. The egg nucleus becomes convoluted along the surface nearest the advancing male nucleus. Immediately prior to nuclear fusion, many egg mitochondria accumulate in the vicinity of the male nucleus (Brawley et al., 1976).

The spherical zygote germinates by forming a primary rhizoid from one side, while the rest of the zygote gives rise to the embryo. There are a number of stages in the determination of which part of the zygote will develop into the rhizoid and which will develop into the embryo (Love et al., 1997):

1 **Apolar** – In the early stages of development, the fucoid zygotes are apolar and have no inherent protoplasmic order.
2 **Axis induction** – A potential polar axis is generated within the zygote. The axis is determined by a number of environmental stimuli, including the direction of incident light, the position of neighboring zygotes, water currents, chemical or ionic gradients and electrical fields. During axis induction, the polarity is labile and can be re-oriented by the subsequent exposure to a stimulus in a different direction.
3 **Axis fixation** – This stage begins when the axis is fixed and no longer susceptible to re-orientation by the environment. The polarity is stabilized by interactions between cell wall components, cortical vesicles and microfilaments (Kropf et al., 1992; Henry et al., 1996).
4 **Rhizoid germination** – The spherical symmetry of the zygote is broken by the emergence of a polar bulge representing the emergence of the rhizoid. The direction of the rhizoid germination defines the direction of the polar axis. Unequal cytokinesis follows, perpendicular to the direction of rhizoid germination, yielding an embryo consisting of two freshly differentiated cells, a large rounded thallus cell, which is the precursor of

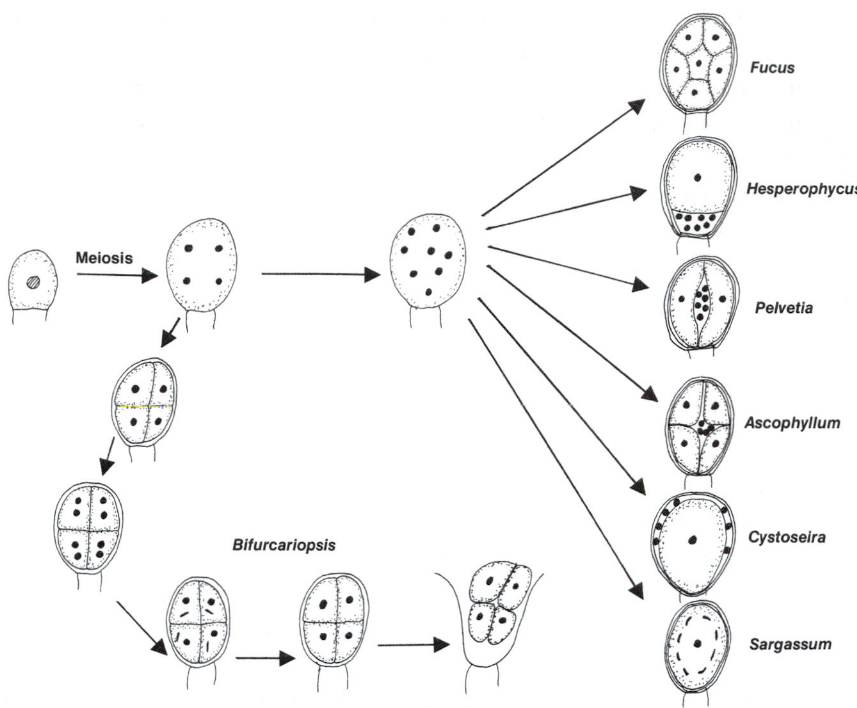

Fig. 17.40 Diagram showing the different types of development of the female reproductive structure in the Fucales. (After Smith, 1955; Jensen, 1974.)

the frond and receives most of the chloroplasts, and a smaller rhizoid cell, which generates the stipe and holdfast.

The embryo then divides to form a minute cylindrical plant with at least one apical cell with trichothallic growth, producing a hair above and the thallus below. This feature is probably an indication of the derivation of the Fucales from an ectocarpalean ancestor with trichothallic growth, such as *Splachnidium* (Fig. 17.10) or *Notheia* (Nizamuddin and Womersley, 1960; Saunders and Kraft, 1995). Eventually the apical cell ceases production of the hair and becomes the three-sided apical cell, which later becomes four-sided. It is only after the initiation of the apical cell that the thallus begins to assume its mature flattened shape.

The most primitive type of fucalian female reproductive structure is that in *Bifurcariopsis* of the Cystoseiraceae (Jensen, 1974) (Fig. 17.40) *Biburcariopsis* produces a unilocular sporangium containing four oogonia, each surrounded by two walls, the inner being the oogonial wall and the outer the "tetraspore" wall. Each oogonial nucleus divides to form a nucleus that functions as the female gamete and a second nucleus, which soon degenerates, that represents the remains of the female gametophyte. From this primitive condition, only one oogonium is formed, which takes a number of different forms (Fig. 17.40): (1) the *Fucus* type, in which there is a cleavage into eight uninucleate eggs; (2) the *Ascophyllum* type, in which four uninucleate eggs are formed, and four degenerate nuclei are extruded between them; (3) the *Pelvetia* type, with two uninucleate eggs and six degenerate nuclei; (4) the *Hesperophycus* type, with one large uninucleate egg and a small seven nucleate spore; (5) the *Cystoseira* type, in which seven nuclei are extruded; and (6) the *Sargassum* type, in which all but one of the nuclei degenerate after fusion of gametes.

In determining phylogeny within the Fucales, five main characteristics are used: (1) A three-sided apical cell is more primitive than a four-sided one; (2) radial branching is more primitive than branching in one plane; (3) a sperm without a proboscis is more primitive than a sperm with one (Manton et al., 1953; Manton, 1964); (4) chloroplasts with pyrenoids are more primitive than those without them (Evans, 1966, 1968); and (5) sperm with an anterior flagellum that is longer than the posterior are more primitive than the reverse. Taking all of the above characteristics into consideration, the Fucaceae are seen as the most advanced family and the Cystoseiraceae as the most primitive.

The Fucales show a periodicity in the formation of receptacles and conceptacles. In *Fucus* and *Ascophyllum* in the North Atlantic, receptacle initiation is a short-day phenomenon, with receptacles being initiated in a 8 : 16 and 12 : 12 light–dark photoperiod and inhibited under a 16 : 8 and continuous light photoperiod (Bird and McLachlan, 1976; Terry and Moss, 1980). White light in the dark period inhibits the short-day response. In the field, conceptacle development commences during September and October, and continues during the following spring until gametes are mature and are liberated during April and May. Following gamete discharge, the conceptacle-bearing receptacles and the rest of the lateral shoot are shed. *Halidrys* (Fig. 17.42) (Moss and Sheader, 1973) has a different periodicity, with vegetative growth in the spring and summer followed by initiation of the conceptacles. The gametes are then shed during the winter. Even though this is the darkest period of the year, the gametes are able to germinate and secure themselves to a substrate with their rhizoids. These germlings are then able to sustain themselves during the periods of little light and subsequently grow normally as soon as there is sufficient daylight. Many of the Fucales show a temperature tolerance similar to that of the Laminariales, with 20 °C being the highest temperature at which eggs of *Halidrys* will germinate.

Ecology

Both the geographical distribution and the location on the shore of a member of the Fucales depend on the ability of the fertilized egg to settle and germinate under the environmental conditions present (Chapman,

1995). Embryos of *Pelvetia fastigata* will almost all survive if they settle under adult *Pelvetia* thalli. Those that settle on exposed rock will almost all die. Within red-algal tufts, most of the younger embryos survive, with survival declining with the increasing age of the settling embryos (Brawley and Johnson, 1991). Movement of water influences the attachment process. It has been shown that a particular direction of current flow will have a marked effect on the subsequent distribution of *Fucus* germlings. In addition, the newly attached zygotes also fall prey to browsing mulluscs – in particular, the limpets and species *Littorina* or periwinkles. The mollusc pressure on fucoid development is one of the key factors in determining the number of plants present.

Lodge (1948) carried out an experiment on the Isle of Man to determine the rate of recolonization of a shore with primarily fucoid algae. A strip of shore 5 m wide was cleared of all macroscopic biological life. In the first spring after the clearance, green algae (*Enteromorpha*, *Urospora*, and *Chaetomorpha*) covered the shore, along with diatoms. *Fucus* germlings then developed beneath the green algae, starting first at the high-tide mark and then establishing themselves toward the low-water mark. The main *Fucus* species during the first year was *F. vesiculosis*, whereas during the second year *F. serratus* became prominent. After the fucoids became established, the green algae gradually disappeared, and a sparse undergrowth of red algae appeared (*Dumontia*, *Laurencia*). After two years the *Fucus* plants dominated the shore, covering a more diverse undergrowth. By this time the limpets had started to recolonize the shore, slowing down colonization by algae. Seven years after the beginning of the experiment, the shore had returned to its original condition.

The zonation of fucalian species on the shore may be determined by the ability of the zygotes to germinate in the dark. *Fucus* spp. and *Ascophyllum nodosum* zygotes can germinate in dark or light, and therefore will germinate in rock crevices and under other light-shading algae. *Pelvetia canaliculata* and *Halidrys elongata* zygotes, however, do not germinate in the dark and are more likely found on exposed rock surfaces and ledges where light is not limiting.

The littoral Fucales are adapted to the difficult conditions in which they live, being able to withstand freezing temperatures and summer temperatures up to 34 to 36°C. The zonation of the different fucoid species is due partly to their ability to photosynthesize better when exposed to air (Madsen and Maberly, 1990) and partly to withstand desiccation during both germination and growth. The plants that live higher up in the littoral zone have thicker walls, more fucoidin, and a higher water content, and reach their dry weight on evaporation later than those lower in the littoral zone. The proportion of polysaccharides also reflects the fucoid position in the littoral zone. *Fucus spiralis* and *Pelvetia canaliculata*, which grow highest in the littoral zone, contain the highest amount of fucoidin, 18% to 24% of the dry weight. *Fucus serratus*, which grows near the low-tide mark, has much less fucoidin, about 13% on a dry-weight basis (Black, 1954b).

Generally the life-span of most shore fucoids is about 2 to 3 years (Boney, 1966). The only exception to this being *Ascophyllum* (Fig. 17.47), which has an average age from 12 to 15 years. The average age of the *Ascophyllum* plants will vary according to their position in the littoral zone. On the Welsh coast, plants from the top of the littoral zone were found to be 4 to 5 years old, whereas those from the bottom of the zone were 5 to 15 years old (David, 1943). Although being long-lived, *Ascophyllum* produces a paucity of sporelings and it takes a couple of decades for recolonization of a denuded area. This has resulted in severely depleted populations in area where the alga is commercially exploited (Bacon and Vadas, 1991).

Morphology of fucoid plants will vary with environment. Moss and Sheader (1973) showed that in *Halidrys siliquosa* the germlings at 10 °C in total darkness produced long rhizoids and a short thallus. If the same plants were grown at a high light intensity (5936 lux), the thalli were long and unbranched, whereas the rhizoids remained short but were pigmented. If the thalli were grown at 20 °C under high light intensity, branched thalli were obtained. Vesiculation will also vary with the environment. *Fucus vesiculosus* growing in areas subjected to very severe wave action will lack vesicles, whereas those growing in calmer areas have gas vesicles. A minimum branch length seems necessary before vesicle formation can begin, and vesiculation is postponed to the following year if growth has not attained this minimum length (Boney, 1966).

Although the members of the order are normally lithophytes, they are also widely represented by unattached growth-forms lacking holdfasts and propagating mostly vegetatively. These free-living forms arise by vegetative growth of detached branches of the normal attached form that have been transported to a sheltered habitat, or by the development of zygotes in a quiet environment into unattached plants. These unattached plants are referred to as **ecads**, a term for a plant whose morphology has been altered by growth in an unusual environment. Most of these ecads are found in sheltered habitats such as bays and salt marshes.

In discussing salt marsh ecads, Boney (1966) states that the fucoids show the following characteristics: (1) vegetative propagation as the main means of propagation; (2) absence of a holdfast; (3) dwarf habit; (4) spiral twisting of the thallus; and (5) profuse branching. Many of the salt marsh forms grow embedded (but not attached by holdfasts) in a muddy substratum, and some are entwined around the stem bases of the dominant angiosperms. The protection afforded by the canopy of angiosperms enables the fucoids to survive at high levels in the marshes.

Ascophyllum nodosum ecad *mackaii* is an unattached form common in Scotland (Gibb, 1957; Moss, 1971). Normally if a plant of *Ascophyllum* becomes detached from rocks in the intertidal zone, it is cast up by the waves and soon disintegrates. In the unusually calm waters at the head of some Scottish lochs, the detached thalli are gently covered and uncovered as the tide advances and recedes, but they never dry out and disintegrate. The external forms of the attached plant and of the ecad are in complete contrast. The attached plant is flattened in one plane and generally dichotomizes once a year in the spring after differentiation of an air bladder. During the summer a series of lateral nodes are developed, from which receptacles are produced as laterals the following year. This yearly cycle of differentiation is completely lacking in the marsh form of the ecad. Here there are no air bladders and no regular dichotomy to mark off one year's growth from another. Instead, apical branching is frequent and in all planes, giving rise to the characteristic cushion form of the ecad. Also there are no lateral nodes and thus no lateral meristems to give rise to lateral branches. Instead, receptacles are sometimes differentiated behind the apices of any branch, apparently at random.

The form of the ecad is caused by the destruction of the apical meristem and the lateral nodes of the ecad. The branches are eventually regenerated from wound-healing tissue to give rise to the reduced ecad form. Growth of the ecad is slow, and large tufts on the shore may have taken several years to grow. The ecad often still has a prominent apical cell, but there is little meristematic activity when compared to the attached plant. This results in a thallus with little cortex.

The receptacles of the ecad are small when compared to those of the attached plant. The eggs are also small and of variable size. The number of germlings produced from the ecad eggs is low. The time of gamete discharge in the ecad approximates that of nearby attached plants.

Classification

The four most prominent families in the order are as follows:

Family 1 Cystoseiraceae: thallus with a three-sided apical cell; no specialized branch systems.

Family 2 Sargassaceae: thallus with a three-sided apical cell; specialized branch systems present.

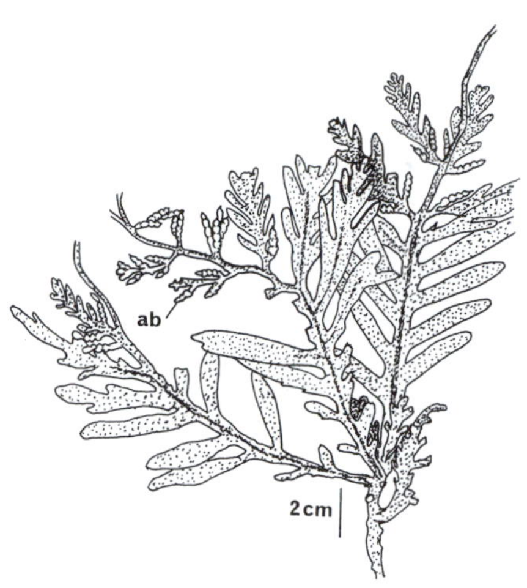

Fig. 17.41 *Cystoseira osmundacea*. (ab) Air bladder. (After Smith, 1969.)

Family 3 Himanthaliaceae: thallus differentiated into a basal cone-shaped sterile portion and the fertile straplike portion.
Family 4 Fucaceae: thallus with a four-sided apical cell in mature plants.

Cystoseiraceae

These organisms are usually monopodially branched, and because branching is copious, they often have a bushy habit. Branching can be either radial or in one plane. It is probable that those forms with radial branching are more primitive than those with branching in one plane. Most of the genera occur in warmer waters and are not able to withstand much desiccation, occurring in deep rock pools or in the sublittoral region.

Cystoseira (Fig. 17.41) exhibits radial branching, with the laterals showing extensive monopodial branching. Air bladders are common on the lower parts of lower laterals, usually occurring in short series, one on top of the other. *Halidrys* (Fig. 17.42) has branching in one plane, with the lower laterals developing into long shoots, some as long as the main axis. Miocene diatomite deposits in California contain a greater number of fossil species of the Cystoseiraceae than the number in existence today (Parker and Dawson, 1966).

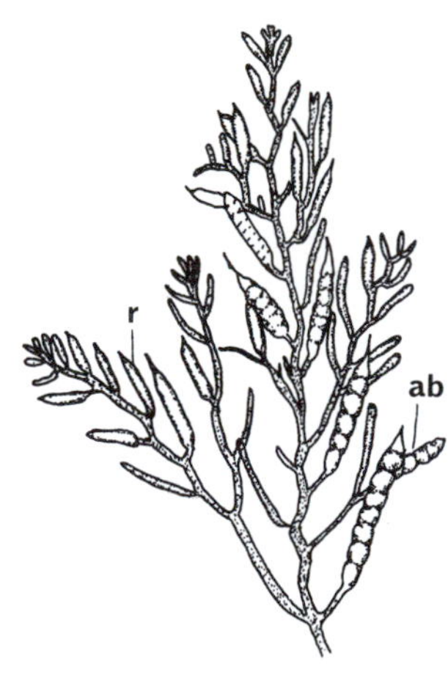

Fig. 17.42 *Halidrys siliquosa*. (ab) Air bladder; (r) receptacle.

Sargassaceae

The branching in this family is always monopodial, and all the members have lateral branch systems. The lateral branch systems have at their base two leaflike structures, with the remainder of the branch system consisting of air bladders and receptacles. *Sargassum* (Figs. 17.43 and 17.44) is the principal genus, and is usually found in warmer waters. *Sargassum* plants make up the huge floating masses of Gulf-weed that compose the Sargasso

Fig. 17.43 *Sargassum muticum* or the wireweed. (After Scagel, 1971.)

Sea, off the African continent between 20° and 35° north latitude. The immense tracts of *Sargassum* are formed by continuous vegetative reproduction of the plants in the area. Holdfasts are never found. Occasionally, conceptacles can be seen, but they are always functionless, and the maintenance of the population is entirely by vegetative means.

Some of the species of *Sargassum* occur as attached seaweeds in the littoral and subtidal zone. *Sargassum muticum* is a Pacific Ocean species that has established itself in Atlantic waters off Europe and North America and has caused some concern among phycologists interested in the balance of the marine environment (Boalch and Potts, 1977; Norton, 1977). In *Sargassum muticum* (Fig. 17.43), most of the fertilized eggs are not released until they are well-developed germlings that sink very rapidly (0.6 mm s^{-1}). The germlings stick to the substratum on contact, although the ability to stick varies with the age of the germling, being optimum between 2 and 18 days and being lost after 49 days. Even in violent water motion, most of the germlings settle close to the parent plant (Norton, 1980).

Himanthaliaceae

This family contains a single genus, *Himanthalia* (Fig. 17.45), which is present near low-tide level on relatively exposed shores and is composed of

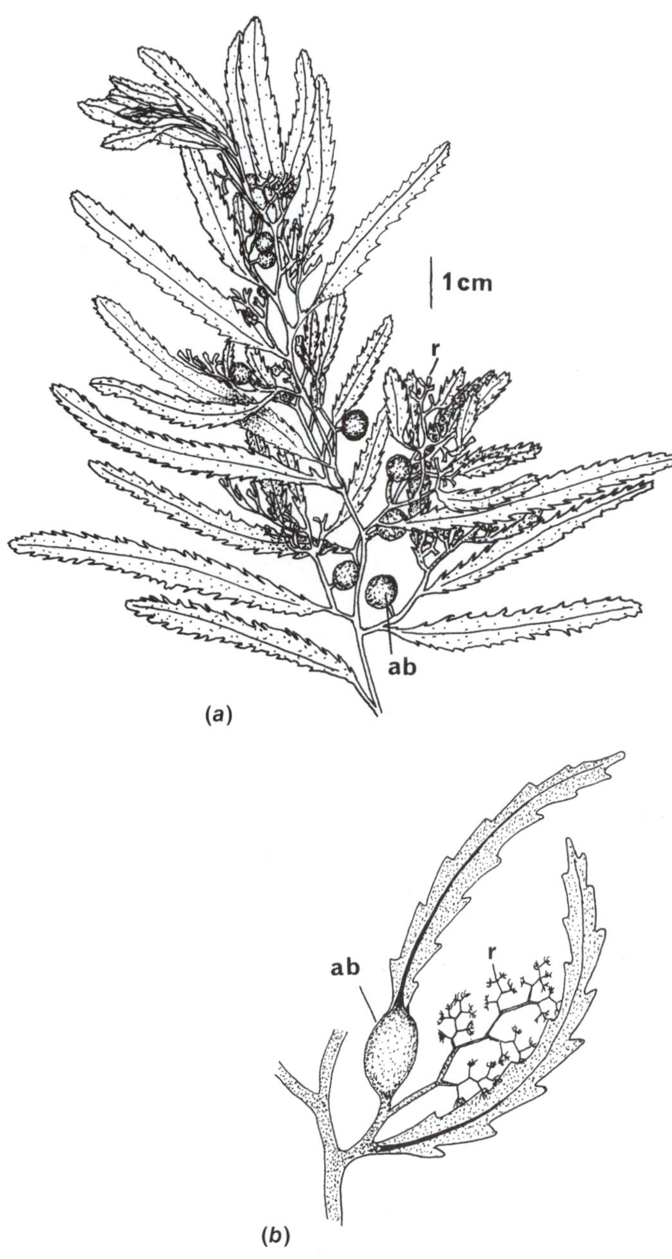

Fig. 17.44 *Sargassum filipendula*, a part of a main branch. (*b*) *S. longifolium*. (ab) Air bladder; (r) receptacle. ((*a*) after Taylor, 1960.)

sterile and fertile portions. The sterile portion resembles a deflated vesicle mounted on a stalk and is several centimeters in width. During the second year of its life, the plant produces from one to four long, strap-shaped, dichotomously branched receptacles from the top of the thallus. These receptacles bear conceptacles, and the whole plant usually dies after the release of the gametes.

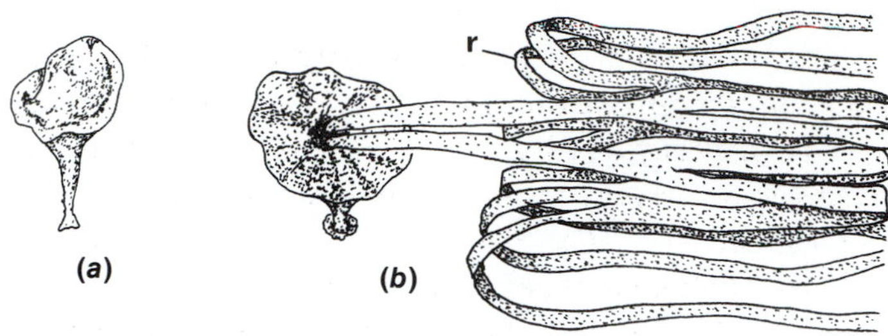

Fig. 17.45 *Himanthalia elongata*. (*a*) Young sterile thallus. (*b*) Mature thallus with straplike receptacle (r).

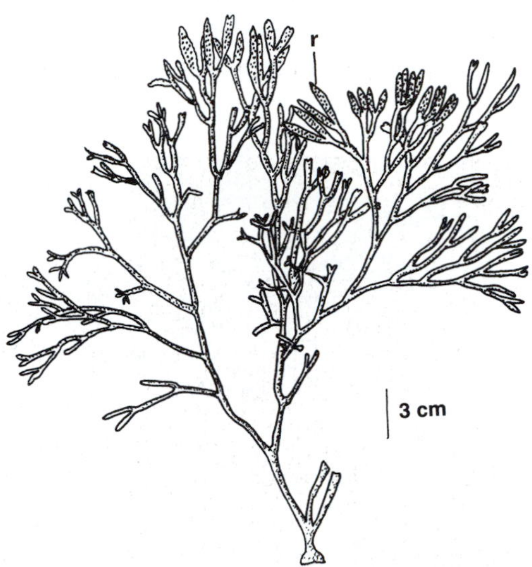

Fig. 17.46 *Pelvetia fastigata*. (r) Receptacle. (After Smith, 1969.)

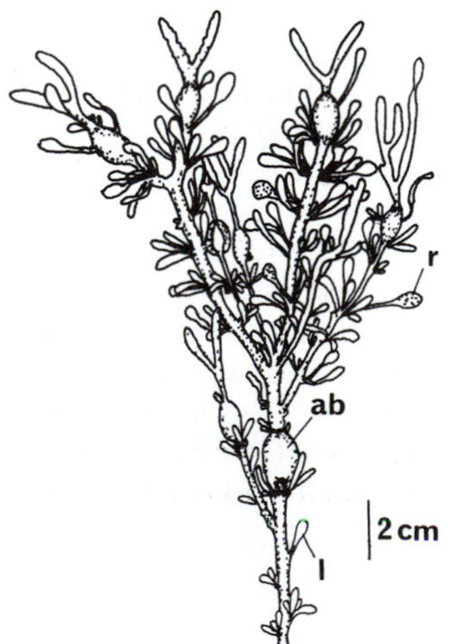

Fig. 17.47 *Ascophyllum nodosum*. Small portion of the distal end of a plant in late summer condition. (ab) Air bladder; (l) lateral; (r) receptacle. (After Taylor, 1957.)

Fucaceae

Algae in this family are the only ones with a four-sided apical cell at maturity. The branching is basically dichotomous and in one plane although some genera show monopodial branching. The three prevalent genera – *Pelvetia, Fucus,* and *Ascophyllum* – are found in the colder waters of the Northern Hemisphere. *Pelvetia* (Fig. 17.46) is a specialized form adapted to the extreme environment near the high-tide mark where it grows. *Ascophyllum nodosum* (Fig. 17.47) is abundant on somewhat protected shores, where it occurs with *Fucus* in the upper part of the littoral region. The mature plants have strap-shaped axes up to 1.5 m long. They do not have a midrib and are usually dilated at frequent intervals by conspicuous vesicles, so that at high water the thallus spreads out on the water.

Fig. 17.48 *Left:* M. J. Dring and *right:* Klaus Lüning.

M. J. Dring Dr Dring received his BSc in Botany from the University of Bristol in 1964, and his PhD from the University of London in 1967. In 1967 he moved to the Queen's University of Belfast where he is currently a Reader. Dr Dring's studies on the responses of algae to light, as well as his book *The Biology of Marine Plants*, have made him a leader in studies of algal interactions with the environment.

Klaus Luning Born in 1941 in Königsberg, Germany. From 1960 to 1968. Dr Lüning studied at the universities of Hamburg and Kiel. In 1968, he worked at the University of Kiel with Professor Fritz Gessner. His dissertation research was carried out at the marine station on Helgoland, where he stayed until 1980, when he moved to the station's central laboratory in Hamburg where he still works. Dr Lüning's publications include *Seaweed Biogeography and Ecophysiology*. His work on light and growth in the algae is well known to phycologists.

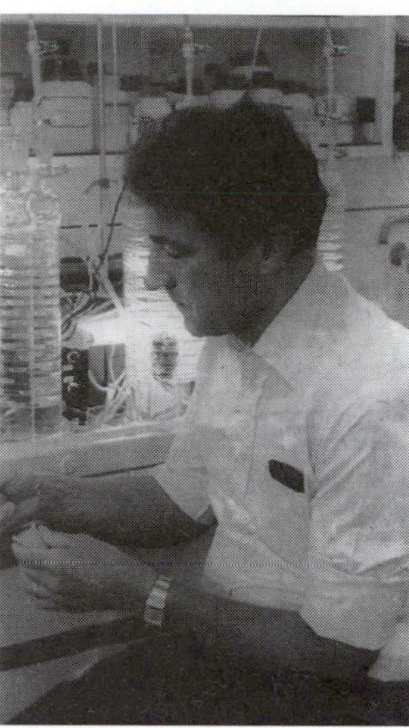

Fig. 17.49 *Left:* J. A. Callow and *right:* Dieter G. Muller.
J. A. Callow Dr Callow obtained both his BSc and PhD from the Botany Department at Sheffield University. In 1969 he accepted a position at Leeds University. Since 1983 he has been Mason Professor of Botany at Birmingham University. From 1988 to 1994 he was Head of the School of Biological Sciences. Dr Callow has performed outstanding research on the reproductive biology of *Fucus*.
Dieter G. Müller Born January 24, 1935, in Stuttgart, Germany. From 1956 to 1961, Dr Müller studied at the Universität Tübingen; from 1961 to 1963, he was a Postdoctoral Fellow at the University of Pennsylvania; from 1962 to 1963 he was a Wissenschaftlicher Assistant at the Botanischen Institut at the Universität Tübingen; from 1964 to 1973, he was a Wissenschaftlicher Mitarbeiter at the Max-Planck-Institut für Zuchtungsforschung at Köln-Vogelsang. In 1973, he received his Habilation at the University of Köln; and since 1973, he has been professor in the Fakultät für Biologie at the Universität Konstanz. In 1964, Dr Müller received strong support from Professor J. Straub, director of the Max-Planck-Institut, Köln, to work out in detail the life history of *Ectocarpus siliculosus*. After evidence for a sexual hormone was discovered, a cooperation scheme was established with the Institut für Biochemie at the University of Köln. These two events led to what is perhaps the most exciting research on algae to have been carried out in the past two decades, that of the characterizations of the sexual hormones of the brown algae.

References

Amsler, C. D., and Neushel, M. (1989). Chemotactic effects of nutrients on spores of the kelps *Macrocystis pyrifera* and *Pterygophora californica*. *Mar. Biol.* (Berlin) 102:557–64.

Andersen, R. J. (1982). The life history of *Desmarestia firma* (C. Ag.) Skottsb. (Phaeophyceae, Desmarestiales). *Phycologia* 21:316–22.

Bacon, L. E., and Vadas, R. L. (1991). A model for gamete release in *Ascophyllum nodosum* (Phaeophyta). *J. Phycol.* 27:166–73.

Bell, G. (1997). The evolution of the life cycle of brown seaweeds. *Biol. J. Linnean Soc.* 60:21–38.

Berkaloff, C., and Rousseau, B. (1979). Ultrastructure of male gametogenesis in *Fucus serratus* (Phaeophyceae). *J. Phycol.* 15:163–73.

Bird, N. L., and McLachlan, J. (1976). Control of formation of receptacles in *Fucus distichus* L., subsp. *distichus* (Phaeophyceae, Fucales). *Phycologia* 15:79–84.

Bisalputra, T. (1966). Electron microscopic study of protoplasmic continuity in certain brown algae. *Can.J. Bot.* 44:89–93.

Bisalputra, T., and Burton, H. (1969). The ultrastructure of the chloroplast of a brown alga *Sphacelaria* sp. II. Association between the chloroplast DNA and the photosynthetic lamellae. *J. Ultrastruct. Res.* 29:224–35.

Bisalputra, T., Shields, M., and Markam, J. W. (1971). *In situ* observations of the fine structure of *Laminaria* gametophytes and embryos in culture. I. Methods and the ultrastructure of the zygote. *J. Microscopie* 10:83–98.

Black, W. A. P. (1954a). Concentration gradients and their significance in *Laminaria saccharina* (L.) Lamour. *J. Mar. Biol. Assoc. UK* 33:49–60.

Black, W. A. P. (1954b). The seasonal variation in the combined L-fucose content of the common British Laminariaceae and Fucaceae. *J. Sci. Food Agric.* 5:445–54.

Boalch, G. T. (1961). Studies on *Ectocarpus* in culture. *J. Mar. Biol. Assoc. UK* 41:287–304.

Boalch, G. T., and Potts, G. W. (1977). The first occurrence of *Sargassum muticum* (Yendo) Fensholt in the Plymouth area. *J. Mar. Biol. Assoc. UK* 57:29–31.

Boney, A. D. (1966). *A Biology of Marine Algae.* Hutchinson, London.

Borowitzka, M. A., Larkum, A. W. D., and Nockolds, C. E. (1974). A scanning electron microscope study of the structure and organisation of the calcium carbonate deposits of algae. *Phycologia* 13:195–203.

Bouck, G. B. (1965). Fine structure and organelle associations in brown algae. *J. Cell Biol.* 26:523–37.

Bouck, G. B. (1969). Extracellular microtubules. The origin, structure and attachment of flagellar hairs in *Fucus* and *Ascophyllum* antherozoids. *J. Cell Biol.* 40:446–60.

Bourne, V. L., and Cole, K. (1968). Some observations on the fine structure of the marine brown alga *Phaeostrophion irregulare. Can. J. Bot.* 46:1369–75.

Brawley, S. H. (1991). The fast block against polyspermy in fucoid algae is an electrical block. *Dev. Biol.* 144:94–106.

Brawley, S.H., and Johnson, L. E. (1991). Survival of fucoid embryos in the intertidal zone depends upon developmental stage and microhabitat. *J. Phycol.* 27:179–86.

Brawley, S. H., Wetherbee, R., and Quatrano, R. S. (1976). Fine structural studies of the gametes and embryo of *Fucus vesiculosus* L. (Phaeophyta). *J. Cell Sci.* 20:233–54.

Buggeln, R. G. (1974). Negative phototropism of the haptera of *Alaria esculenta* (Laminariales). *J. Phycol.* 10:80–2.

Chapman, A. R. O. (1995). Functional ecology of fucoid algae: twenty-three years of progress. *Phycologia* 34:1–32.

Chapman, A. R. O., and Burrows, E. M. (1971). Field and culture studies of *Desmarestia aculeata* (L.) Lamour. *Phycologia* 10:63–76.

Chapman, V. J. (1970). *Seaweeds and Their Uses*, 2nd ed. Methuen, London.

Cheng, T-H. (1969). Production of kelp. A major source of China's exploitation of the sea. *Econ. Bot.* 23:215–36.

Chi, E. Y. (1971). Brown algal pyrenoids. *Protoplasma* 72:101–4.

Clayton, M. N. (1980). Sexual reproduction – A rare occurrence in the life history of the complanate form of *Scytosiphon* (Scytosiphonaceae, Phaeophyta) from southern Australia. *Br. J. Phycol.*15:105–18.

Clayton, M.N., and Ashburner, C. M. (1994). Secretion of phenolic bodies following fertilization in *Durvillaea potatorum* (Durvillaeales, Phaeophyta). *Eur. J. Phycol.* 29:1–9.

Clint, H. B. (1927). The life history and cytology of *Sphacelaria bipinnata Sauv. Publ. Hartley Bot. Lab. Liverpool* 3:1–19.

Cole, K. (1970). Ultrastructural characteristics in some species in the order Scytosiphonales. *Phycologia* 9:275–83.

Colijn, F., and van den Hoek, C. (1971). The life history of *Sphacelaria furcigera* Kütz. (Phaeophyceae). II. The influence of daylength and temperature on sexual and vegetative reproduction. *Nova Hedwigia* 21:899–922.

David, H. M. (1943). Studies in the autecology of *Ascophyllum nodosum* Le Jol. *J. Ecol.* 31:178–98.

Dayton, P. K., and Tegner, M. J. (1984). Catastrophic storms, El Niño, and patch stability in a southern California kelp community. *Science* 224:283–5.

Dayton, P. K., Currie, V., Gerrodette, T., Keller, B. D., Rosenthal, G., and Ven Tresca, D. (1984). Patch dynamics and stability of some California kelp communities. *Ecol. Monogr.* 54:253–89.

Diouris, M. (1989). Long-distance transport of ^{14}C-labelled assimilate in the Fucales: nature of translocated substances in *Fucus serratus*. *Phycologia* 28:504–11.

Druehl, L. D. (1967). Distribution of two species of *Laminaria* as related to some environmental factors. *J. Phycol.* 3:103–8.

Druehl, L. D., and Hsiao, S. I. C. (1977). Intertidal kelp response to seasonal environmental changes in a British Columbia inlet. *J. Fish. Res. Board Can.* 34:1207–11.

Duggins, D. O. (1980). Kelp beds and sea otters: An experimental approach. *Ecology* 61:447–53.

Edelstein, T., Chen, L., and McLachlan, J. (1968). Sporangia of *Ralfsia fungiformis* (Gunn.) Setchell and Gardner. *J. Phycol.* 4:157–60.

Edelstein, T., and Chen, L. C-M., and McLachlan, J. (1970). The life cycle of *Ralfsia clavata* and *R. borneti*. *Can. J. Bot.* 48:527–31.

Edwards, P. (1969). Field and cultural studies on the seasonal periodicity of growth and reproduction of selected Texas benthic marine algae. *Contrib. Mar. Sci. Univ. Texas* 14:59–114.

Emerson, C. J., Buggelin, R. G., and Bal, A. K. (1982). Translocation in *Saccorhiza dermatodea* (Laminariales, Phaeophyceae): Anatomy and physiology. *Can. J. Bot.* 60:2164–84.

Estes, J. A., and Steinberg, P. D. (1988). Predation, herbivory, and kelp evolution. *Paleobiology* 14:19–36.

Evans, L. V. (1965). Cytological studies in the Laminariales. *Ann. Bot.* 29:541–62.

Evans, L. V. (1966). Distribution of pyrenoids from some brown algae. *J. Cell Sci.* 1:449–54.

Evans, L. V. (1968). Chloroplast morphology and fine structure in British fucoids. *New Phytol.* 67:173–8.

Evans, L. V., and Holligan, M. S. (1972a). Correlated light and electron microscope studies on brown algae. I. Localization of alginic acid and sulphated polysaccharides in *Dictyota*. *New Phytol.* 71:1161–72.

Evans, L. V., and Holligan, M. S. (1972b). Correlated light and electron microscope studies on brown algae. II. Physode production in *Dictyota*. *New Phytol.* 71:1173–80.

Evans, L. V., Simpson, M., and Callow, M. E. (1973). Sulphated polysaccharide synthesis in brown algae. *Planta* 110:237–52.

Evans, L. V., Callow, M. E., Callow, J. A., and Bolwell, G. P. (1980). Egg–sperm recognition in *Fucus*. *Br. Phycol. J.* 15:194–5.

Floc'h, J. Y., and Penot, M. (1972). Transport du ^{32}P et du ^{86}Rb chez quelques algues brunes: Orientation des migrations et voies de conduction. *Physiologie Végétale* 10:677–86.

Forster, R.M., and Dring, M. J. (1994). Influence of blue light on the photosynthetic capacity of marine plants from different taxonomic, ecological and morphological groups. *Eur. J. Phycol.* 29:21–27.

Fulcher, R. G., and McCully, M. E. (1969). Histological studies on the genus *Fucus*. IV. Regeneration and adventure embryony. *Can. J. Bot.* 47:1643–9.

Gaines, S. D., and Roughgarden, J. (1987). Fish in offshore kelp forests affect recruitment to intertidal populations. *Science* 235:479–81.

Gibb, D. C. (1957). The free-living forms of *Ascophyllum nodosum* (L.) Le Jol. *J. Ecol.* 45:49–83.

Gibson, G., and Clayton, M. N. (1987). Sexual reproduction, early development and branching in *Notheia anomala* (Phaeophyta) and its classification in the Fucales. *Phycologia* 26:363–73.

Grenville, D.J., Peterson, R. L., Barrales, H. L., and Gerrath, J. F. (1982). Structure and development of the secretory cells and duct system in *Macrocystis pyrifera* (L.) C.A. Agardh. *J. Phycol.* 18:232–40.

Guignard, L. (1892). Observations sur l'appareil mucifere des Laminariacées. *Ann. Sci. Nat. Bot. VII* 15:1–46.

Henley, W. J., and Dunton, K. H. (1997). Effects of nitrogen supply and continuous darkness on growth and photosynthesis of the arctic kelp *Laminaria solidungula. Limnol. Oceanog.* 42:209–16.

Henry, C. A., Jordan, J. R., and Kropf, D. L. (1996). Localized membrane-wall adhesions in *Pelvetia* zygotes. *Protoplasma* 190:39–52.

Henry, E. C. (1987). Primitive reproductive characters and a photoperiodic response in *Saccorhiza dermatodea* (Laminariales, Phaeophyceae). *Br. Phycol. J.* 22:23–31.

Henry, E. C., and Cole, K. M. (1982). Ultrastructure of swarmers in the Laminariales (Phaeophyceae). I. Zoospores. *J. Phycol.* 18:550–69.

Hollenberg, G. J. (1969). An account of the Ralfsiaceae (Phaeophyta) of California. *J. Phycol.* 5:290–301.

Hsiao, S. I. C. (1969). Life history and iodine nutrition of the marine brown alga, *Petalonia fascia* (O. F. Müll.) Kuntze. *Can. J. Bot.* 47:1611–16.

Hsiao, S. I. C. (1970). Light and temperature effects on the growth, morphology, and reproduction of *Petalonia fascia. Can. J. Bot.* 48:1359–61.

Hsiao, S. I. C., and Druehl, L. D. (1973). Environmental control of gametogenesis in *Laminaria saccharina*. IV. In situ development of gametophytes and young sporophytes. *J. Phycol.* 9:160–4.

Hurd, C. L., Galvin, R. S., Norton, T. A., and Dring, M. J. (1993). Production of hylaine hairs by intertidal species of *Fucus* (Fucales) and their role in phosphate uptake. *J. Phycol.* 29:160–5.

Jaenicke, L., Müller, D. G., and Moore, R. E. (1974). Multifidene and aucantene, C_{11} hydrocarbons in the male attracting essential oil from the gynogametes of *Cutleria multifida* (Smith) Grev. (Phaeophyta). *J. Am. Chem. Soc.* 96:3324–5.

Jensen, J. B. (1974). Morphological studies in Cystoseiraceae and Sargassaceae (Phaeophyceae). *Univ. Calif. Publ. Bot.* 68:1–61.

Jones, J. L., Callow, J. A., and Green, J. R. (1988). Monoclonal antibodies to sperm antigens of the brown alga *Fucus serratus* exhibit region-, gamete-, species- and genus-preferential binding. *Planta* 176:298–306.

Kain, J. M. (1966). The role of light in the ecology of *Laminaria hyperborea. Br. Ecol. Soc. Symp.* 6:319–34.

Kain, J. M. (1976). The biology of *Laminaria hyperborea*. VIII. Growth on cleared areas. *J. Mar. Biol. Assoc. UK* 56:267–90.

Kapraun, D. F., and Boone, P. W. (1987). Karyological studies of three species of Scytosiphonaceae (Phaeophyta) from coastal North Carolina. *J. Phycol.* 23:318–22.

Kawai, H., Kubota, M., Kondo, T., and Watanabe, M. (1991). Action spectra for phototaxis in zoospores of the brown alga *Pseudochorda gracilis. Protoplasma* 161:17–22.

Kawai, H., Nakamura, S., Mimuro, M., Furuya, M., and Watanabe, M. (1996). Microspectrofluorometry of the autofluorescent flagellum in phototactic algal zoids. *Protoplasma* 191:172–7.

Kiirikki, M. (1996). Experimental evidence that *Fucus vesiculosus* (Phaeophyta) controls filamentous algae by means of the whiplash effect. *Eur. J. Phycol.* 31:61–6.

Klinger, T., and DeWeede, R. E. (1988). Stipe rings, age and size in populations of *Laminaria setchelli* Silva (Laminariales, Phaeophyta) in British Columbia. *Phycologia* 27:234–40.

Kooareg, D., and Quatrano, R. S. (1988). Structure of the cell walls of marine algae and ecophysiological functions of the matrix polysaccharides. *Oceanogr. Mar. Biol. Rev.* 26:250–315.

Knight, M. (1929). Studies in the Ectocarpaceae. II. The life-history and cytology of *Ectocarpus siliculosus* Dillw. *Trans. R. Soc. Edinburgh* 56:307–32.

Kropf, D. L., Coffman, H. R., Kloareg, B., Glenn, P., and Allen, V. W. (1992). Cell wall and rhizoid polarity in *Pelvetia* embryos. *Dev. Biol.* 160:303–14.

Kuckuck, P. (1899). Über den generationwechsel von *Cutleria multifida* (Engl. Bot.) Grev. *Wiss. Meeresunters. Helgoland N.F.* 3:95–116.

Kuckuck, P. (1929). Fragmente einer Monographie der Phaeosporeen. *Wiss. Meeresunters. Abt. Helgoland* 17, No. 4.

Kumke, J. (1973). Beiträge zur Periodizität der Oogon-Entleerung bei *Dictyota dichotoma* (Phaeophyta). *Z. Pfl. Physiol.* 70:191–210.

Kylin, H. (1934). Zur Kenntnis der Entwicklungsgeschichte einiger Phaeophyceen. *Lunds Univ. Årsskr. N.F. Avd.* 30:1–19.

Kylin, H. (1940). Die Phaeophyceenordnung Chordariales. *Lunds Univ. Årsskr. N.F. Avd.* 36:1–67.

Larkum, A. W. D. (1972). Frond structure and growth in *Laminaria hyperborea*. *J. Mar. Biol. Assoc. UK* 52:405–18.

Lobban, C. S. (1978). The growth and death of the *Macrocystis* sporophyte (Phaeophyceae, Laminariales). *Phycologia* 17:196–212.

Lodge, S. M. (1948). Algal growth in the absence of *Patella* on an experimental strip of seashore. *Proc. Trans. Liverpool Biol. Soc.* 61:78–83.

Loiseaux, S. (1970). *Streblonema anomalum* S. et G. and *Compsonema sporangiiferum* S. et G. stages in the life history of a minute *Scytosiphon*. *Phycologia* 9:185–91.

Loiseaux, S. (1973). Ultrastructure of zoidogenesis in unilocular zoidocysts of several brown algae. *J. Phycol.* 9:277–89.

Loiseaux, S., and West, J. A. (1970). Brown algal mastigonemes: Comparative ultra-structure. *Trans. Am. Microsc. Soc.* 89:524–32.

Love, J., Brownlee, C., and Trewavas, A. J. (1997). Ca^{2+} and calmodulin dynamics during photopolarization in *Fucus serratus* zygotes. *Plant Physiol.* 115:249–61.

Lüning, K. (1981). Egg release in gametophytes of *Laminaria saccharina*: Induction by darkness and inhibition by blue light and U.V. *Br. Phycol. J.* 16:379–93.

Lüning, K. (1986). New frond formation in *Laminaria hyperborea* (Phaeophyta): A photo-periodic response. *Br. Phycol. J.* 21:269–73.

Lüning, K., and Dring, M. J. (1972). Reproduction induced by blue light in female gameto-phytes of *Laminaria saccharina*. *Planta* 104:252–6.

Lüning, K., and Dring, M. J. (1973). The influence of light quality on the development of the brown algae *Petalonia* and *Scytosiphon*. *Br. Phycol. J.* 8:333–8.

Lüning, K., and Dring, M. J. (1975). Reproduction, growth and photosynthesis of gameto-phytes of *Laminaria saccharina* grown in blue and red light. *Mar. Biol.* 29:195–200.

Lüning, K., and Müller, D. G. (1978). Chemical interaction in sexual reproduction of several Laminariales (Phaeophyceae): Release and attraction of spermatozoids. *Z. Pflanzenphysiol.* 89:333–41.

Lüning, K., Schmitz, K., and Willenbrink, J. (1972). Translocation of ^{14}C-labelled assimi-lates in two *Laminaria* species. *Proc. VII Int. Seaweed Symp.* 420–5.

McCully, M. E. (1965). A note on the structure of the cell wall of the brown alga *Fucus*. *Can. J. Bot.* 43:1001–4.

McLachlan, J., and Craigie, J. S. (1964). Algal inhibition by yellow ultraviolet-absorbing substances from *Fucus vesiculosus*. *Can. J. Bot.* 42:287–92.

McLachlan, J., and Craigie, J. S. (1966). Antialgal activity of some simple phenols. *J. Phycol.* 2:133–5.

Madsen, T. V., and Maberly, S. C. (1990). A comparison of air and water as environments for photosynthesis by the intertidal alga *Fucus spiralis* (Phaeophyta). *J. Phycol.* 26:24–30.

Maier, I. (1982). New aspects of pheromone-triggered spermatozoid release in *Laminaria digitata* (Phaeophyta). *Protoplasma* 113:137–43.

Maier, I. (1984). Culture studies of *Chorda tomentosa* (Phaeophyta, Laminariales). *Br. Phycol. J.* 19:95–106.

Maier, I. (1997a). The fine structure of the male gamete of *Ectocarpus siliculosus* (Ectocarpales, Phaeophyceae). I. General structure of the cell. *Eur. J. Phycol.* 32:241–53.

Maier, I. (1997b). The fine structure of the male gamete of *Ectocarpus siliculosus* (Ectocarpales, Phaeophyceae). II. The flagellar apparatus. *Eur. J. Phycol.* 32:255–66.

Maier, I., and Müller, D. G. (1986). Sexual pheromones in algae. *Biol. Bull.* 170:145–75.

Maier, L., Müller, D. G., Gassmann, G., Boland, W., Marner, F-J., and Jaenicke, L. (1984). Pheromone-triggered gamete release in *Chorda tomentosa. Naturwissenschaften* 71:48–9.

Mann, K. H., and Chapman, A. R. O. (1975). Primary production of marine macrophytes. In *Photosynthesis and Productivity in Different Environments*, ed. J. P. Cooper, *Int. Biol. Prog.* 3:307–33. Cambridge: Cambridge University Press.

Manton, I. (1959). Observations on the internal structure of the spermatozoid of *Dictyota. J. Exp. Bot.* 10:448–61.

Manton, I. (1964). A contribution towards understanding of "the primitive fucoid." *New Phytol.* 63:244–54.

Manton, I., and Clarke, B. (1950). Electron microscope observations of the spermatozoid of *Fucus. Nature* 166:973–4.

Manton, I., and Clarke, B. (1951). Electron microscope observations on the zoospores of *Pylaiella* and *Laminaria. J. Exp. Bot.* 23:242–6.

Manton, I., and Clarke, B. (1956). Observations with the electron microscope on the internal structure of the spermatozoid of *Fucus. J. Exp. Bot.* 7:416–32.

Manton, I., Clarke, B., and Greenwood, A. D. (1953). Further observations with the electron microscope on spermatozoids in the brown algae. *J. Exp. Bot.* 4:319–29.

Mariani, P., Tolomio, C., and Braghetta, P. (1985). An ultrastructural approach to the adaptive role of the cell wall in the intertidal alga *Fucus virsoides. Protoplasma* 128:208–17.

Marner, F-J., Müller, B., and Jaenicke, L. (1984). Lamoxirene, the *Laminaria*-pheromone: Structural proof of the spermatozoid releasing and attracting factor of Laminariales. *Z. Naturforsch.* 39c:689–91.

Moe, R. L., and Silva, P. C. (1977). Antarctic marine flora: Uniquely devoid of kelps. *Science* 196:1206–8.

Moe, R. L., and Silva, P. C. (1981). Morphology and taxonomy of *Himantothallus* (including *Phaeoglossum* and *Phyllogigas*), an Antarctic member of the Desmarestiales (Phaeophyceae). *J. Phycol.* 17:15–29.

Moestrup, Ø. (1982). Flagellar structure in algae: A review with new observations particularly on the Chrysophyceae, Phaeophyceae (Fucophyceae), Euglenophyceae, and *Reckertia. Phycologia* 21:427–528.

Moss, B. (1965). Apical dominance in *Fucus vesiculosus. New Phytol.* 64:387–92.

Moss, B., (1967). The apical meristem of *Fucus. New Phytol.* 66:67–74.

Moss, B. (1970). Meristems and growth control in *Ascophyllum nodosum* (L.) Le Jol. *New Phytol.* 69:253–60.

Moss, B. (1971). Meristems and morphogenesis in *Ascophyllum nodosum* ecad *mackii* (Cotton). *Br. Phycol. J.* 6:187–93.

Moss, B. (1983). Sieve elements in the Fucales. *New Phytol.* 93:433–7.

Moss, B., and Sheader, A. (1973). The effect of light and temperature upon the germination and growth of *Halidrys siliquosa* (L.) Lygnb. (Phaeophyceae, Fucales). *Phycologia* 12:63–8.

Motomura, T. (1991). Immunofluorescence microscopy of fertilization and parthenogenesis in *Laminaria angusta* (Phaeophyta). *J. Phycol.* 27:248–57.

Motomura, T. (1994). Electron and immunofluorescence microscopy on the fertilization of *Fucus distichus* (Fucales, Phaeophyceae). *Protoplasma* 178:97–110.

Motomura, T., Ichimura, T., and Melkonian, M. (1997). Coordinative nuclear and chloroplast division in unilocular sporangia of *Laminaria angustata* (Laminariales, Phaeophyceae). *J. Phycol.* 33:266–71.

Müller, D. (1962). Über jahres und lunarperiodische Erscheinungen bei einigen Braunalgen. *Bot. Mar.* 4:140–155.

Müller, D. (1974). Sexual reproduction and isolation of a sex attractant in *Cutleria multifida* (Smith) Grev. (Phaeophyta). *Biochem Physiol. Pflanz.* 165:212–15.

Müller, D. (1978). Locomotive responses of male gametes to the species specific sex attractant in *Ectocarpus siliculosus* (Phaeophyta). *Arch. Protistenk.* 120:371–77.

Müller, D. (1979). Genetic affinity of *Ectocarpus siliculosus* (Dillw.) Lyngb. from the Mediterranean, North Atlantic and Australia. *Phycologia* 18:312–18.

Müller, D. (1982). Sexuality and sex attraction. In *The Biology of Seaweeds* eds. C. S. Lobban, and M. J. Wynne, pp. 661–674. Univ. Calif. Press, Los Angeles and Berkeley.

Müller, D., and Gassmann, G. (1980). Sexual hormone specificity in *Ectocarpus* and *Laminaria. Naturwissenschaften* 67:462–3.

Müller, D., and Jaenicke, L. (1973). Fucoserraten, the female sex attractant of *Fucus serratus* L. (Phaeophyta). *FEBS Lett.* 30:137–9.

Müller, D., and Lüthe, N. M. (1981). Hormonal interaction in sexual reproduction of *Desmarestia aculeata* (Phaeophyceae). *Br. Phycol. J.* 16:351–6.

Müller, D., Jaenicke, L., Donike, M., and Akintobi, T. (1971). Sex attractant in a brown alga: Chemical structure. *Science* 171:815–17.

Müller, D., Gassmann, G., and Lüning, K. (1979). Isolation of a spermatozoid-releasing and -attracting substance from female gametophytes of *Laminaria digitata. Nature* 279:430–1.

Müller, D., Gassmann, G., Boland, W., Marner, F., and Jaenicke, L. (1981). *Dictyota dichotoma* (Phaeophyceae): Identification of sperm attractant. *Science* 212:1040–1.

Müller, D., Peters, A., Gassmann, G., Boland, W., Marner, F-J., and Jaenicke, L. (1982). Identification of a sexual hormone and related substances in the marine alga *Desmarestia. Naturwissenschaften* 69:290–1.

Müller, D., Clayton, M. W., and Germann, I. (1985a). Sexual reproduction and life history of *Perithalia caudata* (Sporochnales, Phaeophyta). *Phycologia* 24:467–73.

Müller, D., Maier, I., and Gassmann, G. (1985b). Survey on sexual pheromone specificity in Laminariales (Phaeophyceae). *Phycologia* 24:475–84.

Nakahara, H., and Nakamura,Y. (1973). Parthenogenesis, apogamy and apospory in *Alaria crassifolia* (Laminariales). *Mar. Biol.* 18:327–32.

Nakamura, Y. (1965). Development of zoospores in *Ralfsia*-like thallus, with special reference to the life cycle of the Scytosiphonales. *Bot. Mag. Tokyo* 78:109–10.

Nakamura, Y. (1972). A proposal on the classification of the Phaeophyta. In *Contributions* to the Systematics of Benthic Marine Algae of the North Pacific, ed. I. A. Abbott, and M. Kurogi, pp. 147–56. New York: Academic Press.

Nakazawa, S. (1962). Polarity. In *Physiology and Biochemistry of the Algae*, ed. R. E. Lewin, pp. 653–61. New York: Academic Press.

Nicholson, N. L. (1970). Field studies on the giant kelp *Nereocystis*. *J. Phycol.* 6:177–82.

Nienberg, W. (1931). Die Entwicklung der Keimlinge von *Fucus vesiculosus* und ihre Bedeutung für die Phylogenie der Phaeophyceen. *Wiss. Meeresunters. Abt. Kiel N.F.* 21:49–63.

Nizamuddin, M., and Womersley, H. B. S. (1960). Structure and systematic position of the Australian brown alga, *Notheia anomala*. *Nature* 187:673–4.

Norton, T. A. (1969). Growth form and environment in *Saccorhiza polyschides*. *J. Mar. Biol. Assoc. UK* 49:1025–45.

Norton, T. A. (1972). The development of *Saccorhiza dermatodea* (Phaeophyceae, Laminariales) in culture. *Phycologia* 11:81–6.

Norton, T. A. (1977). Ecological experiments with *Sargassum muticum*. *J. Mar. Biol. Assoc. UK* 57:33–43.

Norton, T. A. (1978). The factors influencing the distribution of *Saccorhiza polyschides* in the region of Lough Ine. *J. Mar. Biol. Assoc. UK* 58:527–36.

Norton, T. A. (1980). Sink, swim or stick: The fate of *Sargassum muticum* propagules. *Br. Phycol. J.* 15:197.

Norton, T. A., and Burrows, E. M. (1969). Studies on marine algae of the British Isles. 7. *Saccorhiza polyschides* (Lightf.) Batt. *Br. Phycol. J.* 4:19–53.

Novaczek, I. (1981). Stipe growth rings in *Ecklonia radiata* (C. Ag.) J. Ag. (Laminariales). *Br. Phycol. J.* 16:363–71.

O'Kelley, C. J. (1989). The evolutionary origin of the brown algae: information from studies of motile cell ultrastructure. In *The Chromophyte Algae: Problems and Perspectives*, ed. J. C. Green, B. S. C. Leadbeater, and W. L. Diver, pp. 255–78. Oxford: Oxford University Press.

Oliveira, L., Walker, D. C., and Bisalputra, T. (1980). Ultrastructural, cytochemical and enzymatic studies on the adhesive "plaques" of the brown alga, *Laminaria saccharina* (L.) Lamour. and *Nereocystis leutkeana* (Nert.) Post. et Rupr. *Protoplasma* 104:1–15.

Oltmanns, F. (1889). Beitrage zur Kenntniss der Fucaceen. *Bibl. Bot.* 3(14):1–100.

Papenfuss, G. F. (1934). Alternation of generations in *Sphacelaria bipinnata* Sauv. *Bot. Not.* 437–44.

Papenfuss, G. F. (1935). Alternation of generations in *Ectocarpus siliculosus*. *Bot. Gaz.* 96:421–46.

Papenfuss, G. F. (1951). Phaeophyta. In *Manual of Phycology*, ed. G. M. Smith, pp. 119–38. Waltham, MA: Chronica Botanica.

Parke, M. (1948). Studies on the British Laminariaceae. I. Growth in *Laminaria saccharina* (L.) Lamour. *J. Mar. Biol. Assoc. UK* 27:651–709.

Parker, B. C. (1963). Translocation in the giant kelp *Macrocystis*. *Science* 140:891–2.

Parker, B. C. (1965). Translocation in the giant kelp *Macrocystis*. I. Rates, direction, quantity of C^{14}-labelled products and fluorescein. *J. Phycol.* 1:41–6.

Parker, B. C., and Dawson, E. Y. (1966). Non-calcareous marine algae from California Miocene deposits. *Nova Hedwigia* 10:273–95.

Percival, E., and McDowell, R. H. (1967). *Chemistry and Enzymology of Marine Algal Polysaccharides*. London: Academic Press.

Phillips, J. A., and Clayton, M. N. (1993). Comparative flagellar morphology of spermatozoids of the Dictyotales (Pheophyceae). *Eur. J. Phycol.* 28:123–7.

Pollock, E. G. (1970). Fertilization in *Fucus*. *Planta* 92:85–99.

Postels, A., and Ruprecht, F. (1840). *Illustrationes Algarium, Oceani Pacifici, Inprimis*

Septemtrionalis.

Price, I. R., and Ducker, S. C. (1966). The life history of the brown alga *Splachnidium rugosum. Phycologia* 5:261–73.

Rawlence, D.J. (1973). Some aspects of the ultrastructure of *Ascophyllum nodosum* (L.) Le Jolis (Phaeophyceae, Fucales) including observations on cell plate formation. *Phycologia* 12:17–28.

Reed, D. C., Laur, D. R., and Ebeling, A. W. (1988). Variation in algal dispersal and recruitment: the importance of episodic events. *Ecol. Monogr.* 58:321–55.

Reed, R. H., Davison, I. R., Chudek, J. A., and Foster, R. (1985). The osmotic role of mannitol in the Phaeophyta: An appraisal. *Phycologia* 24:35–47.

Reinke, J. (1903). *Studien zur vergleichenden Entwicklungsgeschichte der Laminariaceen,* pp. 1–67. Kiel.

Rousseau, F., Leclerc, M-C., and de Reviers, B. (1997). Molecular phylogeny of European Fucales (Phaeophyceae) based on partial large-subunit rDNA sequence comparisons. *Phycologia* 36:438–46.

Russell, G., and Fletcher, R. L. (1975). A numerical taxonomic study of the British Phaeophyta. *J. Mar. Biol. Assoc. UK* 55:763–83.

Russell, G., and Garbary, D. (1978). Generic circumscription in the family Ectocarpaceae (Phaeophyceae). *J. Mar. Biol. Assoc. UK* 58:517–25.

Sargent, M. C., and Lantrip, L. W. (1952). Photosynthesis, growth and translocation in giant kelp. *Am. J. Bot.* 39:99–107.

Saunders, G. W., and Druehl, L. D. (1992). Nucleotide sequences of the small-subunit ribosomal RNA genes from selected Laminariales (Phaeophyta). Implications for kelp evolution. *J. Phycol.* 28:544–9.

Saunders, G. W., and Kraft, G. T. (1995). The phylogenetic affinities of *Notheia anomala* (Fucales, Phaeophyceae) as determined from small-subunit rRNA sequences. *Phycologia* 34:383–9.

Savaugeau, C. (1899). Les Cutlériacées et leur alternance de générations. *Ann. Sci. Nat. Bot. Ser. 8* 10:265–62.

Savaugeau, C. (1900–14). Remarques sur les Sphacélariacées. Bordeaux. 634 pp. in *J. Bot.* 14–18.

Scagel, R. F., (1971). *Guide to Common Seaweeds of British Columbia.* British Columbia Prov. Museum, Victoria.

Scagel, R. F. Bandoni, R. J., Rouse, G. E., Schofield, W. B., Stein, J. R., and Taylor, T. M. C. (1969). *Plant Diversity: An Evolutionary Approach.* Wadsworth, Belmont, CA: Wadsworth.

Schaffelke, B., and Lüning, K. (1994). A circannual rhythm controls seasonal growth in the kelps *Laminaria hyperborea* and *L. digitata* from Helgoland (North Sea). *Eur. J. Phycol.* 29:49–56.

Schloesser, R. E., and Blum, J. L. (1980). *Sphacelaria lacustris* sp. nov., a freshwater brown alga from Lake Michigan. *J. Phycol.* 16:201–7.

Schreiber, E. (1932). Über die Entwicklungsgeschichte und die systematische Stellung der Desmarestiaceen. *Z. Bot.* 25:561–82.

Shih, M. L., Floch, J-Y., and Srivastava, L. M. (1983). Localization of ^{14}C-labeled assimilates in sieve elements of *Macrocystis integrifolia* by histoautoradiography. *Can. J. Bot.* 61:157–63.

Sideman, E. J., and Scheirer, D. C. (1977). Some fine structural observations on developing and mature sieve elements in the brown alga *Laminaria saccharina. Am. J. Bot.* 64:649–57.

Smith, G. M. (1955). *Cryptogamic Botany,* Vol. 1, 2nd ed. New York: McGraw-Hill.

Smith, G. M. (1969). *Marine Algae of the Monterey Peninsula, California,* 2nd ed. Stanford, Calif.: Stanford University Press.

Stafford, C. J., Green, J. R., and Callow, J. A. (1992). Organization of glycoproteins into plasma membrane domains on *Fucus serratus* eggs. *J. Cell Sci.* 101:437–48.

Sundene, O. (1962a). Growth in the sea of *Laminaria digitata* sporophytes from cultures. *Nytt Mag. Bot.* 9:5–24.

Sundene, O. (1962b). The implications of transplant and culture experiments on the growth and distribution of *Alaria esculenta. Nytt Mag. Bot.* 9:155–74.

Sundene, O. (1963). Reproduction and ecology of *Chorda tomentosa. Nytt Mag. Bot.* 10:159–71.

Tan, I. H., and Druehl, L. D. (1996). A ribosomal DNA phylogeny supports the close evolutionary relationships among the Sporochnales, Desmarestiales, and Laminariales (Phaeophyceae). *J. Phycol.* 32:112–18.

Tatewaki, M. (1966). Formation of a crustaceous sporophyte with unilocular sporangia in *Scytosiphon lomentaria. Phycologia* 6:62–6.

Taylor, W. R. (1957). *Marine Algae of the Northeastern Coast of North America.* University of Michigan Press, Ann Arbor.

Taylor, W. R. (1960). *Marine Algae of the Eastern Tropical and Subtropical Coasts of the Americas.* Ann Arbor: University of Michigan Press.

Terry, L. A., and Moss, B. L. (1980). The effect of photoperiod on receptacle initiation in *Ascophyllum nodosum* (Lo) Le Jol. *Br. Phycol. J.* 15:291–301.

Terry, L. A., and Moss, B. L. (1981). The effect of irradiance and temperature on the germination of four species of Fucales *Br. Phycol. J.* 16:143–51.

Thuret, G. (1854). Recherches sur la fécondation des Fucacées. *Ann. Sci. Nat. Bot. IV* 2:197–214.

Thuret, G., and Bornet, E. (1878). *Études Phycologiques.* Paris.

tom Dieck (Bartsch), I. (1991). Circannual growth rhythm and photoperiodic sorus induction in the kelp *Laminaria setchelli* (Phaeophyta). *J. Phycol.* 27:341–50.

Tovey, D. J., and Moss, B. L. (1978). Attachment of the haptera of *Laminaria digitata* (Huds.) Lamour. *Phycologia* 17:17–22.

Tseng, C. K. (1981). Commercial cultivation. In *The Biology of Seaweeds,* ed. C. S. Lobban, and M. J. Wynne, pp. 680–725. Berkeley and Los Angeles: Univ. Calif. Press.

Vadas, R. L. (1972). Ecological implications of culture studies on *Nereocystis luetkeana. J. Phycol.* 8:196–203.

van den Hoek, C. (1982). The distribution of benthic marine algae in relation to the temperature regulation of their life histories. *Biol. J. Linn. Soc.* 18:81–144.

van den Hoek, and Flinterman, A. (1968). The life history of *Sphacelaria furcigera* Kütz. (Phaeophyceae). *Blumea* 16:193–242.

Vreeland, V. (1972). Immunocytochemical localization of the extracellular polysaccharide alginic acid in the brown seaweed. *Fucus distichus. J. Histochem. Cytochem.* 20:358–67.

Walker, F. T. (1956). Periodicity of the Laminariaceae around Scotland. *Nature* 177:1246.

Williams, J. L. (1898). Reproduction in *Dictyota dichotoma. Ann. Bot.* 12:559–60.

Williams, J. L. (1904). Studies in the Dictyotaceae. I. The cytology of the gametophyte generation. *Ann. Bot.* 18:183–204.

Williams, J. L. (1905). Studies in the Dictyotaceae. II. The periodicity of sexual cells in *Dictyota dichotoma. Ann. Bot.* 19:531–60.

Wright, P. J., Callow, J. A., and Green, J. R. (1995a). The *Fucus* (Phaecophyceae) sperm receptor for eggs. II. Isolation of a binding protein which partially activates eggs. *J. Phycol.* 31:592–600.

Wright, P. J., Green, J. A., and Callow, J. A. (1995b). The *Fucus* (Phaeophyceae) sperm receptor for eggs. I. Development and characteristics of a binding assay. *J. Phycol.* 31:584–91.

Wynne, M. J. (1969). Life history and systematic studies of some Pacific North American Phaeophyceae (brown algae). *Univ. Calif. Publ. Bot.* 50:1–88.

18 · Prymnesiophyta

Prymnesiophyceae

The Prymnesiophyta are a group of uninucleate flagellates characterized by the presence of a haptonema between two smooth flagella. The Prymnesiophyta have two membranes of chloroplast endoplasmic reticulum, as do the Cryptophyta and the Heterokontophyta, but differ in having flagella without mastigonemes. Molecular data also show that the Prymnesiophyta are distinct from the Cryptophyta and Heterokontophyta (Bhattacharya and Ehlting, 1995; Medlin et al., 1994). Until 1962, the organisms were considered part of the Chrysophyceae, at which time Christensen split them off into a separate class, the Haptophyceae (named after the presence of the haptonema). The name Haptophyceae was a descriptive name and not based on a genus in the class; thus the name was later changed to Prymnesiophyceae, based on the genus *Prymnesium* (Hibberd, 1976). The fossil record of the Prymnesiophyceae is known from the Carboniferous (approximately 300 000 000 years ago) (Faber and Preisig, 1994; Jordan and Chamberlain, 1997).

The cells are commonly covered with scales. In many cases, the scales are calcified, thereby producing coccoliths. The chloroplasts lack girdle lamellae and most contain chlorophylls a and c_1/c_2, β-carotene, diadinoxanthin and diatoxanthin (Jeffery and Wright, 1994). The storage product is chrysolaminarin in vesicles in the posterior end of the cell (Janse et al., 1996). The anterior end of the cell has a large Golgi apparatus and sometimes a contractile vacuole.

The Prymnesiophyceae are primarily marine organisms, although there are some freshwater representatives. They make up a major part of the marine nannoplankton and constitute about 45% of the total phytoplankton cells in the middle latitudes of the South Atlantic. They decrease in frequency toward the poles although some still occur in polar waters (Manton et al., 1977).

Cell structure

FLAGELLA

Most of the Prymnesiophyceae have two smooth flagella of approximately the same length (Figs. 18.1 and 18.2(a)). The Pavlovales is the exception,

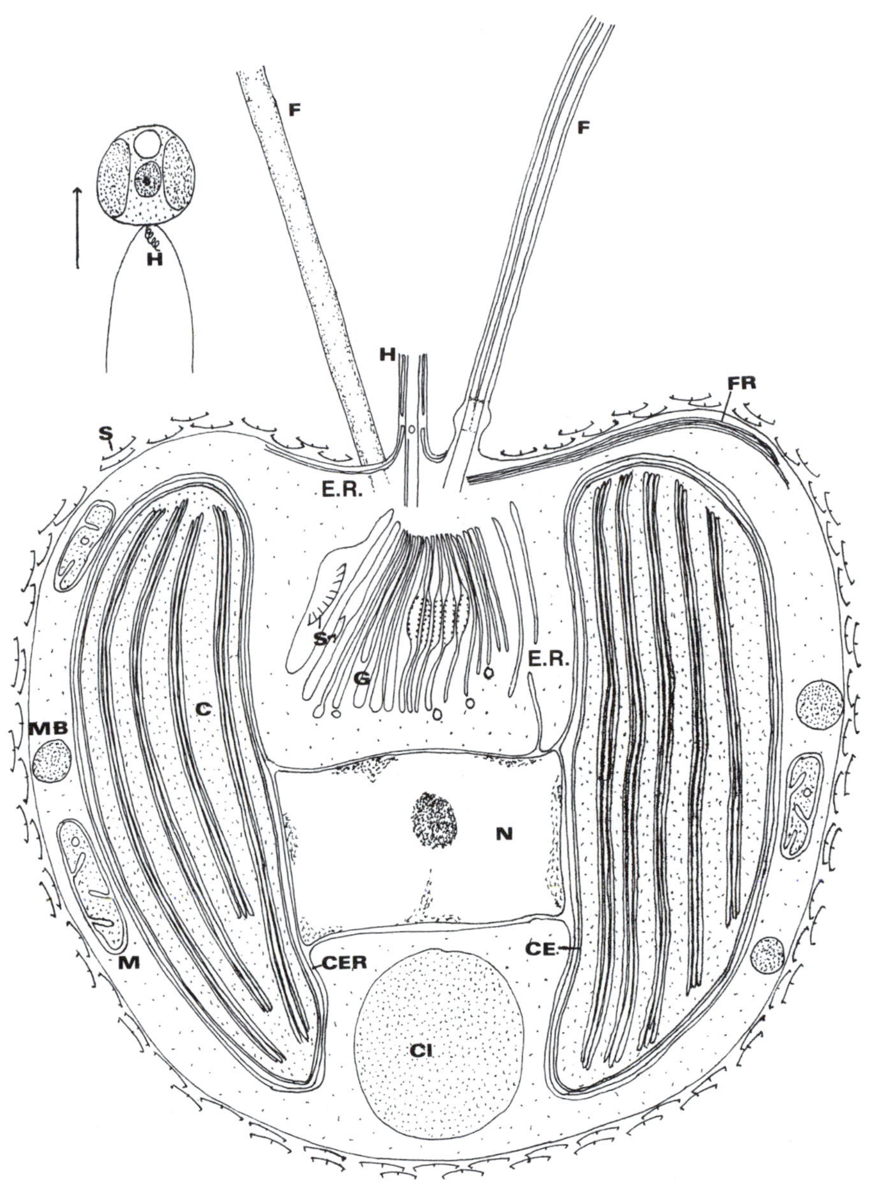

Fig. 18.1 A light and electron microscopical drawing of a cell of a typical member of the Prymnesiophyceae, *Chrysochromulina* sp. A rapidly swimming individual is shown, with the arrow indicating the direction of movement. (C) Chloroplast; (CE) chloroplast envelope; (CER) chloroplast endoplasmic reticulum; (Cl) chrysolaminarin vesicle; (E.R.) endoplasmic reticulum; (F) flagellum; (FR) flagellar root; (G) Golgi body; (H) haptonema; (M) mitochondrion; (MB) muciferous body; (N) nucleus; (S) scale. (Adapted from Hibberd, 1976.)

where one flagellum is longer than the other and is usually covered by small cylindrical to club-shaped hollow scales 70 nm long and 20 nm wide (Fig. 18.2(*b*)) (van der Veer, 1969; Green and Manton, 1970). Because the class is characterized by two more or less equal, smooth flagella, a number of genera, such as *Diacronema* (Fig. 18.3), *Isochrysis* (Fig. 18.16(*b*)), and *Dicrateria*, which have rudimentary or no haptonema, are grouped in the Prymnesiophyceae (Green and Pienaar, 1977). There is usually no flagellar swelling associated with an eyespot as occurs in many other golden-brown

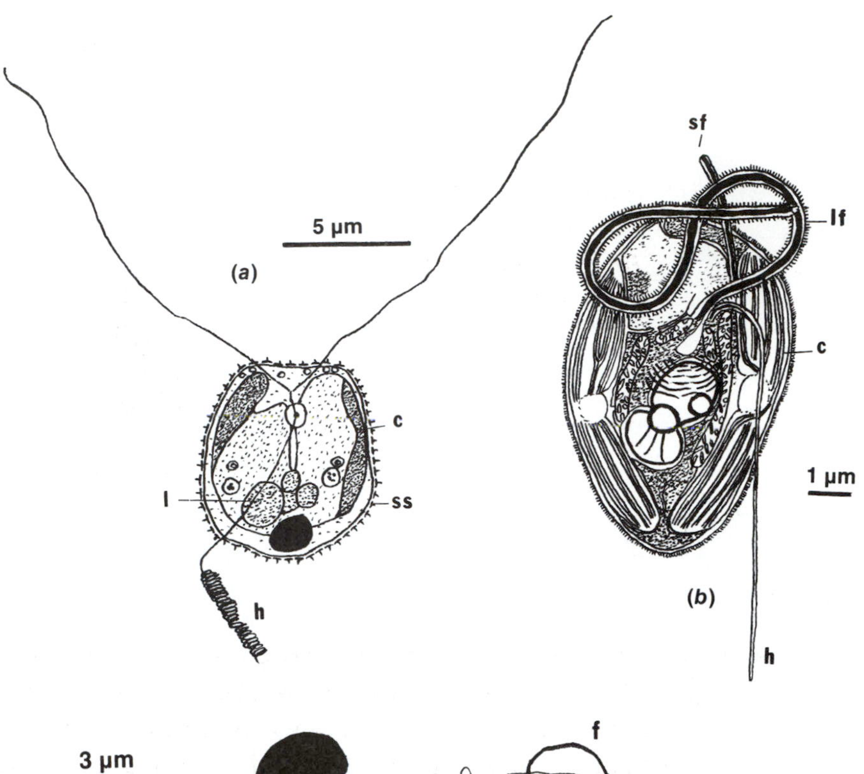

Fig. 18.2 (*a*) Ventral view of a saddle-shaped cell of *Chrysochromulina ephippum* with the haptonema loosely coiled. (*b*) *Pavlova mesolychnon*. (c) Chloroplast; (h) haptonema; (l) leucosin vesicle; (lf) long flagellum; (ss) spined scale; (sf) short flagellum. ((*a*) after Parke et al., 1956; (*b*) after van der Veer, 1969.)

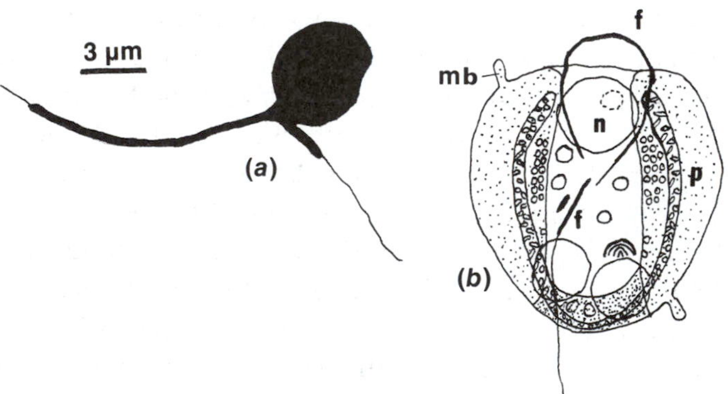

Fig. 18.3 (*a*) Drawing of a direct preparation of *Diacronema vlkianum* showing hair points at the tip of the flagella and the absence of lateral hairs. (*b*) A resting cell of *D. vlkianum*. (f) Flagellum; (mb) muciferous body (discharged); (n) nucleus; (p) plastid. (After Fournier, 1969.)

flagellates with two membranes of chloroplast E.R. (Hibberd, 1976), although there are exceptions (Green and Hibberd, 1977).

During swimming, the flagellar end of the cell can be forward with the flagella sweeping outward and backward down the sides of the body (Fig. 18.4(*a*)), or the flagellar end may be directed backward (Fig. 18.4(*b*)). Movement is usually rapid, the cells swimming only for a short distance in one direction, after which they rapidly change the position of the flagella and move off in the opposite direction. In *Pavlova* sp., the flagellar action is a little different, with the longer flagellum directed forward and the shorter flagellum trailing or directed outward.

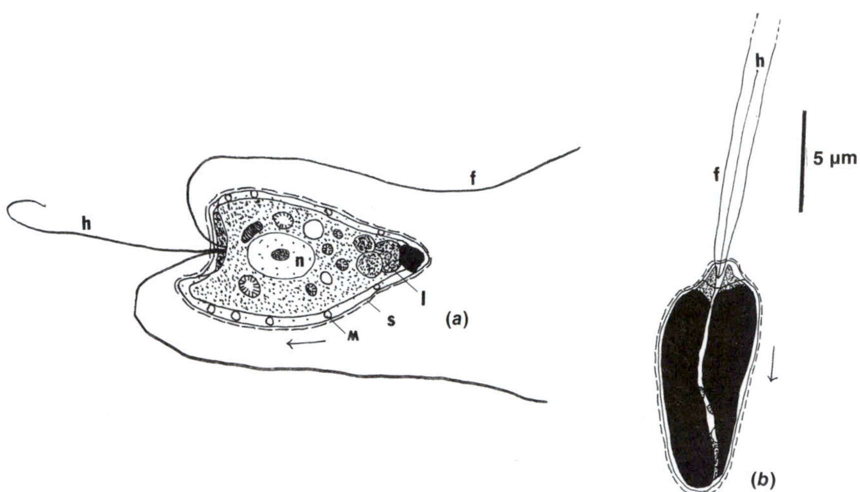

Fig. 18.4 *Chrysochromulina polylepis*. (*a*) Cell with flagella in position for swimming with flagellar pole forward. (*b*) Cell swimming with flagellar pole to the rear. (f) Flagellum; (h) haptonema; (l) leucosin vesicle; (m) muciferous body; (n) nucleus; (s) scale. (After Manton and Parke, 1962.)

HAPTONEMA

A **haptonema** is a filamentous appendage arising near the flagella but thinner and with different properties and structure. The haptonema ranges from a few profiles of E.R. that represent a reduced haptonema in *Imantonia rotunda* (Fig. 18.16(*d*)) (Green and Pienaar, 1977) to a short bulbous structure in *Hymenomonas roseola* (Fig. 18.5(*a*)) to the 80-μm-long whiplike structure in *Chrysochromulina parva* (Fig. 18.5(*b*)) (Manton, 1967a). The haptonema of *Prymnesium parvum* has an internal structure similar to the haptonema of other Prymnesiophyceae and will be used as an example of haptonemal structure (Fig. 18.6) (Manton, 1964). In transverse section the haptonema is composed of three concentric membranes surrounding a core containing seven microtubules. The core is covered by the innermost of the three membranes so that there is no contact between the core microtubules and the outer portion of the haptonema. The space between the innermost and middle membranes is a vesicle continuous over the tip of the core. The haptonema is commonly covered with small body scales.

The microtubules in the haptonema slide, relative to one another, to produce two basic movements, coiling and bending (Greyson et al., 1993).

1 **Coiling** is a sensory response to obstacles (Kawachi and Inouye, 1994). The haptonema coils instantly when forward-swimming cells encounter obstacles. The flagella are thrown backward and generate propulsive forces, resulting in backward swimming.
2 **Bending** by the haptonema occurs during food capture by the cells (Inouye and Kawachi, 1994; Kawachi et al., 1991). Prey particles adhere to

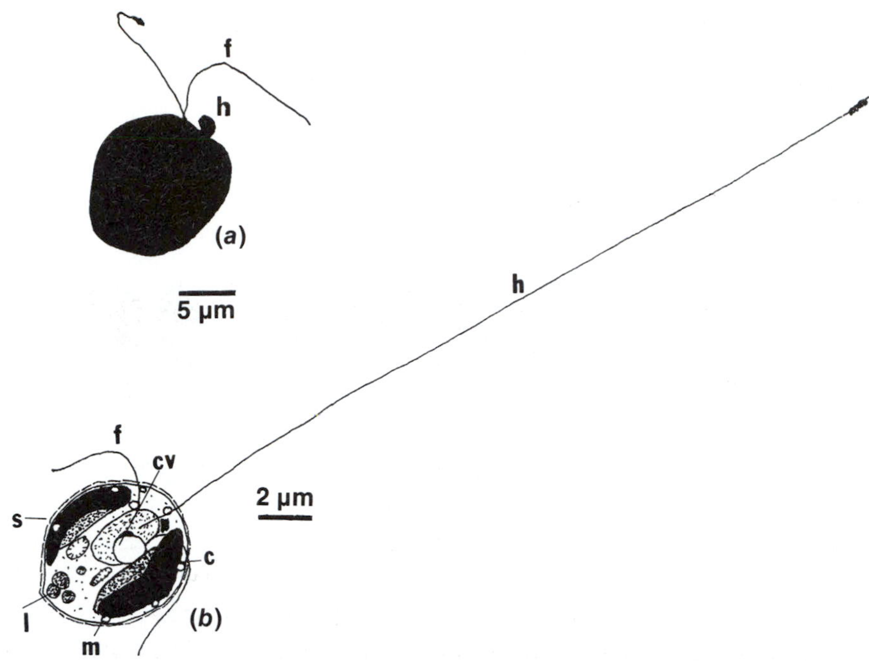

Fig. 18.5 (*a*) *Hymenomonas roseola*, a dried cell showing the flagella and the short bulbous haptonema (which actually arises between the two flagella). (*b*) A slowly gliding cell of *Chrysochromulina parva*. (c) Chloroplast; (cv) contractile vacuole; (f) flagellum; (h) haptonema; (l) leucosin; (m) muciferous body; (s) scale. ((*a*) after Manton and Peterfi, 1969; (*b*) after Parke et al., 1962.)

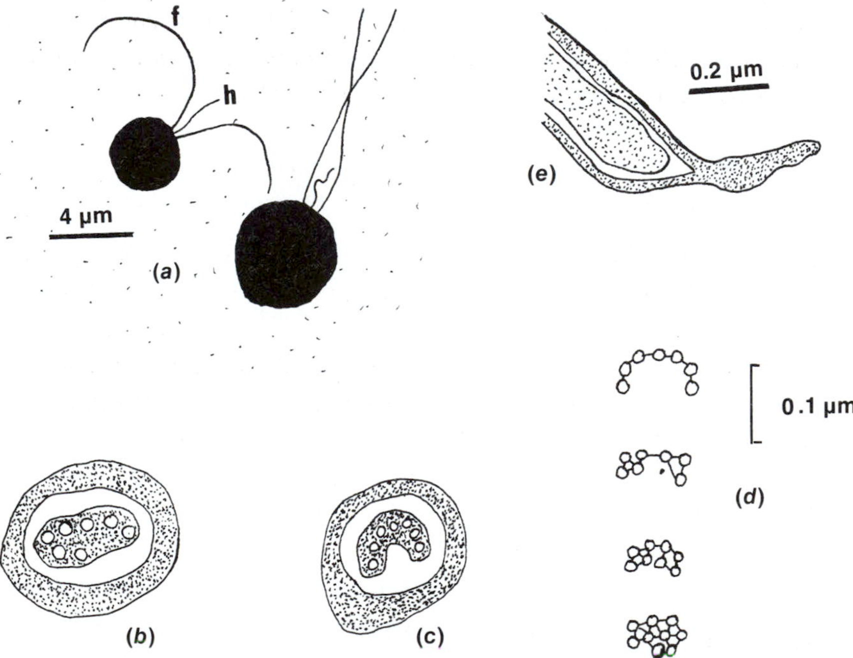

Fig. 18.6 *Prymnesium parvum*. (*a*) Drawing of two cells dried and shadow-cast showing the two smooth flagella (f) and the short haptonema (h) on each cell. (*b*) Transverse section of a haptonema showing the structure of the middle region, seven microtubules in the core surrounded by the membrane-bounded cavity and the haptonemal wall bounded by the plasmalemma externally. (*c*) Transverse section of a haptonema near the point of union with the cell showing the crescentic shape of the core characteristic of this region. (*d*) Progressive levels of haptonema tubules from immediately below the plasmalemma to the distal points of the microtubules. (*e*) Longitudinal section of a tip of a haptonema. ((*a*) after Manton, 1966; (*b*)–(*e*) after Manton, 1968.)

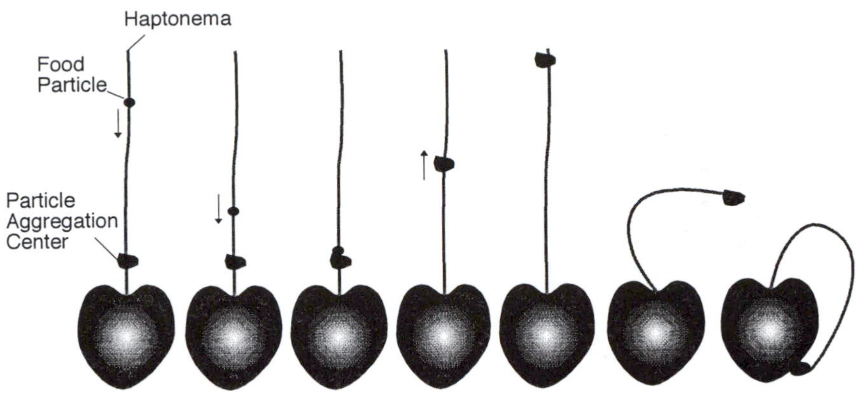

Fig. 18.7 The sequence of events by which food particles adhere to a haptonema and are moved toward the particle aggregation center. The mass of particles at the particle aggregation center then moves to the tip of the haptonema, the haptonema bends to deposit the mass of particles at the posterior end of the cell where the particles are taken up in food vacuoles (Adapted from Kawachi et al., 1991.)

the haptonema as the cells swim with the haptonema projecting ahead of the cell, and the flagella beating alongside. The adhering particles are transported down to a particular point on the haptonema, about 2 μm distal from the base, called the **particle-aggregating center** (Fig. 18.7). The particles accumulate at the particle-aggregating center, resulting in the production of a massive aggregate. In the aggregate, individual particles tightly adhere to one another, suggesting that some sort of cementing material is secreted. After reaching a certain size, the aggregate moves to the tip of the haptonema. The haptonema bends into a sigmoid shape and eventually delivers the aggregate to the posterior end of the cell, where the aggregate is injected into a food vacuole.

CHLOROPLASTS

Usually there are two elongate discoid chloroplasts in each cell (Fig. 18.1). Each chloroplast is surrounded by four membranes: the two membranes of the chloroplast envelope, and outside of them the two membranes of the chloroplast E.R. The thylakoids are aggregated into bands of three. A girdle band of thylakoids is usually absent (Hibberd, 1976). A pyrenoid is commonly present in the center of the chloroplast or as a bulge to one side. Eyespots are not common in the Prymnesiophyceae although *Pavlova* has an eyespot, which consists of a group of lipid droplets inside the anterior end of a chloroplast (Green, 1973) (Fig. 18.19).

OTHER CYTOPLASMIC STRUCTURES

Two types of membrane-bounded vesicles are in the cytoplasm, the first containing lipids and the second the storage product. The storage product is usually stated as being chrysolaminarin (leucosin) (Janse et al., 1996),

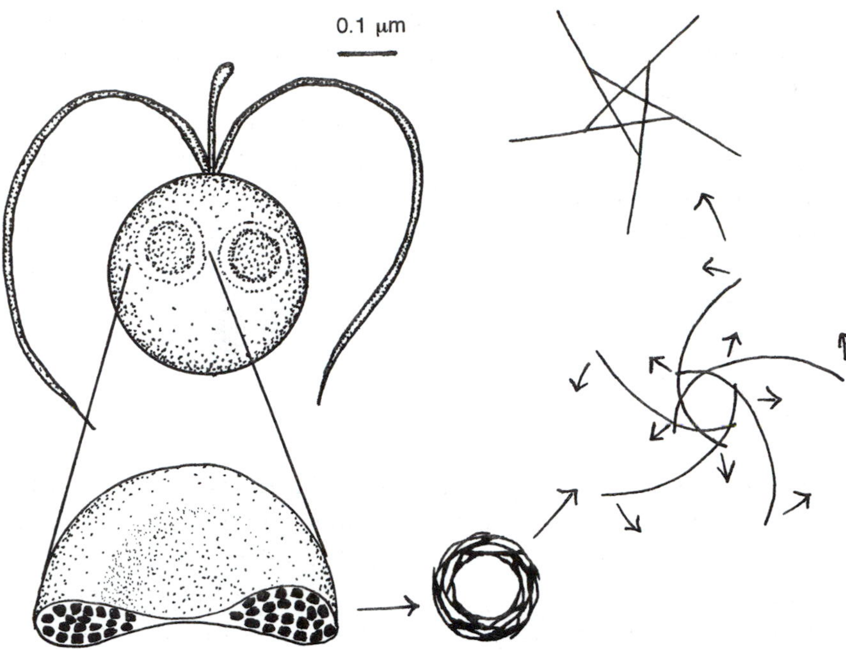

Fig. 18.8 *Phaeocystis globosa* showing a cell with two vesicles containing coiled filaments made of chitin. The coiled filaments unwind when they are discharged outside the cell, forming a 5-ray star structure. (After Chrétiennot-Dinet et al., 1997.)

although in a study of *Pavlova mesolychnon* it was found that one of the storage products present in the cells was a β-1,3 linked glucan similar to paramylon in the Euglenophyceae (Kreger and van der Veer, 1970).

In some of the Prymnesiophyceae and particularly the genus *Chrysochromulina*, muciferous bodies are under the plasma membrane (Figs. 18.1, 18.4(*a*) and 18.9(*b*)). They have the same structure as the muciferous bodies in the Raphidophyceae, Dinophyceae, and Chrysophyceae, and consist of a single-membrane-bounded vesicle filled with semi-opaque contents. When they are discharged outside the cell, they appear as substantial cylinders of opaque material of uniform diameter. The function of muciferous bodies is not known.

Phaeocystis globosa has vesicles that contain tightly-wound filaments of chitin (Chrétiennot-Dinet et al., 1997). Each vesicle contains five chitin filaments that attach near the base of another chitin filament (Fig. 18.8). The chitin filaments produce a five-sided star at their base when they are released from their vesicles.

Surface protrusions containing cytoplasm are common in the Prymnesiophyceae; they are called **pseudopodia**. Also, cells can extrude slender trailing filaments from their surfaces, which can be straight or branched. These filaments, called **filopodia**, eventually become segmentally constricted and break up into droplets. Many of the organisms are phagocytic and consequently have food vesicles in the cytoplasm in which they digest bacteria and other small algae. They are not selective in taking up material into the food vesicles, and will take up indigestible detritus as rapidly as bacteria and other algae.

Like many of the other algal groups, the Prymnesiophyceae participate in symbiotic events. Invertebrate radiolaria can harbor prymnesiophyte algal symbionts (Anderson et al., 1983). The symbionts are held in the rhizopodial network surrounding the central capsule of the radiolarian. The algal symbionts fix carbon dioxide, and some of the photosynthate passes to the radiolarian.

Scales and coccoliths

Most of the members of the Prymnesiophyceae have a cell covering consisting of a number of elliptical scales (Fig. 18.1). The elliptical scales are embedded in a mucilaginous substance, and in some organisms a layer of calcified coccoliths is outside the scales. The structure of the scales can vary considerably, but all of them appear to have an elliptical organic scale constituting either the whole structure of the scale or just the base plate on which the rest of the structure is based. The elliptical organic scale (plate scale) has radiating ridges extending to the edge (Fig. 18.9(*a*)) (Parke and Manton, 1962). *Chrysochromulina minor* (Fig. 18.9(*b*)) (Parke et al., 1955) and *Chrysochromulina parva* (Fig. 18.5(*b*)) (Parke et al., 1962) have a cell covering of only plate scales of the type just described. In other Prymnesiophyceae thc rims of the elliptical base plate have become turned up to form a variety of different types of scales ranging from very shallow tub-shaped scales with the edges of the scales barely turned up, to very long spined structures with the spine being the very elongated upturned rim of the organic scale (Fig. 18.9(*c*)) (Manton, 1972). A number of different types of scales can be present on the same organism. *Chrysochromulina kappa* (Fig. 18.9(*d*)) has plate scales around most of the body with a few short-spined scales near the flagella (Parke et al., 1955). *Chrysochromulina ericina* (Fig. 18.9(*e*)) has a cell covering consisting of plate scales with about 30 long-spined scales intermixed with the plate scales (Parke et al., 1956). *Chrysochromulina pringsheimii* (Fig. 18.9(*f*)) has an even more complex cell covering consisting of four types of scales. Long-spinned scales are at either end of the cell with small-spined scales covering the rest of the cell. Underneath this spined layer is a layer of plate scales, and, finally, near the base of the flagella are a number of small plate scales (Parke and Manton, 1962).

The organic scales originate in the Golgi apparatus (Fig. 18.1). A scale, when first formed in a Golgi vesicle, is closely enveloped by the vesicle

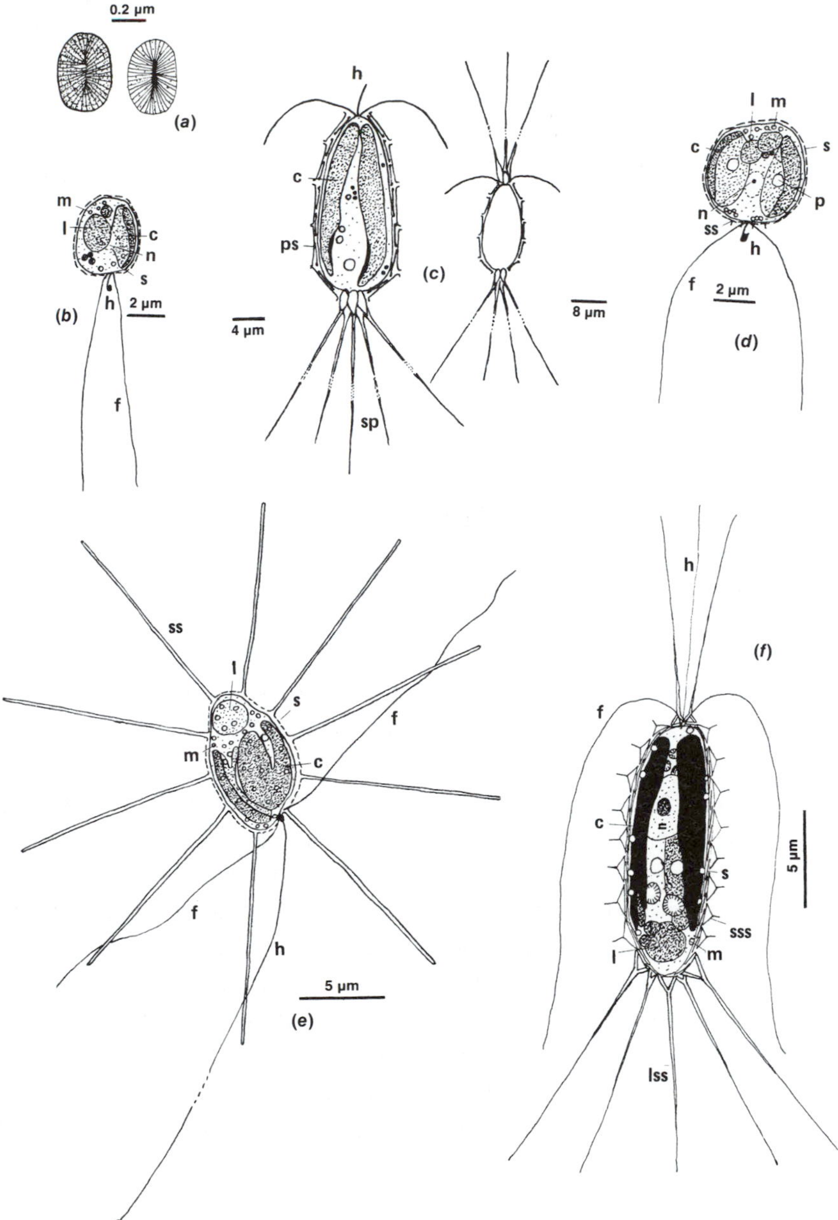

Fig. 18.9 (*a*) Small plate scales of *Chrysochromulina pringsheimii*, the upper face on the left and the lower face on the right. (*b*) *Chrysochromulina minor*, individual swimming with the flagella and haptonema behind the body in the position characteristic of the species during rapid swimming. (*c*) *Chrysochromulina parkae*, two forms.
(*d*) *Chrysochromulina kappa*, swimming with the flagella and haptonema behind the body, the characteristic position of species during rapid swimming.
(*e*) *Chrysochromulina ericina*, individual with dividing chloroplasts anchored by a haptonema which is partially extended; the flagella are in the characteristic position when the cells are stationary.
(*f*) *Chrysochromulina pringsheimii*, individual swimming with the flagella and haptonema in front of the body, with the haptonema fully extended. (c) Chloroplast; (f) flagellum; (h) haptonema; (l) leucosin vesicle; (lss) long-spined scale; (m) muciferous body; (n) nucleus; (p) pyrenoid; (ps) plate scale; (s) scale; (sp) spine; (ss) spined scale; (sss) small-spined scale. ((*a*),(*f*) after Parke and Manton, 1962; (*b*),(*d*) after Parke et al., 1955; (*c*) after Green and Leadbeater, 1972; (e) after Parke et al., 1956.)

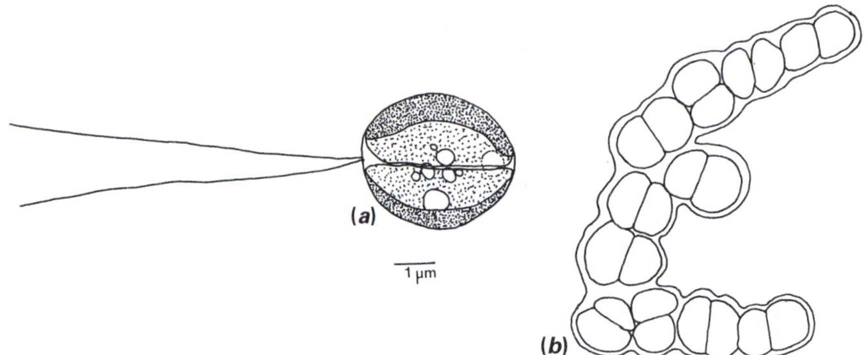

Fig. 18.10 *Pleurochrysis scherffelii*. (*a*) Motile cell. (*b*) Filamentous stage. (After Pringsheim, 1955.)

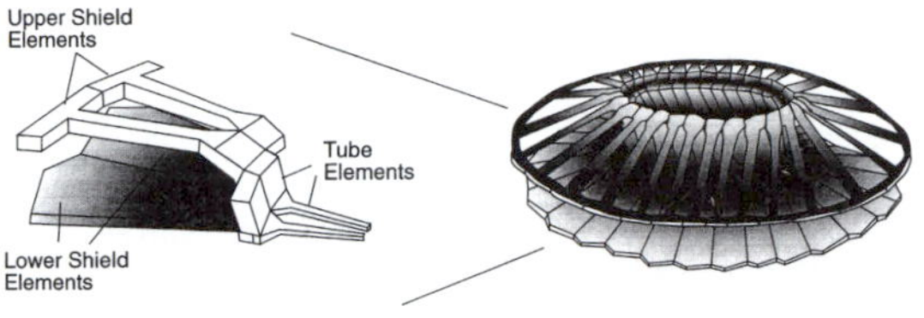

Fig. 18.11 Construction of a coccolith of *Emiliania huxleyi*. The heterococcolith is composed of a tube joining the upper and lower shields, each of which is composed of subunits. (Adapted from Young et al., 1992.)

membrane no matter how elaborate the shape of the scale is (long-spined, etc.); but immediately before the scale is liberated to the outside of the cell, this close contact is lost (Manton, 1967b). There is a diurnal rhythm in the production of scales. Manton and Parke (1962) showed that in *Chrysochromulina polylepis* the greatest production of scales is in the late afternoon with the least in the early morning hours. The time of nuclear division is the reverse, with the most mitotic figures appearing in the early morning.

Even in the non-motile filamentous stages of the Prymnesiophyceae, the cell wall is composed of scales embedded in a gelatinous matrix. In *Pleurochrysis scherffelii* (Fig. 18.10) the cells of the filamentous stage are covered in cellulosic scales produced by the single Golgi apparatus. By the use of the cinephotography, Brown (1969) showed that during wall formation the whole protoplast revolves so that the scale secretions of the Golgi apparatus are received more or less evenly by all portions of the cell wall.

Coccoliths are calcified scales of the Prymnesiophyceae. They were originally described as minute carbonate discs in Cretaceous deposits and thought to be of inorganic origin. Later they were found in sea bottom oozes brought up by the first Atlantic cable survey in 1858. Their algal nature was not recognized until 1898. Coccoliths are basically organic scales that have calcium carbonate ($CaCO_3$) deposited on one surface in a characteristic pattern depending on the species of the alga. Under ordinary conditions

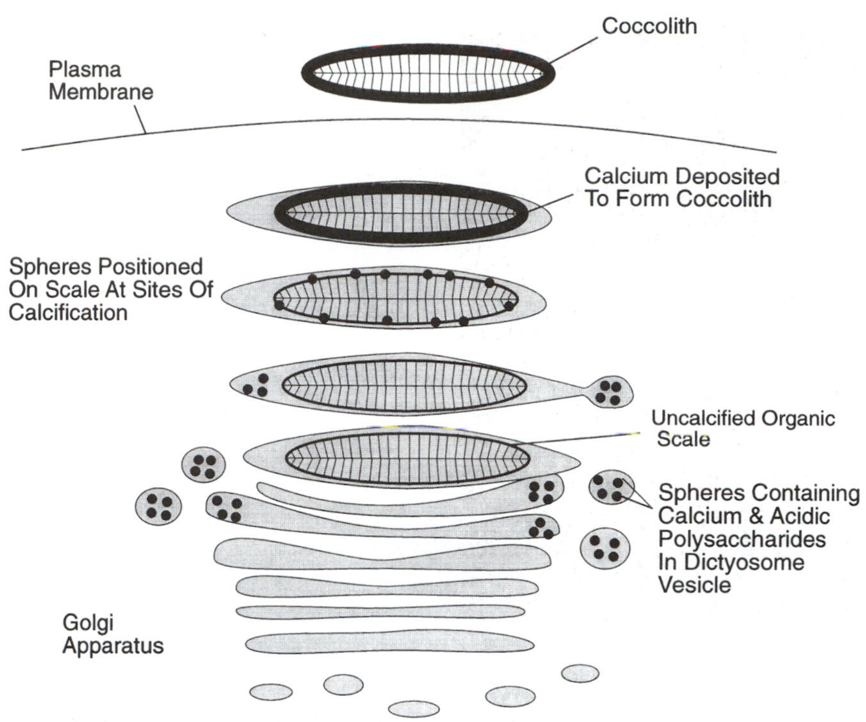

Fig. 18.12 Coccolith formation in *Pleurochrysis*. Spheres containing calcium bound to acidic polysaccharides are produced in vesicles by the Golgi apparatus. These vesicles fuse with Golgi vesicles containing uncalcified organic scales. The calcium-containing spheres become positioned on the scales at sites of calcification and the calcium is released. The calcified scales are carried to the plasma membrane, releasing the coccoliths outside the cell. (Adapted from Marsh, 1994.)

anhydrous $CaCO_3$ exists in nature in two crystalline forms, **calcite** (rhombohedral) and **aragonite** (orthorhombic), which differ in structure, hardness, specific gravity, and solubility. In coccoliths the form of a $CaCO_3$ is usually calcite. The calcite is attached to the outer surface of a plate scale, which has a pattern of radiating ridges (Fig. 18.13) whereas the inner side of the scale (toward the cell) is virtually patternless. The coccoliths of *Emiliania huxleyi* are relatively simple and can serve as an example of coccolith structure although it should be realized that there is a multitude of different coccolith forms, some very complex (Fig. 18.15). The coccoliths of *E. huxleyi* (Figs. 18.11 and 18.15(*b*)) are composed of a number of hollow crystals of calcite arranged around the periphery of a plate scale (Young et al., 1992). Each coccolith consists of an upper and lower shield joined by a tube. The tube and the shields are composed of subunits.

Coccolith formation begins with the production of an uncalcified organic plate-scale in the center of Golgi cisternae (Figs. 18.12, 18.13) (Marsh, 1994, 1996). Highly acidic polysaccharides are produced as 25 nm diameter spheres in the peripheral part of Golgi cisternae. The highly acidic polysaccharides are extremely negative polyanions that sequester large numbers of calcium ions. The 25 nm diameter spheres are pinched off the

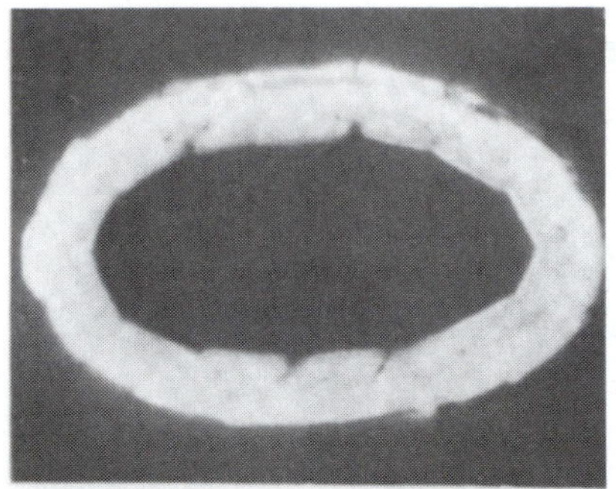

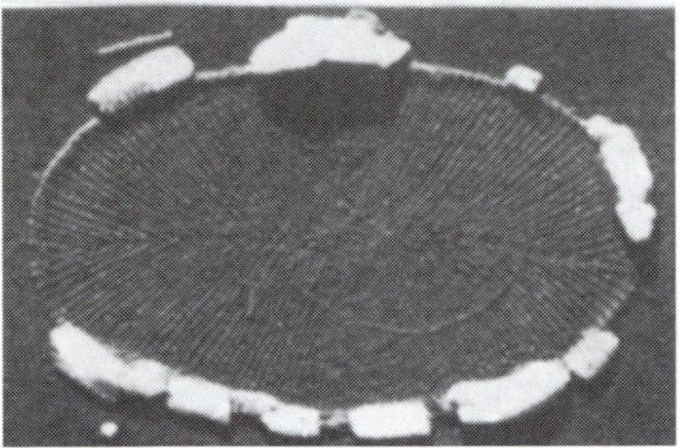

Fig. 18.13 Intact coccolith of *Hymenomonas carterae* and one with the calcium carbonate making up the coccolith partially dissolved away, revealing the scale on which the calcium carbonate is deposited. (From Outka and Williams, 1971.)

Golgi cisternae as vesicles, which eventually fuse with a vesicle containing an uncalcified organic plate-scale. The 25 nm diameter spheres aggregate on the scales in the area of future calcium deposition. Calcium is deposited in the area of the 25 nm diameter spheres to produce calcified coccoliths. The remaining material in the spheres is reorganized into an amorphous coat that surrounds the mature crystals of calcium carbonate that make up the coccoliths. The coccoliths are carried in the vesicles to the plasma membrane where the coccoliths are deposited outside the cell. Two to seven coccoliths are produced per day by *Emiliania huxleyi* (Balch et al., 1993; Linschooten et al., 1991).

Coccoliths are detached from cells in layers at the same time as other coccoliths are produced. During logarithmic growth, about the same number of coccoliths are detached as are produced. In stationary growth, however, the rate of coccolith detachment increases about threefold, while coccolith production drops off (Balch et al., 1993).

Coccolithophorids can have an outer covering of **heterococcoliths**, which contain assemblages of morphologically complex and diverse $CaCO_3$ elements. Conversely, the coccolithophorid may be covered with **holococcoliths**, which are composed of similar (often rhombohedral) calcite crystals. In *Coccolithus*, holococcoliths occur in one part of the life cycle (the motile *Crystallolithus* stage) (Figs. 18.14, 18.15), whereas heterococcoliths occur in the non-motile stage. The base plates of heterococcoliths are pro-

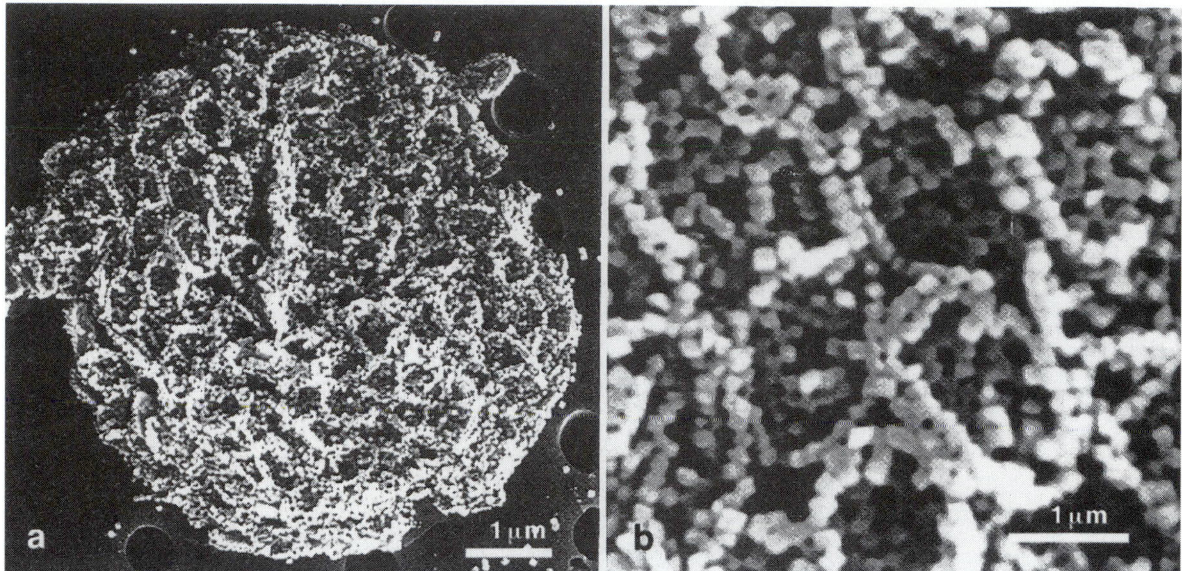

Fig. 18.14 Scanning electron micrographs of holococcoliths covering a cell of *Crystallolithus hyalinus*. (From Faber and Preisig, 1994).

duced in Golgi cisternae followed by calcification in the same cisternae. However, only the base plates of holococcoliths are produced in Golgi cisternae. These base plates are discharged outside the cell where calcification occurs within an outer vestment (Rowson et al., 1986).

The coccolithophorids (algae with coccoliths) (Fig. 18.15) are common in tropical waters because these warm waters have a low partial pressure of carbon dioxide and are usually saturated or supersaturated with calcium carbonate, the concentrations being especially high in the upper layers. Supersaturation of calcium carbonate is favorable for the formation of coccoliths, and the distribution of coccolithophorids shows a close correlation with the degree of saturation of seawater by calcium carbonate. In the seas of polar regions, the degree of saturation does not even reach 90%.

The calcification reaction is:

$$2\,HCO_3^- + Ca^{2+} \rightleftharpoons CaCO_3 + CO_2 + H_2O$$

Carbon dioxide is released in calcification, and this may make coccolithophorids more competitive by increasing the amount of CO_2 available for photosynthesis inside the cell (Nielsen, 1995). This may be particularly important in seawater where the pH of 8.2 results in a very low concentration of dissolved CO_2 of about 10 μM compared to a total dissolved carbonate (mostly HCO_3^-) of 2000 μM.

The external coccoliths can be removed by lowering the pH of the culture medium. Cells of *Coccolithus huxleyi* decalcified in this manner may acquire a complete coccolith envelope (about 15 coccoliths) within 15

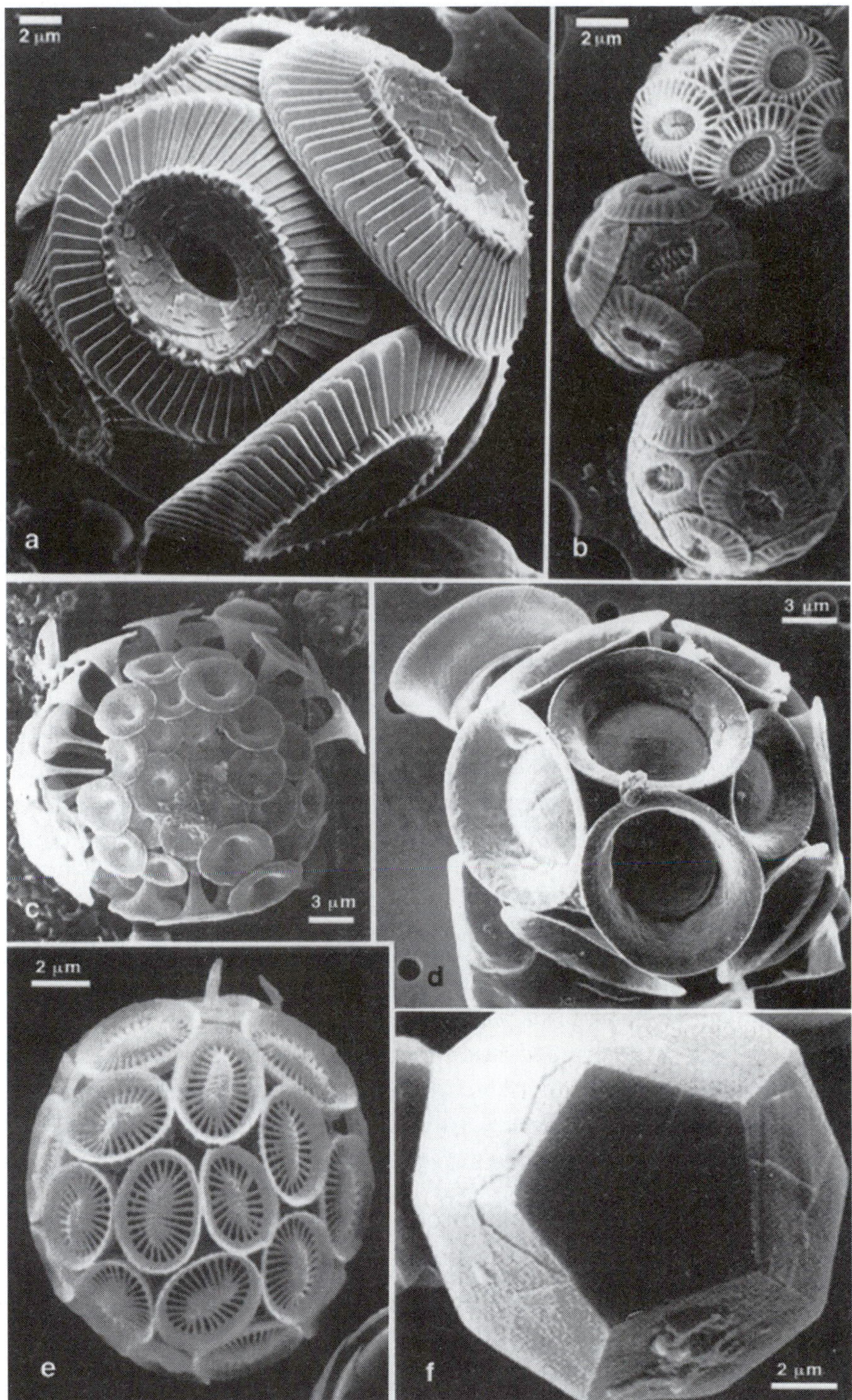

Fig. 18.15 Scanning electron micrographs of coccolithophorid cells. (*a*) *Coccolithus pelagicus*. (*b*) *Emiliania huxleyi*. (*c*) *Discosphaera tubifera*. (*d*) *Pontosphaera syracusana*. (*e*) *Syracosphaera nodosa*. (*f*) *Braarudosphaera bigelowii*. (From Faber and Preisig, 1994.)

hours of their being transferred back to a normal medium in the light. Complete recalcification in *Cricosphaera* sp. may require 40 hours. In both instances, cell division is not a prerequisite for the formation of new coccoliths. If the organisms are grown in artificial seawater, the coccoliths dissolve if the product of the concentrations of calcium and carbonate is appreciably smaller than the solubility product of calcite. In *C. huxleyi*, coccoliths are still formed inside cells even when the calcium content of the medium is reduced to levels where external coccoliths dissolve, although coccolith production is retarded at calcium concentrations less than half that of normal seawater. A photochemical process is apparently directly associated with coccolith production because when light is turned off, there is a sharp drop in coccolith production.

Pleurochrysis carterae has biflagellate cells surrounded by a coccosphere consisting of a single layer of 100 to 200 coccoliths. The cells incorporate calcium into extracellular coccoliths at a more or less constant rate throughout a 16-hour light: 8-hour dark cycle. The cells divide during the dark periods with a concomitant decrease in cell size during the dark period followed by an increase in cell size during the light period. The cells form coccoliths in the light as well as in the dark at a similar rate (van der Wal et al., 1987) (although *Emiliania huxleyi* (Fig. 18.15(*b*)) produces coccoliths only during the light period (Linschooten et al., 1991)).

Although coccolithophorids constitute a minor part of recent **calcareous oozes** (bottom sediments composed of calcified remains of organisms) in the ocean, in the Cretaceous they dominated the calcareous nannoplankton, an indication that during this time they were more abundant in the oceans than they are today (Tasch, 1973). Coccolithophorids provided the major constituent of Mesozoic (Jurassic and Cretaceous) and Tertiary chalks and marls. The abundance of coccolithophorids in these chalks can be demonstrated by taking a piece of ordinary blackboard chalk, pulverizing it, mixing it with distilled water in a test tube, and letting it stand for 20 minutes. Draw some of the solution into a pipette, dispose of the first four to five drops, and place the next few drops on a slide. Place a cover slip on the slide, and view it at a magnification of 400 to 500 ×. Many coccoliths and other remains will be seen.

Coccoliths in sedimentary rocks can be used as markers in the discovery and mode of deposition of oil deposits. For example, the oil shales of the Kimmeridge Clays in England are sandwiched between limestone bands that are composed mostly of coccoliths of one species, *Ellipsagelosphaera britannica* (Gallois, 1976). Other oil-bearing rocks have similar characteristic coccoliths. Therefore petroleum geologists know that when a drill core shows certain coccoliths that are associated with petroleum, there is a good chance of finding oil in that stratum of rock.

Toxins

The prymnesiophycean alga *Prymnesium parvum* (Fig. 18.6) forms a potent exotoxin that causes extensive fish mortalities in brackish water conditions in many countries in Europe and in Israel (Shilo, 1967). The toxin is most effective against aquatic gill-breathing animals, such as fish and molluscs. In Amphibia, only the gill-containing tadpole stage is sensitive to immersion in solutions containing the ichthytoxin. The rapidity of the action of *Prymnesium* toxin on immersed fish suggests that the immediate target must be an exposed organ, probably the gill. Experiments have shown that the toxins affect the permeability of the gill, resulting in the increased sensitivity of the fish. In fish removed promptly from such toxin solutions, the gill damage is repaired within hours. In Israel, Shilo (1967) has found that it is possible to control *P. parvum* in fish breeding ponds by adding small amounts of ammonium salt, which causes the algal cells to lyse.

In cultures of *P. parvum*, the synthesis of the toxins is greatest during the late stages of the logarithmic phase of growth and continues into the stationary phase. Initially the toxins are within the cell, but later they are excreted and found in the medium. For formation of the toxin, light is essential; under heterotrophic conditions in the dark, cell multiplication continues but no toxin is formed. Phosphate-starved cells show a large increase in toxins. The toxins require a cofactor to be toxic: Calcium, magnesium, spermine, or streptomycin can be substituted for this cofactor.

Some species of *Chrysochromulina* produce toxins that kill fish, mussels and ascidians (Hansen et al., 1995; Moestrup, 1994; Simonsen and Moestrup, 1997). The best documented fish kills have occurred off the coast of Norway and Sweden. The large blooms of *Chrysochromulina* causing the fish kills have been associated with a lack of predation by the normal ciliate grazers of *Chrysochromulina*. It appears that the long spines on the surface of the *Chrysochromulina* cells make them too large to be taken up by the ciliates (Hansen et al., 1995).

In the North Sea of Europe, blooms of the prymnesiophyte *Phaeocystis* (Fig. 18.8 and 18.16(*a*)) occur as macroscopic lobed colonies or "bladders" in the spring and fall, coating the fishing nets with a slime. The *Phaeocystis* colors the water yellow-brown, and the bloom in the spring is greater than the one in the fall. Herring specifically avoid the ocean areas where there are *Phaeocystis* blooms because of its unpalatability to the fish (Savage, 1930). Also, for a number of years biologists were puzzled by the near sterility of the intestines of certain Antarctic birds in regard to Protozoa and microorganisms. This condition was shown to be due to the large quantities of *Phaeocystis* in their diet, with *Phaeocystis* producing large amounts of

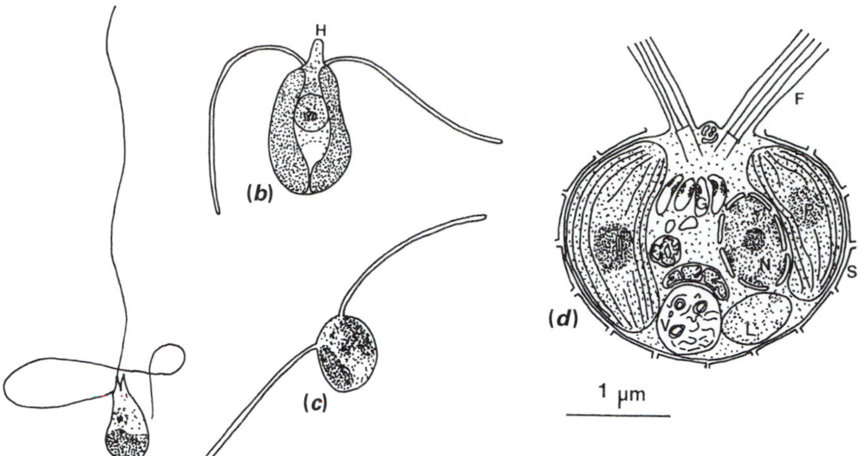

Fig. 18.16 (*a*) Motile cell of *Phaeocystis poucheti*. (*b*) *Isochrysis galbana*. (*c*),(*d*) *Imantonia rotunda*. (F) Flagellum; (H) haptonema, (L) leucosin vesicle; (N) nucleus; (P) pyrenoid; (S) scale. ((*d*) after Reynolds, 1974.)

acrylic acid which has a strong bactericidal action. *Phaeocystis* secretes 16% to 64% of the carbon assimilated in photosynthesis as polysaccharides of varying molecular weight. As much as 7 μg liter^{-1} of acrylic acid, and at least 0.3 mg liter^{-1} of polysaccharides can be released in the formation of a dense bloom (Guillard and Hellebust, 1971; van Rijssel et al., 1997).

Every spring from 1978 to 1983, large areas (10 to 100 square miles) of water off the Atlantic Coast of France and southern England gave a strong reflectance of visible light to the Nimbus space satellite. Samples showed that the water contained mostly the coccolithophorid *Emiliania huxleyi* (Fig. 18.15(*b*)). The large coccoliths of this organism resulted in the high light reflectance which was picked up by the satellite (Holligan et al., 1983).

Growth

Generally, the Prymnesiophyceae seem to require either thiamine or vitamin B_{12} for growth, or, in some cases, both; so the algae are not completely autotrophic but auxotrophic. *Emiliania huxleyi* needs only thiamine, and this lack of a vitamin B_{12} requirement is thought to be ecologically important in that it enables the organism to grow regularly in the vitamin B_{12}-deficient offshore waters of the Sargasso Sea. The coccolithophorids are among the fastest-growing algae, with the fastest growth rates for *Cricosphaera elongata* being 2.25 divisions per day and those of *C. huxleyi* being 1.85 divisions per day.

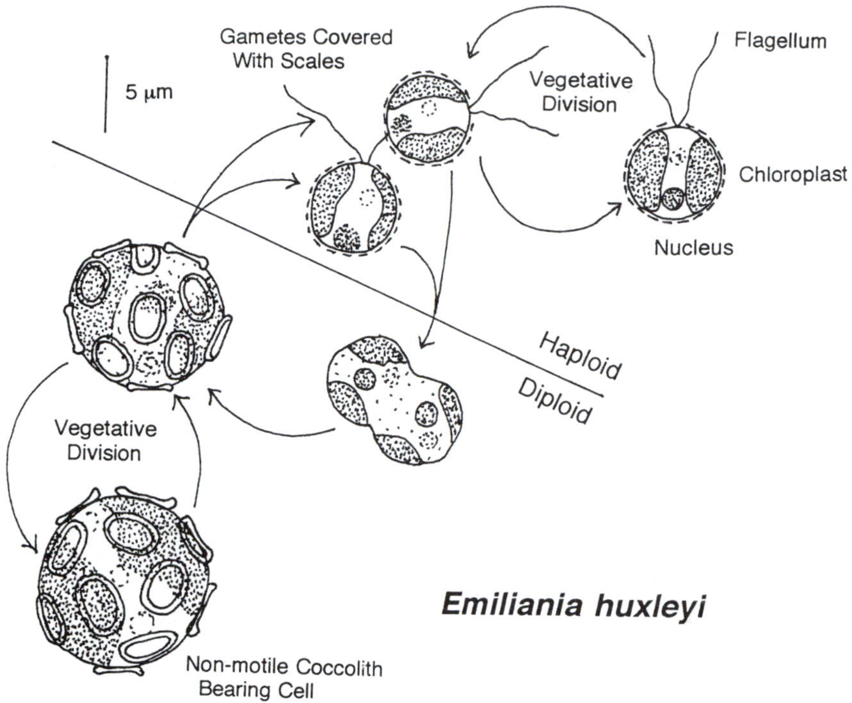

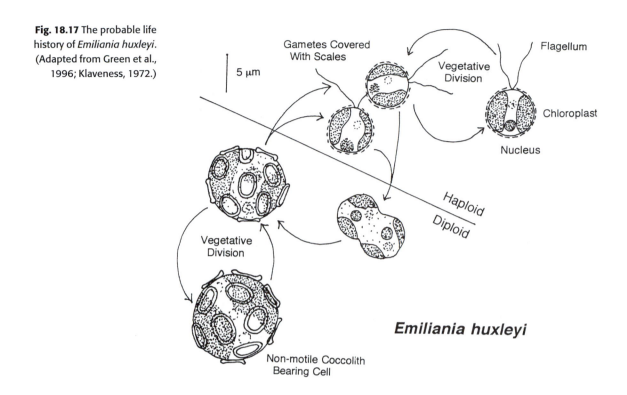

Fig. 18.17 The probable life history of *Emiliania huxleyi*. (Adapted from Green et al., 1996; Klaveness, 1972.)

Classification

The Prymnesiophyceae can be divided into two orders (Green and Jordan, 1994):

Order 1 Prymnesiales: cells with two equal smooth flagella, no eyespot, scales commonly covering the cell body.

Order 2 Pavlovales: cells with two unequal flagella often covered with hairs and deposits, eyespots may be present.

Molecular studies have shown the Prymnesiales and the Pavlovales to be two distinct and monophyletic groups (Fujiwara et al., 1994; Simon et al., 1997).

PRYMNESIALES

Emiliania huxleyi (Fig. 18.15(*b*)) is typical of the order with motile cells having two equal flagella. The life cycle of *Emiliania huxleyi* (Fig. 18.17) probably involves a diploid, non-motile phase with coccoliths, that alternates with a haploid phase that has scales but no coccoliths (Green et al., 1996; Klaveness, 1972).

The life cycle of the coccolithophorid *Hymenomonas carterae* is more

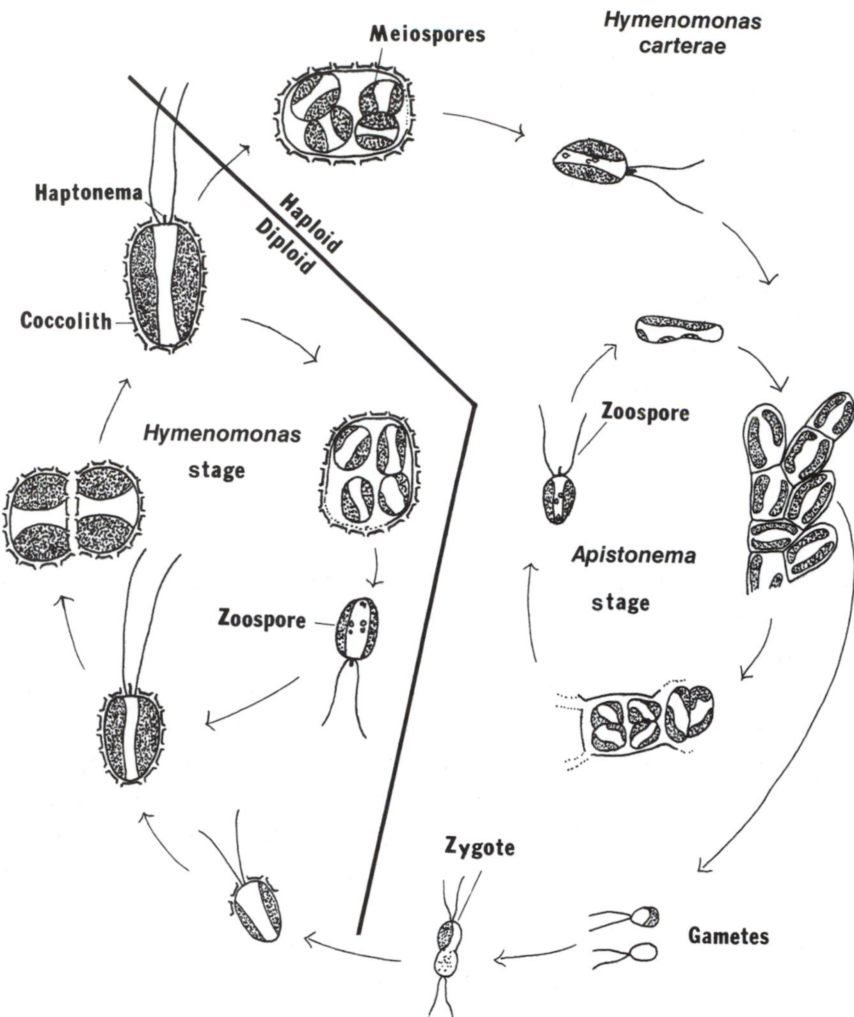

Fig. 18.18 The life history of *Hymenomonas carterae*.

complex. *H. carterae* has a haploid *Apistonema* stage (21 ± 1 chromosomes) that consists of a filamentous branched system of spherical or elongate cells (Fig. 18.18). Cells of the *Apistonema* stage can give rise either to asexual swarmers, which form new *Apistonema* thalli, or to motile gametes. Each asexual swarmer has two long flagella and a cup-shaped chloroplast, and the cell is covered with uncalcified scales. Gametes can be distinguished from asexual swarmers by their smaller size and the reduced number or absence of chloroplasts. After fusion of two gametes, the resulting zygote develops coccoliths. The diploid *Hymenomonas* stage (42 ± 1 chromosomes) has an outer layer of coccoliths under which are several layers of organic scales. A short haptonema is present between the flagella. The coccolith-bearing stage can perpetuate itself asexually by mitotic divisions.

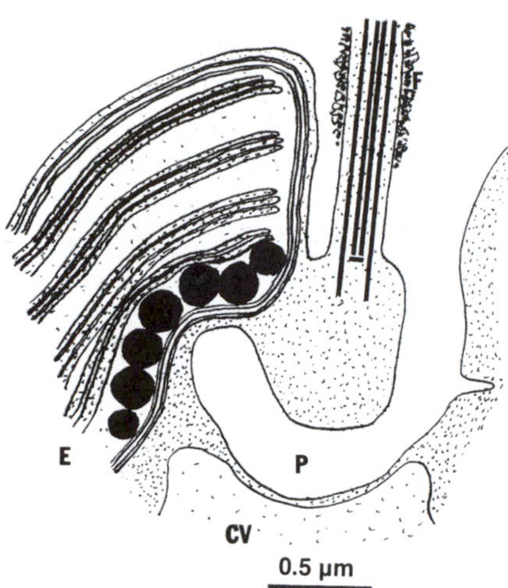

Fig. 18.19 A drawing of a section through the apical area of *Pavlova granifera* showing the contractile vacuole (CV), eyespot (E), long flagellum (F), and pit (P). (After Green, 1973.)

Under certain circumstances the protoplasts of *Hymenomonas* divide meiotically to produce four meiospores, which form haploid *Apistonema* thalli (von Stosch, 1967; Leadbeater, 1970; Parke, 1971).

The non-motile benthic stages of this order are resistant to fairly wide variations in the environment. The benthic stage of *Cricosphaera* sp. can survive temperatures of 35 to 40 °C for an hour or more as well as deep freezing for periods up 4 days. The same benthic phase is able to withstand a salinity up to the crystallization of the salt in solution. Normally benthic stages of *Cricosphaera* sp. grow near the high-water mark where they are presumably exposed to great variations in the environment. The motile phase of *Cricosphaera* sp. will tolerate salinities only as low as 0.4% to 0.8% salt and as high as 4.5% to 5.0% salt (see Paasche, 1968, for a review).

PAVLOVALES

The Pavlovales is the most primitive order in the Prymnesiophyceae (Green, 1980). The cells have two unequal flagella, with a haptonema arising between the flagella. In *Pavlova*, the two unequal flagella are attached some distance below the cell apex (Fig. 18.2(*b*)) (Green, 1967, 1976; van der Veer, 1969, 1976; Green and Manton, 1970). The longer flagellum is directed forward during swimming and is covered with fine hairs and dense bodies, whereas the short flagellum projects outward and can have fibrillar hairs on it. The flagella and haptonema are attached at the bottom of a depression (Green, 1973). A pit or canal passes from the bottom of the depression, under the base of the long flagellum, and terminates at the inner face of the eyespot (Fig. 18.19). The shape of the cells is variable, and there is a single two-lobed chloroplast with a pyrenoid. An eyespot is present inside the chloroplast. No sexual reproduction is known, and the cells propagate by longitudinal division in the motile state.

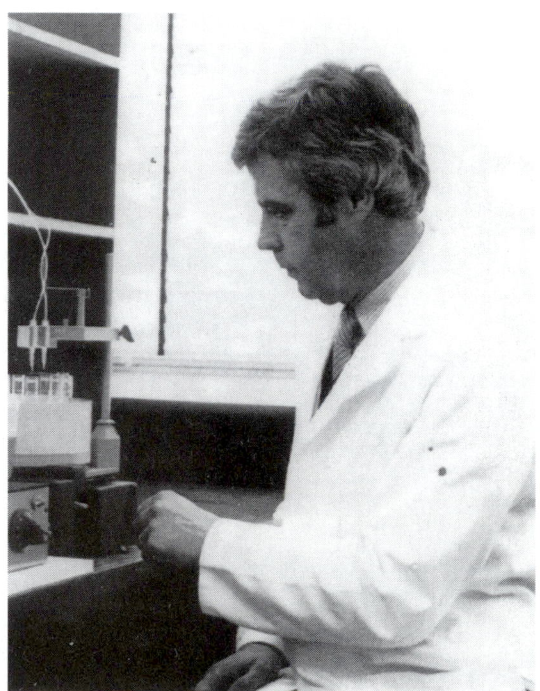

Fig. 18.20 *Left:* Irene Manton and *right:* Barry S. C. Leadbeater.
Irene Manton 1904–1988. Dr Manton was a graduate of Cambridge University. She was a lecturer in botany at the University of Manchester from 1929 to 1945 and a professor of botany at the University of Leeds from 1945 to 1969. Dr Manton was one of the foremost cytologists during the 1950s and 1960s when the new science of electron microscopy was adding vast amounts of new information to the understanding of the algae. The work of Dr Manton and her colleagues led to the discovery of the distinctive features of the haptophytes and the recognition of these organisms as a distinct group of algae.
Barry S. C. Leadbeater Dr Leadbeater graduated in Biology from the University of London in 1963 and obtained his PhD from the same institution in 1967. As a research student under Professor John Dodge, he studied the ultrastructure of dinoflagellates. From 1967 to 1969, Dr Leadbeater worked with Professor Irene Manton, FRS, on the ultrastructure and life cycles of microscopic prymnesiophytes. Since 1969, he has been at the University of Birmingham where he is currently senior lecturer in Plant Biology. His research has centered on the ultrastructure, life cycles, and physiology of algae and protozoa comprising the nannoplankton.

References

Anderson, O. R., Swanberg, N. R., and Bennett, P. (1983). Fine structure of yellow-brown symbionts (Prymnesiida) in solitary radiolaria and their comparison with similar Acantharian symbionts. *J. Protozool.* 30:718–22.

Balch, W. M., Kilpatrick, K., Holligan, P. M., and Cucci, T. (1993). Coccolith production and detachment in *Emiliania huxleyi* (Prymnesiophyceae). *J. Phycol.* 29:566–75.

Bhattacharya, D., and Ehlting, J. (1995). Actin coding regions: gene family evolution and use as a phylogenetic marker. *Arch. Protistenk.* 145:155–64.

Brown, R. M. (1969). Observations on the relationship of the Golgi apparatus to wall formation in the marine chrysophycean alga, *Pleurochrysis scherfferlii* Pringsheim. *J. Cell Biol.* 41:109–23.

Chrétiennot-Dinet, M-J., Giraud-Guille, M. M., Vaulot, D., Putaux, J-L., Saito, Y., and Chanzy, H. (1997). The chitinous nature of filaments ejected by *Phaeocystis* (Prymnesiophyceae). *J. Phycol.* 33:666–72.

Christensen, T. (1962). *Alger Botanik*, Vol. 2, No. 2. Copenhagen: Munksgaard.

Faber, W. W., and Preisig, H. R. (1994). Calcified structures and calcification in protists. *Protoplasma* 181:78–105.

Fournier, R. O. (1969). Observations on the flagellate *Diacronema vlkianum* Prauser (Haptophyceae). *Br. Phycol. J.* 4:185–90.

Fujiwara, S., Sawada, M., Someya, J., Minaka, N., Kawachi, M., and Inouye, I. (1994). Molecular phylogenetic analysis of *rbc*L in the Prymnesiophyta. *J. Phycol.* 30:863–71.

Gallois, R. W. (1976). Coccolith blooms in the Kimmeridge Clay and origin of the North Sea Oil. *Nature* 259:473–5.

Green, J. C. (1967). A new species of *Pavlova* from Madeira. *Br. Phycol. Bull.* 3:299–303.

Green, J. C. (1973). Studies in the fine structure and taxonomy of flagellates in the genus *Pavlova*. II. A freshwater representative, *Pavlova granifera* (Mack) comb. nov. *Br. Phycol. J.* 8:1–12.

Green, J. C. (1976). Notes on the flagellar apparatus and taxonomy of *Pavolva mesolychnon* van der Veer, and on the status of *Pavlova* Butcher and related genera within the Haptophyceae. *J. Mar. Biol. Assoc. UK* 56:595–602.

Green, J. C. (1980). The fine structure of *Pavlova pinguis* Green and a preliminary survey of the order Pavlovales (Prymnesiophyceae). *Br. Phycol. J.* 15:151–91.

Green, J. C., and Manton, I. (1970). Studies in the fine structure and taxonomy of flagellates in the genus *Pavlova*. I. A revision of *Pavlova gyrans*, the type species. *J. Mar. Biol. Assoc. UK* 50:1113–30.

Green, J. C., and Leadbeater, B. S. C. (1972). *Chrysochromulina parkae* sp. nov. (Haptophyceae) a new species recorded from S.W. England and Norway. *J. Mar. Biol. Assoc. UK* 52:469–74.

Green, J. C., and Hibberd, D. J. (1977). The ultrastructure and taxonomy of *Diacronema vlkianum* (Prymnesiophyceae) with special reference to the haptonema and flagellar apparatus. *J. Mar. Biol. Assoc. UK* 57:1125–36.

Green, J. C., and Pienaar, R. N. (1977). The taxonomy of the order Isochrysidales (Prymnesiophyceae) with special reference to the genera *Isochrysis* Parke, *Dicrateria* Parke and *Imantonia* Reynolds. *J. Mar. Biol. Assoc. UK* 57:7–17.

Green, J. C., and Jordan, R. W. (1994). Systematic history and taxonomy. In *The Haptophyte Algae*, ed. J. C. Green, and B. S. C. Leadbeater, Systematics Assn. Special Vol. 51, pp. 1–22. Oxford: Clarendon Press.

Green, J. C., Course, P. A., and Tarran, G. A. (1996). The life-cycle of *Emiliania huxleyi*: A brief review and a study of relative ploidy levels analyzed by flow cytometry. *J. Marine Systems* 9:33–44.

Greyson, A. J., Green, J. C., and Leadbeater, B. S. C. (1993). Structure and physiology of the haptonema in *Chrysochromulina* (Prymnesiophyceae). II. Mechanisms of haptonemal coiling and the regeneration process. *J. Phycol.* 29:686–700.

Guillard, R. R. L., and Hellebust, J. A. (1971). Growth and production of extracellular substances by two strains of *Phaeocystis poucheti*. *J. Phycol.* 7:330–8.

Hansen, P. J., Nielsen, T. G., and Kaas, H. (1995). Distribution and growth of protists and mesozooplankton during a bloom of *Chrysochromulina* spp. (Prymnesiophyceae, Prymnesiales). *Phycologia* 34:409–16.

Hibberd, D. J. (1976). The ultrastructure and taxonomy of the Chrysophyceae and Prymnesiophyceae (Haptophyceae): A survey with some new observations on the ultrastructure of the Chrysophyceae. *Bot. J. Linn. Soc.* 72:55–80.

Holligan, P. M., Viollier, M., Habour, D. S., Camus, P., and Champagne-Phillipe, M. (1983). Satellite and ship studies of coccolithophore production along a continental shelf. *Nature* 304:339–42.

Inouye, I., and Kawachi, M. (1994). The haptonema. In *The Haptophyte Algae*, ed. J. C. Green, and B. S. C. Leadbeater, Systematics Assn. Special Vol. 51, pp. 73–89. Oxford: Clarendon Press.

Janse, I., van Rijssel, M., van Hall, P. J., Gerswig, G. J., Gottschel, J. C., and Prins, R. A. (1996). The storage glucan of *Phaeocystis globosa* (Prymnesiophyceae) cells. *J. Phycol.* 32:382–7.

Jeffrey, S. W., and Wright, S. W. (1994). Photosynthetic pigments in the Haptophyta. In *The Haptophyte Algae*, ed J. C. Green, and B. S. C. Leadbeater, Systematics Assn. Special Vol. 51, pp. 111–52. Oxford: Clarendon Press.

Jordan, R. W., and Chamberlain, A. H. L. (1997). Biodiversity among haptophyte algae. *Biodiversity and Conservation* 6:131–52.

Kawachi, M., and Inouye, I. (1994). Ca^{2+} mediated induction of the coiling of the haptonema in *Chrysochromulina hirata* (Prymnesiophyceae = Haptophyta). *Phycologia* 33:53–7.

Kawachi, W., Inouye, I., Maeda, O., and Chihara, M. (1991). The haptonema as a food-capturing device: observations on *Chrysochromulina hirata* (Prymnesiophyceae). *Phycologia* 30:563–73.

Klaveness, D. (1972). *Coccolithus huxleyi* (Lohm.) Kamptn. II. The flagellate cell, aberrant cell types, vegetative propagation and life cycles. *Br. Phycol. J.* 7:309–18.

Kreger, D. R., and van der Veer, J. (1970). Paramylon in a Chrysophyte. *Acta Bot. Neerl.* 19:401–2.

Leadbeater, B. S. C. (1970). Preliminary observations on differences of scale morphology at various stages in the life cycle of "*Apistonema-Syracosphaera*" *sensu* von Stosch. *Br. Phycol. J.* 5:57–69.

Linschooten, C., van Bleijswijk, J. D. L., van Emburg, P., de Vrind, J. P. M., Kempers, E. S., Westbroek, P., and de Vrind-de Jong, E. W. (1991). Role of light-dark cycle and medium composition in the production of coccoliths by *Emiliania huxleyi* (Haptophyceae). *J. Phycol.* 27:82–6.

Manton, I. (1964). Further observations on the fine structure of the haptonema in *Prymnesium parvum*. *Arch. Mikrobiol.* 49:315–30.

Manton, I. (1966). Further observations on the fine structure of *Chrysochromulina chiton*, with special reference to the pyrenoid. *J. Cell Sci.* 1:187–92.

Manton, I. (1967a). Further observations on the fine structure of *Chrysochromulina chiton* with special reference to the haptonema, "peculiar" Golgi and aspects of scale production. *J. Cell Sci.* 2:265–72.

Manton, I. (1967b). Further observations on scale formation in *Chrysochromulina chiton*. *J. Cell Sci.* 2:411–18.

Manton, I. (1968). Further observations on the microanatomy of the haptonema in *Chrysochromulina chiton* and *Prymnesium parvum*. *Protoplasma* 66:35–53.

Manton, I. (1972). Preliminary observations on *Chrysochromulina mactra* sp. nov. *Br. Phycol. J.* 7:21–35.

Manton, I., and Parke, M. (1962). Preliminary observations on scales and their mode of origin in *Chrysochromulina polylepis* sp. nov. *J. Mar. Biol. Assoc. UK* 42:565–78.

Manton, I., and Peterfi, L. S. (1969). Observations on the fine structure of coccoliths, scales and the protoplast of a freshwater coccolithophorid. *Hymenomonas roseola* Stein, with supplementary observations on the protoplast of *Cricosphaera carterae*. *Proc. R. Soc. Lond.* [*B*] 172:1–15.

Manton, I., Sutherland, J., and Oates, K. (1977). Arctic coccolithophorids: *Wigwammia arctica* gen. et sp. nov. from Greenland and arctic Canada. *W. annulifera* sp. nov. from South Africa and S. Alaska and *Calciarcus alaskensis* gen. et sp. nov. from S. Alaska. *Proc. R. Soc. Lond.* [*B*] 197:145–68.

Marsh, M. E. (1994). Polyanion-mediated mineralization – assembly and reorganization of acidic polysachharides in the Golgi system of a coccolithophorid alga during mineral deposition. *Protoplasma* 177:108–22.

Marsh, M. E. (1996). Polyanion-mediated mineralization – a kinetic analysis of the calcium-carrier hypothesis in the phytoflagellate *Pleurochrysis carterae*. *Protoplasma* 190:181–8.

Medlin, L. K., Barker, G. L. A., Baumann, M., Hayes, P. K., and Lange, M. (1994). Molecular biology and systematics. In *The Haptophyte Algae*, ed. J. C. Green, and B. S. C. Leadbeater, Systematics Assn. Special Vol. 51, pp. 393–411. Oxford: Clarendon Press.

Moestrup, Ø. (1994). Economic aspects: "blooms", nuisance species, and toxins. In *The Haptophyte Algae*, ed. J. C. Green, and B. S. C. Leadbeater, Systematics Assn. Special Vol. 51, pp. 265–85. Oxford: Clarendon Press.

Nielsen, M. V. (1995). Photosynthetic characteristics of the coccolithophorid *Emiliania huxleyi* (Prymnesiophyceae) exposed to elevated concentrations of dissolved inorganic carbon. *J. Phycol.* 31:715–19.

Outka, D. E., and Williams, D. C. (1971). Sequential coccolith morphogenesis in *Hymenomonas carterae*. *J. Protozool.* 18:285–97.

Paasche, E. (1968). Biology and physiology of coccolithophorids. *Annu. Rev. Microbiol.* 22:71–86.

Parke, M. (1971). The production of calcareous elements by benthic algae belonging to the class Haptophyceae (Chrysophyta). *Proc. II Plank. Conf.*, pp. 929–38.

Parke, M., and Manton, I. (1962). Studies on marine flagellates. VI. *Chrysochromulina pringsheimii* sp. nov. *J. Mar. Biol. Assoc. UK* 42:391–404.

Parke, M., Lund, J. W. G., and Manton, I. (1962). Observations on the biology and fine structure of the type species of *Chrysochromulina* (C. *parva* Lackey) in the English Lake District. *Arch. Mikrobiol.* 42:333–52.

Parke, M., Manton, I., and Clarke, B. (1955). Studies on marine flagellates. II. Three new species of *Chrysochromulina*. *J. Mar. Biol. Assoc. UK* 34:579–609.

Parke, M., Manton, I., and Clarke, B. (1956). Studies on marine flagellates. III. Three further species of *Chrysochromulina*. *J. Mar. Biol. Assoc. UK* 35:387–414.

Pringsheim, E. G. (1955). Kleine Mitteilungen über Flagellaten und Algen. I. Algenartige Chrysophyceen in Reinkultur. *Arch. Mikrobiol.* 21:401–10.

Reynolds, N. (1974). *Imantonia rotunda* gen. et sp. nov., a new member of the Haptophyceae. *Br. Phycol. J.* 9:429–34.

Rowson, J. D., Leadbeater, B. S. C., and Green, J. C. (1986). Calcium carbonate deposition in the motile (*Crystallolithus*) phase of *Coccolithus pelagicus*. *Br. J. Phycol.* 21:359–70.

Savage, R. E. (1930). The influence of *Phaeocystis* on the migration of the herring. *Fish. Invest., Lond., Ser. II* 12:5–14.

Shilo, M. (1967). Formation and mode of action of algal toxins. *Bacteriol. Rev.* 31:180–93.

Simon, N., Brenner, J., Edvardsen, B., and Medlin, L. K. (1997). The identification of *Chrysochromulina* and *Prymnesium* species (Haptophyta, Prymnesiophyceae) using fluorescent or chemiluminescent oligonucleotide probes: a means for improving studies on toxic algae. *Eur. J. Phycol.* 32:393–401.

Simonsen, S., and Moestrup, Ø. (1997). Toxicity tests in eight species of *Chrysochromulina* (Haptophyta). *Can. J. Bot.* 75:129–36.

Tasch, P. (1973). *Paleobiology of the Invertebrates.* New York: John Wiley.

van der Veer, J. (1969). *Pavlova mesolychnon* (Chrysophyta), a new species from the Tamar Estuary, Cornwall, *Acta Bot. Neerl.* 18:496–510.

van der Veer, J. (1976). *Pavlova calceolata* (Haptophyceae), a new species from the Tamar Estuary, Cornwall, England. *J. Mar. Biol. Assoc. UK* 56:21–30.

van der Wal, P., deVrind, J. P. M., deVrind-deJong, E. W., and Borman, A. H. (1987). Incompleteness of the coccosphere as a possible stimulus for coccolith formation in *Pleurochrysis carterae* (Prymnesiophyceae). *J. Phycol.* 23:218–21.

van Rijssel, M., Hamm, C.E., and Gieskes, W. W. C. (1997). *Phaeocystis globosa* (Prymnesiophyceae) colonies: hollow structures built with small amounts of polysaccharides. *Eur. J. Phycol.* 32:185–92.

von Stosch, H. A. (1967). Haptophyceae. In Vegetative Forplanzung, Parthenogenese und Apogamie bie Algen, ed. W. Ruhland, *Encyclopedia of Plant Physiology* 18:646–56.

Young, J. R., Didymus, J. M., Bown, P. R., Prins, B., and Mann, S. (1992). Crystal assembly and phylogenetic evolution in heterococcoliths. *Nature* 356:516–18.

GLOSSARY

abaxial: located in a position away from the plant.
acronematic flagellum: a flagellum without hairs.
acropetal: toward the apex.
adaxial: located on the side toward the main axis.
adelphoparasite: parasite that is closely related to the host.
aerobic: needing oxygen.
aeroplankton: air-borne microscopic organism.
agar: one or more polysaccharides containing sulfated galactose obtained from the walls of some red algae.
agarophyte: red algae used in the production of agar.
agglutin: chemical substance involved in the recognition of a gamete of the opposite strain.
agglutination: adherence of gametes of different mating types by their flagella tips.
akinete: thick-walled resting spore.
alkalinity: in water chemistry, the total quantity of base in equilibrium with carbonate or bicarbonate that can be determined by titration with strong acid. Alkaline waters have a high pH.
alginates: salts of polysaccharides composed of D-mannuronic and L-guluronic acids obtained from brown algae.
alloparasites: parasites not closely related to their host.
allophycocyanin: a blue biliprotein obtained from cyanobacteria and red algae.
alpha granule: protoplasmic structure containing myxophycean starch in the cyanobacteria.
alveoli: membrane-bounded flattened vesicles or sacs underlying the plasma membrane in certain algae, particularly the dinoflagellates.
amphiesma: the plasma membrane and underlying flattened vesicles of dinoflagellates which, in some species, contain plates.
amylopectin: the storage polysaccharide in the cyanobacteria, composed of α-1,4 glucoside linkages, with 1,6 linked side chains.
amyloplast: a colorless plastid containing starch.
amylum star: star-shaped aggregate of cells filled with starch that forms new plants in the stoneworts (Charales).
anaerobic: without oxygen.
androsporangium: sporangium that forms androspores.
androspore: spore that forms a dwarf male filament in the Oedogoniales (green algae).
anhydrobiotic: organisms that can withstand the removal of the bulk of their intracellular water for extended periods of time.
anisogamy: fusion of gametes that are unequal in size or physiology.
antapical: opposite from the apex.
anterior: the front end, forward.
antheridium: the sex organ in which the male gametes are formed.
antherozoid: male gamete.

anthropomorphic: related to man.
anticlinal: perpendicular to the circumference of the thallus.
apical growth: growth by means of an apical cell dividing to form the thallus beneath it.
aplanogamete: non-motile gamete.
aplanospore: non-motile spore.
apochlorotic: colorless.
aragonite: orthorhombic crystals of calcium carbonate.
areolae: the chambers in the honeycomb arrangement of some diatom valves.
asexual: reproduction without the fusion of gametes.
autospore: aplanospore with the same shape as the parent cell.
autotroph: not needing an external source or organic compounds as an energy source. Energy is obtained from light or inorganic chemical reactions.
auxiliary cell: a cell that receives a nucleus from the zygote in the red algae.
auxospore: a resting cell in the diatoms that is commonly formed from the zygote.
axenic culture: a culture containing only one species.
axoneme: an axial array of (usually) nine outer doublet and two central microtubules.

bacteriocin: antibiotic secreted by cyanobacteria that kills related strains of cyanobacteria.
baeocyte: endospore produced by cyanobacteria.
basal apparatus: flagellar or ciliary apparatus exclusive of flagella/cilia.
basal body (kinetosome): cylindrical structure (*ca.* 0.2 μm diameter) found at the base of a flagellum/cilium consisting of a continuation of the nine outer axonemal doublets (A,B) but with the addition of a C-microtubules to form triplets.
basipetal: toward the base.
bathal zone: ocean water over continental slope.
benthic: pertaining to any part of a lake or ocean bottom.
benthos: organisms living on, and attached to, the bottom of aquatic habitats.
bilaterally symmetrical: a plane through an object divides it into mirror image halves.
biliprotein: a red or blue pigment and attached bile (linear tetrapyrrole) chromophore in the cyanobacteria, cryptophytes and red algae.
bioluminescence: emission of light by a living organism.
biomass: the amount at any one time of living water in a habitat.
bisexual: both sexes produced on the same individual.
bisporangium: a sporangium forming two meiospores in the red algae.
bloom: heavy growth of planktonic algae in a body of water.
brackish: saline water with a salinity less than that of seawater.
budding: exospore formation in the cyanobacteria.
bulbil: small plant formed on rhizoids in the stoneworts (charophytes).

calcareous ooze: bottom sediment in oceans composed of calcified remains of organisms.
calcification: deposition of calcium carbonate, usually in association with smaller amounts of other carbonates.
calcite: rhombohedral crystals of calcium carbonate.
callose: polysaccharide associated with pores in sieve cells.
canal: rigid opening in some flagellates.
capsule: extracellular mucilage.
carotene: oxygen-free, unsaturated, hydrocarbon carotenoid.
carotenoid: yellow, orange, or red hydrocarbon fat-soluble pigment.
carpogonium: female gametangium in the red algae.

carposporangium: carpospore-producing sporangium derived directly, or indirectly, from the zygote in the red algae.

carpospore: usually diploid spore produced by the carposporangium in the red algae.

carposporophyte: usually diploid generation in the red algae derived from the zygote, it forms carpospores.

carrageenan: red algal polysaccharide (phycocolloid) similar to agar, but needing higher concentrations to form a gel.

cellulose: polysaccharide composed of β-1,4 linked glucose molecules that forms the main skeletal framework of most algal cells.

cell wall: A mostly rigid, often multilayered structure consisting of discrete microfibrillar polysaccharides embedded in an amorphous matrix composed of polysaccharides, lipids, and proteins, which together comprise an outermost layer of the cell proper.

centric: type of ornamentation arranged around a central point in the diatoms.

centriole: equivalent to a flagellar basal body.

chemoautotroph: an organism that obtains energy from oxidation of reduced inorganic compounds, and cell carbon primarily from carbon dioxide.

chemotaxis: the movement of a whole cell in response to a concentration gradient of a chemical substance. If it is toward higher concentration, it is positive; if away from a higher concentration, it is negative chemotaxis.

chitin: polysaccharide made up of repeating units of *N*-acetylglucosamine.

chlorophyll: fat-soluble, green, porphyrin-type pigment.

chloroplast: plastid with chlorophyll.

chloroplast endoplasmic reticulum (chloroplast E.R.): one or two membranes surrounding the chloroplast membrane.

chromatic adaption: change in the proportions of different photosynthetic pigments enabling optimum absorption of the available wavelengths of light.

chromoplast or **chromatophore:** a chloroplast with some other color than green.

chrysolaminarin or **leucosin:** a liquid polysaccharide storage product composed principally of β-1,3 linked residues of glucose.

ciliary/flagellar matrix: the cytosol of the flagellum/cilium, often lacking structural detail.

cilium (flagellum): a long, cylindrical extension of a eukaryotic cell, bounded by the plasma membrane and containing an axoneme. A flagellum/cilium is a motility organelle that is mainly involved in cell movement by means of water propulsion, but can perform additional functions, such as feeding, mating and sensory perception.

cingulum or **girdle:** transverse furrow in the dinoflagellates containing the transverse flagellum.

circadian rhythm: repeated sequence of events that occur about 24-hour intervals.

circein: female hormone secreted by oogonia in the green alga *Oedogonium*.

cirri: curled appendages on a zygote.

clone: the group of individuals derived from a single individual.

coccoid: spherical.

coccolith: calcareous scale or plate-like particle deposited on the surface membrane of some prymnesiophytes (the coccolithophorids), varying in complexity and surface decoration according to species.

coenobium: colony of algal cells in a specific arrangement and number that is fixed at the time of origin and is not subsequently augmented.

coenocyte: large multinucleate cell without cross walls, except where reproductive bodies are concerned.

colony: a group of unicells which cohere and remain together as a unit.
compensation depth: depth of water where there is sufficient light so photosynthesis equals respiration over a 24-hour period.
compensation point: light intensity at which respiration equals photosynthesis over a 24-hour period.
conceptacle: cavity in the thallus where gametangia are produced.
conjugation: fusion of two non-flagellated protoplasts.
connecting band: part of the girdle in diatoms.
contractile vacuole: vacuole fed by smaller vesicles that expels water and solutes rhythmically outside the cell.
coralline: reference to calcified algae.
cornuate process: hornlike wall extension.
corona: crown.
Corps de Maupas: a vesicular body in the cryptophytes used in digestion of unwanted cell components.
cortex: the outer portion or layer(s) of a protist cell, including the plasma membrane, but excluding secreted non-living structures that may lie outside the plasma membrane. The outer portion of an algal thallus.
cosmopolitan: occurring in many diverse places.
costa: an elongated, solid thickening of a diatom valve.
crenulate: wavy with small teeth or scallops.
cribellum: small sieve plate covering the pores in a cribrum in the diatoms.
cribrum: silicified plate with tiny pores covering the holes of the valve in the diatoms.
cross fertilization: the union of gamates from different thalli.
cruciate: cross shaped.
cryoplankton: plankton of polar or cold regions.
cryptobiotic crust: accumulations of cyanobacteria, lichens, fungi, and/or mosses in desert soils.
cryptoendolith: organism that lives inside rocks.
cryptostomata or **cryptoblast:** flasklike opening in the thallus, with hairs.
cuticle: a thin hydrophobic layer deposited on the outside surface of the cell wall.
cyanelle: endosymbiotic cyanobacterium.
cyanobacterocin: an antibiotic produced by a cyanobacterium that inhibits growths of related cyanobacteria.
cyanoglobin: myogloblin-like molecule capable of scavenging oxygen in heterocysts of cyanobacteria.
cyanome: host cell containing a cyanelle.
cyanophage: virus in cells of the cyanobacteria.
cyanophycin granule: polypeptide storage granule in the cyanobacteria.
cyst: a non-motile, often dehydrated, resistant, inactive dormant stage in the life cycle.
cystocarp: in the red algae, the carposporophyte and surrounding gametophytic tissue (pericarp).
cytokinesis: division of the cytoplasm usually right after karyokinesis (nuclear division).
cytoskeleton: intracellular network of protein filaments that is insoluble in non-ionic detergents.
cytosome: cell opening used for ingestion of food particles.

dendricule: short protoplasmic extension with no organelles.
dendroid: a type of non-motile colony that produces mucilage in one area, usually forming a stalk.
detritus: particulate organic matter.

dexiotropic: clockwise when seen from the cell apex (the opposite is *leiotropic*).
diastole: filling of a contractile vacuole (*systole* is the emptying).
diatomaceous earth: a mineral consisting of the remains of the silicified frustules of diatoms.
diazotroph: organism capable of fixing atmospheric N_2 into ammonium.
dichotomy: division of a thallus into two equal branches.
dictyosome: stack of vesicles in a Golgi apparatus.
diel: occurring over a 24-hour period.
diffuse growth: type of growth where most of the cells in a thallus are capable of division.
dioecious: an organism that has male and female gametes borne on separate plants.
diplobiontic: having two separate stages in the life cycle.
diplohaplont: an organism having a separate multicellular diploid and multicellular haploid stage.
diploid: possessing two sets of chromosomes.
diplont: an organism in which the haploid stage of the life cycle consists of only the gametes.
disc or **thylakoid:** membrane-bound sac in the chloroplast.
discobolocyst: type of projectile in the Chrysophyceae.
distichous: arranged in two equal rows.
distromatic: thallus that is two cells thick.
diurnal: daily.
dorsiventral: flattened in one plane.
dulse: a preparation of *Rhodymenia* (red alga) used as food.

ecad: a form of a plant species produced in response to a particular habitat, the modifications not being inheritable.
ecdysis: shedding of the theca in the dinoflagellates.
ecotype: a locally adapted variant.
ectoderm: outer protoplasm next to the plasma membrane in the Charales.
egg: large non-motile female gamete.
ejectisome: type of projectile found in green algae and cryptomonads.
electron-dense or **electron-opaque:** term used to describe a material that absorbs electrons and appears dark in electron micrographs.
electron-transparent: term used to describe a material that does not absorb electrons and appears light in electron micrographs.
endemic: occurring in a specific area.
endochite: inner wall of the oogonium in the Fucales (Phaeophyceae).
endoderm: inner protoplasm next to the vacuole in the Charales, capable of cytoplasmic streaming.
endolithic: living inside rock.
endophyte: plant living inside another plant.
endosome: nucleus in the euglenoids.
endospore: asexual spore produced by some cyanobacteria.
endosymbiotic: term that describes an organism living inside a host in a symbiosis.
endozoic: living inside an animal.
enucleate: without a nucleus.
envelope: a general term used variously for such structures as plasma membranes, pellicles, walls, sheaths and gelatinous coverings.
epicone: part of the cell above the girdle in the dinoflagellates.
epidermis: outer layer of cells.
epipelic: growing on mud.

epiphyte: one plant living on another plant.
epitheca: larger of the two halves of a diatom frustule.
epizoic: living on an animal.
epontic: living on the bottom of ice.
estuary: the mouth of a river where tidal effects are evident, and where freshwater and seawater mix.
euendolith: an organism that bores into rock.
eukaryotic or **eucaryotic:** cells with a membrane-bounded nucleus.
euphotic or **photic zone:** water above the compensation depth.
eurysaline (euryhaline): tolerant of a wide salinity range.
eurythermic: tolerant of a wide temperature range.
eutrophic: term that described a body of water that receives large amounts of nutrients, usually resulting in a large growth of algae.
exocite: outer wall of an oogonium of the Fucales (Phaeophyceae).
exotoxin: toxin secreted into the medium.
extant: living today.
extinct: not living today.
extracellular matrix: mucilaginous glycoproteins external to the plasma membrane.
eyespot: red to orange area in a cell, composed of lipid droplets.

facultative heterotroph or **autotroph:** organism that is able to live as a heterotroph or an autotroph.
facultative parasite or **saprophyte:** organism that is able to live as a parasite or a saprophyte.
false branching: in the cyanobacteria, breakage of a trichome through a sheath, giving the appearance of a branch.
fibulae: in diatoms, small nodules in rows on the inside of the valve.
filament: in the cyanobacteria, one or more trichomes enclosed in a sheath.
flagellar apparatus (kinetid): an organellar complex consisting of one or more basal bodies/kinetosomes that may bear flagella/cilia, may have microtubular and fibrous roots associated with their bases, and may function in locomotion, feeding, sensation, and reproduction.
flagellar/ciliary matrix: the cytosol of the flagellum/cilium, often lacking structural detail.
flagellar hairs: filamentous appendages usually arranged in one or more rows but not covering the entire surface of a flagellum/cilium. There are two types of flagellar hairs. (1) Tubular flagellar hairs consisting of at least a hollow shaft (>15 nm diameter), often with a cylindrical shaft and one or more terminal filaments. (2) Non-tubular flagellar hairs, assembled in the Golgi apparatus and consisting of primarily of carbohydrates, forming two rows along the length of the flagellum and attached through the flagellar membrane to specific outer doublets.
flagellar/ciliary roots: fibrous, microtubular or amorphous structures originating at or near basal bodies/kinetosomes and terminating somewhere else in the cell but not at nearby basal bodies/kinetosomes.
flagellar scales: organic structures of discrete size and shape, often covering the whole surface of the flagellum/cilium and generally assembled in the Golgi apparatus.
flagellate: a unicell having at least one flagellum.
flagellum (cilium): a long, cylindrical extension of a eukaryotic cell, bounded by the plasma membrane and containing an axoneme. A flagellum/cilium is a motility organelle that is mainly involved in cell movement by means of water propulsion, but can perform additional functions, such as feeding, mating and sensory perception.

floridean starch: red algal storage product composed of α-1,4 and α-1,6 linked glucose residues.
floridoside: primary product of photosynthesis in the Rhodophyta.
foliose: leaflike.
foramen: chamber or hole.
fragmentation: type of asexual reproduction where a thallus breaks into two or more parts, each of which forms a new thallus.
frustule: silicified cell wall in the diatoms.
fucoidin: polysaccharide in the cell wall and mucilage of the brown algae composed of sulfated fucose units.
fucosan or **phaeophycean tannin:** a colorless acidic fluid in brown algae, giving a characteristic red color with vanillin hydrochloride.
fucoserraten: a sex attractant secreted by the eggs of *Fucus* brown algae.
funori: a commercially produced phycocolloid of red algae used as a glue.
fusiform: spindle-shaped.

gametangium: structure forming gametes.
gamete: cell capable of fusion with another to form a zygote.
gametogenesis: the formation of gametes.
gametophyte: plant generation that forms the gametes, usually haploid.
gamone: a chemical involved in the attraction of one gamete to another.
gas vacuole: a collection of gas vesicles.
gas vesicle: hollow cylindrical gas-filled structures in the Cyanophyta.
generative auxiliary cell: cell in the Rhodophyta forming the gonimoblast filaments.
geotaxis: movement of a cell away (negative geotaxis) or toward (positive geotaxis) gravity.
girdle: the transverse groove containing the transverse flagellum in the Dinophyta.
girdle band: part of a diatom wall where the theca overlap; a band of thylakoids running parallel under the chloroplast envelope.
gliding: active movement of an organism in contact with a solid substrate where there is neither a visible organ responsible for the movement nor a distinct change in the shape of the organism.
globule: male reproductive structure in the Charales.
glycocalyx: sticky polysaccharide secreted by the zygote in the Fucales (Phaeophyceae).
glycogen: a storage polysaccharide related to amylopectin and stains red-purple with iodine.
glycopeptide or **glycoprotein:** polysaccharide composed of sugars and amino acids or peptides.
glycoside: a polysaccharide composed of glucose.
gonidium: a cell that divides to form a daughter colony.
gonimoblast: usually diploid cells that form the carposporangia in the Rhodophyta.
gonoid: type of ornamentation that is dominated by angles in the diatoms.
granum: stack of thylakoids in a chloroplast.
gullet: anterior of invagination in euglenoids and cryptomonads.
gyrogonite: fossilized nucule of the Charales.

hair: appendage on a flagellum; colorless elongate cell.
haplobiontic: having only one multicellular stage.
haploid: having one complete set of chromosomes.
haplont: an organism in which the only diploid stage is the zygote.
hapteron or **holdfast:** bottom part of an alga that attaches the plant to the substrate.

haptonema: appendage that arises between the flagella in the Prymnesiophyta.
heleoplankton: plankton that grow in marshy areas or in small ponds.
hematochrome: red or orange lipid bodies occurring outside the chloroplast.
hermaphroditic or **homothallic:** producing both male and female gametangia on the same thallus.
heterocyst: thick-walled, hollow-looking enlarged cell in the cyanobacteria.
heterokont: having flagella of unequal length.
heteromorphic alternation of generations: having haploid and diploid generations of different morphology.
heterothallic: producing male and female gametangia on different plants.
heterotrichous: term used to describe division of a plant into an erect and a prostrate part.
heterotrophic: needing an external source of organic compounds as an energy source.
histone: basic protein.
holdfast: part of an alga that attaches a plant to a substrate.
holophytic or **autotrophic:** needing only light and inorganic substances for growth.
holozoic or **phagocytic:** absorbing food particles whole into food vesicles for digestion.
homothallic: producing male and female plants on the same plant.
hormogonium: short pieces of a trichome in the cyanobacteria that become detached from the parent filament and move away by gliding, subsequently developing into new filament.
hyaline: transparent.
hydrophilic: water-attracting.
hydrophobic: water-repelling.
hypersaline: greater than normal salinity.
hypha: long slender cell in the medulla of Laminariales (brown algae).
hypnospore: aplanospore with a greatly thickened cell wall.
hypocone: lower part of the cell in the Dinophyta, usually having a longitudinal sulcus.
hypogenous cells: in the Rhodophyta, those cells under the carpogonium.
hypolimnion: water beneath the thermocline in thermally stratified water bodies.
hypothallus: lower part of the thallus composed of large cells in the coralline reds.
hypotheca: smaller half of a diatom frustule.
hystrichospore or **hystrichosphaerid:** fossilized resting spore in the Dinophyta.

ichthytoxin: toxin that kills gill-bearing animals.
intercalary: in-between two cells or tissues.
intercalary bands: bands between the valve and girdle band in the diatoms.
internode: part of axis between nodes.
interstitial water: that water trapped between particles of soil or mud.
intertidal: occurring between the low- and high-tide marks.
inversion: phenomenon in the green algae in which a colony turns itself inside out through a pore.
iridescence: the play of colors caused by refraction and interference of light waves at the surface.
isoagglutination: adhesion of flagella of the same sex when a mating-type substance of the opposite sex is added.
isoenzyme: enzymes having the same function but of a somewhat different structure.
isogamy: fusion of similar gametes.
isokont: cell with flagella of the same length.
isomorphic alternation of generations: generations that are morphologically alike.
isthmus: a passage connecting two bodies.

karyogamy: fusion of two gamete nuclei.
karyokinesis: division of the nucleus.
keel: in pennate diatoms, an extension of the valve running lengthwise, similar to the keel of a ship.
kelp: a member of the Laminariales (brown algae); also used for the burnt ash of plants of the Laminariales.
kerogen: yellow-brown amorphous organic matter in sedimentary deposits.
kinetid (flagellar apparatus): an organellar complex consisting of one or more basal bodies/kinetosomes that may bear flagella/cilia, may have microtubular and fibrous roots associated with their bases, and may function in locomotion, feeding, sensation, and reproduction.
kombu (Japanese): vegetable made from Laminariales (Phaeophyta).

labiate process: in diatoms, an appendage at the periphery of the valve through which mucilage may be secreted, functions in the movements of centric diatoms.
laminarin: food storage polysaccharide in the brown algae composed principally of β-1,3 linked glucose residues.
laminate: flat.
laver or **laver bread:** similar to Japanese nori, vegetable made from dried *Porphyra* (Rhodophyta).
leiotropic: counterclockwise when seen from the cell apex (opposite is dexiotropic).
lentic: related to a pond or lake.
leucosin or **chrysolaminarin:** food storage polysaccharide of golden-brown algae composed mostly of β-1,3 linked glucose residues.
leucoplast: colorless plastid usually having a large number of starch grains and few thylakoids.
list: extension of the theca in the Dinophyta.
lithophyte: plant growing on rock.
lithotrophic or **autotrophic:** needing only light and/or inorganic compounds for growth.
littoral zone: zone from the water's edge to a water depth of about 6 m or the maximum depth of rooted vegetation, if any exists.
loculus: hexagonal chamber in the wall of diatoms.
log phase of growth: growth phase characterized by rapidly dividing cells.
lorica: an envelope around the protoplast, not attached to the protoplast as the wall is.
lotic: related to rivers or streams.
luciferase: enzyme that oxidizes luciferin.
luciferin: compound responsible for bioluminescence.
lysosome: single-membrane-bounded cytoplasmic particle containing destructive enzymes.

macrandous: species in the Oedogoniales (green algae) that do not produce dwarf male filaments.
macroplankton: plankton larger than 75 μm.
mäerl: coralline red algae applied to soil to increase the pH of the soil.
mannan: polysaccharide composed of mannose residues.
mannitol: sugar alcohol, $C_6H_{14}O_6$; primary product of photosynthesis in the brown algae.
mantle or **valve jacket:** part of a valve in the diatoms that is bent inward.
marl: deposits of calcium and magnesium carbonate.
mastigoneme or **hair:** filamentous appendage of a flagellum.
mating-type reaction: flagella adhesion between gametes of different sexes.

mating structure: a dense plate in the anterior part of the protoplasm of a gamete that determines its sex.
medulla: inner part of algal thallus, usually composed of packed colorless filaments.
meiocyte: a cell which undergoes meiosis.
meiosis: cell division in which the chromosome number is halved.
meiosporangium: structure in which spores are produced by meiosis.
meiospore: spore formed by meiosis.
meristem: dividing tissue that forms new cells.
meristoderm: dividing layer of cells in the brown algae.
mesochite: middle wall of an oogonium of the Fucales (brown algae).
metachromatic granule: protoplasmic body containing stored polyphosphate in the cyanobacteria.
microaerophilic: with small amounts of oxygen.
microfibril: crystalline anhydrous cellulose found in many algal cell walls.
microfilament: submicroscopic solid filament in protoplasm.
micrometer (μm): 10^{-6} m, 1 μm equals 1 micron.
microtubule: submicroscopic tubule in the protoplasm.
mitosis: nuclear division resulting in two daughter nuclei which are genetically identical to their parent.
mitosporangium: a sporangium in which spores are produced by mitosis.
mitospore: a spore produced as a direct result of mitosis.
mixotroph or **facultative heterotroph:** photosynthetic organism capable of using organic compounds in the medium.
monoecious or **homothallic:** having male and female gametangia borne on the same plant.
monopodial: having one main axis of growth.
monosporangium: a sporangium that forms a monospore in the Rhodophyta.
monospore: asexual spore that germinates to re-form the parent in the Rhodophyta.
monostromatic: a thallus only one cell thick.
muciferous body: a body, usually in the outer protoplasm of a cell, that discharges mucilage, usually explosively.
mucilage canal: canal present in some brown algae, composed of elongated cells in the cortex area that secrete mucilage.
mucopeptide: polysaccharide of the walls of the cyanobacteria, composed of sugars and amino acids.
multiaxial: having an axis with a number of apical cells that give rise to a number of nearly parallel filaments.
multicellular: composed of many cells.
multiseriate: with more than one row of cells.
myxophycean starch: storage polysaccharide of the cyanobacteria, similar to glycogen.

nannandrous: in the Oedogoniales (green algae), producing dwarf males.
nannoplankton or **nanoplankton:** plankton smaller than 75 μm but larger than 2 μm.
nanometer (nm): 10^{-9} meter, 1 nm equals 10 angstrom units.
necridium or **separation disc:** a cell that dies in a trichome of the cyanobacteria resulting in the formation of a hormogonium from part of the trichome.
nemathecium: wartlike surface elevation containing the reproductive structures in the Rhodophyta.
neritic region: ocean water over the bottom that extends from the high-tide mark to a depth of 200 m.
net plankton or **macroplankton:** plankton larger than 75 μm.
nitrogen fixation: the intracellular fixation of nitrogen gas from the atmosphere to

ammonia in cyanobacteria.
node: part of thallus that bears branches.
nori or **laver:** vegetable made from dried *Porphyra* (Rhodophyceae) in Japan, similar to laver bread.
nucule: female reproductive structure in the Charales.

oceanic region: open seas beyond 200-m bottom depth.
oligotrophic: term describing a body of water low in nutrients.
ooblast or **connecting filament:** a filament produced by the zygote that fuses with an auxiliary cell in the Rhodophyta.
oogamy: fusion of a large non-motile egg with a small motile sperm.
oogonium: single-celled female gametangium.
oospore or **zygospore:** thick-walled zygote with food reserves.
organelle: a membrane-bounded part of a cell.
osmotrophic: term describing a heterotrophic organism that absorbs organic molecules in a soluble form.
ostiole: an opening to the outside in a conceptacle.
overturn: phenomenon in a body of water in which the surface water becomes colder than the bottom water, causing the surface water to sink and resulting in a mixing of the water column.
ovoid: spherical (0.2–2 mm in diameter), concentrically laminated, carbonate grains that form by carbonate accretion in aggitated, shallow tropical marine environments.
ovum or **egg:** non-motile large female gamete.

pallium: a pseudopod projection used for feeding in thecate heterotrophic dinoflagellates.
palmelloid: term describing a colony of an indefinite number of single, non-motile cells in a mucilaginous matrix.
pantonematic or **tinsel flagellum:** flagellum with hairs attached to the surface.
papilla: a small rounded protuberance.
paraflagellar or **paracrystalline body:** photoreceptor in the Eugleophyta consisting of a crystalline swelling in one of the flagella.
paramylon: storage polysaccharide composed of β-1,3 linked glucose molecules.
paraphysis: sterile structure found with sporangia or gametangia.
parasite: heterotrophic organism that derives nutrients from a living host.
parasporangium: sporangium producing more than one asexual spore in the Rhodophyta.
paraxonemal body: proteinaceous structure restricted to a certain area along the flagellum/cilium.
paraxonemal rod: long cylindrical structure (solid or hollow) that extends nearly the entire length of a flagellum/cilium, located between the axoneme and flagellar membrane, and usually connected to the axoneme and flagellar membrane by specific links.
parenchyma: a tissue formed of thin-walled living cells produced by division in three planes.
parietal: peripheral.
parthenogenetic: germination of an egg without fertilization to form a new plant.
pedicel: supporting structure of a reproductive tissue.
peduncle: tail-like process with a central core of microtubules.
pelagic: living in the open ocean or oceanic region; in some definitions, living at or near the surface of the open sea.
pellicle: proteinaceous outer covering in the Euglenophyta.

pennate: term describing type of ornamentation arranged on either side of a central line in the diatoms (bilateral symmetry).
peptidoglycan: polysaccharide composed of sugars and amino acids in the walls of the cyanobacteria.
pericentral cell: a small cell formed around a central axis.
periclinal: parallel to the circumference of the surface.
peridinin: a xanthophyll in dinoflagellate cells.
periphyton: organism attached to submerged vegetation.
periplast: outer cell covering in the Cryptophyta.
perithallus: outer part of the thallus in the coralline Rhodophyta.
perizonium: siliceous wall of auxospore in diatoms, consisting of multiple overlapping bands.
phaeophycean tannin or **fucosan:** colorless, acidic fluid in physodes in the brown algae.
phagotrophic or **holozoic:** ingesting solid food particles into a food vesicle for digestion.
pheromone: hormonal substance secreted by one individual that stimulates a physiological or behavioral response in another individual of the same species.
phialopore: hole in an inverted daughter colony in the Volvocales (Chlorophyta).
photic or **euphotic zone:** depth of water above the compensation depth.
photoautotrophic: term describing autotrophic plant that obtains energy from photosynthesis.
photoheterotroph: organism capable of using organic compounds as a source of carbon in the light but not in the dark.
photoreceptor: the part of the cell that receives the stimulus in phototaxis, usually a dense area in a flagellar swelling.
photosynthate: organic product of photosynthesis.
phototaxis: movement of a whole organism toward (positive) or away from (negative) light.
phragmoplast: wall formation by the coalescence of Golgi vesicles between spindle microtubules.
phycobiliprotein or **phycobilin:** water-soluble blue-green or pink pigment in the cyanobacteria, Rhodophyta, and Cryptophyta.
phycobilisome: an aggregation of phycobiliproteins on the surface of a thylakoid.
phycobiont: algal partner in a lichen.
phycocolloid: polysaccharide colloid formed by an alga.
phycocyanin: blue-green-colored phycobiliprotein.
phycoerythrin: pink-colored phycobiliprotein.
phycomata: walled cyst produced by unicellular green algae.
phycophaein: black oxidized phaeophycean tannins.
phycoplast: type of cell division in which the mitotic spindle disperses after nuclear division with the two daughter nuclei coming close together, another set of microtubules arising perpendicular to the former position of the microtubules of the mitotic spindle, and the new cell wall forming along these microtubules.
physode: vesicle containing phaeophycean tannins in the brown algae.
phytochrome: photoperiod-regulating chemical.
phytoplankton: plants that float aimlessly or swim too feebly to maintain a constant position against a water current.
picoplankton: plankton that will pass through a filter with pores 2 μm in diameter but not through a filter with pores of 0.2 μm in diameter.
pit connection: a continuous area between two red algal cells consisting of an aperture in a cross wall, a plug, and a plug cap.
placoderm desmids: desmids which have two semicells joined by a narrow isthmus (contrasted to a saccoderm desmid without semicells).

plakea: flat plate of cells in the Volvocales (Chlorophyta).
plankton: organisms that float aimlessly or swim too feebly to maintain a constant position against a water current.
planogamete: motile gamete.
planospore: motile spore.
planozygote: motile zygote.
plasma membrane (plasmalemma): the outermost living membrane of a cell.
plasmodesma (plural **plasmodesmata**): the minute cytoplasmic threads that extend through openings in cell walls and connect the protoplasts of adjacent living cells.
plasmogamy: fusion of protoplasm without fusion of nuclei (karyogamy).
plastid: double-membrane-bounded organelle usually containing the photosynthetic apparatus or some part of it.
pleomorphic: having more than one shape.
plethysmothallus: a stage composed of filaments or compacted filaments that can multiply itself by spores.
plurilocular sporangium: many-chambered sporangium in the brown algae, each chamber forming one swarmer.
pneumatocyst or **air bladder:** expanded part of thallus containing gases.
polar nodule: wall swelling near the end of a cell in diatoms.
polyglucan granule: protoplasmic structure containing the storage product in the cyanobacteria.
polyhedral body: protoplasmic structure in the cyanobacteria associated with DNA microfibrils; it may contain the carbon dioxide-fixing enzyme ribulose-1,5-bisphosphate carboxylase.
polymorphic: having more than one shape.
polyol: sugar alcohol.
polyphosphate body: protoplasmic structure containing stored phosphate in the cyanobacteria.
polysiphonous: term describing thallus made up of vertical files of parallel cells.
polystichous: a type of parenchymatous growth in the brown algae.
pore: a single hole.
pore membrane: part of wall over a loculus in diatoms.
poriod: pore occluded by a plate in diatoms.
primary producer: photosynthetic plant.
proboscis: microtubules in the anterior part of a sperm, probably associated with the chemotactic response.
procarp: association of carpogonium and auxiliary cells in the Rhodophyta.
process: extension of a wall.
productivity: change in biomass per unit time.
profundal zone: part of a lake beneath the compensation depth.
prokaryotic cell: a type of cell lacking membrane-bounded organelles.
prolamellar body: a body composed of membrane-bounded connected tubules in dark-grown chloroplasts.
promeristem: a non-dividing apical cell controlling the division of a number of smaller promeristematic cells beneath it.
propagulum: branchlets that fall off and form new plants in the Sphacelariales (brown algae).
proplastid: a small plastid that usually matures to a chloroplast or amyloplast.
pseudocilia: non-functional flagella.
pseudoparenchyma: densely packed filaments resembling parenchyma tissue.
pseudoraphe: unornamented area on a theca of diatoms where a raphe would occur.
psychrophile: organism able to grow at temperature less than 15 °C.

psychrotroph: organism able to tolerate severe winter conditions and then grow in the warmer summer months.
puncta: opening in the frustule in diatoms, either a pore or a loculus.
pusule: a structure in the Dinophyta that is associated with the expulsion of excess protoplasmic water.
pyrenoid: proteinaceous area of the chloroplast associated with the formation of storage product.
pyriform: shaped like a pear.

radial: occurring around a central point.
radially symmetrical: when an object is cut in half, the two halves are superimposable by folding over at the plane of section.
ramulus: reproductive branch.
raphe: longitudinal slit in the valve of some diatoms.
receptacle: swollen tip of thallus containing conceptacles in the Fucales (brown algae).
red tide: water with a large number of dinoflagellates or other organisms that color the water red.
reservoir: large empty space at the bottom of a canal in some flagellates.
resting cell: cell with the same morphology as vegetative cells in diatoms, but with a large amount of lipid and reduced size of organelles.
resting spore: thick-walled cell resistant to unfavorable environmental conditions.
reticulate: netlike.
rhabdosome: an inclusion in dinoflagellates of the order Dinophysales that resembles a trichocyst but does not function as a projectile.
rhizoid: rootlike filament without vascular tissue.
rhizoplast: flagellar root composed of microfibrils that are often contractile.
rhizopodium: long delicate cytoplasmic protrusion.
rhizostyle: flagellar root composed of microtubules and associated structures.
rib: extension of the cell wall.

saccoderm desmid: a desmid without semicells (contrasted to a placcoderm desmid with semicells).
saprophyte: heterotrophic organism living off dead material.
scale: an organic or inorganic cell-surface element of variable geometry, distributed individually or arranged in a pattern sometimes forming an envelope around the cell.
scintillon: particle associated with bioluminescence.
scytonemin: pigment that accumulates in the sheaths of cyanobacteria acting as a sunscreen to reduce the amount of near ultraviolet (370–384 μm) reaching the protoplasm.
separation disc or **necridium:** a cell that dies in a trichome of the cyanobacteria, resulting in the separation of a hormogonium from part of the thallus.
septum: cross wall.
serrate: toothed as in a saw.
sessile: lacking a stalk.
seta or **awn:** elongated hollow wall extension.
sheath: extracellular mucilage.
siderophore: secondary hydroxamate secreted by some cyanobacteria to solubilize external iron compounds.
sieve cell: cell in the medulla of the Laminariales (brown algae), involved in active transport of organic molecules.
sieve membrane: wall structure covering a loculus in diatoms.

sieve plate: end wall in a sieve cell with pores through which the cytoplasm is continuous (plasmodesmata).
silica deposition vesicle: vesicle in which silica is deposited.
silicalemma: membrane of a silica deposition vesicle.
sinus: the incision in the midregion of a desmid cell (green algae).
siphonaceous or **siphonous** or **coenocytic cells:** large multinucleate cells without cross walls except when reproductive bodies are formed.
sirenine: a sex attractant in *Ectocarpus* (brown algae).
skeleton: a hardened non-living, protective or supportive structure, enclosed by, or attached to, cytoplasmic structures.
slime layer: extracellular mucilage.
somatic: vegetative.
sorus: cluster of reproductive bodies.
sperm: male gamete.
spermatangium: male gametangium forming one spermatium in the Rhodophyta.
spermatogenesis: formation of sperm.
spermocarp: zygote plus an enclosing layer of cells in *Coleochaete* (Chlorophyta).
spicule: a rod-like, spindle-shaped, stellate, or variously curved and ornamental structure, usually siliceous or calcareous, with blunt or tapered tips deposited individually on the cell surface, or distributed throughout the peripheral cytoplasm.
spine: a non-living, rod-like or tapered elongate structure attached to a scale, wall or skeletal framework.
spinulax: very small wall extension.
sporangium: spore-producing structure.
spore: cell that germinates without fusing to form a new individual.
spore mother cell: a cell that divides to produce spores.
sporeling: young plant arising from a spore.
sporocyte: a cell that divides and gives rise to spores.
sporogenesis: the process of spore production.
sporophyte: diploid plant that forms spores.
stalk: an elongate structure specifically formed to attach an organism to a substrate.
standing crop: the amount of biomass present at a specific time.
starch: storage polysaccharide composed of α-1,4 and α-1,6 linked glucose residues.
statospore or **cyst:** resting spore.
stellate: star-shaped.
stenohaline: able to tolerate only small changes in salinity.
stenothermic: able to tolerate only a small variation in temperature.
stephanokont: cell with a ring of flagella at one pole.
stichidia: specialized reproductive branches in *Polysiphonia* (red alga).
stigma or **eyespot:** group of pigmented lipid bodies that are associated with phototaxis.
stipe: organ between a holdfast and a blade.
stria: row of punctae (pores or loculi) in diatoms.
stroma: non-membranous part of a plastid.
stromatolite: rocklike deposition of carbonates and trapped sediments formed by cyanobacteria and diatoms.
stud process: short stubby wall extension.
sublittoral zone: in a lake the zone from the end of rooted vegetation (about 6 m) to the compensation depth; in the sea the zone from the lowest low-tide mark to 200-m depth.
suffultory cell: cell to which the dwarf male filament attaches in *Oedogonium* (Chlorophyta).
sulcus: longitudinal groove in the hypocone of Dinophyta.

supporting cell: a cell that bears the carpogonial branch in some Rhodophyta.
supralittoral zone: zone above the high-tide mark in the ocean and above the standing-water mark in lakes, which receives splash during windy periods.
suture: area of fusion of two adjacent structures.
swarmer: motile cell.
symbiosis or **reciprocal parasitism:** two organisms living together to the mutual benefit of each.
sympodial axis: axis formed from successive dichotomous branches in which one branch is shorter than the other, giving the appearance of a simple stem.
syncyanosis: the symbiotic association between a cyanome and a cyanelle.
syngamy: fusion of gametes.
systole: contraction of a contractile vacuole (opposite is diastole).

taxon (plural **taxa**)**:** a taxonomic group.
terrestrial: growing on soil.
test: a hardened cell covering typically secreted by the organism, or built up of particles gathered from the environment, forming a protective barrier around the cell.
tetrasporangium: a sporangium producing four tetraspores, usually by meiosis.
tetraspore: spore formed in a tetrasporangium, usually by meiosis.
tetrasporophyte: usually diploid plant forming tetraspores in the Rhodophyceae.
thallophyte: plant lacking roots, stem, and leaves.
theca: outer covering of the Dinophyceae and some Chlorophyceae.
thecal plate: a plate in a vesicle under the plasmalemma in the Dinophyceae.
thermal stratification: phenomenon in a body of water in which the water is progressively colder with depth, resulting in no interchange of water between the bottom and top.
thermocline: layer in a thermally stratified lake where the temperature changes suddenly with depth.
thermophiles: organisms that grow at high temperatures.
thylakoid: membrane-bound sac in a plastid.
tinsel flagellum: flagellum with hairs.
trabeculae: wall ingrowths in some coenocytic green algae.
transitional region: the most proximal (basal) part of a flagellum/cilium adjacent to the basal body/kinetosome, comprising matrix, axoneme, and flagellar/ciliary membrane.
trellisoid: type of uniformly arranged ornamentation without reference to a point or line in diatoms.
trichoblast: uniseriate, usually colorless hair, bearing sex organs in the Rhodophyceae.
trichocyst: projectile in the Dinophyceae and Raphidophyceae.
trichogyne: long colorless part of a carpognium that receives the spermatium in the Rhodophyceae.
trichome: a row of cells without the sheath in the cyanobacteria.
trichothallic: term describing intercalary meristem producing a hair in one direction and the thallus in the other direction.
trophocyte: vegetative cell.
trumpet hyphae: drawn-out sieve cells wider at the cross walls than in the middle of the cells (Laminariales).

uniaxial: having a main axis consisting of a single row of usually large cells.
unilocular sporangium: sporangium composed of a single cell producing zoospores usually by meiosis.

uniseriate: having a single row of cells.
unisexual: having only one type of gametangium formed on one plant.
upwelling: an area of the ocean where nutrient-rich bottom water rises to the surface.
uronic acid: a type of monosaccharide.
utricle: inflated branchlet.

valve: part of the cell wall in diatoms, a valve plus a connecting band making up a theca.
valve jacket or **mantle:** part of the valve in diatoms that is bent inward.
velum: wall extension over a loculus in diatoms.
volutin granule: protoplasmic body containing stored polyphosphate.

water column: a vertical section of a body of water.
whiplash flagellum: flagellum without hairs on its surface.

xanthophyll: a carotenoid composed of an oxygenated hydrocarban.
xylan: polysaccharide composed of xylose sugar residues.

zoochlorellae: Chlorophyta living inside invertebrate animals.
zooplankton: animal plankton.
zoosporangium: sporangium that forms zoospores.
zoospore: flagellated planospore.
zoosporogenesis: formation of zoospores.
zooxanthellae: non-green algal cells, usually Dinophyta, living inside invertebrates.
zygospore: thick-walled resting spore.
zygote: product of the fusion of two gametes.

INDEX

The most important page references are in **bold**, and page references that contain figures are in *italics*.